Electric Motor Controls

Electric Motor Controls

Rex Miller
State University College
Buffalo, New York

Mark R. Miller
Texas A&M University
College Station, Texas

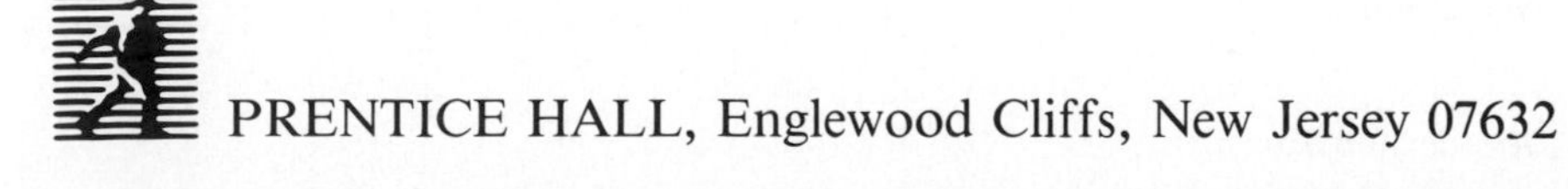

Library of Congress Cataloging-in-Publication Data
Miller, Rex
Electric motor controls / Rex Miller, Mark R. Miller.
p. cm.
Includes index.
ISBN 0-13-249376-4
1. Electric controllers. 2. Electric motors—Electronic controllers. I. Miller, Mark Richard. II. Title.
TK2851.M488 1992
621.46—dc20 91-3833
CIP

Editorial/production supervision
and interior design: *Julie Boddorf*
Cover design: *Wanda Lubelska*
Page layout: *C. Hart Pelletreau*
Prepress buyers: *Mary McCartney and Ilene Levy*
Manufacturing buyer: *Ed O'Dougherty*
Editorial assistant: *Elizabeth O'Brien*
Supplements editor: *Judy Casillo*

NOTE: This book is intended for use by persons familiar with the operation and installation of control systems, as well as with electricity and electric theory and operation. The publisher and authors disclaim any and all liability, loss or risk, incurred as a consequence, directly or indirectly, of the use and application of any of the contents of this book. The reader is urged to adhere to all safety practices associated with the activities described in this book and to consult competent practitioners in this field.

Printed in the United States of America

10 9 8 7 6 5 4 3 2 1

ISBN 0-13-249376-4

PRENTICE-HALL INTERNATIONAL (UK) LIMITED, *London*
PRENTICE-HALL OF AUSTRALIA PTY. LIMITED, *Sydney*
PRENTICE-HALL CANADA INC., *Toronto*
PRENTICE-HALL HISPANOAMERICANA, S.A., *Mexico*
PRENTICE-HALL OF INDIA PRIVATE LIMITED, *New Delhi*
PRENTICE-HALL OF JAPAN, INC., *Tokyo*
SIMON & SCHUSTER ASIA PTE. LTD., *Singapore*
EDITORA PRENTICE-HALL DO BRASIL, LTDA., *Rio de Janeiro*

Dedicated to Betty Ann Curren

Contents

4

Control Circuits and Diagrams 39

5

Switches 50

6

Magnetism and Solenoids 70

7

Relays 77

12

Motor Starting Methods 159

13

Solid-State Reduced-Voltage Starters 182

14

Speed Control and Monitoring 192

15

Motor Control and Protection 203

16

Three-Phase Controllers 221

Drives 233

18

Transformers 243

19

Power Generation 264

Power Distribution Systems 279

21

Programmable Controllers 305

22

Troubleshooting and Maintenance 317

Preface

Electric Motor Controls has been designed for use by construction electricians and apprentices who install and operate control systems. Others concerned with motor controls are journeyman electricians, technicians, engineers, electrical contractors and others, such as drafters and designers of control systems. We assume that the reader has a working knowledge of basic electricity and electric motor theory and operation. The book builds on that base to provide a working knowledge of the many aspects of motor control systems.

All the information needed for a thorough understanding of motor controllers and their theory, operation, installation, and maintenance is provided. The book is designed primarily for use by apprentice training programs, journeyman training, vocational–technical schools, and two-year colleges.

The glossary addresses those terms used in various technical areas encompassed by the control technician. The subject of motor control has some very specialized terminology which has been highlighted in the glossary, with a particular emphasis on applications.

The appendices have been separated from the text so that the information presented does not get in the way of learning about a topic within a chapter organization.

The study questions at the ends of chapters are designed to cause students to summarize their reading and put it into an organized format for use later when pressure of a job requires that information be forthcoming in a hurry.

The *National Electrical Code® Handbook* is a necessity for anyone working in this field. A copy of the latest edition should be in every toolbox.

The many illustrations included show a variety of parts and techniques used in present-day practice in the field. Obviously, not all problems can be presented here; a great deal of on-the-job ingenuity is required.

As you know, it is not possible to learn to swim without first getting a feel for the water. Neither is it possible to learn all there is to know about electrical controls by reading about them. You must be willing to get practical experience and devote time to the development of skills related to the job.

It is the authors' hope that the book will serve as an appetizer to those really interested in going into this exciting, yet demanding field.

Rex Miller
Mark R. Miller

Acknowledgments

No author works without being influenced and aided by others. Every book reflects this fact and this book is no exception. A number of people cooperated in providing technical data and illustrations. For this we are grateful. We thank especially the people at Allen-Bradley, Square D, and Westinghouse, who contributed both technical information and illustrations.

The following list indicates the willingness of industry to aid in the preparation of textbooks of quality for the trade.

Allen-Bradley Company: Jim Olson, Anita Fisher
Amprobe Instrument Division of Core Industries, Inc.
Automatic Switch Company
James G. Biddle Company
B&K Precision
Bodine Electric Company
Bussman Division of Cooper Industries, Inc.
Honeywell, Minneapolis, Minnesota
Ideal Industries, Inc.
Klein Tools, Inc.
The Lincoln Electric Company
National Association of Fire Equipment Distributors
Reliance Electric Company
Simpson Electric Company
Square D Company: Dave Stachowicz
Tecumseh Products Company
Westinghouse Electric Corporation

Electric Motor Controls

1

Tools

Objectives

After studying this chapter, you will be able to:

1. Identify the hand tools needed for motor control work.
2. Operate a VOM, scope, voltage tester, and digital probe.
3. Identify wire wrap/unwrap tools, conduit benders, PVC cutters, cable benders, and conduit reamers.

Tools are an extension of the person. They allow a human being to do things that are impossible using the human hands alone. Tools are also expensive and represent a sizable investment in money and in the time necessary to keep them in working condition. Two types of tools are needed by electricians who work with motors and motor controls: hand tools, and electrical instruments for measurement. We discuss both types in this chapter.

HAND TOOLS

Tools may be placed in an electrician's pouch (Fig. 1–1) or in a toolbox (Fig. 1–2). Keeping tools organized saves time and money. If tools do not have a place to call their own, you will spend too much time looking for them. Not only do you need to have a definite place for a tool, but you should also develop the habit of returning it to that place after you have used it. Whenever your tools are used on a workbench or around a piece of equipment, you need a toolbox to hold the tools and make them useful at the work site.

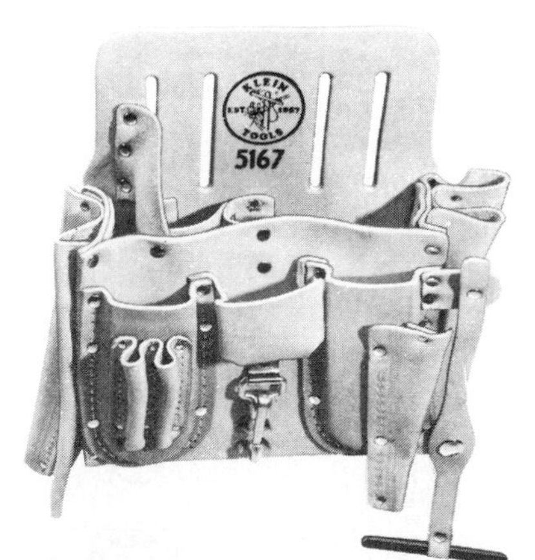

FIGURE 1–1 *Electrician's pouch. (Klein)*

FIGURE 1–2 *Portable toolbox. (Klein)*

Some tools can be mounted on a silhouetted outline on a board in front of the workbench. That way, if a tool is missing, you know which one it is, and when using it, you know where to put it when you are finished with it. Pegboard can be used or a piece of plywood with cutouts attached and pegs on which to hang the tools.

Screwdrivers

Always use the right tool for the job. Screwdrivers are the tool most often misused. Two types of screwdrivers are encountered most often: the Phillips-head and the standard or slot type. Screwdrivers come in thousands of variations.

Generally, screwdrivers are available with either wooden or plastic handles, but plastic is most common. Plastic handles are very helpful when working around electricity. Plastic is supposed to be shockproof if the handle is kept clean. The blade tip may vary in size from $\frac{1}{8}$ to $\frac{1}{4}$ in. The shaft is from 4 to 8 in. long and is usually made of nickel-plated chrome-vanadium steel. The tips or points must withstand the force applied when a screw sticks or is difficult to remove. The main thing to remember when using a screwdriver is to get a good fit between the tip of the screwdriver and the slot in the screw. This will prevent damage to both the screwhead and the screwdriver.

Other type of screwdrivers are also available. They are usually adapted for a particular purpose. The main reason for changing the screwhead from the slot to some other type is to increase efficiency in getting the driver into the screwhead for positive mating and more positive driving force. See Fig. 1–3 for other types of screwhead configurations.

Pliers

There are a number of pliers available for special jobs. The pliers shown in Fig. 1–4 are indicative of the variety available for work in the electrical and electronics field. Each is designated for a particular job.

1. 4-in. midget for close work.
2. 4-in. pliers for fast, clean tip cutting. Has a tapered nose and nearly flush cutting edges and will cut to the tip. Produces burr-free cuts.
3. 7-in. diagonal pliers for heavy-duty cutting.

FIGURE 1–3 *Screwhead configurations. (Klein)*

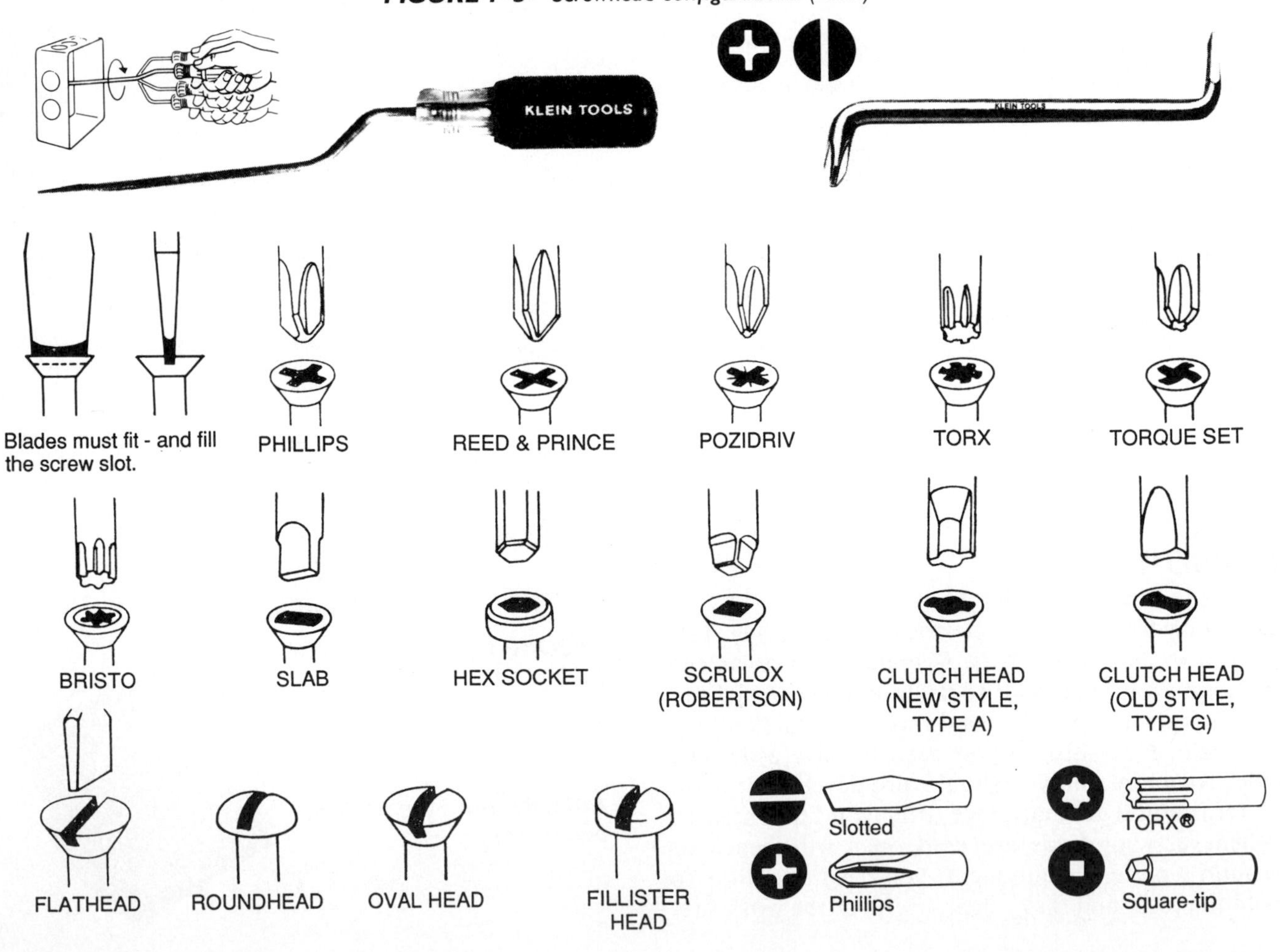

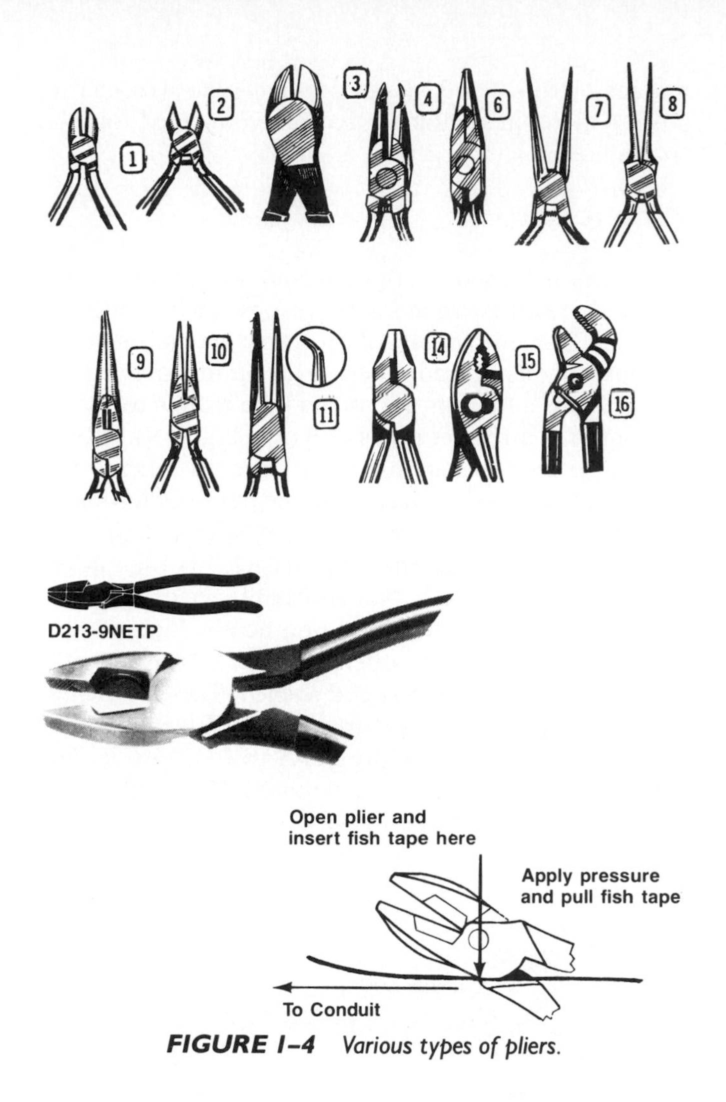

FIGURE 1–4 *Various types of pliers.*

4. $4\frac{1}{2}$-in. thin needle-nose pliers with cutter at the tip.
6. 5-in. thin chain-nose pliers, whose smooth jaws are slightly beveled on the inside edges.
7. $5\frac{1}{2}$-in. pliers with fine serrated jaws for firm gripping or looping wire.
8. Slim serrated jaws (6 in. long); permit entry in areas inaccessible to regular long-nose pliers.
9. Long-nose pliers ($6\frac{1}{2}$ in. long) with side cutter.
10. Long-nose pliers ($6\frac{1}{2}$ in. long) without side cutter.
11. Thin bent-nose pliers (5 in.) with fine serrated jaws: 60° angle thin bent-nose for thin wire.
14. 8-in. serrated upper and lower jaws with side cutter.
15. 8-in. chrome-plated combination pliers for general use.
16. Four-position 10-in. utility pliers with forged rib and lock design with serrated jaws.

Hammers

There are a number of types of hammers. The three most common are best suited for use by the electrician or motor control technician.

The claw hammer is used to drive nails and to work mainly with wood; it has claws with which to pull out nails. It is the most common type of hammer, but it is used only occasionally in motor control work—when a ball-peen hammer is not available. The claw hammer most often used is shown in Fig. 1–5.

The ball-peen hammer (Fig. 1–6) has a rounded top and a flat, large-diameter bottom surface. It is used in most work around machinery.

In some cases a mallet is needed to force a connection or to make a slight adjustment. Mallets can be obtained with either a soft face or a hard face. Leather, plastic, rubber, wood, and lead mallets are

CAUTION
Hammers are perhaps the most abused—and misused—of all hand tools. Improper use of hammers can cause injury. Use of damaged hammers can cause injury. And use of hammers for jobs other than those for which they were specifically designed can cause injury.

These basic rules for proper use apply to all hammers:

(1) Strike square blows only. Avoid glancing blows.
(2) Be sure striking face of hammer is at least 3/8″ larger than tools to be struck (chisels, punches, wedges, etc.). Face should be slightly crowned with edges beveled.
(3) Never use one hammer to strike another hammer.
(4) Always use a hammer of the right size and weight for the job.
(5) Never strike with the side or "cheek" of a hammer.
(6) Never use a hammer with a loose or damaged handle. Replace the handle.
(7) Discard any hammer having chips, cracks, dents, mushrooming, or excessive wear. Replace the hammer. DO NOT TRY TO REPAIR IT.
(8) Always wear safety goggles to protect your eyes.

For instructions on safe use, see label on each hammer.

FIGURE 1–5 *Claw hammer. (Klein)*

FIGURE 1–6 *Ball-peen hammer. (Klein)*

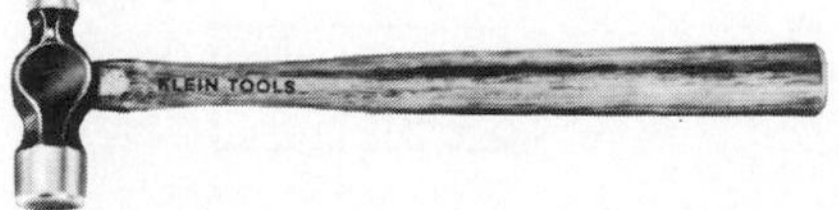

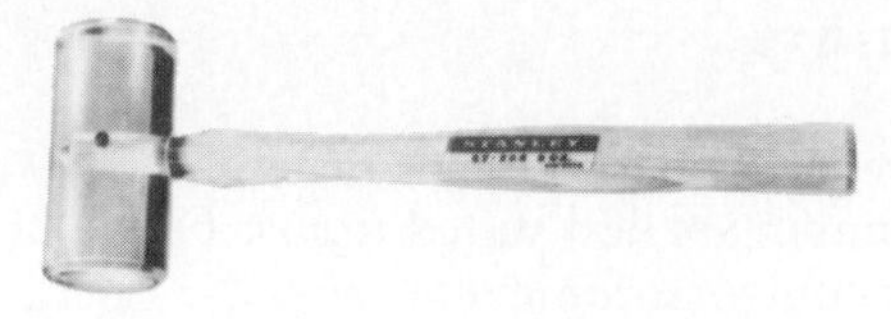

FIGURE 1–7 Plastic-tip mallet. (Klein)

used for various jobs as the need arises (Fig. 1–7). In most instances a rubber mallet is used to enable you to do the job without marring the surface of the metal being hit. Keep in mind that you should be careful with your hammer or mallet blows. Use eye protection.

Hacksaws

Hacksaws are very useful for cutting metal. Figure 1–8 shows a hacksaw blade being installed. A blade will break if it is not properly inserted and tightened; so once the blade is pointing in the right direction—away from the handle—tighten it. Note that the cutting takes place when the hacksaw is pushed away from the operator. Lift up on the saw when bringing it back to the starting position. Riding the blade as it is drawn back through the metal ruins the blade.

Blades come in 8-, 10-, and 12-in. lengths. The hacksaw is usually adjustable to fit any of the three blade lengths. Hacksaw blades also come in a number of tooth sizes. Use 14-teeth-per-inch blades for cutting 1-in. or thicker sections of cast iron, machine steel, brass, copper, aluminum, bronze, or slate. Use 18-teeth-per-inch blades for cutting $\frac{1}{4}$- to 1-in.-thick sections of annealed tool steel, high-speed steel, rail, bronze, aluminum, light structural shapes, and copper. The 24-teeth-per-inch blade is used for cutting material that is $\frac{1}{8}$ to $\frac{1}{4}$ in. thick in section. It is usually best for iron, steel, brass, and copper tubing, wrought-iron pipe, drill rod, conduit, light structural shapes, and metal trim. The 32-teeth-per-inch blade is used for cutting material similar to that recommended for 24-tooth blades.

FIGURE 1–8 Inserting a hacksaw blade.

With blade horizontal, flat-headed retaining pins permit cutting flush to surface.

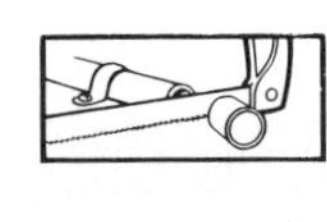

Flush handle design and absence of projecting wing nuts permit cuts all the way through to flat surfaces.

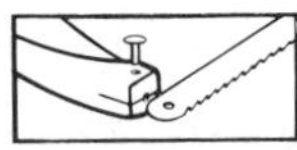

Snap-in retaining pins hold blade in accurately machined slots, assuring perfect alignment full length of blade.

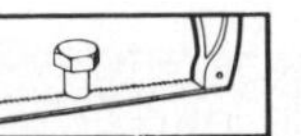

Blade can be turned to cut flush in either direction against flat surface.

Wrenches

Wrenches are used to tighten and loosen nuts and bolts. There are two general types of wrenches: adjustable and nonadjustable. Adjustable wrenches have one jaw that can be adjusted to accommodate different-size nuts and bolts and may range from 4 to 18 in. in length for different types of work (Fig. 1-9). When adjustable wrenches are used, there are two rules to remember: (1) Place the wrench on the nut or bolt so that the force will be placed on the fixed jaw, and (2) tighten the adjustable jaw so that the wrench fits the nut or bolt snugly (Fig. 1–10). Nonadjustable wrenches have fixed openings to fit nuts and bolt heads. Figure 1–11 shows a nonadjustable open wrench, and Fig. 1–12 shows a nonadjustable box-end wrench. These wrenches are available in sets. They are also available in metric sizes. Openings are actually 0.005 to 0.015 in. larger than the size marked on the wrench. This is to allow the wrench to be slipped easily over the nut or bolt head. Make sure, however, that the wrench fits the nut

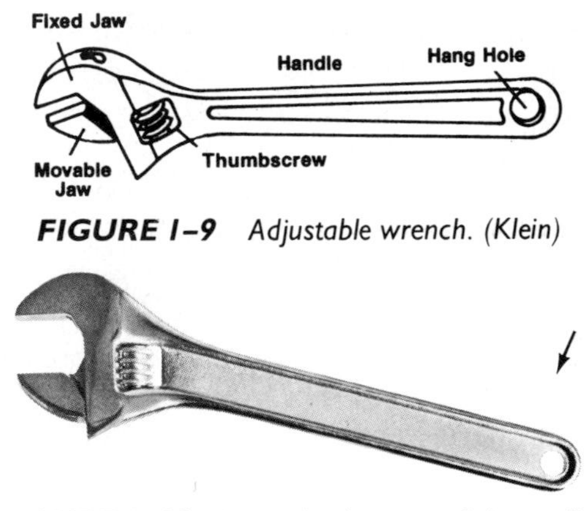

FIGURE 1–9 Adjustable wrench. (Klein)

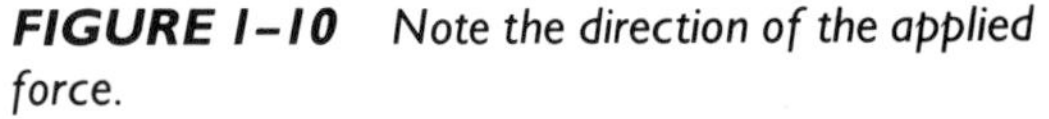

FIGURE 1–10 Note the direction of the applied force.

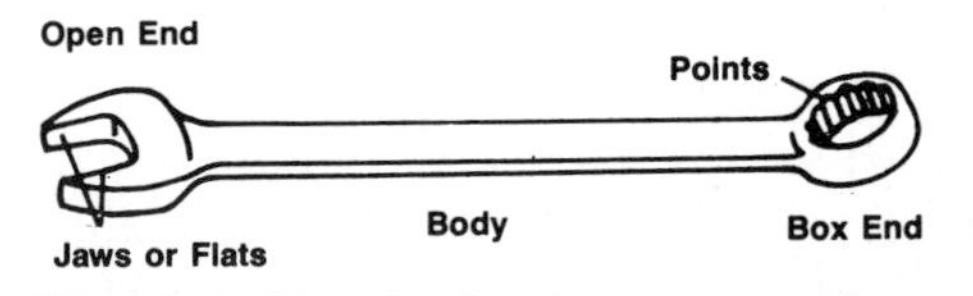

FIGURE 1–11 Combination open-end box-end wrench. (Klein)

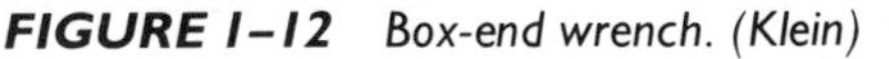

FIGURE 1–12 Box-end wrench. (Klein)

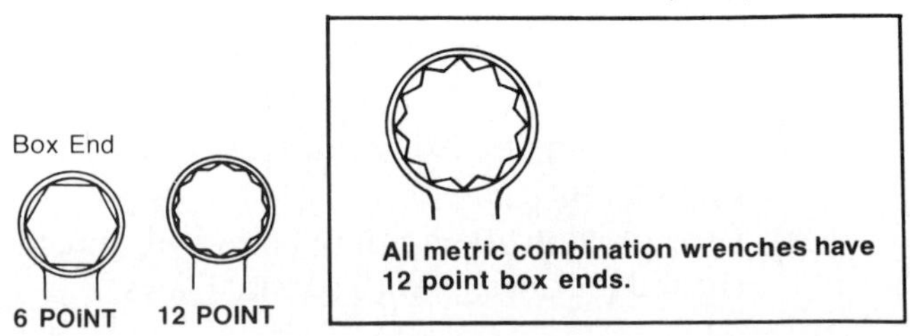

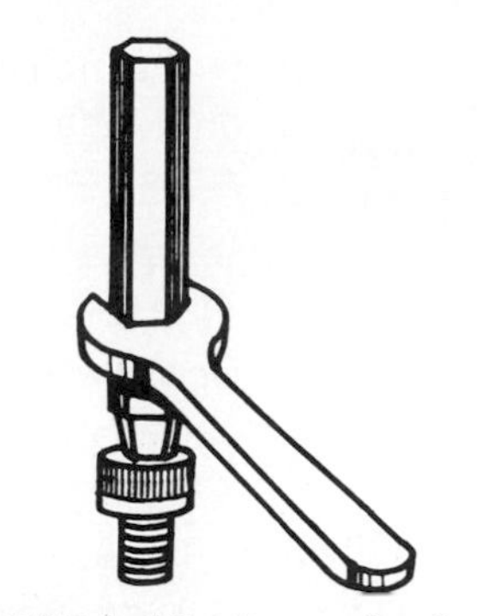

FIGURE 1–13 *Proper fit of wrench.*

or bolt head properly (Fig. 1–13). If it is does not, it may damage the nut or bolt head. It is safer to pull than to push a wrench. If you exert pressure on a wrench and the nut or bolt head suddenly breaks loose, there is a chance that you will injure your hand. If the wrench must be pushed rather than pulled, use the palm of the hand so that the knuckles will not be injured if there is a slip.

Allen Wrenches. Allen wrenches are designed to be used with headless screws. These screws are used in many devices as setscrews (Fig. 1–14). Allen wrenches come in a variety of sizes to fit any number of setscrews (Fig. 1–15). A complete set is especially helpful in motor control work.

Another type of recessed setscrew head has an Allen-type hole, but ridges in each flat side make it difficult for an Allen wrench to fit. Newer hex-socket key sets are made in $\frac{3}{32}$- to $\frac{1}{4}$-in. sizes with eight blades in a set (Fig. 1–16).

Socket Wrenches. Socket wrenches may be used in locations that are not easily accessible to the box-end or open-end wrench. Sockets are easily taken off the ratchet and replaced with another size. Sockets come in both a 12-point and a 6-point arrangement. Make sure that you use the proper size for each nut or bolt head. This type of wrench is also available in metric sizes. Figure 1–17 shows some of the sockets, extensions, flexible handles, and universal joints that make sockets effective in almost any location.

Torque Wrenches. Torque wrenches, designed to apply the proper torque to various bolts and nuts, are made to fit various socket drives. Two popular sizes of torque wrenches are the $\frac{3}{8}$-in. drive and the $\frac{1}{2}$-in. drive. Torque wrenches are made to measure in pound-inches (lb-in.) and in lb-ft. Use the proper wrench for the torque that has to be applied. The wrenches come in various handle lengths. Normally, the longer the wrench, the greater the torque the wrench will measure. A typical wrench is shown in Fig. 1–18.

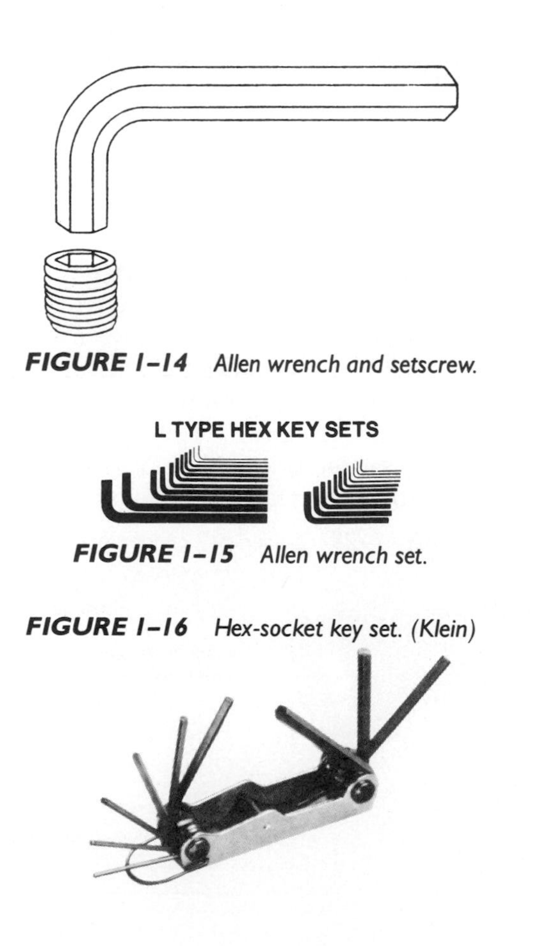

FIGURE 1–14 *Allen wrench and setscrew.*

FIGURE 1–15 *Allen wrench set.*

FIGURE 1–16 *Hex-socket key set. (Klein)*

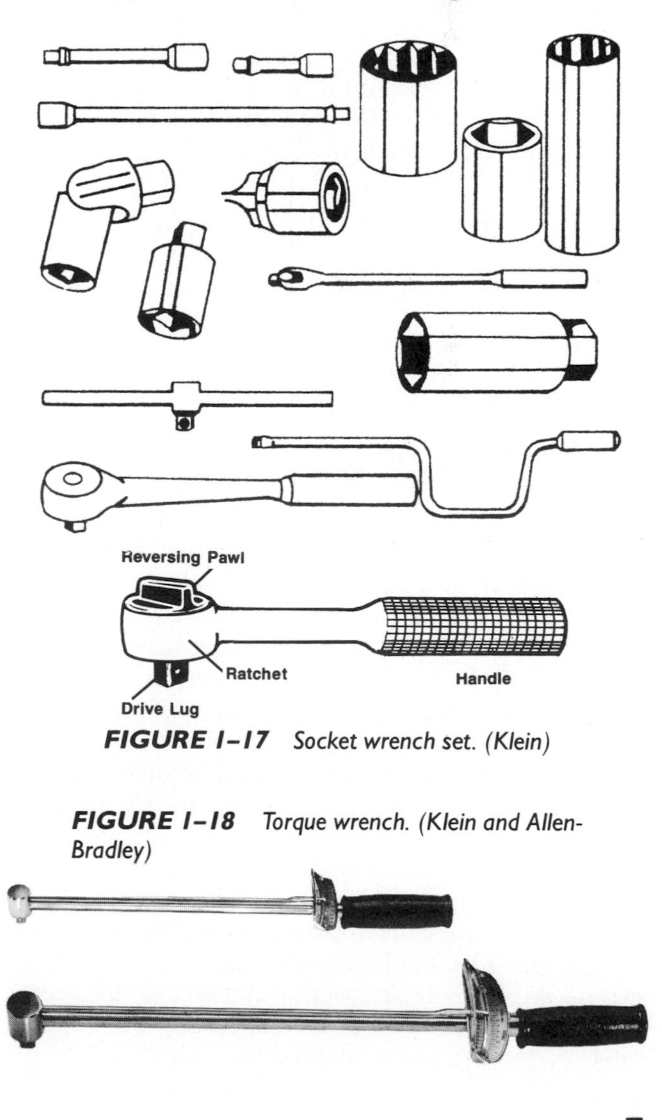

FIGURE 1–17 *Socket wrench set. (Klein)*

FIGURE 1–18 *Torque wrench. (Klein and Allen-Bradley)*

Nut Drivers

The nut driver is nothing more than a socket attached to a screwdriver handle. It is an excellent tool for most on-the-bench work. Nut drivers come in a variety of sizes and usually have the size stamped on the handle.

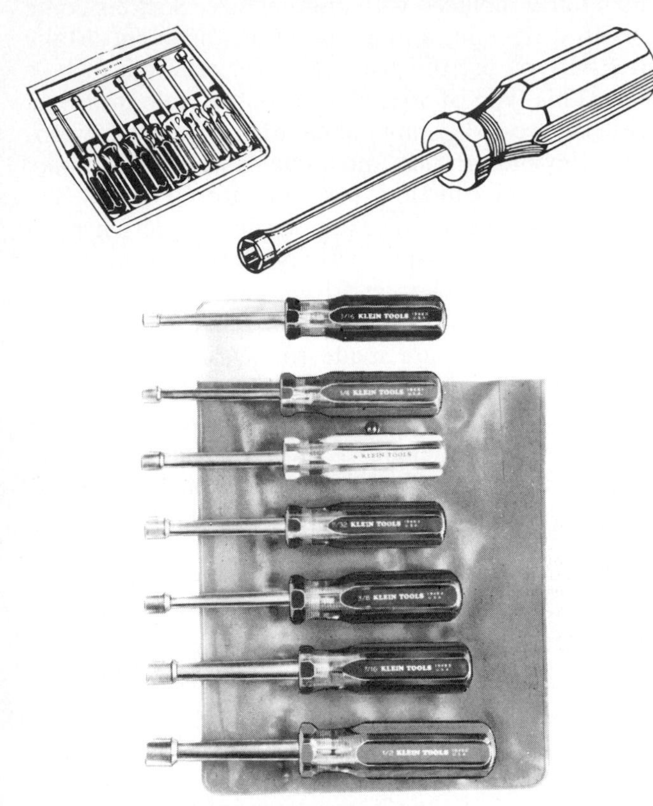

FIGURE 1–19 Nut driver set.

Sometimes they are color-coded according to size. Figure 1–19 shows a set of nut drivers.

Tools for Bearings and Bushings

In motor control and repair it is sometimes necessary to remove a bearing from the end bell of a motor. A bearing tool (Fig. 1-20) makes this task somewhat easier. The set illustrated has nine adapters that easily and quickly remove or insert any sleeve motor bearing or bushing with a $\frac{1}{2}$- to 1-in. inside diameter. A bearing tool eliminates the chance of broken bearings or end bells.

Some bearing removals need a different approach. The pulley or gear puller can, in some instances, be used to remove a bearing that has stuck onto the motor shaft. These pullers come in a variety of sizes and styles (Fig. 1-21).

Bushing tools have been designed for removing or inserting bushings in motors. They are handy timesavers. A complete set usually consists of 20 pieces: the box, three drivers, and 16 adapters that cover a range of $\frac{3}{8}$ to $1\frac{1}{2}$ in. (Fig. 1–22).

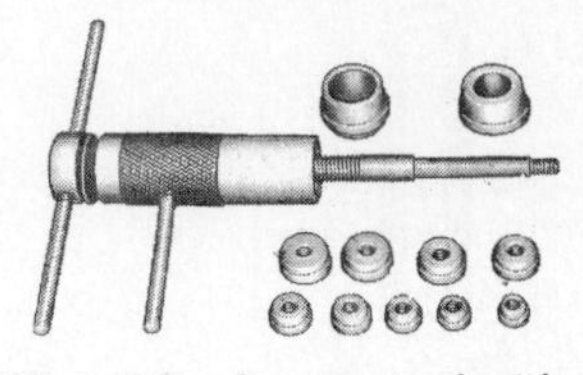

FIGURE 1–20 Bearing tool with adapters.

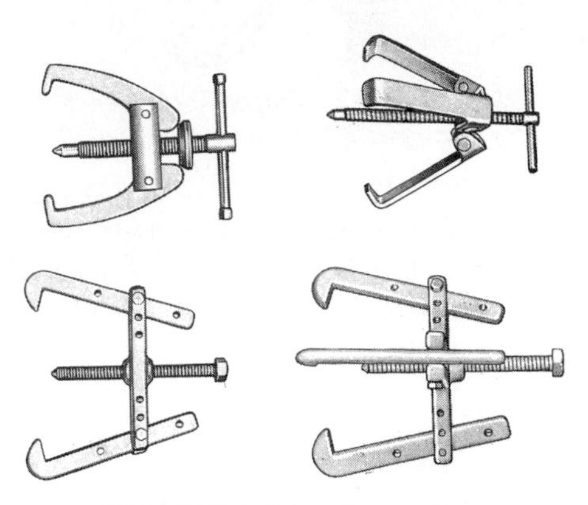

FIGURE 1–21 Gear pullers.

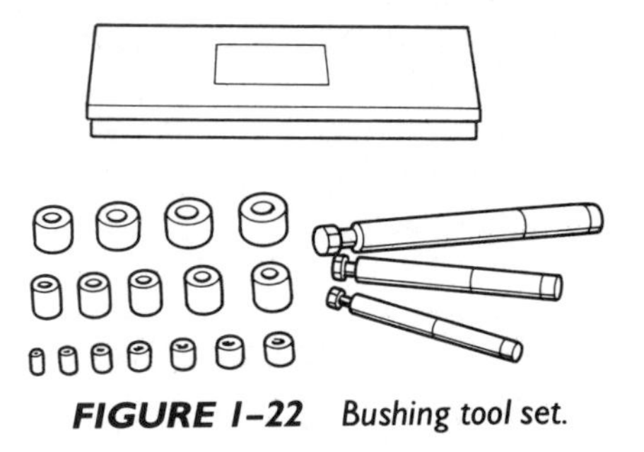

FIGURE 1–22 Bushing tool set.

Solderless Connector Crimper

The solderless connector crimper (Fig. 1–23) is very useful in motor work. It takes a good connection to withstand the vibration of a motor. A number of connectors have been designed for electrical work. The tool and kit of connectors and lugs of various sizes are available at most electrical supply houses.

FIGURE 1–23 Solderless connector–crimper. (Klein)

Soldering Iron

The soldering gun (Fig. 1–24) comes in handy when it is necessary to make a solder connection that will take vibration and withstand corrosion. Soldering irons are available from about 15 watts (W) to over 600 W.

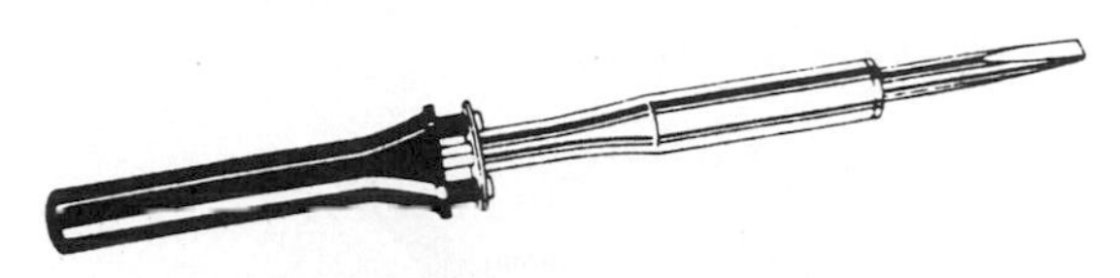

FIGURE 1–24 *Soldering iron.*

The best all-purpose soldering iron for use in the shop is about 100 W. This will do the job in most cases where larger wires are involved. The small 15-W irons are very useful in electronics work on printed circuit boards.

Soldering Gun

The soldering gun (Fig. 1–25) is very handy for making quick disconnects of soldered joints. Cold solder joints result when the operator heats the tip and places solder on it, then lets the solder cool on the joint. The wires being soldered or the metal surface and the wire being connected to it must be of sufficient temperature to melt the solder. This means that the gun must be left in one spot long enough to cause the joint to be heated to the temperature needed to melt the solder. The secret is to heat the material, not the solder.

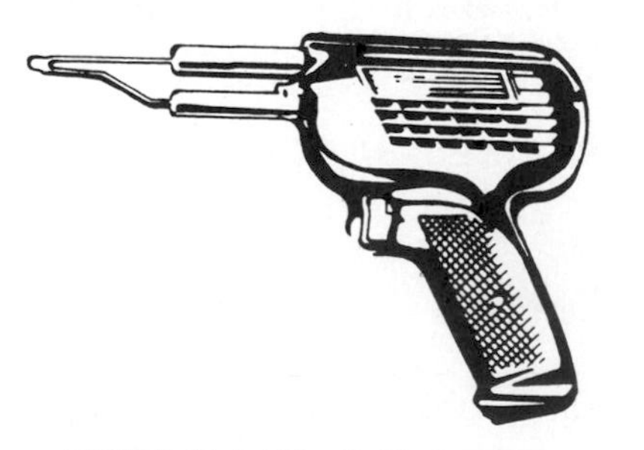

FIGURE 1–25 *Soldering gun.*

Wire Gages

Wire gages are needed to measure wire size. The numbers on the gage (Fig. 1–26) tell you the size of the wire. Keep in mind, however, that wire with Formvar insulation will read one size larger. Keep in mind also that the wire is moved through the slit in the gage. The hole is there to pass the wire through. The slot does the measuring. Pull the wire free of the slot and through the hole. Decimal equivalents are usually stamped on the metal disk on the opposite side of the gage numbers. Every toolbox needs a wire gage.

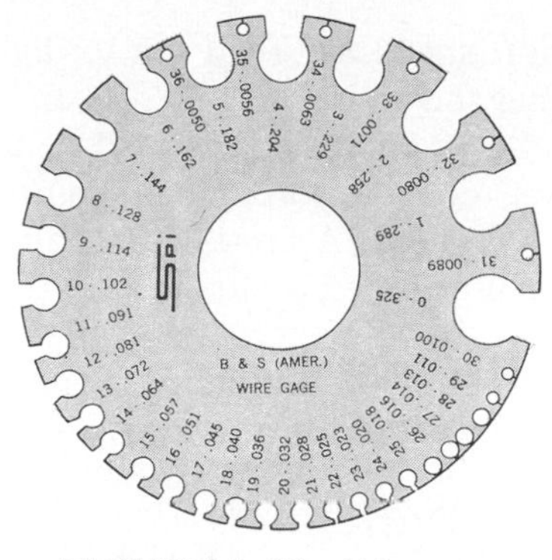

FIGURE 1–26 *Wire gage.*

Fuse Puller

Fuse pullers are made of phenolic material that has been shaped so that you can pull at least two sizes of fuses with it. It catches the round body of the fuse and allows you to extract it from its holder without coming in contact with the live circuit (Fig. 1–27).

FIGURE 1–27 *Fuse puller.*

Tachometer

To check the speed of rotation of an electric motor, it is best to use a hand-held tachometer, available in both analog and digital form. They represent speed in revolutions per minute (rpm) (Fig. 1–28). The tachometer is a very useful device for measuring motor speed. It can help locate possible problems and can indicate if a motor is operating as it should after it has been repaired.

FIGURE 1–28 *Tachometer. (Biddle)*

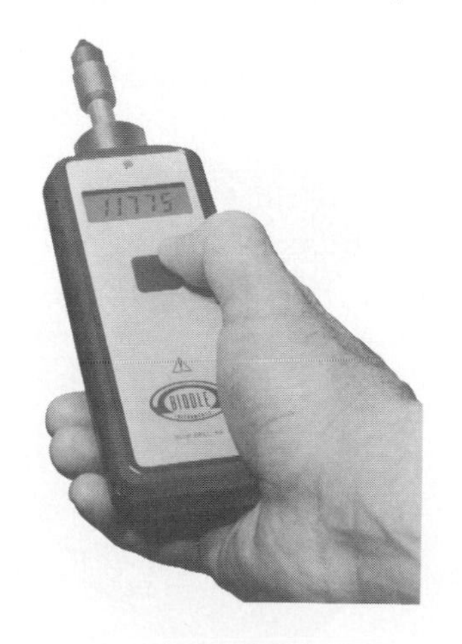

The shaft speed will read out in digital form in the tachometer shown in Fig. 1–28. It can be placed on the open end of a motor shaft, or it can be used on motors, saws, compressors, fans, pumps, grinders, and other equipment. A cone-shaped tip is used for shafts with center holes; a cup-shaped tip is used for flat-end shafts. Tachometers are available that use a strobe light to detect the number of revolutions per minute, but they are somewhat more expensive.

Knock-Out Punch

Knock-out punches come in many sizes. They are used to make holes in metal boxes for enclosures of various control devices. They range from a simple punch that uses wrenches to large hydraulic units utilized to punch holes in very heavy gage metal (Fig. 1–29).

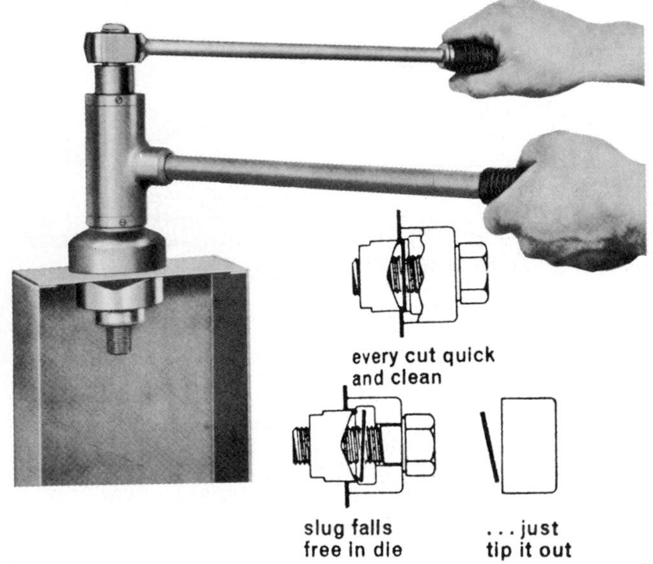

FIGURE 1–29 *Knockout punch in operation.*

Fish Tape

For pulling wires through conduit, walls, or into junction boxes, fish tape is essential (Fig. 1–30).

FIGURE 1–30 *Fish tape and reel. (Ideal Industries)*

Cable Stripper

Larger-gage wire (cable) has to be stripped for a T-tap or midspan strip. In most instances a cable stripper is used for stripping insulation so that placing lugs on the end of the wire or cable is much easier. Figure 1–31 shows how it can be used to strip insulation from a conductor without scoring.

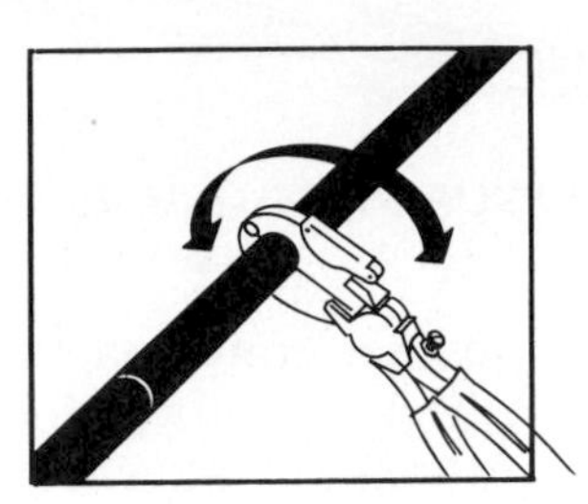

1. Make two cuts around circumference of cable, rotating jaws around cable for a clean, accurate insulation cut.

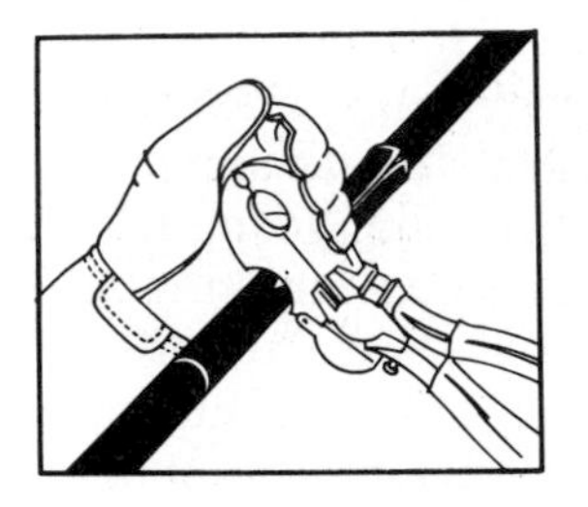

2. Slit insulation between cuts, parallel to the cable. Integral steel slitter, guided by tool, makes it safe and easy. No knife needed.* Won't damage cable.

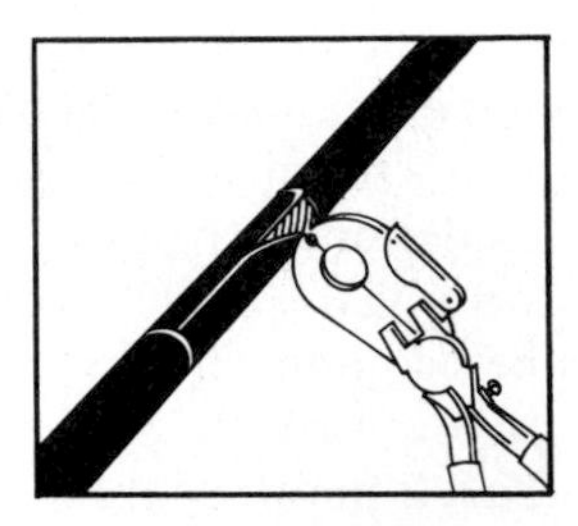

3. Pinch edge of slitted insulation between the built-in grippers at tip of tool and peel it away.

FIGURE 1–31 *Cable stripper. (Klein)*

Cable Cutter

Hand-operated cable cutters are used to make a shear-type cut for large-size wire. Long fiberglass handles give leverage. A cable cutter can cut cable up to 750 MCM (750,000 circular mils)* (Fig. 1–32). The clean cut that results makes it easier to fit cable ends into lugs.

FIGURE 1–32 *Cable cutter. (Klein)*

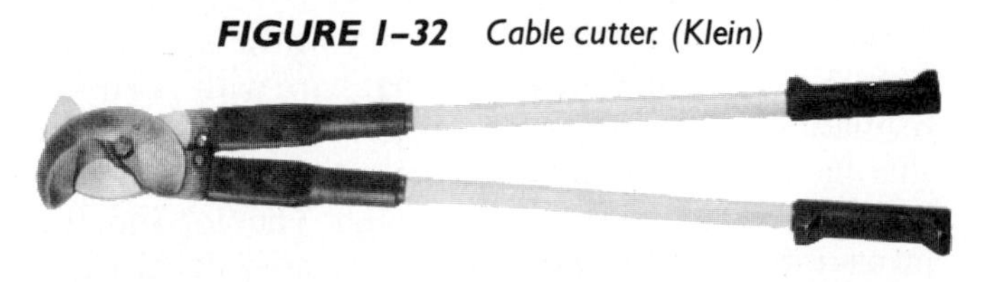

*M is the Roman numeral for 1,000. CM stands for circular mil. A mil is 0.001 in. A circular mil is a 0.001 in. in cross-sectional area. MCM is 1000 CM.

Electrician's Knives

The jack knife (Fig. 1–33) is used to cut insulation. Every toolbox should have handy a jack knife with very sharp blades. Another knife that comes in handy for the electrician or anyone working around electrical equipment is the skinning knife (Fig. 1–34). This knife fits in the toolbox or in the tool pouch with the proper cover.

FIGURE 1–33 *Jack knife. (Klein)*

FIGURE 1–34 *Skinning knife. (Klein)*

Other Tools

Other tools that may be useful on the job are a flashlight, Polaroid camera, polarized receptacle tester, wire markers, and various wire-pulling apparatus and threading tools for rigid conduit. A heavy-duty electric drill or one with hammer action to drill through concrete is also handy. Various tools will come to mind as you develop your workshop or toolbox. As different situations arise on the job you will be better equipped to know which tools to invest in. Keep in mind that your tools are an investment. Mark them with your name or the name of your company.

ELECTRICAL TOOLS

There are a number of electrical instruments that are needed for the motor repair technician to do the job of installation and maintenance of motors and their controls. Everything from a simple VOM to an oscilloscope is needed to do various tests and to increase the efficiency of the troubleshooter.

Portable Ammeter and Voltmeter

The clamp-on type of ammeter and voltmeter has extended leads with replaceable probe tips that make

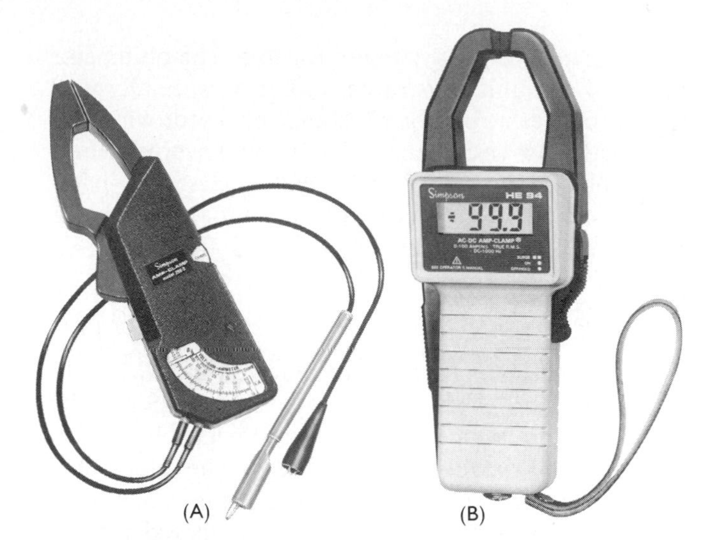

FIGURE 1–35 *(A) Analog clamp-on ammeter; (B) digital clamp-on ammeter. (Simpson)*

voltage reading easy and fast. The clamp-on jaw goes around a conductor of 1 in. diameter/500 MCM or 2 in. diameter/2000 MCM to take the current reading without interrupting service. One hand operation is possible: Open the jaws, change ranges, and read the current. The meter comes in a handy, sturdy case to protect it while not in use (Fig. 1–35).

Megohmmeter

The megohmmeter is usually referred to as a *megger*. The megger shown in Fig. 1–36 has ranges for measuring insulation resistances of motors, compressors, conductors, or anything else that needs measuring. It can also read continuity and can measure low resistances of motor windings.

Test voltage is generated by a hand crank. This means that it does not require any other power source

FIGURE 1–36 *Megger. (Biddle)*

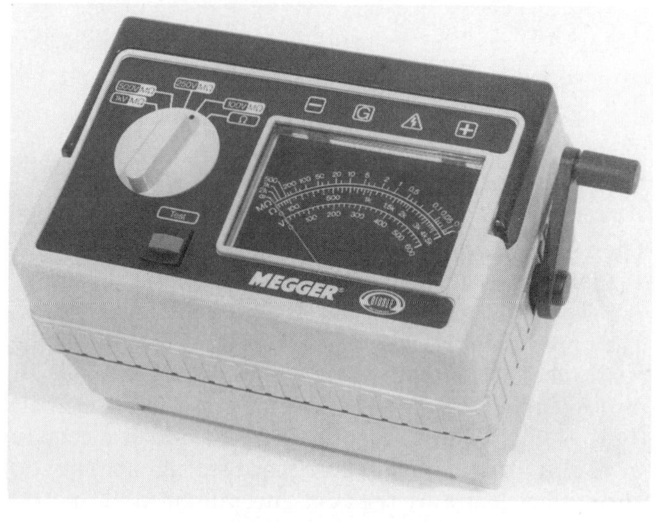

and is therefore always ready for use. The ohmmeter range of the unit illustrated makes it especially well suited for measuring the resistance of motor windings and other low resistances. The guard terminal eliminates the effect of any surface leakage that may influence a reading. It comes with a manual that shows how to use it for various purposes.

Voltage Tester

The voltage tester is a handy device that checks 10 ac/dc voltage levels. It fits in a shirt pocket. The lighted windows indicate the voltage level. This makes it easily read in dimly lighted areas. The coiled lead cord extends to 50 in. A test button distinguishes normal readings from those due to distributed capacitance or high-resistance leakage. It is also helpful in checking out 115-V ac grounded convenience outlets, and it will operate on 25 to 800 Hz (Fig. 1–37).

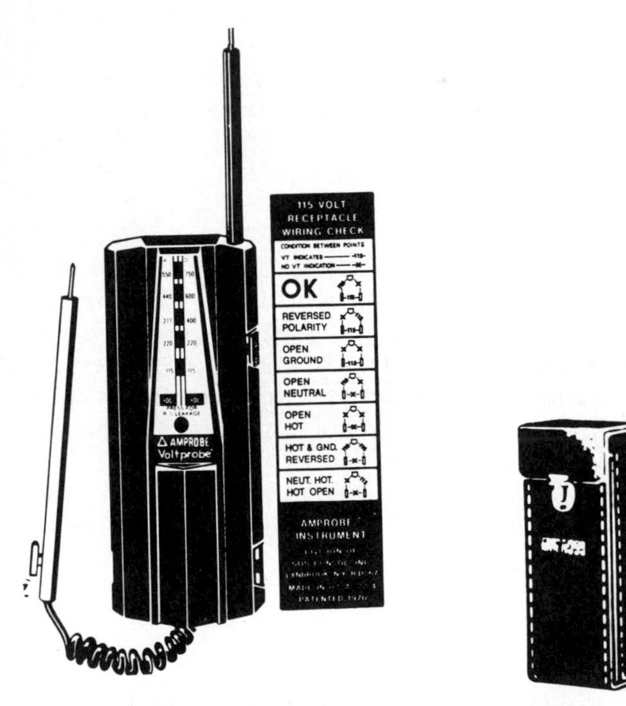

FIGURE 1–37 Voltage tester. (Amprobe Instrument)

VOM

The volt-ohm-milliammeter, usually called a VOM, is available in many sizes designed to read many ranges. VOMs check out voltages, currents, and resistances. They are portable and contain a battery to power the resistance checking ranges. Most are also capable of testing capacitors and inductances as well as decibels (Fig. 1–38). A flip-open carrying case makes it easy to store and provides a place for leads storage so that you know where they are the next time you need the meter.

FIGURE 1–38 Analog VOM. (Simpson)

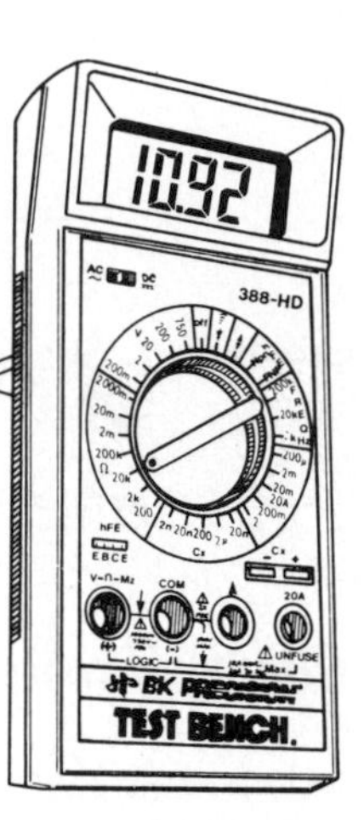

FIGURE 1–39 Digital VOM.

This type of meter is also available in the digital form, which reads out in numbers instead of your having to read the scale (Fig. 1–39). It is possible to get more accurate readings with a digital meter than with an analog meter. It will read down to a tenth of an ohm (0.1 Ω), so it comes in handy when testing motor windings and continuity (See Chapter 22.) Other digital meters include the clamp-on type shown in Fig. 1–35B.

Digital Logic Probe

The digital logic probe has become a necessity in troubleshooting programmable controllers and other equipment that uses computer logic to do switching or sequencing. It is a quick way to peek inside TTL, LSI, and CMOS digital circuits.* The probe shown in Fig. 1–40 has color-coded light-emitting diodes (LEDs) to indicate high, low, or pulsed logic states (up to 10 MHz). It puts out a tone that really speeds up the testing.

*TTL, transistor-transistor logic; LSI, large-scale integrated; CMOS, complementary metal-oxide semiconductor.

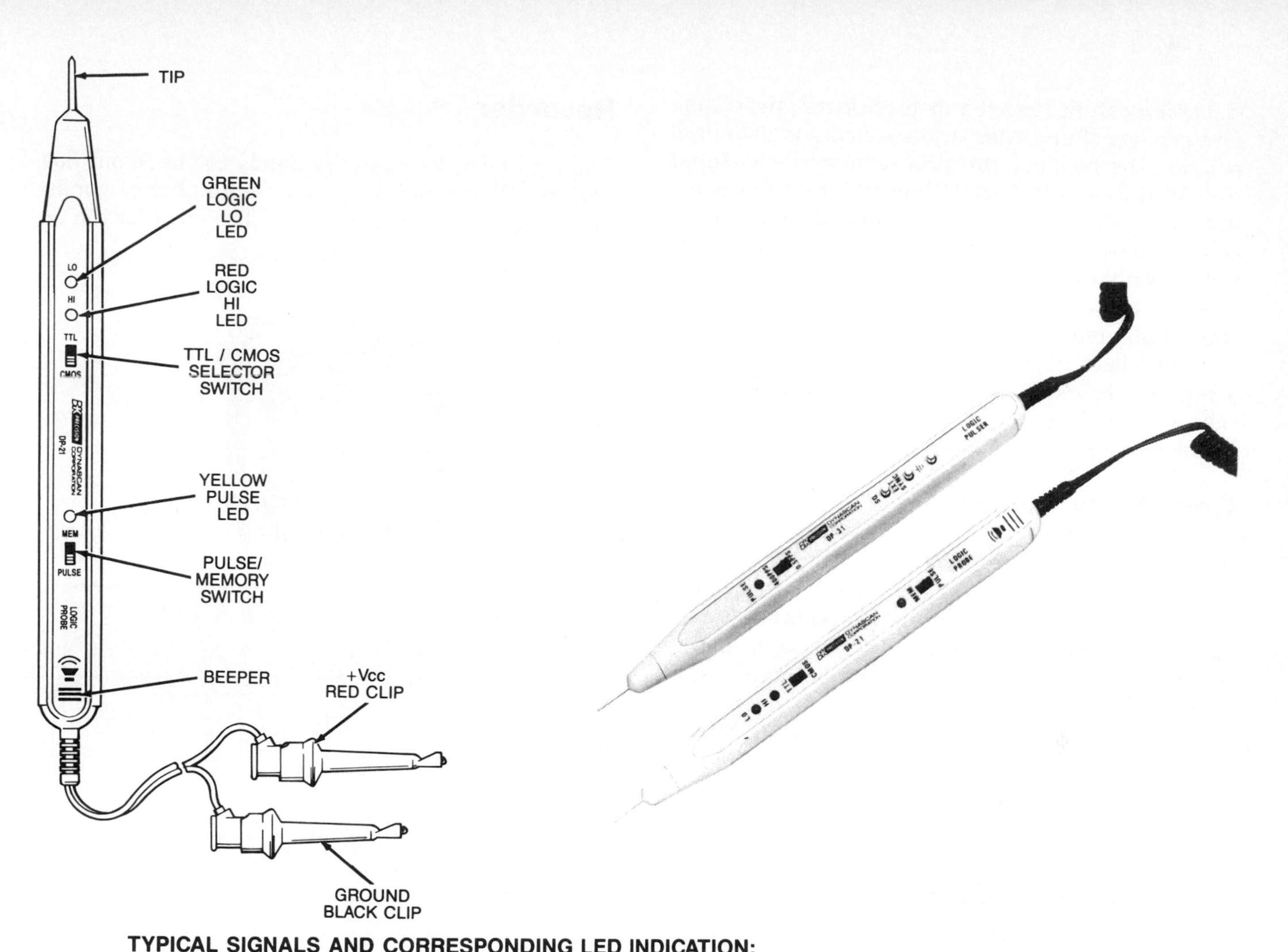

TYPICAL SIGNALS AND CORRESPONDING LED INDICATION:

ITEMS	WAVEFORM	LED INDICATIONS LEVEL RED	GREEN	PULSE YELLOW	BEEPER
Logic "1" no pulse activity	1 / 0	●	○	○	High tone
Logic "0" no pulse activity	1 / 0	○	●	○	Low tone
Signal level between "1" & "0"	1 / 0	○	○	○	
Logic "1" with pulse	1 / 0	●	○	☆	Intermittent high tone.
Logic "0" with pulse	1 / 0	○	●	☆	Intermittent low tone.
Pulse train with freq. < 200KHZ	1 / 0	●	●	☆	1. Alternate and Intermittent sound. 2. Mixed and Intermittent sound.
Pulse train with freq. > 200KHZ	1 / 0	○	○	☆	

● LED On ○ LED Off ☆ LED blinks ----------- ref. level ———— signal

FIGURE 1–40 *Digital logic probe. (B&K)*

The next order of business involves the operation of the probe and what it indicates at various points in the test procedure. Typical signals and corresponding LED indications are shown in Fig. 1–40. To operate the probe, apply power to it by connecting the black clip to GND (−). Connect the red clip to the V_{cc} (+). Make sure that V_{cc} is less than 20 V.

The TTL/CMOS switch can be switched to TTL mode for use in TTL circuits. The TTL logic 1 threshold is 2.3 ± 0.2 V; the logic 0 threshold is 0.8 ± 0.2 V. When switching to CMOS mode, the CMOS logic 1 threshold is 70% V_{cc}; the logic 0 threshold is 30% V_{cc}.

The logic probe can also detect and memorize the level transition. Either positive or negative level transition can be detected or memorized, depending on the mode (Pulse/Memory) selected. On the PULSE position the memory function is inoperative. Input state transition is "0" to "1"; or "1" to "0" will activate the pulse indicator (flicker for 500 milliseconds) and will generate a chop sound for the beeper. On the MEMORY position the memory function is activated. The pulse indicator lights and a beeper generates sound until reset if any pulse or level transition occurs.

The beeper generates sound when the red or green LED lights. Some typical combinations of indications from the probe are shown in the table of typical signals in Fig. 1-40.

Continuity Tester

Whenever an ohmmeter is not available or handy, it is convenient to use the continuity tester, since it is usually carried in the toolbox or on the toolbelt. This device is a simple battery, lamp, and probe, with another lead to complete the circuit and light the lamp to indicate a complete path for current flow. It can be used for testing for opens, shorts, or continuity (Fig. 1-41).

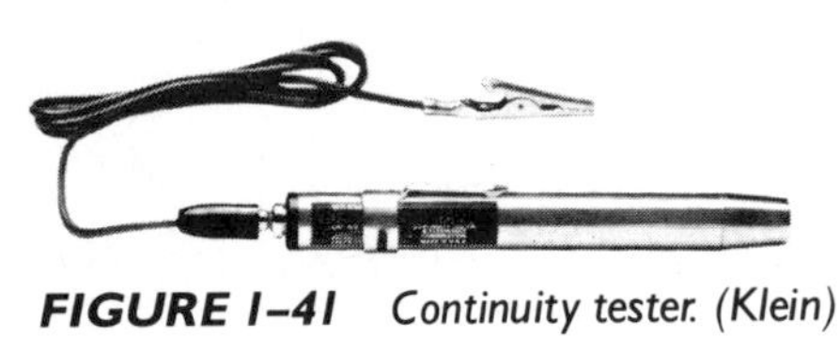

FIGURE 1-41 *Continuity tester. (Klein)*

Polarized Plug Tester

There are a number of plug-in devices to indicate if a circuit has the proper ground and if the hot side of the receptacle is where it is supposed to be. Figure 1-42 shows one of these plug-in types and how the readout in the LEDs indicates the various conditions of the circuit.

FIGURE 1-42 *Polarized receptacle tester.*

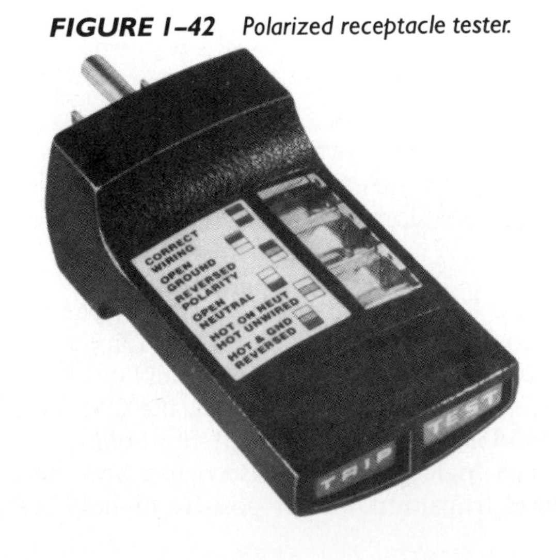

Recorder

The strip recorder comes in handy to check out line voltages. These may vary under a number of conditions that are not easily checked. The recorder can be

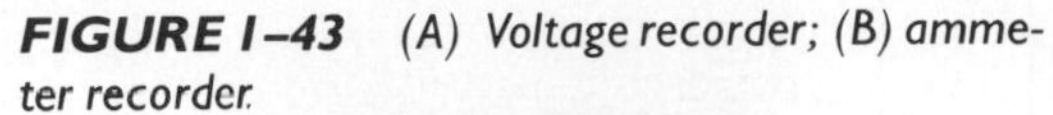

FIGURE 1-43 *(A) Voltage recorder; (B) ammeter recorder.*

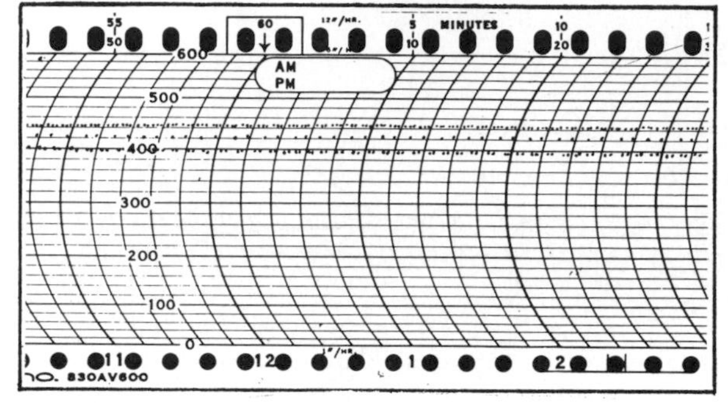

Strip Chart

(A)

(B)

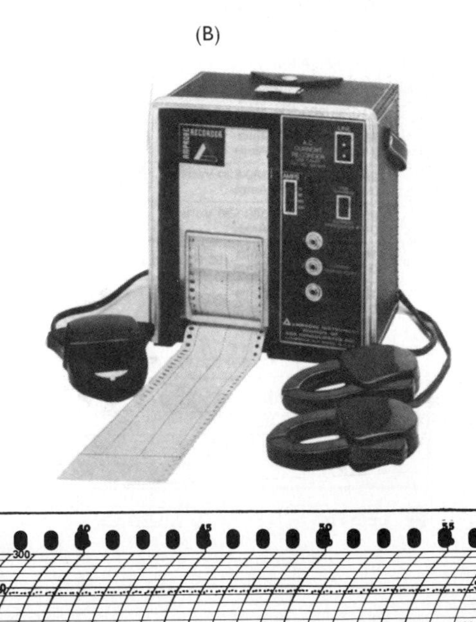

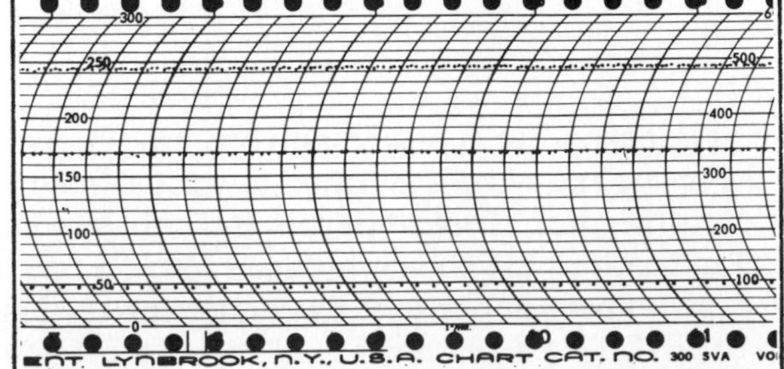

Strip Chart

placed on the line and will record variations in line voltages over a long period of time. Knowing when the variations occur aids in locating a number of types of problems. The recorder shown in Fig. 1–43A monitors voltage conditions over a wide variety of ranges. It helps pinpoint overvoltage conditions, which shorten lamp and motor life, and undervoltage conditions, which affect oven, tool, and lighting output. The ammeter type of recorder (Fig. 1–43B), which uses the split-core transformer principle for clamp-on transducers, simply clamps around the conductor in which the current is to be measured. There is no need to interrupt the service. With this type of device you can monitor all three phases, and in conjunction with a voltmeter recorder it can be utilized to check out when various power surges occur and can aid in regulating placement of various machines on the line at the most economical time for power consumption.

Oscilloscope

The oscilloscope (Fig. 1–44A) is also a voltage-indicating device. It shows the shape of the power being used. It can aid in tracing pulses on the line that may cause timing problems with programmable controllers or other computer-operated machines. By using a function reference signal generator, such as that shown in Fig. 1–44B, with an oscilloscope, it is possible to properly adjust and tune the stability circuit of a high-gain motor drive and regulators. The appropriate stability circuits can be optimized by using a step function into the regulator and observing the feedback loop output with an oscilloscope or chart recorder.

FIGURE 1–44 *(A) Oscilloscope; (B) function reference signal generator.*

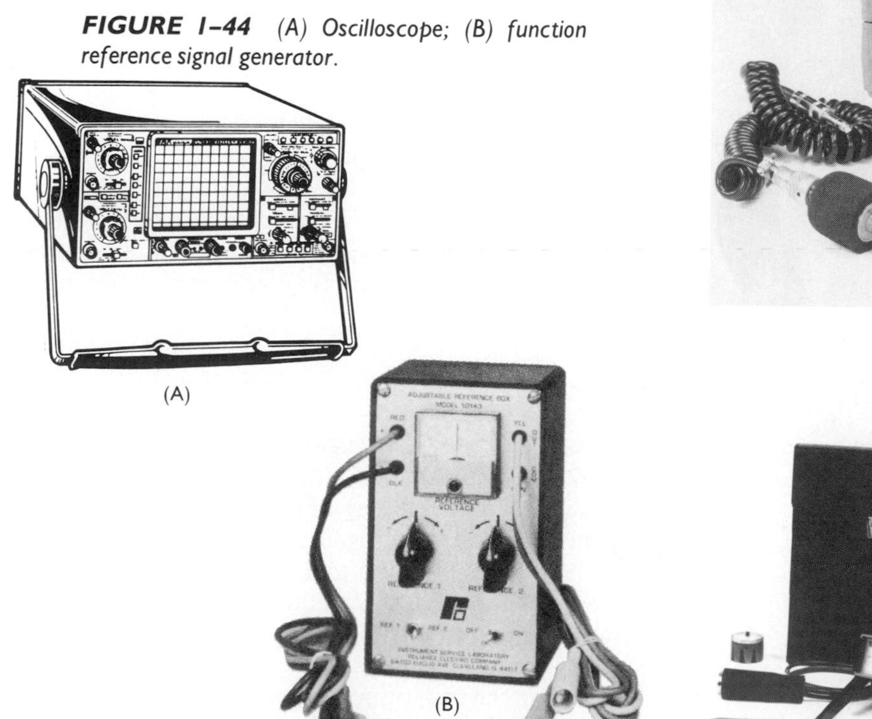

Phase Sequence Adapter

The phase sequence adapter (Fig. 1–45) is used in conjunction with any Amprobe volt/ammeter (or other voltmeter with appropriate ac range). It permits you to determine the phase sequence of any electrical equipment using three-phase lines up to 550 V, 25 to 60 Hz. The carrying case and manual are included.

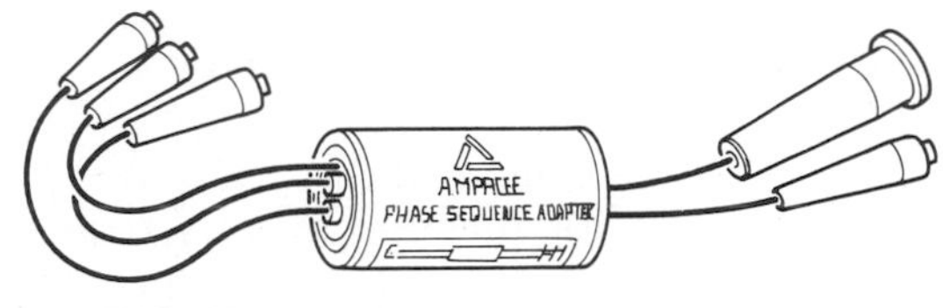

FIGURE 1–45 *Phase sequence adapter.*

Balance Analyzer

The balance analyzer (Fig. 1–46) has a digital meter that provides readings of displacement, velocity, and acceleration in the vibration mode. The strobe mode

FIGURE 1–46 *(A) Balance analyzer; (B) vibration meter and bearing tester. (Vitec)*

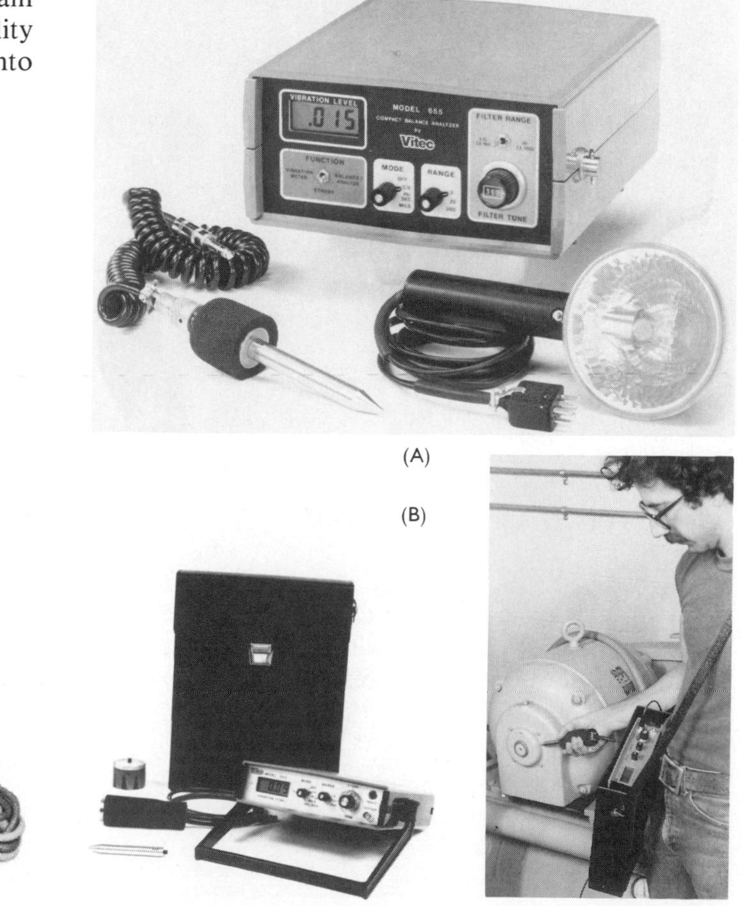

allows the operator to select the strobe and find the rate for stop-motion analysis. The balance/analyze mode allows the model to be tuned to specific frequencies for signature analysis and balancing purposes. Although this is a very expensive instrument, it is usually worth its price in locating and eliminating problems caused by vibration and off-balance loads.

SPECIAL TOOLS FOR SPECIFIC JOBS

Printed Circuit Board Puller

Motor controls are sometimes incorporated into printed circuit boards and mounted into racks. Extreme caution must be observed when installing or removing boards because some boards are extremely sensitive to static electricity discharges. The printed circuit board puller shown in Fig. 1–47 clamps firmly onto the circuit board. It is covered with plastic to prevent physical damage to the board, and its use prevents you from touching the board while removing it.

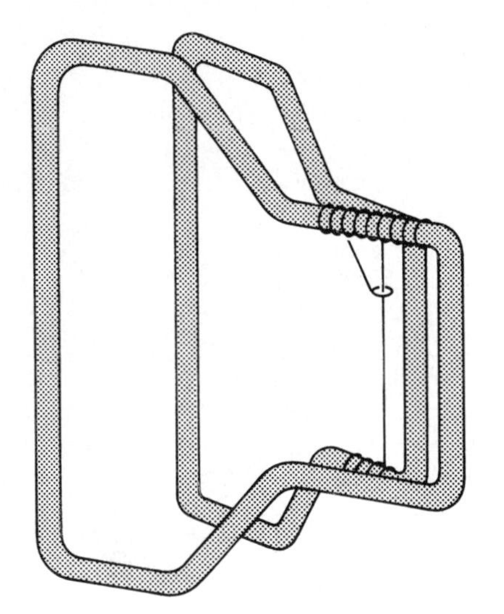

FIGURE 1–47 Printed circuit board puller. (Reliance)

Wire Wrap/Unwrap Tool

Soldering is not the only way that wire terminations are made at present. Wire wrapping provides fast, secure electrical connections without soldering. Wire wrapping and unwrapping tools (Fig. 1–48) are made of case-hardened steel for long life. They are hand operated, so there is no need for air or electrical supply. This method of termination eliminates wire crystallization due to soldering heat and subsequent fracturing from on-the-job vibration. The tool shown can handle 18- to 22-gage wire.

FIGURE 1–48 Manual wire wrapper and unwrapper. (Reliance)

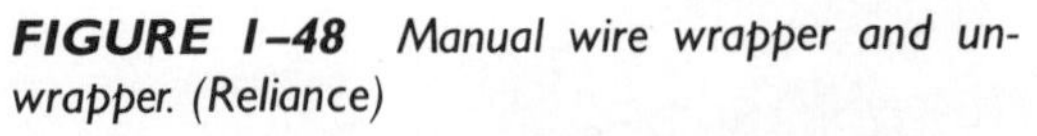

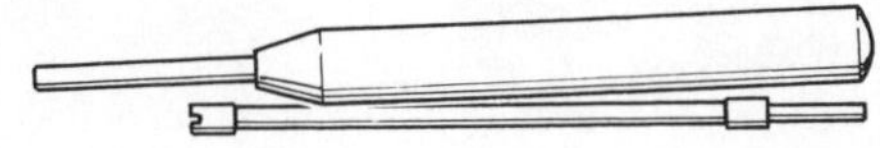

Conduit Benders

Conduit benders take the guesswork out of this job. Benders have built-in benchmark symbols, degree scales, and multiplier scales. An arrow points to the beginning of a bend; a star indicates back-of-bend locations. A teardrop symbol indicates the exact center of a 45° bend.

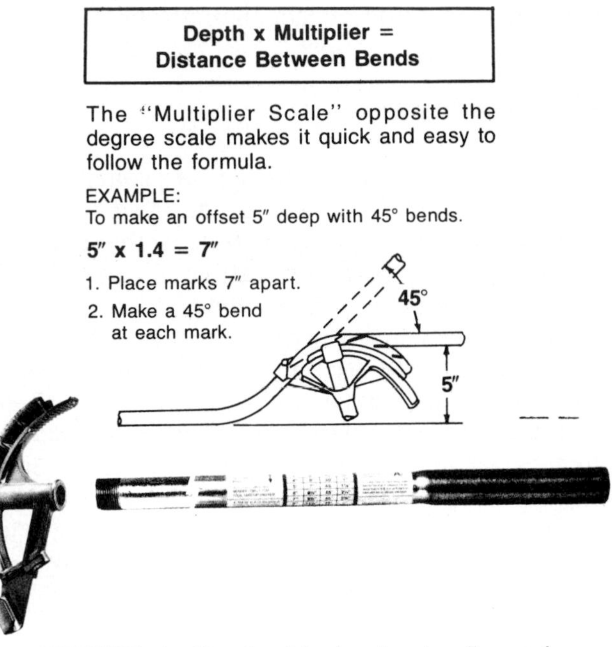

FIGURE 1–49 Conduit bender, handle, and bending formula. (Klein)

Benders for $\frac{1}{2}$-in. EMT also have a cast-in offset formula and multiplier, providing instant, on-the-spot information for making accurate offset angle bends (Fig. 1–49).

PVC Cutters

Cutting PVC plastic has traditionally been done using a backsaw. This saw cut can be rather ragged when finished, because the kerf of the saw blade is designed for cutting wood. Making clean, burr-free cuts through PVC is a quick, simple task with specially designed cutters. They use ratchet action for maximum cutting power with very little effort (Fig. 1–50). The one shown is only 9 in. long, so it can fit in the toolbox or be hung on the tool pouch. It will cut $\frac{1}{4}$- to $1\frac{1}{4}$-in. PVC. Just open the handles completely, place the conduit in the lower jaw hook, and alternately squeeze and re-

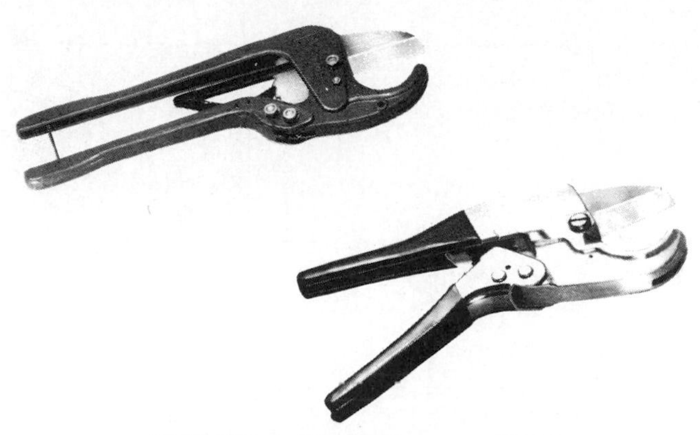

FIGURE 1–50 PVC cutters. (Klein)

lease the handles, letting the ratchet action do the work until the shear is complete. By using a cutter with longer handles it is possible to cut $\frac{1}{2}$- to 2-in. PVC as easily as the short-handled cutter cuts smaller diameters. The advantage of this type of cutter is the reduction in pipe deformation and the possibility of PVC conduit cracking during cutting. A clean cut is also obtained. The blades can easily be replaced when they become dull.

Cable Benders

Some cables require a great deal of effort to bend to fit their intended placement. A tool has been designed to aid in the bending operation (Fig. 1–51). The forged, one-piece bender has been made of steel to meet the demands for such a tool. The head is angled 22° for ease in bending the cable in tight, hard-to-access places. A $\frac{7}{8}$-in. opening will allow inserting cables up to 300 MCM capacity. A $1\frac{1}{64}$-in. opening is used for cable

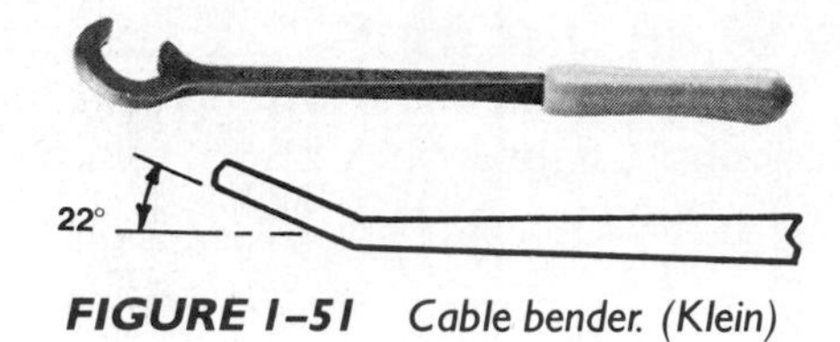

FIGURE 1–51 Cable bender. (Klein)

up to 500 MCM. The handle is plastic coated and contoured for good gripping action. For easier bends, two benders should be used.

Conduit Reamer

The conduit reamer (Fig. 1–52) locks onto a screwdriver and reams and deburrs $\frac{1}{2}$-, $\frac{3}{4}$-, and 1-in. thin-wall conduit ends. The smooth ends protect wire being pulled through the conduit, which permits correct installation of fittings. It reams inside and outside at the same time. Two setscrews hold it tightly on round or square screwdriver shanks. The reamer can be left on the screwdriver for normal use. The steel cutting blade is replaceable to keep it cutting easily.

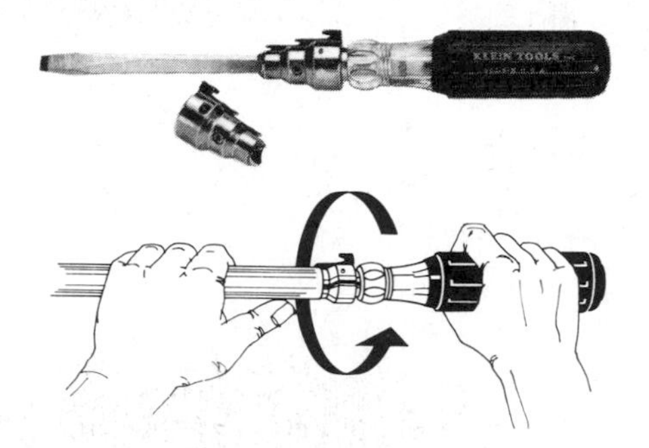

FIGURE 11–52 Conduit reamer fits onto standard screwdriver. (Klein)

QUESTIONS

1. List at least five types of pliers that may be put to good use by an electrician.
2. What is the difference between a ball-peen hammer and a claw hammer?
3. Why do hacksaw blades have different amounts of teeth?
4. Describe an Allen wrench.
5. Why are torque wrenches necessary?
6. What is a nut driver, and how is it used?
7. What is the wattage rating of the best all-purpose soldering iron?
8. What are fuse pullers made of? Why?
9. What is another name for an electrician's knife?
10. What does a megger do?
11. Why do you need a digital logic probe?
12. What does *CMOS* stand for?
13. How is a continuity tester used?
14. What function does an oscilloscope serve?
15. What is a balance analyzer?
16. How is PVC conduit cut?
17. What does *MCM* stand for?
18. Why is a conduit reamer needed?

REVIEW PROBLEMS

One of the most important tools that an electrician can have is a good grasp of *Ohm's law:*

$$E = I \times R \qquad I = \frac{E}{R} \qquad R = \frac{E}{I}$$

1. What drop in voltage exists across a 5000-Ω resistor if it carries 100 mA?
2. What voltage is necessary to cause a current of 5 A to flow through 150 Ω of resistance?
3. How much electromotive force (EMF) is needed to cause 10 mA to flow through a resistance of 1000 Ω?
4. A resistor has a resistance of 60 Ω. What is the voltage drop across it when 100 mA flows through it?
5. A magnetic brake coil requires 6.5 A to load a motor. If the coil has a resistance of 0.5 Ω, what voltage is required?
6. An automobile headlight pulls 7.5 A from a 12.6-V car battery. What is the resistance of the lamp?
7. A solenoid coil pulls a current of 4.5 A when connected to a 12-V supply. What is the resistance of the coil?
8. Find the missing values using Ohm's law:

	Voltage (V)	Current (A)	Resistance (Ω)
a.	120	10	
b.	240	5	
c.	660	10	
d.	440	5	
e.	277	2	
f.	120		10
g.	240		5
h.	660		100
i.	440		75
j.	277		25
k.	12	2	
l.		2	10
m.		1	5
n.		2.5	100
o.		3	2

2

Safety in the Workplace

Objectives

After studying this chapter, you will be able to:

1. List safety measures for working safely on the job.
2. List values associated with the use of GFCIs.
3. Identify conditions that lead to overloads, short circuits, and blown fuses.
4. Identify fuses by their voltage rating and ampere rating.
5. Describe how electrical codes promote safety.
6. Describe the role of UL and CSA in electrical safety.
7. Identify the OSHA safety color code colors.
8. Choose the proper fire extinguisher for electrical fires.
9. Use hammers properly.
10. Wear the correct working clothes.
11. Use motors and generators safely.
12. Select proper equipment for doing a job safely.
13. Explain the need for proper grounding of electrical equipment.

Most fatal electrical shocks happen to people who should know better. Below we provide some electromedical facts that should make you think twice before taking chances. For a safe workplace you have to think before you act. Working safely involves a number of considerations. Before working around motors that require large currents and usually operate on very high voltages, there are a number of things you should be aware of.

ELECTRICAL SHOCK

It is not the voltage but the current that kills. People have been killed by 110 V in the home and with as little as 42 V direct current (dc). The real measure of a shock's intensity lies in the amount of current (in milliamperes) forced through the body, not in the voltage. Any electrical device used on a house wiring circuit can, under certain conditions, transmit a fatal current.

Since you do not know how much current went through the body in an accident, it is necessary to perform artificial respiration to try to get the person breathing again, or if the heart is not beating, CPR.

Note: A heart that is in fibrillation cannot be restricted by closed-chest cardiac massage. A special device called a defibrillator is available in some medical facilities and by ambulance services.

Muscular contractions are so severe with 200 mA and above that the heart is forcibly clamped during the shock. Clamping prevents the heart from going into ventricular fibrillation, making the victim's chances for survival better.

ELECTRICAL SAFETY MEASURES

Working with electricity can be dangerous. However, electricity can be safe if properly respected.

Using Ground-Fault Circuit Interrupters

Some dangerous situations have been minimized by using ground-fault circuit interrupters (GFCIs; see Fig. 2–1). Since 1975 the *National Electrical Code*® (NEC) has required installation of GFCIs in outdoor and bathroom outlets in new construction, but most homes built before then have no GFCI protection.

FIGURE 2–1 *Ground-fault protection system. (Westinghouse)*

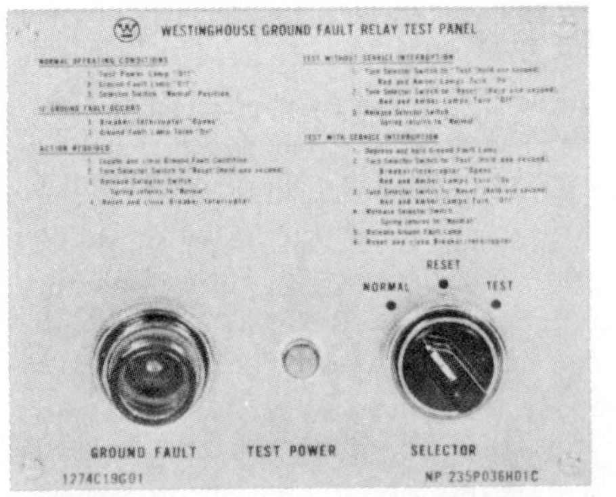

Ground-fault protection is also available for installation in motor control centers and high-voltage starters.

The ground-fault protection shown in Figs. 2–1 and 2–2 that consists of a ground-fault relay and a current sensor. Two types of relays are available from Westinghouse. Type GR operates instantaneously and fulfills the basic need for fast, sensitive ground-fault detection. Type GRT incorporates a time delay and is adjustable from instantaneous to 36 cycles (0.6 second at 60 Hz). Both relays employ reliable solid-state circuits and are self-powered by fault current.

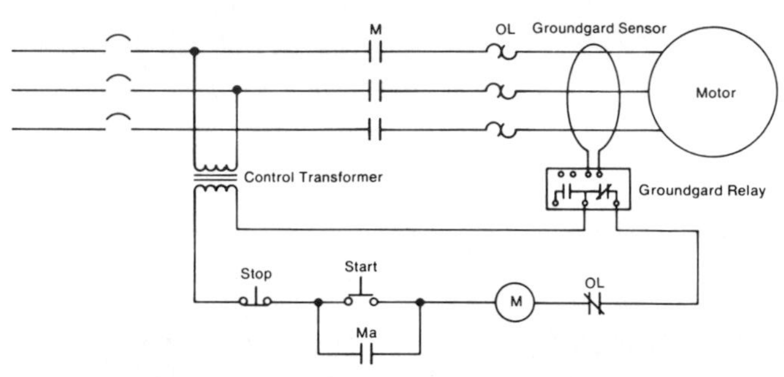

FIGURE 2–2 *Typical Groundgard application. (Westinghouse)*

Current sensors are window current transformer-type devices through which the cable or bus of all phases is run. Various "window" sizes are available for a wide variety of applications. Relay contacts are available on relays both tripped open for application in standard contactor holding coil circuits and tripped closed for mechanically latched contactors.

Safety Devices

Electricity can create conditions almost certain to result in bodily harm, property damage, or both. It is important for those who work with or around electricity to understand the hazards involved when they are working around electrical power tools, maintaining electrical equipment, or installing equipment for electrical operation. Available safety devices are essentially overcurrent devices: specifically, fuses and circuit breakers as well as GFCIs.

Circuit Protection

Electrical distribution systems are often quite complicated. They cannot be absolutely fail-safe. Circuits are subject to destructive overcurrents. Harsh environments, general deterioration, accidental damage or damage from natural causes, excessive expansion, or overloading of the electrical distribution system are factors that contribute to the occurrence of overcurrents. Reliable protective devices prevent or minimize costly damage to transformers, conductors, motors, and the many other components and loads that make up the complete distribution system. Reliable circuit protection is essential to avoid the severe monetary losses that can result from power blackouts and prolonged downtime of facilities. It is the need for reliable protection, safety, and freedom from fire hazards that has made the fuse a widely used protective device.

Overcurrents

An overcurrent is either an overload current or a short-circuit current. The overload current is an excessive current relative to normal operating current but one that is confined to the normal conductive paths provided by the conductors and other components and loads of the distribution system. As the name implies, a short-circuit current is one that flows outside the normal conducting paths.

Overloads

Overloads are most often between one and six times the normal current level. Usually, they are caused by harmless temporary surge currents that occur when motors are started or transformers are energized. Such overload currents or transients are normal occurrences. Since they are of brief duration, any temperature rise is trivial and has no harmful effect on the circuit components. It is important to ensure that protective devices do not react to them.

Continuous overloads can result from defective motors (such as worn motor bearings), overloaded equipment, or too many loads on one circuit. Such sustained overloads are destructive and must be cut off by protective devices before they damage the distribution system or system loads. However, since they are of relatively low magnitude compared to short-circuit currents, removal of the overload current within a few seconds will generally prevent equipment damage. A sustained overload current results in overheating of conductors and other components and will cause deterioration of insulation, which may eventually result in severe damage and short circuits if not interrupted.

Short Circuits

Overload currents usually occur at rather modest levels; the short-circuit or fault current can be many hundreds of times larger than the normal operating current. A high-level fault may be 50,000 A or larger. If not cut off within a matter of a few milliseconds, damage and destruction can become rampant. There can be severe insulation damage, melting of conductors, vaporization of metal, ionization of gases, arcing, and fire. Simultaneous high-level, short-circuit currents can develop huge magnetic field stresses. The magnetic forces between busbars and other conductors can be many hundreds of pounds per lineal foot; even heavy bracing may not be adequate to keep them from being warped or distorted beyond repair.

Fuses

The fuse is a reliable overcurrent protective device. The fundamental element of the basic fuse is a fusible link or links encapsulated in a tube and connected to contact terminals. The link's electrical resistance is so low that it simply acts as a conductor. However, when destructive currents occur, the link very quickly melts and opens the circuit to protect the conductor and other circuit components and loads. Fuse characteristics are stable. Fuses do not require periodic maintenance or testing. Fuses have three unique performance characteristics:

1. They are safe. Fuses have a high interrupting rating, which means that they can withstand very high fault currents without rupturing.
2. Properly applied, fuses prevent blackouts. Only a fuse nearest a fault opens without upstream fuses (feeders or mains) being affected—fuses thus provide selective coordination.
3. Fuses provide optimum component protection by keeping fault currents to a low value. They limit current.

Voltage Rating. Most low-voltage power distribution fuses have 250- or 600-V ratings (others have ratings of 125 V and 300 V). The voltage rating of a fuse must at least equal the circuit voltage. It can be higher but can never be lower. The voltage rating determines the ability of the fuse to suppress the internal arcing that occurs after a fuse link melts and an arc is produced. If a fuse is used with a voltage rating lower than the circuit voltage, arc suppression will be impaired, and under some fault current conditions, the fuse may not safely clear the overcurrent.

Ampere Rating. Fuses have ampere ratings. In selecting the ampacity of a fuse, consideration must be given to the type of load and code requirements. The ampere rating of a fuse should *normally* not exceed the current-carrying capacity of the circuit. There are some specific circumstances where the ampere rating is permitted to be greater than the current-carrying capacity of the fuse. A typical example is a motor circuit. The dual-element fuse generally is permitted to be sized up to 175% and non-time-delay fuses up to 300% of the motor full-load amperes (Fig. 2–3). Generally, the ampere rating of a fuse and switch combi-

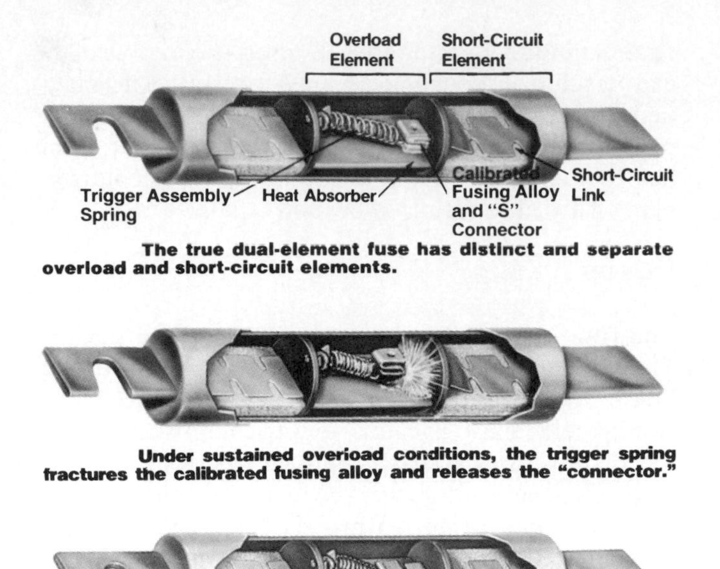

The true dual-element fuse has distinct and separate overload and short-circuit elements.

Under sustained overload conditions, the trigger spring fractures the calibrated fusing alloy and releases the "connector."

The "open" Dual Element fuse after opening under an overload condition.

Like the single element fuse, a short-circuit current causes the restricted portions of the short-circuit elements to melt and arcing to burn back the resulting gaps until the arcs are suppressed by the arc quenching material and increased arc resistance.

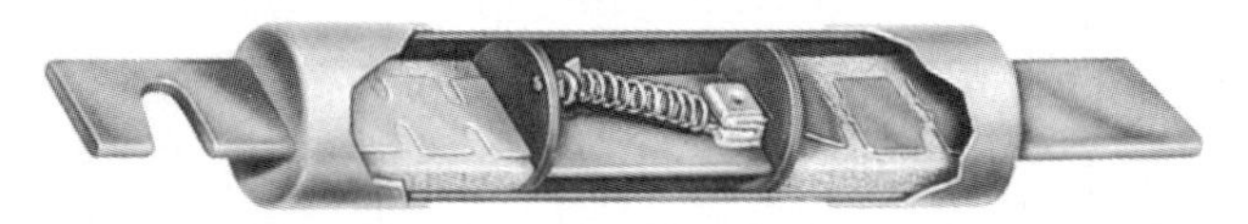

The "open" Dual Element fuse after opening under a short circuit condition.

FIGURE 2–3 *Dual-element fuse operation. (Bussmann Division, Cooper Industries, Inc.)*

nation should be selected at 125% of the load current (this usually corresponds to the circuit capacity, which is also selected at 125% of the load current). There are exceptions, such as when the fuse–switch combination is approved for continuous operation at 100% of its rating.

ELECTRICAL CODES

Thomas Edison's first electric station could transmit electricity only 5000 feet, while modern power pools enable a customer to use electricity produced in a power plant many states away. Early generators had a power-producing capacity of only 100 kW, but generators with the ability to produce millions of kilowatts are now in operation. Since 1900 transmission voltages have increased from 30,000 V to 500,000 V, and lines of up to 765,000 V are now in operation.

In the early 1880s the New York Board of Fire Underwriters were concerned with the new method of electric lighting proposed by Thomas Edison. Although he did not realize the danger of the giant force that he was helping to create, the New York Board recognized that unless proper precautions were followed, the new method of lighting could prove to be as hazardous as the open flame that it was replacing. In 1881, one person was appointed to inspect every electrical installation before power was turned on. This was the beginning of the Electrical Department of the New York Board. It was necessary for the inspector to check not only the installation within the building but to carry the investigation back to the power station, then only one or two blocks distant. At that time, the board investigated the safety of the entire power system and required the power companies to make weekly tests for grounds and open circuits and report to the board the results of their tests. In 1882, the Committee of Surveys drew up a *set of safeguards* for arc and incandescent lighting, which was the forerunner of the present *National Electrical Code®*.

National Electrical Code®

The *National Electrical Code®* is the most widely adopted code in the world. The combined sales of its handbooks are over 1 million copies each time it is published, which is once every three years. The *National Electrical Code®* is a nationally accepted guide for the safe installation of electrical conductors and equipment and is, in fact, the basis for all electrical codes used in the United States. It is also used extensively outside the United States, particularly where American-made equipment is installed. No electrician should be caught without a copy in the toolbox.

The *National Electrical Code® Handbook* is published by the National Fire Protection Association to assist those concerned with electrical safety in understanding the intent of the 1990 edition of the Code. A verbatim reproduction of the 1990 *National Electrical Code®* is included, and added where necessary are comments, diagrams, and illustrations that are intended to clarify further some of the intricate requirements of the Code.

The Code is purely advisory as far as the NFPA (National Fire Protection Association) and ANSI (American National Standards Institute) are concerned but is offered for use in law and for regulatory purposes in the interest of life and property protection. Anyone noticing any errors is asked to notify the NFPA executive office, the chairman, and the secretary of the committee.

Underwriters' Laboratories

Underwriters' Laboratories, Inc. (UL) was founded in 1894 by William Henry Merrill. He came to Chicago to test the installation of Thomas Edison's new incandescent electric light at the Columbian Exposition. He later started UL as a laboratory where insurance companies could test products for electrical and fire hazards. It continued as a testing laboratory for insurance underwriters until 1917. It then became an independent, self-supporting, safety-testing laboratory. The National Board of Fire Underwriters (now the American Insurance Association) continued as sponsors of UL until 1968. At that time, sponsorship and membership were broadened to include representatives of consumer interests, governmental bodies or agencies, education, public safety bodies, public utilities, and the insurance industry, in addition to safety and standardization experts. UL has expanded its testing services to more than 13,000 manufacturers throughout the world. Over 1 billion UL labels are used each year on products listed by Underwriters' laboratories.

UL is chartered as a not-for-profit organization without capital stock, under the laws of the state of Delaware, to establish, maintain, and operate laboratories for the examination and testing of devices, systems, and materials. Its stated objectives are: "By scientific investigation, study, experiments and tests, to determine the relation of various materials, devices, products, equipment, constructions, methods and systems to hazards appurtenant thereto or to the use thereof, affecting life and property, and to ascertain, define and publish standards, classifications and specifications for materials, devices, equipment, construction, methods and systems affecting such hazards, and other information tending to reduce and prevent loss of life and property from such hazards."

The corporate headquarters, together with one of the testing laboratories, is located on West Street in Northbrook, Illinois; Melville, New York; Santa Clara, California; and Tampa, Florida.

Underwriters' Laboratories has a total staff of over 2000 employees. More than 700 persons are engaged in engineering work, and of this number, approximately 425 are graduate engineers. Supplementing the engineering staff are more than 500 factory inspectors. The engineering functions are divided among these six departments:

- Burglary protection and signaling
- Casualty and chemical hazards
- Electricity
- Fire protection
- Heating, air conditioning, and refrigeration
- Marine equipment

The electrical department is the largest of the six engineering departments. Safety evaluations are made on hundreds of different types of appliances for use in homes, commercial buildings, schools, and factories. The scope of the work in this department includes electrical construction materials that are used in buildings to distribute electrical power from the meter location to the electrical load.

Underwriters' Laboratories publishes annual lists of manufacturers whose products have met UL safety requirements. These lists are kept up to date by quarterly supplements. Eleven lists are published each year. UL presently publishes more than 300 *Standards for Safety* for materials, devices, constructions, and methods. Copies of these are available to interested persons, and a free catalog is available.

FIGURE 2–4 *UL stickers and label.*

For your own safety, the products you are using to wire a house, building, or installation of any kind should be marked "UL" (Fig. 2-4).

Canadian Standards Association

The Canadian organization that parallels UL is the *Canadian Standards Association* (CSA). However, CSA has more authority to remove from the market products that do not meet standards; the UL program is strictly voluntary. If an electrical product (or in some cases, another type of product) used in Canada is connected in any way with the consumption of power from the electrical sources owned by the provinces, that product must have CSA approval. This is a measure in the interest of public safety.

Products certified by CSA are eligible to bear the CSA certification mark. Misuse of the mark may result in suspension or cancellation of certification. CSA

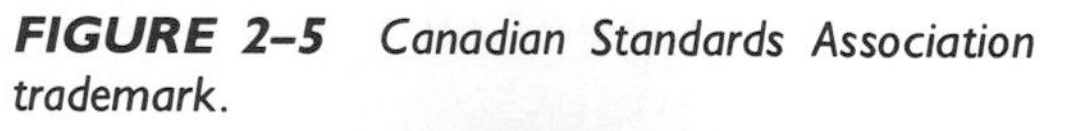

FIGURE 2–5 *Canadian Standards Association trademark.*

may resort to legal action to protect its registered trademark in the event of abuse. Figure 2–5 shows the CSA mark. In addition, CSA information tags and other markings are made available for certified products and their containers, to supplement the CSA mark.

Standards in Other Countries

Other countries also have standards and testing laboratories. Figure 2–6 shows some of the marking used by other countries.

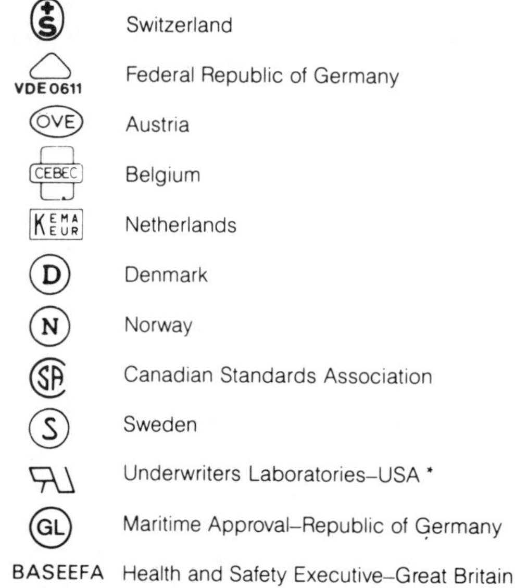

Products recognized under the component program are identified by the recognized marking . Recognized components are suitable for factory installation on other equipment where their use and limitations are determined by UL.

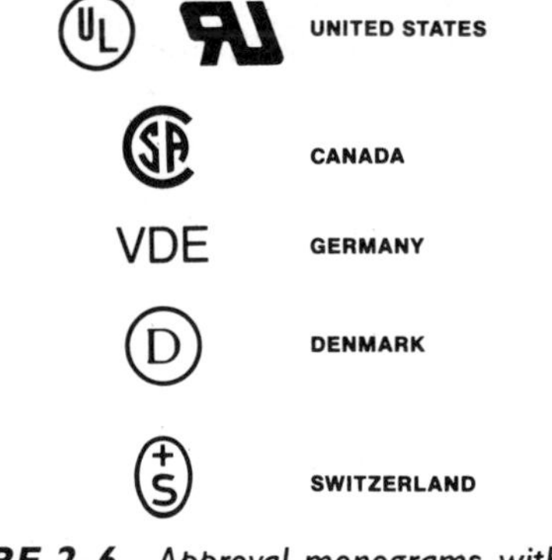

FIGURE 2–6 *Approval monograms with country of origin.*

OSHA

The Occupational Safety and Health Act (OSHA) of 1970 sets uniform national requirements for safety in the workplace—anywhere that people are employed. Originally, OSHA adopted the 1971 NEC as rules for electrical safety. The involved process for modifying a federal law such as OSHA made it impossible for the act to adopt each new NEC revision, as the Code is amended every three years. To avoid this problem, the OSHA administration in 1981 adopted its own code, a condensed version of the NEC containing only those provisions considered related to occupational safety. OSHA was amended to adopt this code, based on the NFPA standard 70E, Part 1, which is now federal law.

Equipment Standards

Many equipment standards have been established by the National Electrical Manufacturers Association (NEMA) and the American National Standards Institute (ANSI). Underwriters' Laboratory (UL) has standards that equipment must meet before UL will list or label it. Most jurisdictions and OSHA require that where equipment listed as safe by a recognized laboratory is available, unlisted equipment may not be used. UL is by far the most widely accepted national laboratory, although Factory Mutual Insurance Company lists some equipment, and a number of other testing laboratories have been recognized and accepted. The Institute of Electrical and Electronics Engineers (IEEE) publishes a number of books (the "color book" series) on recommended practices for the design of industrial buildings, commercial buildings, emergency power systems, grounding, and the like. Some of these IEEE standards have been adopted as ANSI standards. They are excellent guides, although they are not in any way mandatory.

OSHA Color Code

In order to establish a safe environment for everyone, OSHA has standardized some colors for specific applications. Red, yellow, orange, purple, and green have been chosen to indicate various locations, machines, and devices.

Red	Fire protection equipment and apparatus. Portable containers of flammable liquids. Emergency stop buttons and switches.
Yellow	Caution, and for marking physical hazards. Waste containers for explosive or combustible materials. Caution against starting, using, or moving equipment under repair. Identification of starting point or power source of machinery.
Orange	Dangerous parts of machines. Safety starter buttons. The exposed parts (edges) of pulleys, gears, rollers, cutting devices, and power jaws.
Purple	Radiation hazards.
Green	Safety. Location of first-aid equipment (other than firefighting equipment).

FIRE EXTINGUISHERS

Fire extinguishers are limited in their application. They are used to control small fires that are identified properly by the person selecting the extinguishers. Extinguishers are made for various uses. Needless to say, the water type is *not* useful for electrical fires. Electrical fires are classified as class C. Class C fires are described as those dealing with energized electrical equipment, where the electrical nonconductivity of the extinguishing medium is of importance. Electrical fires call for carbon dioxide, dry chemicals, multipurpose dry chemical, and Halon 1211 types of extinguishers (Fig. 2–7).

SAFE WORKING PRACTICES

Safety equipment is available for use by the electrician or anyone working with electric motors. In most instances a person working in a commercial or industrial location will need steel-toed shoes, a hard hat, goggles, or a face shield. Fuse pullers are insulated to protect the user, but must be handled properly. Using pliers and cutting equipment can be harmful if certain commonsense rules are not followed.

Tool Safety

Some precautions to be observed when using pliers:

1. Never cut any wire or metal unless your eyes are protected. Safety goggles or other protective equipment are an absolute necessity. It is easy to forget to wear them. It is a big bother to put them on for "just one cut." You have heard all the reasons and excuses. But none of them make any sense. They are all part of the lazy person's way, not the professional's way—the safe way.
2. Never cut any wire or metal unless your fellow workers' eyes are also protected. Although you may not have heard that precaution before, you'll see that it makes sense. The wire that does not get you may get someone else. So think about others as well as yourself.
3. Never depend on plastic-dipped handles to insulate you from electricity. Plastic-dipped handles are for comfort and a firmer grip. They are not intended for protection against electric shock.
4. Always wear protective goggles!

Using Hammers Properly

There are some precautions that should be observed when using hammers on the job. Hammers are perhaps the most abused—and misused—of all hand tools. Improper use of hammers can cause injury. Use of damaged hammers can cause injury, as can use of hammers for jobs other than those for which they were specifically designed.

There are some basic rules for the proper use of hammers:

1. Strike square blows only. Avoid glancing blows.
2. Make sure that the striking face of the hammer is at least $\frac{3}{8}$ in. larger than tools to be struck (chisels, punches, wedges, etc.). The face should be slightly crowned with the edges beveled.
3. Never use a hammer to strike another hammer.
4. Always use a hammer of the right size and weight for the job.
5. Never strike with the side or "cheek" of a hammer.
6. Never use a hammer with a loose or damaged handle. Replace the handle.
7. Discard any hammer having chips, cracks, dents, mushrooming, or excessive wear. Replace the hammer. Do not try to repair it.
8. Always wear safety goggles to protect your eyes (Fig. 2–8).

The professional takes both the job and tools seriously. The right tool for a job means saved time and professional results. Using tools safely also means saving time and getting the job done properly. There is only one way to do a job right and that is the *safe* way.

Working Clothes

Clothes worn on the job should be selected to protect you while you work. Of course, everyone will not work under the same conditions; therefore, some general safety tips will be mentioned here. Others may present themselves as you work in a specific area.

1. Wear head protection—a hard hat if the job requires it. Long hair should be concealed and not allowed to move freely. Short hair is preferred for personal safety.
2. Wear goggles or tempered glasses or both.
3. Do not wear a tie.
4. Wear long-sleeved shirts with the cuffs tightly buttoned near the wrists.

Extinguisher Classifications †	A		AB	BC				ABC			D
Extinguishing Agent	**Water Types (includes antifreeze)**		**AFFF Foam and FFFP**	**Carbon Dioxide**	**Dry Chemical Types** Purple K Super K Monnex Potassium Bicarb. Urea based		**Halogenated Types** 1211 1301 1211/1301	**Multipurpose Dry Chemical**		**Halogenated Types** 1211 1211/1301	**Dry Powder**
Discharge Method	Stored Pressure	Pump Tank	Stored Pressure	Self Expelling	Stored Pressure	Cartridge Operated	Stored Pressure	Stored Pressure	Cartridge Operated	Stored Pressure	Cartridge Operated
Sizes Available	2 ½ Gal.	2 ½-5 Gal.	2 ½ Gal. (33 Gal.)	5-20 lb. (50-100 lb.)	2 ½-30 lb. (50-350 lb.)	4-30 lb. (125-350 lb.)	1-5 lb.	2 ½-20 lb. (50-350 lb.)	5-30 lb. (125-350 lb.)	5 ½-22 lb. (50-150 lb.)	30 lb. (150-350 lb.)
Horizontal Range (Approx.)	30-40 ft.	30-40 ft.	10-25 ft. (30 ft.)	3-8 ft. (3-10 ft.)	10-15 ft. (15-45 ft.)	10-20 ft. (15-45 ft.)	10-16 ft.	10-15 ft. (15-45 ft.)	10-20 ft. (15-45 ft.)	9-16 ft. (20-35 ft.)	5 ft. (15 ft.)
Discharge Time (Approx.)	1 Min.	1-3 Min.	50-65 Sec. (1 Min.)	8-15 Sec. (10-30 Sec.)	8-25 Sec. (25-60 Sec.)	8-25 Sec. (25-60 Sec.)		8-25 Sec. (20-60 Sec.)	8-25 Sec. (25-60 Sec.)	10-18 Sec. (30-45 Sec.)	20 Sec. (150 lb., 70 Sec., 350 lb., 1 ¾ Min.)
Operating Precautions and Agent Limitations	Conductor of electricity. Needs protection from freezing. (except antifreeze). Use on flammable liquids and grease will spread fire.		Conductor of electricity. Needs protection from freezing. Not effective on water-soluble flammable liquids such as alcohol, unless otherwise stated on nameplate. AFFF not effective on pressurized flammable liquid/gas fires.	Smothering occurs in high concentrations. Avoid contact with discharge horn. Limited effectiveness under windy conditions. Severely reduced effectiveness at sub-zero (F) temperatures.	Extensive cleanup, particularly on delicate electronic equipment. Obscures visibility in confined spaces.		Avoid high concentrations and unnecessary use.	Extensive cleanup. Damages electronic equipment. Obscures visibility in confined spaces. Limited penetrating ability on deep-seated Class A fires.		Avoid high concentrations and unnecessary use.	Not listed.
	NOTE: Protection required below 40° F and above 120° F.							NOTE: Only dry chemical types are effective on pressurized flammable gases and liquids; for deep fat fryers, multipurpose ABC dry chemicals are not acceptable.			
NOTE: These photos are not proportional in relation to one another.											

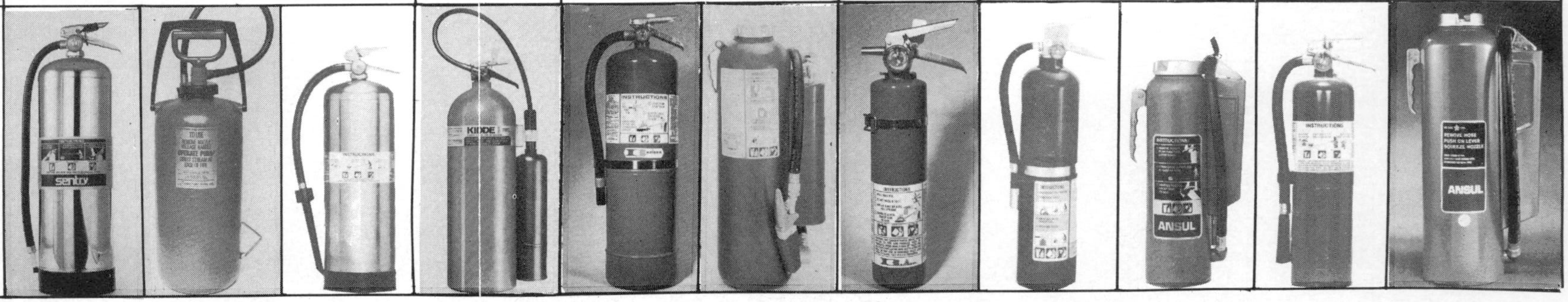

FIGURE 2–7 *Fire extinguishers. (National Association of Fire Equipment Distributors)*

How to Select the Proper Fire Extinguisher

Different types of extinguishers are used to put out different kinds of fires. For example, a Halon 1211 extinguisher is one of the types recommended for putting out an electrical fire (halon does not conduct electricity). A water type extinguisher should never be used on such a fire (water does conduct electricity).

To help you select the proper extinguisher for the type of hazard you are most likely to encounter in the home, car, or workplace, the charts here classify both fires and extinguishers into types. Types of fires are classified as: A — ordinary combustible materials, such as wood, cloth, paper, and many plastics; B — flammable liquids, gases, and grease; and C — energized electrical equipment (such as computers). There is a class D — combustible metals, but this hazard is rare and found usually in industrial situations.

Unlike fires, extinguishers have more classifications than simply A, B, C, or D. They are A, AB, BC, ABC, and D; this is because some extinguishers can put out more than just one type of fire. For example, a BC dry chemical fire extinguisher can put out a grease fire (Class B), as well as an electrical fire (Class C).

You might wonder, if that is so, why doesn't everyone just use ABC extinguishers, which put out virtually all types of fires? The answer is, although an ABC extinguisher is capable of putting out Class A, B, and C fires, it is not always as effective as another extinguisher designed solely for putting out only certain kinds of fires.

For example, there are *multipurpose dry chemical* extinguishers, classified as ABC, and simply *dry chemical* extinguishers classified as BC. An ABC multipurpose dry chemical extinguisher is prohibited for use on deep fat fryers in restaurants. Only BC dry chemical types are effective on deep fat fryers. The BC dry chemical reacts with the hot grease to create a kind of soapy foam that suffocates the fire (this process if called saponification). The multipurpose dry chemical of ABC fire extinguishers breaks down this foam once it is formed and allows the fire to re-ignite.

For general home use, however, the ABC extinguishers are fine for all applications. We will, however, show you what other kinds of extinguishers may be used in different areas of the home, car, or office.

All extinguishers are labeled with picture symbols.* These symbols depict the kinds of fires on which the extinguisher is effective. A slash through a picture means the extinguisher is not to be used on that class fire. If it is used, it may actually spread the fire rather than put it out.

* Picture symbols for Class D fires are not practical.

Choosing the Best Size Extinguisher

We have learned that extinguishers are classified by the type of fire they put out. They are also *rated* with a number that tells you the *size* of fire they can put out.

For example, a five pound *multipurpose* dry chemical extinguisher (ABC) has a rating of 2-A:10-B:C. That means the extinguisher can put out approximately twice as much fire as a 1-A extinquisher. A 1-A rated extinguisher is required to put out a blazing wood crib consisting of 50 pieces of 20 inch long 2′ x 2′s and an 8′ x 8′ wood panel. It can put out five times the size of a Class B fire that consists of 3.25 gallons of liquid fuel burning in a 2.5 square foot pan.

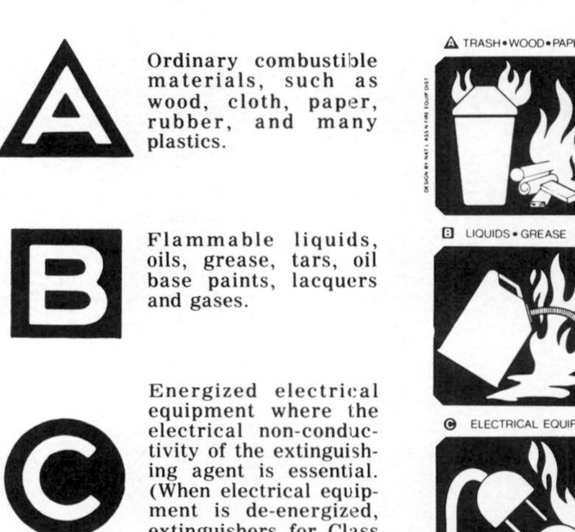

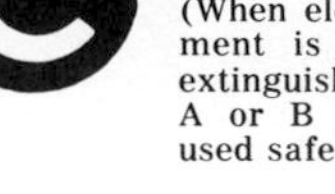

Classification of Fires		Picture Symbols	Recommended Extinguishers
A	Ordinary combustible materials, such as wood, cloth, paper, rubber, and many plastics.	TRASH • WOOD • PAPER	WATER AFFF ABC Dry Chemical Halon 1211 LIQUIDS: AFFF Regular
B	Flammable liquids, oils, grease, tars, oil base paints, lacquers and gases.	LIQUIDS • GREASE	Dry Chemical ABC Dry Chemical Purple K Dry Chemical Halon 1211 CO_2 GASES: Regular & Purple K Dry Chemical
C	Energized electrical equipment where the electrical non-conductivity of the extinguishing agent is essential. (When electrical equipment is de-energized, extinguishers for Class A or B fires may be used safely.)	ELECTRICAL EQUIP	Regular Dry Chemical ABC Dry Chemical Purple K Dry Chemical Halon 1211 CO_2

Recommendations

- Kitchen — 2½-5 lb. dry chemical, UL rating of 5B:C, and/or halon type
- *Car/garage* — 1-5 lb. Halon 1211 or multipurpose dry chemical, UL rating of 1A:10B:C
- *Personal computer* — 1-2½ lb. Halon 1211, UL rating of 2B:C

The National Association of Fire Equipment Distributors

FIGURE 2–7 *continued*

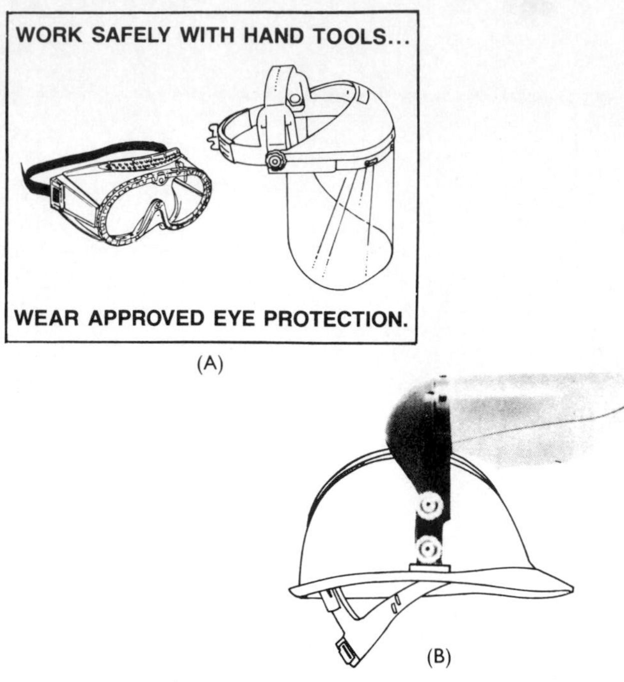

FIGURE 2–8 (A) General-purpose cover-type protective goggles are made of soft vinyl plastic. This type protects the eyes from all angles. (B) For more protection you may want to add a hard hat and face shield as a combination safety step while working in hazardous locations. (Klein)

5. Wear your trousers tucked into your safety shoes or tightly fastened near the ankles.
6. Do not wear rings, earrings, necklaces, or bracelets.
7. Wear safety shoes. These should have an arch support and steel-tipped or safety toes. Soles should be such as not to slip on oily or wet surfaces. Wear rubber boots in wet or damp locations.

Disconnects

Electrical shock can be prevented if the circuit is not "live" when you are working on it or any equipment connected to the circuit. Remove the power. Remove the fuse or disconnect. If it has a lock-out feature, make sure that it is used—that the lock is locked and you have the key.

Report to your supervisor any unsafe conditions observed. Keep in mind not only your own safety, but that of others.

Using Motors and Generators Safely

The use of electric motors and generators, like that of all other utilization of concentrated power, is potentially hazardous. The degree of hazard can be greatly reduced by proper design, selection, installation, and use, but hazards cannot be completely eliminated. The reduction of hazards is the joint responsibility of the user, the manufacturer of the driven or driving equipment, and the manufacturer of the motor or generator.*

Most manufacturers make their equipment to meet the NEMA standards for safe operation. However, even well-built equipment can be installed or operated in a hazardous manner. It is important that safety considerations be observed by the user. The user must properly select, install, and use the apparatus with respect to load and environment.

Selection of Apparatus. Before selecting a piece of equipment you should study the recommendations of the manufacturer of the apparatus, generally available in catalogs. Manufacturers usually maintain engineering departments to assist you in making sure that you have the right equipment for the job. Local sales representatives or factory representatives can usually arrange for more details, if needed.

Installation of Apparatus. The equipment manufacturer or the person installing the apparatus must take care in the installation. The *National Electrical Code®* (NEC), sound local electrical and safety codes, and when applicable, the Occupational Safety and Health Act (OSHA) should be followed when installing the apparatus to reduce hazards to persons and property.

Use of Apparatus. The chance of electric shocks, fires, or explosions can be reduced by giving proper consideration to the use of grounding, thermal and overcurrent protection, type of enclosure, and good maintenance procedures.

Safety Considerations

1. All motors, gearmotors, and controls must be grounded adequately and securely. Failure to ground properly may cause serious injury to personnel.

2. Do not insert objects into ventilation openings, motors, or other apparatus.

3. Sparking of starting switches in ac motors so equipped, and of brushes in commutator-type motors, can be expected during normal operation. In addition, enclosures may eject flame in the event of an insulation or component failure. Therefore, avoid, protect from, or prevent the presence of flammable or combustible materials in the environment of motors, gearmotors, and controls.

4. When dealing with hazardous locations (flammable or explosive gas, vapor, dust) an explosion-proof or dust-ignition-proof product is the recommended approach. Exceptions are allowed by the

*See Standards Publication No. ANSI C51.1/NEMA MG2 *Safety Standard for Construction and Guide for Selection, Installation and Use of Electric Motors and Generators,* available from National Electrical Manufacturers Association, 2101 L Street N.W., Washington, DC 20037.

National Electrical Code®. The NEC and NEMA safety standard should be studied thoroughly before exercising this option.

5. Ventilated products are suitable for clean, dry locations where cooling air is not restricted. Enclosed products are suitable for dirty, damp locations. For outdoor use, washdown, and so on, enclosed products must be protected by a cover while allowing adequate air flow.

6. Moisture will increase the electrical shock hazard of electrical insulation. Therefore, consideration should be given to the avoidance of (or protection from) liquids in the area of motors and controls. Use of totally enclosed motors/gearmotors will reduce the hazard if all openings are sealed.

7. Chassis controls should be properly guarded or enclosed to prevent possible human contact with "live" circuitry.

8. Proper consideration should be given to rotating members. Before starting, make sure that keys, pulleys, and so on, are securely fastened. Proper guards should be provided for rotating members to prevent hazards to personnel.

9. Before servicing or working on equipment, disconnect the power source (this applies especially to equipment using automatic restart devices instead of manual restart devices (Fig. 2–9).

FIGURE 2–9 *Specially designed switch and panel boxes have double interlocking hasp of tempered steel with interlocking tabs for extra security. For safety reasons the box should be locked whenever someone is working on the circuit or equipment. In this way another person is unable to apply power while someone else is working on equipment. (Klein)*

10. In selecting, installing, and using equipment, the hazard of mechanical failure should be considered in addition to electrical hazards. Mechanical considerations include proper mounting and alignment of apparatus and safe loads on shafting and gearing. Do not depend on gear friction to hold loads.

11. Ambient temperatures around apparatus should not exceed 40°C (104°F) unless the nameplates specifically permit higher values.

12. Power supply to all equipment must be that shown on the nameplate.

Grounding

In industrial installations, the effect of a shutdown caused by a single ground fault could be disastrous. An interrupted process could cause the loss of all materials involved, often ruin the process equipment itself, and sometimes create extremely dangerous situations for operating personnel.

Grounding encompasses several different but interrelated aspects of electrical distribution system design and construction, all of which are essential to the safety and proper operation of the system and equipment supplied by it. Among these are equipment grounding, system grounding, static and lightning protection, and connection to earth as a reference (zero) potential.

Equipment Grounding. Equipment grounding is essential to the safety of personnel. Its function is to ensure that all exposed non-current-carrying metallic parts of all structures and equipment in or near the electrical distribution system are at the same potential, and that this is the zero reference potential of the earth. Grounding is required by both the *National Electrical Code*® (Article 250) and the National Electrical Safety Code. Equipment grounding also provides a return path for ground-fault currents, permitting protective devices to operate. Accidental contact of any energized conductor of the system with an improperly non-current-carrying metallic part of the system (such as a motor frame or panelboard enclosure) would raise the potential of the metal object above ground potential. Any person coming in contact with such an object while grounded could be seriously injured or killed. In addition, current flow from the accidental grounding of an energized part of the system could generate sufficient heat (often arcing) to start a fire.

The equipment grounding system must be bonded to the grounding electrode at the source or service. However, it may also be connected to ground at many other points. This will not cause problems with safe operation of the electrical distribution system.

Where computers, data processing, or microprocessor-based industrial process control systems are installed, the equipment grounding system must be designed to minimize interference with their proper operation. Often, isolated grounding of this equipment, or a completely isolated electrical supply system, is required to protect microprocessors from power system "noise" that does not in any way affect motors or other electrical equipment.

System Grounding. System grounding connects the electrical supply—from the utility, transformer secondary windings, or a generator—to ground. A system can be solidly grounded (no intentional impedance to ground), impedance grounded (through a resistance or reactance), or ungrounded (with no intentional connection to ground).

Solidly grounded three-phase systems are shown in Fig. 2–10. They are usually wye-connected, with the neutral point grounded. Less common is the "red-leg" or high-leg delta, a 240-V system supplied by some utilities with one winding center-tapped to proved 120 V to ground for lighting. This 240-V, three-phase, four-wire system is used where the 120-V lighting load is small compared to the 240-V power load, because the installation is low in cost to the utility.

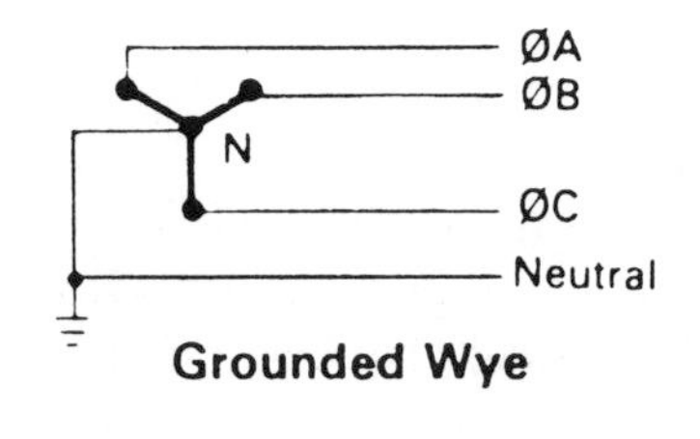

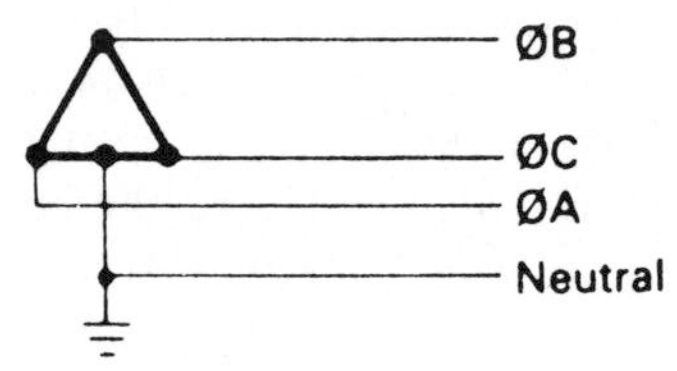

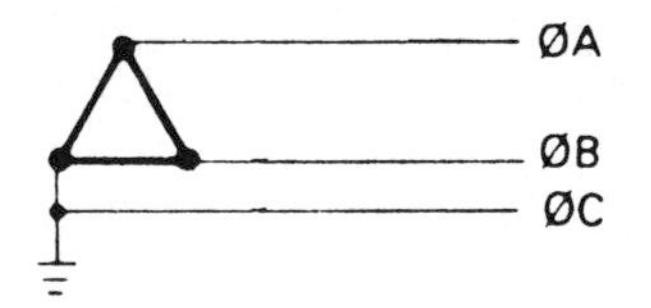

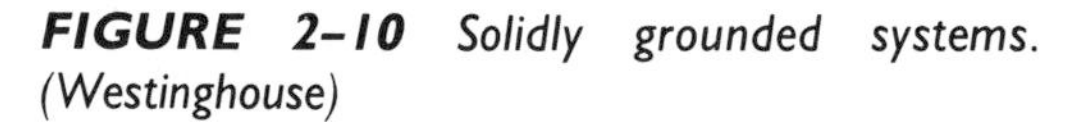

FIGURE 2–10 *Solidly grounded systems. (Westinghouse)*

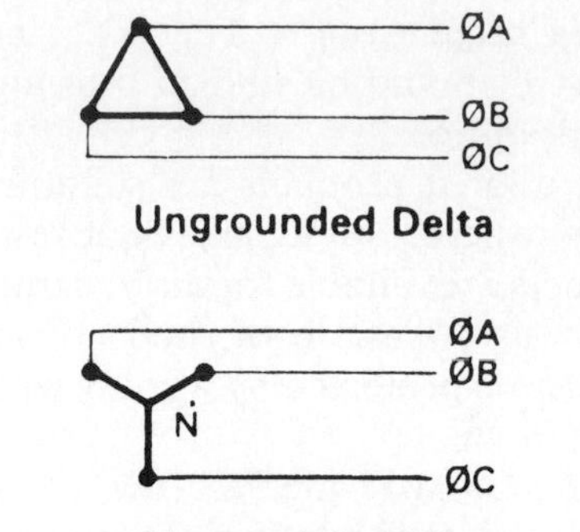

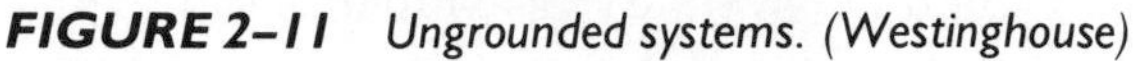

FIGURE 2–11 *Ungrounded systems. (Westinghouse)*

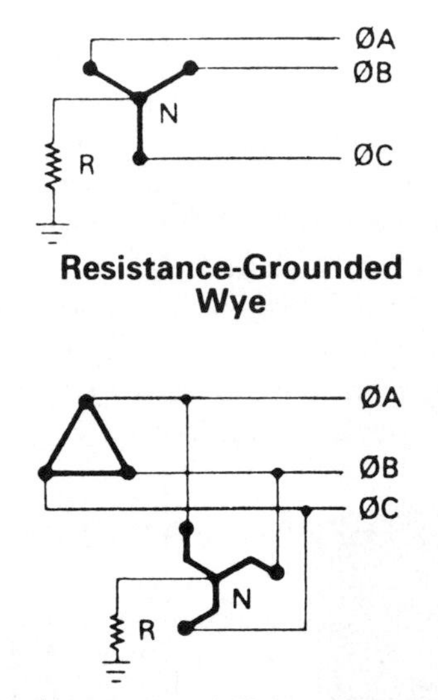

FIGURE 2–12 *Resistance-grounded systems. (Westinghouse)*

A corner-grounded three-phase delta system is sometimes used, with one phase grounded to stabilize all voltages to ground. Better solutions are available for new installations. Ungrounded systems (Fig. 2–11) can be either wye or delta, although the ungrounded delta system is far more common.

Resistance-grounded systems (Fig. 2–12) are simplest with a wye connection, grounding the neutral point directly through the resistor. Delta systems can be grounded by means of a zigzag or other grounding transformer. This derives a neutral point, which can be either solidly or impedance grounded. If the grounding transformer has sufficient capacity, the neutral created can be solidly grounded and used as part of a three-phase, four-wire system. Most transformer-supplied systems are either solidly grounded or resistance grounded. Generator neutrals are often grounded through a reactor, to limit ground-fault (zero sequence) currents to values the generator can withstand. Medium-voltage resistance-grounded neutrals are often grounded through a grounding transformer, although many 2400-V and some 4160-V systems are grounded directly, like low-voltage systems.

Which System Is Best? There is no best distribution system for all applications. That is why you may find any one of these systems in any plant. In choosing among solid-grounded, resistance-grounded, or ungrounded power distribution, the characteristics of the system must be weighed against the requirements of power loads, lighting loads, continuity of service, safety, and cost.

QUESTIONS

1. If it is not the voltage that kills, what does?
2. How much current does it take to kill?
3. What is ventricular fibrillation?
4. What does *GFCI* stand for?
5. What does *NEC* stand for?
6. Give two examples of two overcurrent devices.
7. How is an overload described?
8. What is a fusible link?
9. How is a dual-element fuse sized?
10. What is the highest transmission voltage used today?
11. What is the *National Electrical Code®*?
12. What is the purpose of the Underwriters' Laboratories?
13. What does the Canadian Standards Association do?
14. What does *OSHA* stand for?
15. What do *IEEE* and *ANSI* stand for?
16. What does the OSHA color orange denote?
17. Why isn't a water-type fire extinguisher used on electrical fires?
18. How can the chance of electrical shocks be reduced?
19. What is grounding? What is a system ground?
20. Where do you encounter resistance-grounded systems?

REVIEW PROBLEMS

Electrical energy can be converted to heat energy by the use of resistances. Heat and power are closely related in the electrical field, especially with large resistors. To help you handle and locate such heat-producing devices safely, a review of the power formulas will come in handy.

$$P = E \times I \qquad P = \frac{E^2}{R} \qquad P = I^2 R$$

1. How much power is required to operate a solenoid that draws 4.5 A when connected to a 12-V supply?
2. How much power is drawn when a spotlight draws 2.5 A at 120 V?
3. What is the current needed when a 120-V, 150-W bulb is connected to a power source?
4. A 24-V transformer delivers 1500 W of power. What is the current?
5. If a resistor of 500 Ω is connected across a source of 277 V, what is the power consumed?
6. How much current will a 240-V $\frac{1}{2}$-hp electric motor pull?
7. If a 5-hp motor is connected to a 240-V line, what will the current requirements be?
8. What current do you need for a $\frac{1}{2}$-hp motor on a 120-V line?

3

Symbols

Objectives

After studying this chapter, you will be able to:

1. Draw the symbols used in electrical wiring diagrams and ladder diagrams.
2. Use the resistor color code to select the proper-size resistors.
3. Identify relay contact symbols.
4. Identify electronic schematic symbols.

Electrical wiring diagrams and ladder diagrams, which are the main means of communicating between engineer and technician in the motor controls field, are made up of nothing more than lines connecting certain symbols.

Symbols are used as a sort of shorthand that allows many electrical functions to be identified and sketched out in a small area. People familiar with the symbols used in the electrical field are able to identify certain components and figure out how they are wired and how they operate.

Symbols are the heart of any wiring diagram. Because they are so important a part of the job, it is best to become acquainted with them at an early stage in the learning of a trade.

ELECTRICAL SYMBOLS

Electrical symbols are used on blueprints and in wiring diagrams for buildings and plants. Figure 3–1 shows the standardized symbols utilized by drafters who draw wiring diagrams for the placement of electrical equipment and control devices.

Common Switch (Button) Symbols

Motor controls technicians work with switches (buttons) and their associated components to make a motor do what it is supposed to do when it is supposed to do it. Some of the symbols used for switching are shown in Figure 3–2. Note the pressure and vacuum switches, liquid-level switch, temperature-actuated switch, and flow switch. These will be utilized by the motor control technician in their many applications. Switches may also be referred to as *buttons*.

Standard Wiring Diagram Symbols

Some basic symbols used in wiring diagrams that you will be studying are shown in Fig. 3–3. All of these will be encountered in a normal wiring diagram for the control of electric motors in an industrial or commercial installation.

1. Lighting Outlets

Ceiling *Wall*

1.1 Surface or Pendant Incandescent, Mercury-Vapor, or Similar Lamp Fixture

1.2 Recessed Incandescent, Mercury-Vapor, or Similar Lamp Fixture

1.3 Surface or Pendant Individual Fluorescent Fixture

1.4 Recessed Individual Fluorescent Fixture

1.5 Surface or Pendant Continuous-Row Fluorescent Fixture

1.6 Recessed Continuous-Row Fluorescent Fixture

1.7 Bare-Lamp Fluorescent Strip

1.8 Surface or Pendant Exit Light

1.9 Recessed Exit Light

1.10 Blanked Outlet

1.11 Junction Box

1.12 Outlet Controlled by Low-Voltage Switching When Relay Is Installed in Outlet Box

2. Receptacle Outlets

Grounded *Ungrounded*

2.1 Single Receptacle Outlet

2.2 Duplex Receptacle Outlet

2.3 Triplex Receptacle Outlet

2.4 Quadruplex Receptacle Outlet

2.5 Duplex Receptacle Outlet—Split Wired

2.6 Triplex Receptacle Outlet—Split Wired

2.7 Single Special-Purpose Receptacle Outlet

2.8 Duplex Special-Purpose Receptacle Outlet

2.9 Range Outlet (typical)

2.10 Special-Purpose Connection or Provision for Connection

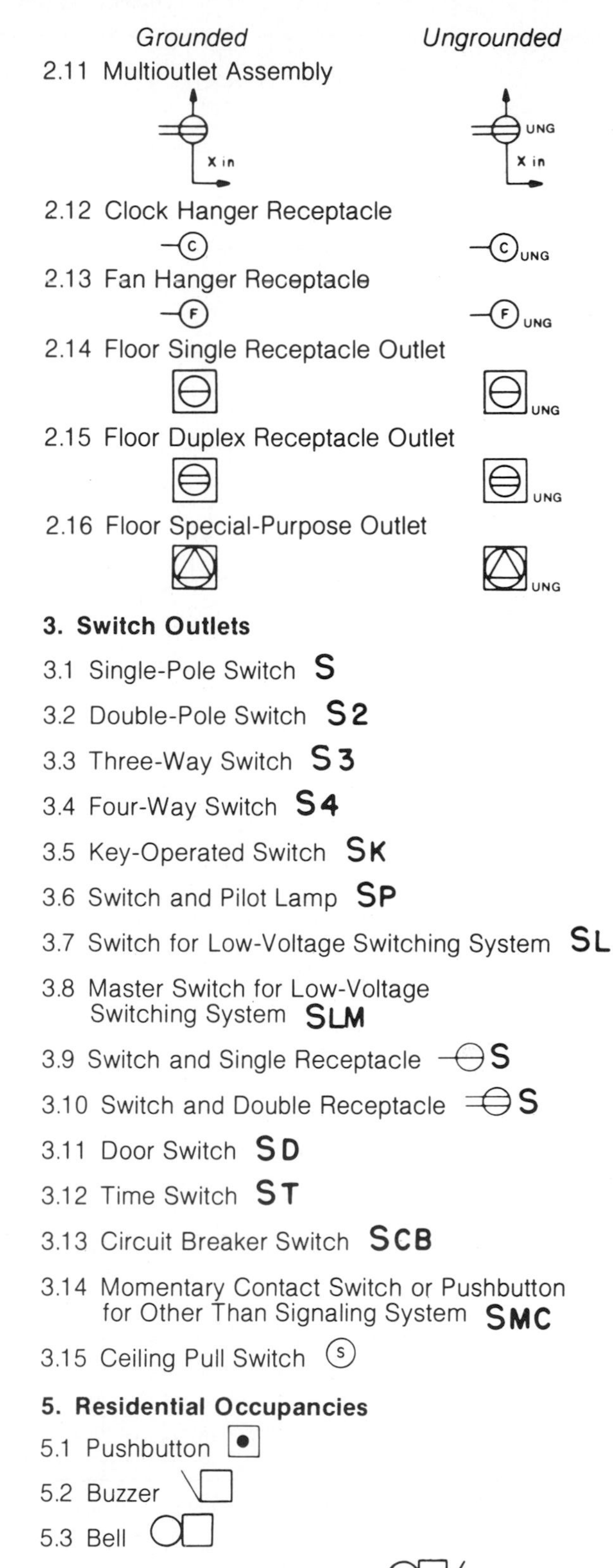

FIGURE 3–1 *Electrical symbols for architectural plans.*

COMMON SWITCH SYMBOLS

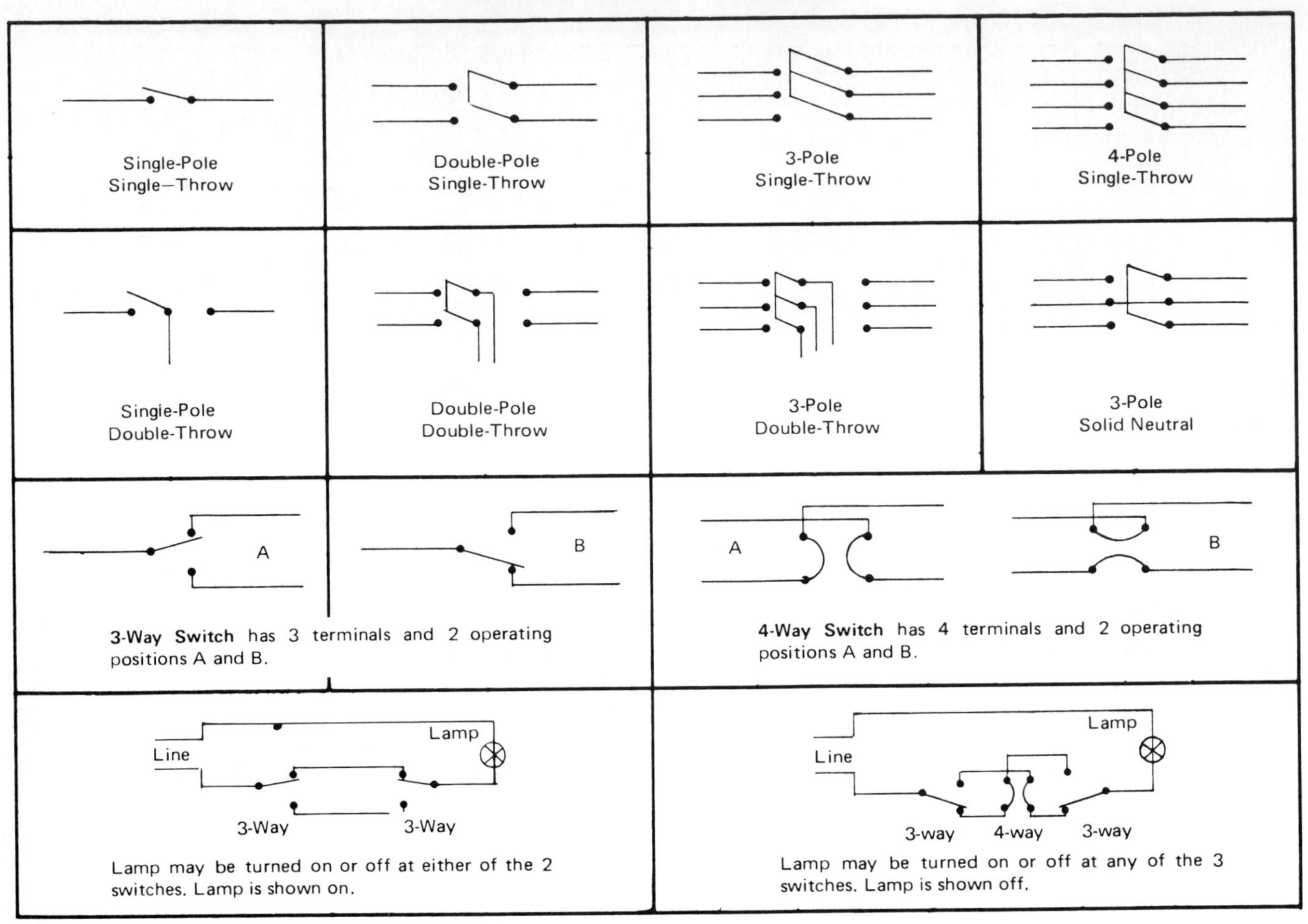

OTHER SWITCHES

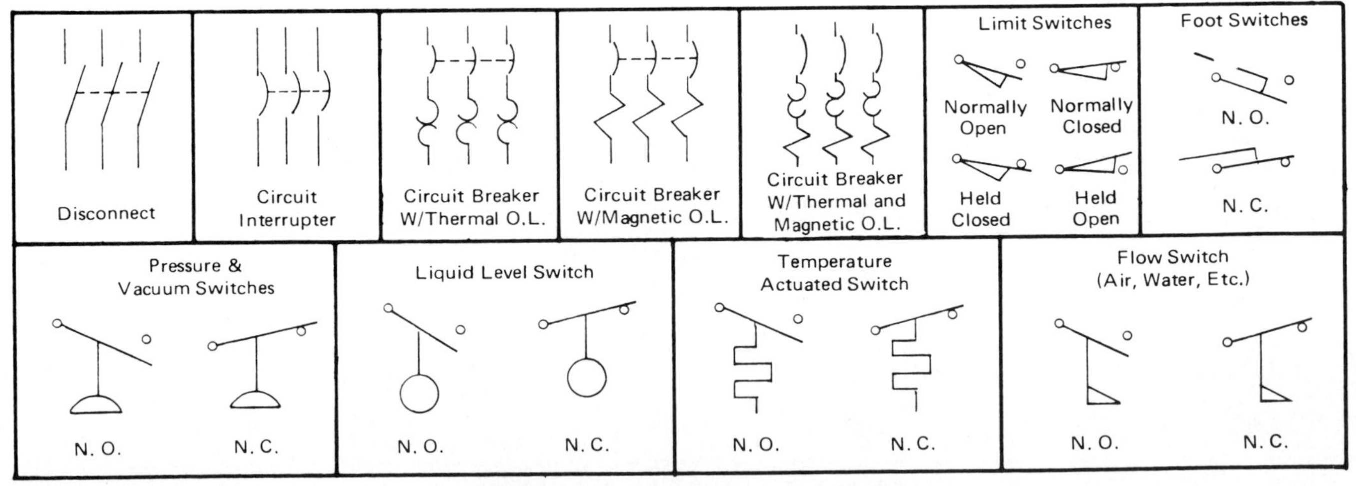

FIGURE 3–2 *Common switch symbols.*

The diagram symbols shown below have been adopted by the Square D Company and conform where applicable to standards established by the National Electrical Manufacturers Association (NEMA).

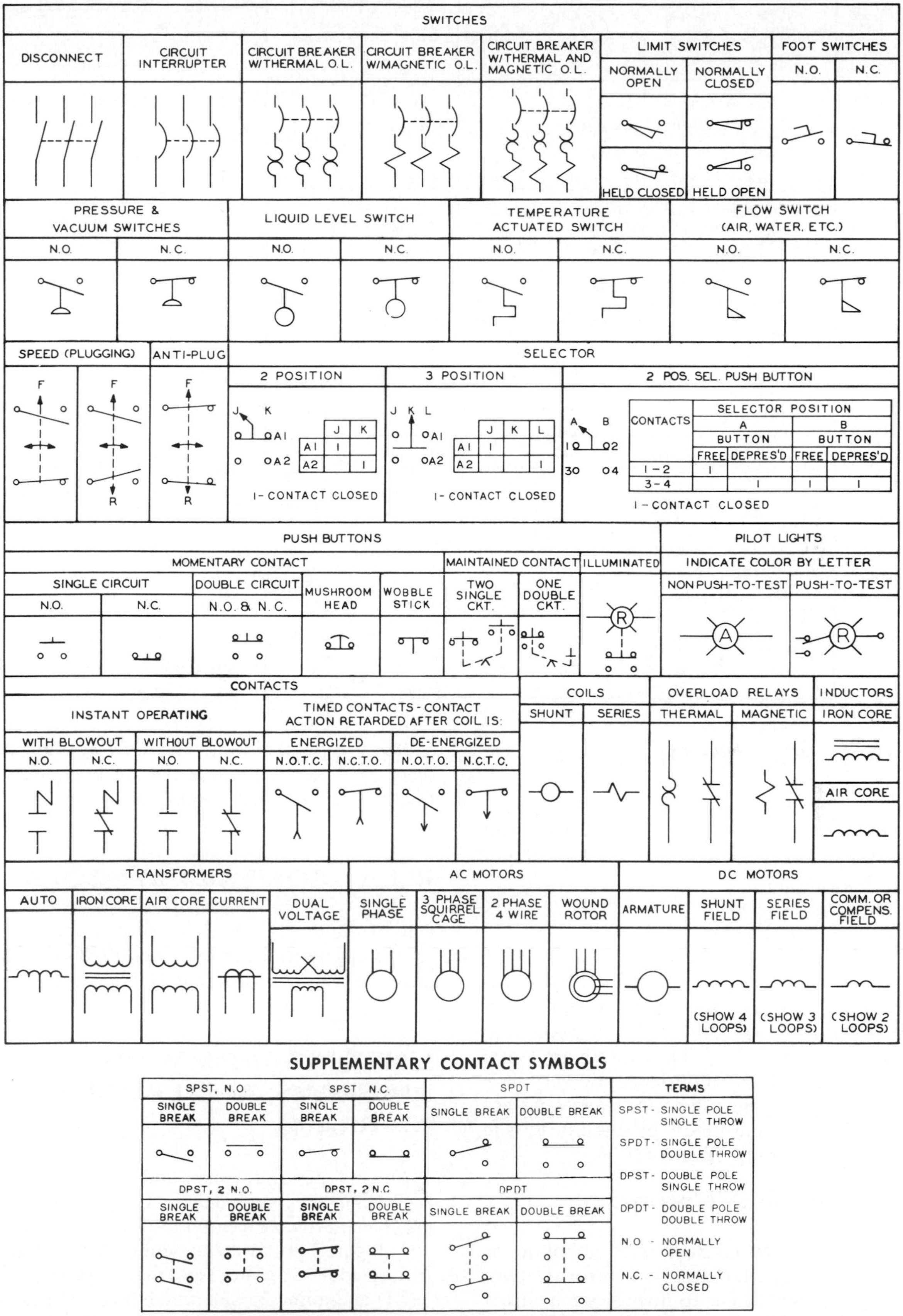

SWITCHES

- DISCONNECT
- CIRCUIT INTERRUPTER
- CIRCUIT BREAKER W/THERMAL O.L.
- CIRCUIT BREAKER W/MAGNETIC O.L.
- CIRCUIT BREAKER W/THERMAL AND MAGNETIC O.L.
- LIMIT SWITCHES: NORMALLY OPEN; NORMALLY CLOSED; HELD CLOSED; HELD OPEN
- FOOT SWITCHES: N.O.; N.C.
- PRESSURE & VACUUM SWITCHES: N.O.; N.C.
- LIQUID LEVEL SWITCH: N.O.; N.C.
- TEMPERATURE ACTUATED SWITCH: N.O.; N.C.
- FLOW SWITCH (AIR, WATER, ETC.): N.O.; N.C.
- SPEED (PLUGGING): F, R
- ANTI-PLUG: F, R

SELECTOR

2 POSITION

J K; A1; A2

	J	K
A1	1	
A2		1

1 - CONTACT CLOSED

3 POSITION

J K L; A1; A2

	J	K	L
A1		1	
A2			1

1 - CONTACT CLOSED

2 POS. SEL. PUSH BUTTON

A B; 1 2; 3 4

CONTACTS	SELECTOR POSITION A BUTTON FREE	A BUTTON DEPRES'D	B BUTTON FREE	B BUTTON DEPRES'D
1-2	1			
3-4		1	1	1

1 - CONTACT CLOSED

PUSH BUTTONS

- MOMENTARY CONTACT
  - SINGLE CIRCUIT: N.O.; N.C.
  - DOUBLE CIRCUIT: N.O. & N.C.
  - MUSHROOM HEAD
  - WOBBLE STICK
- MAINTAINED CONTACT: TWO SINGLE CKT.; ONE DOUBLE CKT.
- ILLUMINATED (R)

PILOT LIGHTS

INDICATE COLOR BY LETTER

- NON PUSH-TO-TEST (A)
- PUSH-TO-TEST (R)

CONTACTS

- INSTANT OPERATING
  - WITH BLOWOUT: N.O.; N.C.
  - WITHOUT BLOWOUT: N.O.; N.C.
- TIMED CONTACTS - CONTACT ACTION RETARDED AFTER COIL IS:
  - ENERGIZED: N.O.T.C.; N.C.T.O.
  - DE-ENERGIZED: N.O.T.O.; N.C.T.C.

COILS

- SHUNT
- SERIES

OVERLOAD RELAYS

- THERMAL
- MAGNETIC

INDUCTORS

- IRON CORE
- AIR CORE

TRANSFORMERS

- AUTO
- IRON CORE
- AIR CORE
- CURRENT
- DUAL VOLTAGE

AC MOTORS

- SINGLE PHASE
- 3 PHASE SQUIRREL CAGE
- 2 PHASE 4 WIRE
- WOUND ROTOR

DC MOTORS

- ARMATURE
- SHUNT FIELD (SHOW 4 LOOPS)
- SERIES FIELD (SHOW 3 LOOPS)
- COMM. OR COMPENS. FIELD (SHOW 2 LOOPS)

SUPPLEMENTARY CONTACT SYMBOLS

- SPST, N.O.: SINGLE BREAK; DOUBLE BREAK
- SPST N.C.: SINGLE BREAK; DOUBLE BREAK
- SPDT: SINGLE BREAK; DOUBLE BREAK
- DPST, 2 N.O.: SINGLE BREAK; DOUBLE BREAK
- DPST, 2 N.C.: SINGLE BREAK; DOUBLE BREAK
- DPDT: SINGLE BREAK; DOUBLE BREAK

TERMS

- SPST - SINGLE POLE SINGLE THROW
- SPDT - SINGLE POLE DOUBLE THROW
- DPST - DOUBLE POLE SINGLE THROW
- DPDT - DOUBLE POLE DOUBLE THROW
- N.O. - NORMALLY OPEN
- N.C. - NORMALLY CLOSED

FIGURE 3–3 *Standard wiring diagram symbols. (Square D)*

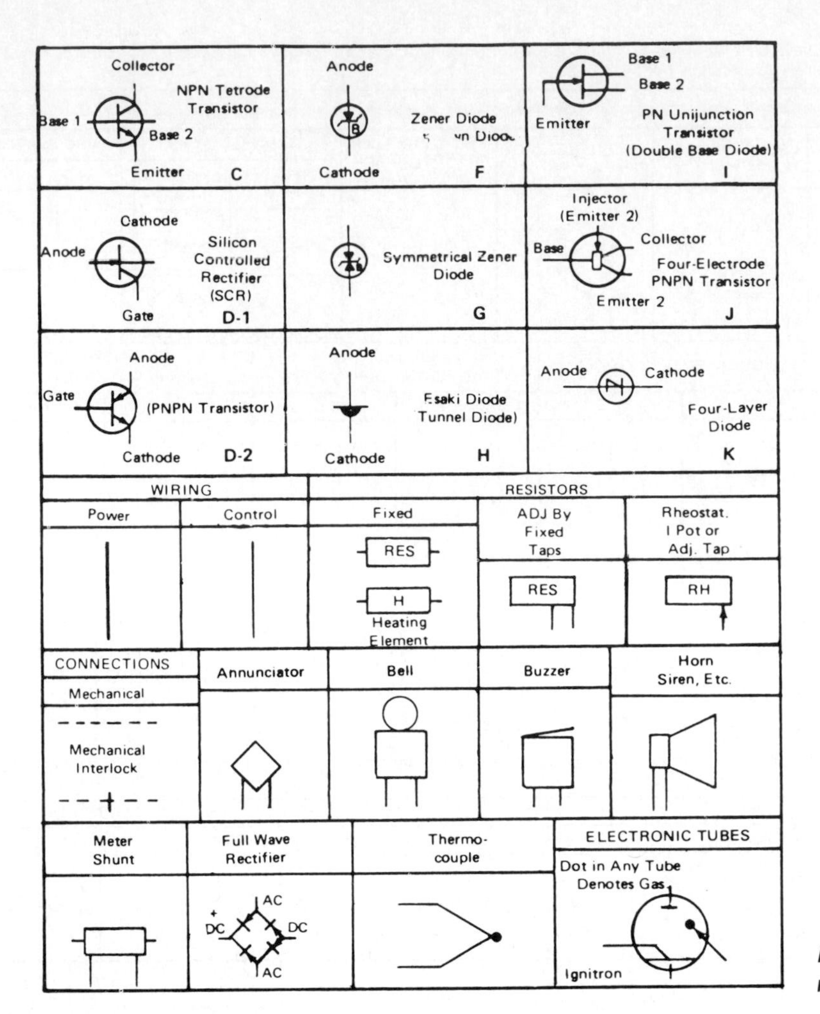

FIGURE 3–4 Electronic schematic symbols.

ELECTRONIC SYMBOLS

The field of electronics has its own symbols that are used by those who draw wiring diagrams. More and more electronics is being utilized in the control of motors (Figs. 3–4 and 3–5).

Resistor Color Code

It is a good idea to learn how the resistor color code is read. Many electronic controls have resistors identified by the color code. The color code is used on resistors from 0.1 to 2 W. The small resistors are useful primarily in electronic circuits that draw very little current. They usually have large resistances, as opposed to the larger wire-wound types used directly in series with motor windings (Fig. 3–6).

Electronic Symbols Compared

Figure 3–7 shows the usual electronic symbol on the left and compares it with industrial control symbols. This will give you an idea as to how they may differ when you are looking at a wiring diagram. In most instances, the capacitor symbol is a straight line with a curved line parallel to it. The older symbol shown is there for those schematics or wiring diagrams you may find hidden away in a closet from bygone days.

RELAY CONTACT SYMBOLS

One of the things you do work with a great deal in controlling motors is relays. Figure 3–8 shows some of the basic relay contact symbols that may be used.

LINE DIAGRAMS, WIRING DIAGRAMS, AND LADDER DIAGRAMS

Different types of drawings and diagrams are covered in Chapter 4 in greater detail. However, at this point it may be well to take a closer look at symbols in action.

Figure 3–9 shows the symbols put to work in the typical ladder diagram. Note that there are both a line 1 (L1) and a line 2 (L2), where the power is brought into the circuit. Then the various devices are connected to both sides of the power line. At this time you

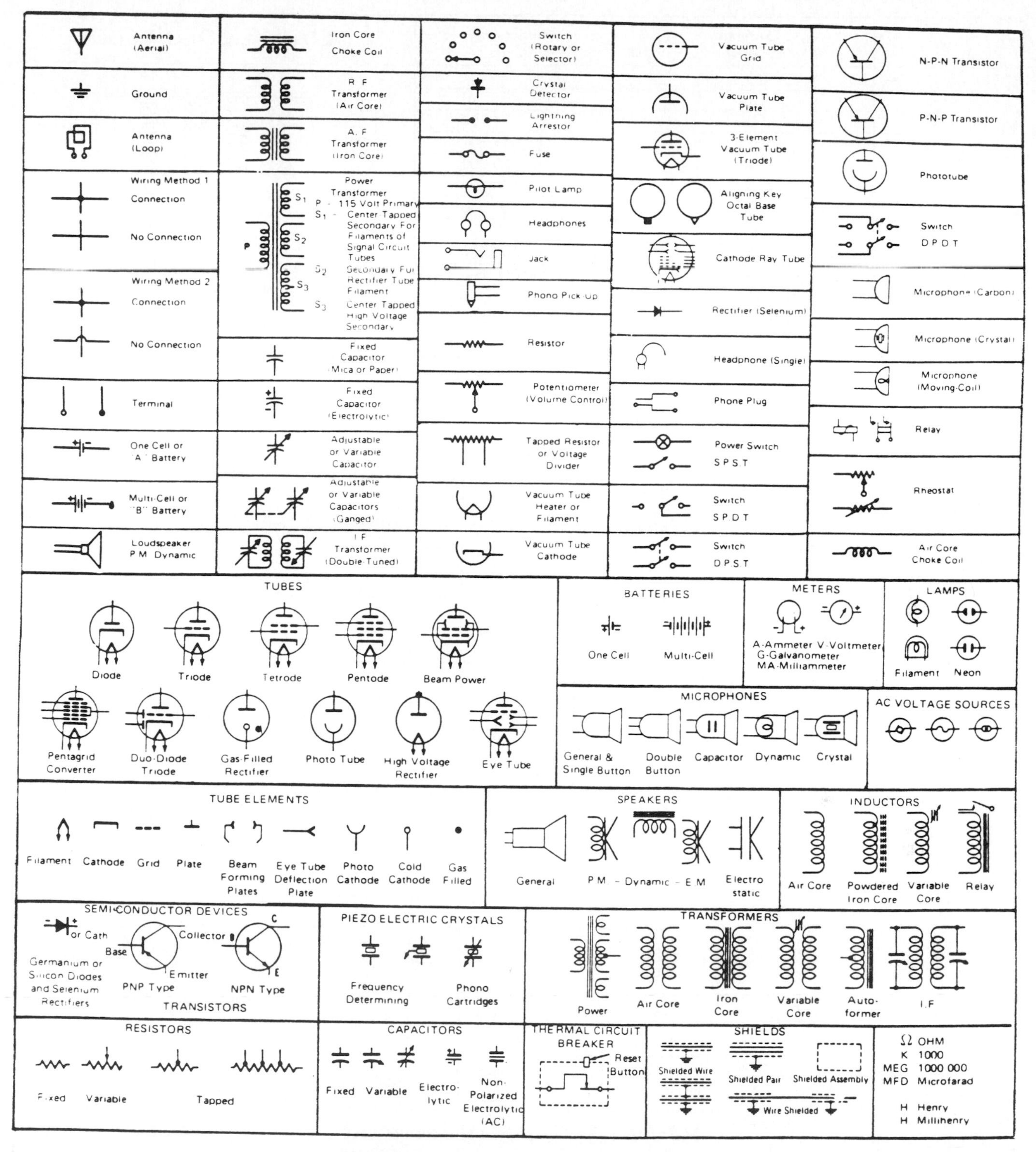

FIGURE 3–5 *Electronic schematic symbols.*

should focus on how the contacts are drawn in the circuit in reference to what device is being controlled by the normally open (NO) and normally closed (NC) contacts. Also take a close look at the START and STOP switches. As soon as the START switch (button) is closed, M, in the circle, will energize. This causes the M contacts (located across the START switch), which are normally open (NO), to close. That completes the circuit from L1 through the STOP switch and the START switch through coil M and the overload contacts (O.L.) to the other side of the line (L2). Once the coil is energized, the NO contacts across the START switch stay closed and the START switch can be released. However, when the STOP switch is opened, it causes

Color	Significant Figure	Multiplying Value
Black	0	1
Brown	1	10
Red	2	100
Orange	3	1,000
Yellow	4	10,000
Green	5	100,000
Blue	6	1,000,000
Violet	7	10,000,000
Gray	8	100,000,000
White	9	1,000,000,000
Gold	±5% tolerance	0.1
Silver	±10% tolerance	0.01
No color	±20% tolerance	

EIA/MIL COLOR CODE

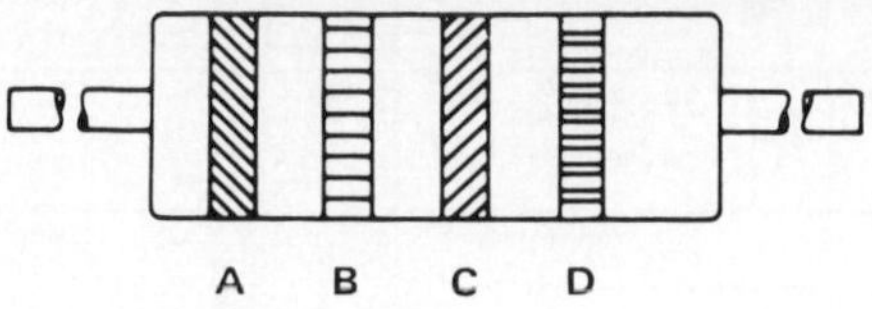

A B C D

Band A – Ist significant figure
Band B – 2nd significant figure
Band C – Number of zeros or decimal multiplier
Band D – Tolerance

FIGURE 3–6 *Resistor color code.*

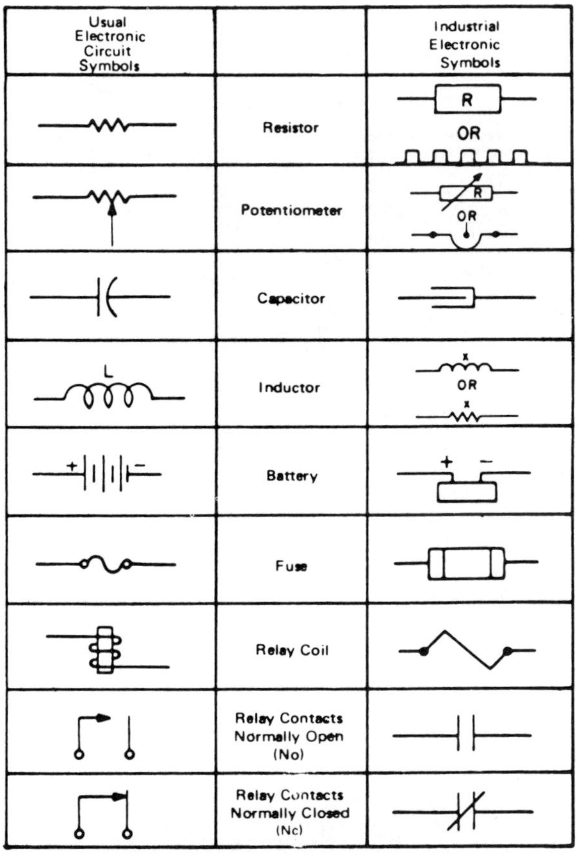

Usual Electronic Circuit Symbols		Industrial Electronic Symbols
	Resistor	R OR
	Potentiometer	R OR
	Capacitor	
L	Inductor	x OR x
+ –	Battery	+ –
	Fuse	
	Relay Coil	
	Relay Contacts Normally Open (No)	
	Relay Contacts Normally Closed (Nc)	

FIGURE 3–7 *Comparison of industrial electricity circuit symbols with the more common electronic circuit symbols.*

the circuit to deenergize and the contacts M to open again until the START button is again closed.

Now drop down to the next line in the diagram. Note that the NC contacts have an M above them. This means that they are connected to the M coil or relay. When the M coil is energized, these contacts open and turn off the pilot light connected in series with the contacts.

The pushbutton station wiring diagram is a representation of the physical station, showing the relative positions of units, the suggested internal wiring, and connections with the starter.

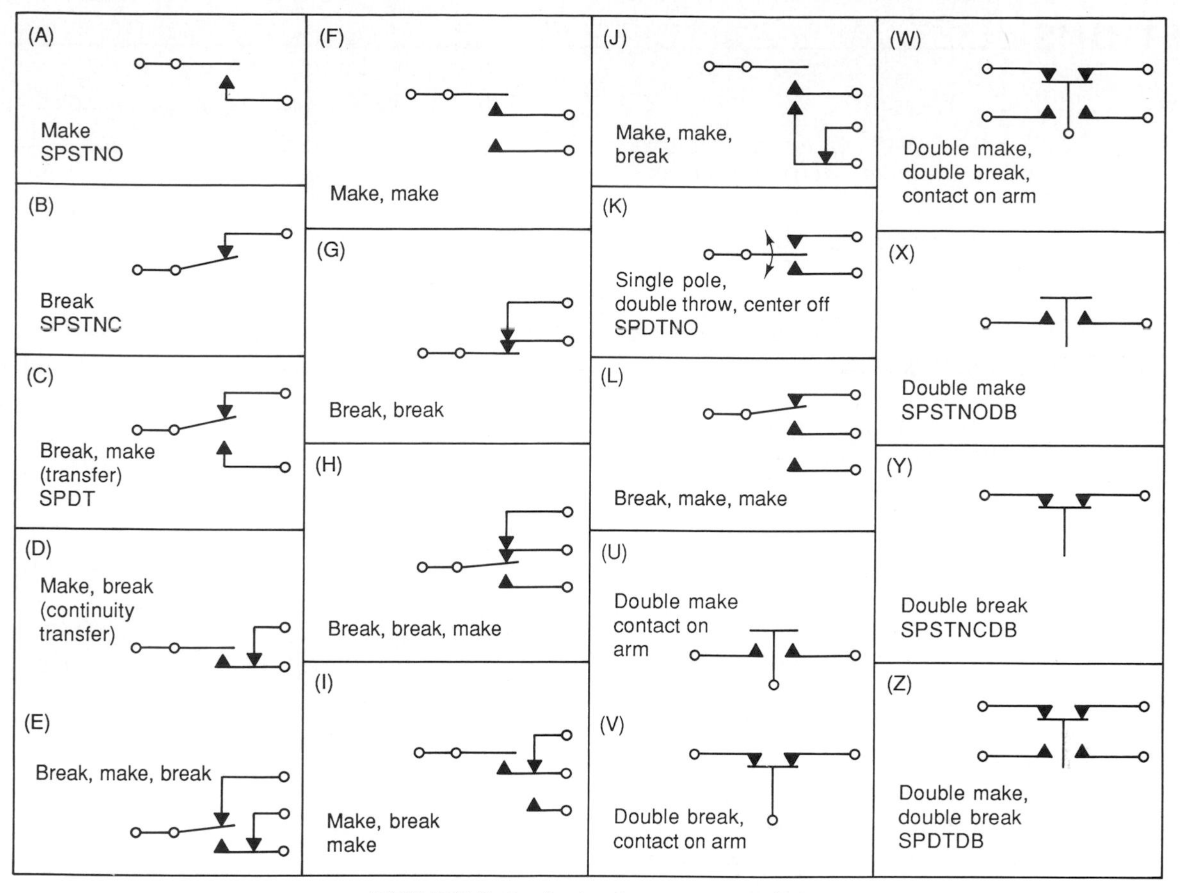

FIGURE 3–8 *Basic relay contact symbols.*

FIGURE 3–9 *Explanation of control circuit symbols.*

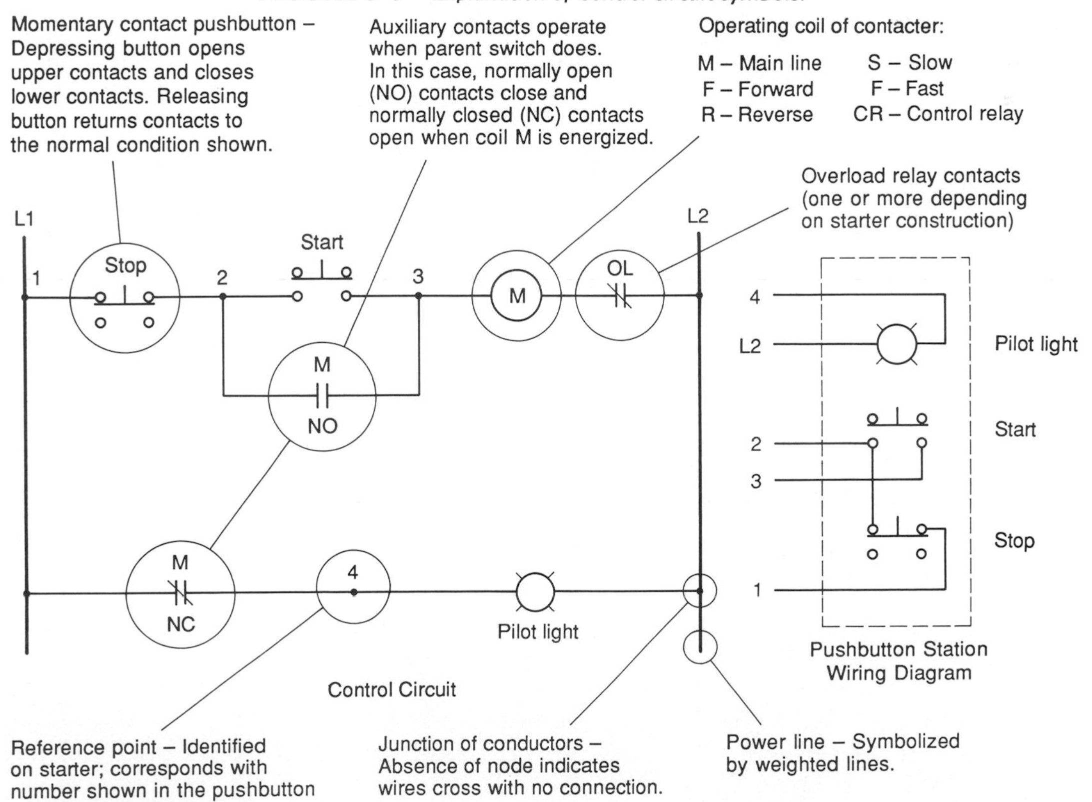

QUESTIONS

1. Where are electrical symbols used?
2. Draw five basic symbols used in wiring diagrams that you will need in studying electric motor controls.
3. What are the resistor color code colors? What do they mean?
4. What are the basic relay symbols used on diagrams?
5. What is a ladder diagram?
6. What is the difference between a schematic and a ladder diagram?

REVIEW PROBLEMS

The use of symbols is important in electrical circuitry. Symbols make easier the task of drawing circuits and analyzing them. Series circuits utilize symbols at the beginning level. A review of series circuits is presented here in the form of problems.

$$R_T = R_1 + R_2 + R_3 + \cdots$$

1. Three resistors of 10, 100, and 200 Ω are connected in series. What is the total resistance?
2. If the resistance in problem 1 is connected to 620 V, what is the current drawn?
3. A set of eight old-style Christmas tree lights are connected in series and each draws 0.15 A. What is the voltage drop across each lamp if the applied voltage is 120? What is the total current of the circuit?
4. A series resistor combination of 6, 12, and 18 Ω is connected to a power source of 72 V. What is the current flowing through an ammeter inserted between the 12- and 18-Ω resistors?
5. What is the resistance of a third resistor if the first two are 10 Ω and 20 Ω and the total resistance of the circuit measures 50 Ω?
6. The total resistance of a series circuit is 220 Ω. It has two resistors of 100 Ω each. What is the resistance of the third resistor?
7. If one lamp burns out in a series string of ten 12-V lamps and you short the burned-out filament, what will be the voltage drop across each lamp?
8. What happens to the current in a series circuit if the voltage stays the same at the source but the resistance is reduced by having one resistor short?

Control Circuits and Diagrams

Objectives

After studying this chapter, you will be able to:

1. Draw a ladder diagram.
2. Explain how to use symbols in diagrams.
3. Describe undervoltage release and protection.
4. Identify a two-wire control circuit.
5. Identify a three-wire control circuit.
6. Describe the difference between a line drawing and a wiring diagram.
7. Define the role of a thermal protector.

A wiring diagram shows, as closely as possible, the actual location of all of the component parts of the device. The open terminals (marked by an open circle) and arrows represent connections made by the user. Figure 4–1 shows a circle with *motor* written in it with three arrows leaving the circle and labeled T1, T2, T3; this represents a three-phase motor. The T stands for the terminals on the motor and are to be connected by the electrician.

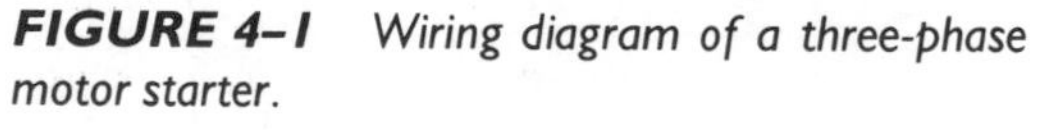

FIGURE 4–1 *Wiring diagram of a three-phase motor starter.*

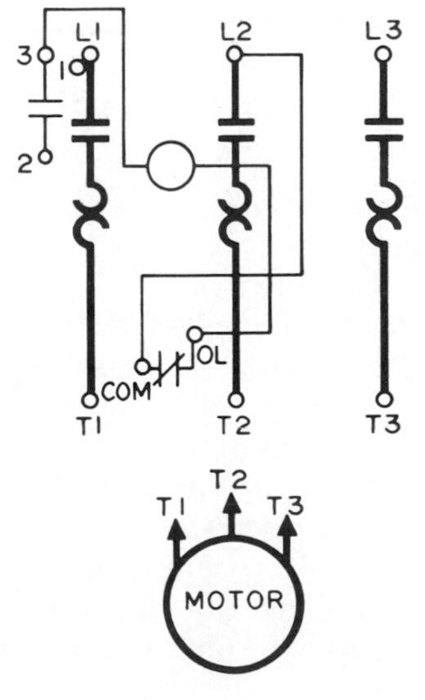

Since wiring connections and terminal markings are shown, this type of diagram is helpful when wiring the device or in tracing wires when troubleshooting. Note that bold lines denote the power circuit, and thin lines are used to show the control circuit. In most instances, ac magnetic equipment uses black wires for power circuits and red wiring for control circuits. It is to your advantage to learn how to read both wiring diagrams and line diagrams.

WIRING DIAGRAMS

A wiring diagram gives the necessary information for actually wiring up a group of control devices or for physically tracing wires when troubleshooting is necessary.

Wiring diagrams or connection diagrams include all of the devices in the system and show their physical relation to each other. All poles, terminals, coils, contacts, and switches are shown in their proper place on each device. These diagrams are helpful in wiring-up systems, because connections can be made exactly as they are shown on the diagram. In following the electrical sequence of any circuit, however, the wiring diagram does not show the connections in a manner that can easily be followed. For this reason a rearrangement of the circuit elements to form a line diagram is desirable.

Start Circuits

Manual starting switches are designed for starting and protecting small ac and dc motors rated at 1 hp or less where undervoltage protection is not needed. They are operated by a toggle lever mounted on the front of the switch (Fig. 4-2). Wiring diagrams do not show the operating mechanism since it is not electrically controlled. These motor starters consist of an on–off snap switch combined with a thermal overload device operating on the eutectic alloy ratchet principle. Terminal markings corresponding to those shown on the diagrams in Figs. 4–3 to 4–6 will be found on each switch.

Other simple wiring diagrams are used to show how to connect manual starters operated by START–

FIGURE 4–2 *Manual starting switch. On–off snap switch with thermal overload. (Allen-Bradley)*

FIGURE 4–3 *Single-pole switch used to start–stop a motor. (Allen-Bradley)*

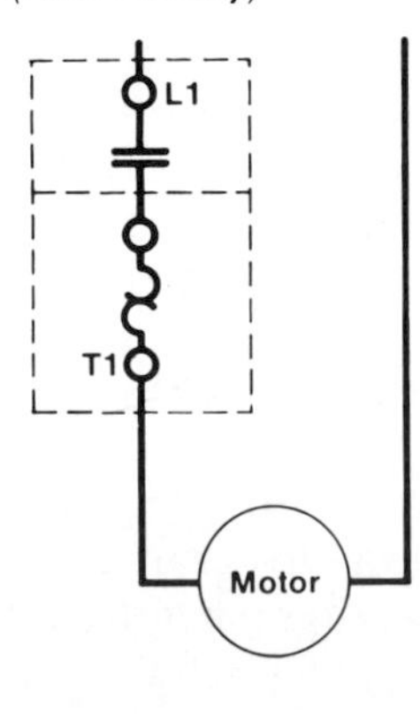

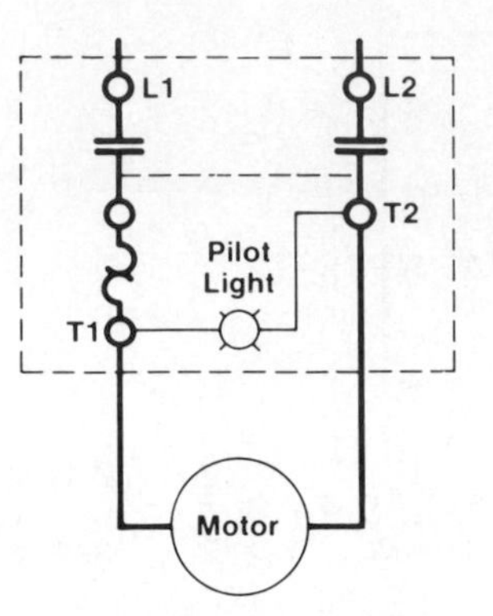

FIGURE 4–4 *Double-pole switch with built-in neon pilot light. (Allen-Bradley)*

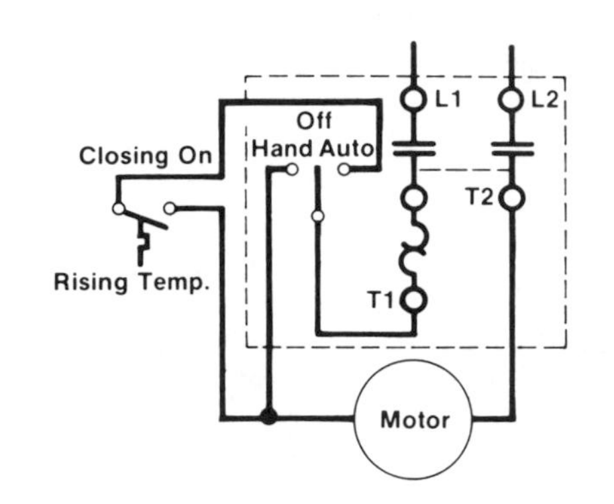

FIGURE 4–5 *HAND–OFF–AUTO selector switch circuit. (Allen-Bradley)*

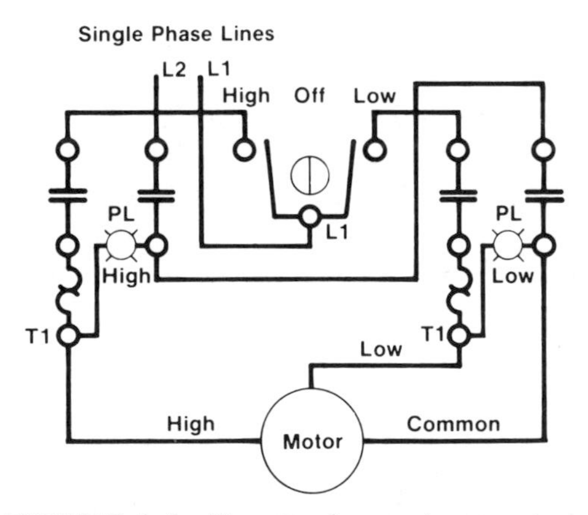

FIGURE 4–6 *Two-speed manual motor starter. (Allen-Bradley)*

STOP pushbuttons mounted on the front of the starter. Figure 4–7 shows what the pushbutton looks like when properly packaged with the cover removed for ease in viewing the terminals. This type of starter is used where undervoltage protection is not required. Wiring diagrams do not show the operating mechanism since it is not electrically controlled. Pushing the START button mechanically closes the contacts, connecting the motor to the line. The contacts are opened

FIGURE 4–7 *Start–stop manual starter. (Allen-Bradley)*

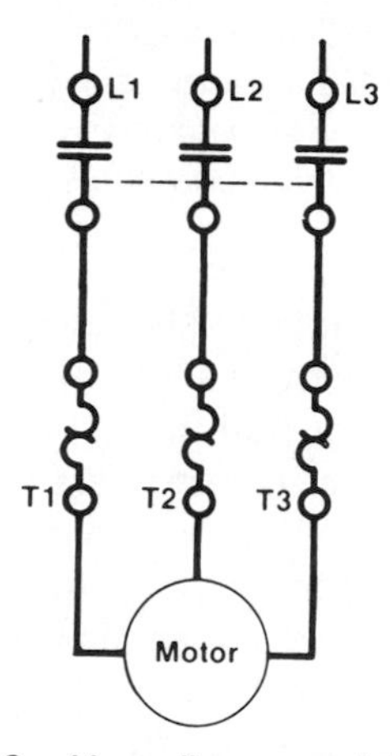

FIGURE 4–8 *Manually operated three-phase or two-phase three-wire starter. (Allen-Bradley)*

FIGURE 4–9 *Direct-current motor starter. (Allen-Bradley)*

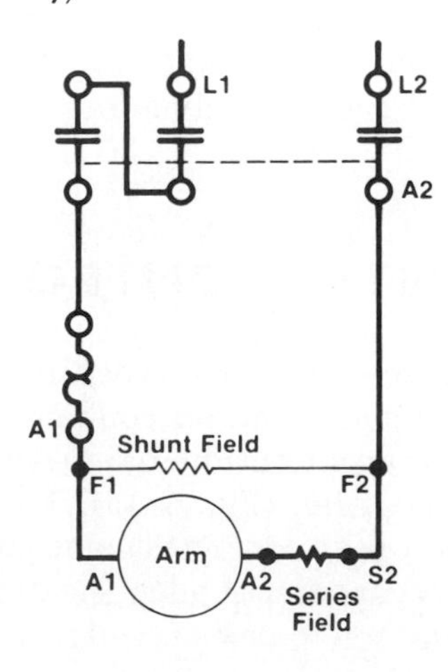

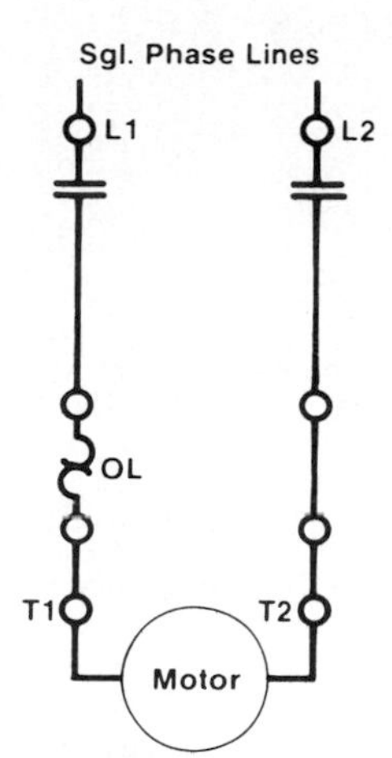

FIGURE 4–10 *Single-phase, two-contact manual starter. (Allen-Bradley)*

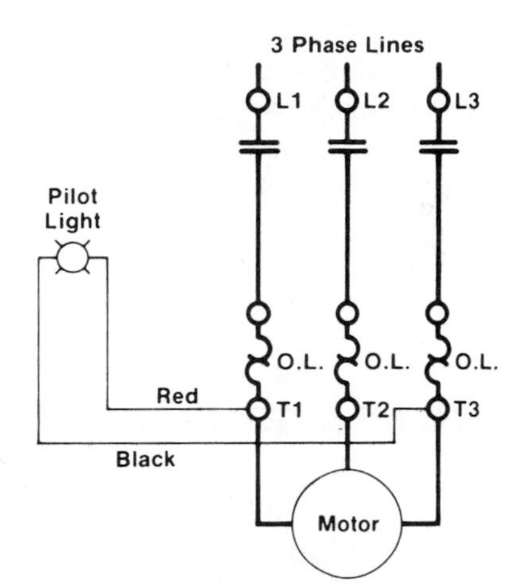

FIGURE 4–11 *Three-phase manual starter with pilot light. (Allen-Bradley)*

by pressing the STOP button. Terminal markings corresponding to those shown on the diagrams (Figs. 4–8 to 4–11) will be found on each switch.

LINE DIAGRAM OR LADDER DIAGRAM

A line diagram gives the necessary information for easily following the operation of the various devices in the circuit. It is very helpful in troubleshooting, for it shows, in a simple way, the effect that opening or closing various contacts has on other devices in the circuit.

A line diagram is sometimes referred to as an *elementary diagram* or a *schematic diagram.* No attempt is made to show the various devices in their actual positions. All control devices are shown between vertical lines which represent the source of control power, and circuits are shown connected as directly as possible from one of these lines to the other. All con-

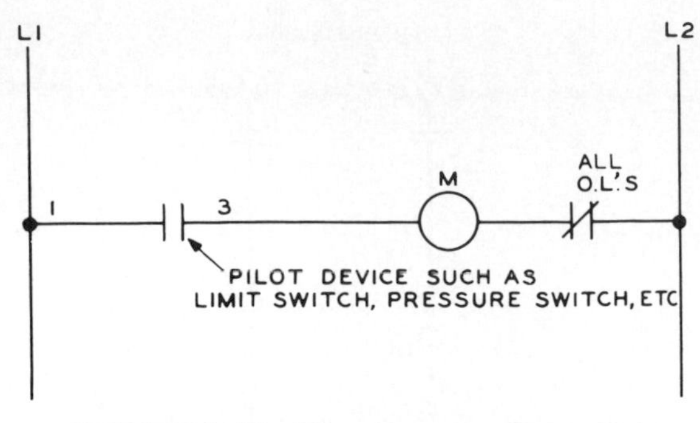

FIGURE 4–12 *Elementary two-wire control.*

nections are made in such a way that the functioning of the various devices can easily be traced (Fig. 4–12).

This is a simple or elementary diagram of a switch in series with a magnetic coil (M) and overload contacts. Note how the L1 is to the left and the number 1 is placed near the junction point of L1 with the horizontal line. Each line or point where a horizontal line is taken off the power line is numbered. This comes in handy later when the programmable controller is utilized to do the work of switching. This type of drawing is also referred to as a *ladder diagram,* due to its design with parallel power lines and rungs that make up the circuits.

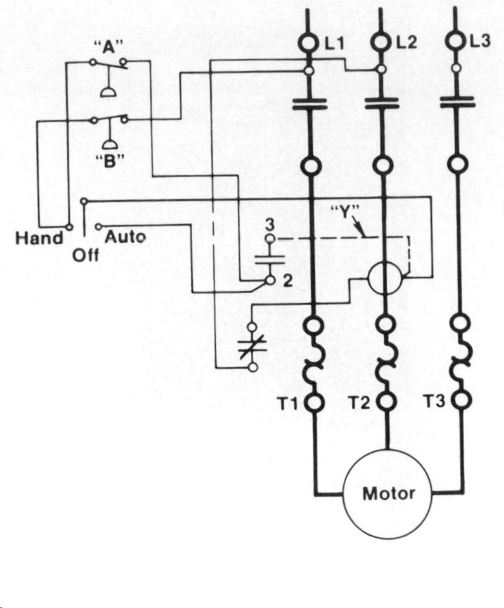

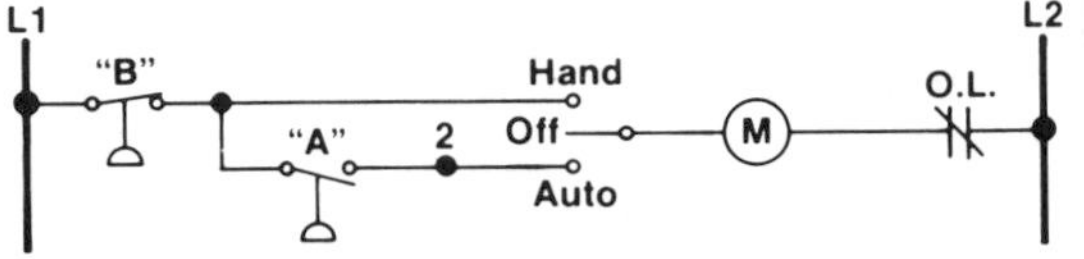

FIGURE 4–13 *Starter operated by pressure switch or thermostat with manual control provided by a selector switch. (Allen-Bradley)*

UNDERVOLTAGE RELEASE

Another name for undervoltage release is *low-voltage release—two-wire control.* These terms mean that the starter will drop out when there is a voltage failure and will pick up again as soon as power is restored (Fig. 4–12). This low-voltage release is a two-wire control using a *maintained contact* pilot device (limit switch, pressure, or float switch) in series with the starter coil. This arrangement is used when a starter is required to function automatically without the attention of an operator. If a power failure occurs while the contacts of the pilot device are closed, the starter will drop out. When the power is restored, the starter will pick up automatically through the closed contacts of the pilot device. The term *two-wire control* arises from the fact that in the basic circuit, only two wires are required to connect the pilot device to the starter.

TWO-WIRE CONTROL CIRCUITS

There are a number of circuits that can be drawn and used to good advantage in specific situations. Figure 4–13 shows a starter operated with manual control provided by a selector switch. A high-pressure cutout switch can be added. Both the wiring diagram and the line drawing are shown to illustrate how things are connected and arranged. The selector switch makes it possible to operate the starter manually for testing in case of failure of the automatic pilot control. When a standard across-the-line starter without pushbuttons is used, connection Y is removed and the wiring follows the solid lines of the diagram.

Note that in the ladder diagram only L1 and L2 are used. The control circuitry uses these two for power, while the magnetic starter M has three sets of contacts that will interrupt the power to the three-phase motor in all three legs or phases.

If a high-pressure cutout switch is added, it should be inserted in the line leading from L1 to the HAND terminal of the selector switch. A low-pressure switch or thermostat can be used where A is located (Fig. 4–13). B represents the high-pressure switch or safety cutout. Note how the HAND–OFF–AUTO switch operates in both the wiring diagram and the ladder diagram.

THERMOSTAT CONTROL

Since the contacts of a gage-type thermostat cannot handle the current to a starter coil, a thermostat relay must be used as an intermediate step between the thermostat and the starter (Fig. 4–14). The relay is energized when the CLOSE contact is made and deenergized when the OPEN contact is made. The OPEN contact bypasses the relay coil to deenergize it. A resistor (Res.)

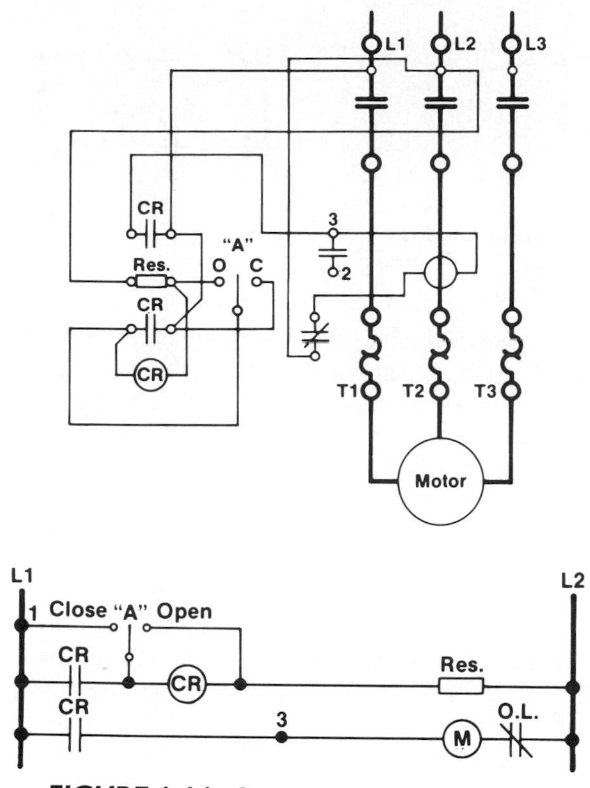

FIGURE 4-14 Starter controlled by a gage-type thermostat. (Allen-Bradley)

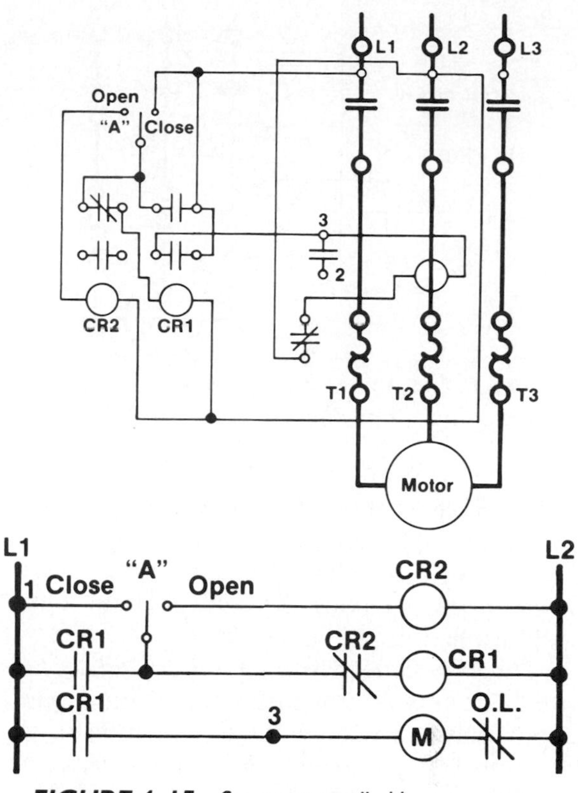

FIGURE 4-15 Starter controlled by a gage-type thermostat whose contacts can switch only a small amount of current. (Allen-Bradley)

is built into the relay to guard against a short circuit when this is done.

The thermostat contacts must not overlap or be adjusted too close to one another, since this may burn out the resistance unit. It is also advisable to check the inrush current of the relay against the current rating of the thermostat. This arrangement can also be used with pressure controls. "A" represents the three-wire gage-type temperature control device. CR is the thermostat relay, which in this case has two sets of contacts also labeled CR.

Another method of protecting thermostat contacts from high currents can be seen in Fig. 4-15. This diagram shows the use of a thermostat whose contacts can handle even less current than the ones in the preceding example. It is advisable in such cases as this to use relays having coils that operate at a very small value of volt-amperes. This will reduce the burden on the thermostat contacts and is especially advisable where frequent operation is required. "A" represents the three-wire sensitive gage-type control. CR1 and CR2 are the low coil current relays.

When the CLOSE position is indicated it means that relay CR1 is energized. Energizing CR1 causes the contacts CR1 to close. Closing CR1 contacts in series with the control coil (M) causes it to energize, starting the motor. It also keeps itself energized by the other set of CR1 contracts. However, when OPEN is indicated by the thermostat it completes the circuit through the closed contacts of CR1 to cause relay CR2 to energize. Once CR2 is energized it opens its contacts (CR2), which are normally closed and in series with CR1 relay coil. This causes CR1 to deenergize and open the CR1 contacts, one of which is in series with the motor control coil (M); this stops the motor. Once CR2 is energized it is quickly deenergized by its contacts opening CR1's circuit and opening the CR1 contacts, which deenergize the motor starter coil.

Note the difference between the line drawing and the wiring diagram. The line drawing (ladder diagram) makes it easier to see how the relays operate in sequence to control the magnetic starter coil and thereby the motor.

UNDERVOLTAGE PROTECTION

Undervoltage protection is also referred to as *low-voltage protection, three-wire control.* Both terms mean the same and indicate that the starter will drop out when there is a voltage failure but will not pick up automatically when voltage returns. The control cir-

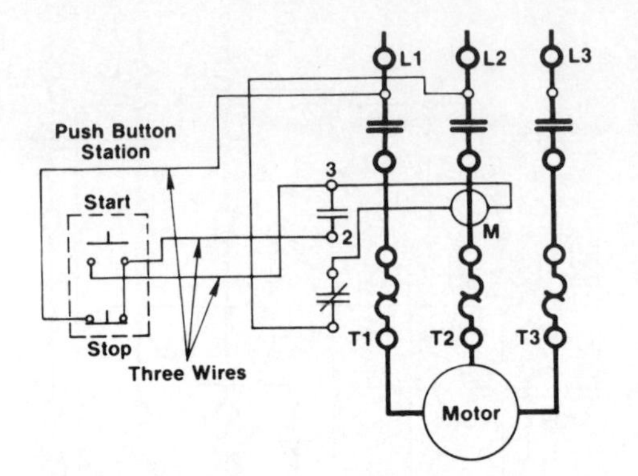

FIGURE 4–16 *Low-voltage protection: three-wire control. (Allen-Bradley)*

cuit is completed through the STOP button and also through a holding contact. See contact 2–3 on the starter in Fig. 4–16. When the starter drops out, these contacts open, breaking the control circuit until the START button is pressed once again.

Three wires lead from the pilot device to the starter. The START–STOP pushbutton station is usually thought of when undervoltage protection or three-wire control is mentioned. This is the most common means of providing this type of control.

The main distinction between the two-wire and three-wire controls is that with undervoltage release (two-wire) and undervoltage protection (three-wire) is the fact that one will cause the motor to start again when the power is on again and the other will not start until the START button is pressed. The designations *two-wire* and *three-wire* are used only to describe the simplest applications of the two types. Actually, in other systems, there might be more wires leading from the pilot device to the starter, but the principle of two-wire or three-wire control would still be present.

THREE-WIRE CONTROL CIRCUITS

One of the methods used to improve your ability to read wiring diagrams and ladder diagrams is to practice. Figures 4–17 through 4–26 are three-wire control circuits that can do a variety of things. Each will be analyzed to show how it operates. Knowing how they operate will aid in troubleshooting since troubleshooting is nothing more than determining why a device is not performing the way it should normally.

One of the simplest types of three-wire control circuitry is shown in Fig. 4–17. This three-wire control has a pilot light in the circuit to indicate when the motor is running. The pilot light is wired in parallel with the starter coil to indicate when the starter is energized and the motor is running or at least has power applied to its terminals.

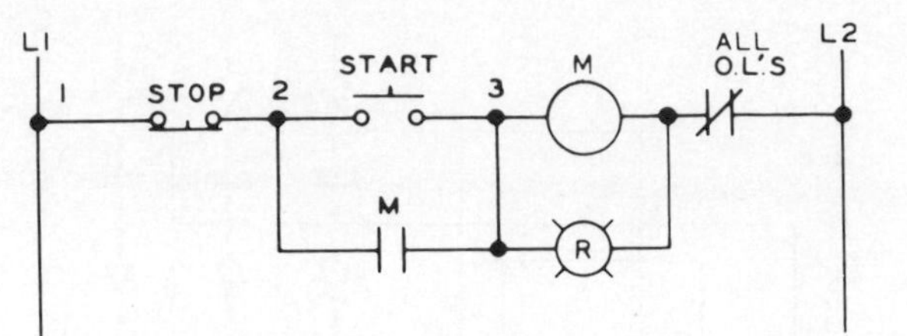

FIGURE 4–17 *Three-wire control with pilot light. Motor running.*

Figure 4–18 takes the three-wire control circuitry a step further. It shows a pilot light used to indicate when the motor is stopped. A pilot light may be required to indicate when the motor is stopped in some cases. This can be done by wiring a normally closed auxiliary contact located on the starter in series with the pilot light. When the starter is deenergized, the pilot light is on. When the starter picks up, the auxiliary contacts open, turning off the light.

Figure 4–19 uses a push-to-test pilot light to indicate when the motor is running. When the motor-running pilot light is not glowing, there may be doubt as to whether the circuit is open or whether the pilot light bulb is burned out. The push-to-test pilot light enables testing of the bulb simply by pushing on the switch cap. This indicates only that the bulb is or is not working; it does not indicate other problems that may be in the coil or wiring up to the motor.

The illuminated pushbutton combines a start button and a pilot light (see Fig. 4–20). Pressing the pilot light lens operates the start contacts. Space is saved by requiring only a two-unit pushbutton station instead of three.

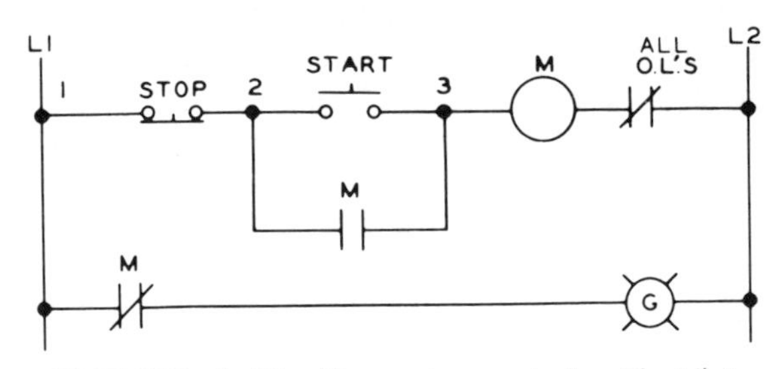

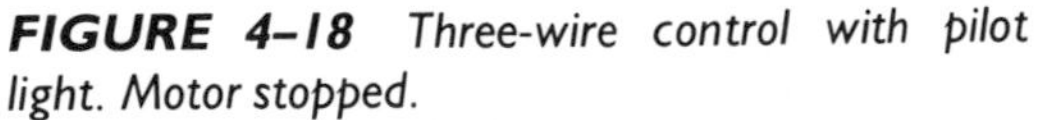

FIGURE 4–18 *Three-wire control with pilot light. Motor stopped.*

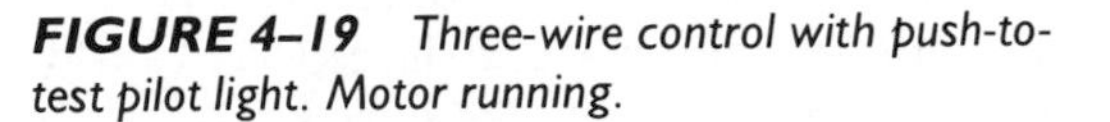

FIGURE 4–19 *Three-wire control with push-to-test pilot light. Motor running.*

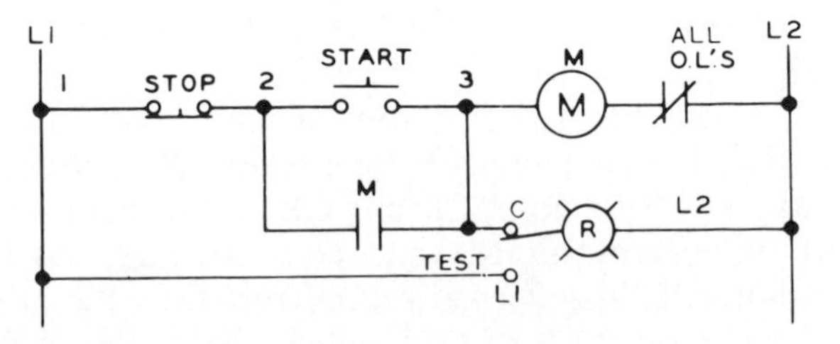

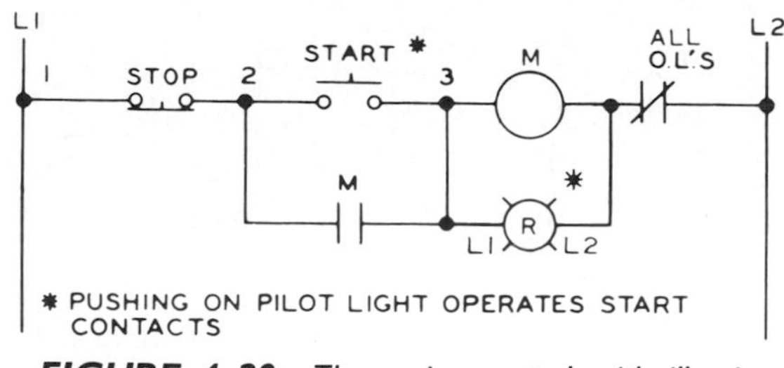

FIGURE 4–20 *Three-wire control with illuminated pushbutton. Motor running.*

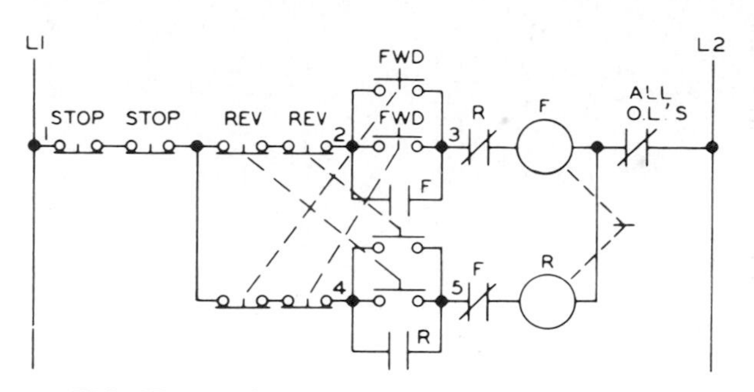

FIGURE 4–23 *Three-wire control. Reversing starter. Multiple-pushbutton station. (Allen-Bradley)*

When one START–STOP station is required to control more than one starter, the arrangement shown in Fig. 4–21 can be used. A maintained overload on any one of the motors will drop out all three starters. Note how M1 contacts control M2 and M2's contacts control M3. Once the STOP button is depressed it opens the circuit to M1, and its dropout causes the other two coils to drop out also. The START switch is paralleled with contacts from M3.

So far the circuits have been utilized to keep the motor running or stopped when unsafe conditions occur. Now it is time to look at what can be done to reverse the direction of motor rotation. Three-wire control of a reversing starter can be accomplished with a FORWARD–REVERSE–STOP pushbutton station, as shown in Fig. 4–22. Limit switches can be added to stop the motor at a certain point in either direction. Jumpers 6 to 3 and 7 to 5 must then be removed. The circle with F and the circle with R above are the coils whose contacts actually do the switching of the leads to the motor in order to make it reverse direction.

In some cases more than one pushbutton station is needed. That calls for an arrangement similar to Fig. 4–23. Note how the two switches (forward and reverse) are paralleled but arranged so that pressing FORWARD in either location will cause the F coil to energize and pushing REVERSE will cause the R coil to energize from either position. Note how the F contacts are in series with the R coil and the R contacts are placed in series with the F coil. This deenergizes the R coil when the F coil is energized.

Sometimes it is necessary to know in which direction the motor is rotating. A pilot light can be connected to indicate the direction of rotation. Pilot lights can be connected in parallel with the forward and reverse contactor coils to indicate which contactor is energized and thus in which direction the motor is running (Fig. 4–24).

Start–stop and reversing of rotation have been examined up to this point. Now it is time to introduce the two-speed starter. Three-wire control of a two-speed starter with a HIGH–LOW–STOP pushbutton station is shown in Fig. 4–25. The diagram allows the operator to start the motor from rest at either speed or to change from low to high speed. The STOP button must be operated before it is possible to change from high to low speed. This arrangement is intended to prevent excessive line current and shock to the motor and the driven machinery. Shock to the motor can result when

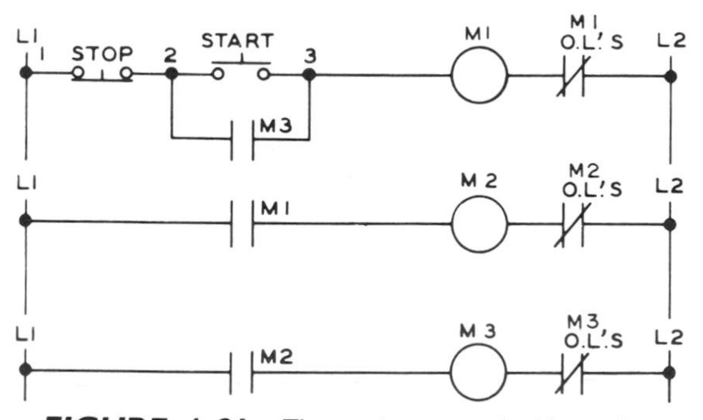

FIGURE 4–21 *Three-wire control. More than one starter. One pushbutton station controls all. (Allen-Bradley)*

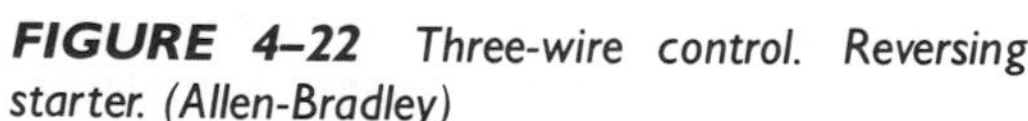

FIGURE 4–22 *Three-wire control. Reversing starter. (Allen-Bradley)*

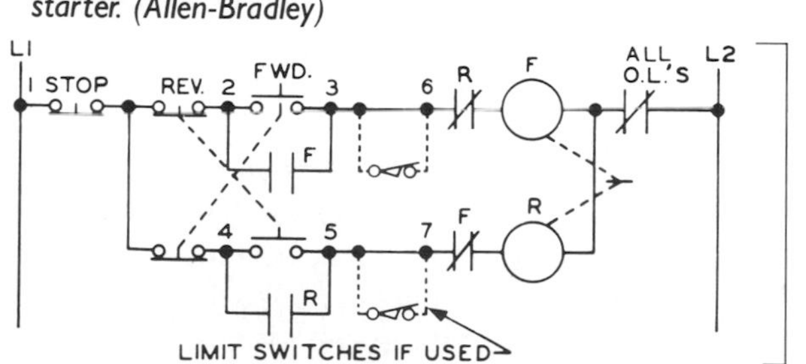

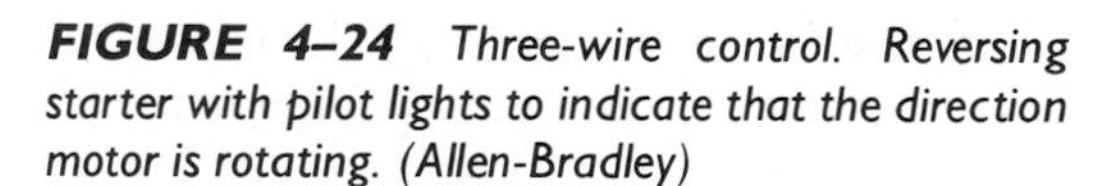

FIGURE 4–24 *Three-wire control. Reversing starter with pilot lights to indicate that the direction motor is rotating. (Allen-Bradley)*

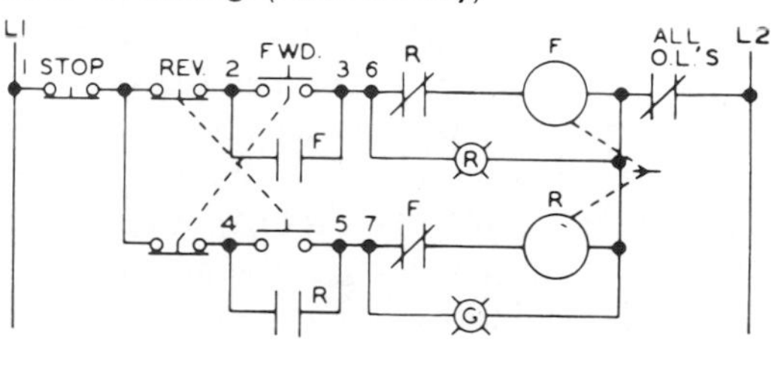

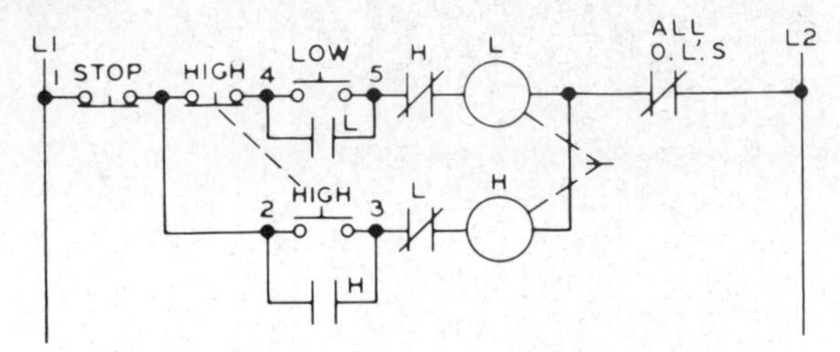

FIGURE 4–25 *Three-wire control, two-speed starter. (Allen-Bradley)*

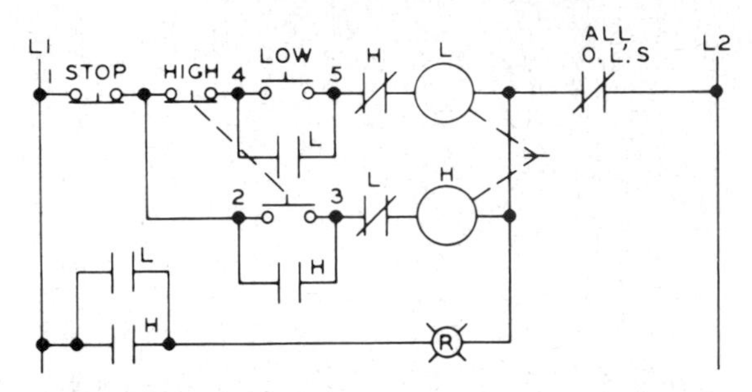

FIGURE 4–26 *Three-wire control. Two-speed starter with one pilot light to indicate motor operation at each speed. (Allen-Bradley)*

motors running at high speed are reconnected for a lower speed.

Once the two-speed motor has been controlled, it is, in most instances, preferable to have some indication as to which speed is engaged. Figure 4–26 shows a three-wire two-speed starter with one pilot light to indicate that the motor is operational at both speeds. One pilot light can be used to indicate operation at *both* low and high speeds. One *extra* normally open interlock on each contactor is required. Two pilot lights, one for *each* speed, could be used by connecting pilot lights in parallel with high and low coils, such as shown in Fig. 4–24.

OVERCURRENT PROTECTION FOR CONTROL CIRCUITS

A high-quality electric motor, properly cooled and protected against overloads, can be expected to have a long life. The goal of proper motor protection is to prolong motor life and postpone the failure that ultimately takes place. Good electrical protection consists of providing both proper overload protection and current-limiting, short-circuit protection. Ac motors and other types of high-inrush loads require protective devices with special characteristics. Normal, full-load, running currents of motors are substantially less than the currents that result when motors start or are subjected to temporary mechanical overloads. This characteristic is shown in Fig. 4–27.

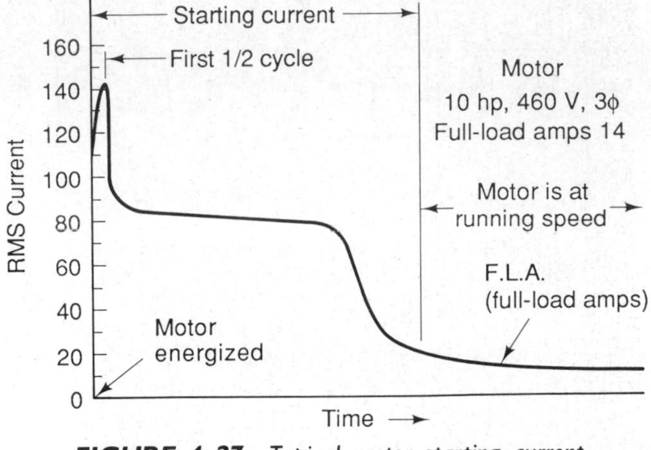

FIGURE 4–27 *Typical motor starting current characteristics. (Bussmann Division, Cooper Industries, Inc.)*

At the moment an ac motor circuit is energized, the starting current rapidly rises to many times normal current and the rotor begins to rotate. As the rotor accelerates and reaches running speed, the current declines to the normal running current. Thus, for a period of time, the overcurrent protective devices in the motor circuit must be able to *tolerate* the rather substantial temporary overload.

Motor starting currents can vary substantially depending on the motor type, load type, starting methods, and other factors. For the initial half-cycle, the momentary transient *rms current* can be as high as 11 times or more. After the first half-cycle, the starting current subsides to four to eight times (typically, six times) the normal current for several seconds. This current is called the *locked rotor current.* When the motor reaches running speed, the current then subsides to its normal running level.

The special requirements for protection of motors require that the motor overload protective device withstand the temporary overload caused by motor starting currents, and at the same time, protect the motor from continuous or damaging overloads.

Overload protection is provided by three main types of devices:

1. *Overload relays in motor controllers.* Usually, the melting alloy or bimetallic type, which are designed to simulate motor damage curve characteristics. Correctly sized, overload relays in a good-quality controller, properly maintained, provide good protection.
2. *Dual-element, time-delay fuses and low-peak dual-element fuses.* The time-delay element in these fuses provides a minimum of 10 seconds delay at 500% load and will yield good running protection when sized cor-

rectly. In addition, short-circuit protection is afforded by the short-circuit element.

3. *Thermal protectors.* Sensitive to heat and internally embedded in small fractional-horsepower motors and hermetic compressor motors of integral sizes; resetting can be automatic or manual.

Circuit breakers are not generally recommended for overload protection. They usually do not have sufficient time delay to permit close sizing (typically, sizing must be 200 to 250%) of motor full-load amperes (FLA) and are prone to trip out magnetically under starting conditions. The instantaneous trips of circuit breakers have to be set high enough to overcome the motor momentary transient current. Fuses other than dual-element are not recommended because they must be substantially oversized (typically, 300 to 400% of motor FLA) to permit starting.

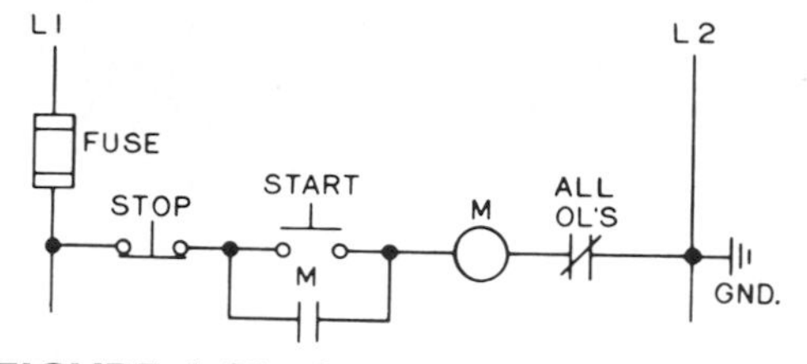

FIGURE 4–28 *Overcurrent protection for control circuit. One fuse. (Allen-Bradley)*

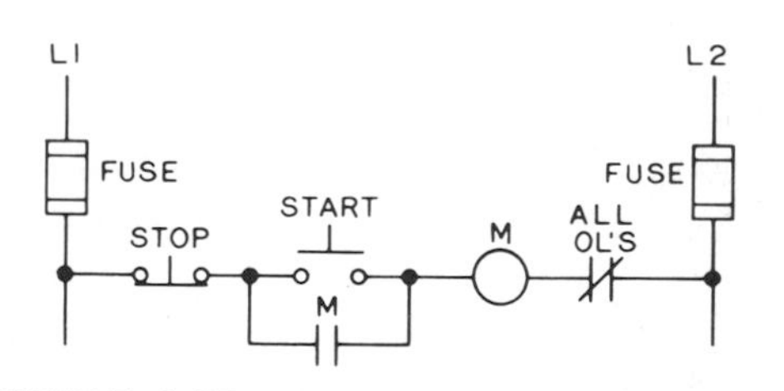

FIGURE 4–29 *Overcurrent protection for control ciruit. Two fuses. (Allen-Bradley)*

Common control with fusing in one line only and with both lines ungrounded or, if user's conditions permit, with one line grounded are shown in Fig. 4–28. Note the fuse symbol and that it is in *series* with L1. L2 is grounded and not fused. Common control with fusing in both lines and with both lines ungrounded is shown in Fig. 4–29.

TRANSFORMERS IN CONTROL CIRCUITS

Transformers can be used to step down or step up voltages. A step-down transformer is usually employed in control circuits where the equipment utilized to control the main operating equipment is connected to high-voltage lines. Low voltage can be used to control high voltages and high currents by energizing starter coils that have contacts that can handle the higher currents and voltages. This lower voltage allows for smaller wire to be used and for control stations to be placed in remote areas far from the actual operating device or motor.

There are a number of control circuits that use fuses for protection. Note the fusing employed in Fig. 4–30. One fuse is used to protect the transformer in this circuit, where it is placed in the secondary circuit of the transformer. However, Fig. 4–31 uses two fuses to protect the secondary transformer circuitry, while Fig. 4–32 shows how both the primary and the second-

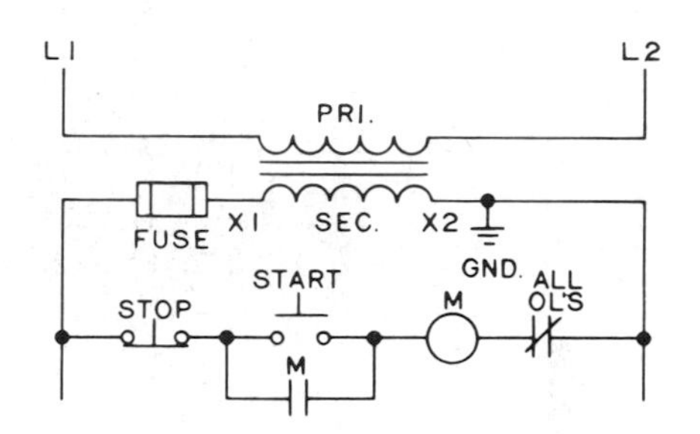

FIGURE 4–30 *Control circuit transformer with one-fuse protection. (Allen-Bradley)*

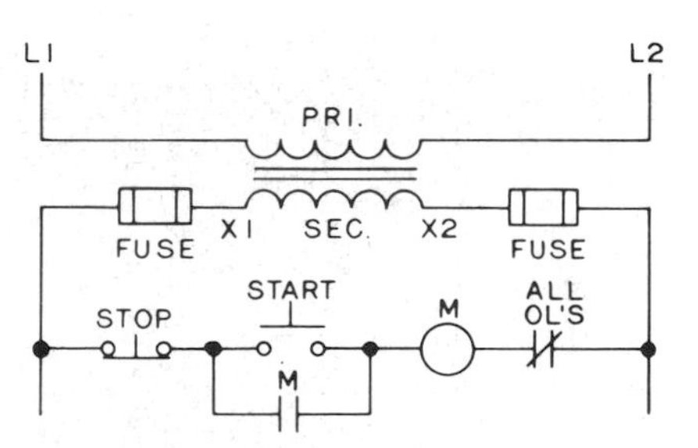

FIGURE 4–31 *Control circuit transformer with two-fuse protection.*

FIGURE 4–32 *Control circuit transformer with one primary and one secondary fuse. (Allen-Bradley)*

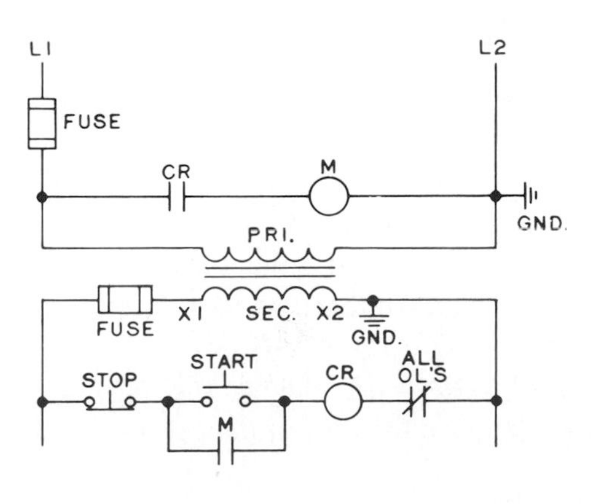

ary of the transformer is fused with one of the primary lines being grounded. In Fig. 4–33 the primary circuit is not grounded, so two fuses are used, one in L1 and one in L2, as well as two fuses in the secondary circuitry. Figure 4–34 shows the control circuit transformer with fusing in both primary lines, with no secondary fusing and with all lines ungrounded.

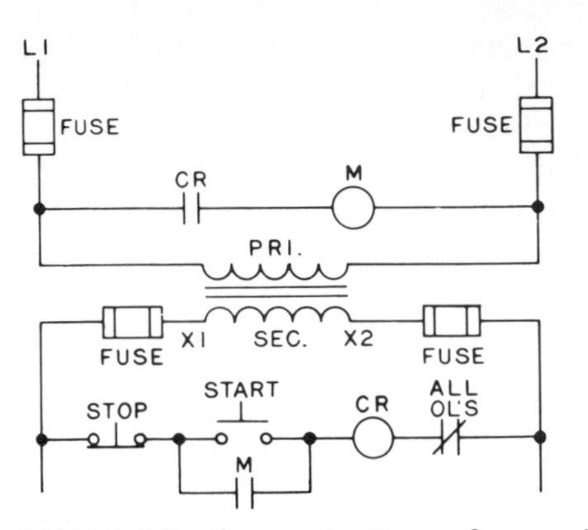

FIGURE 4–33 Control circuit transformer with both primary and secondary lines fused. (Allen-Bradley)

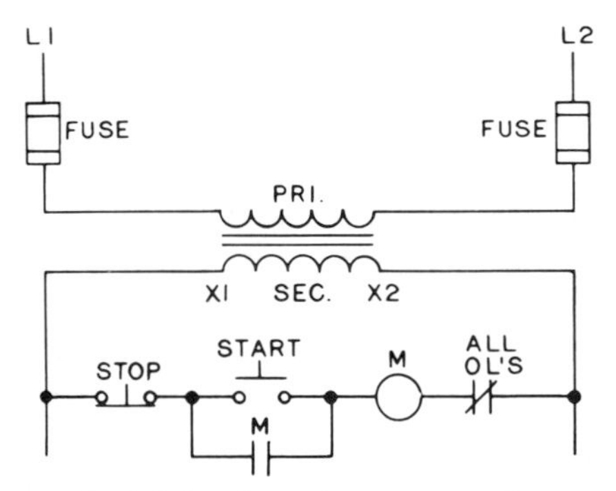

FIGURE 4–34 Control circuit transformer with fusing of both primary lines. (Allen-Bradley)

A wiring diagram and a ladder diagram are shown in Fig. 4–35, where the step-down transformer provides low voltage for the control circuit, which is wired for three-wire control. The starter coil is operated on a voltage lower than line voltage. This is usually done for safety reasons. This also requires the use of a step-down transformer in the pilot circuit. The startcr is operated from a START–STOP pushbutton station. When a control circuit step-down transformer is used with this type of starter, wiring connection X must be removed. Note that a fuse is added to the transformer secondary.

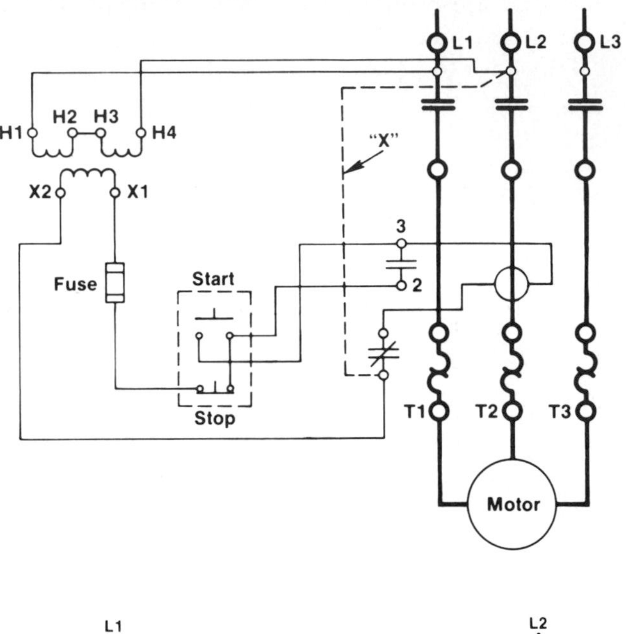

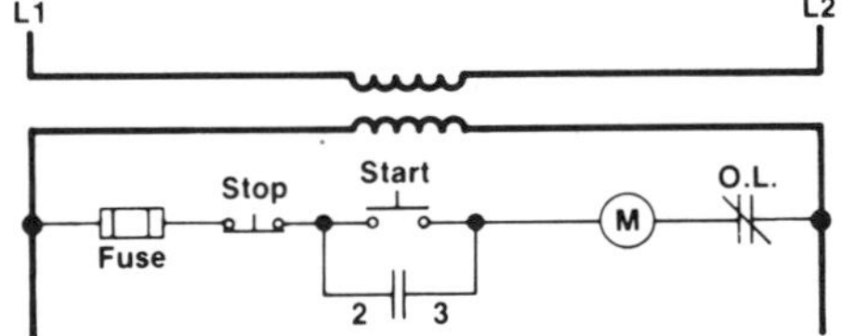

FIGURE 4–35 Step-down transformer in a control circuit. Wiring diagram and ladder diagram. (Allen-Bradley)

QUESTIONS

1. Why are wiring connections and terminal markings useful to the electrician?
2. What is the purpose of a manual start switch?
3. Describe a ladder diagram.
4. What is a maintained contact pilot device?
5. What is the difference between a two-wire and a three-wire control system?
6. What does *deenergized* mean with a relay?
7. What is meant by *undervoltage protection?*
8. What is a jumper used for?
9. How high can momentary transient rms current go?
10. What is locked rotor current?
11. What is a thermal protecιor?
12. Why are fuses needed in control circuits?
13. What are manual starting switches designed for?
14. What type of information does a line diagram provide?
15. What is another name for a line diagram?

16. What does L1 stand for on a diagram?
17. What does the term *two-wire* mean?
18. Why is a resistor built into the relay?
19. What do CR1 and CR2 mean?
20. What is the difference between a line drawing and a wiring diagram?

REVIEW PROBLEMS

The parallel circuit is utilized in all phases of electrical work. A review of the properties of such circuits can

$$I_T = I_{R_1} + I_{R_2} + I_{R_3} + \cdots$$

$$R_T = \frac{R_1 \times R_2}{R_1 + R_2}. \quad \frac{1}{R_T} = \frac{1}{R_1} + \frac{1}{R_2} + \frac{1}{R_3} + \cdots$$

benefit anyone working with electricity.

1. Two resistors, one of 4 Ω and one of 8 Ω, are connected in parallel. What is the resistance of the circuit?
2. A resistance of 36 Ω is connected in parallel with a resistor of 18 Ω. What is the total resistance?
3. A 5400-, 78,000-, and 112,000-Ω resistor are connected in parallel. If 10 mA flows through the 112,000-Ω resistor, what is the current flow in the other two resistors?
4. A 10,000-, 5000-, and a 15,000-Ω resistor are connected in parallel. If 30 mA flows in the 15,000-Ω resistor, what is the current flow in the other two resistors?
5. What is the total current in problem 3?
6. What is the total current in problem 4?
7. How many 100,000-Ω resistors must be connected in parallel to give a combined resistance of 10,000 Ω?
8. Two resistors of 3.3 and 4.7 kΩ are connected in parallel across a power source of 9 V. What is the current drain on the 9-V battery?

5

Switches

Objectives

After studying this chapter, you will be able to:

1. Explain how a drum switch operates.
2. Describe float switch action.
3. Tell how a joystick is used.
4. Tell why interlock switches are utilized for safety reasons.
5. Describe pushbutton interlocking.
6. Identify the various types of limit switches.
7. Explain pushbutton, pressure, and selector switch operation.
8. Explain the use and action of snap switches.
9. Explain start–stop switch actions.
10. Explain how toggle switches, temperature switches, and vacuum switches operate.

Switches come in many sizes and shapes. They have different types of contacts for various applications. In the case of motor controls the switches have to have contacts that are able to withstand the onrush of current at the start and the inductive kickback produced by the collapsing magnetic field of the motor when it is turned off. In this chapter we discuss those switches used to turn motors on and off and to reverse their direction of rotation. The switches have been arranged alphabetically by name.

DRUM SWITCH

One of the most commonly used switches for motor control is the drum switch (Fig. 5–1). It has the capability to reverse a motor or turn it off. Drum switches may be used for across-the-line starting and the reversing of ac polyphase, ac single phase, or dc motors. They are compact and inexpensive but ruggedly constructed.

Easy Conversion

Drum switches are field convertible from MAINTAINED to MOMENTARY operation. This conversion consists of removing the handle screw and handle, turning the shaft 180°, then replacing the handle and handle screw.

A handle is used to move contacts that are mounted on an insulated rotating shaft. Moving contacts make and break contact with stationary contacts within the controller as the shaft is rotated. This drumlike rotation or moving of the contacts causes the switch to be called a *drum switch*. It is also sometimes called a *cam switch* since it uses a cam action to accomplish its switching action.

Figure 5–1 shows that the drum switch is fully

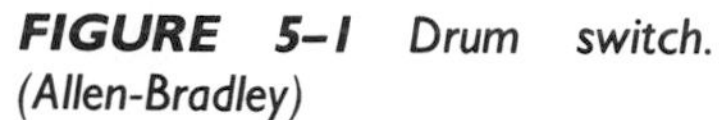

FIGURE 5-1 *Drum switch. (Allen-Bradley)*

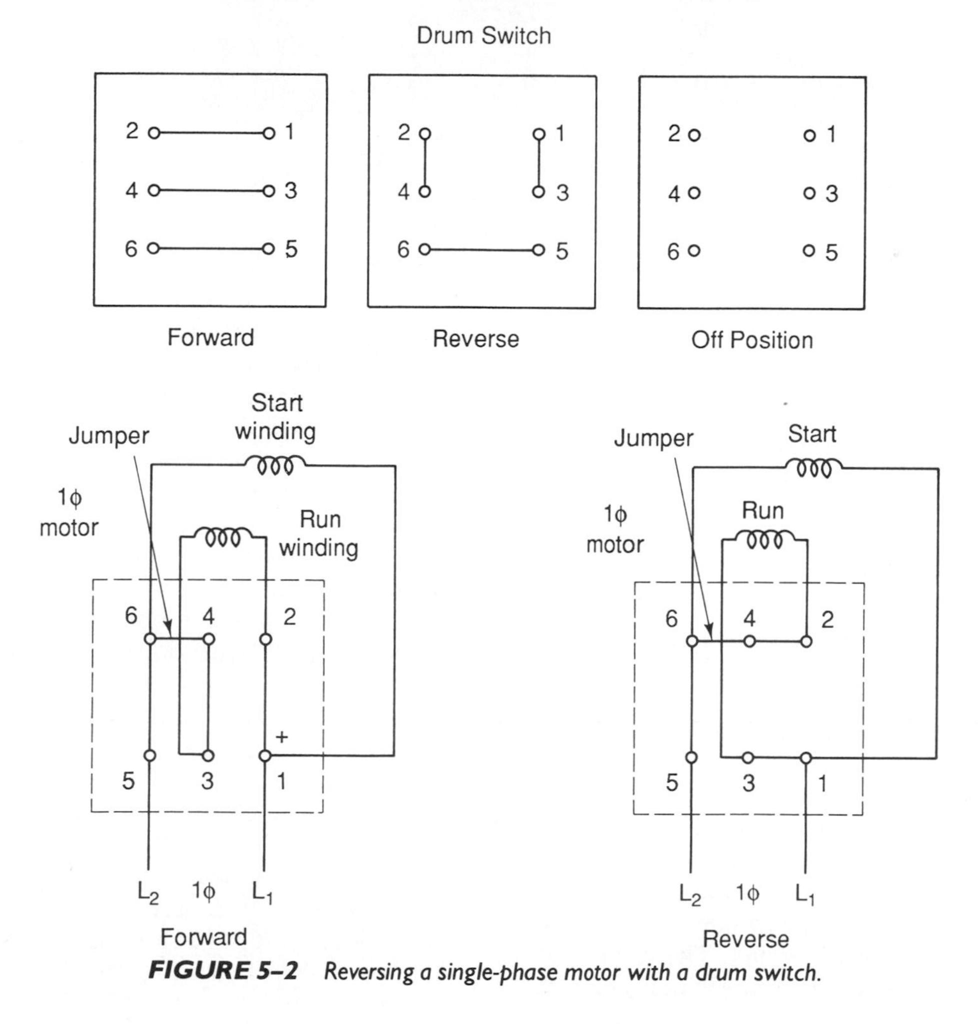

FIGURE 5-2 *Reversing a single-phase motor with a drum switch.*

enclosed and its insulated handle is used to move the contacts from point to point. Figure 5-2 shows how the six contacts of the switch are connected for off–reverse–forward positions. Note how the start and run windings connections are changed by the switching action to cause a reverse rotation.

Three-Phase Switching

Three-phase motors can also be reversed by using the drum switch. Keep in mind that only two of the phases have to be changed to cause a three-phase motor to reverse its direction of rotation (Fig. 5-3).

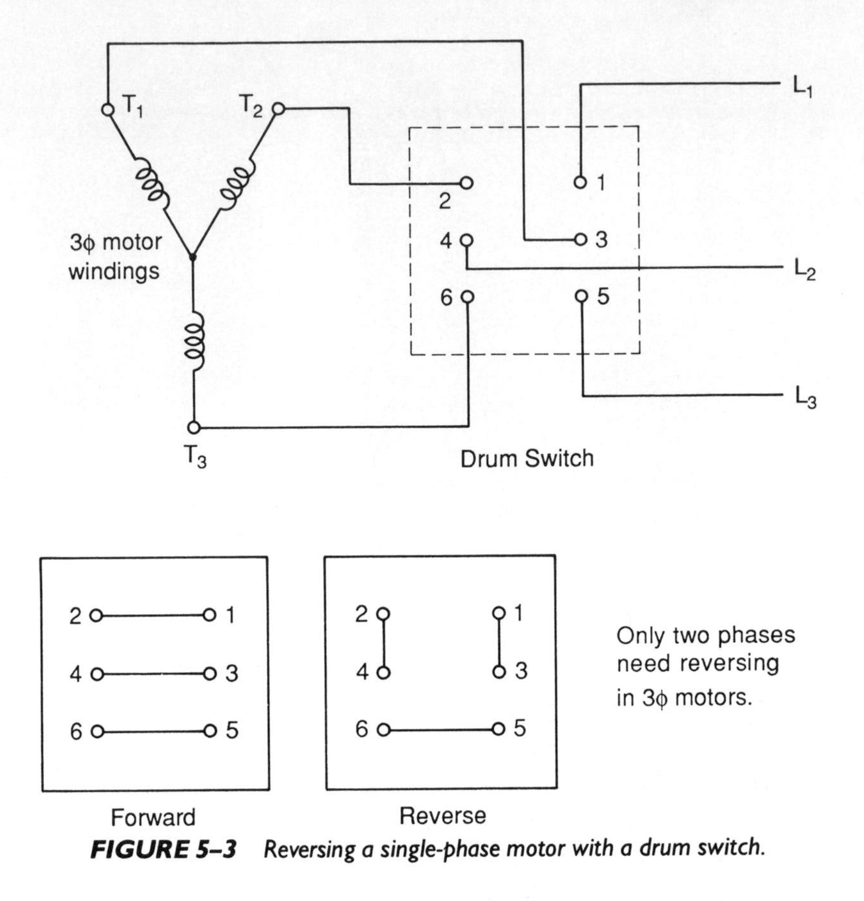

FIGURE 5–3 *Reversing a single-phase motor with a drum switch.*

DC Switching

Dc motor control can also be provided by the drum switch (Fig. 5-4). The reversal is accomplished by changing the polarity. Note how the switching action causes the armature and field coil windings of the series motor to be placed in series in the forward position. Then trace it out for the reversal of polarity in the REVERSE position of the drum switch. The compound dc motor is also reversed by changing the polarity provided to the armature. The field coil maintains the same polarity in both positions of the switch. Note how the coils are connected to − and + in the same way for both positions of the switch. Then take a look at how the armature has − and + switched to make it rotate in the opposite direction. That is accomplished by 5 and 6 on the switch being the same in both the forward and reverse positions.

Some dc motors use interpoles (commutating windings) to suppress brush arcing. Therefore, it is a good idea to keep in mind that these windings are part of the armature circuit.

FLOAT SWITCHES

Float switches provide automatic control for motors operating tank or sump pumps. They are built in five styles and can be supplied with accessories to provide rod or chain operation, wall, or floor mounting (Fig. 5-5). Float switches are designed for automatic control of ac or dc pump motor magnetic starters, or for direct control of light motor loads.

A great variety of operating accessories are available to make float switches operate by many methods, depending on the position, depth of a tank, or sump (Fig. 5-6). Stainless steel floats, rods, chains, and stop collars are available for use in certain corrosive liquids. They are made in both normally open and normally closed configurations or both. Figure 5–7 shows how the float switch is connected in the circuit. Note the normally open and normally closed operation of the switch. This is one of the most popular means of pumping water from a basement automatically and prevents damage from seepage or flooding of any type. Once the level of the water causes the switch to close, it turns on the pump motor. If it is a small fractional-horsepower motor, the switch can control the motor directly. If it is a larger-horsepower pump motor, the float switch is used to energize a magnetic motor starter.

FLOW SWITCHES

A flow switch is used to detect the flow of liquids, water, oil, or other gases in a pipe or duct. It is nothing more than a normally open or normally closed

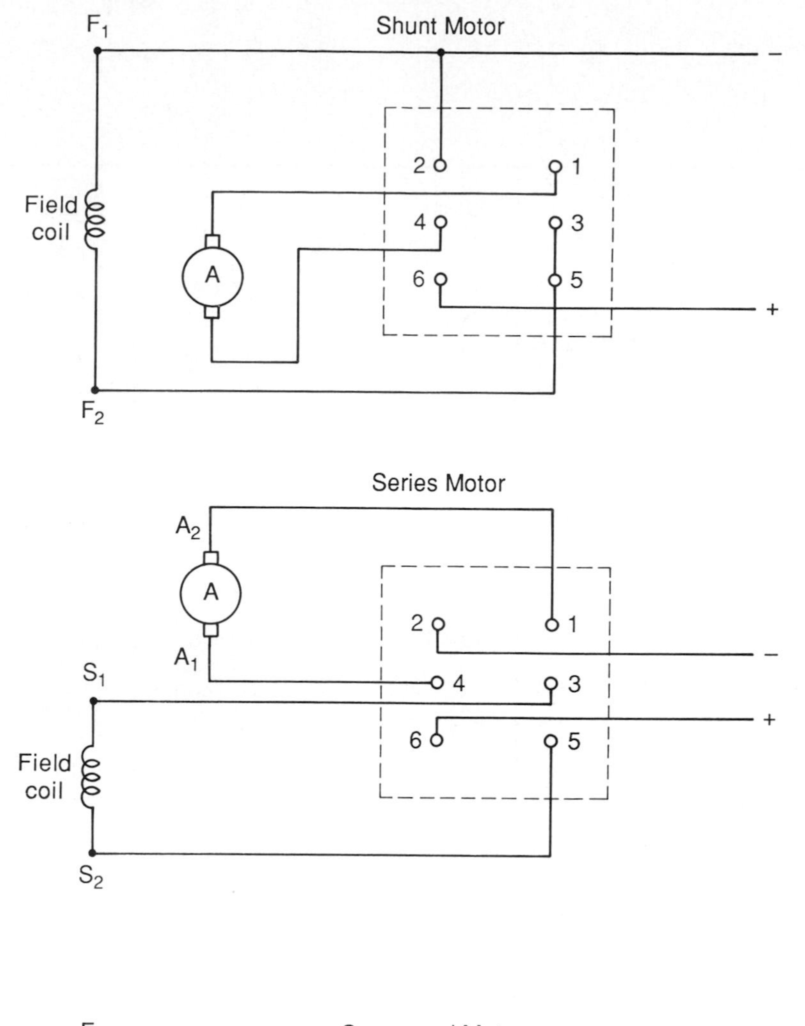

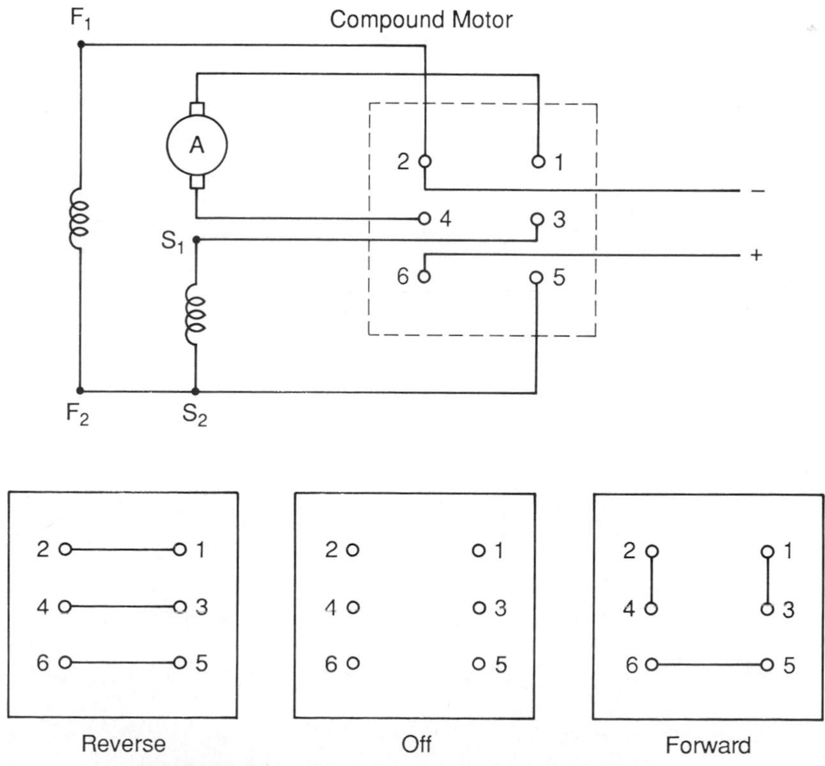

FIGURE 5–4 *Dc motor control by drum switch.*

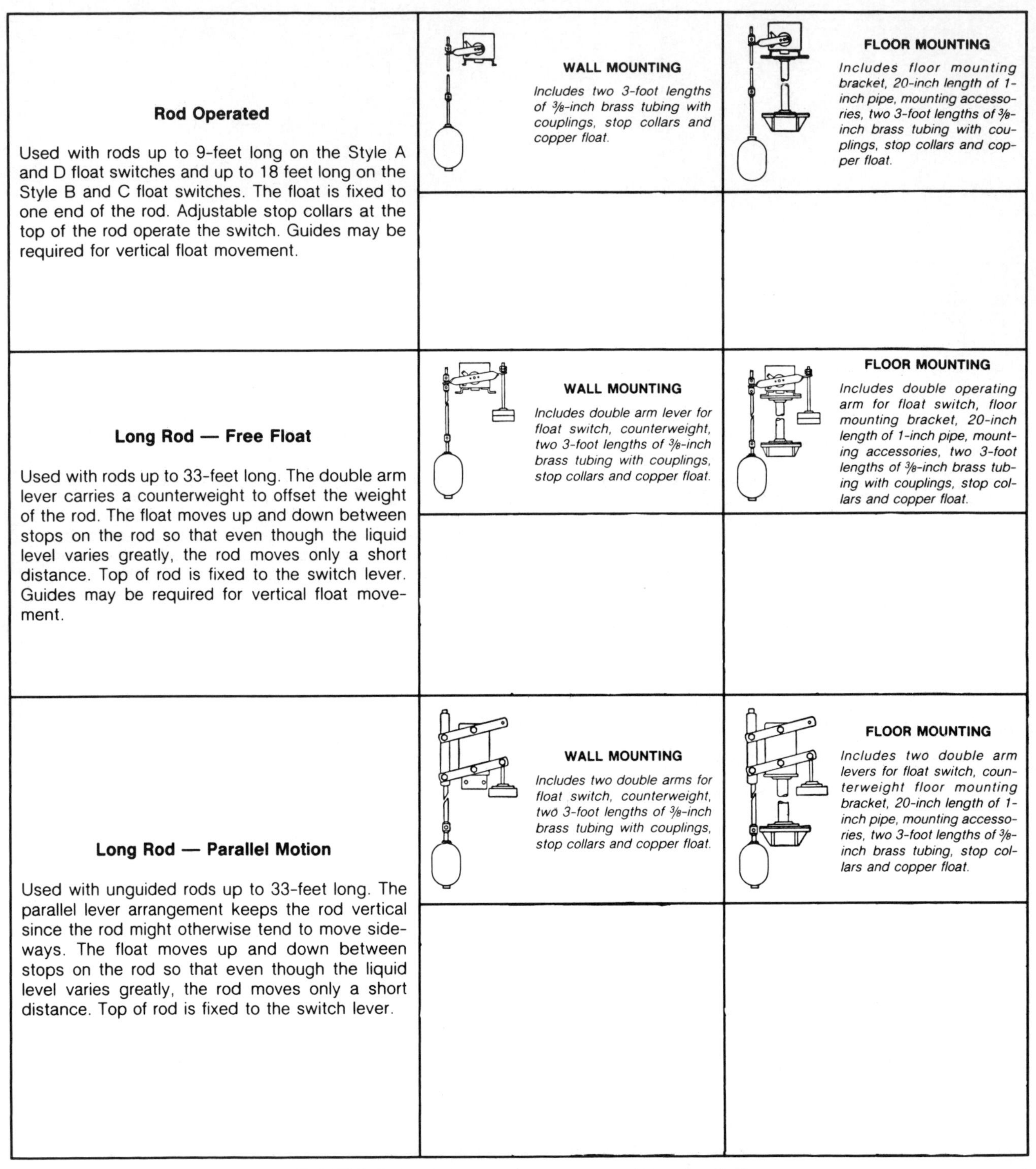

FIGURE 5–5 *Automatic float switches. (Allen-Bradley)*

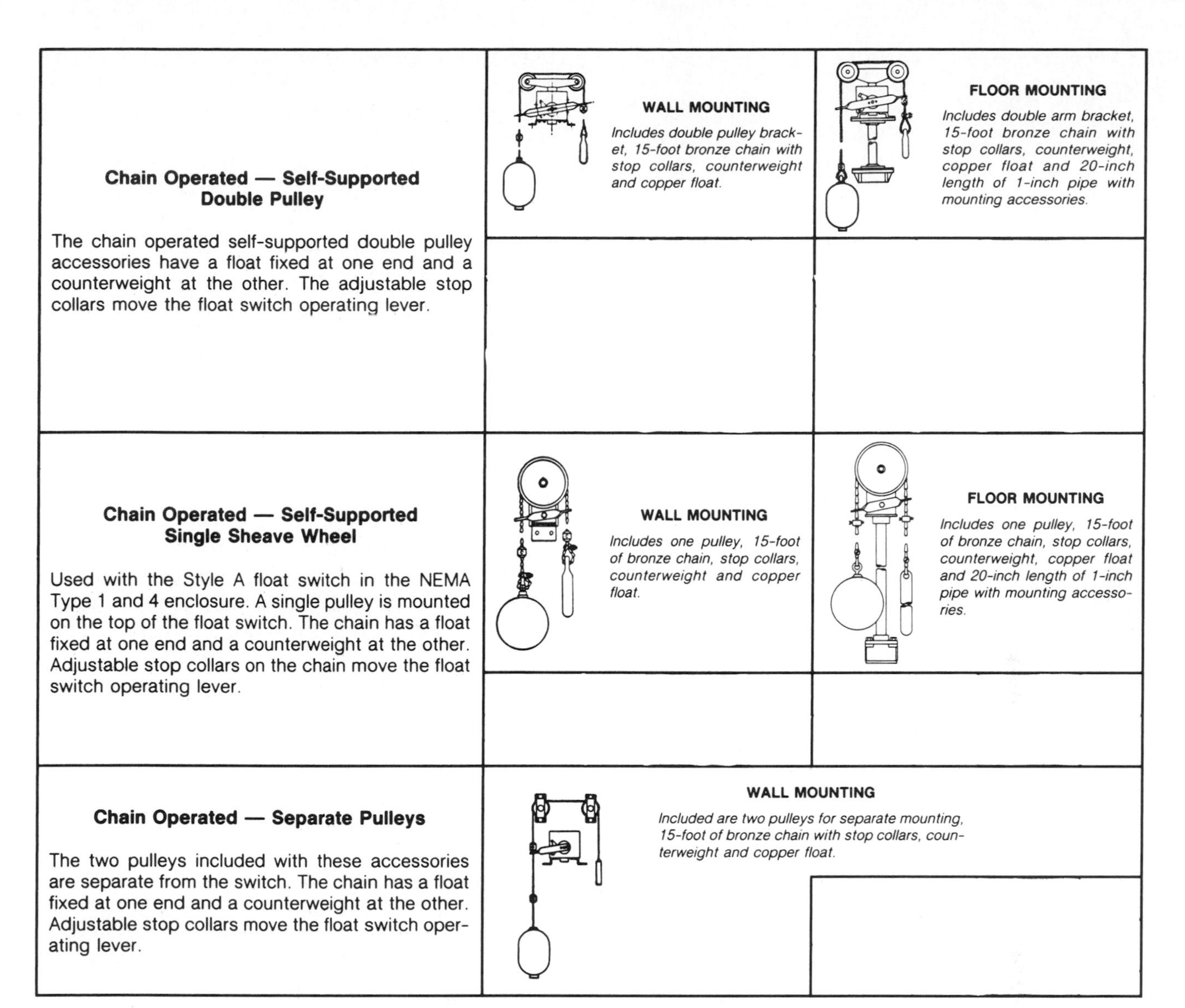

Chain Operated — Self-Supported Double Pulley The chain operated self-supported double pulley accessories have a float fixed at one end and a counterweight at the other. The adjustable stop collars move the float switch operating lever.	**WALL MOUNTING** *Includes double pulley bracket, 15-foot bronze chain with stop collars, counterweight and copper float.*	**FLOOR MOUNTING** *Includes double arm bracket, 15-foot bronze chain with stop collars, counterweight, copper float and 20-inch length of 1-inch pipe with mounting accessories.*
Chain Operated — Self-Supported Single Sheave Wheel Used with the Style A float switch in the NEMA Type 1 and 4 enclosure. A single pulley is mounted on the top of the float switch. The chain has a float fixed at one end and a counterweight at the other. Adjustable stop collars on the chain move the float switch operating lever.	**WALL MOUNTING** *Includes one pulley, 15-foot of bronze chain, stop collars, counterweight and copper float.*	**FLOOR MOUNTING** *Includes one pulley, 15-foot of bronze chain, stop collars, counterweight, copper float and 20-inch length of 1-inch pipe with mounting accessories.*
Chain Operated — Separate Pulleys The two pulleys included with these accessories are separate from the switch. The chain has a float fixed at one end and a counterweight at the other. Adjustable stop collars move the float switch operating lever.	**WALL MOUNTING** *Included are two pulleys for separate mounting, 15-foot of bronze chain with stop collars, counterweight and copper float.*	

FIGURE 5–5 *continued*

FIGURE 5–6 *Float switch accessories. (Square D)*

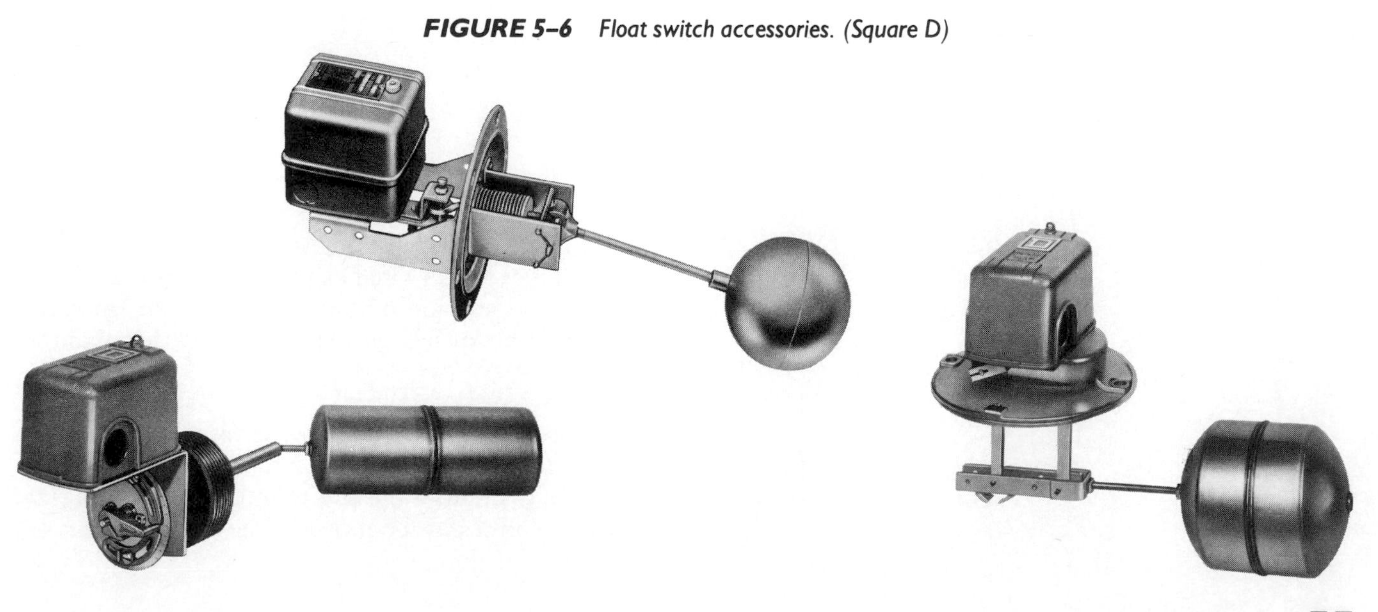

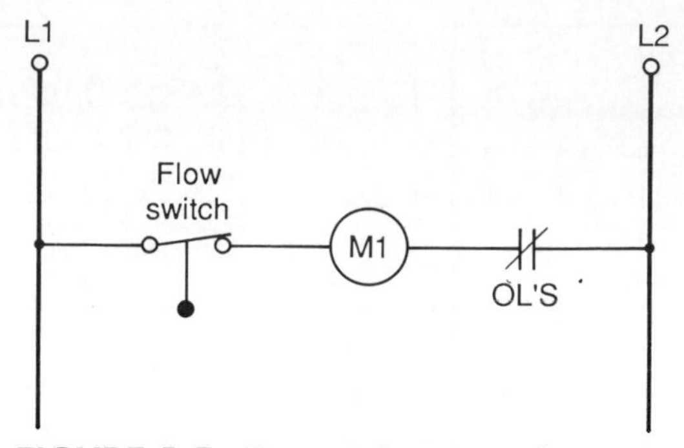

FIGURE 5–7 *Flow switch connected in a circuit.*

FIGURE 5–8 *Oil-tight limit switch with nylon rod. (Allen-Bradley)*

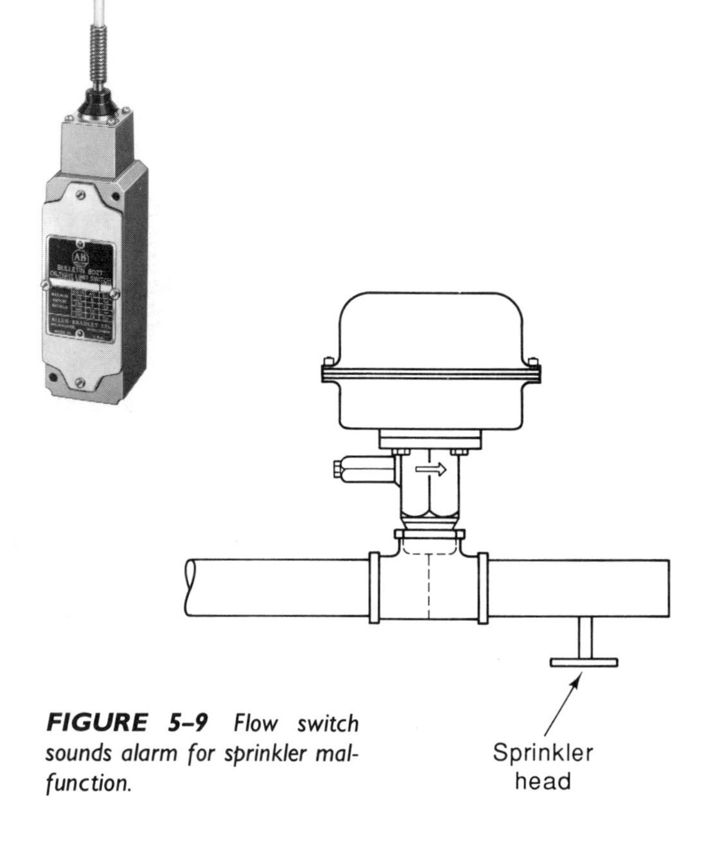

FIGURE 5–9 *Flow switch sounds alarm for sprinkler malfunction.*

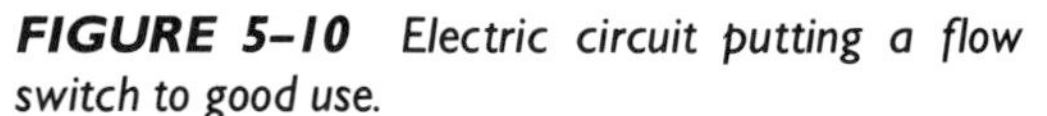

FIGURE 5–10 *Electric circuit putting a flow switch to good use.*

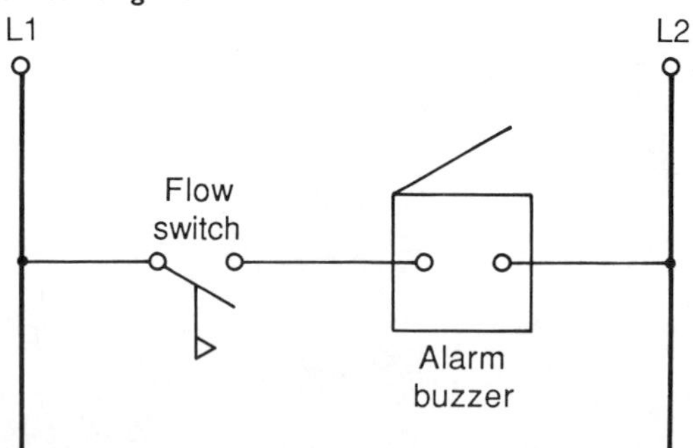

switch with an operator extending from the switch enclosure (Fig. 5–8). Most are adjustable so that allowances can be made for buildup or changes in the application of the switch.

These switches are very useful in a number of locations. They may be used to detect the flow of air in a wind tunnel or in air-conditioning ductwork, and they may be used to set off alarms or start motors.

They are used to indicate the flow of water in a sprinkler system. This can be used to alert the person in charge in case a sprinkler system begins to operate without being triggered by fire. The sounding of the alarm can allow the operator to shut off the sprinkler system before too much water damage occurs (Fig. 5–9).

Clogged air filters can be detected by flow switches. They can be adjusted to indicate when a filter is clogged and is not allowing enough air through to keep a motor or other device properly cooled. Airflow detection is important in heating systems. Electrical heating systems are especially prone to damage if the proper amount of air is not allowed to circulate over the heating coils. The coils can burn out if not properly cooled by a steady flow of air. See Fig. 5–10 for the electrical circuit that puts the flow switch to work.

FOOT SWITCHES

Foot switches are just what the name suggests. They are operated by pressure directed by a human foot. They can be found on stapling machines, older-model automobile headlight dimmers, and any number of machine operations that require a person to use both hands and thus operate the on–off switch by foot. This type of switch is usually a heavy-duty type and comes in a variety of sizes, voltages, and shapes. Many limit switches can be converted to foot operation.

JOYSTICK

The joystick is a rather recent device and is used most commonly in conjunction with computers, although it can be used to control a number of operations (Fig. 5–11). This type of switch can be used to good advantage in the operation of a crane, where the operator has to keep his or her eyes on the load rather than on the switch.

Two types of joysticks are available: the standard model, where the operator is free to move from position to position, and the latched lever type. The latched lever requires the operator to lift a locking ring before the lever can be moved. The switching action

FIGURE 5–11 *Joystick switch. (Square D)*

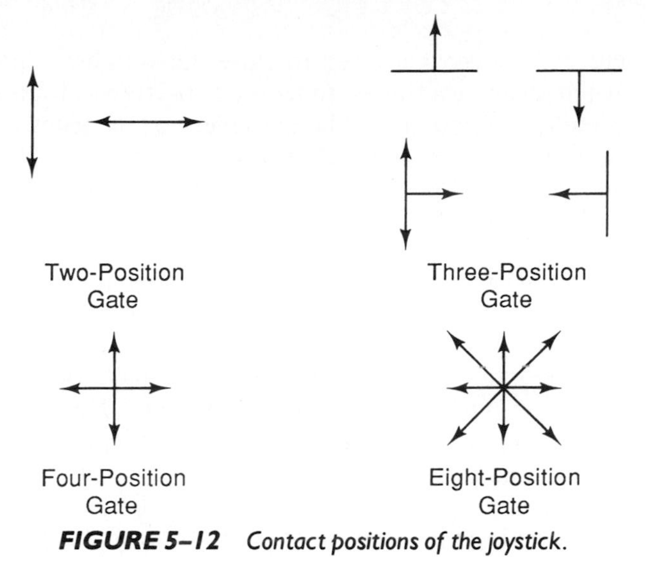

FIGURE 5–12 *Contact positions of the joystick.*

can be momentary or maintained contact or a combination of both.

Note the different combinations of contact positions available in Fig. 5–12. The center position on the joystick is used for the off or stop control. Figure 5–13A shows how the joystick is wired into a circuit and can operate at least four loads, depending on

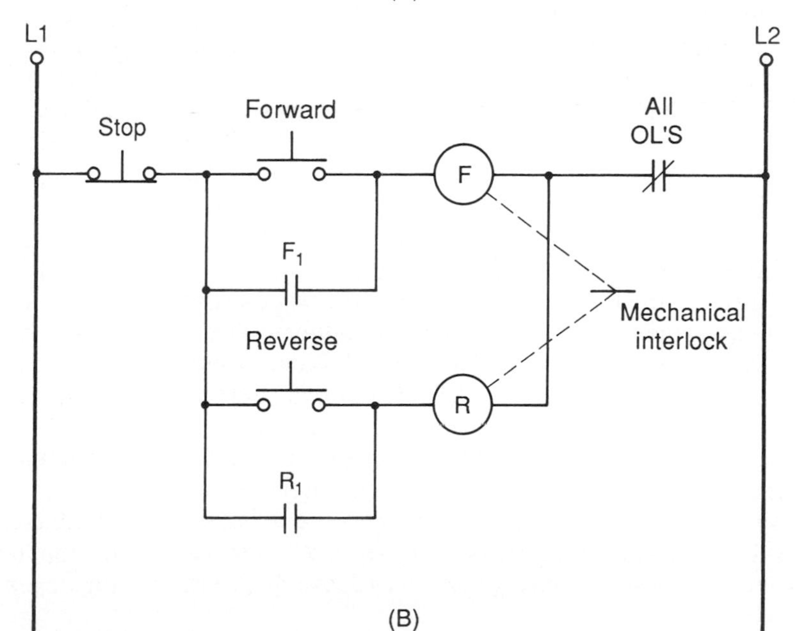

FIGURE 5–13 *(A) Joystick contacts. Note the black dot that indicates left, right, up, and down positions of the handle. (B) Mechanical interlocking in reverse and forward with a two-pushbutton station..*

where the lever is located to close the switches. The switches can be adjusted to have a two-, three-, four-, or eight-position gate. The placement of the gates in the switch causes it to change its possibilities. The gates are metal plates that can be changed as needed by the person who usually makes repairs and installations.

Whenever a job requirement dictates the utilization of moving an object from left to right or up and down, the joystick can be made to control the movement needed. A good example is operating a foundry ladle or a hoist for any number of jobs. There are some advantages to joystick operation of a load. Inasmuch as no two contacts can be closed at the same time, there is a slight delay between the break of one circuit and the make of another by the switching action. This means that the load will not have to change directions quickly.

INTERLOCK SWITCHES

Interlock switches are used for safety purposes. The idea is to make sure that the forward motion of the motor has been stopped before the reversing action is started by changing the electrical path through the motor windings. There are three ways to incorporate interlocking and the protection needed. Auxiliary contact interlocking, mechanical interlocking, and pushbutton interlocking are the three methods most commonly used.

Mechanical Interlocking

A ladder diagram can be used to illustrate the way a mechanical interlock functions (Fig. 5-13B). The dashed lines from the forward coil to the reverse coil are used to indicate a mechanical interlock. This type of interlock switching is usually provided by the manufacturer of the starter. A piece of material is usually inserted so that when one pushbutton is depressed, the other cannot be depressed. The FORWARD pushbutton is depressed, completing the circuit to the F coil. The coil is energized, but in energizing, it closes contacts F1. This means that the pushbutton can be released and the circuit stays energized by completing the circuit through the F1 contacts instead of the pushbutton contacts. This is often referred to as *memory*. The F coil stays energized until the STOP switch is depressed. Once it is depressed, the coil deenergizes and contacts (F1) open. The STOP button returns to its normally closed position and the REVERSE button can be depressed, energizing the R coil. The coil energizes and closes contacts (R_1) that are across the REVERSE pushbutton. The circuit stays energized until the STOP switch is depressed, causing the coil to deenergize. The

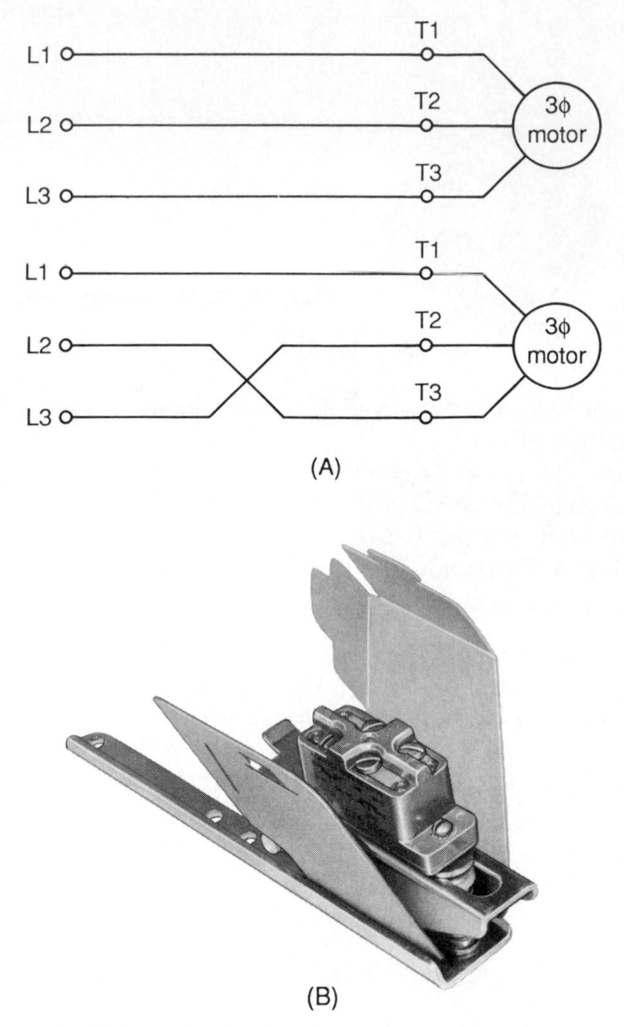

FIGURE 5–14 *(A) Reversing a three-phase motor means changing two phases. (B) A one- or two-pole electrical interlock can be added to the disconnect switch or circuit breaker. (Square D)*

mechanical interlock prevents both forward and reverse from being energized if the buttons were to be depressed at the same time. It also prevents the reverse power being applied while forward power is still being applied to the motor.

Keep in mind that the contacts for the mechanical interlock are not shown in a line or ladder diagram. The assembly of the switches causes one to be disabled when the other is being depressed. In this way, short circuits and burnouts are prevented. Take a look at Fig. 5-14. The three-phase motor can be reversed by changing any two of the phases. Reversing motor starters are available in vertical or horizontal construction.

Pushbutton Interlocking

Double-circuit pushbuttons can be used for pushbutton interlocking (Fig. 5-15). This type of interlocking depends on the electrical circuit for its operation.

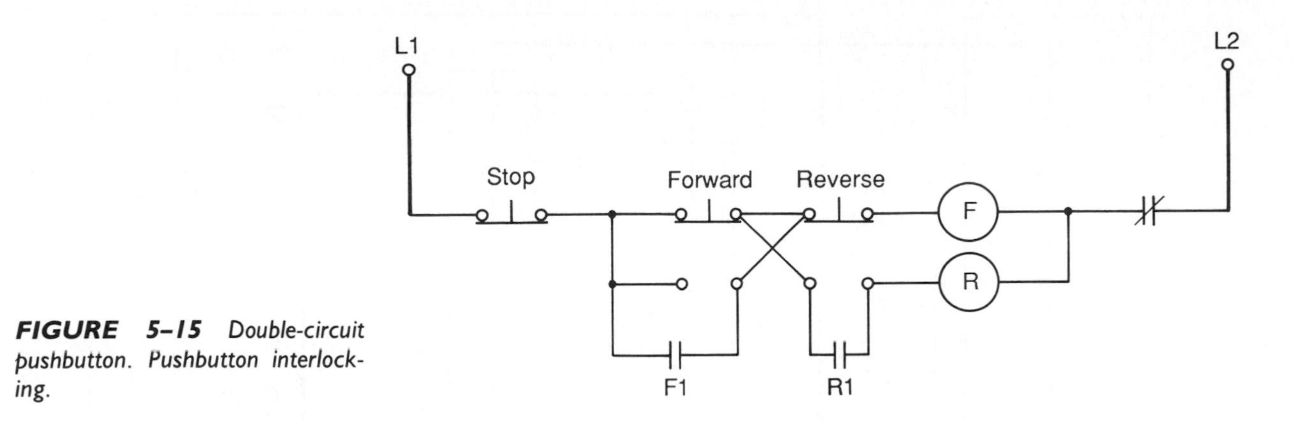

FIGURE 5-15 Double-circuit pushbutton. Pushbutton interlocking.

The idea is to make sure that both coils are not energized at the same time.

Note that the STOP button is normally closed in Fig. 5-15. That means that when the FORWARD button is pressed, coil F is energized and the normally open (NO) contact F1 closes to hold in the forward contactor. Because the normally closed (NC) contact are used in the forward and reverse pushbutton units, there is no need to press the STOP button before changing direction of rotation. If the REVERSE button is pressed while the motor is running in the forward direction, the forward control circuit is deenergized and the reverse contactor is energized and held closed.

If reversing is repeated quickly, it may cause the motor to overheat and overload relays and fuses to function as designed. In most applications for motors it is best for the motor to coast to a stop before reversing. However, if plugging is called for, the starter has to be rated to handle the reversing, especially if it occurs at a rate of more than five times per minute.

Auxiliary Contact Interlocking

Normally closed (NC) auxiliary contacts on the starter can be used for interlocking. This is done by a separate set of contacts on the reversing starter (Fig. 5-16). The three-phase motor shown has a forward (F) and reverse (R) coil that energize to close contacts in L1, L2, and L3 to complete the power to the motor through its overload (OL) contactors, inserted inside the motor windings. The forward pushbutton is depressed. This completes the circuit to the F coil because the R normally closed contacts (R_2) are closed. Once coil F is energized, it holds the F_1 contacts across the forward switch contacts closed, keeping coil F energized. When coil F energized, it opened contacts F_2, making

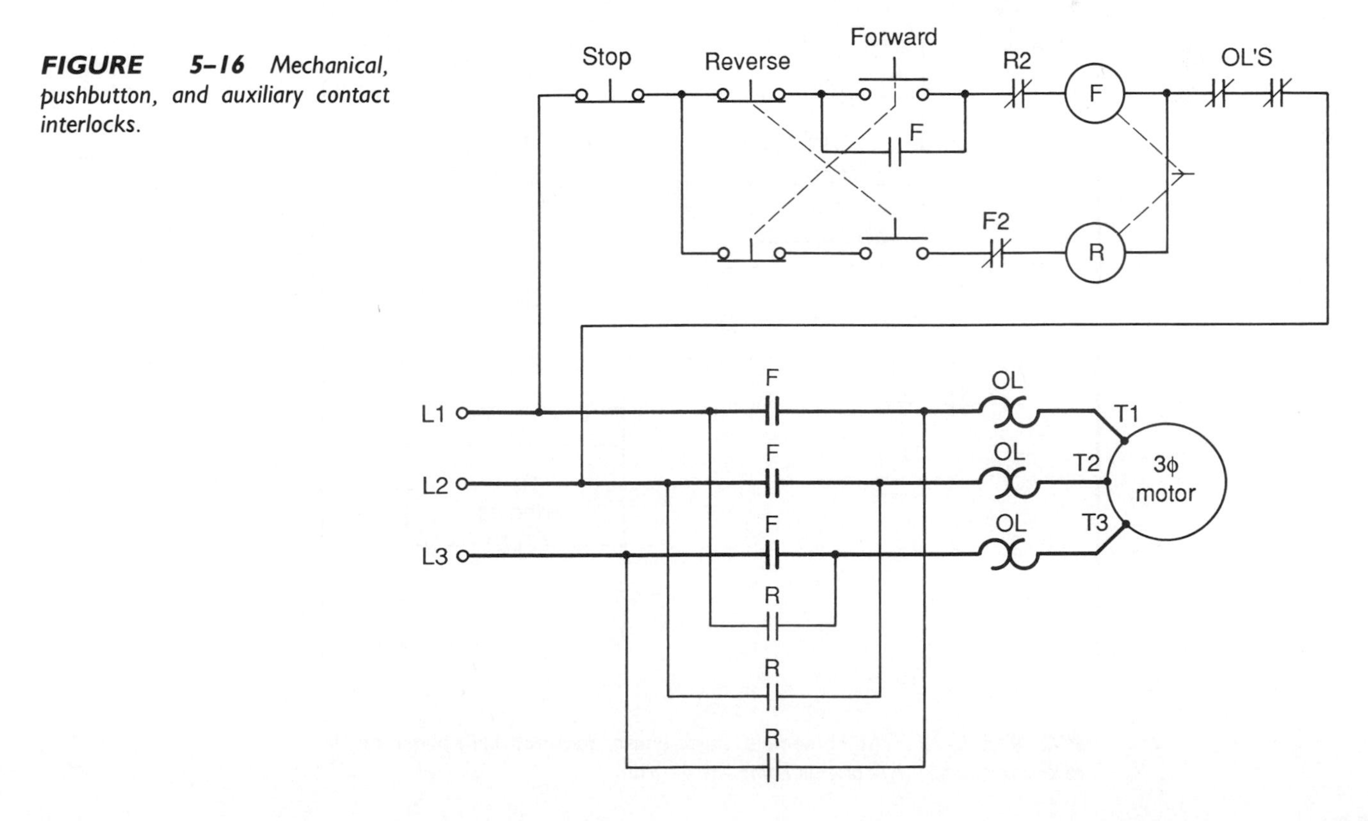

FIGURE 5-16 Mechanical, pushbutton, and auxiliary contact interlocks.

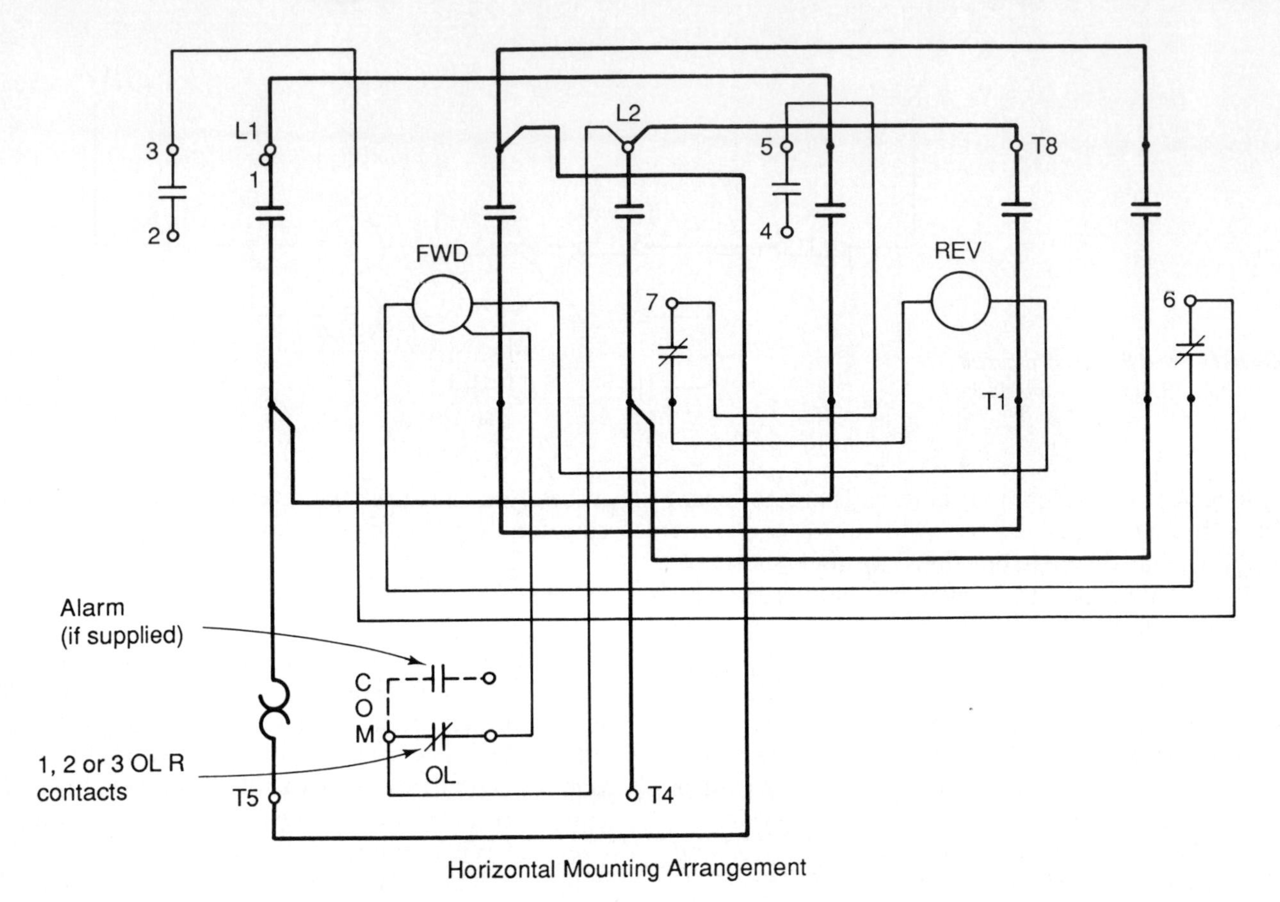

(A)

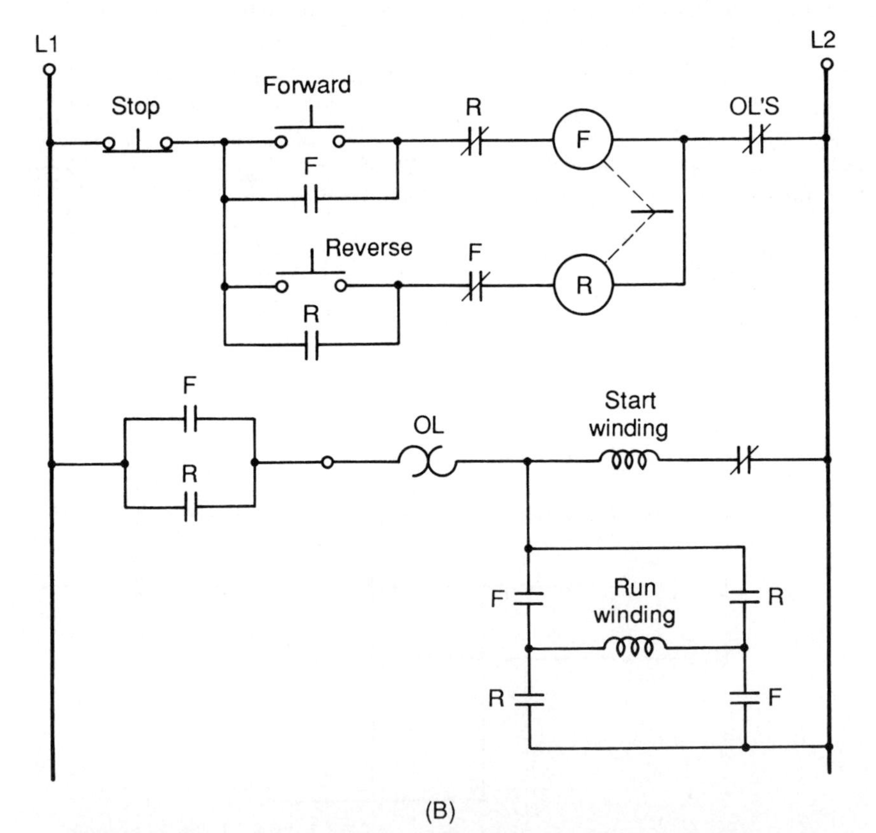

(B)

FIGURE 5–17 *(A) Three-pole, single-phase, four-lead split-phase motor reversing starter; (B) corresponding line diagram.*

it impossible for the R coil to energize. Contacts F_2 and R_2 are auxiliary contacts and are sometimes mounted on the outside of the reversing starter. In reversing the single-phase motor, keep in mind that either the running or the starting winding leads are reversed, not both. Figure 5–17 shows how this is done with a reversing starter using single-phase.

LIMIT SWITCHES

As the name implies, the limit switch limits the operation of a machine that is connected in line with the switching action (Fig. 5-18). Many industrial production lines use this type of switch to limit the travel of various devices on the line. A good example of the use of a limit switch is a garage door opener. When the door is lifted it must stop before it hits the motor. A limit switch turns off the motor before the door crashes into the motor that is pulling it up. Then, too, the motor must be turned off when the door is lowered, or else the motor keeps running after the door hits the floor. Limit switches do the job and make it possible for these operations to be semiautomatic, or once the action has been initiated, it is limited by limit switches to conform to the physical conditions.

Limit switches can be used to start, stop, forward, reverse, recycle, slow down, or speed up various operations. Inasmuch as they can do all these things, a variety of sizes and shapes are needed.

Limit switches are made up of two parts, the electrical contacts and the mechanical device that operates the on-off function of the contacts. The actuating mechanism may take a number of forms.

Types of Limit Switches (Fig. 5–19)

Heavy-duty precision oil-tight (type C) provides long electrical and mechanical life together with easy installation and wiring. This type is available in a variety of head and body styles, including an explosion-proof version that is also watertight and submersible. Also available are standard, logic reed, and power reed contacts, as well as many special features.

Miniature enclosed reed switch (type XA) is a small, inexpensive die-cast zinc switch utilizing a her-

FIGURE 5–18 *Limit switches: (A) sealed contact switch; (B) oil-tight limit switch; (C) time-delay oil-tight limit switch.*

(A)

(B)

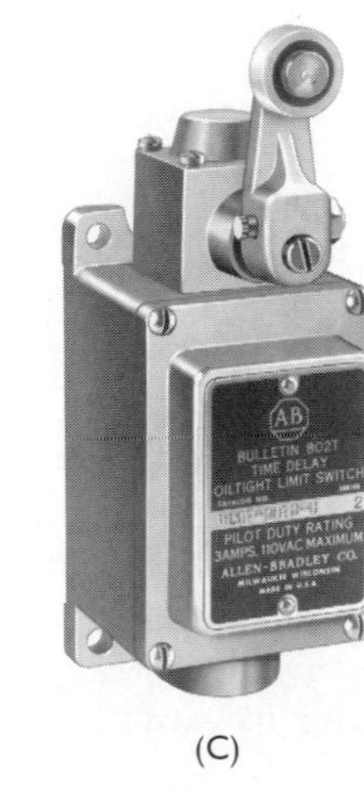

(C)

FIGURE 5–19 *Limit (position) switches. (Square D)*

HEAVY DUTY PRECISION OILTIGHT **TYPE C**	The Type C provides superior electrical and mechanical life along with easy installation and wiring. Available in a variety of head and body styles, including an explosion proof version that is also watertight and submersible. Also available are standard, logic reed and power reed contacts, as well as many special features. The Type C should be the first choice for all applications.
HEAVY DUTY PRECISION OILTIGHT **TYPE B**	Most Type B devices are obsolete — Select the Type C equivalent for new installations.
MINIATURE ENCLOSED REED SWITCH **TYPE XA**	The Type XA is a small, inexpensive die cast zinc switch utilizing a hermetically sealed reed for the contact mechanism. Prewiring and potting combined with the sealed reed make this switch a good choice where contact reliability and environmental immunity are required along with small size and low cost.
HEAVY DUTY OILTIGHT **TYPE T** FOUNDRY **TYPE FT**	If load exceeds Type C contact ratings, if a required operating sequence is not available on the Type C or if high trip and reset forces are required, use the Type T. Use FT in foundries or mills where one or more of the above listed Type T features are required and where hot, falling sand or similar foreign material could cause jamming of other limit (position) switches.
PRECISION OILTIGHT **TYPE AW**	The Type AW is suitable for applications requiring a precision, oiltight switch. It is suggested especially for applications requiring a lever type switch for use in extremely low ambients or micrometer adjustment on a plunger type switch.
INTERNATIONAL POSITION SWITCHES **TYPE BD**	New Product — Meets International standards, CENELEC EN 50041. Uses same lever arms as the Type C and AW.
SNAP SWITCHES	Use snap switches on applications requiring a basic contact mechanism with or without operator where an enclosure is furnished separately or not required.

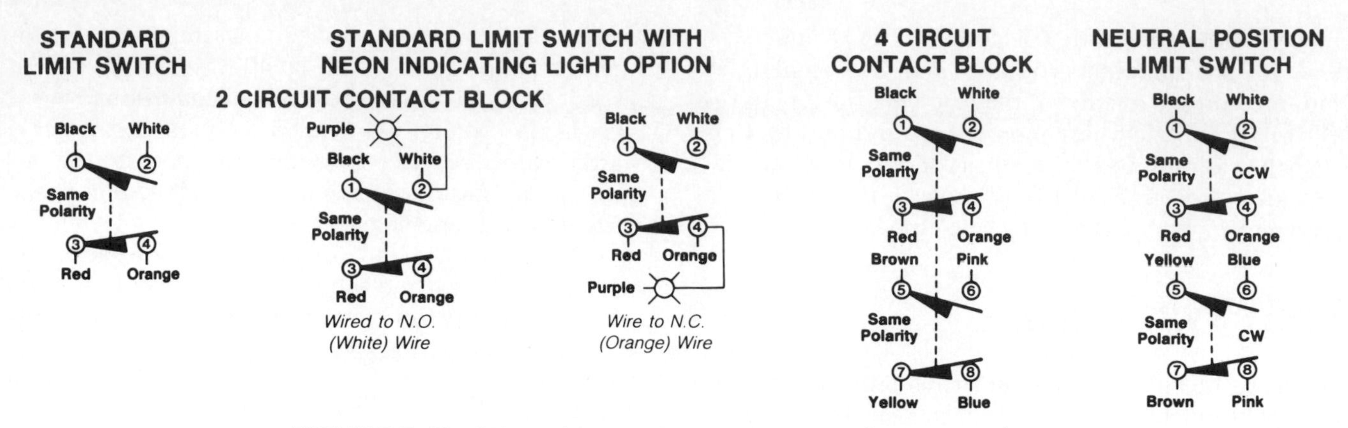

FIGURE 5–20 *Wiring diagrams showing contact configurations and terminal wire color code. (Allen-Bradley)*

metically sealed reed for the contact mechanism. Prewiring and potting combined with the sealed reed make this switch a good choice where contact reliability and environmental immunity are required along with small size and low cost.

Heavy-duty oil-tight foundry (type FT) is used in foundries or mills where one or more of the type T features are required and where hot, falling sand or similar foreign material could cause jamming of other limit (position) switches.

Snap switches are used on applications requiring a basic contact mechanism with or without operator where an enclosure is furnished separately or not required.

FIGURE 5–21 *Gravity return limit switches: (A) slotted shaft to aid in adjustment; (B) limit switch with steel operating lever; (C) example of clockwise operation.*

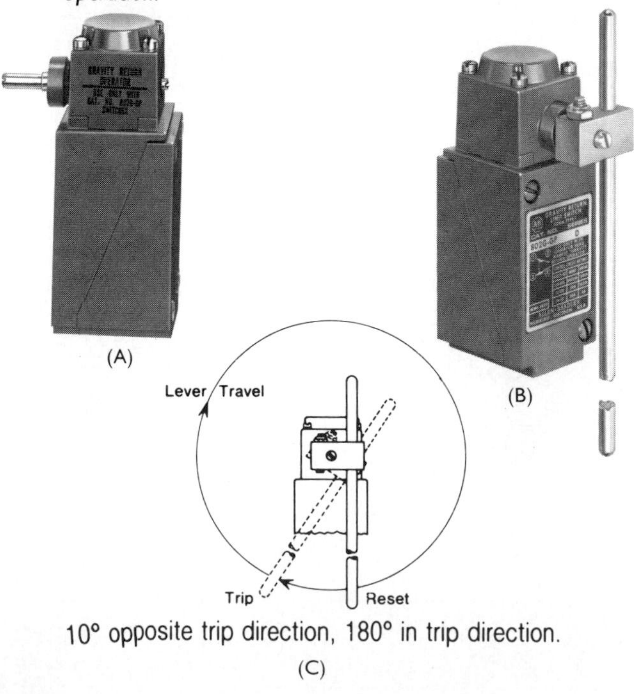

There are limit switches designed for special applications or found to be particularly useful under certain conditions. Snap-action limit switches are designed to snap over (trip instantly) once the mechanism that operates the limit switch has traveled the required distance to trip the limit switch, regardless of the speed with which it travels. This type should be used whenever machine operation acts at a slow rate of speed or where machine motion is short. Figure 5–20 shows some of the wiring diagrams and contact configuration and terminal wire color code for limit switches.

Gravity return limit switches are designed for conveyor-type operations with small, lightweight moving objects. They have an extremely light operating torque and use the action of gravity on the lever arm to reset the contacts (Fig. 5–21). Limit switches are also described according to how they work. They may be top push, spring return, roller type, rod type, side push, lever type, and maintained contact.

Limit Switch Circuits

Figure 5–22 shows the limit switch in a single-station, maintained-contact configuration. The START button mechanically maintains the contacts that take the place of hold-in contacts. Depressing the START button maintains the circuit. By depressing the STOP button it breaks the circuit by opening the start contacts. If the contactor is deenergized by a power failure or overload operation, the start contacts are unaffected. The motor restarts automatically.

PRESSURE SWITCHES

The control of pumps, air compressors, welding machines, lube systems, and machine tools requires control devices that respond to the pressure of a medium such as water, air, or oil. The control device that does

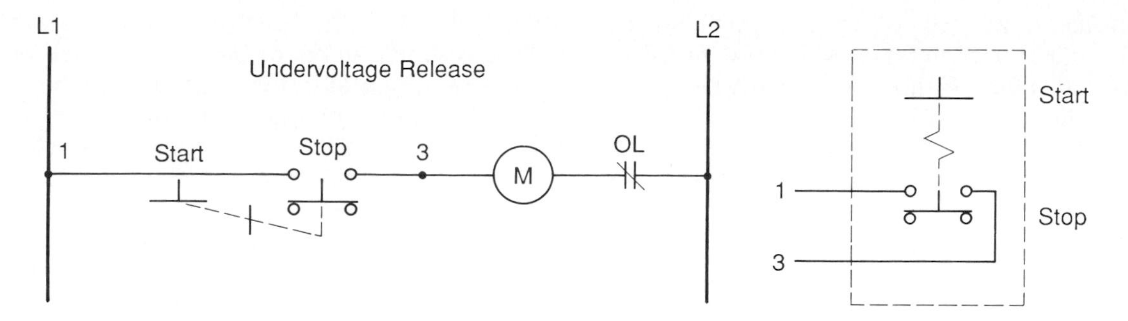

The start button mechanically maintains the contacts that take the place of hold-in contacts. Depressing the start button maintains the circuit; depressing the stop button breaks the circuit by opening the start contacts. If the contactor is deenergized by a power failure or overload operation, the start contacts are unaffected. The motor restarts automatically.

FIGURE 5–22 *Single-station, maintained-contact buttons. (Allen-Bradley)*

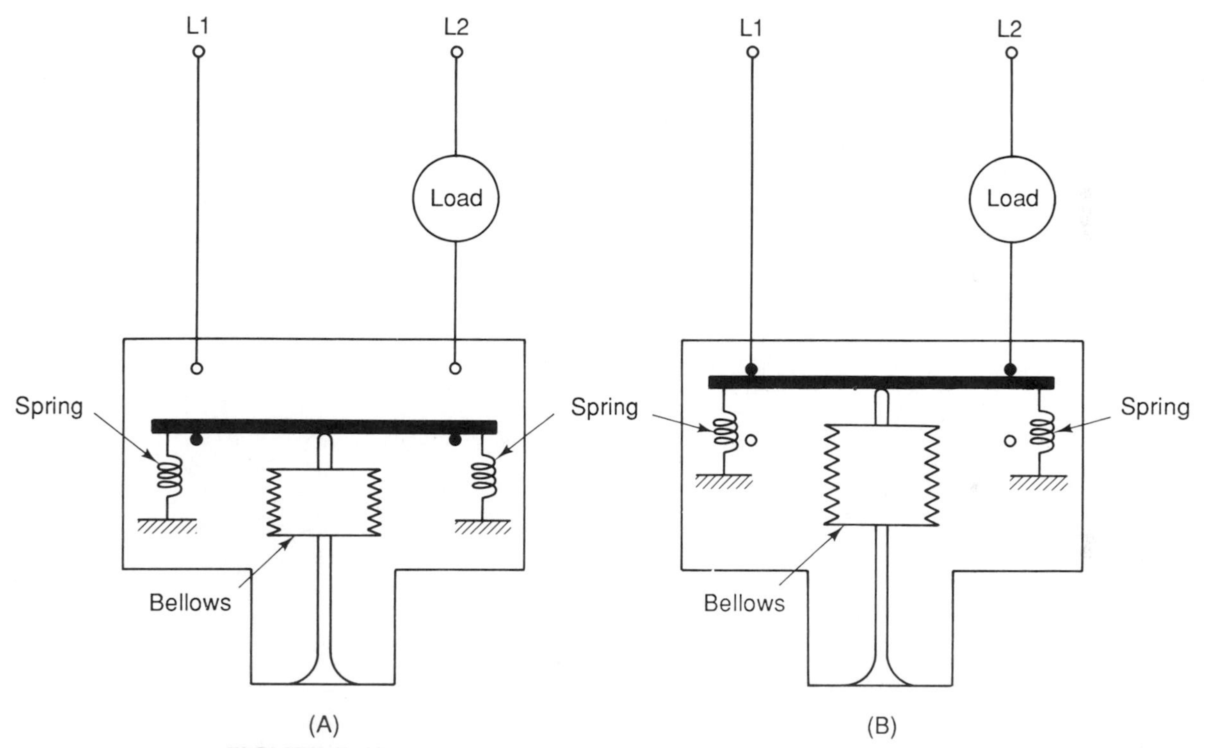

FIGURE 5–23 *Pressure switch operation: (A) open switch, bellows contracted; (B) closed switch, bellows expanded.*

this is a pressure switch. It has a set of contacts that are operated by the movement of a piston, bellows, or diaphragm against a set of springs. The spring pressure determines the pressures at which the switch closes and opens it contacts (Fig. 5–23).

Industrial pressure switches are designed for use in pneumatic and hydraulic systems in a wide variety of applications. Open types are suitable for panel mounting where permitted. They can be used on air, water, oil, or any pressure media that is compatible with the actuator material. However, piston-actuated devices used on dry gas media will have reduced seal life due to lack of seal lubrication.

Pressure controls are accurate, rugged, compact, and adaptable to a wide variety of pressure applications. The control has a snap-action precision switch equipped with silver contacts. Contact force is maintained at a high value up to the instant of snap-over. This avoids dead-center conditions. Normal industrial vibration has little effect on the positive opening and closing of the contacts. Straight in-line and relatively friction-free construction provides accurate and consistent operation regardless of the angle at which the control is mounted.

Long-life copper-alloy bellows are designed for use with air, water, oil, noncorrosive liquids, vapors, or gases in a series of pressure ranges from 30 in. vacuum (Hg) to 900 psi. Stainless steel bellows are available for use with many of the more corrosive liquids or gases at pressures up to 375 psi.

Piston-type controls are designed for use on oil only and must not be used on air or gases, water, or liquids that will rust cast iron. All pistons are provided with a drain, which should be connected to an oil return line with the reservoir vented to atmosphere. Drains should never be plugged, as back pressure on the diaphragm assembly may change the setting and could result in forcing fluid into the enclosure. Controls with seal rings generally do not require return lines, provided that a limited amount of leakage is not objectionable. Phosphate ester–based hydraulic fluids require a special diaphragm assembly. Figure 5–24 shows the symbols for the various types of pressure-operated switches.

FIGURE 5–24 *Pressure controls symbols for automatic operation and manual reset. (Allen-Bradley)*

CONTACT BLOCKS

Symbol		Description
Pressure Controls	Temperature Controls	
AUTOMATIC OPERATION		
		Single pole double throw — automatically opens or closes on rise or fall.
		Single pole double throw — slow acting contact with no snap action. Contacts close on rise and close on fall with an open circuit between contact closures.
		Single pole single throw, normally open — closes on rise.
		Single pole single throw, normally closed — opens on rise.
		Single pole single throw, normally open — closes on rise.
		Single pole single throw, normally closed — opens on rise.
		Two circuit, single pole single throw, normally open — a common terminal is connected to 2 separate contacts which close on rise.
		Two circuit, single pole single throw, normally closed — a common terminal is connected to 2 separate contacts which open on rise.
MANUAL RESET		
		Single pole single throw, normally open — contacts open at a predetermined setting on fall and remain open until system is restored to normal run conditions at which time contacts can be manually reset.
		Single pole single throw, normally closed — contacts open on rise and remain open until system is restored to normal run conditions at which time contacts can be manually reset.
		Single pole double throw, one contact normally closed — contact opens on rise and remains open until system is restored to normal run condition at which time contact can be manually reset. A second contact closes when the first contact opens.
		Single pole single throw, normally closed — contacts close on fall and remain closed until system is restored to a higher predetermined setting.

PUSHBUTTON SWITCHES

Pushbutton switches are used in stations to control motor operation. Figure 5–25 shows pushbutton stations fitted with two, three, and six switches. They can be used to start, stop, forward, and reverse action, as well as to reset and test or cause up-and-down movement. Pushbutton stations are used in the control cir-

FIGURE 5–25 *Pushbutton stations: (A) pendant station; (B) cavity-mounted station without pilot light; (C) cavity-mounted station with pilot light; (D) six-unit custom-built station. (Allen-Bradley)*

(A)

(B)

(C)

(D)

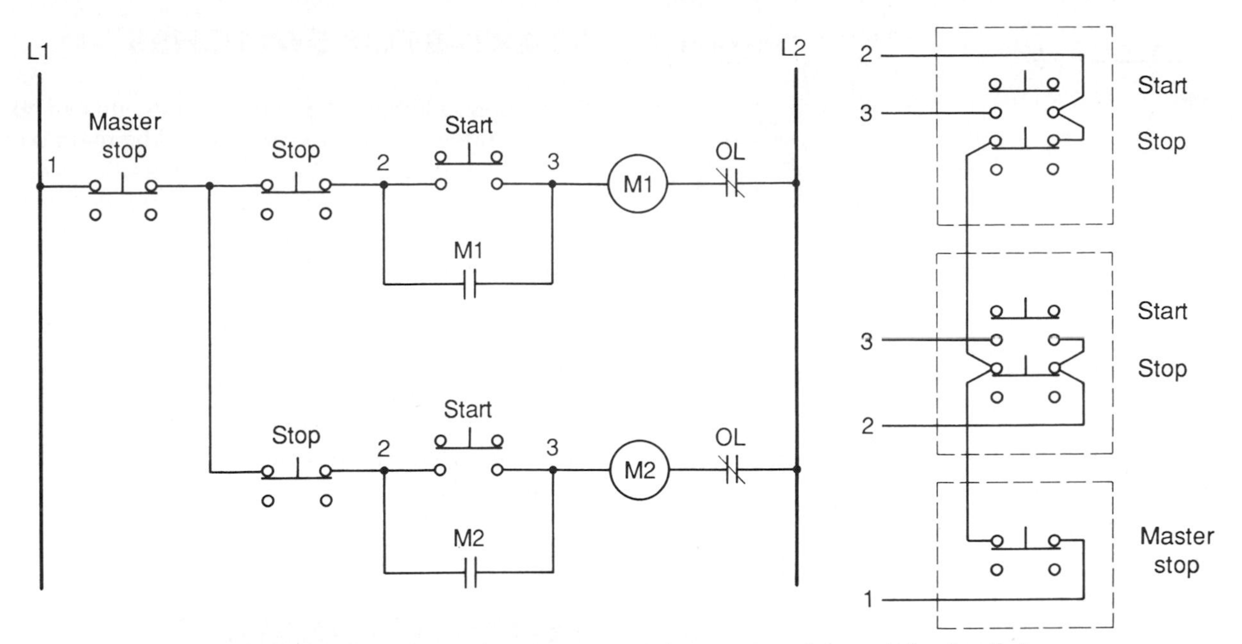

FIGURE 5–26 *Group of single stations with master* STOP *button. (Allen-Bradley)*

cuits ofmagnetic starters. They are usable on the great majority of applications where compact size and dependable performance is needed. Assembled stations are available with any combination of pushbuttons, selector switches, pilot lights, or special-purpose devices.

Push buttons can be used in a number of ways to control motors and manufacturing processes, for instance: start–stop control, reversing control, two-speed control, jogging control, thermostat-controlled motor, and ground detection with push-to-test pilot lights.

Figure 5–26 shows a group of single stations with a MASTER STOP button. Note the difference between the ladder or line diagram and the wiring diagram of the switch. Notice how the numbers on the wiring diagram are connected in the circuit for proper operation. This circuit has a momentary contact MASTER STOP that is connected in series with a group of parallel-connected circuits. Depressing the MASTER STOP button deenergizes all the circuits.

The circuit in Fig. 5–26 is the basic start–stop circuit. Two-wire control or undervoltage release circuits are not applicable here because they would be reenergized as soon as the MASTER STOP button is released.

SAFETY SWITCHES

A safety switch can be used for motors that can be placed directly across the line for starting (Fig. 5–27). This type of switch is operated manually and usually has a handle on the outside of the enclosure to operate the switching blades. The switch enclosures are made of steel or fiberglass-reinforced polyester especially formulated to withstand attack from almost any corrosive atmosphere found in industrial applications. Electrical interlocks are available for safety purposes. A pivot arm operated from the switch mechanism breaks the control circuit before the main switch blades break. Slow-blow fuses are usually installed inside the enclosure to handle the current surge caused by a starting motor.

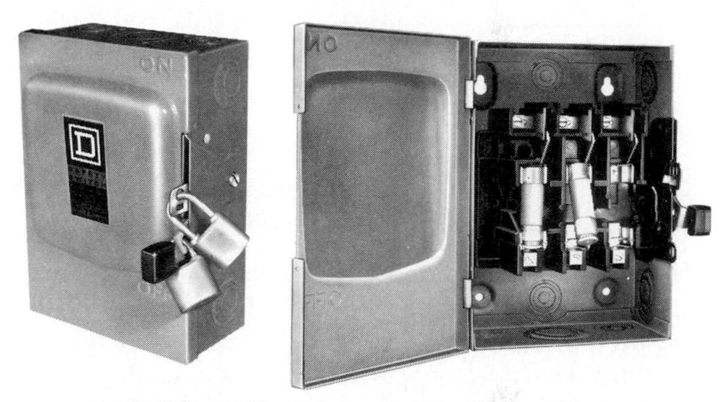

FIGURE 5–27 *Safety switches. (Square D)*

Manual transfer switches are designed to transfer loads from one supply to another. They are not fusible. They are available from 30 to 600 A. Of course, the higher current switches with four-pole switching are over $4000.

SELECTOR SWITCHES

Selector switches can be used for determining the direction of motor rotation, for starting–stopping and jogging. They are available in two-, three-, or four-

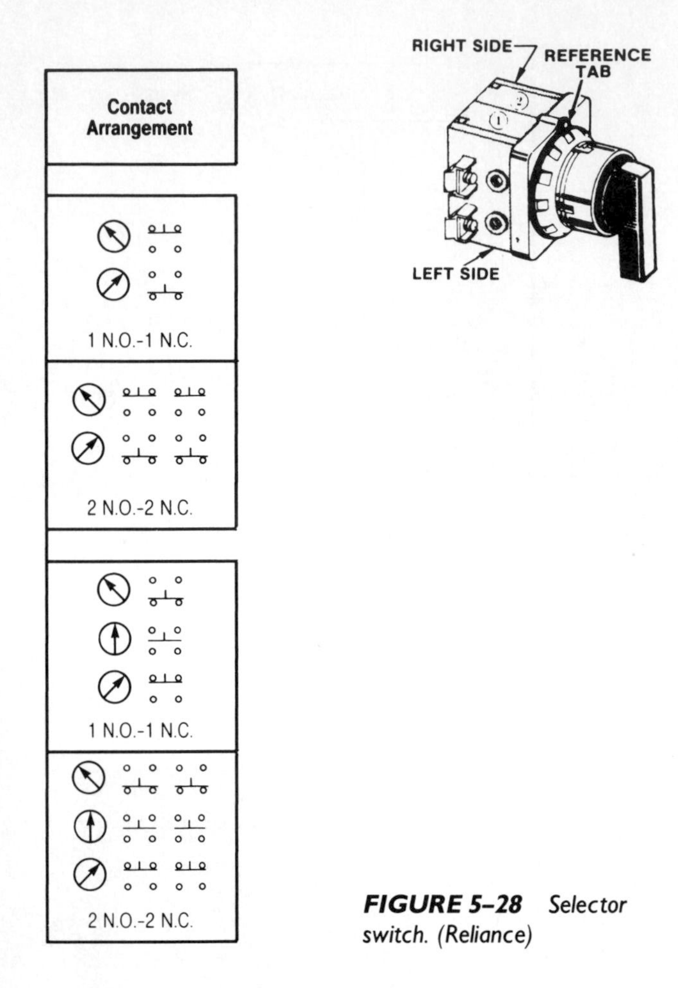

FIGURE 5–28 Selector switch. (Reliance)

position configurations. Figure 5–28 shows the switch and the contact arrangements. Selector switches are available with a lever operator, coil operator, slot operator, knob operator, or key operator. They can also be obtained with a momentary contact where testing a circuit is necessary.

SINGLE-POLE SWITCHES

Single-pole switches are simple on–off types. They are made in thousands of different shapes and styles and can be utilized as a simple on–off switch for a small motor or can be made into larger motor controllers by making the switch contacts of different materials. Many pushbutton switches are single-throw. They use two contacts, as shown in Fig. 5–29.

FIGURE 5–29 Single-pole switch circuit. (Allen-Bradley)

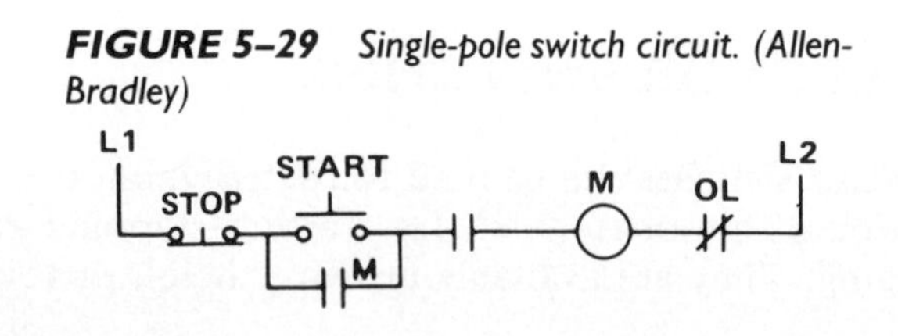

START–STOP SWITCHES

Start–stop switches take many shapes and can be used for a number of control purposes. The main concern in a switch used to turn a motor on and off is its current rating; that is, will it be able to handle the onrush current? Figure 5–30 shows a circuit where the START button maintains the contacts that take the place of hold-in contacts. Depressing the START button maintains the circuit. Depressing the STOP button breaks the circuit by opening the start contacts. If the contactor is deenergized by a power failure or overload operation, the start contacts are unaffected. The motor restarts automatically. Note the symbol for the maintained contact switch.

TEMPERATURE SWITCHES

Temperature-operated switches are usually referred to as thermostats. Figure 5–31 shows how the high-temperature cutout is placed in the circuit. Then note the location of the thermostat. This particular circuit has undervoltage release and uses a selector switch to change from automatic operation with the thermostat in the circuit to one where the thermostat is taken out of the circuit or the circuit is deenergized by the selector switch being placed in the no-contact center position.

TOGGLE SWITCHES

Toggle switches are available in many sizes and shapes and are made for special purposes as well as for motor control circuits (Fig. 5–32). They are used to turn various devices on and off, or to switch from one device to another. They usually have a metal handle and are mounted through a round hole. Screw terminals are usually provided for attaching wires. However, some may have wire leads furnished. A good example of a toggle switch is located in the home where it is mounted on the wall and turns on and off the overhead light.

TRANSISTOR SWITCHING

One of the functions of the transistor is switching. It is a very efficient device when it comes to switching. The other function for the transistor is amplifying. These functions are discussed in other chapters.

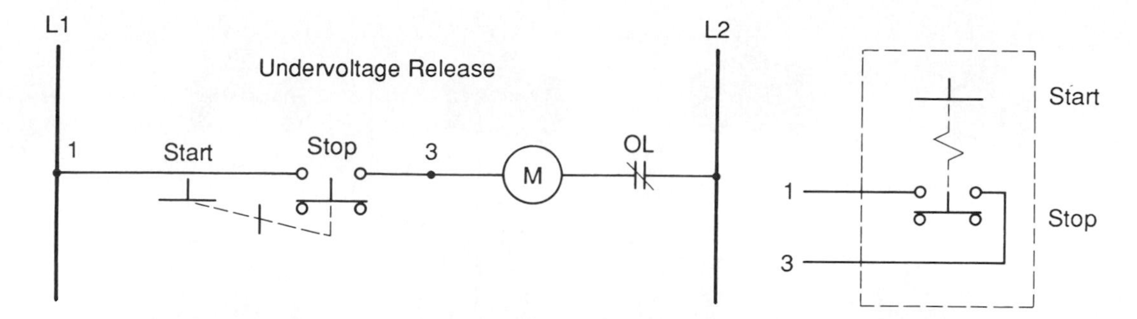

The start button mechanically maintains the contacts that take the place of hold-in contacts. Depressing the start button maintains the circuit; depressing the stop button breaks the circuit by opening the start contacts. If the contactor is deenergized by a power failure or overload operation, the start contacts are unaffected. The motor restarts automatically.

FIGURE 5–30 *START button with maintained contacts. (Allen-Bradley)*

FIGURE 5–31 *(A) Temperature switches. (Square D) (B) Thermostat-controlled motor with selector switch. (Allen-Bradley)*

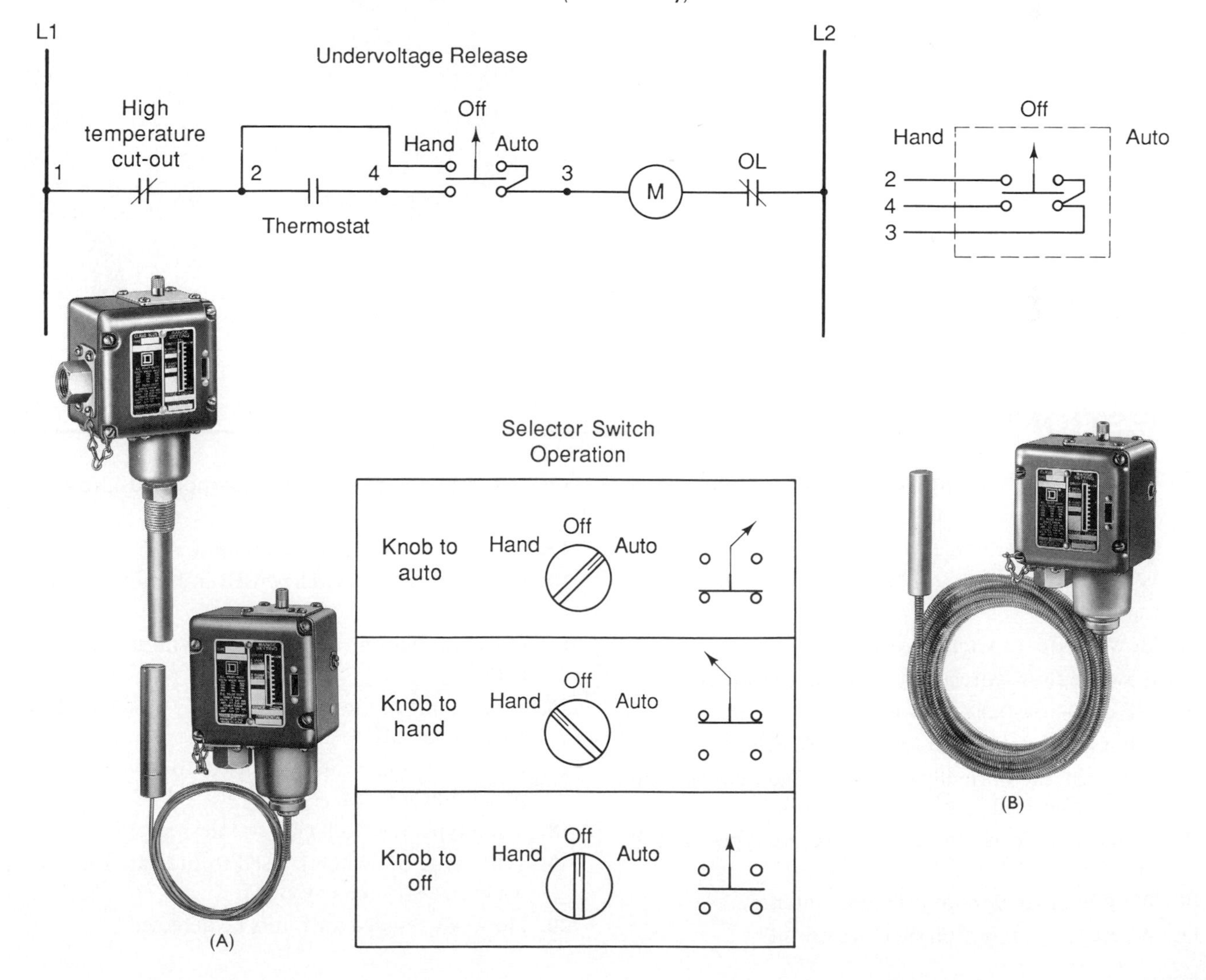

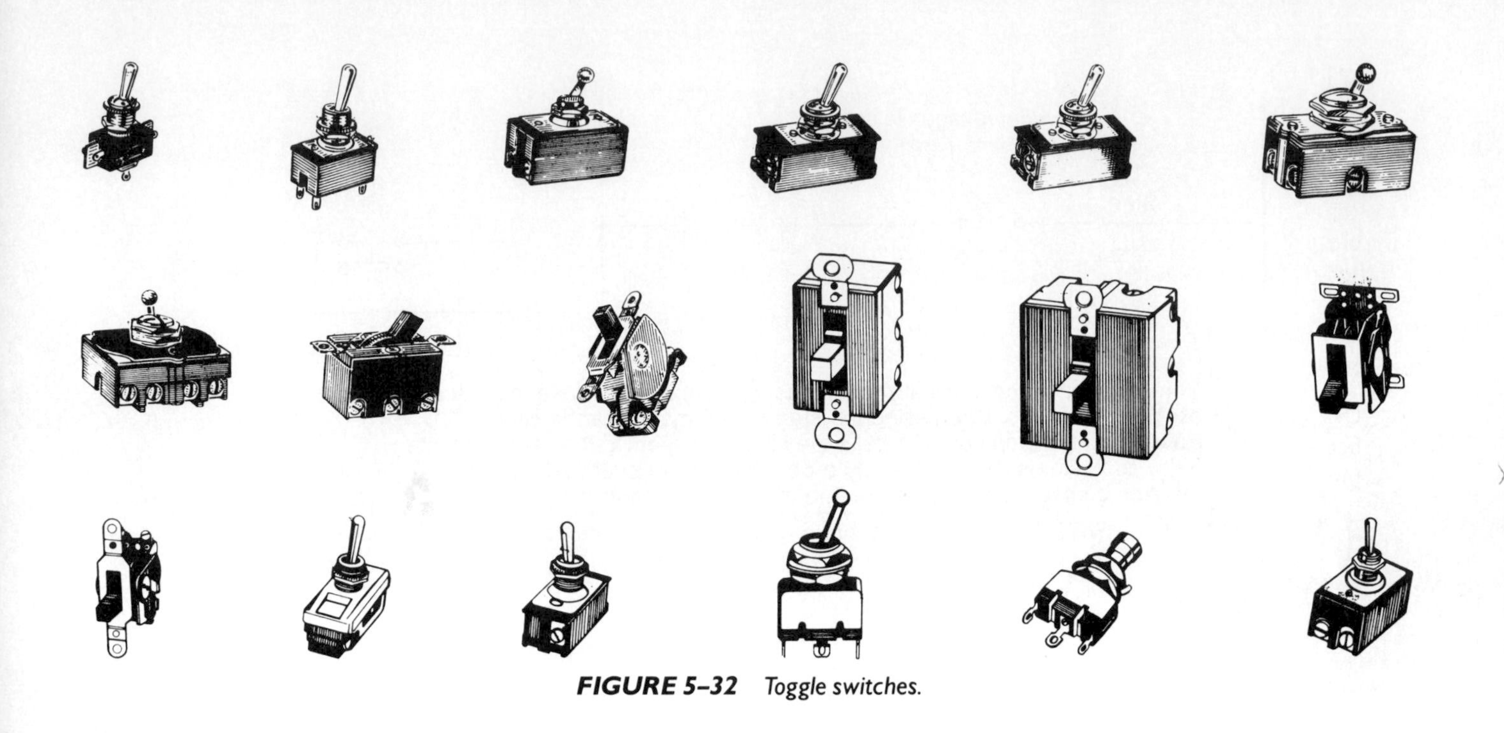

FIGURE 5–32 *Toggle switches.*

VACUUM SWITCHES

These types of switches use a vacuum for proper operation (Fig. 5-33). They may have double-throw contacts that are normally open and normally closed so they can be used to control forward and reverse operation of a motor.

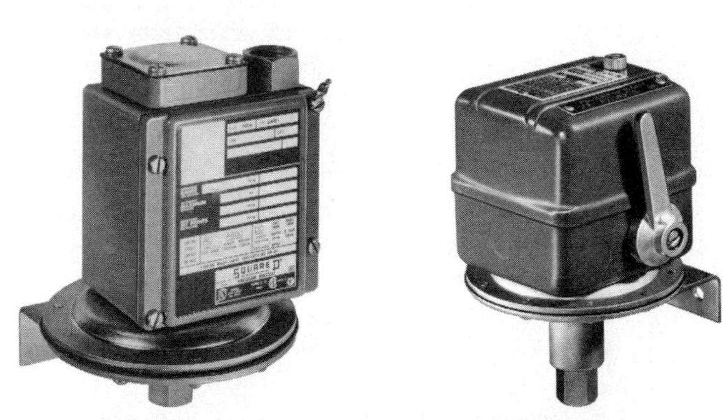

FIGURE 5–33 *Vacuum switches. (Square D)*

QUESTIONS

1. How is a drum switch converted from maintained to momentary operation?
2. What is another name for a drum switch?
3. What do some dc motors use to suppress brush arcing?
4. How are float switches used?
5. How are flow switches used in alarm systems?
6. What is a joystick? How is it used?
7. What is the purpose of an interlock switch?
8. What can the normally closed auxiliary contacts on a starter be used for?
9. How are limit switches used? List five types of limit switches.
10. What are two types of pressure switches?
11. Where are pushbutton switches useful?
12. How are safety switches incorporated into various types of equipment?
13. Where are selector switches used?
14. How are toggle switches used in motor control circuits?
15. How are vacuum switches used in motor control circuits?
16. What is one of the most commonly used switches for motor control?
17. Can the drum switch be used to reverse a three-phase motor?
18. What type of switch uses stainless steel floats?
19. What type of switch is used to indicate the flow of water in a sprinkler system?
20. How can clogged air filters be detected?

REVIEW PROBLEMS

Switches make up one of the most often used means of controlling electricity. Being able to complete the circuit or interrupt the circuit is of great importance to any electrician at one time or another. A good grasp of the drawings associated with switching improves your comprehension of circuit operation and enhances your ability to visualize the operation of switches.

1. Using Fig. P–1, draw lines to complete the circuit between the switch terminals to cause the three-phase motor to run in the *forward* direction.

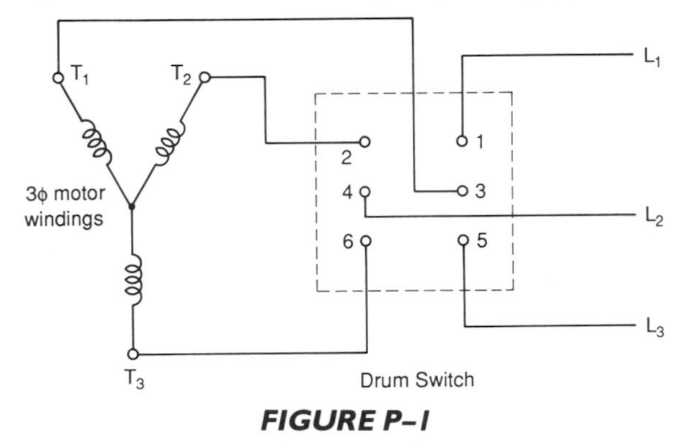

FIGURE P–1

2. Using Fig. P–2, draw lines to complete the circuit between the switch terminals to cause the three-phase motor to run in the *reverse* direction.

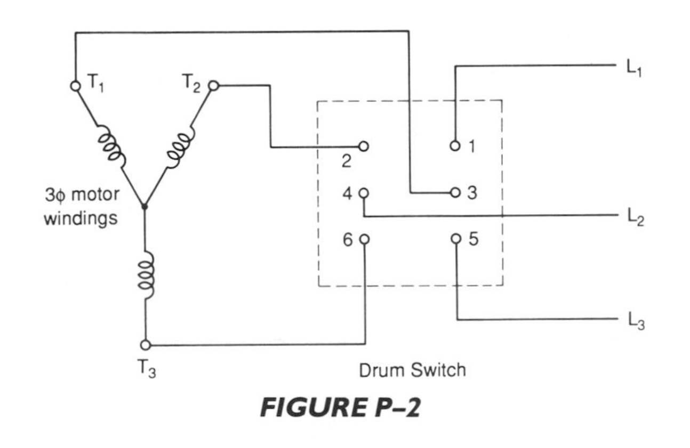

FIGURE P–2

3. Using Fig. P–3, draw lines to complete the circuit between the switch terminals to cause the shunt motor to run in the *forward* direction.

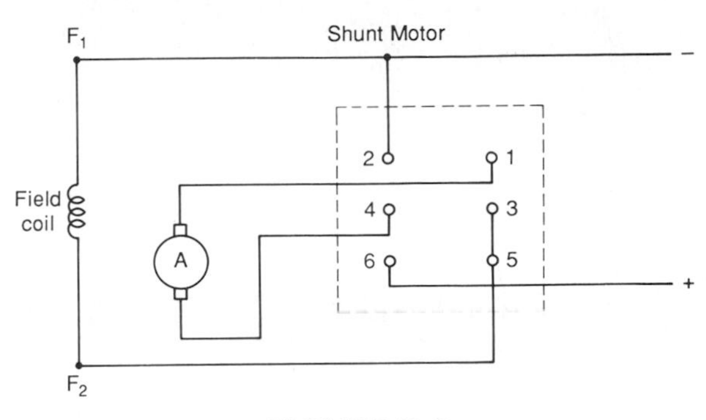

FIGURE P–3

4. Using Fig. P–4, draw lines to complete the circuit between the switch terminals to cause the series motor to run in the *reverse* direction.

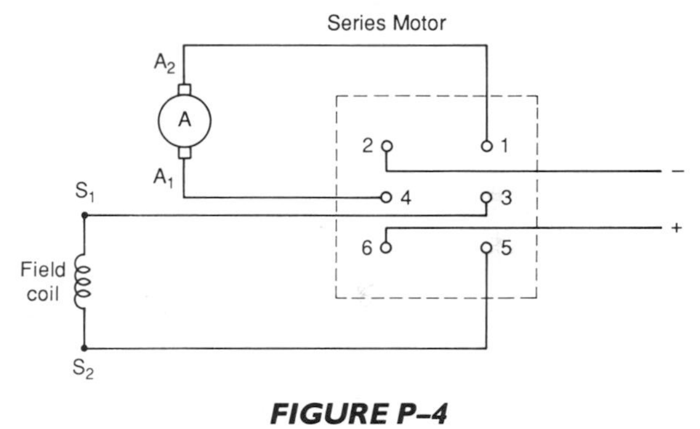

FIGURE P–4

5. Using Fig. P–5, draw lines to cause the circuit between the switch terminals to cause the compound motor to run in the *forward* direction.

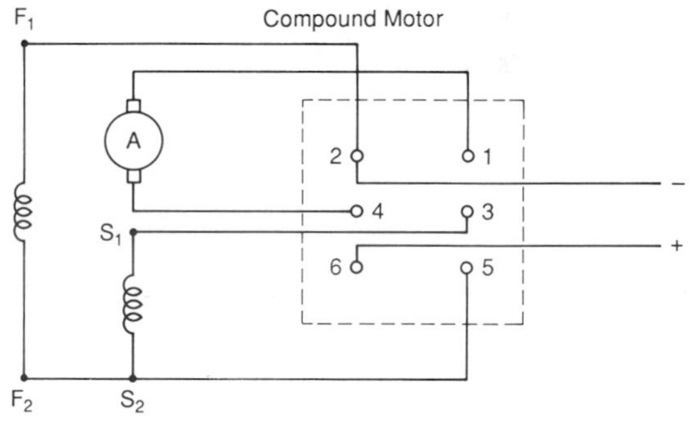

6. Draw the switching circuit in Fig. P–6 so that the switches complete the circuit and cause the lamp to light.

FIGURE P–6

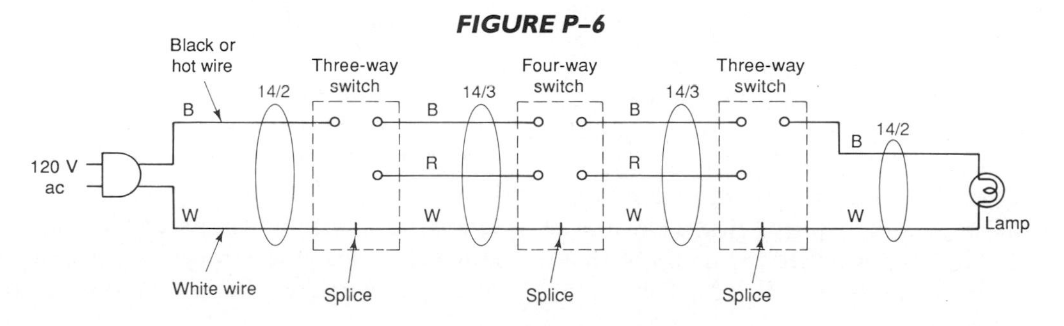

6

Magnetism and Solenoids

Objectives

After studying this chapter, you will be able to:

1. Understand how magnetism was discovered and put to work.
2. Describe a solenoid.
3. Explain the use and operation of a solenoid valve.
4. Explain the importance of the air gap and the armature in a solenoid's operation.
5. Define the role of shading coils in solenoids.
6. Identify a coil by its sealed current rating.
7. Differentiate between pickup voltage and seal-in voltage.
8. Differentiate between pickup voltage and dropout voltage.
9. Explain the role of hum in magnetic devices.
10. Identify various problems associated with magnetic coil devices and their causes.

Observation of the effects of magnetism predate the discovery of static electricity. Magnetism was discovered about 2600 B.C. Some say it was first observed by the Chinese; others believe that the Greeks were the first to observe it. There is little doubt that primitive man also noticed its effect but lacked the knowledge to use it. Certain heavy stones or rocks had the power to attract and lift similar stones as well as pieces of iron. The material in these stones was called *magnetite,* named by the Greeks for the province of Magnesia in Asia Minor, where some of the stones were found. Its properties were later referred to as possessing *magnetism.* Magnetism is a property of stones made of magnetite (Fig. 6–1).

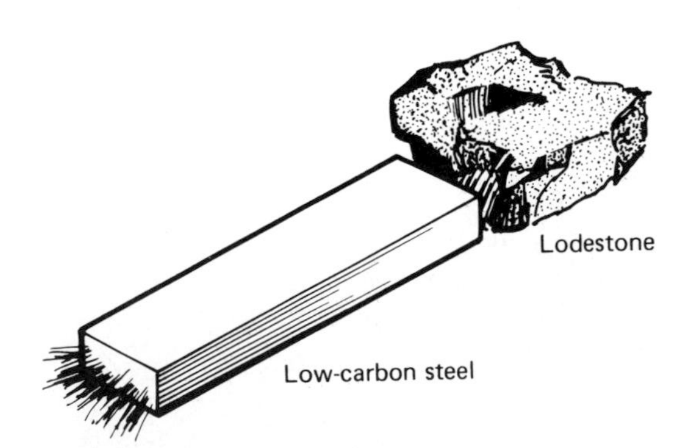

FIGURE 6–1 *The lodestone is a natural magnet with no definite shape but a north–south pole alignment.*

The Chinese were among the first to record observations of the stone and its properties. They hung magnetite on a string and watched it line up so that one end pointed north and the other south. This led to later development of the magnetic compass. Since this stone could lead the way for travelers, it was called the *leading stone* (later changed to *lodestone*).

Many years passed with no developments in the field of magnetism. Then William Gilbert, a physician to Queen Elizabeth I of England, began studying the mystery of the lodestone and formed the basic principles of magnetism. He also experimented with static electricity and established the use of the word *electron*. Others were also experimenting with the properties of the lodestone at the same time. Many added knowledge to the field and some have contributed their names to units of measurement in both electricity and magnetism. These pioneering workers include Alesandro *Volta,* Charles Augustin de *Coulomb,* André Marie *Ampère,* Luigi *Galvani,* Hans Christian *Oersted,* and Georg Simon *Ohm*.

Benjamin Franklin caused some interest in electricity in the United States by flying his kite with metal objects attached during a thunderstorm. About 80 years later Michael Faraday of England and Joseph Henry of the United States each discovered the relationship between magnetism and electron flow. This discovery became one of the most important contributions to the field of electricity and electronics.

After the establishment of this relationship, progress in the field was quite rapid. Thomas Edison did a lot of work with the electric motor and the electric light and developed a generator to produce electricity. The advent of the transformer led to developments in the modern era.

SOLENOIDS

One device that grew out of the development of electric power and its transmission was the solenoid. The solenoid is a current-carrying coil used as a magnet. In a solenoid, there is a tendency for the core to move so that it encloses the maximum number of magnetic lines of force, each having the shortest possible path. In Fig. 6-2 the core is shown outside the coil. Because it is made of a ferromagnetic material, it presents a low-reluctance path to the magnetic lines of force at the north end of the coil. These lines of force concentrate the soft-iron core and then complete their paths back to the south pole of the electromagnet.

A movable iron core tends to be pulled to the center of a solenoid. The electromagnetic lines of force passing through the core material have thus magnetized the core. The direction of the magnetic lines has produced a south pole in the core at the north end of the electromagnet. The electromagnetic lines of force leave the opposite end of the iron core, and this end is the north pole of the magnetized core.

An attraction between the north pole of the coil and the south pole of the iron core tends to pull the core into the coil. Magnetic lines of force fan outward from the north pole of the magnetized core. They have a shorter magnetic path back to the south pole of the coil. As the iron core is pulled into the coil, this path becomes increasingly shorter. The magnetic lines of force travel the shortest possible distance when the core centers itself in the coil. This movement of the core into the coil can be harnessed to cause the closing of switch contacts or the opening of a valve that controls air, gas, hydraulic fluid, or any other flowing medium.

Solenoid Valves

Solenoid valves are used on many refrigeration systems. They are electrically operated. A solenoid valve, when connected as in Fig. 6-3, remains open when current is supplied to it. It closes when the current is turned off. In general, solenoid valves are used to control the liquid refrigerant flow into the expansion valve, or the refrigerant gas flow from the evaporator when it or the fixture it is controlling reaches the temperature desired. The most common application of the solenoid valve, in the liquid line, operates with a thermostat. With this hookup, the thermostat may be

FIGURE 6-2 *A solenoid with an electric current has a tendency to pull the core into the coil.*

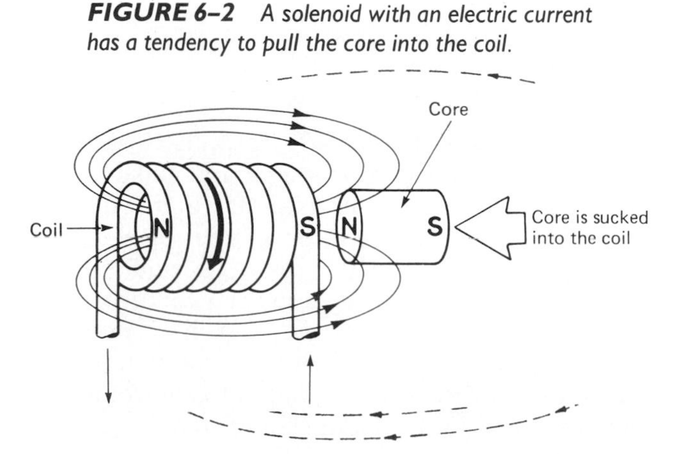

FIGURE 6-3 *Solenoid valve connected in the suction and liquid evaporator lines of a refrigeration unit.*

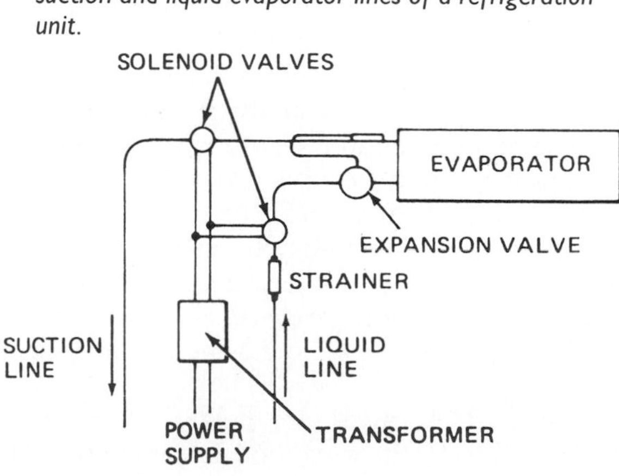

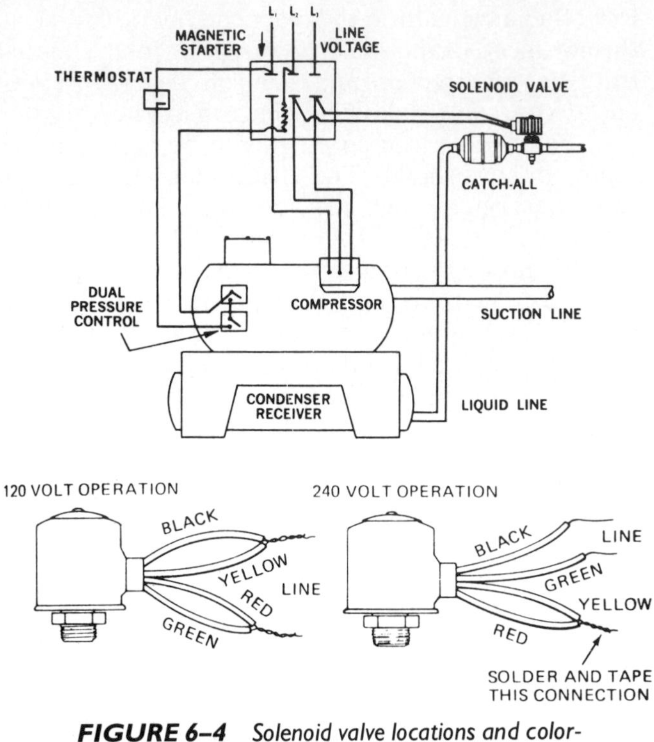

FIGURE 6–4 *Solenoid valve locations and color-coded wires.*

set for the desired temperature in the fixture. When the temperature is reached, the thermostat will open the electrical circuit and shut off the current to the valve. The solenoid valve then closes and shuts off the refrigerant supply to the expansion valve. The condensing unit operation should be controlled by a low-pressure switch. In other applications, where the evaporator is to be in operation for only a few hours each day, a manually operated snap switch may be used to open and close the solenoid valve.

The solenoid valve shown in Fig. 6–4 is operated with a normally closed status. A direct-acting metal ball and seat assure tight closing. The two-wire, class W coil is supplied standard for long life on low-temperature service or sweating conditions. Current failure or interruption will cause the valve to fail-safe in the closed position. The solenoid cover can be rotated 360° for easy installation. Explosion-proof models are available for use in hazardous areas.

Uses. This solenoid valve is usable with all refrigerants except ammonia. It can also be used for air, oil, water, detergents, butane or propane gas, and other noncorrosive liquids or gases. A variety of temperature control installations can be accomplished with these valves. Such installations include bypass, defrosting, suction line, hot gas service, humidity control, alcohols, unloading, reverse cycle, chilled water, cooling tower, brine, and liquid-line stop installations and ice makers.

Operation. The valves are held in the normally closed position by the weight of the plunger assembly and the fluid pressure on top of the valve ball. The valve opens by energizing the coil and magnetically lifting the plunger and allowing full flow by the valve ball. Deenergizing the coil permits the plunger and valve ball to return to the closed position.

Automatic Gas Furnace Solenoid

The solenoid shown in Fig. 6–5 is used to control the flow of natural gas into a furnace. When heat is desired, the thermostat in the room goes on, making contact and completing the circuit through the power supply to the valve coil (3). The energized coil creates a magnetic field that lifts the plunger (4). This allows pressurized gas to flow through the inlet at the left side of the valve body (9), beneath the plunger, and through the outlet at the right. When the room reaches the desired temperature, the thermostat breaks the circuit and the plunger drops back into the valve seat (11), shutting off the flow of gas.

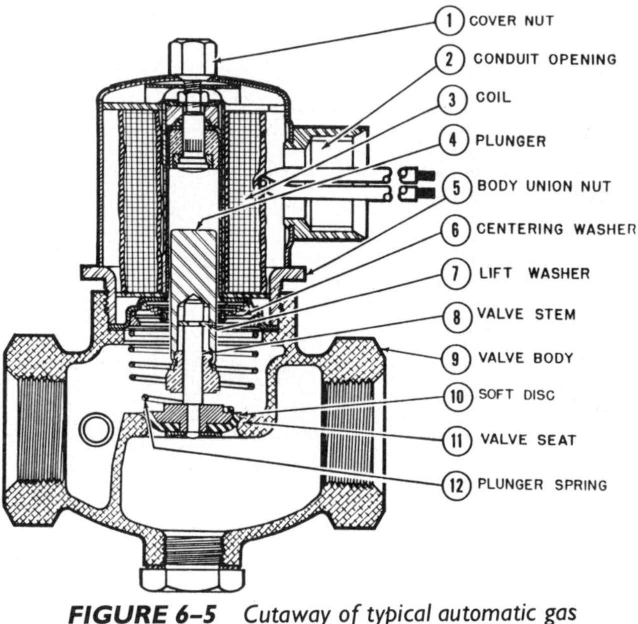

FIGURE 6–5 *Cutaway of typical automatic gas furnace control.*

CONSTRUCTION OF SOLENOIDS

The manufacture of solenoids warrants closer attention. Some parts and their interaction make for certain considerations by their manufacturer. Eddy currents can affect the operation of the solenoid. The armature gap and the coil's ability to handle current and voltage are part of the designer's concerns. They are also part of the maintenance person's concern, inasmuch as these factors contribute to the malfunctioning, in time, of the solenoids used to control motors and processes in industry and commerce.

Eddy Currents in the Armature and Core

Eddy currents are generated any time a varying magnetic field is brought near a piece of metal (Fig. 6–6).

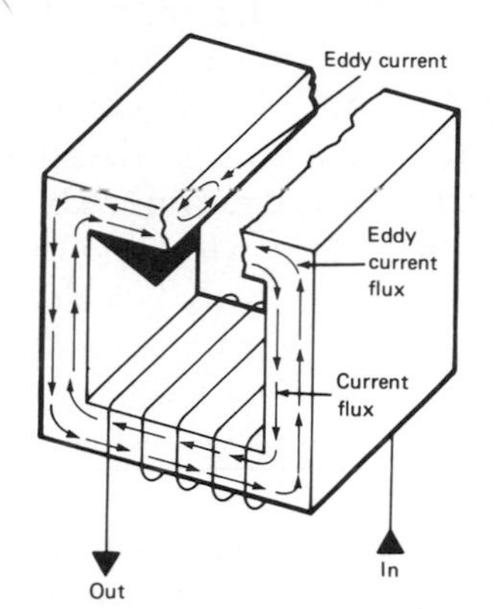

FIGURE 6–6 *Generation of eddy currents in an open core.*

Eddy currents can create heat and cause problems. In the ac solenoid note that the core and armature are made of laminated steel. Lamination of the steel core and armature reduce the buildup of heat produced by eddy currents. Eddy currents meet with high opposition in a lamination. Laminations are sheets of metal rather than a solid piece. Dc solenoids are made with a solid core, inasmuch as the magnetic field varies only when the coil is energized and deenergized.

Air Gap and the Armature

The armature completes the path of the magnetic lines of force. It is important that the armature match or fit

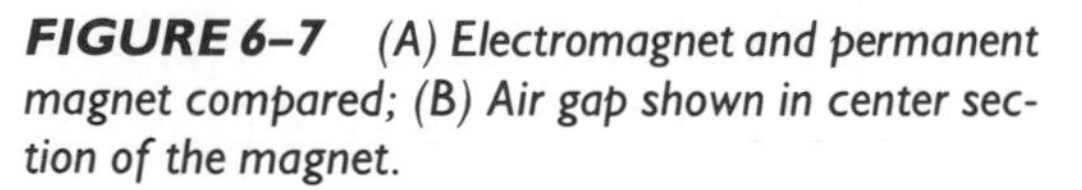

FIGURE 6–7 *(A) Electromagnet and permanent magnet compared; (B) Air gap shown in center section of the magnet.*

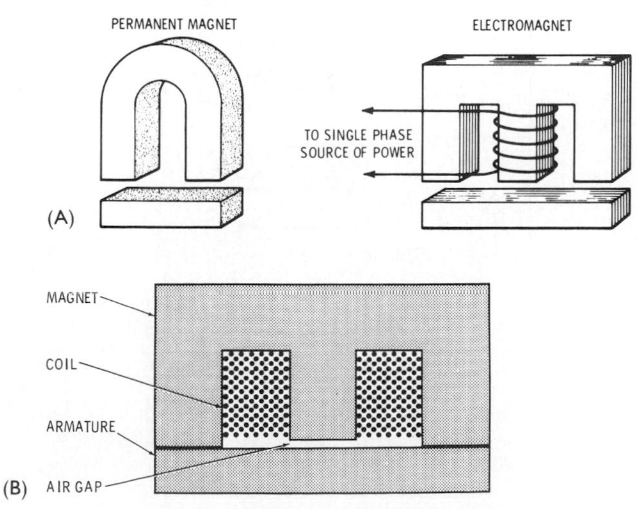

the rest of the magnetic assembly of the solenoid. A close fit leads to less chattering. Very close tolerances are required in the fitting of the surfaces. However, some problems are associated with the armature being too closely mated to the rest of the magnetic (iron) circuit. Inasmuch as residual magnetism is generated by the electric current producing the magnetic field, it has a tendency not to let go when the current is turned off in the coil. The residual magnetism holds the armature closely even after the power has been removed. That is why a small air gap is left in the iron circuit. This allows the magnetic field to be broken and the armature to drop away after the coil is deenergized (Fig. 6–7).

SHADING COILS

Shading coils are used on ac solenoids to reduce the chatter produced by the varying electric current through the coil and thus the varying magnetic field produced in the iron circuit. Figure 6–8 shows how the shading coil is placed on the magnetic assembly. The shading coil is a single ring of heavy-duty wire. A moving magnetic field causes current to be induced into this single turn in much the same way as in the

FIGURE 6–8 *(A) Magnet assembly and armature. Note the shading coils. (B) Shading coil current versus magnet coil current. (Square D)*

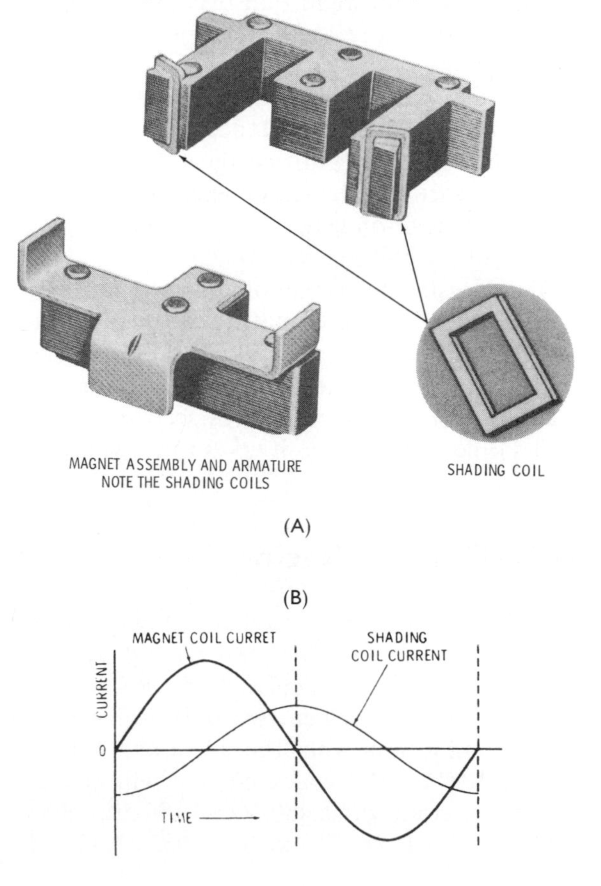

secondary of a transformer. The induced current is slightly out of phase with the current that is flowing in the coil circuit. The result is a magnetic field, even though slightly less than the original field, that is slightly behind the rise and fall of the original field produced by the coil. By being slightly behind in its peak, the induced field created by the shading coil has a tendency to fill in when the original drops to zero. This has a tendency to hold the armature close to the magnetic circuit and reduces chatter that would otherwise result from a varying magnetic field.

SOLENOID COILS

The coils of solenoids have special needs. They are the source of the magnetic field when energized. Therefore, the amount of current and voltage needed for their operation is a result of their intended use. More current is needed to energize the coil than to keep it energized, especially when the coil is used on alternating current. A coil has inductive reactance that is the product of the inductance of the coil and the frequency of the current times a factor of a 2π. This inductive reactance is measured in ohms (Ω) and in some instances is referred to as impedance (Z). This means that the coil has a tendency to hold back the flow of current after it is energized and pulls in the core material to change its inductance. The inrushing current produces a magnetic field that extends past the first coil of wire into those on top of it, thus inducing an EMF that is in the direction opposite to that which caused it. Inasmuch as alternating current is constantly changing, in most instances 120 times per second (60 Hz), it does not always meet the physical limitations imposed by the inductance of the coil and there is an opposition generated (X_L) greater than the resistance of the coil's copper wire.

The inrush current is about 6 to 10 times as great as is needed to hold the armature to the magnetic assembly. Some heat is generated when the coil is energized for a period of time. The coil is rated in terms of volt-amperes (VA). Volt-amperes are the product of the volts times the amperes it takes to cause the coil to energize.

Sealed Current Rating

If a coil has a VA rating of 300 and 30 VA sealed, the inrush current of the 120-V coils is 300 VA/120 V. This means that it takes a current of 2.5 A to produce the 300 VA. If the sealed current of the same coil (that is, the current it pulls when the solenoid is completely closed) is 30/120, the sealed current will be 0.25 A. The same solenoid with a 480-V coil will draw 300/480 or 0.625 A inrush and 30/480 or 0.0625 A sealed. As you can see, the VA rating of the coil can be useful in determining the amount of current drawn at the start and at the hold-in condition of the solenoid.

Coil Voltage

Voltage ratings of coils are determined by the manufacturer. The number of turns of wire and the size of the wire make a difference in the voltage it takes to energize the coil. Keep in mind also that the amount of current drawn by the coil is determined by the number of turns (dc resistance of the wire) and the core material. Core material and the turns as well as the length and diameter of the coil determine its inductance. The inductance is an important factor when considering ac operation of the coil.

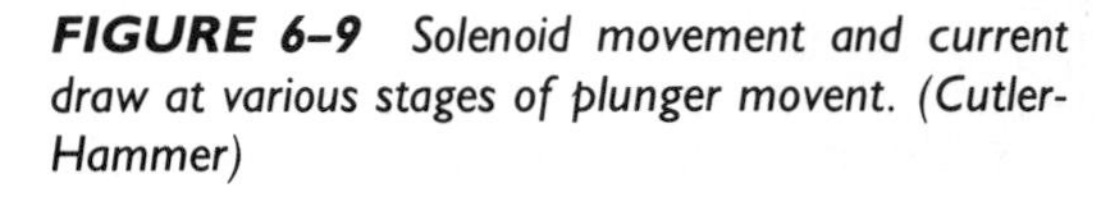

FIGURE 6–9 Solenoid movement and current draw at various stages of plunger movent. (Cutler-Hammer)

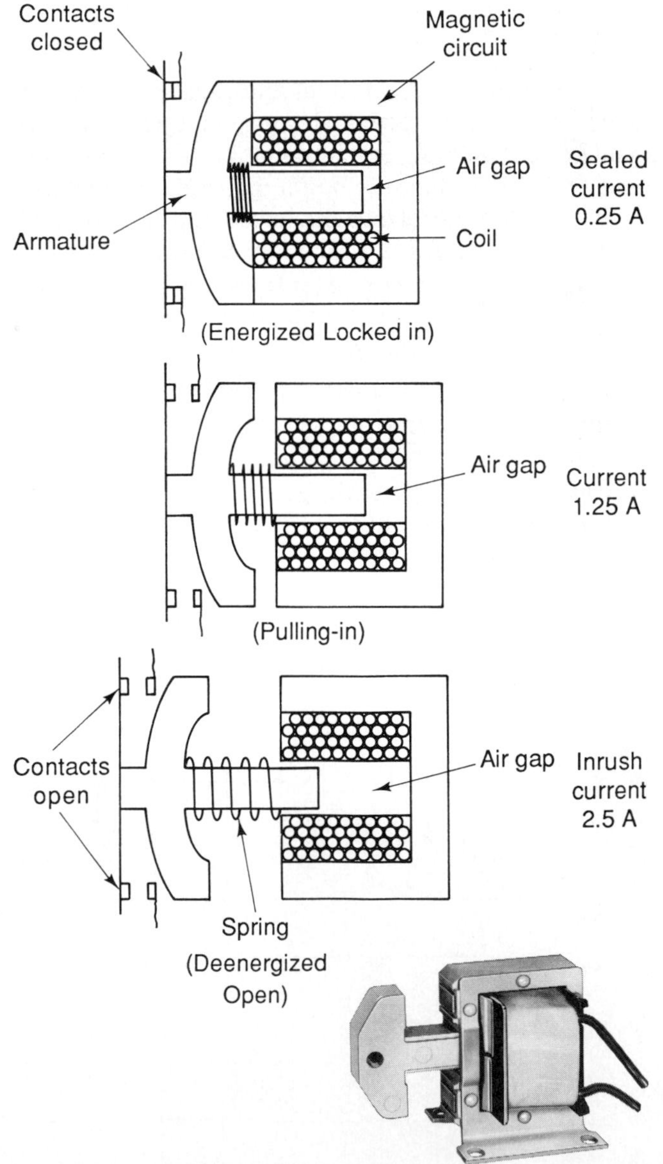

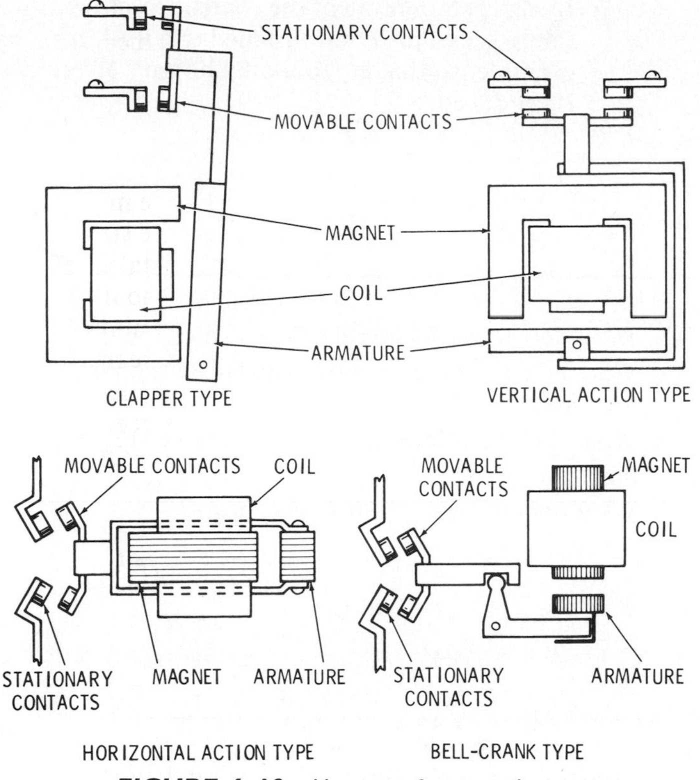

FIGURE 6–10 *Magnetic frame and armature assemblies. (Square D)*

Pickup voltage is the lowest voltage at which the coil armature will start to move. *Seal-in voltage* is the lowest voltage required to cause the armature to seat against the pole faces of the magnet. *Dropout voltage* is that voltage at which the armature will start to lose its attraction to the magnet and fall back to its open position (Fig. 6–9).

Pickup voltage is usually higher than seal-in voltage for most solenoids. However, there are some exceptions to this general statement. For instance, the bell-crank armature solenoid and magnetic assembly is an exception. Figure 6–10 shows the types of solenoid construction.

Effects of Voltage Variation

Low control voltage produces low coil currents and reduced magnetic pull. On devices with vertical action assemblies, if the voltage is greater than pickup voltage but less than seal-in voltage, the controller may pick up but will not seal. With this condition, the coil current will not fall to the sealed value. As the coil is not designed to carry continuously a current greater than its sealed current, it will quickly get very hot and burn out. The armature will also chatter. In addition to the noise, wear on the magnet faces results (Fig. 6–11).

AC HUM

All ac devices that incorporate a magnetic effect produce a characteristic hum. This hum or noise is due mainly to the changing magnetic pull (as the flux changes) inducing mechanical vibrations. Contactors, starters, and relays could become excessively noisy as a result of some of the following operating conditions:

1. Broken shading coil
2. Operating voltage too low
3. Wrong coil
4. Misalignment between the armature and magnetic assembly—the armature is then unable to seat properly

FIGURE 6–11 *Magnetic starter power circuit. (Square D)*

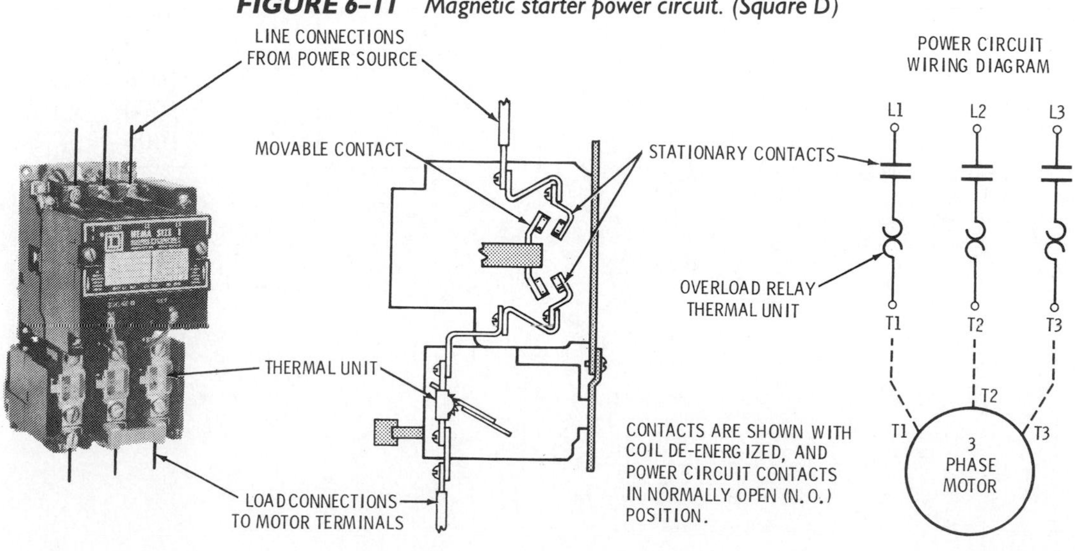

5. Dirt, rust, filings, and so on, on the magnetic faces—the armature is unable to seal-in completely
6. Jamming or binding of moving parts (contacts, springs, guides, yoke bars) so that full travel of the armature is prevented
7. Incorrect mounting of the controller, as on a thin piece of plywood fastened to a wall, for example, so that a "sounding board" effect is produced.

QUESTIONS

1. When was magnetism discovered?
2. What is a solenoid? How does it operate to control switching action?
3. What does a solenoid valve do?
4. How are eddy currents generated? How are eddy currents minimized?
5. Why are shading coils used on ac solenoids?
6. What is inductive reactance? In what unit is it measured?
7. What does a sealed current rating mean?
8. What effect does low control voltage have on magnetic assemblies?
9. What five conditions may cause excessive hum in a magnetic assembly?
10. What does *sounding board effect* mean?

REVIEW PROBLEMS

A solenoid has a coil. The coil has inductance and inductive reactance when used on alternating current. Impedance (Z) is also a factor in some instances when inductive reactance and resistance are both present. A quick review of these two important factors makes you aware of their presence and the need to understand their role in the proper operation of solenoids on alternating current.

$$X_L = 2\pi FL \qquad Z = \sqrt{R^2 + X_L^2}$$

1. What is the reactance of an inductor whose inductance is 6 H when the frequency of the applied voltage is 60 Hz?
2. What is the inductive reactance of a coil on 60 Hz if its inductance is 6 H?
3. What is the inductance of a coil whose resistance is 120 Ω, the reactance is 1721.7184 Ω, and the 120 V has a frequency of the applied voltage is 60 Hz?
4. What is the maximum dc current in the inductor in problem 3?
5. What is the inductive reactance of a coil if the frequency of the applied voltage is 50 Hz and the inductance is 10 H?
6. Find the inductive reactance and the impedance of the following combinations connected in series:

	F (H_z)	L (H)	R (Ω)	X_L	Z
a.	100	10	5000		
b.	200	10	5000		
c.	300	10	5000		
d.	400	10	5000		
e.	500	10	5000		

7. Find the inductive reactance and the impedance of the following combinations connected in series:

	F (H_z)	L (H)	R (Ω)	X_L	Z
a.	60	5	1000		
b.	50	5	1000		
c.	25	5	1000		

7

Relays

Objectives

After studying this chapter, you will be able to:

1. Define a relay.
2. Identify uses for relays.
3. Describe solid-state relay actions.
4. Identify the condition of relay contacts.
5. Explain how transistors are used for switching circuits.
6. Differentiate between a triac and a diac.
7. Explain how triacs work.
8. Explain how SCRs work.
9. Understand how phase failure relays operate.
10. Describe the actions of a solid-state relay.
11. Define zero-current turn-off and zero-voltage turn-on.
12. Identify various types of solid-state relay switching.
13. Differentiate between load detector and load converter relays.
14. Explain how thermal overload relays operate.
15. Identify the various types of thermal overload relays.

A relay is a device designed for remote control of another device. The word *relay* is defined as *to pass on.* That is what the electromagnetic device does. It uses low voltage and low current to cause switching of high voltage or high current, usually at a distant or remote location.

RELAY SOLENOIDS

The relay puts to work the solenoid and its ability to generate an attracting magnetic force. A practical example of the solenoid principle is the relay (Fig. 7-1). One or more sets of contacts are associated with the moving armature of a relay. These serve as electrical contacts and make or break when the relay current is switched on and off. The relay contacts can be arranged for a variety of functions, such as SPST (single-pole single-throw), SPDT (single-pole double-

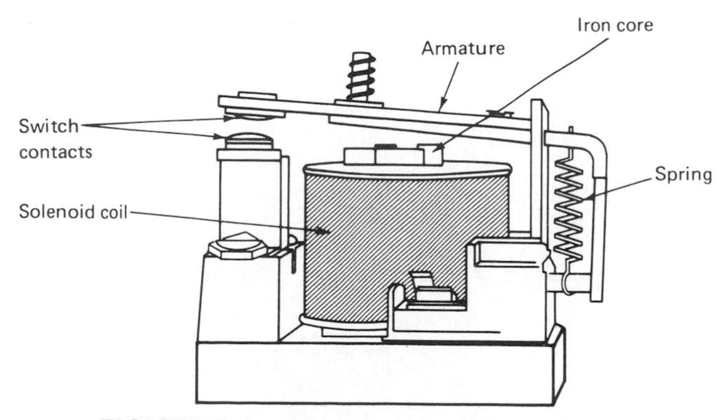

FIGURE 7–1 *Solenoid used to operate a relay.*

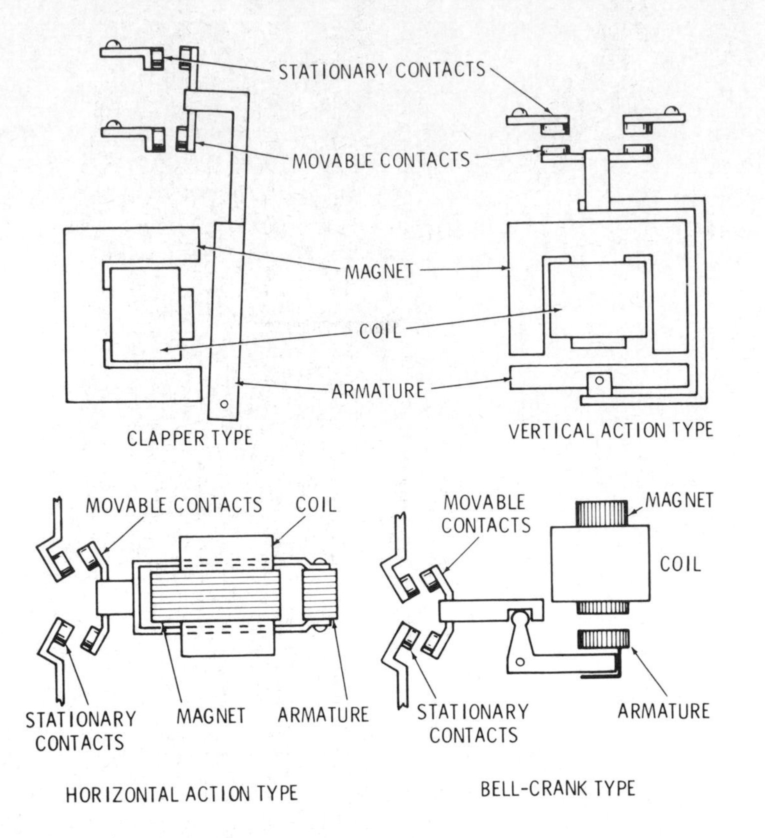

FIGURE 7–2 *Magnetic frame and armature assemblies.*

throw), DPDT (double-pole double-throw), or other desired combinations.

The advantage of the relay is the substantial pulling power that can be developed with a small coil current. The contacts themselves can be made quite large and can handle and switch high values of electrical power. An extremely small amount of control power can be used to switch much higher voltage and currents in a safe manner.

Uses for Relays

Relays are used to control the start and stop operations of small electric motors. Figure 7–2 shows how magnetic frames and armature assemblies are used to control the large currents used in the motor circuits. Figure 7-3 shows a double-pole double-throw (DPDT) relay that can be used to remotely control an electric motor or other electrical devices.

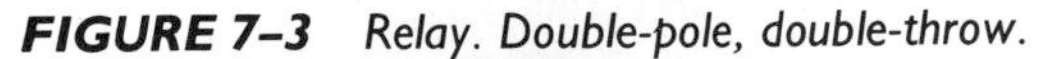

FIGURE 7–3 *Relay. Double-pole, double-throw.*

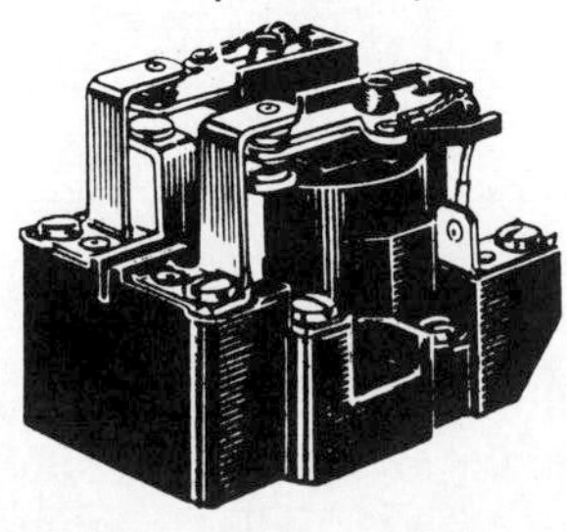

Relay Armature

Note how the armature of the relay is connected to a set of contacts so that when the armature moves to its closed position, the contacts also closed. The electromagnet makes the difference between a remote control possibility and a manual control. The electromagnet consists of a coil of wire placed on an iron core. When current flows through the coil, the iron bar, called the *armature,* is attracted by the magnetic field created by the current in the coil. To this extent, both will contact the iron bar. The electromagnet can be compared to the permanent magnet; however, the electromagnet does have the ability to be turned on and off electrically. Interrupting the current flow to the coil or electromagnet causes it to deenergize. This means that it will cause the armature to open and return to its original position, with the contacts no longer making contact. That means that the electrical circuit connected to the contacts is then open or broken. A low voltage and small current can be used to energize the coil. This, in turn, causes the contacts to close and turn on the device connected to the relay contacts circuit. By making certain modifications, it is possible to make some rather useful motor controllers.

Relay Contacts

If the contacts of a relay are pitted or burned, it calls for a burnishing (polishing) so that they will close

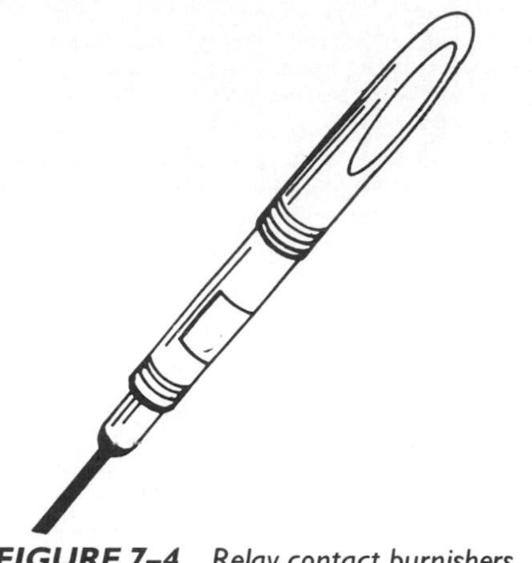

FIGURE 7–4 *Relay contact burnishers.*

tightly against one another and complete the circuit. This can be done with the power off. Place a piece of sandpaper (very fine grain) between the contacts and hold the contacts closed. Then move the sandpaper until it sands down the high points on the contacts. Use an even-finer grade and polish the points further. Make sure that you do not get the contacts to the point where they no longer "mate" properly. It may be sufficient to use a tool such as that shown in Fig. 7–4. It should be noted that some manufacturers do not recommend sanding contacts, but simply replacing with a new set.

SOLID-STATE RELAYS

Solid state refers to relays made with silicon or germanium materials that operate on the same basic principles as transistors and diodes. In most instances the relay is nothing more than a transistor, either PNP or NPN type. In other instances the solid-state relay is a silicon-controlled rectifier (SCR). Of course, the circuit arrangements are such as to allow the switching needed for the relay action. Other features, not readily available in electromechanical relays, are also available in this type of relay. As with everything, there are advantages and some limitations or disadvantages. In most instances the manufacturer will point to the advantages and you have to become aware of the limitations by closely examining the information provided by the manufacturer of the device.

One of the first differences noted between electromechanical and solid-state relays is the absence of a coil and no contacts. The solid-state relay needs very low voltage and current to cause it to do its job of switching. The transistor or SCR does the actual switching and the change in control voltage causes the semiconductor device to conduct or not conduct according to the control voltage applied to its elements. Figure 7–5 is an example of a solid-state relay.

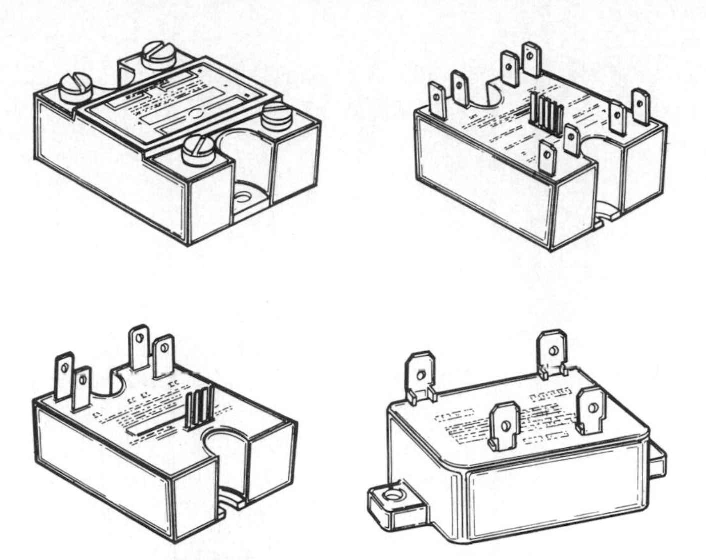

FIGURE 7–5 *Solid-state relays.*

Transistor

Inasmuch as the transistor is the device used to do the switching in a solid-state relay, it may be a good idea at this point to take a closer look at its operation (Fig. 7–6). The transistor is the key element in this type of relay. There are three elements in a transistor: base, collector, and emitter. The *base-to-emitter* voltage of the transistor can control the current flow between the emitter and the collector. In this PNP type of transistor, a negative voltage on the base allows emitter–base current to flow. This is due to the properties of the silicon-doped material at the junction of the emitter and base. The emitter–base voltage can then cause the transistor to conduct current from the emitter to the collector. A positive voltage on the base and negative on the emitter prevents emitter–base current from flowing, and the transistor stops conducting. This means that it behaves as a closed contact in the first

FIGURE 7–6 *Transistors in various case configurations.*

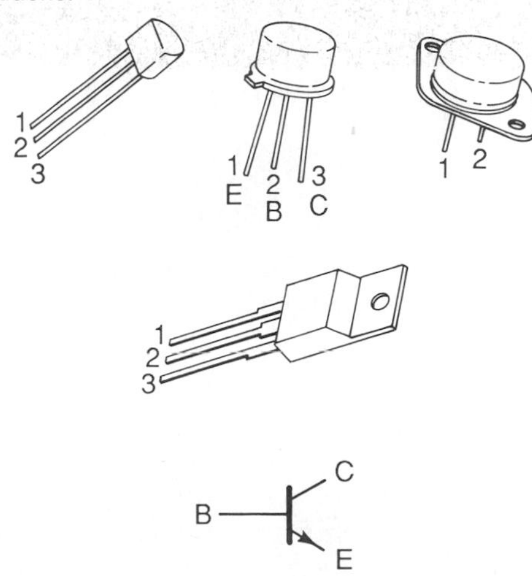

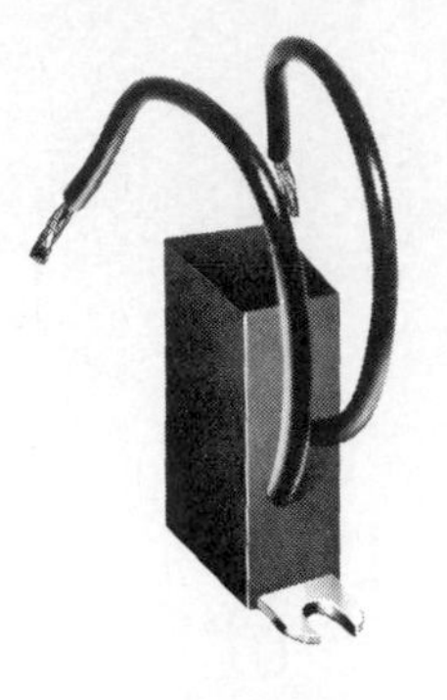

	**MANUAL TEST TOOL** — Provides a means of manually switching the contacts of a basic relay or timing relay and holding all contacts in their switched state until the tool is removed. This simplifies the checking of control circuits without power on the coil or contacts..............................
	TRANSIENT SUPPRESSOR — Consists of an R-C circuit designed to suppress coil generated transients to approximately 200 percent of peak voltage. It is particularly useful when switching the TYPE X relay near solid state equipment. It is designed for use on 120VAC coils only.............

FIGURE 7–7 *Surge suppressors.*

state and as an open contact in the second. This means that the current flow from emitter to collector can be controlled by a small voltage change in the base-emitter connection. There are no moving parts and no contacts to be concerned with at this time. However, there are limitations on how much current the transistor will conduct. The fact that there are no moving contacts, wear or arcing, deterioration or vibration, or dust and dirt damage makes the solid-state relay very much in demand.

Surge Protection

Solid-state devices are sometimes used with magnetic switches. This means that a voltage transient suppressor may be needed to prevent some of the harmful electrical impulses generated by electromagnetic devices on the line. Figure 7–7 shows a transient suppressor, and Fig. 7–8 shows a transient suppressor mounted on a magnetic relay.

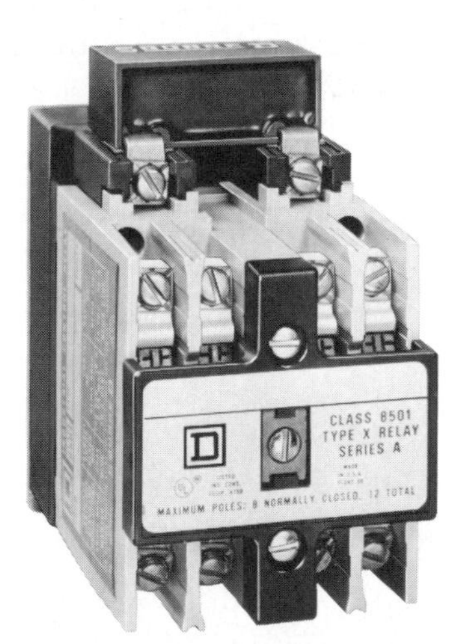

FIGURE 7–8 *Surge suppressor mounted on top of magnetic relay. (Square D)*

Triacs

Triacs are also used in solid-state switching. Since the SCR can control current in only one direction, it is limited in its applications. The triac can conduct in both directions. The triac has the same characteristics as the SCR. This means that the triac can be thought of as two SCRs placed in parallel but connected in the opposite direction.

Triac Construction. Examine the drawing of the triac in Fig. 7–9. The triac has three terminals: main terminal 1 (MT_1), main terminal (MT_2), and gate (G). By examining the PN structure of the triac it can be seen that current can pass through a PNPN layer, or it can pass through a NPNP layer. The device can be described as having an NPNP layer in parallel with a PNPN layer. This arrangement of four-layer material gives the triac a connection of two SCRs in parallel. This connection is shown in Fig. 7–10. The connection in Fig. 7–10 is not how the triac operates. That is because the triac gate voltage responds differently than the SCR gate voltage. Figure 7–11 shows the schematic diagram of the triac. Because the triac can con-

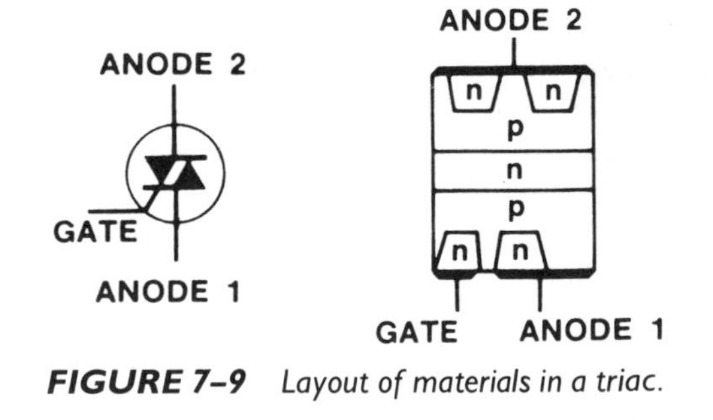

FIGURE 7–9 *Layout of materials in a triac.*

FIGURE 7–10 *Two SCRs connected in parallel.*

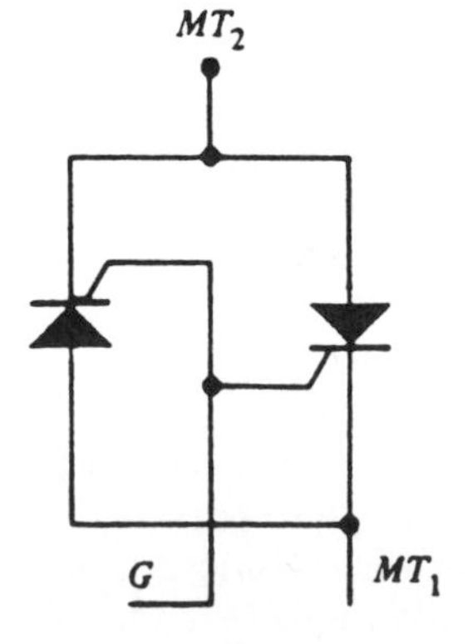

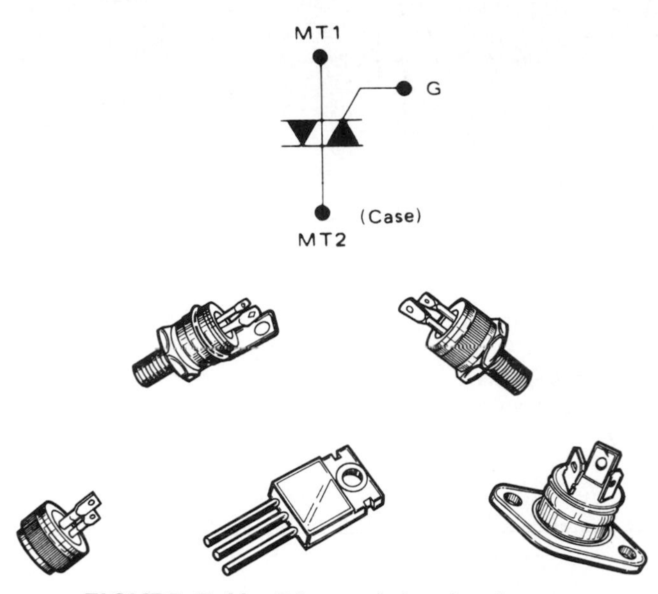

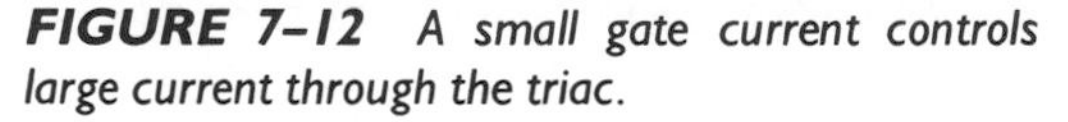

FIGURE 7–11 Triac symbol and various case configurations.

duct current in both directions, the schematic diagram contains two diodes facing in opposite directions.

Triac Applications. Because the triac can conduct in both directions, it is best when used to control ac power. Take a look at the diagram in Fig. 7–12. In this circuit, full power is applied to the load when the gate is triggered on. When S_1 is open, the triac cannot conduct. This is because the voltage applied to the triac is below the breakover point. When S_1 is closed, the triac is triggered on, and both halves of the ac power are applied to the load. This differs from SCR operation. The SCR can apply only half of the power to the load because it conducts in only one direction. The advantage of all thyristors is that small gate currents can control large load currents.

The triac conducts in both directions and requires a small current to operate. It does, however,

FIGURE 7–12 A small gate current controls large current through the triac.

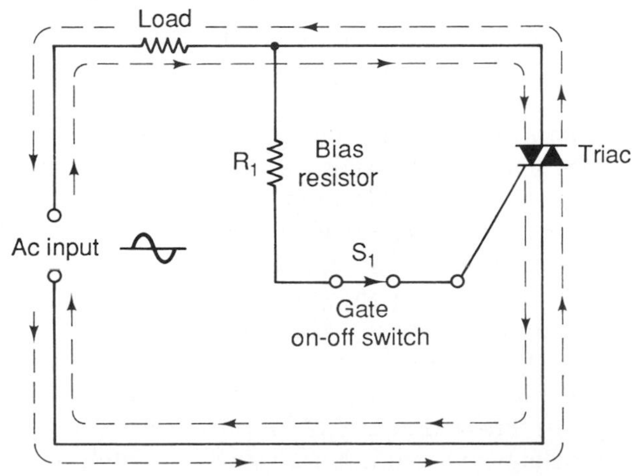

have some disadvantages compared to the SCR. The SCR has higher current ratings than the triac. The triac can handle currents up to 25 A, whereas the SCR can safely handle currents of around 800 A. That means when large currents are required the SCR is the better choice.

There are some differences in the frequency-handling abilities of the SCR and triac. The triac is usually slower in turning on when used with an inductive load, such as when it is used to control a motor. The triac is also designed to operate mainly in the low-frequency range of 30 to 400 Hz. The SCR can safely handle frequencies up to 30 kHz.

Silicon-Controlled Rectifiers

The silicon-controlled rectifiers (SCR) is a three-junction semiconductor device that can operate as a switch (Fig. 7-13). It is basically a rectifier. It will conduct current in only one direction. The best part of SCR operation is its ability to be turned on and off. The on–off action makes the SCR very useful in controlling current.

Construction of the SCR. Construction details of a component provide information about how the component will operate. Solid-state devices are made by joining P and N material into junction or junctions. Bipolar transistors, diodes, and FETs are all constructed this way. The SCR is made by joining four alternating layers of P and N material. Most SCRs are made of silicon, but germanium is also used (Fig. 7–14).

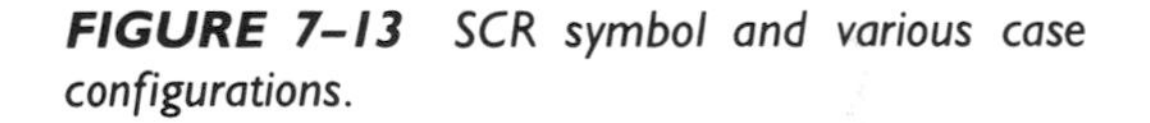

FIGURE 7–13 SCR symbol and various case configurations.

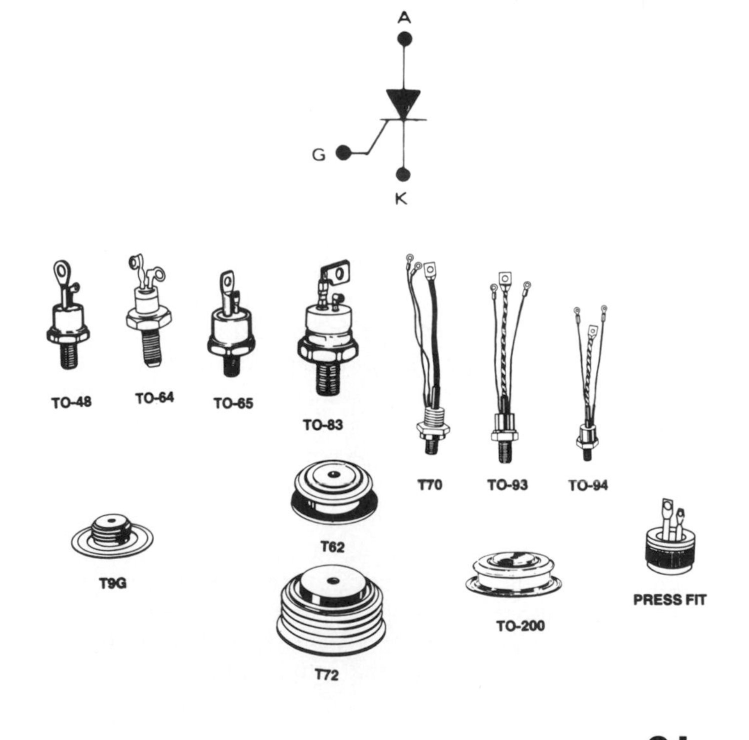

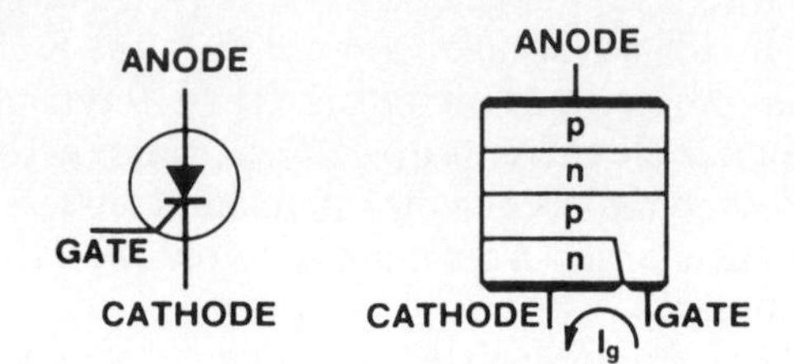

FIGURE 7-14 *Layer construction of the SCR.*

In Fig. 7-14, note how the layers of P and N material are sandwiched together. Also note the three junctions. Leads for external connections are attached to these layers. These three connections are called the anode (A), cathode (K), and gate (G).

Now take a look at the schematic diagram of the SCR shown in Fig. 7-13. The schematic symbol is about the same as that for a rectifier diode. The main difference is the gate. In some cases the circle around the symbol is not shown. The leads may not be identified on a schematic drawing. When the leads are marked, they are identified with the letters A, K, and G.

Operation of the SCR. Inasmuch as the SCR is a semiconductor device, it requires a biasing voltage to cause it to turn on. Figure 7-15 shows a simplified arrangement that causes the SCR to operate. A switch is used in the gate circuit to apply voltage to the gate.

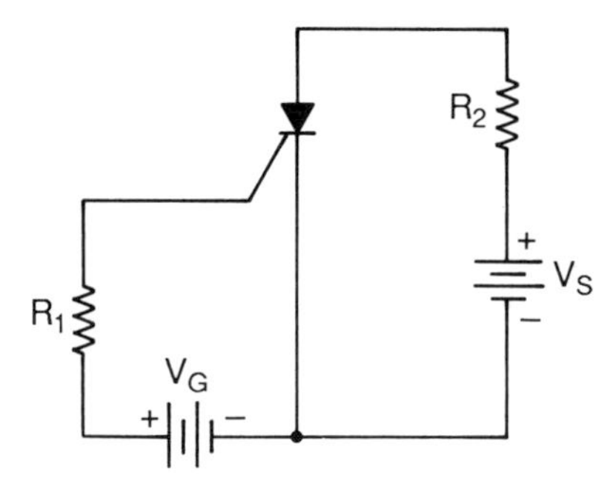

FIGURE 7-15 *Biasing circuit for an SCR.*

Resistor R_1 is used to limit the current flow in the gate circuit. A second voltage source supplies the needed forward bias to the anode and cathode. A resistor is in series with the anode-cathode circuit. This resistor is also used as a current-limiting resistor. It prevents high currents from causing damage to the SCR. Without the resistor, the SCR conducts hard in forward bias and burns out after a short operating period. A specific gate current must be reached before and SCR will become a conductor. Each SCR has its own breakover voltage. This means that each SCR must have the proper forward bias applied and the proper gate current in order to operate effectively as a switch.

AC Operation of the SCR. The SCR can be used to control dc or ac. Since it is a rectifier, it operates on only one ac alternation. The SCR conducts only when the input cycle makes the anode positive and the cathode negative (Fig. 7-16). By closing S_1 a positive voltage is developed that turns the SCR on. The series resistor is in the gate circuit for current-limiting purposes. The diode is in the circuit to protect the anode and cathode during reverse voltage operation.

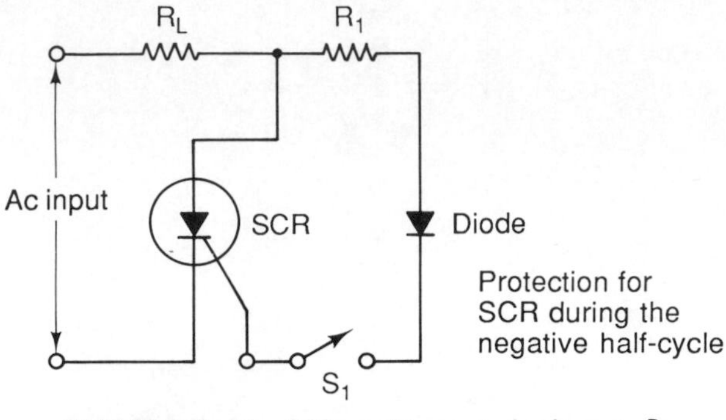

FIGURE 7-16 *SCR used to control voltage to RL (load resistor).*

If S_1 is closed, the SCR conducts when the proper polarity appears at the anode. If the gate switch is opened, the SCR continues to conduct until the voltage between the anode and cathode falls below the breakover voltage. Once the voltage falls below this level the SCR remains off until S_1 is closed again.

There are many advantages to the use of an SCR over an electrical switch. For instance, the SCR will not wear out. It will not develop contact arcing, nor will it stick in one position. This means that the SCR is a more reliable component than the mechanical switch, especially in high current applications. The SCR may be controlled by a switch, or it may be controlled by an electrical pulse from a computer. The most important characteristic is that a small amount of power applied to the gate controls large amounts of current to a load.

PHASE-FAILURE RELAYS

Interchanging any two phases of a three-phase induction motor power source causes the motor to reverse its direction of rotation. This is called *phase reversal.* In elevators and other industrial operations, equipment damage and personnel injuries may result when a phase reversal occurs unexpectedly. This can happen if a fuse blows or a wire to a motor breaks while the motor is running. The motor continues to operate on single phase but that causes serious overheating. Phase failure and phase-failure reversal relays are used to protect motors against these situations.

Both voltage-sensing and current-sensing phase-failure relays are available. Voltage-sensing relays may be connected at any point on the lines, but only detect abnormal conditions ahead of the point of con-

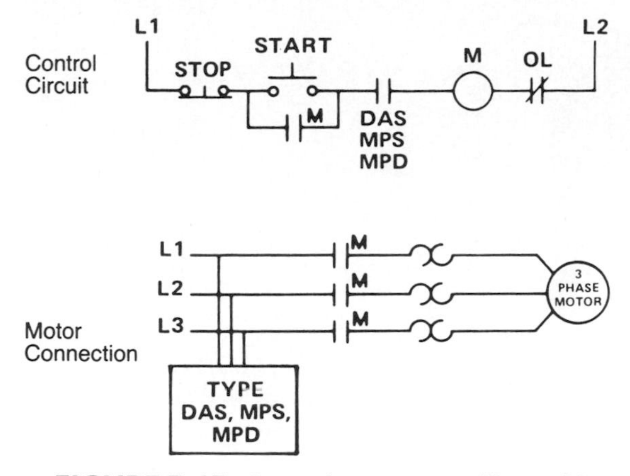

FIGURE 7–17 *Line side monitoring. (Square D)*

nection. Voltage sensing offers the advantage of being able to detect abnormal conditions independent of motor running status. They are also easy to apply since motor voltage is all that is required to select the relay. Figure 7–17 shows line-side monitoring. With the relay connected before the starter, the motor cannot be started in the reverse direction. However, the motor is unprotected against phase failures between the relay and the motor.

Phase-failure relays are often used to control a shunt trip circuit breaker. When this is done, care must be taken to ensure that the shunt trip circuit always has an adequate source available. This can be accomplished by using the diagram shown in Fig. 7–18.

When a phase failure occurs on L2 or L3, the shunt trip coil draws power from L1 through the control relay (CR) contacts and phase-failure relay contacts (which change state on detecting a phase failure). If a phase failure occurs on L1, the control relay (CR) contacts change state. The shunt trip coil draws power from L2 through the CR contacts and phase-failure relay contacts.

If the control relay contacts, the phase-failure relay contacts, or the shunt trip coil does not have the same voltage rating as the motor, control transformers may be interposed where needed.

Load-side monitoring is shown in Fig. 7–19. With the relay connected directly to the motor, the total feed lines are monitored. The motor may sustain a momentary bump in the reverse direction with this connection.

Current-sensing relays require three externally connected current transformers. These must be sized to match the motor full-load current. Current sensing offers the advantage of being able to detect imbalances more precisely by monitoring currents. Relay selection is independent of motor voltage and requires a separate 120-V source or supply.

Three-wire control is necessary with current control to prevent the relay from "cycling" on and off when an open phase or phase reversal occurs (Fig. 7–20). Due to the current-sensing features of this

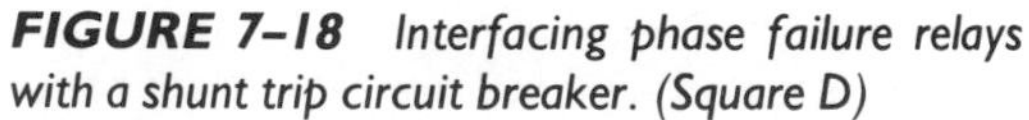

FIGURE 7–18 *Interfacing phase failure relays with a shunt trip circuit breaker. (Square D)*

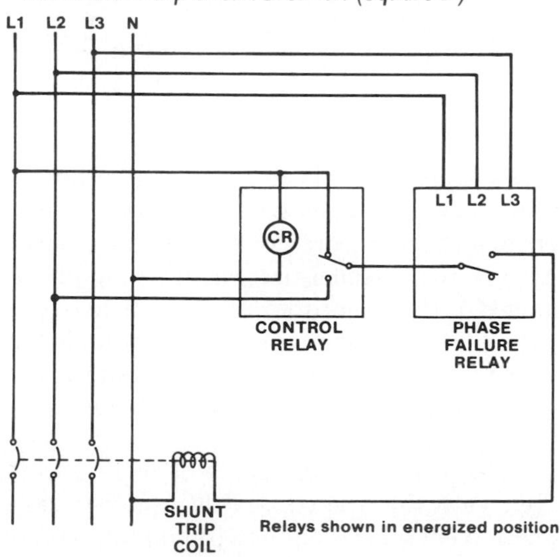

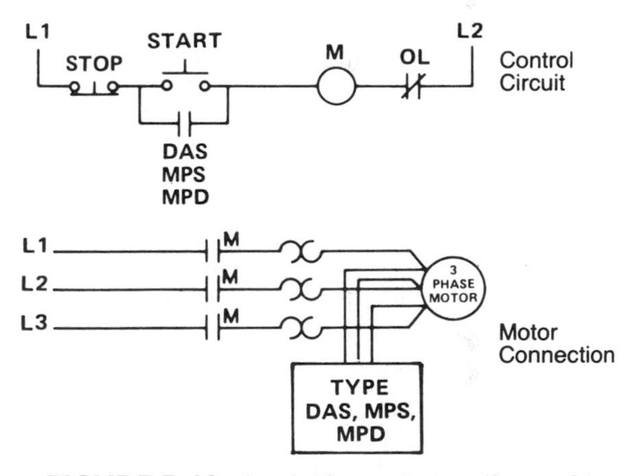

FIGURE 7–19 *Load-side monitoring. (Square D)*

FIGURE 7–20 *Current monitoring. (Square D)*

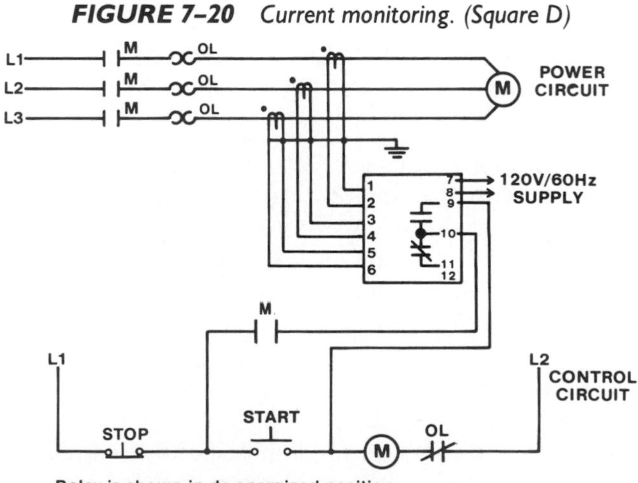

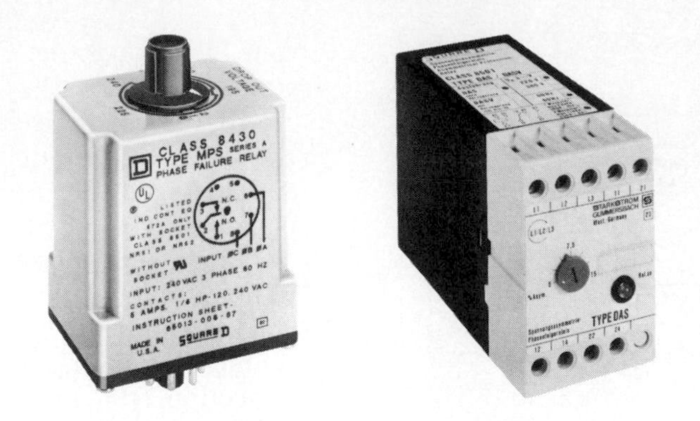

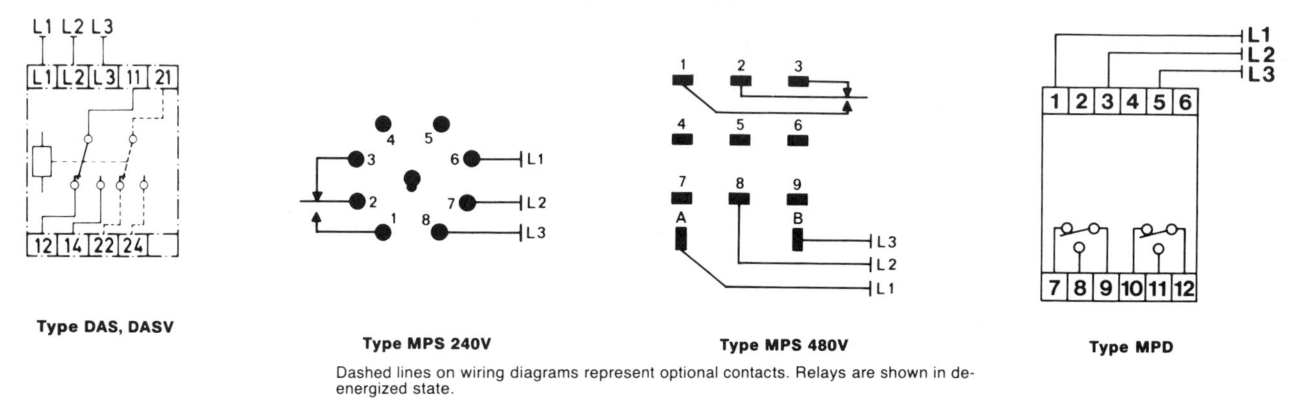

FIGURE 7–21 *Phase failure relays with wiring diagrams and contact connections. (Square D)*

relay, the load must first draw current before a reverse-phase condition can be sensed. This means that this relay is not used for protecting motors driving equipment unable to tolerate a momentary "bump" in the reverse direction.

Figure 7–21 shows three types of phase-failure relays. These offer the reliability and accuracy of solid-state sensing circuitry with the isolation of hard output contacts. MPS is a phase-failure and undervoltage relay. DAS is a phase-failure relay. *DAS* and *MPS* are used by Square D to designate this type of relay.

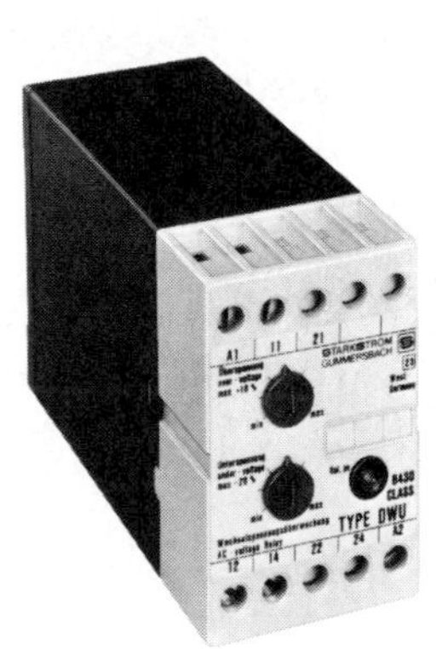

FIGURE 7–22 *Solid-state monitoring relay. (Square D)*

SOLID-STATE MONITORING RELAYS

Solid-state monitoring relays use solid-state sensing circuitry with the isolation of hard output contacts. This type of monitoring is both accurate and reliable. Figure 7–22 shows the enclosure of several types of monitoring relays. All of these relays have an LED indication when the relay is energized.

Voltage Relay

The relay monitors ac single-phase and dc (independent of polarity) voltages. The relay has independent adjustable controls for both pickup and dropout voltage. The relay energizes when the supply voltage is present and the monitored voltage is above the pickup setting. The relay deenergizes when the supply voltage is removed or the monitored voltage is below the dropout setting. The dropout voltage is adjustable from 50 to 95% of pickup voltage. Figure 7–23 shows how the relay is wired in the circuit and how its contacts operate.

In electromechanical types of relays the pull-in voltage means that the voltage necessary to cause the relay coil to energize is enough to pull the armature to

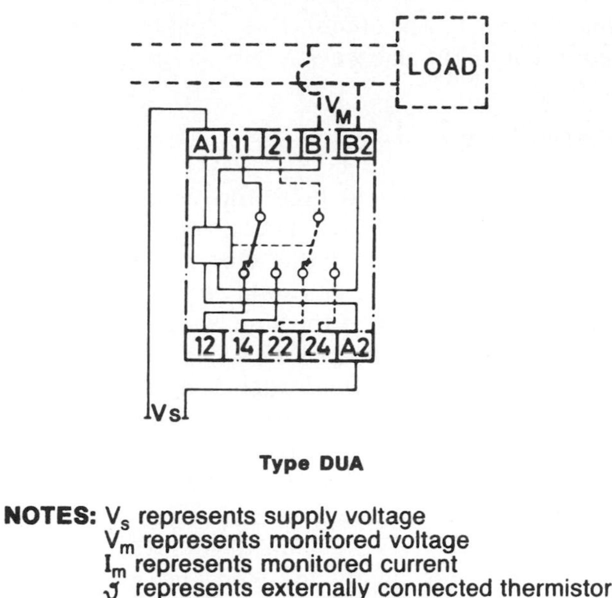

NOTES: V_s represents supply voltage
V_m represents monitored voltage
I_m represents monitored current
ϑ represents externally connected thermistor
S_k represents monitored contacts
Dashed lines represent optional contacts
Relays are shown in de-energized position

FIGURE 7–23 *Voltage relay. (Square D)*

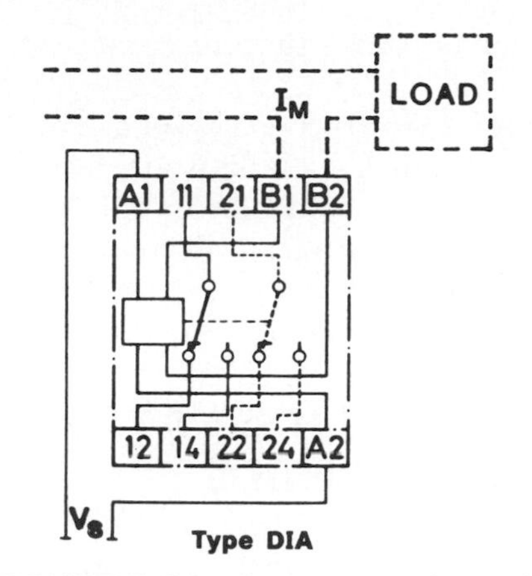

FIGURE 7–24 *Current relay. (Square D)*

the coil core and thereby close the contacts attached to the armature. In the semiconductor or solid-state type of relay it refers to the voltage needed to cause proper biasing of the transistor or semiconductor device to energize or conduct, thereby lowering its emitter–collector resistance or forward conduction resistance of the SCR or triac.

The term *dropout voltage* means the same in both electromechanical and semiconductor solid-state relays. The dropout voltage is lower than the voltage it takes for the relay to energize. In other words, the relay will continue to be energized between the pull-in and dropout voltages once it is energized. However, once the voltage has reached a point where there is not enough current flowing to cause a sufficient magnetic field to hold the armature to the core, it will open. In the case of the solid-state relay, it is the point at which the device stops conducting or turns off. The solid-state relay has no armature to pull in or drop out. It takes very little difference in voltage to cause the voltage monitoring relay to drop out or stop conducting. It also makes this type of relay very useful in operating with computer signals. It is very useful in digitally controlled logic circuits.

Current Relay

This relay is almost identical to the voltage monitoring relay except that it monitors the variations in current. It is also hooked up differently (Fig. 7–24). The relay monitors ac single-phase and dc (independent of polarity) currents. The relay has independent adjustable controls for both pickup and dropout current. The relay energizes when the supply voltage is present and the monitored current is above the pickup setting. The relay deenergizes when the supply voltage is removed or the monitored current is below the dropout setting. The dropout current is adjustable from 50 to 95% of the pickup current.

Over/Under Relay

This type of relay monitors single-phase voltages and requires no additional supply voltage (Fig. 7–25). Overvoltage is adjustable from 100 to 110% of nominal voltage. Undervoltage is adjustable from 80 to 100% of nominal voltage. The relay energizes when voltage is between these two settings.

Zero-Current Turn-Off. One of the unique features of a semiconductor or solid-state relay is its ability to automatically turn off when the ac load current sine wave is crossing its zero axis. This property of a solid-state device is very important when switching inductive loads, since the turn-off occurs when the current is at its minimum point on the sine wave. This reduces the inductive kickback that often sustains an arc in the electromechanical relay contacts. Inductive kickback

FIGURE 7–25 *Over/under voltage relay. (Square D)*

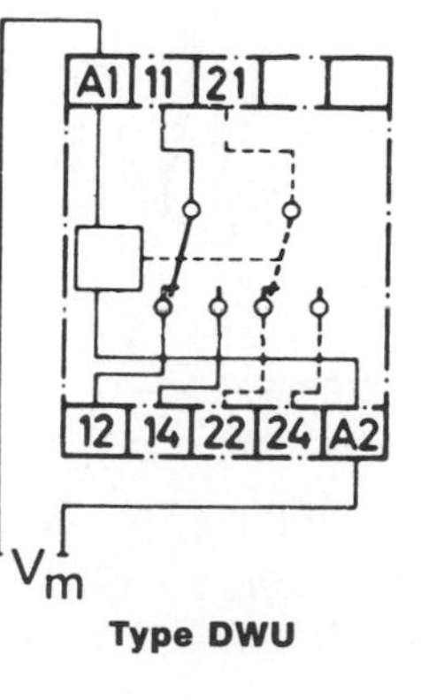

increases voltage spikes that can damage semiconductors.

Zero-Voltage Turn-On. Zero-voltage turn-on is not necessarily available on solid-state relays. It is one condition that can be added to a solid-state relay to provide certain features that will greatly extend the life of some types of loads.

TYPES OF SOLID-STATE RELAY SWITCHING

The three types of relay switching that can be accomplished by the solid-state relay are the instant on (IO), the universal (US), and the zero switch (ZS). Each of these relays is designed to switch on at different locations on the sine wave and for good reasons. The type of load—resistive, inductive, lamp, or a combination of these—determines the type of relay to select for the job.

The instant-on (IO) type turns on *immediately* as soon as the control input is switched. That means that it can turn on at any point on the sine wave. However, the turn-off of the instant-on type is always at the zero point where the current crosses the zero point of the sine wave (Fig. 7–26).

The universal-type solid-state relay turns on within a given *window* on the ac sine wave (Fig. 7–26). The universal relay does not turn on at the zero point or at the peak of the sine wave. This type of relay can turn on only inside the given window. The universal relay turns off at zero current, just as in the other two basic types of solid-state relays.

The zero switch has an advantage in handling the turn-on surges so often associated with a circuit. This type of relay monitors the load circuit to make sure that it turns off when the sine wave passes through the zero point on the sine wave.

FIGURE 7–26 *Zero points and universal relay turn on windows.*

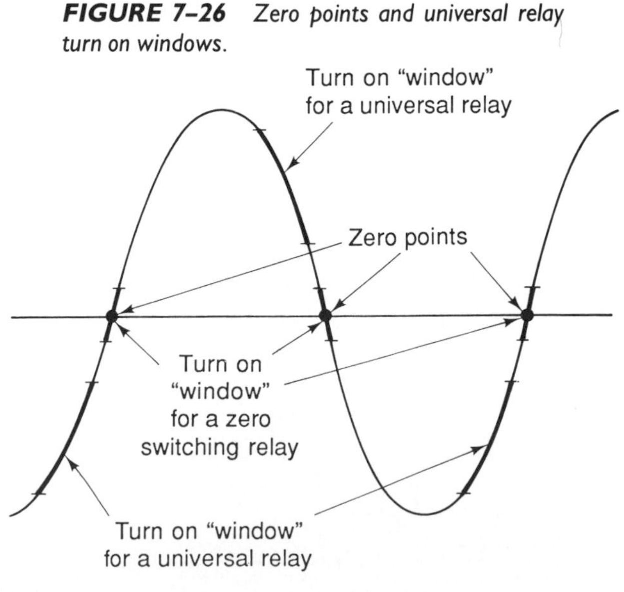

Switching Relay Loads

Relays are designed to take into consideration the characteristics of inductive loads, resistive loads, and variations of these two types. One of the advantages of the solid-state relay is its ability to turn on and turn off at the zero point of the sine wave. Some devices benefit more than others if switched at these points. For instance, *resistive loads* such as light bulbs have a low resistance when cold and higher resistance when hot. If they are turned on cold, they can draw large amounts of current, at least more than normal operating current, and shorten their life. When turned on at the zero point it gives the filaments time to heat up and change resistance slightly and keep the surge down to a manageable point, and in turn extends the life of the bulb or lamp. Zero-point switching also eliminates noise spikes. This means that a zero-switching relay should be used with lamps and other types of resistive loads.

Other types of load do not necessarily respond well to zero switching. *Inductive loads* are one type that need an instant-on type of relay. This is due to the nature of an inductive load presented by a motor, transformer, or solenoid coil. With an inductive load the current and voltage are nearly 90° out of phase. This means that the current is lagging the voltage by this amount. It also means that there is no inductive reactance to increase the opposition to current flow at the instant when starting voltage is applied. This means that there is a large current draw when the voltage is at the zero-crossing point of the sine wave. Switching at any point other than zero is preferable for an inductive load.

Loads consisting of lamps are also resistive, but the inrush current is high (up to 15 times normal operating currents) due to the low resistance of the cold filaments. This high inrush current can do great damage to relays. The main concern here is the inrush current at the start of operation. That means that a zero-switching relay can be of some advantage with this type of load. However, the inrush currents should be taken into consideration when choosing the proper relay for the job.

When there is a combination of resistive and inductive loads in a circuit to be switched, it is best to use the universal relay. This type of relay does not switch at the zero point or at the peak point of the sine wave. Therefore, it is often selected for a combination load.

Thermistor Relay

The thermistor relay operates from the signal of externally connected thermistors (Fig. 7–27). The relay en-

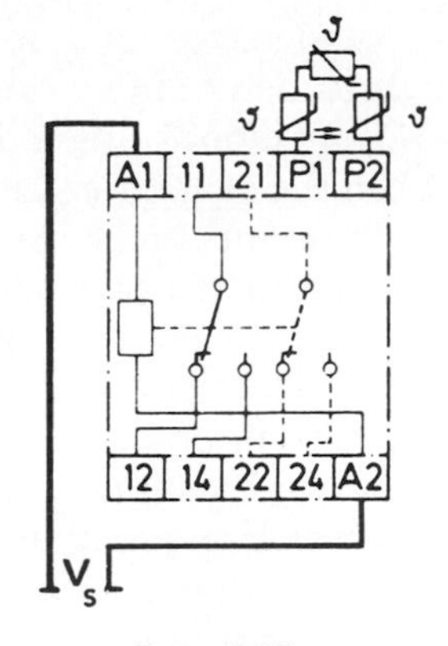

FIGURE 7–27 Thermistor relay. (Square D)

ergizes when the supply voltage is present and the thermistor resistance is below 1.9 kΩ (range of 1.5 Ω to 2.3 kΩ). The relay deenergizes when the supply voltage is removed or the thermistor resistance increases above 3 kΩ (range 2.5 to 3.6 kΩ).

Contact Amplifier Relay

Contact amplifier relays are used in cases where contacts do not have sufficient current and voltage ratings to switch loads such as coils, solenoids, or small motors. Typical examples are manometer contacts or supervisory relays for broken wire at wire-producing machines (Fig. 7–28).

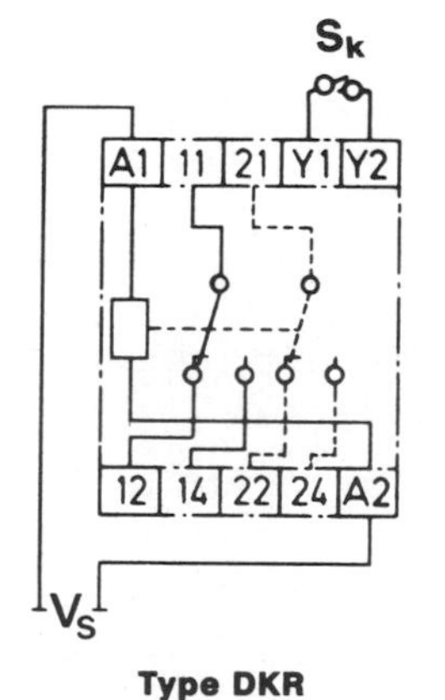

FIGURE 7–28 Contact amplifier relay. (Square D)

Load Detector and Load Converter Relays

The load on an induction motor can be monitored with the right choice of solid-state circuitry. The load detector relay and the load converter relay shown in Fig. 7–29 are used to control equipment and processes by monitoring the load of an induction motor. The devices simulate the true output power of the motor

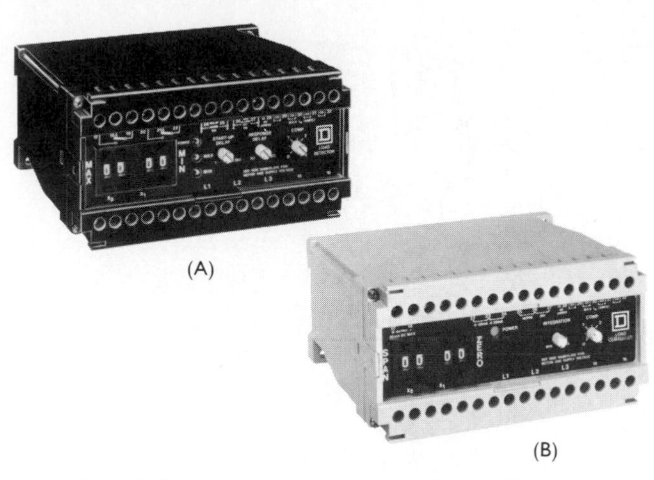

FIGURE 7–29 (A) Load detector and (B) Load converter relays. (Square D)

by measuring the input voltage, current, power factor, and compensating for internal losses of the motor. Using these parameters, the output power is computed continuously. In an induction motor, the current varies as the input voltage varies. By computing power, the load relays are not affected by these variations and can provide much greater accuracy over a wider range of load than is possible by measuring current alone. Three relay models are available. One type provides both a maximum and a minimum trip point, a second type provides two maximum trip points, and a third type provides an analog current output.

Three types of load detectors are available. The type V load detector monitors the motor load and has two separate output relays to indicate when a trip point has been exceeded. On the type V3, the two output relays correspond to a maximum and minimum trip point. On the type V4, the two output relays correspond to two maximum trip points. Two separate pushbutton thumbwheel switches, located on the face of the device, are used to select the trip points. The load detector has two SPDT relay outputs, to indicate maximum or minimum trip points exceeded. The unit is also adjustable for startup delay, response delay, and compensation of motor losses. LEDs indicate power on, maximum trip, or minimum trip.

Typical Applications

Either the type V3 load detector or the type G load converter is used in a crusher–conveyor situation (Fig. 7-30). The load converter can monitor the crusher load and feed the output signal to an adjustable-frequency drive controlling the feeder conveyor motor. In this way the load on the crusher is be kept constant by varying the speed of the conveyor. If an adjustable drive is not used, the load detector starts and stops the conveyor motor to keep the load on the crusher within preset limits.

The type V load detector has a monitoring resolution fine enough to determine worn tools on machinery (Fig. 7–31). By monitoring a drill motor, the drill is stopped when the drill bit is dull. This extends the life of tools and prevents breakage problems.

By continuously monitoring the power consumption of a fan motor, an accurate indication of the system status can be obtained. The V load detector monitors the loading on three-phase motors and controls and protects fan systems (Fig. 7–32).

The load detector can simultaneously monitor four common problems that may occur in heating, ventilating, and air-conditioning systems: broken or loose fan belts, closed dampers, blocked filters, and mechanical wear on the motor (such as bearings).

Figure 7–33 shows how the load detector and the load converter are connected in a circuit. Figure 7–34 shows the outputs of the type V3 unit that can be used to disconnect a motor when load limits are exceeded. In a three-wire control scheme, normal operating mode, with an alarm signal, the relay is wired as shown.

The typical load converter monitors the motor load and generates a 0- to 20-mA or 4- to 20-mA output signal proportional to the load on the motor. An integration adjustment on the face of the device smooths the output signal from short-term variations

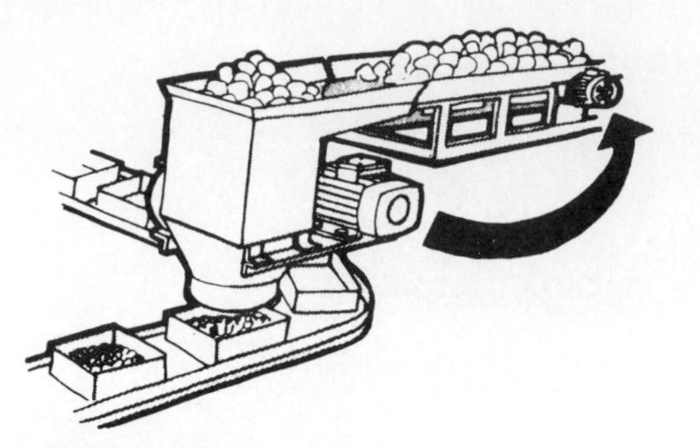

FIGURE 7–30 *Crusher–conveyor with load detector. (Square D)*

FIGURE 7–31 *Machine tool bit load detector senses dull bit. (Square D)*

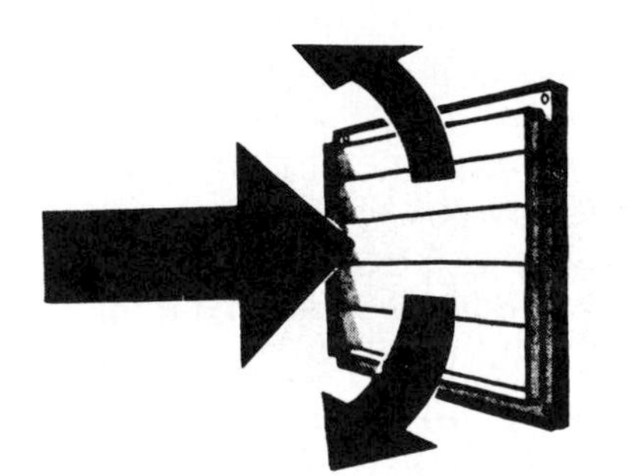

FIGURE 7–32 *Load detector senses closed dampers. (Square D)*

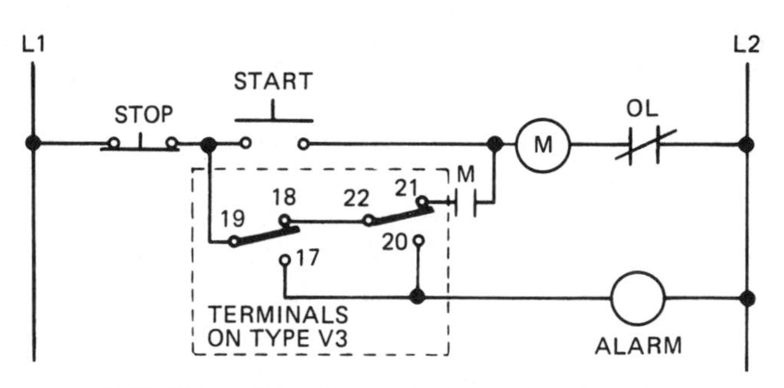

FIGURE 7–34 *Three-wire control and load detector circuit. (Square D)*

FIGURE 7–33 *Load detector and load converter hookups. (Square D)*

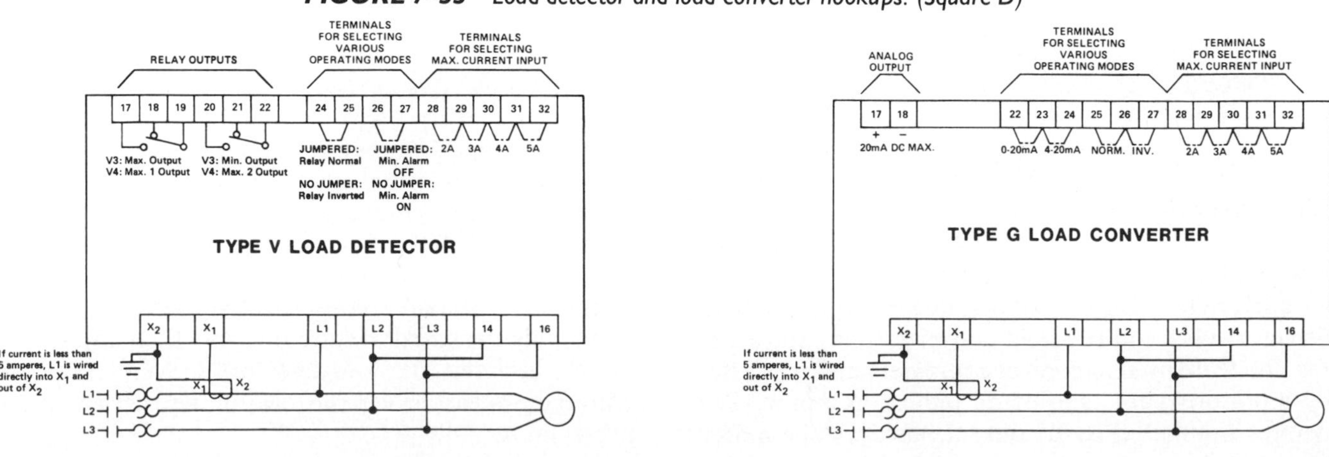

in the input signals. Two separate pushbutton thumbwheel switches on the face of the device are used to select the zero point of the range and the span. The output signal can be selected to increase with load or decrease as load increases. An LED indicates when power is applied to the device.

THERMAL OVERLOAD RELAYS

Thermal overload relays sense motor current by converting this current to heat in a resistance element. The heat generated is used to open a normally closed contact in series with a starter coil causing the motor to be disconnected (Fig. 7–35).

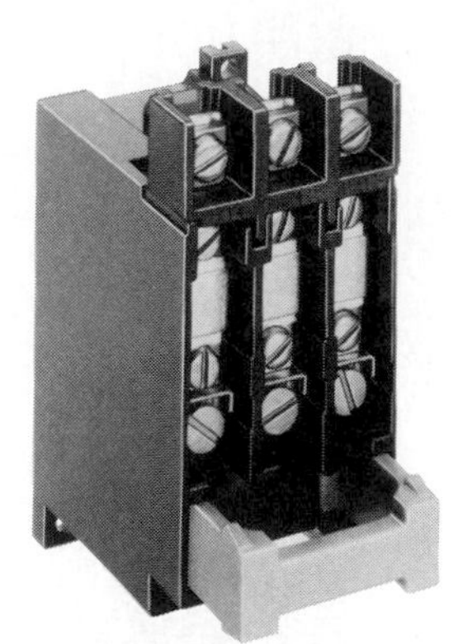

FIGURE 7–35 *Thermal overload relays with replaceable contacts. (Square D)*

The thermal overload relay is simple and inexpensive. It is very effective in providing motor-running overcurrent protection. This is possible because the most vulnerable part of most motors is the winding insulation, and this insulation is very susceptible to damage by excessively high temperature.

Being a thermal model for a motor, the thermal overload relay produces a shorter trip time at higher current similar to the way in which a motor reaches its temperature limit in a shorter time at a higher current. In a high ambient temperature, a thermal overload relay trips at a lower current, or vice versa, allowing the motor to be used to its maximum capacity in its particular ambient temperature (if the motor and overload are in the same ambient).

After it is tripped, the thermal overload relay does not reset until it has cooled. This allows the motor to cool before it can be restarted.

Types of Thermal Overload Relays

There are two types of thermal overload relays: bimetallic and melting alloy. In some types the bimetallic is available in both noncompensated and ambient-temperature-compensated versions. In both melting alloy and bimetallic, single-element and three-element overloads are available. See Fig. 7–36 for a cutaway view of a standard trip melting alloy thermal unit.

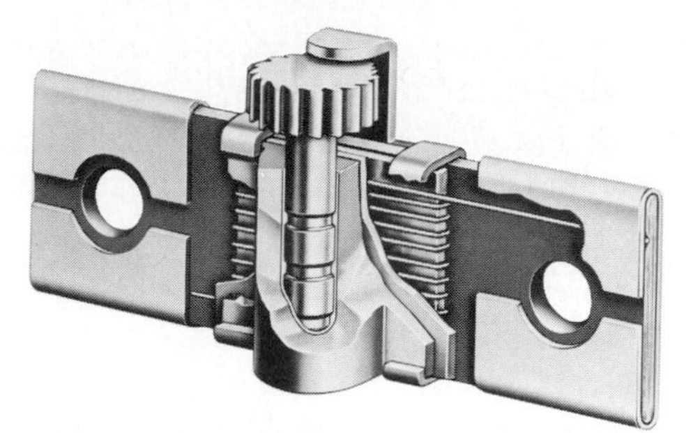

FIGURE 7–36 *Cutaway view of standard trip melting-alloy thermal unit. (Square D)*

With the exceptions of a few types, all thermal overloads incorporate a trip-free reset mechanism that allows the relay to trip on an overload even though the reset lever is blocked or held in the reset position. This mechanism also prevents the control circuit contact from being reclosed until the overload relay and the motor have cooled.

Hand-Reset Melting Alloy (NEMA Style)

Hand-reset melting alloy overload relays use a eutectic alloy solder that responds to the heat produced in a heater element by the motor current. When tripped, the overload relay is reset manually after allowing a few minutes for the motor and the relay to cool and the solder to solidify.

Repeated tripping does not affect the original calibration. Melting-alloy thermal units are available in three designs: quick trip, standard trip, and slow trip. *Quick trip* (class 10) units are used to protect hermetically sealed, submersible pump, and other motors that can endure locked rotor current for a very short time, or motors that have a low ratio of locked rotor to full-load current. *Standard trip* (class 20) units provide trip characteristics for normal motor acceleration up to approximately 7 seconds on a full-voltage start. *Slow trip* (class 30) units provide trip characteristics for motor acceleration up to approximately 12 seconds on a full-voltage start. The motor should be suitable for extended starting periods.

Overload Relay Class Designation

Class 10. Relay will trip in 10 seconds or less at a current equal to 600% of its current rating. Used with hermetic motors, submersible pumps, or motors with short locked-rotor time capability.

Class 20. Relay will trip in 20 seconds or less at a current equal to 600 times its current rating.

Class 30. Relay will trip in 30 seconds or less at a current equal to 600 times its current rating.

General applications call for a class 20 thermal relay.

Bimetallic Overloads (NEMA Style)

Bimetallic overload relays are used where the controller is remote or difficult to reach. Three-wire control is recommended when automatic restarting of a motor could be hazardous to personnel.

Normally, bimetallic relays are used on automatic reset. They are supplied from the factory on hand reset, but can be adjusted for either hand or automatic reset in the field. When used on hand reset, allow the motor and thermal units a few minutes to cool before resetting.

Temperature Compensation

Ambient temperature compensation is available on some overload relays (Fig. 7–37). These relays have all the features of the noncompensated bimetallics. In addition, an extra bimetal element maintains a nearly constant trip current in relay temperatures from −20 to +165°F for one type and −4 to 131°F for the other type. Trip current is adjustable from 85 to 115% of the trip current ratings. A SPDT contact is standard on the 25- and 45-A sizes. The NO contact can be used in an alarm circuit and must be wired on the same polarity as the NC contact. Contacts are not replaceable.

Protection Level. Protection level is the relationship between trip current rating and full-load current. Protection level is in percent and is the trip current rating divided by the motor full-load current times 100. Check with the manufacturer of the thermal relay (usually listed with instructions in tables in the back of the catalog) to make sure that the correct unit is selected for the job the motor is expected to do.

Thermal units are not included with any overload relays. They are selected and ordered and priced separately. Ideally, thermal units should be selected from the instruction sheet that is included with every starter or overload relay. If it is desirable to order thermal units along with the controller, they should be selected from a catalog with specifications listed and based on the type of controller being ordered and the nameplate full-load current of the motor. If the motor full-load current is not known at the time thermal units are ordered, an approximate selection can be made using Table 7–1 as follows:

1. Locate motor horsepower and voltage.
2. Determine approximate full load current from the table.
3. Use approximate full-load current in place of actual nameplate full-load current.

Ambient-temperature correction curves for thermal overload relays are given in Fig. 7–38.

FIGURE 7–37 *Bimetallic overloads.*

TABLE 7–1 APPROXIMATE THERMAL UNIT SELECTION BASED ON HORSEPOWER AND VOLTAGE[a]

Motor Horse-power	*Motor Full-Load Current*					
	Three Phase				*Single-Phase*	
	200 V	*230 V*	*460 V*	*575 V*	*115 V*	*230 V*
$\frac{1}{20}$	0.39	0.34	0.17	0.14	1.30	0.65
$\frac{1}{12}$	0.55	0.48	0.24	0.19	1.90	0.95
$\frac{1}{8}$	0.74	0.64	0.32	0.26	2.60	1.30
$\frac{1}{6}$	0.90	0.78	0.39	0.31	3.24	1.62
$\frac{1}{4}$	1.22	1.06	0.53	0.42	4.40	2.20
$\frac{1}{3}$	1.52	1.32	0.66	0.53	5.47	2.74
$\frac{1}{2}$	2.07	1.80	0.90	0.72	7.45	3.73
$\frac{3}{4}$	2.88	2.50	1.25	1.00	10.1	5.07
1	3.68	3.20	1.60	1.28	12.6	6.31
$1\frac{1}{2}$	5.18	4.50	2.25	1.80	17.2	8.59
2	6.67	5.80	2.90	2.32	21.4	10.7
3	9.66	8.40	4.20	3.36	29.1	14.5
5	15.4	13.4	6.68	5.35	42.9	21.4
$7\frac{1}{2}$	22.6	19.6	9.82	7.86	58.4	29.2
10	29.7	25.8	12.9	10.3		36.3
15	43.6	38.0	19.0	15.2		49.4
20	57.4	49.9	24.9	20.0		
25	70.9	61.7	30.8	24.7		
30	84.3	73.3	36.7	29.3		
40	111.	96.4	48.2	38.5		
50	137.	119.	59.6	47.6		
60	163.	142.	70.8	56.6		
75	201.	175.	87.6	70.0		
100	265.	230.	115.	92.0		
125	327.	284.	142.	114.		
150	389.	338.	169.	135.		
200	511.	445.	222.	178.		

Source: Courtesy of Square D

[a]Use only when motor full-load current is not known. Thermal units selected using approximate full-load currents from the table will provide a trip current between 100 and 125% of full-load current for many four-pole, single-speed, normal-torque, 60-Hz motors. Since the full-load current rating of different makes and types of motors vary so widely, these selections may not be suitable. Thermal units should be selected on the basis of motor nameplate full-load current and service factor. Thermal unit sizes originally selected on an approximate basis should always be rechecked and corrected at the time of installation if required.

Note: These currents should not be used for selection of fuses, circuit breakers, or wire sizes. See NEC tables 430–148 through 430–150. For motors rated 208 to 220 V, use the 230 V column. For motors rated 440 to 550 V, use 460 and 575 V columns, respectively.

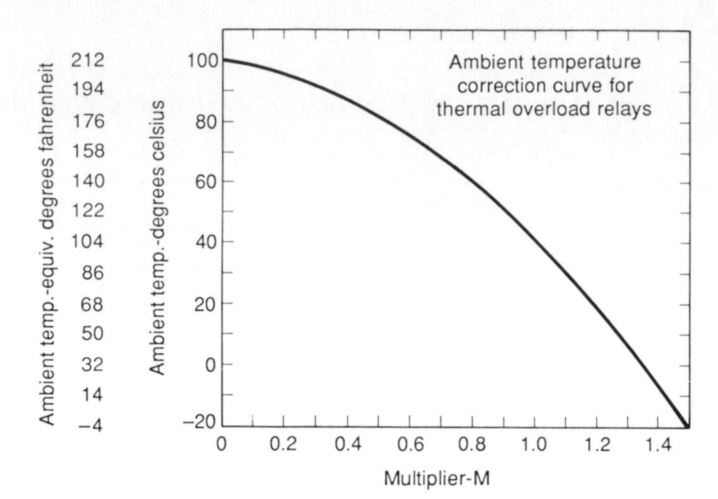

FIGURE 7–38 *Ambient temperature correction curve for thermal overload relays.*

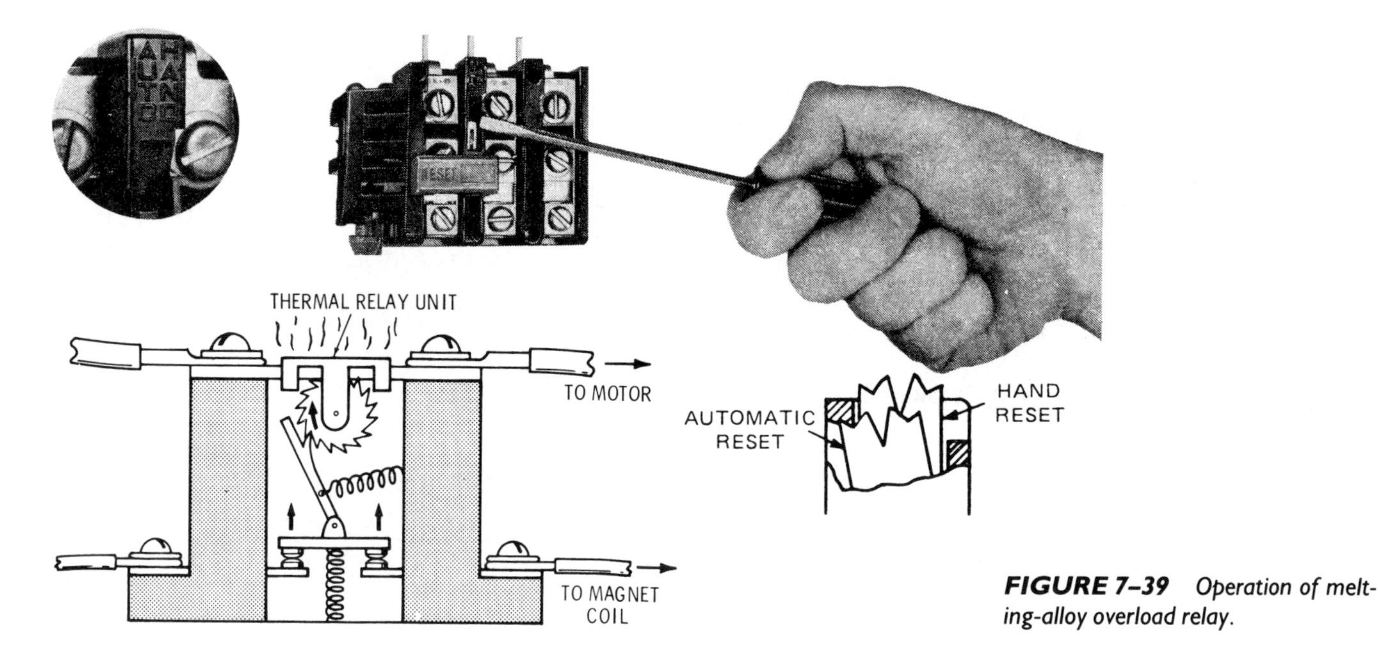

FIGURE 7–39 *Operation of melting-alloy overload relay.*

Mounting Thermal Units. Always be certain the correct thermal units are installed in the starter before operating a motor. Thermal units should always be mounted so that their type designation can be read from the front of the starter (Fig. 7-39). Melting-alloy thermal units should be mounted so that the tooth of the pawl assembly can engage the teeth of the ratchet wheel when the reset button is pushed. Mounting surfaces of the starter and thermal units should be clean. Make sure that the thermal unit mounting screws are securely fastened.

ELECTROMAGNETIC RELAYS AND MOTORS

Relays are a necessary part of many control and pilot light circuits. They are similar in design to contactors, but are generally lighter in construction, so they carry smaller currents.

Magnetic contactors are normally used for starting polyphase motors, either squirrel-cage or single-phase. Contactors may be connected at any convenient point in the main circuit between the fuses and the motor. Small-diameter control wires may be run between the contactor and the point of control.

Protection of the motor against prolonged overload is accomplished by time-limit overload relays that are operative during the starting period and running period. Relay action is delayed long enough to take care of heavy starting currents and momentary overloads without tripping.

Motors for commercial condensing units on refrigeration or air-conditioning systems are normally protected by a metallic switch operated on the *thermo,* or heating principle. This is a built-in motor overload

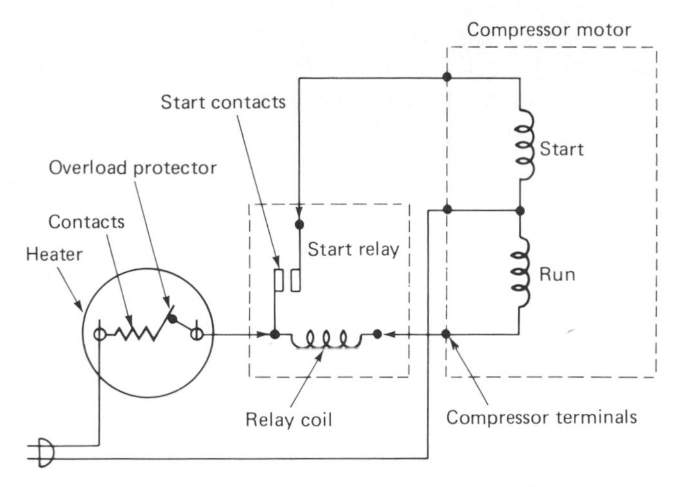

FIGURE 7-40 *Circuit for a domestic refrigerator with start relay.*

protector. It limits the motor winding temperature to a safe value. In its simplest form, the switch or motor protector consists essentially of a bimetal switch mechanism that is permanently mounted and connected in series with the motor circuit (Fig. 7-40).

When the motor becomes overloaded or stalled, excessive heat is generated in the motor winding due to the heavy current produced by this condition. The protector located inside the motor is controlled by the motor current passing through it and the motor temperature. The bimetal element is calibrated to open the motor circuit when the temperature, as a result of excessive current, rises above a predetermined value. When the temperature decreases, the protector automatically resets and restores the motor circuit.

Motor Winding Relays

Following the thermal overload protector in the circuit (Fig. 7-40) is the motor start relay. A motor winding relay is usually incorporated in single-phase motor-compressor units. This relay is an electromagnetic device for making and breaking the electrical circuit to the start winding. A set of normally closed contacts is placed in series with the motor start winding (Fig. 7-40).

The electromagnetic coil is in series with the auxiliary winding of the motor. When the motor start and run windings are energized a fraction of a second later, the motor comes up to speed and sufficient voltage is induced in the auxiliary winding to cause current to flow through the relay coil. The magnetic force created by the current through the coil is sufficient to attract the spring-loaded armature, which mechanically opens the relay starting contacts. With the starting contacts open, the start winding is out of the circuit. The motor continues to run only on the run winding. When the control contacts open, power to the motor is interrupted. This allows the relay armature to close the starting contacts. The motor is now ready to start a new cycle when the control contacts again close.

A hermetic compressor motor relay is an automatic switching device designed to disconnect the motor start winding after the motor has attained a running speed. There are two types of motor relays used in refrigeration and air conditioning compressors: the *current-type* relay and the *potential-type* relay.

Potential-Type Relay. The potential-type relay is generally used with large commercial air-conditioning compressors. The motors may be capacitor start–capacitor run types up to 5 hp. Relay contacts are normally closed. The relay coil is wired across the start winding. It senses voltage change. Start winding voltage increases with motor speed. As the voltage increases to the specific pickup value, the armature pulls up, opening the relay contacts and deenergizing the start winding. After switching, there is still sufficient voltage induced in the start winding to keep the relay coil energized and the relay starting contacts open. When power is turned off, the voltage drops to zero. That means that the coil is deenergized and the start contacts reset (Fig. 7-41).

Many of these relays are extremely *position sensitive.* When changing a compressor relay, care should be taken to install the replacement in the same position as the original. Never select a replacement relay solely by horsepower or other generalized rating. Select the correct relay from the parts guidebook furnished by the manufacturer.

Current-Type Relay. The current-type relay is generally used with small refrigeration compressors up to $\frac{3}{4}$ hp. When power is applied to the compressor motor, the relay solenoid coil attracts the relay armature upward. This causes the bridging contact and stationary contact to engage (Fig. 7-42). This energizes the motor start winding. When the compressor motor at-

FIGURE 7-41 *Potential-type relay used in refrigeration circuits. (Tecumseh Products)*

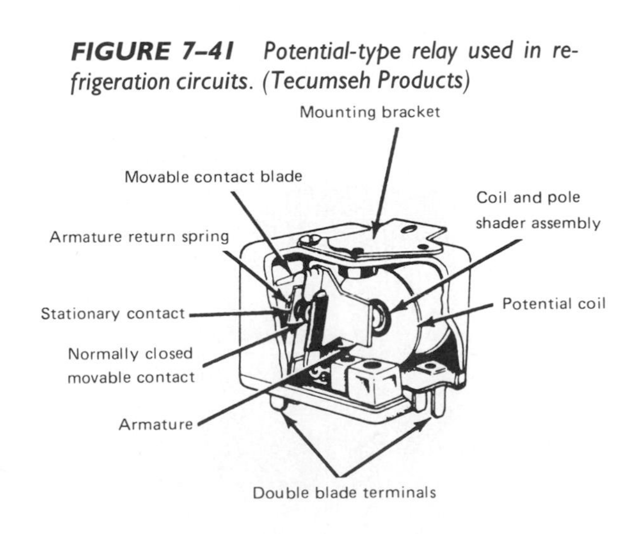

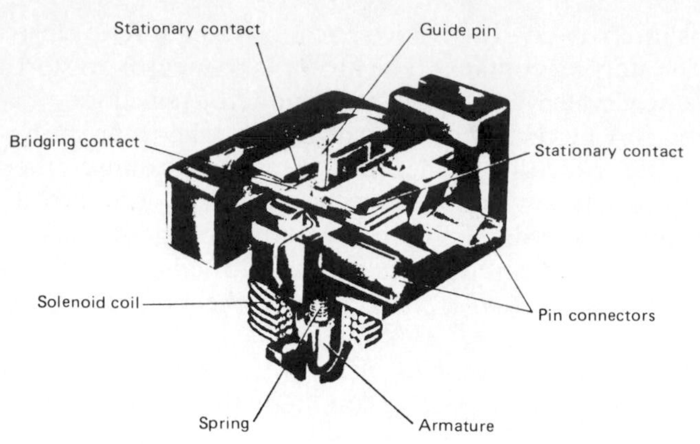

FIGURE 7–42 *Current-type relay used in refrigeration circuits. (Tecumseh Products)*

tains running speed, the motor main winding current is such that the relay solenoid coil deenergizes. This allows the relay contacts to drop open, thereby disconnecting the motor start winding.

ELECTROMECHANICAL RELAYS

The electromechanical relay is available in a number of different arrangements and have any number of names to indicate their purpose. The general-purpose relay can perform a number of operations in a variety of locations.

The control relay is designed specifically for use as a machine tool relay (Fig. 7–43). This relay is designed to handle the logic switching requirements of machine tools, conveyors, hoists, elevators, cranes, tire machines, and practically every type of motor-driven machinery.

The *control relay* is electromagnetically operated and held. Energization of the magnet coil causes the normally open contacts to close and the normally closed contacts to open. Deenergization of the coil causes the contacts to switch back to their original state.

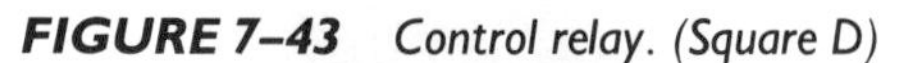

FIGURE 7–43 *Control relay. (Square D)*

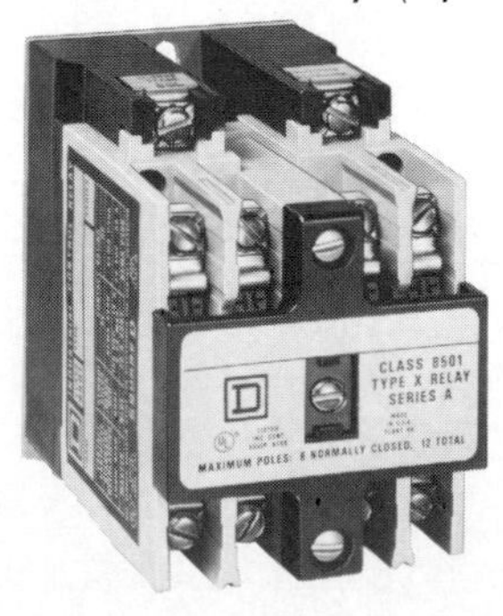

The latching relay, the control relay, and the timing relay work together to form a logic system for making automated plants possible. The control relay is designed to switch inductive and resistive loads in both ac and dc circuits. By far the greatest number of applications involve the switching of inductive loads in ac circuits. Typical loads are operating coils on such devices as other relays, timers, starters, contactors, and solenoids.

The *latching relay* is electromagnetically operated and is held by means of a mechanical latch. Energizing the latching coil causes the normally open contacts to close and the normally closed contacts to open. The mechanical latch holds all contacts switched, *even after power is removed* from the latching coil. Energizing of a second coil, the unlatching coil, results in all contacts switching back to their original state. Both coils are continuously rated and require no coil clearing contacts.

If the latching and unlatching coils are energized at the same time, the latching coil will override the contacts and the contacts will go to, and remain in, their switched state. Figure 7–44 shows the conventional two-coil circuit.

The latch attachment has its own coil and is mounted on the control relay (Fig. 7–45). The mechanism fits on any two- to eight-pole relay and can be mounted on the relay in the field even after the relay

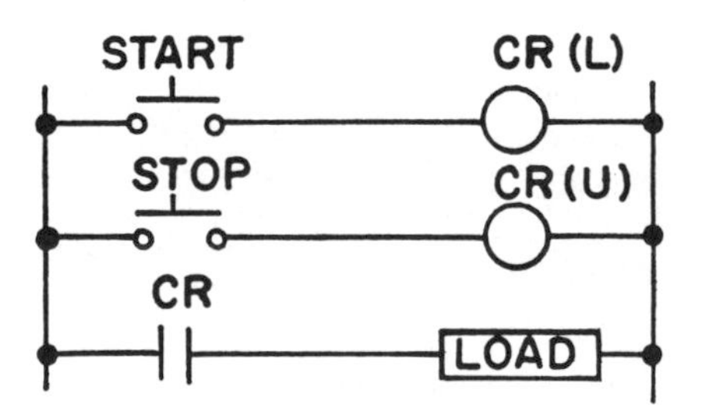

FIGURE 7–44 *Two-coil circuit. (Square D)*

FIGURE 7–45 *Latching relay. Note the latch attachment on the front of the control relay. (Square D)*

contacts and coil are wired. The latching attachment has a self-adjusting feature that adapts the stroke of the latch to the stroke of the relay on which it is mounted to provide optimum performance.

The *solid-state timer* is recommended for use in control system work, with the latching relay. The solid-state timer is useful in extremely high duty cycles. This type of relay is discussed in Chapter 9.

RELAY OPERATING CHARACTERISTICS

Ratings of relays are important. The *make* rating, the *break* rating, and the *continuous* rating have to be taken into consideration when the relay is specified.

Resistive Ratings. This indicates the resistive load that the contacts can make, break, or carry continuously. Resistive ratings are based on 75% power factor.

Inductive Ratings. The inductive rating refers to loads, such as coils of contactors, starters, relays, and solenoids, that the contacts can make, break, and carry continuously. Inductive rating tests are run with 35% power factor loads.

Make Ratings. This rating applied to the current that can be handled by the contact at the time of contact closure. In inductive ac circuits, the momentary inrush current is often 10 times the sealed current, and a relay must be able to handle this inrush current as well as be able to break it in an emergency.

Break Ratings. This rating applies to the current that can be interrupted successfully by the contact. The inductive break rating is always less than the resistive or continuous ratings. When contacts break an inductive circuit, the inductance in the load tends to maintain the current. The result is an arc across the contact that causes heating and erosion of the contacts. Because of the extra heat generated, the allowable inductive current must be less than the resistive current for equal contact life.

Continuous Ratings. Continuous rating indicates the load that the contacts can carry continuously without making or breaking the circuit and without exceeding a certain temperature rise.

Contact Life

The life of control relay contacts depends on the magnitude and characteristics of the electrical load, inductance, duty cycle, mechanical properties of the device in which they are used, voltage fluctuations, and environment.

When control circuit relays are operated at maximum rated load, the life of the contacts is usually less than that of the remainder of the device. If the application requires a large number of operations during the life of the contacts, the contacts must be applied at values less than their maximum make and break ratings.

NEMA standards recommend that control relays for automatically operated sequencing systems be utilized with loads of less than 25% of the 60-A make and 6-A break ratings. NEMA standards do not recommend using a relay at its maximum ampere rating where the number of operations are expected to exceed substantially the 6000 operations required by the NEMA endurance test.

Contact Construction

The relay uses a double-break contact (Fig. 7–46). This places, for practical purposes, two single breaks in series, so that two arcs occur when the contact interrupts the current flow. This division of the energy in the arc materially extends the electrical life of the contact when compared to devices using single-break contacts. The stationary movable contacts are made of silver cadmium oxide material. This choice of material is important because of its resistance to welding when closing on the inrush currents normally associated with inductive loads. It also helps to reduce the contact erosion associated with repeated interruption of inductive circuits. Note how the contacts are constructed in Fig. 7–47. They provide two parallel paths per pole. The fact that both halves of the movable finger are not rigidly connected assures that all four contact points are held closed with nearly equal force. A conductive crossover saddle (Fig. 7–48) straddles the two fingers to provide a crossover path for even greater reliability.

An ohmmeter test on a contact is unreliable because the most common voltage source for an ohmme-

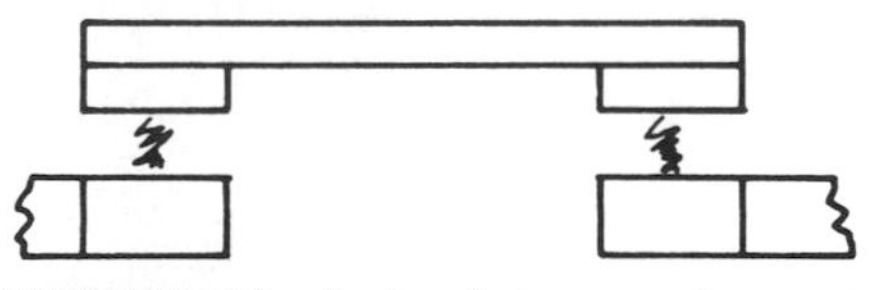

FIGURE 7–46 Arcing between relay contacts. (Square D)

FIGURE 7–47 Relay contacts. (Square D)

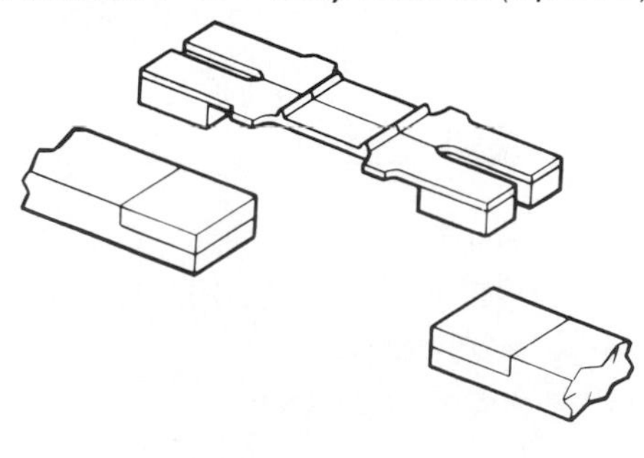

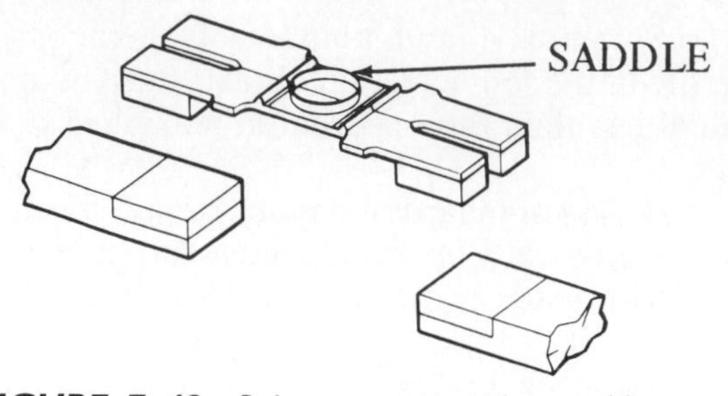

FIGURE 7–48 *Relay contacts with a saddle. (Square D)*

ter is a 1.5-V dry cell, and to contacts being tested, this is a low-energy circuit load. Several ohmmeter readings of the same contacts may vary from a few hundredths of an ohm to several ohms, yet the contacts may work perfectly well with a relay coil load.

Relay Coils

Coils for relays are designed to operate satisfactorily on voltages varying as much as 15% below and 10% above the nominal values. The nominal value is that value stamped on the coil.

Most coils are stamped with a part number, operating voltage, frequency, and date code. In most relays the external part of the coil is pressure-molded epoxy (Fig. 7–49). This forms a dense protective cover that provides high strength and resistance to mechanical damage. Construction of this type also provides good heat transfer for better cooling. Moisture absorption is also reduced. Coils for control relays and latch relays are designed so that they may be operated continuously without overheating.

FIGURE 7–49 *Exploded view of control relay. (Square D)*

Transients. The coil is relatively unaffected by switching transients. That is, because the pickup time of the relay is about 9 milliseconds, a short transient voltage is very unlikely to cause false switching of the output contacts. High-quality coil insulation plus low-voltage differential between turns makes failure due to shorted turns, resulting from transients, also quite unlikely. However, when a relay coil is deenergized, a transient is generated by the relay coil that may interfere with proper operation of nearby solid-state equipment. An optional transient suppressor is available to suppress transients to approximately 200% of peak voltage (Fig. 7–7).

Shock and Vibration

Shock and vibration show up in the form of contact bounce, false switching of contacts due to armature travel, or mechanical breakage. Most relays are designed with vibration and shock in mind, inasmuch as this is part of the operating environment of the relay. They are given consideration and compensated for by high contact pressure along with low-mass movable contacts.

Relays and Altitude

Higher altitudes produce an atmosphere with less pressure. The low pressure in itself does not affect the relay directly. However, the higher altitudes tend to reduce the insulation value of the air as well as its cooling effect. As a result it is necessary to consider the possible need to derate electrical equipment when it is used at high altitude. The relay should be derated when used at altitudes above 6000 feet but not more than 15,000 feet. This calls for derating the relay to 75% of its normal rating.

QUESTIONS

1. Describe a relay.
2. What do SPST, SPDT, and DPDT mean?
3. List the parts of a relay.
4. What does *burnishing* mean?
5. What are two types of solid-state relays?
6. What are some of the advantages of solid-state relays?
7. What is a triac? Draw the symbol for a triac.

8. What is the difference between an SCR and a triac?
9. Which can handle higher currents, an triac or an SCR?
10. Draw the symbol for an SCR and label the leads.
11. Describe, briefly, the operation of an SCR.
12. What is phase reversal?
13. What is dropout voltage?
14. What does an over/under relay do?
15. Why is zero-current turn-off an advantage to the solid-state relay?
16. What makes the universal-type solid-state relay different?
17. Why does an inductive load need an instant on relay?
18. Explain how the thermistor relay operates.
19. Where are contact amplifier relays useful?
20. Where are load detectors utilized?
21. What are the two types of thermal overload?
22. What are the three designs of thermal trip overload units?
23. What does the overload relay trip class designation mean?
24. What is the difference between the operation of the current relay and the potential relay?
25. What three types of relays make it possible to have a logic system in automated plants?
26. Describe the make rating and break rating of relay contacts.
27. Why are transient suppressors needed on relay coils?
28. How does high temperature affect relay operation?
29. Does the latching relay hold all contacts closed even after power is removed?
30. What does *inductive rating* mean?
31. What does *resistive rating* mean?
32. What does *continuous rating* mean?
33. What determines relay contact life?

REVIEW PROBLEMS

A review of series–parallel circuits can be of assistance in many electrical applications. Relays are no exception. Sometimes it is necessary to add relays in circuits with resistances in various combinations.

1. Find the current through R_1 in Fig. P–1.
2. Find the current through R_4 in Fig. P–1.
3. Find the voltage drop across R_3 in Fig. P–1.

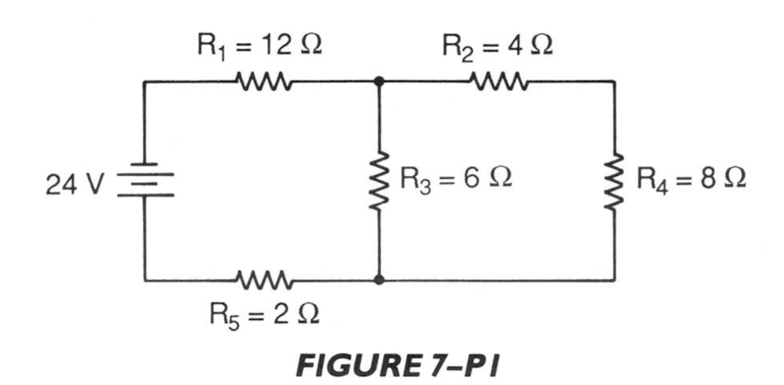

FIGURE 7–P1

4. Find the voltage drop across R_2 in Fig. P–1.
5. Find the current through R_5 in Fig. P–1.
6. Find the correct values from Fig. P–2 and place them in the following table:

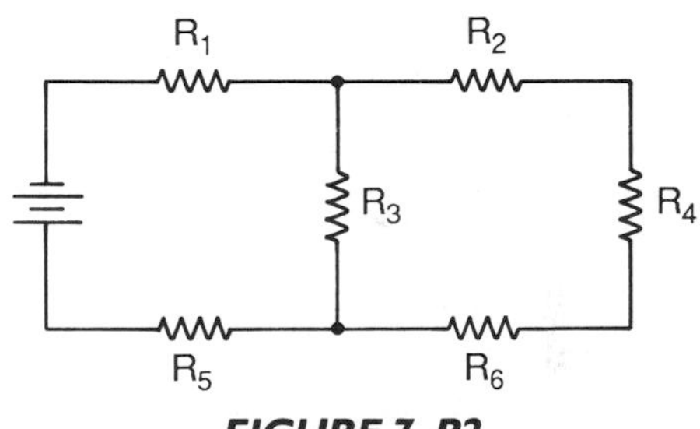

FIGURE 7–P2

Resistor	Resistance (Ω)	Voltage Drop (V)	Current (A)
R_1	3		2
R_2	5		1.00
R_3			
R_4	3		
R_5		8	
R_6	8		1.00

8

Motors

Objectives

After studying this chapter, you will be able to:

1. Identify and classify types of motors.
2. Define *CEMF*.
3. Explain the advantages and disadvantages of the various types of dc motors.
4. Describe the reversing procedure for ac motors.
5. Understand reasons for compensating windings and interpoles.
6. Explain how fields rotate in ac motors.
7. Draw three-phase waveforms.
8. Explain how synchronous motors are started.
9. List the characteristics of squirrel-cage motors.
10. Define *slip* in a motor.
11. Understand single-phase motor operation.
12. Determine how the number of poles affects motor speed at various frequencies.
13. Calculate the horsepower of a motor given certain values.
14. Describe the classification of insulation systems used for motors.

There are three types of motors when classified into large groups. The dc motor runs on direct current only, the ac motor runs on alternating current only, and the universal motor can run on ac or dc. Before you learn to control motors, it is best to know what makes them run and why they were chosen to do a particular job. The job they are designed to do and their ability to do the job are important factors in putting a motor to work at its greatest efficiency.

DC MOTORS

The dc motor is a mechanical workhorse. Many large pieces of equipment depend on a dc motor for power to move. The speed and direction of rotation of a dc motor are easily controlled. Just reverse the polarity and you reverse the direction of rotation. Change the voltage and the speed changes. This makes it especially useful for operating equipment such as winches, cranes, missile launchers, and elevators.

Operating Principles

The operation of a dc motor is based on the principle that *a current-carrying conductor placed in a magnetic field, perpendicular to the lines of flux, tends to move in a direction perpendicular to the magnetic lines of flux* (Fig. 8–1). The relationship between the direction of the magnetic field, the direction of current in the conductor, and the direction in which the conductor

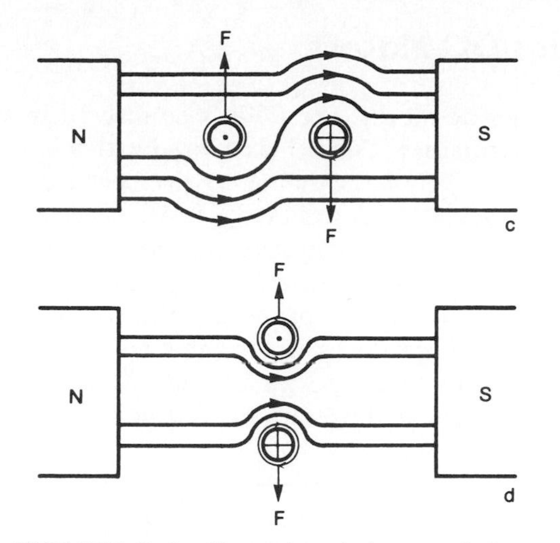

FIGURE 8–1 *Upward and downward forces created by interaction of field and armature flux.*

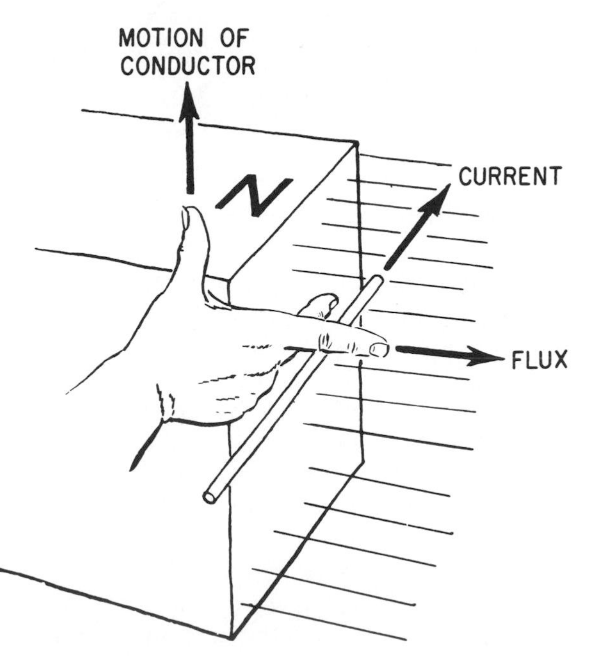

FIGURE 8–2 *Right-hand rule for motors.*

tends to move is called the *right-hand rule for motors* (Fig. 8–2).

The right-hand rule can be used to find the direction of rotation of a motor. If the motion of a conductor is unknown, it can be found by extending the thumb, index finger, and middle finger of your right hand so that they are at right angles to each other. If the forefinger is pointed in the direction of magnetic flux (north to south) (check the polarity of the power source to determine this), and the middle finger is pointed in the direction of current flow in the conductor, the thumb will point in the direction the conductor will move.

Keep in mind that a dc motor rotates as a result of two magnetic fields interacting with one another. The armature of a dc motor acts through its coils. Since the armature is located within the magnetic field of the field poles, these two magnetic fields interact. Like magnetic poles repel each other, and unlike magnetic poles attract each other. The dc motor has field poles that are stationary and an armature that turns on bearings in the space between the field poles. The armature of a dc motor has windings that are connected to commutator segments. Figure 8-3 shows how the simple dc motor operates.

Note how the brushes fitting on the commutator segments create a magnetic field in the conductor according to the polarity of the battery voltage. The changes in direction of current flow through the armature loop, caused by the switching action of the commutator segments, change the polarity of the magnetic field around the conductor. The magnetic fields repel and attract each other and the armature continues to turn. The momentum of the rotating armature carries the armature past the position where the unlike poles are exactly lined up. However, if these fields are exactly lined up when the armature current is turned on, there is no momentum to start the armature moving. In this case the motor will not rotate or start. It would be necessary to give the motor a spin to start it.

This disadvantage is eliminated when there are more turns on the armature, because then there is more than one armature field. No two armature fields can be exactly aligned with the field from the field poles at the same time.

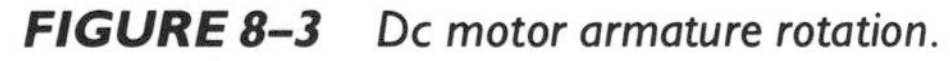

FIGURE 8–3 *Dc motor armature rotation.*

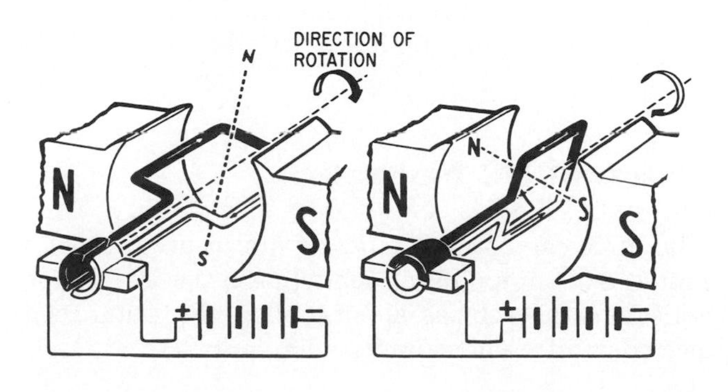

Counter Electromotive Force

While a dc motor is running, it acts somewhat like a dc generator. For instance, there is a magnetic field from the field poles. This means that a loop of wire is turning and cutting this magnetic field. For the moment, disregard the fact that there is current flowing through the loop of wire from the battery. As the loop sides cut the magnetic field, a voltage is induced in them just as it is in the loop sides of a dc generator. This induced voltage causes current to flow in the loop. The current induced is in the opposite direction from that which caused it (the battery). Inasmuch as this current flows in the opposite direction from that which caused it, it is called a *counter electromotive force* (CEMF).

In a dc motor, a counter EMF is always developed. The *counter EMF cannot be equal to or greater than the applied battery voltage* because then the motor would not run. The *counter EMF* is *always a little less.* Counter EMF opposes the applied voltage enough so that it is able to keep the armature current from the battery to a low value. If there were no such thing as counter EMF, much more current would flow through the armature. This is because it would have only its low dc resistance to determine the current draw. This means that the motor would run much faster. There is no way to avoid counter EMF. It makes a dc motor more economical to operate.

Loads

Dc motors are used to turn many mechanical devices: such things as water pumps, grinding wheels, fan blades, and circular saws, to name a few. Keep in mind that the pump or fan blade is the load. It is the mechanical device that the motor must move. This is the motor load. This load can cause the motor to draw more current as the amount of mechanical energy demanded is increased. This, in turn, means that there is more electrical power consumed since the voltage times the current equals the power consumed. The load on a motor affects its speed, current drawn, and its efficiency.

To get the most from a motor or operate it at its most efficient point, the load and the motor characteristics and abilities must be matched. This makes for a better operating condition for the load and the motor.

Types of DC Motors

The three categories for the dc motor are the series, shunt, and compound. Each type has distinct characteristics since each has its field coils and armature connected in different arrangements.

Series DC Motors

In a series dc motor, the field is connected in series with the armature. The field is wound with a few turns of comparatively large diameter wire because it must carry full armature current (Fig. 8–4). There are both advantages and disadvantages in this arrangement. This type of motor develops a very large amount of turning force (torque) from standstill. Because of this characteristic, the series motor can be used to operate electrical appliances, portable electric tools, cranes, winches, hoists, and to start an automobile engine.

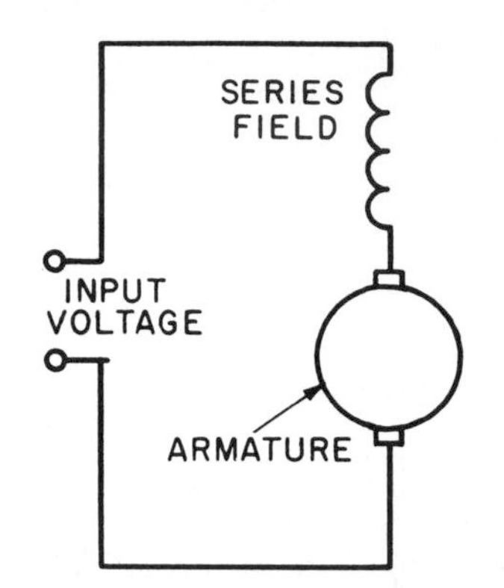

FIGURE 8–4 *Series-wound dc motor diagram.*

Another characteristic is that the speed varies widely between no-load and full-load. Series motors cannot be used where a relatively constant speed is needed for varying loads. A major disadvantage of the series motor is related to the speed characteristic. The speed of a series motor with no load increases to the point where the motor may become damaged. Usually, either the bearings are damaged or the windings fly out of the slots in the armature because the motor keeps increasing in speed until it self-destructs. With large motors such as cranes there is some danger to both the equipment and the personnel around it. A load must *always* be connected to a series motor before it is turned on. This means that there can be no belt-driven loads, since the belt may break or slip off. Small motors, such as those used in electrical hand drills, have enough internal friction (gearbox) to load themselves. Larger motors must be treated with more caution.

The series motor can be operated on ac or dc. This makes it more flexible in its use. However, it operates best on dc. The universal motor (operates on ac or dc) is discussed later.

Shunt DC Motors

A shunt motor is connected with the field windings in parallel (shunt) with the armature windings (Fig. 8–5). Once adjusted, the speed of a dc shunt motor remains

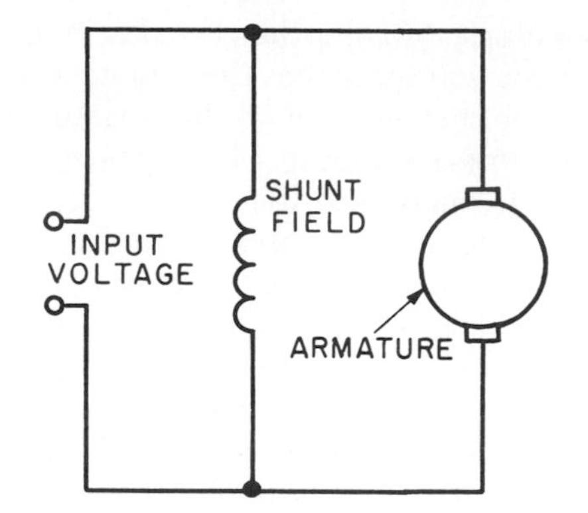

FIGURE 8–5 *Shunt-wound dc motor diagram.*

relatively constant even under changing load conditions. One reason for this is that the field flux remains constant. A constant voltage across the field makes the field independent of variations in the armature circuit.

When the load on the motor is increased, the motor tends to slow down. This slowdown decreases the amount of counter EMF generated in the armature. This decrease in counter EMF decreases the opposition to the flow of battery current through the armature. Armature current then increases. The increased armature current causes the motor to speed up. The conditions that established the original speed are then reestablished, and the original speed is maintained.

Now, if the motor load decreases, the motor tends to increase its speed. However, the counter EMF increases. This means that the armature current decreases and the decrease in armature current causes the speed to decrease. The decrease in load and the decrease in speed cause an almost instantaneous response. This means that the speed has a tendency to appear to be constant or to have so slight a fluctuation as to be unnoticed in most cases.

Compound DC Motors

A compound motor has two *field windings* (Fig. 8–6). One is a shunt field; it is connected in parallel with the armature. The other is a series field; it is connected in series with the armature. The shunt field gives the motor a constant speed advantage. The series field gives it the advantage of being able to develop a large torque when the motor is started under a heavy load. The compound motor has *both* shunt and series motor characteristics.

There are two types of compound motors, the long shunt and the short shunt. In the *long shunt* the shunt field is connected in parallel with the series field and armature (Fig. 8–6A). In the *short shunt* the shunt field is across the armature and the series field is in series with this parallel (shunt) arrangement (Fig. 8–6B).

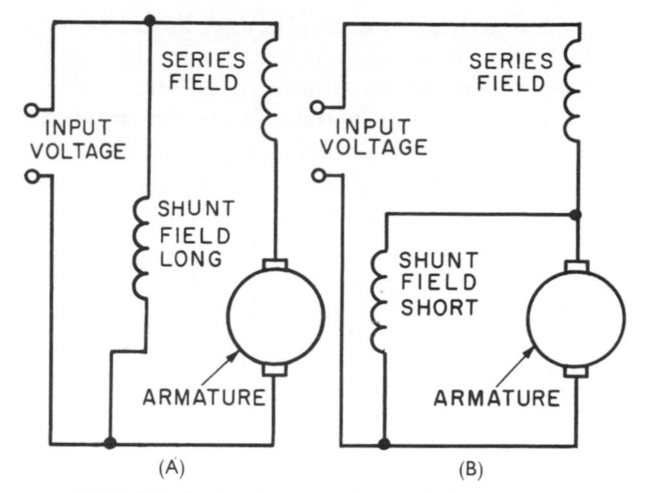

FIGURE 8–6 *Compound-wound dc motor diagram: (A) long shunt; (B) short shunt.*

Types of Armatures

Two types of armatures are used for dc motors. The gramme-ring armature is inefficient and is not necessarily used for any purpose except to get a better understanding of the drum-wound armature. The drum-wound armature is used on ac motors.

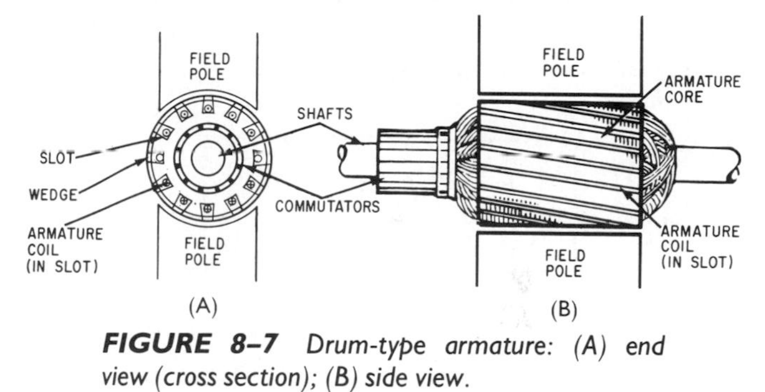

FIGURE 8–7 *Drum-type armature: (A) end view (cross section); (B) side view.*

Figure 8–7A shows an end view of the drum-wound armature as it appears cut through the middle. Current flow through the coils is indicated by a dot to show the current flowing toward you and the + indicates that the current is flowing away from you or that it resembles the tail feathers of an arrow as it goes away from you. Figure 8–7B is a side view of the armature and pole pieces. Notice that the length of each conductor is positioned parallel to the faces of the pole pieces. Each conductor of the armature can then cut the maximum flux of the motor field. The inefficiency of the gramme-ring type of armature is overcome by this positioning and makes it the drum-wound type.

Direction of Rotation

The direction of rotation of a dc motor depends on the direction of the magnetic field and the direction of current flow in the armature. If either the direction of the field or the direction of current flow through the armature is reversed, the rotation of the motor will reverse. However, if both of these factors are reversed at the same time, the motor will continue rotating in the same direction. In actual practice, the field excitation voltage is reversed to reverse the motor direction. This means that if you want a motor that can be connected to a reversing switch, the leads from the field must be brought out for easy access to the switching device.

Motor Speed

Dc motors are variable-speed motors. The speed of a dc motor is changed by changing the *current in the field or by changing the current in the armature.* A decrease in field current causes a decrease in the field flux. That means that the counter EMF decreases. This decrease in CEMF permits more armature current. That means that the motor speeds up. When the field current is increased, the field flux increases. More counter EMF is developed. The increase in CEMF decreases the armature current. The armature current then decreases and the motor slows down. Decreasing the applied voltage to the armature causes the armature current to decrease and the motor slows down. Increasing the armature voltage and current causes the motor to speed up.

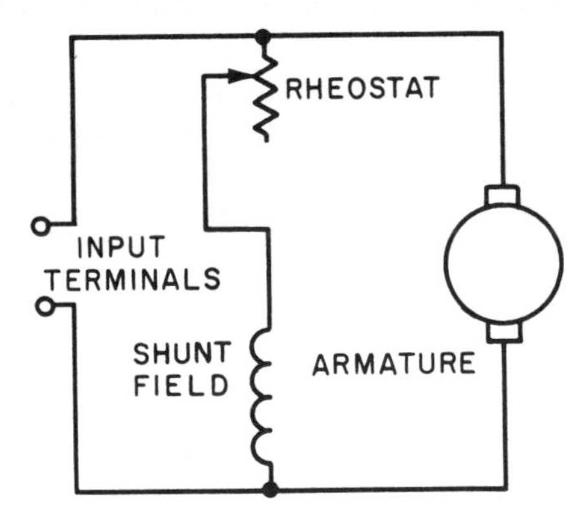

FIGURE 8–8 *Rheostat used as a speed control.*

Shunt motor speed can be controlled by a rheostat connected in series with the field windings (Fig. 8–8). Increasing the resistance of the rheostat causes the current through the field winding to decrease. This decreases the flux momentarily, which, in turn, decreases the counter EMF. The motor then speeds up. This momentary increase in speed increases the counter EMF and keeps the armature current the same. A decrease in rheostat resistance increases the current flow through the field windings and causes the motor to slow down.

In a series motor, the rheostat speed control may be connected either in parallel or in series with the armature windings. Moving the rheostat in a direction that allows the voltage across the armature to be lowered allows the current through the armature to be decreased and slows the motor. Moving the rheostat in a direction that increases the voltage and current through the armature increases the motor speed. One disadvantage of putting a resistor in series with a series motor is that it destroys the torque advantage the motor characteristically possesses. A good example of this is the foot control on a sewing machine. It is a series motor that powers the sewing machine. Inserting the series resistor in the foot-speed control decreases the motor's starting torque, and to start the sewing machine the motor must be given an assist by spinning the wheel. Changing speed controls from rheostats to SCRs gives the advantage of speed control without the loss of starting torque. This type of control is discussed in Chapter 7.

Armature Reaction

Since the armature revolves in a magnetic field it is subject to the same laws of nature that control the generator. This means that it cuts the flux field produced by the field coil. Of course, it also has its own magnetic field, produced by current flowing through its windings. There is a generator reaction produced by the armature revolving in the magnetic field, as well as its own magnetic field, used to give the motor effect. This armature effect has to be compensated for in the design of the motor. There are several ways to do this, one of which is to shift the brushes after the motor has started (Fig. 8–9). Note how the armature field has distorted the flux field between the pole pieces. The effect has shifted to the neutral plane to the left. This is in opposition to the direction of rotation. As the brushes are shifted it causes the neutral plane to shift. The proper location is indicated when there is no sparking from the brushes. Another way is to place compensating windings and interpoles in the motor permanently.

FIGURE 8–9 *Armature reaction.*

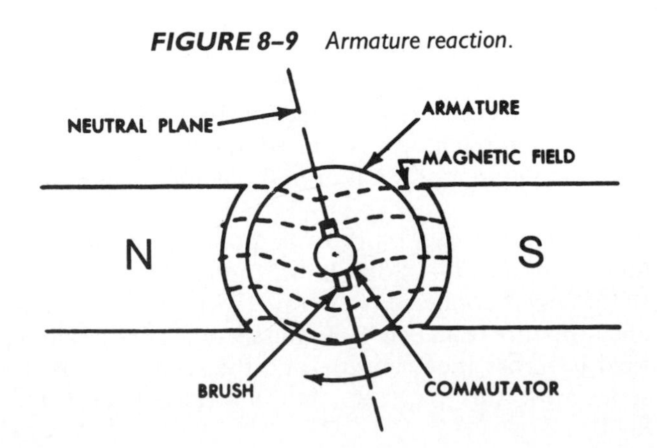

Compensating Windings and Interpoles

Compensating windings and interpoles cancel armature reaction in a motor. Shifting brushes reduces sparking and makes the field less effective. Canceling armature reaction eliminates the need to shift brushes.

Compensating windings and interpoles can be found in both motors and generators. Compensating windings are somewhat expensive. This means that most large dc motors depend on interpoles to correct armature reaction. Compensating windings are the same in motors as in generators. Interpoles, however, are slightly different. The difference is that a generator interpole has the same polarity as the main pole ahead of it in the direction of rotation. In a motor the interpoles have the same polarity as that of the main pole following it (Fig. 8–10).

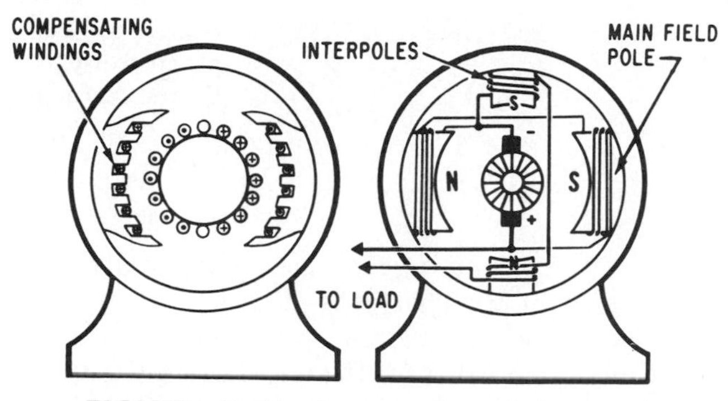

FIGURE 8–10 *Compensating windings and interpoles in a generator.*

Interpoles are connected to carry the armature current. As the load varies, the interpole flux varies, and commutation is corrected automatically as the load changes. This means that it is not necessary to shift the brushes when the load increases or decreases. Brushes are located on the no-load neutral plane and they remain in that position for all conditions of load.

The dc motor is reversed by reversing the direction of current in the armature. This means that the current through the interpole is also reversed. The interpole, then, still has the proper polarity to provide automatic commutation.

DC Motor Starting Resistance

The resistance of most dc motor armatures is low, somewhere between 0.05 and 0.5 Ω. Counter EMF does not exist until the armature begins to turn. This means that it is necessary to use an external starting resistance in series with the armature of a dc motor to keep the initial armature current to a safe level. As the armature begins to turn, counter EMF increases and the applied voltage is opposed by this increase in counter EMF. This means that the armature current is then reduced by its own generator effect. Then, once the motor comes up to normal speed, the external resistance in series with the armature is decreased or eliminated and full voltage can be applied across the armature.

Starting resistance can be controlled either manually, by an operator, or by any of several automatic devices. The automatic devices are usually just switches controlled by motor speed sensors. Automatic starters are covered in later chapters.

DC Motor Characteristics and Applications

A quick reference for checking out the characteristics and applications of dc motors is provided in Table 8-1.

Troubleshooting DC Motors

Methods used for solving problems with dc motors and a *troubleshooting dc motors table* are described in Chapter 22.

AC MOTORS

The alternating-current motor is less expensive than the dc motor of comparable size. The ac motor also requires less maintenance since it has, in most cases, no brushes or commutator to be maintained. Since most commercial power is generated as ac, it is only natural that this type of motor be used for doing the work that needs to be done.

An ac motor is well suited for constant-speed jobs. This is mainly because its speed is determined by the frequency of the power source. However, for some applications, the dc motor is better suited than the ac motor. The dc is more easily varied in its speed. But in some instances the speed of the ac motor can also be controlled within very narrow limits. Inverter-type drives make it easier to vary ac motor speed. It is these limits of control that we discuss here.

Types of AC Motors

In this chapter we discuss three types of ac motors: series, synchronous, and induction. Synchronous motors may be considered as polyphase motors. They have a constant speed and their rotors are energized with dc voltage. Induction motors are commonly used, single-phase or polyphase. Their rotors are energized by induction. The series ac motor is a familiar type of motor. It is very similar to a dc motor that has already been discussed.

TABLE 8–1 DC MOTOR CHARACTERISTICS AND APPLICATIONS

Speed Regulation	*Speed Control*	*Starting Torque*	*Pull-out Torque*	*Application*
Series dc motors				
Varies inversely as the load. Races on light loads and full voltage.	Zero to maximum, depending on control and load.	High. Varies as square of voltage. Limited by the commutation, heating, and line capacity.	High. Limited by commutation, heating, and line capacity.	Where high torque is required and speed can be regulated: cranes, hoists, gates, starters.
Shunt dc motors				
Drops 3 to 5% from no load to full load.	Any desired range, depending on motor design and type of system.	Good. With constant field, varies directly as voltage applied to the armature.	High. Limited by commutation, heating, and line capacity.	Where constant speed is needed and starting conditions are not severe: fans, pumps, blowers, conveyors.
Compound dc motors				
Drops 3 to 20% from no load to full load, depending on amount of compounding.	Any desired range, depending on motor design and type of control.	Higher than for shunt, depending on the amount of compounding.	High. Limited by commutation, heating, and line capacity.	Where high starting torque combined with fairly constant speed is required: Plunger pumps, punch presses, shears, geared elevators, conveyors, hoists.

Series AC Motor

The series ac motor is the same electrically as a dc series motor (Fig. 8-11). Use the left-hand rule for the polarity of coils and you can see that the instantaneous magnetic polarities of the armature and field oppose each other. This means that motor action results. By reversing the current, you reverse the polarity of the input. Note that the field magnetic polarity still opposes the armature magnetic polarity. This is because the reversal affects both the armature and the field. The ac input causes these reversals to take place continuously and the motor continues to rotate in the same direction.

FIGURE 8–11 *Series ac motor.*

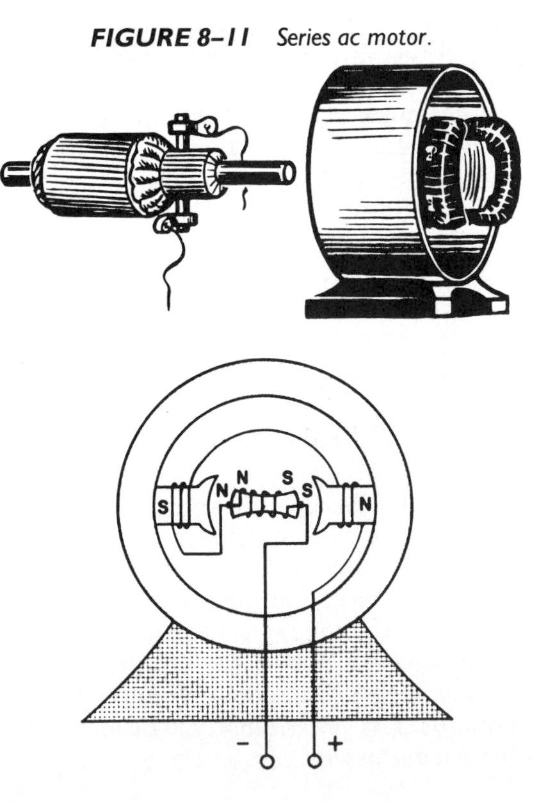

Construction of the ac series motor does vary slightly from the dc series motor. Since ac is being used it calls for special metals in the pole pieces—such as silicon steel, which easily reverses its magnetic polarity without causing residual magnetism to remain after the reversal. It also means that laminations are used in the ac motor to decrease the amount of eddy currents generated in the pole pieces. Dc can be used to power ac series motors efficiently, but putting ac to a series motor made for dc does not produce the same efficiencies.

Characteristically, the ac and dc series motors are similar. They both have a varying speed characteristic. Low speeds are possible for large loads and light loads produce high speeds. Speed varies directly with the size of the load. The larger the load, the slower the speed.

Ac–dc types of series motors, called *universal motors,* are especially designed for use on both power sources. They are usually made in small horsepower sizes, usually less than 1 hp, and are used most frequently in vacuum cleaners. Universal motors cannot be operated on polyphase ac power.

Magnetic Fields in AC Motors

Rotating magnetic fields are the key to the operation of ac motors. This is because the alternating current causes a continuously changing magnetic field as it rotates around a series of stator pole pairs.

The magnetic field in a stator can be made to rotate electrically. This means that it can move around and around. Then another magnetic field in the rotor can be made to chase it. This is done by having the rotor field attracted and repelled by the stator field. By allowing the rotor to turn freely, it is allowed to chase the rotating field in the stator. Rotating magnetic fields are set up in two-phase or three-phase machines. To establish a rotating magnetic field in a motor stator, the number of pole pairs must be the same as (or a multiple of) the number of phases in the applied voltage. The poles must be displaced from each other by an angle equal to the phase angle between the individual phase of the applied voltage.

Rotating Magnetic Field: Two-Phase. A two-phase stator shows the rotating magnetic field most easily. The stator of a two-phase induction motor is made up of two windings (or a multiple of two) placed at right angles to each other around the stator (Fig. 8–12). Note that the voltages applied to phases 1-1A and 2-2A are 90° out of phase; that is, the currents that flow in the phases are displaced from each other by 90°. The magnetic fields generated in the coils are in phase with their respective current. The magnetic fields are also 90° out of phase with each other. The coil axes of these two out-of-phase magnetic fields are at right angles to each other. They also add together at every instant during their cycle. They produce a resultant field that rotates one revolution for each cycle (hertz) of ac.

FIGURE 8–12 *Two-phase motor stator.*

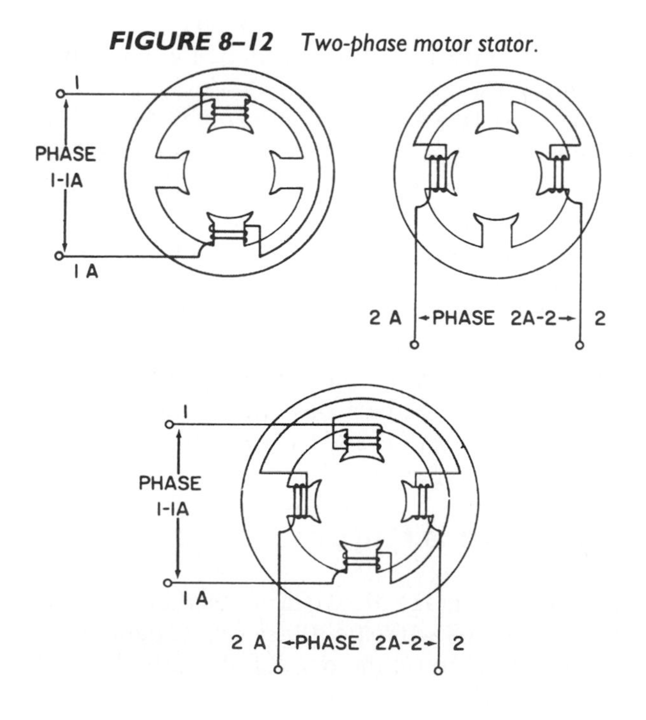

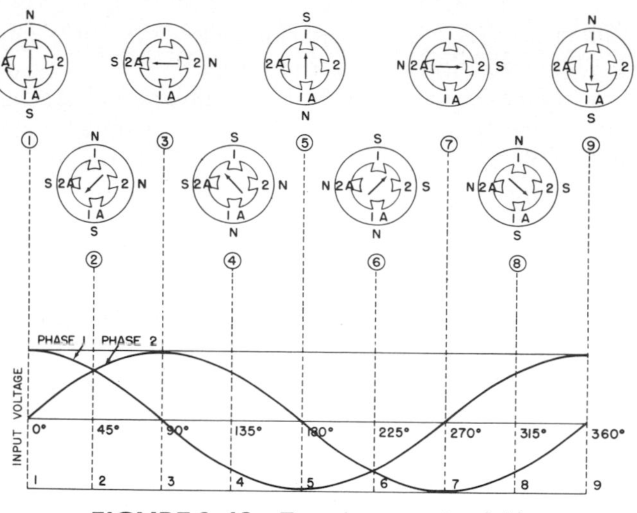

FIGURE 8–13 *Two-phase rotating field.*

Analyzing how the two-phase stator's rotating field works is one way of understanding how all ac motors work. Figure 8–13 shows the two-phase rotating field. By taking a look at the nine different locations on this figure, you will be able to see how the field moves step by step with each of the two phases.

The arrow indicates the rotor. Keeping track of the arrow shows how the rotation of the rotor is accomplished. This chart shows the voltage of each phase. The current flows in a direction that causes the magnetic polarity indicated at each pole piece. Note that from one point to the next, the polarities are rotating from one pole to the next in a clockwise manner. One complete cycle of input voltage produces a 360° rotation of the pole polarities.

Two-Phase Waveforms. The waveforms in Fig. 8–13 are applied to the windings as shown in Fig. 8–12. Note that they are displaced by 90°. That means that when one is at its maximum, the other is at its minimum. Now take a closer look at Fig. 8–13 and position 1, indicated by the circled 1. The current flow and the magnetic field in winding 1-1A are at maximum (because the phase voltage is maximum). The current flow and magnetic field in winding 2-2A are zero (because the phase voltage is zero). The resultant magnetic field is therefore in the direction of the 1-1A axis. At the 45° point (position 2), the resultant magnetic field is midway between windings and 1-1A and 2-2A. The coil currents and magnetic fields are equal in strength. At 90°, position 3, the magnetic field in winding 1-1A is zero. The magnetic field in winding 2-2A is at maximum. Now the resultant magnetic field is along the axis of the 2-2A winding. The resultant magnetic field has rotated clockwise through 90° to get from position 1 to position 3.

Once the two-phase voltages have completed one full cycle and arrived at position 9, the resultant magnetic field has rotated through 360°. This means that by placing two windings at right angles to each other and exciting them with voltages 90° out of phase, a rotating magnetic field can be produced. Note that the arrow representing the rotor is now pointing to where it started in position 1, indicating a complete revolution of the rotor.

Two-phase current is rarely used in this country. However, it is used here to show how similar operation is used to start a single-phase motor with its start winding placed in the circuit to get the rotor turning. Single-phase and three-phase motors use the same principle of rotating magnetic fields to cause their rotors to rotate.

Rotating Magnetic Field: Three-Phase. A three-phase induction motor also operates on the principle of a rotating magnetic field. Figure 8–14 shows how a three-phase operation is connected. This one is Y-connected. The three-phase windings can be connected to a three-phase ac input and have a resultant magnetic field that rotates (Fig. 8–15).

Figure 8–14 shows how three-phase stators are connected in a wye configuration. The dot shows where connections are made to ensure a wye configuration. The pole pieces are placed 120° apart. Note that 3 times 120 equals 360. Now take a closer look at Fig. 8–15. It shows how instantaneous polarities are generated. Current flows toward the terminal number

FIGURE 8–14 *Three-phase, wye connected stator.*

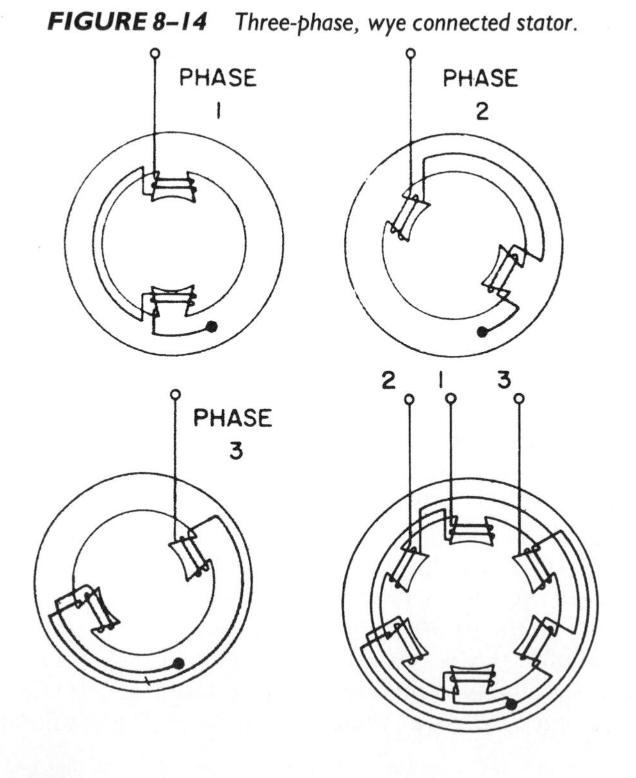

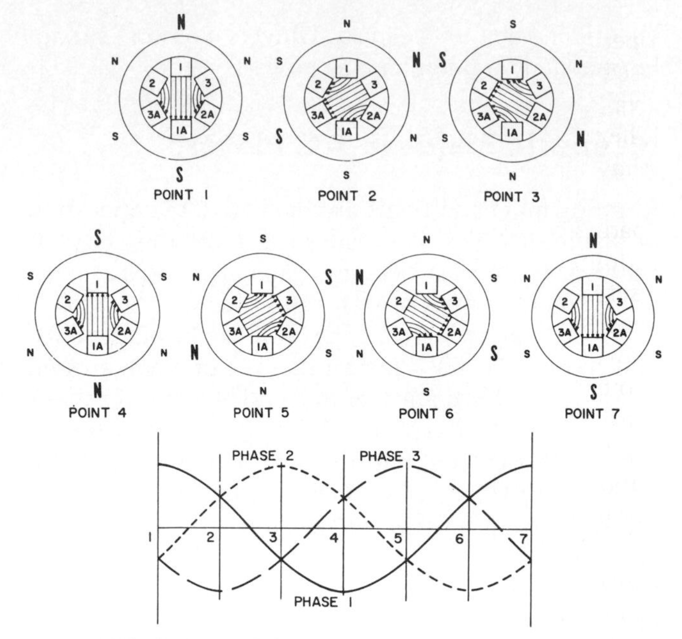

FIGURE 8–15 *Three-phase rotating field polarities and input voltages.*

in Fig. 8–14 for positive voltages and away from the terminal number for negative voltages.

Now refer to Fig. 8–15. Note that at position 1 the magnetic field in coils 1-1A is maximum. Check the polarities. At the same time, negative voltages are present in the 2-2A and 3-3A windings. These negative voltages create weaker magnetic fields that tend to aid the 1-1A field. At position 2, maximum negative voltage is present in the 3-3A windings. This creates a strong magnetic field that is aided by the weaker fields in 1-1A and 2-2A. Now move along each point on the voltage graph. Notice that the resultant magnetic field is rotating in a clockwise direction. This means that when the three-phase voltage completes one full cycle and reaches position 7, the magnetic field has rotated through 360°.

If you place a permanent bar magnet in this rotating magnetic field with a shaft through it and allow the magnet to rotate freely in step with the rotating magnetic field, you will be able to see how the shaft rotates at the same rate as the moving magnetic field. Keep in mind that this simplified explanation of rotating fields is given here to show you how a rotating magnetic field can be utilized to cause a shaft to turn and produce usable mechanical energy from electrical energy. Motors have been designed to use a number of principles and can be utilized to do many jobs efficiently.

Synchronous Motors

The main advantage of synchronous motors is a constant-speed characteristic. They are capable of correcting the low power factor of an inductive load when operated under certain conditions. They are often used to drive dc generators. Synchronous motors are available in sizes up to thousands of horsepower. They may be designed as either single-phase or three-phase machines (Fig. 8–16).

Synchronous motors are often used without a load. They can be used for power factor correction. Adding the synchronous motor to the circuit can be useful in power factor correction. This means that the machine will do the same amount of work as before power correction; however, it will draw less current from the power lines. This is accomplished by the amount of excitation applied to the wound rotor in the form of low-voltage dc. Fixed condensers (capacitors) are often used in place of synchronous motors for power factor correction.

This type of motor is used whenever exact speed must be maintained or for power factor correction. Synchronous motors are more expensive than other types at the lower horsepower ratings, but may possibly be more economical for 100 hp and higher ratings (Fig. 8–16).

If three-phase ac is applied to a synchronous motor, a rotating magnetic field is set up around the rotor. The rotor is then energized with dc. That is, it acts as a bar magnet since the dc produces a fixed north–south polarity for the rotor. The strong rotating magnetic field attracts the strong rotor field activated by the dc. This results in a strong turning force on the rotor shaft. The rotor is therefore able to turn a load as it rotates in step with the rotating magnetic field.

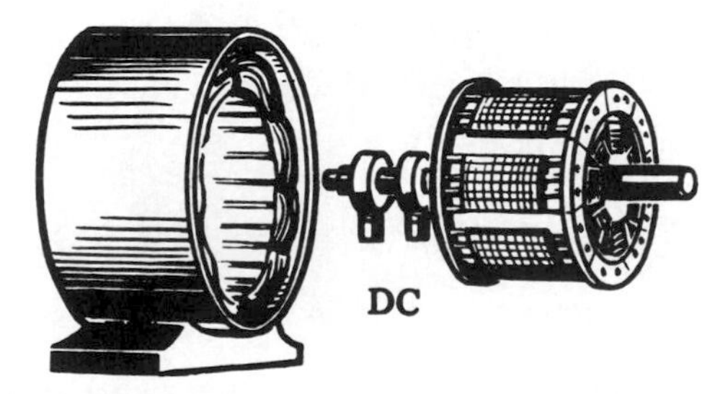

FIGURE 8–16 *Revolving field synchronous motor.*

FIGURE 8–17 *Self-starting synchronous ac motor.*

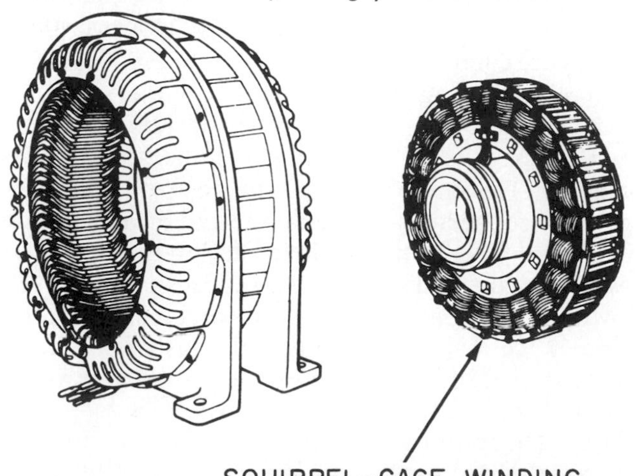

Getting the motor started is accomplished by adding a squirrel-cage winding to the rotor. It cannot start from standstill without the aid of a squirrel-cage rotor. That is because when ac is applied to the stator, a high-speed rotating magnetic effect appears immediately. This rotating field rushes past the rotor poles so quickly that the rotor does not have a chance to get started. In effect, the rotor is repelled first in one direction and then the other. In its purest form, a synchronous motor has no starting torque. It has torque only when running at synchronous speed. A synchronous motor used to correct the power factor has to operate at less than its nominal mechanical load.

Squirrel-Cage Motor. The three-phase current with which the motor is supplied establishes a rotating magnetic field in the stator. This rotating magnetic field cuts the conductors in the rotor, inducing voltages and causing currents to flow. These currents set up an opposite-polarity field in the rotor. The attraction between these opposite stator and rotor fields produces the torque that causes the rotor to rotate. This is, in essence, how a squirrel-cage motor works.

A squirrel-cage winding can be added to the rotor of a synchronous motor to cause it to start. The squirrel cage is shown as part of the rotor in Fig. 8–17. The name comes from the shape—it looks something like a turnable squirrel cage. The windings are heavy copper bars shorted together by copper rings. A low voltage is induced in these shorted windings by the three-phase stator field. Because of the short circuit, a relatively large current flows in the squirrel cage. This causes a magnetic field that interacts with the rotating field of the stator. Because of the interaction the rotor begins to turn, following the stator field and the motor starts. The squirrel cage is also used in single-phase and other three-phase motors.

Starting the Synchronous Motor. To start a practical synchronous motor, the stator is energized. However, the dc supply to the rotor field is not energized. The squirrel-cage windings are allowed to bring the rotor close to synchronous speed. At that point the dc field is energized. This locks the rotor in step with the rotating stator field.

Full torque is developed in the synchronous motor when it comes up to speed. Once it comes up to synchronous speed the load can be applied. A switch operated by centrifugal force is used to apply dc to the rotor as it reaches synchronous speed. The need for a dc power source for the rotor makes it an expensive type of machine to operate and maintain. The dc

TABLE 8–2 SYNCHRONOUS MOTOR CHARACTERISTICS AND APPLICATIONS

Speed regulation:	Constant.
Speed control:	None, except special motors designed for two fixed speeds.
Starting torque:	40% for slow-speed to 160% for medium-speed 80%-power factor designs. Special designs develop higher torques.
Pull-out torque:	Unity of motors, 170%; 80%-power factor motors, 225%. Special designs up to 300%.
Applications:	For constant-speed service, direct connection to slow-speed machines and where power factor correction is required.

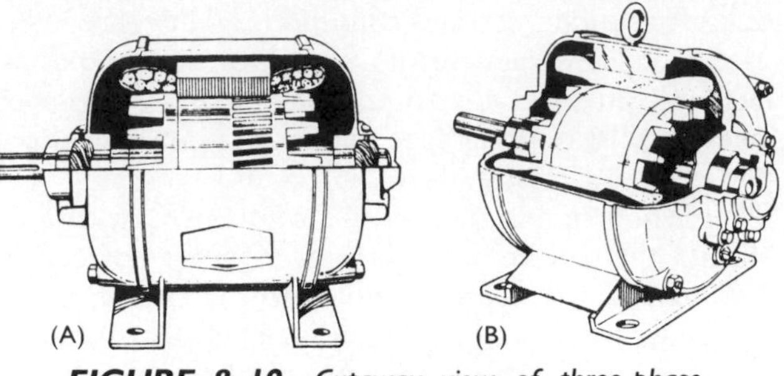

FIGURE 8–19 *Cutaway view of three-phase motor: (A) with a half-etched squirrel-cage rotor; (B) with a cast rotor.*

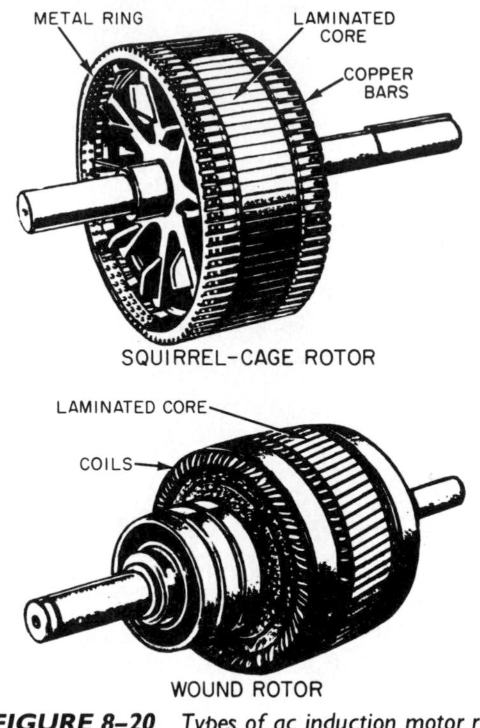

FIGURE 8–20 *Types of ac induction motor rotors.*

source may be part of the motor or it may come from an external generator.

As you can see from this discussion, synchronous motors have advantages and disadvantages, as do other types of motors. If expense is not a factor, the selection of one over another depends on the job and the power available. Table 8–2 summarizes the synchronous motor characteristics and applications.

Induction Motors

The induction motor is the most commonly used ac motor (Fig. 8–18). It is simple in design and rugged in construction. It costs relatively little to manufacture. The induction motor rotor is not connected to an external source of voltage. The induction motor derives its name from the fact that ac voltages are induced in the rotor circuit by the rotating magnetic field in the stator. Induction in this motor is similar to the induction between the primary and secondary of a transformer. The rotor can be thought of as a short-circuited secondary of a transformer that is mounted on a shaft and supported by bearings that allow it to rotate freely as the rotating field in the stator moves from stator pole to stator pole (Fig. 8–19).

Induction motors are often large and permanently mounted. They drive loads at fairly constant speed. They are used in washing machines, refrigerator compressors, bench grinders, and table saws.

FIGURE 8–18 *Induction motor.*

Stator Construction. The stator construction of the three-phase induction motor and the three-phase synchronous motor are almost identical. However, their rotors are completely different (Fig. 8–20). An induction motor is made of a laminated cylinder with slots in its surface. The windings in these slots are one of two types. The most common is the squirrel-cage winding. This entire winding is made up of heavy copper bars connected together at the end by a metal ring made of copper or brass. No insulation is required between the core and the bars. This is because of the very low voltages generated in the rotor bars. The other type of windings contains actual coils placed in the rotor slots. This type of rotor is then called a wound rotor (Table 8–3).

Slip. The rotor of an induction motor cannot turn at the same speed as the rotating magnetic field. If the speeds were the same, there would be no relative mo-

TABLE 8–3 MOTOR CHARACTERISTICS AND APPLICATIONS (Two- and Three-Phase)

	General-Purpose Squirrel-Cage (Class B)
Speed regulation:	Drops about 3% for large to 5% for small sizes.
Speed control:	None, except multispeed types, designed for two to four fixed speeds.
Starting torque:	200% of full load for two-pole to 105% for 16-pole designs.
Pull-out torque:	200% of full load.
Applications:	Constant-speed service where starting torque is not excessive: fans, blowers, rotary compressors, centrifugal pumps.
	High-Torque Squirrel Cage (Class C)
Speed regulation:	Drops about 3% for large to 6% for small sizes.
Speed control:	None, except multispeed types, designed for two to four fixed speeds.
Starting torque:	250% of full load for high-speed designs to 200% for low-speed designs.
Pull out torque:	200% of full load.
Applications:	Constant-speed service where fairly high starting torque is required at infrequent intervals with starting current of about 400% of full load: reciprocating pumps, compressors, crushers.
	High-Slip Squirrel-Cage (Class D)
Speed regulation:	Drops about 10 to 15% from no load to full load.
Speed control:	None, except multispeed types, designed for two to four fixed speeds.
Starting torque:	225 to 300% of full load, depending on speed with rotor resistance.
Pull-out torque:	200%. Will usually not stall until loaded to maximum torque, which occurs at standstill.
Applications:	Constant-speed service and high starting torque, if starting is not too frequent, and for taking high-peak loads with or without flywheels: punch presses, shears, elevators.
	Low-Torque Squirrel Cage (Class F)
Speed regulation:	Drops about 3% for large to 5% for small sizes.
Speed control:	None, except multispeed types, designed for two to four fixed speeds.
Starting torque:	50% of full load for high-speed designs to 90% for low-speed designs.
Pull-out torque:	150 to 170% of full load.
Applications:	Constant-speed service where starting duty is light: fans, blowers, centrifugal pumps, similar loads.
	Wound Rotor
Speed regulation:	With rotor rings short-circuited, drops about 3% for large to 5% for small sizes.
Speed control:	Speed can be reduced to 50% by rotor resistance to obtain stable operation.
Starting torque:	Up to 300%, depending on external resistance to obtain stable operation. Speed varies inversely as load.
Pull-out torque:	200% when rotor slip rings are short circuited.
Applications:	Where high starting torque with low starting current or where limited speed control is required: fans, centrifugal and plunger pumps, compressors, conveyors, hoists, cranes.

tion between the stator and rotor fields. Without relative motion there would be no induced voltage in the rotor. For relative motion to exist between the two, the rotor must rotate at a slower speed than that of the rotating magnetic field. The difference between the speed of the rotating stator field and the rotor speed is called *slip*. The smaller the slip, the closer the rotor speed approaches the stator field speed.

Rotor speed depends on the torque requirements of the load: The greater the load, the stronger the turning force needed to rotate the rotor. Turning force increases only if the rotor-induced EMF increases. This EMF can increase only if the magnetic field cuts through the rotor at a faster rate. To increase the relative speed between the field and rotor, the rotor must slow down. This means that for heavier loads the induction motor turns more slowly than for lighter loads.

Slip is directly proportional to the load on the motor. A slight change in speed is necessary to produce the usual current changes required for normal changes in load. This is because the rotor windings have such a low resistance. Induction motors are referred to as constant-speed motors.

Single-Phase Induction Motors. The single-phase induction motor is the most commonly used of all types of electric motors. It has the lowest initial cost and needs little maintenance. There are a number of descriptive titles for each type of single-phase motor: such names as split-capacitor, capacitor-start, split-phase, shaded-pole, capacitor start–capacitor run, and permanent-split capacitor. Of course, other groupings include the capacitor start and resistance start as part of the overall title "split-phase."

The stator field in the single-phase motor does not rotate. It simply alternates polarity between poles as the ac voltage changes polarity (Fig. 8–21). Magnetic induction causes a voltage to be induced in the rotor, thereby producing a magnetic field in the rotor. This field is in opposition to the stator field because, as Lenz's law states: The induced voltage is in the opposite direction from that which produced it. Thus the

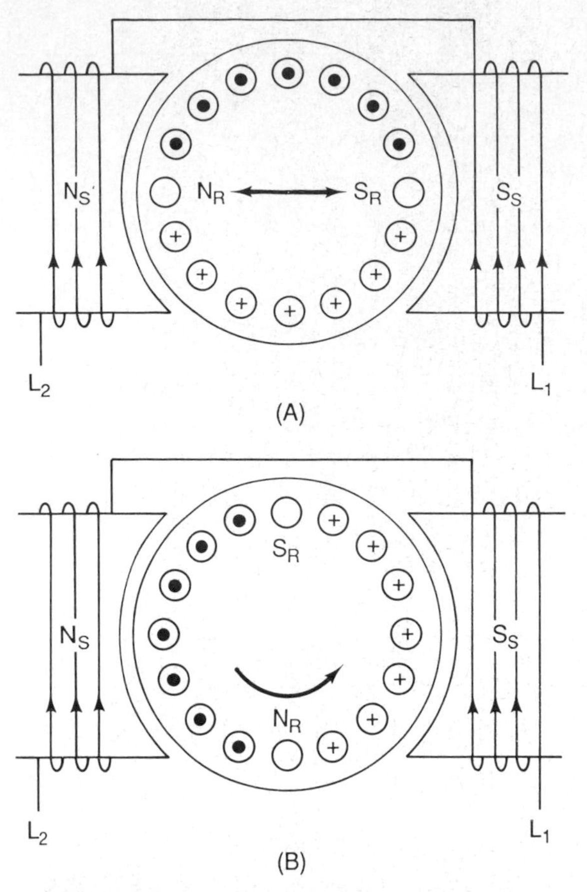

FIGURE 8–21 *Rotor currents in a single-phase ac induction motor: (A) stationary; (B) rotating.* N_{R7} S_{R7} *rotor field;* N_S, R_S, *stator field.*

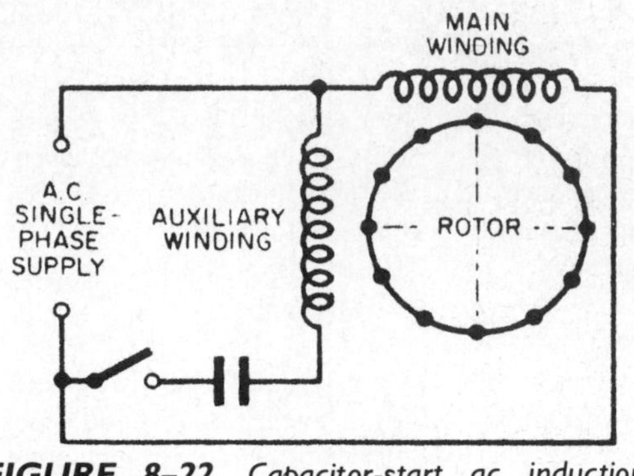

FIGURE 8–22 *Capacitor-start ac induction motor.*

interaction between rotor and stator fields will not produce rotation. Instead of rotation it produces a north pole directly opposite a south pole. Therefore, the rotor stands still until pushed or nudged into motion. This is shown in Fig. 8–21A.

Once the movement is started (Fig. 8–21B) the south pole on the rotor is attracted by the left-hand pole. That is, the north rotor pole is attracted to the right-hand pole. This is a result of the rotor being rotated 90° by an outside force. Once the pull exists between the two fields, it becomes a rotary force, turning the rotor toward the magnetic field of the stator. The two fields alternate continuously. They never actually line up, so the rotor will continue to rotate once started. The next problem is to design a starting method to get this single-phase motor to start without an outside force turning it each time it is started.

The basic operation of the single-phase motor is the same for all types, but the starting methods utilized create a specialized label for each. To become familiar with two of the slightly different methods used to get the single-phase motor started, we will look at the capacitor-start and the resistance-start in addition to the shaded-pole types.

Split-Phase Induction Types. Split-phase motors are designed to use inductance, capacitance, or resistance to develop a starting torque. The capacitor-start type uses, of course, capacitance. Figure 8–22 shows how an auxiliary winding is added with a capacitor in series with it. This winding is placed in parallel with the main (run) winding and is located at right angles (90°) to it. This produces a phase difference of 90 electrical degrees between the two windings. The start or auxiliary winding is connected with a switch that is operated centrifugally. When the motor comes up to about 75% of its rated speed, opening the switch disconnects the start winding.

When the power is first applied, the starting switch is closed. This places the capacitor in series with the auxiliary winding. The capacitor is of such value that the auxiliary circuit is effectively a resistive–capacitive circuit. In this circuit the current leads the line voltage by 45°. That is because X_c equals the resistance in the circuit. The main (run) winding has enough resistance and inductance to cause the current to lag the line voltage by about 45° because X_L about equals the resistance. The currents in each winding are therefore 90° out of phase. The magnetic fields are thus displaced by the same amount. The effect is that the two windings act as a two-phase stator and produce the rotating field required to start the motor.

Once the motor is started, the centrifugal switch opens and takes the start winding out of the circuit; the rotor continues to rotate until power is removed from the main winding. Split-phase motors are available in small sizes, usually less than 1 hp, because they do not have sufficient starting torque to handle large loads.

Another type of split-phase induction motor is the resistance-start (Fig. 8–23). Note that there is a resistor in series with the start winding. The auxiliary circuit, consisting of the winding and the resistor, is switched in the circuit and cut out by a centrifugal switch. The electrical phase shift between the currents in the auxiliary and main windings is obtained by making the impedance of the windings unequal. Note

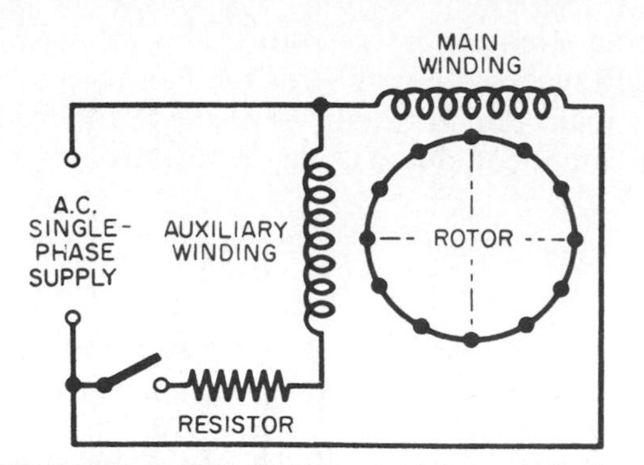

FIGURE 8–23 Resistance-start ac induction motor.

how the main winding has a high inductance (little resistance due to windings) and the auxiliary winding has a low inductance plus a high-value series resistance. The start winding then has a lagging phase angle due to the inductance.

The start winding has a smaller phase angle, so that the current does not lag as much as in the main winding. There is about a 30° difference between the two windings in their phase angles. This is enough of a difference to develop enough torque to get the rotor moving if there is little or no load on the motor. The centrifugal switch opens the circuit once the rotor is up to about 75% of its normal design speed.

Shaded-Pole Induction Motors. Shaded-pole motors are also of the induction type. They use a different type of starting arrangement. The effect of a moving magnetic field is produced by constructing the stator in a special way. These motors have projecting pole pieces similar to some dc motors. Portions of the pole-piece surfaces are surrounded by a copper strap called a *shading coil.* Figure 8–24 shows a pole piece with the strap in place. The strap causes the field to move back and forth across the face of the pole piece. Note the numbered sequence and points on the magnetization curve in the figure. As the alternating stator field starts increasing from zero step 1, the lines of force expand across the face of the pole piece and cut through the strap. A voltage is induced in the strap. The current that results generates a field that opposes the cutting action (and decreases the strength) of the main field. This produces the following:

FIGURE 8–24 Starting a shaded-pole induction motor.

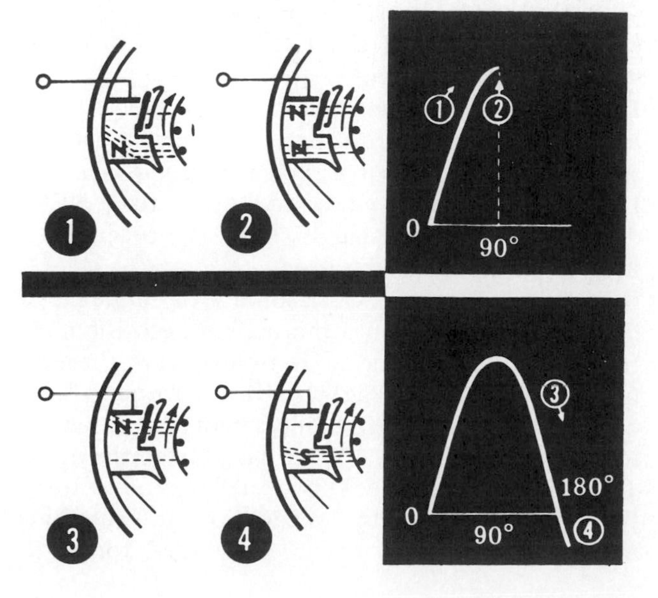

A. As the field increases from zero to a maximum at 90°, the field reaches its maximum value.

B. A large portion of the magnetic lines of force are concentrated in the unshaded portion of the pole ①.

C. At 90° the field reaches its maximum value and stops expanding. No EMF is induced in the strap. No opposition magnetic field is generated, so the main field is uniformly distributed across the poles as shown in ②.

D. From 90 to 180°, the main field starts decreasing or collapsing inward. The field generated in the strap opposes the collapsing field. The effect is to concentrate the lines of force in the shaded portion of the pole face as shown in ③.

E. From 180 to 360° the main field goes through the same change as it did from 0° (180°). However, note that it is in the opposite direction as shown in ④.

The direction of the field does not affect the way the shaded pole works. The motion of the field is the same during the second half-cycle as it was during the first half of the cycle.

The motion of the field back and forth between shaded and unshaded portions produces a weak torque to start the motor. Because of the weak starting torque, shaded-pole motors are built in very small sizes. They drive such things as fans, clocks, blowers, and electric razors. Timer motors on many large pieces of equipment use the shaded-pole motor for the clock mechanism.

Speed and Slip in Squirrel-Cage Motors

The speed of a squirrel-cage motor depends on the frequency of the power source and the number of poles the motor has. The higher the frequency, the faster the motor; the more poles the motor has, the slower it runs. The smallest number of poles ever used in a squirrel-cage motor is two. A two-pole 60-Hz motor

TABLE 8–4 MOTOR SPEED VERSUS NUMBER OF POLES

Number of Poles	*60-Hz Speed*	*50-Hz Speed*
2	3600	3000
4	1800	1500
6	1200	1000
8	900	750
10	720	600
12	600	500

runs at approximately 3600 rpm (Table 8–4). Formula:

$$\text{synchronous speed} = \frac{60 \times 2f}{p}$$

where f is the frequency of power supply and p is the number of poles in the motor. Most standard commercial motors (143T through 445T frame sizes) are wound with a maximum of eight poles.

The actual speed of the motor is somewhat less than its synchronous speed. This difference between the synchronous and actual speeds is defined as *slip.* If the squirrel-cage rotor rotated as fast as the stator field, the rotor conductor bars would be standing still with respect to the rotating field. See Fig. 8–25 for a look at the rotor construction. This means that no voltage would be induced into the bars and no current would be set up to produce torque. Since no torque is produced, the rotor will slow down until sufficient current is induced to develop enough torque to keep the rotor at a constant speed. Therefore, the rotor rotates slower than the rotating magnetic field of the stator. For example, a 3600-rpm motor usually rotates at 3450 rpm and a 1800-rpm turns at 1725 rpm at rated load.

An increased load on the motor causes the rotor to slow down; that is, the rotating field cuts the rotor bars at a higher rate than before. This has the effect of increasing the current in the bars and hence increasing

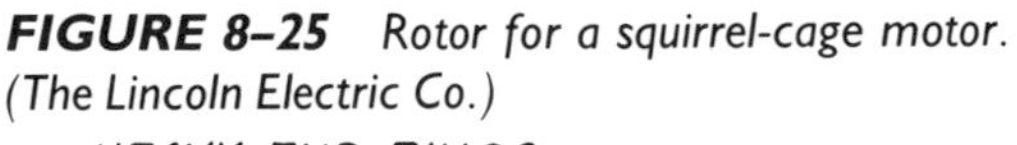

FIGURE 8–25 *Rotor for a squirrel-cage motor. (The Lincoln Electric Co.)*

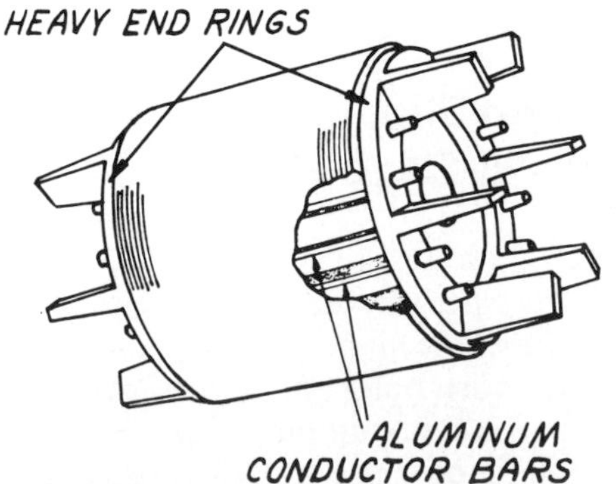

the pole strength of the rotor. This increased pole strength makes it possible for the motor to carry the larger load. Slip is usually expressed in percent and easily can be computed using the formula

$$\text{percent slip} = \frac{\text{synchronous speed} - \text{actual speed}}{\text{synchronous speed}} \times 100$$

Squirrel-cage motors are made with the slip ranging from less than 5% to around 20%. Motors with a slip of 5% or higher are used for hard-to-start applications. A motor with a slip of 5% or less is called a *normal slip* motor. A normal slip motor is often referred to as a constant-speed motor because the speed changes very little with load variations.

Manufacturers usually specify the motor speed on the nameplate as that which is normal at its rated load. Actual speed is, of course, lower than the synchronous speed.

Rotation

The direction of rotation of a polyphase squirrel-cage motor depends on the motor connection to the power lines. Rotation can readily be reversed by interchanging any two input leads.

Torque and Horsepower

Torque and horsepower are two very important motor characteristics that determine the size of the motor for a particular job. *Torque* is the turn effort produced by the motor. Torque is measured in pound-feet (lb-ft). *Horsepower* is a rate of doing work. One horsepower equals 33,000 pounds being lifted for a distance of 1 foot in 1 minute:

$$\text{hp} = \frac{\text{speed in (rpm)} \times 2\pi \times \text{torque}}{33{,}000}$$

It takes 746 watts of electrical power to produce 1 horsepower.

Locked-Rotor Torque. An induction motor is built to supply the extra torque needed to start a load. The speed–torque curve for a typical motor is shown in Fig. 8-26. This curve shows the torque for a NEMA design B motor. The curve shows the torque to be 210% of the rated load torque.

Breakdown Torque. Occasionally, a sudden overload will be placed on a motor. To keep the motor from stalling every time an overload occurs, these motors have what is called *breakdown torque.* Breakdown torque is much higher than the rated-load torque. It takes quite an overload to stall the motor. The speed–torque curve shows the breakdown torque for a typical motor to be 270% of the rated-load torque. Operating a motor overloaded for an ex-

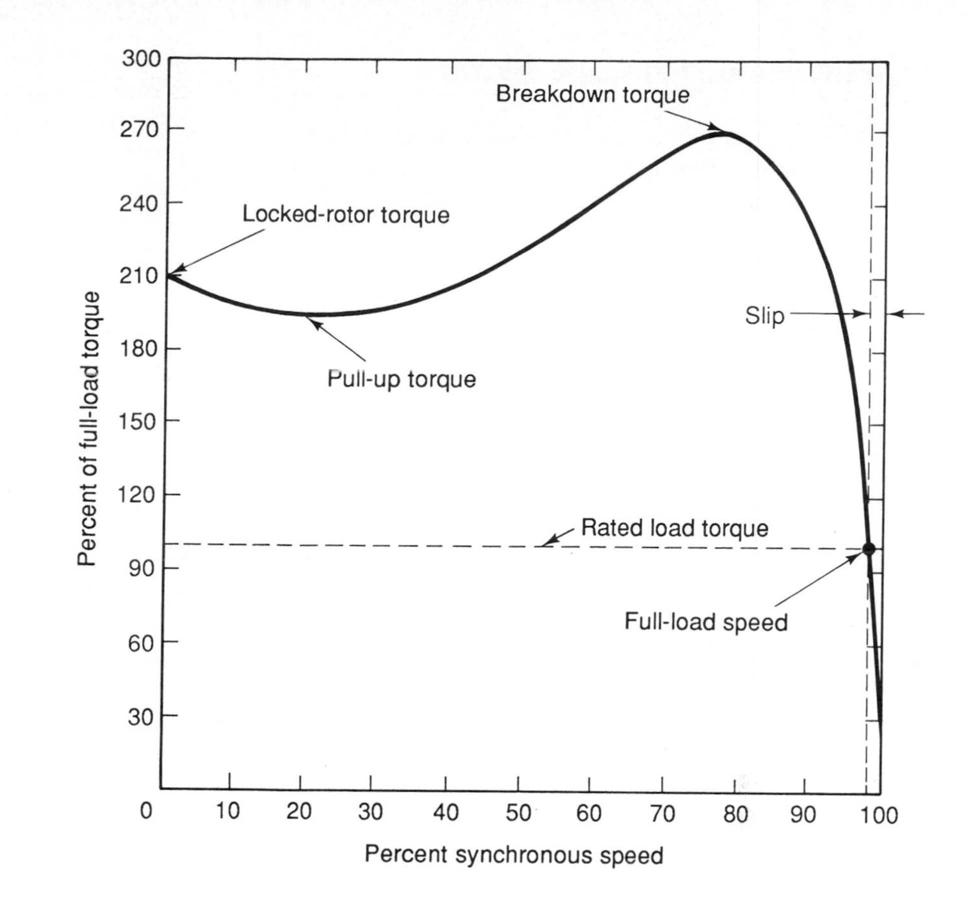

FIGURE 8–26 *Representative speed–torque curve for NEMA design B motors. (The Lincoln Electric Co.)*

tended period of time causes excessive heat buildup in the motor and may eventually burn up the motor windings.

Because of the variety of torque requirements, NEMA has established different *designs* to cover almost every application (Table 8–5). These designs take into consideration starting current and slip as well as torque. These designs should not be confused with various *classes* of insulation, which also are designated by letter (Table 8–6).

Locked-Rotor kVA/hp

Another rating specified on motor nameplates and determined by the motor design is locked-rotor kVA per horsepower. A letter appears on the nameplate corresponding to various kVA/hp ratings (Table 8–5).

These nameplate code ratings give a good indication of the starting current the motor will draw. A code letter at the beginning of the alphabet indicates a low starting current, and a letter at the end of the alphabet indicates a high starting current for the particular horsepower rating of the motor. Starting current can be figured by using the formula

$$\text{locked-rotor amperes} = \frac{1000 \times \text{hp} \times \text{kVA/hp}}{1.73 \times \text{volts}}$$

The starting current is important to the buyer of a motor since it indicates the amount of protection needed in the way of overcurrent devices. The buyer must install power lines big enough to carry the required currents and put in fuses of the proper size.

TABLE 8–5 LOCKED-ROTOR kVA/hp

Code Letter	*kVA/hp*[a]
A	0–3.15
B	3.15–3.55
C	3.55–4.00
D	4.00–4.50
E	4.50–5.00
F	5.00–5.60
G	5.60–6.30
H	6.30–7.10
J	7.10–8.00
K	8.00–9.00
L	9.00–10.0
M	10.0–11.2
N	11.2–12.5
P	12.5–14.0
R	14.0–16.0
S	16.0–18.0
T	18.0–20.0
U	20.0–22.4
V	22.4 and up

[a]The locked-rotor kVA/hp range includes the lower figure up to, but not including, the higher figure.

Starting Squirrel-Cage Motors

Squirrel-cage motors are usually designed for across-the-line starting. This means connecting it directly to the power source by means of a suitable contactor.

In large squirrel-cage motors and in some other types of motors the starting currents are very high. Usually, the motor is built to stand these high currents. However, since these currents are almost six times rated load current, there may be a large voltage drop in the power system. Some method of reducing the starting current has to be used to limit the voltage drop to a tolerable level. Reduced-voltage starting is described in Chapter 12. A summary of the mechanical and electrical formulas used with electric motors is given in Table 8–7.

TABLE 8–6 CLASSIFICATION OF INSULATION SYSTEMS

An insulation system is an assembly of insulating materials in association with the conductors and the supporting structural parts of a motor or generator. Insulation systems are divided into classes according to the thermal endurance of the system for temperature rating purposes. Four classes of insulation systems are used in motors and generators: classes A, B, F, and H. These classes have been established in accordance with IEEE Std 1, *General Principles for Temperature Limits in the Rating of Electric Equipment.*

Insulation systems are classified as follows:

Class A—A class A insulation system is one which by experience or accepted test can be shown to have suitable thermal endurance when operating at the limiting class A temperature specified in the temperature rise standard for the machine under consideration.

Class B—A class B insulation system is one which by experience or accepted test can be shown to have suitable thermal endurance when operating at the limiting class B temperature specified in the temperature rise standard for the machine under consideration.

Class F—A class F insulation system is one which by experience or accepted test can be shown to have suitable thermal endurance when operating at the limiting class F temperature specified in the temperature rise standard for the machine under consideration.

Class H—A class H insulation system is one which by experience or accepted test can be shown to have suitable thermal endurance when operating at the limiting class H temperature specified in the temperature rise standard for the machine under consideration.

TABLE 8–7 SUMMARY OF FORMULAS

Mechanical Formulas

$$\text{Torque lb-ft} = \frac{\text{hp} \times 5252}{\text{RPM}} \qquad \text{hp} = \frac{\text{torque} \times \text{rpm}}{5252}$$

$$\text{Sync. rpm} = \frac{120 \times \text{frequency}}{\text{no. poles}}$$

Rules of Thumb (Approximation)*

At 1800 rpm, a motor develops 3 lb-ft/hp
At 1200 rpm, a motor develops 4.5 lb-ft./ hp
At 575 V, a three-phase motor draws 1 A/hp
At 460 V, a three-phase motor draws 1.25 A/hp
At 230 V, a three-phase motor draws 2.5 A/hp
*Departs on lower hp and rpm motors.

Temperature Conversion

$$°C = (°F - 32) \times \frac{5}{9}$$

$$°F = (°C \times \frac{9}{5}) + 32$$

Electrical Formulas

To Find	Alternating Current: Three-Phase
Amperes when horsepower is known	$\frac{hp \times 746}{1.73 \times E \times Eff \times PF}$
Amperes when kilowatts are known	$\frac{KW \times 1000}{1.73 \times E \times PF}$
Amperes when kVA are known	$\frac{kVA \times 1000}{1.73 \times E}$
Kilowatts	$\frac{1.73 \times I \times E \times PF}{1000}$
Horsepower = (output)	$\frac{1.73 \times I \times E \times EFF \times PF}{746}$

I = current
E = voltage
Eff = efficiency
PF = power factor
kVA = kilovolt-amperes
kW = kilowatts

Source: The Lincoln Electric Co.

QUESTIONS

1. What is the right-hand rule for motors?
2. Describe the armature of a dc motor.
3. What is counter EMF?
4. What does the load on a motor affect in the circuit?
5. What are the three types of dc motors?
6. List the advantages of the series motor.
7. List the advantages of the shunt motor.
8. What is the difference between short-shunt and long-shunt compound motors?
9. What controls the speed of a dc motor? How is speed control accomplished?
10. Describe armature reaction.
11. Describe generator effect in a motor.
12. What are the three types of ac motors?
13. What type of motor can use both ac and dc?
14. What is the main advantage of a synchronous motor?
15. How did the squirrel-cage motor get its name?
16. How is a synchronous motor started?
17. What is the most commonly used ac motor? Why?
18. What is slip?
19. Why does the induction motor turn slower for heavier loads than for light loads?
20. List four types of single-phase induction motors.
21. How does the shaded-pole motor get started?
22. What is the term used to describe the difference between actual speed and synchronous speed?
23. What does the direction of rotation of a polyphase squirrel-cage motor depend on?
24. How many watts of electrical power does it take to equal 1 horsepower?
25. How are motor insulation classes designated on the nameplate?

REVIEW PROBLEMS

Table 8–4 shows some rule-of-thumb math methods utilized to obtain various motor torques and current drain.

1. Using the rule-of-thumb approximations, what is the torque developed by a three-phase motor at 1800 rpm?
2. Using the rule-of-thumb approximations, what is the torque developed by a three-phase motor at 1200 rpm?
3. How much current does a 575-V three-phase motor draw when producing 10 hp?
4. How much current does a 460-V three-phase motor draw when producing 10 hp?
5. How much current does a 230-V three-phase motor draw when producing 10 hp?
6. How many amperes does a 10-hp three-phase, 230-V motor draw when it has an efficiency of 80% and a power factor of 0.8?
7. How many amperes does a 100-kW motor draw if it is connected to a three-phase, 240-V line and has a power factor of 0.85?
8. If a three-phase motor needs 10 kVA on 240 V, what is the current it draws?
9. What is the horsepower of a three-phase motor that draws 10 A on a 240-V line and has an efficiency of 80% and a power factor of 0.75?
10. A six-pole synchronous motor is connected to a 240-V 60-Hz line. How fast does it rotate?

Timers and Sensors

Objectives

After studying this chapter, you will be able to:

1. Describe the difference between synchronous clock timers and solid-state timers.
2. Draw a ladder diagram of a control circuit with an on-delay relay.
3. List the uses of the general-purpose relay.
4. Explain how programmable timers operate.
5. Understand how thumbwheel switches are used to set time counts on electronic counters.
6. Describe a DIP switch.
7. Explain the advantages of a pneumatic timing relay.
8. Explain sequence control.
9. List the uses for sensors in industry.
10. Explain how solid-state level controls work.
11. Determine dielectric constants of various materials.
12. Describe the difference between a thermistor and a thermocouple.
13. Describe strain gage operation.

Time is important in any industrial or commercial operation. Proper timing makes it possible to meet production schedules and demands. Timing the operation of a machine is an important function and needs the proper equipment to do the job. Many types of industrial timers are available to control operations of machines. In most instances it is necessary to control the motor that drives the machine. A close examination of timers shows that there are general-purpose, pneumatic, programmable, and solid-state types. Of course, there are such individual-purpose timers as those required for welders and similar pieces of equipment. Complex operations and production schedules are timed by equipment that has been designed for doing the job and making sure it is done properly and accurately. Timers are called upon to time a fraction of a second up to hours and in some cases a year. Timers can be broken down into three categories: dashpot, synchronous clock, and electronic timers. Electronic timers are gaining more favor as they become less expensive and more reliable than the older mechanical types.

DASHPOT OR PNEUMATIC TIME-DELAY TIMERS

The dashpot timer (Fig. 9-1) operates using a pneumatic chamber with a variable orifice. When a voltage is applied to the relay coil, the armature pushes against a diaphragm and air is forced out of the chamber. The speed at which the air exits the chamber controls the length of time delay.

FIGURE 9–1 *Pneumatic time-delay relay. (Square D)*

SYNCHRONOUS CLOCK TIMERS

A synchronous motor is used to turn a clock mechanism (Fig. 9–2). The synchronous motor is very accurate since it operates on the frequency of the line voltage and is not affected by fluctuations in line voltage. This type of timer has one or more contacts that open or close depending on the position of the clock hands.

FIGURE 9–2 *Synchronous motors used to power motor-driven timers.*

The hands can be set to where they are to make contact. This means that it is possible to obtain from 1 minute up to 12 hours of time delay on the clock. Some can be made to time in seconds. A good example of a clock timer is the one used in darkrooms to time the development of film.

SOLID-STATE TIMERS

The solid-state timer is more accurate and versatile and is becoming the standard in industry. Although the price has a tendency to continue to decrease, it has become the most reliable type of timer. In most instances there are no moving parts to cause trouble or need servicing. The time delay is set by a resistance-capacitance (*RC*) network and the switching is usually done by an SCR. Most units are encapsulated in plastic of some type, usually an epoxy.

TIME-DELAY RELAYS

The pneumatic-type relay has a synthetic rubber bellows for controlling the tripping time (Fig. 9–3). The timer is particularly useful for applications requiring a timer with greater accuracy and a longer timing range than is afforded by the fluid dashpot timer. These pneumatic timing relays have a range of $\frac{1}{20}$ to 180 seconds with a repetitive accuracy of $\pm$ 10%. A minimum reset time of 75 milliseconds must be provided to ensure repetitive accuracy. The times are adjustable for a period longer than 180 seconds (3 minutes) provided that 100% accuracy is not needed.

Relays can be timed so that the delay takes place with the contacts closed or open. When power is applied to the relay coil of an *on-delay* relay, a period of time passes before the relay contacts change state. Figure 9–3 shows the schematic symbols for time-delay relays.

FIGURE 9–3 *(A) On-delay timer. (Allen-Bradley) (B) Schematic symbols for an on-delay and its NO and NC contacts.*

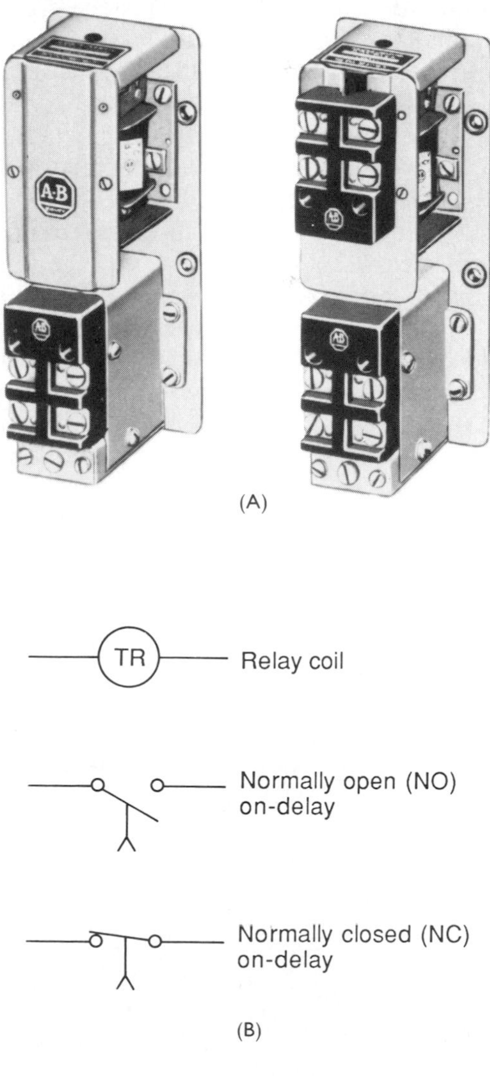

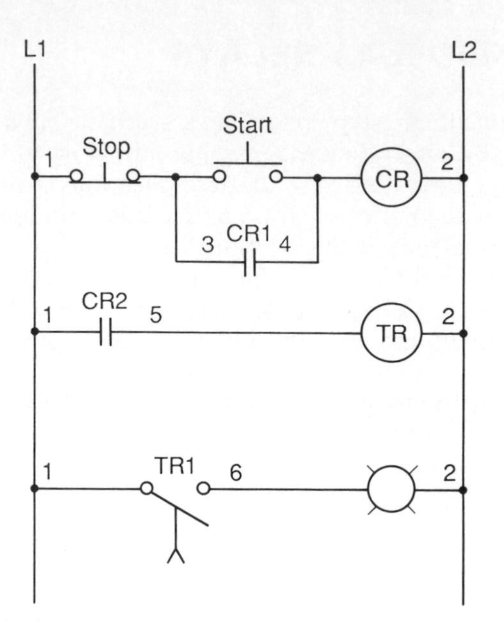

FIGURE 9–4 *Ladder diagram of control circuit with a delay relay.*

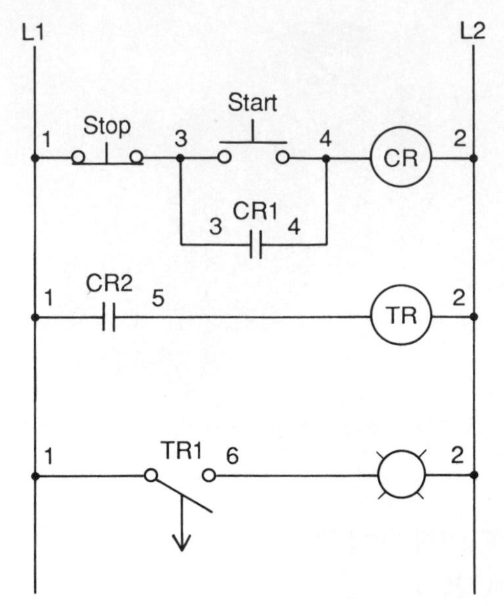

FIGURE 9–6 *Ladder diagram of control circuit with off-delay relay.*

A ladder diagram of a control circuit with an on-delay relay is shown in Fig. 9–4. The start button is pushed and the control relay (CR) is energized; thus all CR contacts close. The control relay (CR) remains energized even after the pushbutton is released. Relay contacts CR-1 and CR-2 are closed, and a voltage is applied to the coil of the time-delay relay (TR). Even though voltage is applied to the timing relay coil, contact TR-1 does not close immediately. After a predetermined period of time, TR-1 closes and the indicator lamp turns on.

This *on-delay relay* gets its name from the fact that a period of time passes before the contact is closed. If the circuit also has an NC contact, the NC contact is also delayed for a period of time before it opens. When the circuit is turned off, all contacts immediately return to their normal deenergized positions.

The *off-delay relay* symbols are shown in Fig. 9–5. This relay has NO and NC contacts. Voltage is applied to the coil of the off-delay relay. The contacts change state immediately as they do in a normal control relay. NO contacts close and remain closed as long as a voltage is applied to the coil. Even after the voltage is removed from the coil, the contacts remain in their activated states for a preset period. The contacts return to their normal states only after the relay *times-out*.

FIGURE 9–5 *Schematic symbols for an off-delay relay and its NO and NC contacts.*

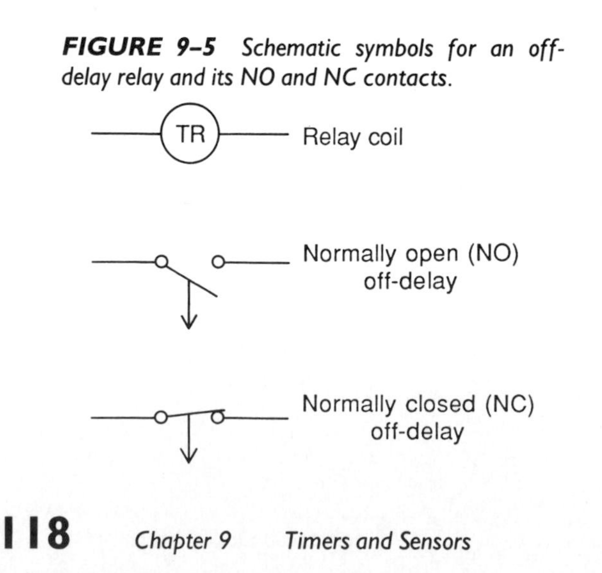

The off-delay relay shown in Fig. 9–6 is a ladder diagram of a control circuit. The START button is pushed and a voltage is applied to the relay coil (CR). Then the START button is released, so CR-1 closes and maintains voltage to coil CR. Contact CR-2 also closes and a voltage is applied to coil TR. Contact TR-1 closes immediately and the indicator lamp turns on. The STOP button is pushed and voltage is removed from coil CR. All CR contacts open. Voltage is also removed from coil TR. However, contact TR-1 remains closed for a preset period of time and the indicator lamp remains on. After the relay times-out, contact TR-1 opens and the light goes out.

GENERAL-PURPOSE TIMING RELAYS

The general-purpose timing relay has many uses. It is used for automatic control of machine tool programming, sequencing controls, heating and cooling operations, and warm-up delays. Most of this type are designed for plug-in housing (Fig. 9–7). The plug-in base resembles the eight-pin base of the older vacuum tubes. In fact, they can fit into some vacuum-tube sockets. Note how the keyway indicates the beginning or pin 1 and the location of pin 8. The pins are

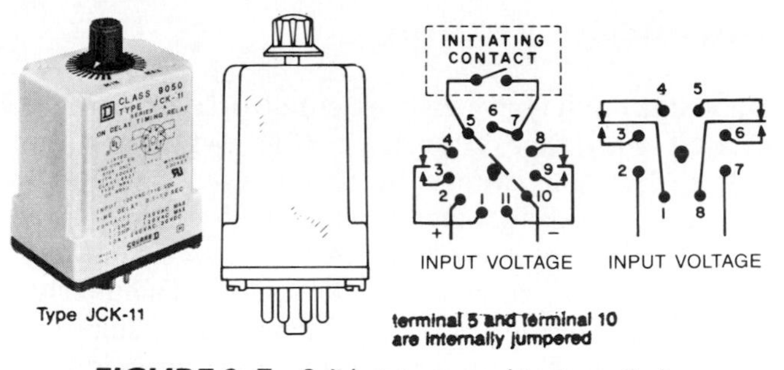

FIGURE 9–7 *Solid-state general-purpose timing relay. (Square D)*

counted from the keyway in a clockwise direction. However, looking at the bottom of the socket, you have to count in the opposite (counterclockwise) direction for making connections to the socket. Some timers have an 11-pin base. An adjustable knob on top of the case allows for the time delay to be adjusted from 0.1 to 10 seconds on some and other time ranges for others (Table 9–1).

CONTROL OF PROGRAMMABLE TIMERS

Programmable timers are microprocessor controlled to provide flexibility with accurate timing. The on-delay timer has five programmable timing ranges:

0.5 to 9.99 seconds
0.1 to 99.9 seconds
1.0 to 999 seconds
0.1 to 99.9 minutes
1.0 to 999 minutes

A five-position rotary switch is used to select the timing ranges. Three pushbutton thumbwheels are used to select the time value (Fig. 9–8). Timing modes are shown in Fig. 9–9. These relays draw about 1.2 VA, making the solid-state relay very low in power consumption.

DIGITAL SOLID-STATE TIMER /COUNTER

There are some extremely flexible microcomputer-based digital timer/counters. They can be field programmed to perform as a timer or counter with a variety of timing or counting modes and ranges. They are available in either a single- or dual-stage model and can be ordered with either relay or solid-state outputs and with ac or dc inputs to make them compatible with most control systems.

TABLE 9–1 TIMING RELAYS

(a) Variable Time-Delay Relays[a]

Range	Description
0.1–10 seconds 0.3–30 seconds 0.6–60 seconds 1.2–120 seconds 1.8–180 seconds	Integrated circuit with CMOS circuitry is used to provide an accurate time delay.
0.1–10 minutes 0.3–30 minutes 0.6–60 minutes 1.2–120 minutes	Integrated circuit with CMOS circuitry is used to provide an accurate time delay.

(b) Fixed Time-Delay Relays

Timing Mode	Timing Range (seconds)
On delay	1–80
	181–3600
Off delay	1–180
	181–3600
Interval	1–180
	181–3600
One Shot	1–180
	181–3600
Repeat cycle	1–180
	181–3600

[a]Fixed repeat cycle timers are supplied with the same on and off times.

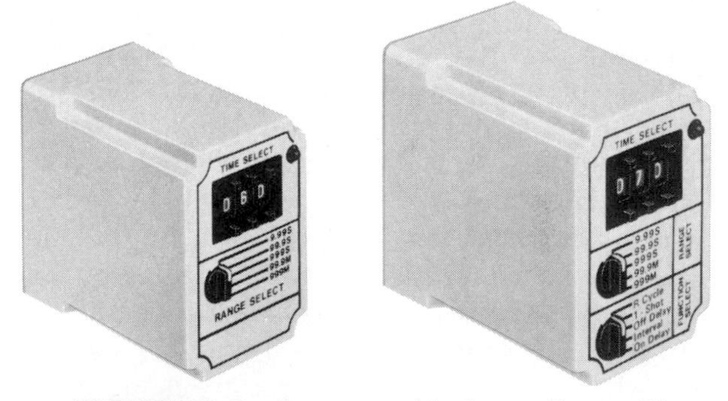

FIGURE 9–8 *Programmable timers. (Square D)*

FIGURE 9–9 *Timing modes for programmable timer. (Square D)*

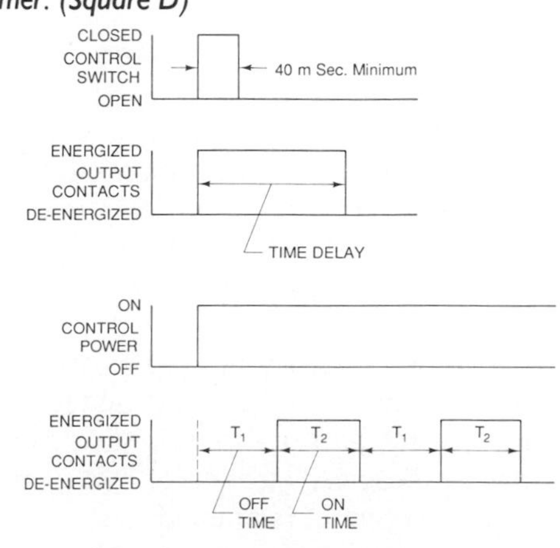

Single-Stage Version

Single-stage units have a selector switch and four pushbutton thumbwheels located on the front face of the device (Fig. 9-10). The selector switch is used to select the time or count range, while the four pushbutton thumbwheels are used to set the time or count value. Operating mode selection is accomplished by programming seven internal DIP switches. The units are available with either a digital readout capable of up or down timing/counting or a LED status indicator.

FIGURE 9–10 Digital solid-state timer/counter—single-stage version: (A) type P; (B) type PM. (Square D)

Time/Count Modes

Relay switches require a great deal of attention, which results in a lot of downtime for the machines they control. In industry, the relay has quickly been replaced by electronics, just as the telephone company uses transistor switching to replace relay switching. There are no moving parts in transistors and no contact points to be cleaned periodically. With relays, the entire circuit had to be rewired to change a program, but with a digital solid-state timer/counter, only the thumbwheels on front of the device are changed.

The four modes for the solid-state timer/counter are on-delay, off-delay, interval, and repeat cycle pulse.

On-Delay Mode. Closing the initiating contact begins the delay period, and the time/count outputs are energized.

Off-Delay Mode. Closing the initiating contact energizes the time/count outputs. Opening the initiating contact begins the delay period. At the end of the delay period, the time/count outputs deenergize.

Interval. Closing the initiating contact energizes the time/count outputs and begins the delay period. At the end of the delay period, the time/count outputs deenergize.

Repeat Cycle Pulse. Closing the initiating contact begins the delay period. At the end of the delay period, the time/count outputs energize for 50 milliseconds and then deenergize. The device repeats this cycle until the initiating contact is opened.

Dual-Stage Version

Dual-stage units have eight pushbutton thumbwheels located on the face of the device (Fig. 9–11). The four left pushbutton thumbwheels are used to set the time/count value for stage 1, while the four right pushbutton thumbwheels are used to set the time/count value for stage 2. A seven-position DIP switch located inside the device is used to set the time or count range and mode. The units come with two LED status indicators, one for stage 1 and one for stage 2.

FIGURE 9–11 Digital solid-state timer/counter—dual-stage version: (A) type P; (B) type PM. (Square D)

Two-Stage (On-Delay) Mode. Closing the initiating contact begins the stage 1 delay period. At the end of stage 1 delay period, the stage 1 time/count outputs energize and the stage 2 delay period begins. At the end of the stage 2 delay period, the stage 2 time/count outputs energize.

Repeat Cycle (On-Delay). Closing the initiating contact begins the OFF stage 1 delay period. At the end of the OFF stage 1 delay period, the stage 2 time/count outputs energize and the ON stage 2 delay period begins. At the end of the ON stage 2 delay period, the stage 2 time/count outputs deenergize and the OFF stage 1 delay period begins. This cycle repeats until the initiating contact is opened. The stage 1 time/count outputs act as instantaneous outputs and are energized when the initiating contact is closed.

Two-Stage (OFF-Delay). Closing the initiating contact energizes the stage 1 and stage 2 time/count outputs. When the initiating contact is opened, the stage 1 delay period begins. At the end of the stage 1 delay period, the stage 1 time/count outputs deenergize and the stage 2 delay period begins. At the end of the stage 2 delay period, the stage 2 time/count outputs deenergize.

Repeat Cycle (OFF-Delay). Closing the initiating contact energizes the stage 1 and stage 2 time/count outputs. When the initiating contact is opened, the stage 1 time/count outputs deenergize and the ON stage 1 delay period begins. At the end of the ON stage 1 delay period, the stage 2 time/delay outputs deenergize and the OFF stage 2 delay period begins. At the

Replaceable Hard Relay Outputs:
Wiring and Contact Arrangement

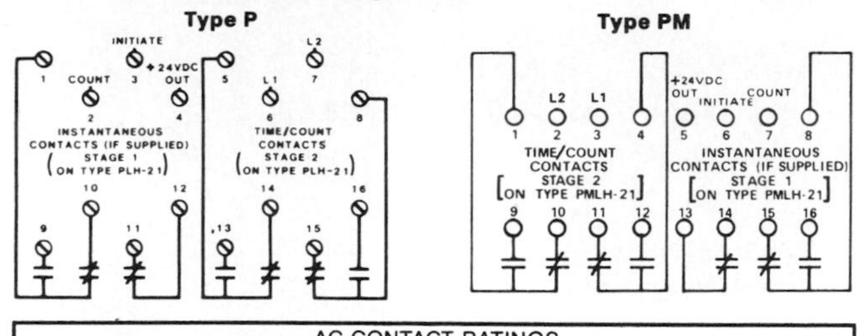

AC CONTACT RATINGS						
	Inductive 35% Power Factor					Resistive 75% Power Factor
	Make		Break		Continuous Amperes	Make, Break and Continuous Amperes
Volts	Amps	VA	Amps	VA		
120	30	3600	3.0	360	7	7
240	15	3600	1.5	360	7	7
DC Contact Rating: 7 amps max. @ 28VDC resistive or inductive						

Output Relay Operating Times:
Pick-up 25 ms
Drop-out 25 ms

Solid State Outputs:
Wiring and Contact Arrangement

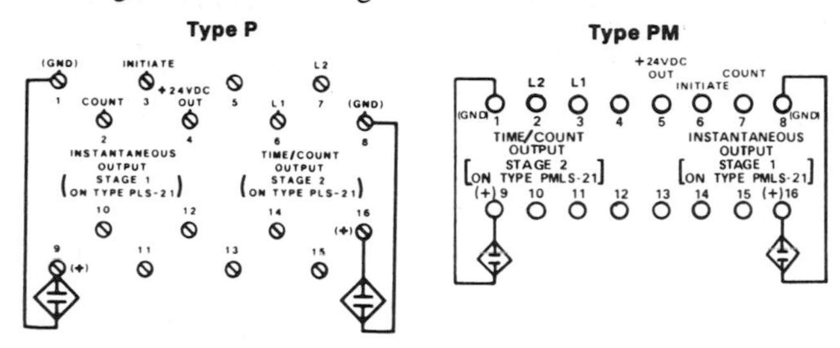

Solid State Contact Ratings:
50 mA @ 30VDC

FIGURE 9–12 *Contact points on timer/counter. (Square D)*

end of the OFF stage 2 delay period, the stage 2 time/count outputs energize and the ON stage 1 delay period begins. This cycle repeats until the initiating contact is closed. The stage 1 outputs act as instantaneous outputs and are energized when the initiating contact is closed.

These timer/counters operate on 120 V 50/60 Hz (102 to 132 V) and 240 V 50/60 Hz (204 to 264 V). Figure 9–12 shows the wiring diagrams for these timers/counters.

THUMBWHEEL SWITCHES

Thumbwheel switches are utilized on many electronic devices to allow for the setting of time or counts. They can be ganged to make more than one. A nutdriver is used to tighten the nuts on the end of the long screw that inserts into the body of the switches to hold the outside cases and any separators together (Fig. 9–13). You can see from those shown that the push-button

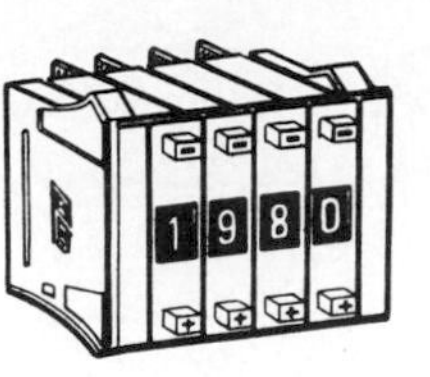

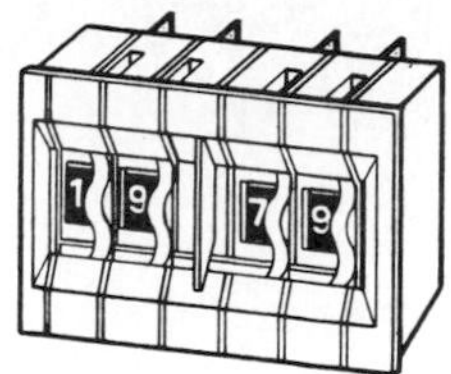

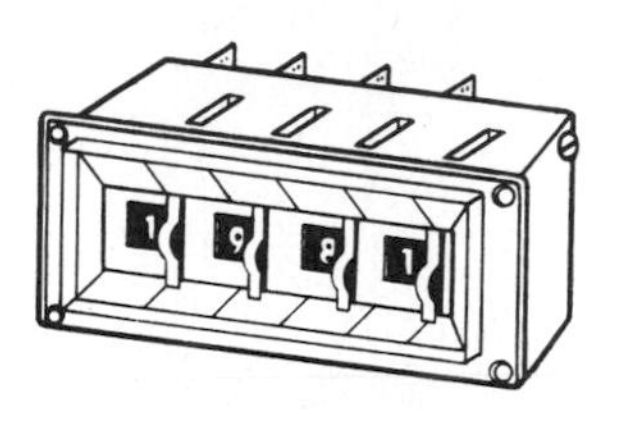

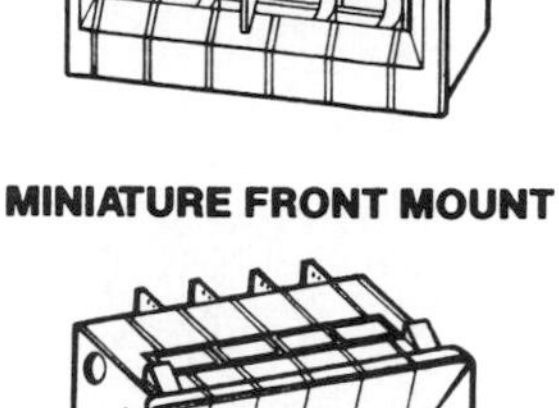

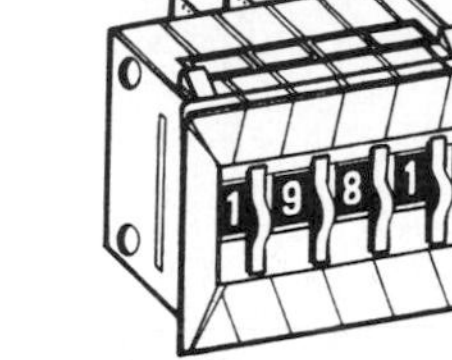

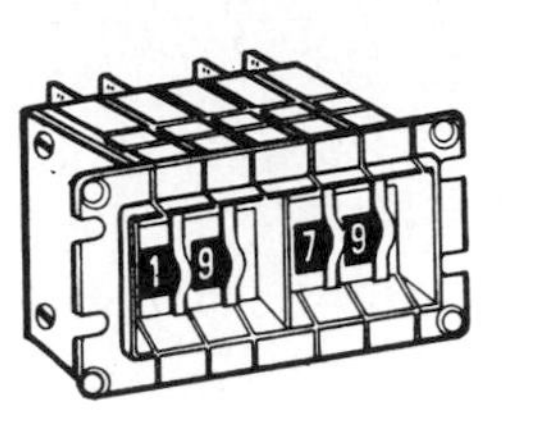

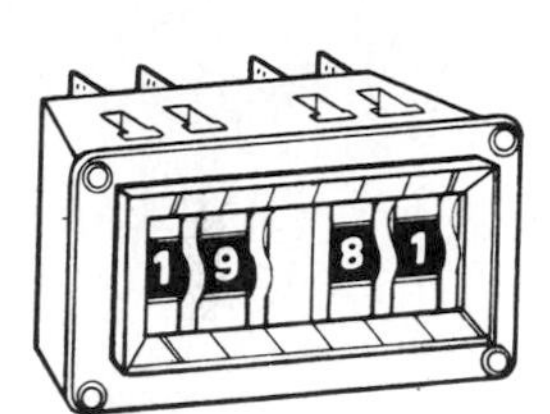

FIGURE 9–13 *Thumbwheel switches.*

front mount switch has a pushbutton for setting each number, whereas the others have the thumbwheel that can be rotated until the correct digit comes up. Components needed to assemble a complete switch are:

1. One pair of end plates
2. Blank bodies to divide switches in a bank
3. Divider plates to separate switch banks into framed sections
4. Hardware

DIP SWITCHES

DIP (dual-in-line packaging) is also used with integrated circuits or chips and their packaging. DIP switches are small and have contacts that resemble

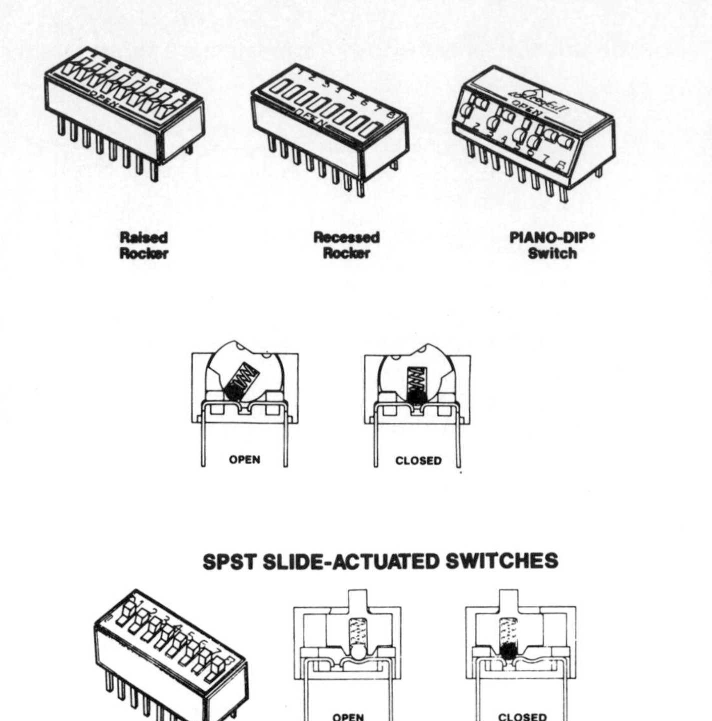

FIGURE 9–14 *DIP switches.*

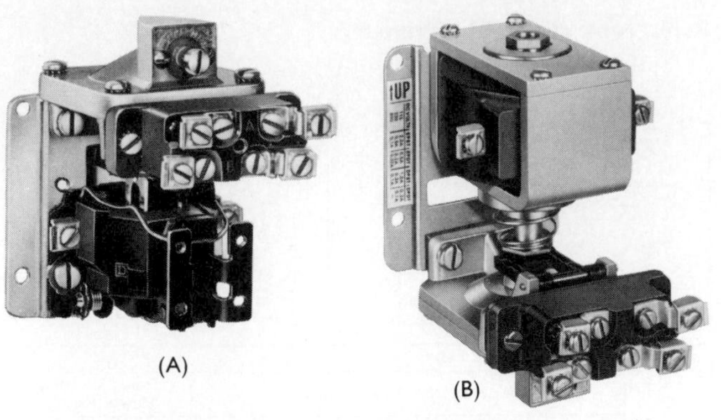

FIGURE 9–15 *Pneumatic timing relays: (A) type AO-10E; (B) type HO-10E. (Square D)*

those of an integrated circuit (IC) or chip. They are designed to be attached to a printed circuit board and soldered in place. As you can see from Fig. 9–14, these are very small and it takes a ballpoint pen or something equally small to cause the switches to be rotated from on-to-off or off-to-on. They also come in raised rocker, recessed rocker, and piano DIP switch. A high-pressure spring-and-ball contact system causes them to make and break contact. The ball makes the complete circuit between the two contacts as shown in the cutaway views.

A high-pressure, spring-and-ball contact system is also used to make and break contact in single-pole, single-throw *slide-actuated switches,* which are very small and need a very small screwdriver tip or ballpoint pen to close or open them. These are also mounted on printed circuit boards and are used to program a circuit or chip to function in such a manner as in the timer/counters discussed in previous paragraphs.

PNEUMATIC TIMING RELAYS

The pneumatic timing relay has some advantages over other types. It has the advantage of not being affected by normal variations in ambient temperature and atmospheric pressure. It is adjustable over a wide range of timing periods and has good repeat accuracy. This type of relay is available with a variety of contact and timing arrangements (Fig. 9–15).

This pneumatic time-delay unit is mechanically operated by a magnet structure. The time-delay function depends on the transfer of air through a restricted orifice. This restriction is done by the use of a reinforced synthetic rubber bellow or diaphragm. The timing range is adjusted by positioning a needle valve to vary the amount of orifice or vent restriction.

Energizing or deenergizing pneumatic timing relays can be controlled by devices such as pushbuttons, limit switches, and thermostatic relays. They draw very small amounts of current, so sensitive control devices are used to control the operating sequence.

This type of timing relay is used for motor acceleration and in automatic control circuits. Automatic control is needed where uses are repetitive and accuracy is required.

Two types of time delay are provided by pneumatic timers: *on-delay,* which means that the relay provides time delay when it is *energized,* and *off-delay,* which means that the relay is *deenergized* when it provides time delay. Figure 9–16 shows the on-delay arrangement in an Allen-Bradley relay. When the operating coil (O) is energized, solenoid action raises the solenoid plunger (A). The pressure on the rod (B) is released. A spring (W) is located inside the bellows (E). The spring then allows the pushrod (B) to move upward. As (B) moves upward, it causes the off-center linking mechanism (C) to move the end of the snap-action mechanism (X) upward. This action, in turn, raises (D) to operate the switch or time-controlled contacts.

The position of the needle valve at the bottom determines how fast the bellows rises. The needle valve setting determines the time interval between the solenoid closing and the rise of the bellows to operate the switch. If the needle valve is almost closed, it takes a considerable length of time for air to pass the valve and cause the bellows to rise. Gravity causes the plunger (B) to drop when the coil is deenergized. The action of the reset spring and the falling plunger resets the timer almost instantaneously.

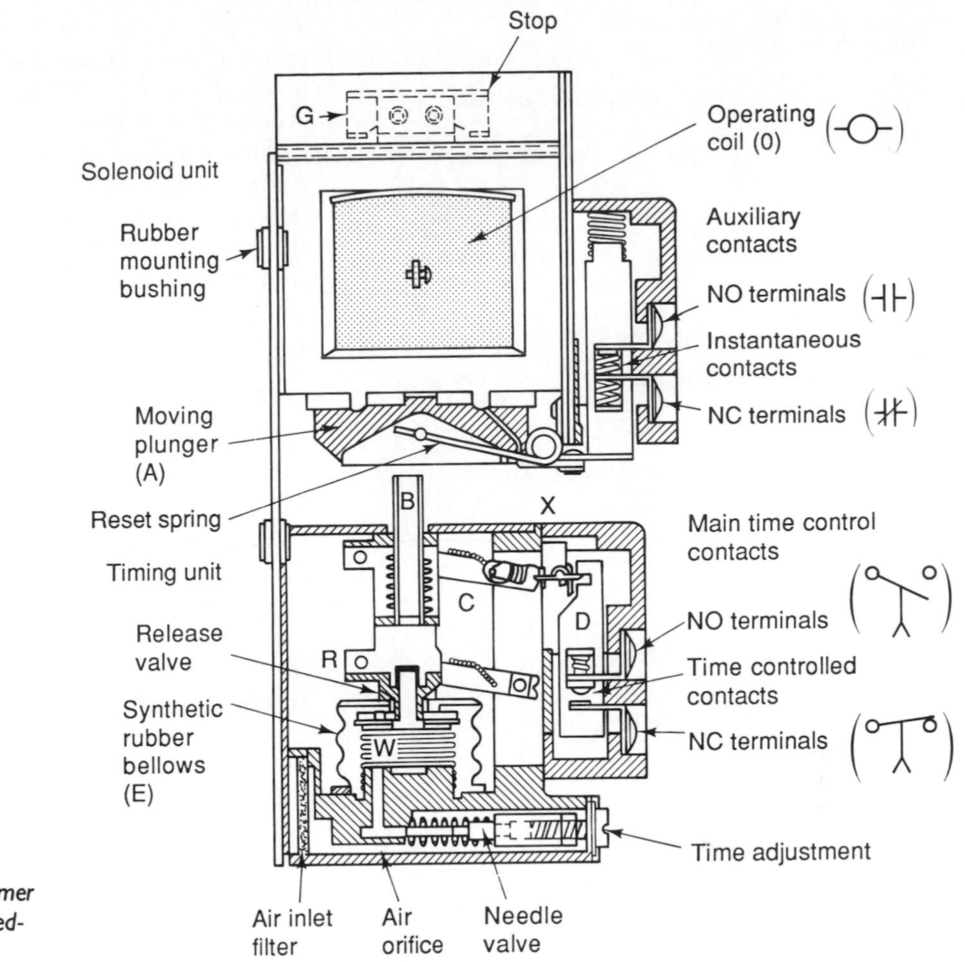

***FIGURE 9–16** Ac on-delay timer with coil energized and timer timed-out. (Allen-Bradley)*

The off-delay mode in this Allen-Bradley relay is shown in Fig. 9–16 with the solenoid rotated 180°. When the operating coil (O) is energized, plunger (A) is held down and (G) pushes against (B). That holds the bellows (E) in a fully compressed position. Deenergizing the coil allows the reset spring to force (A) upward. This also causes (G) to rise. That releases the downward force on (B). With the pressure taken from (B), the bellows (E) slowly expands. This forces (B) upward. As (B) moves up, it trips to toggle; this is the same as in the on-delay mode. Tripping the toggle causes the switch to be tripped. Figure 9–17 shows the symbols used in the standard elementary diagrams for timed contacts.

MOTOR-DRIVEN TIMERS

There are many types of motor-driven timers used for various purposes. Figure 9–18 is only one of the many types utilized in industry and in commerce. This timer is operated from a sustained or momentary contact

***FIGURE 9–17** Timed contacts symbols.*

Contact action is retarded after the coil is:

Energized		Deenergized	
NOTC	NCTO	NOTO	NCTC
On-Delay Mode		Off-Delay Mode	

***FIGURE 9–18** Reset timer.*

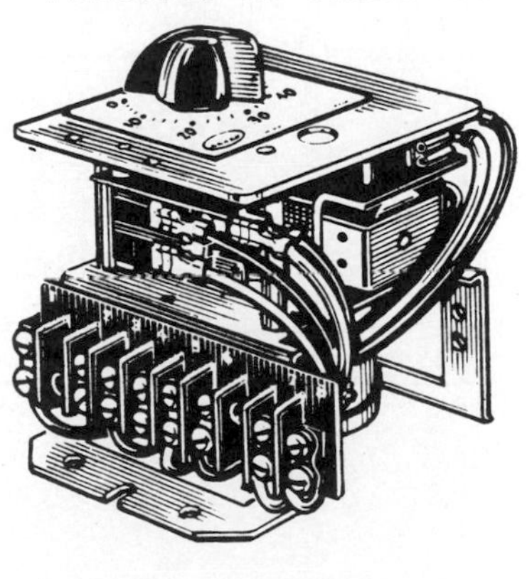

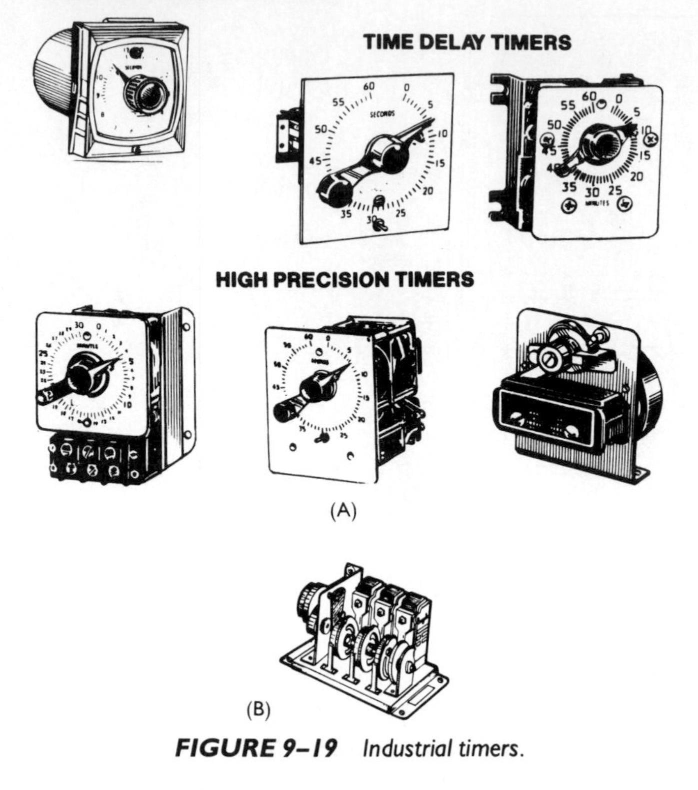

FIGURE 9–19 *Industrial timers.*

switch. In the latter case, the solenoid-actuated switch can energize a load during the time period as an interval timer. Its features are cam-actuated switches that allow programmed timing, reduce switch actuator overtravel, and greatly extend switch life. A differential clutch employs hardened gears for reliable, long-life operation. Power required is 120 V and the motor consumes about 2.5 W. Scale reset is within 0.5 second. Switch ratings are 10 A resistive at 125 V ac or $\frac{1}{3}$ hp at 125 V.

Many timers are easily converted from on-delay to off-delay. Figure 9–19A shows some of the types available for various jobs. Most are synchronous motor driven. These are hysteresis-synchronous motors and are self-starting but very accurate since they synchronize on frequency rather than depend on voltage for constant speed. Figure 9–19B shows a set of cams that can be programmed for their on–off periods. A number of these switches can be ganged to control a number of operations. Figure 9–20 shows how a standard start–stop station controlling an on-delay timer with a set of *timed-closed* contacts is arranged in an elementary diagram.

SEQUENCE CONTROL

The circuit shown in Fig. 9–21 is a sequence control of two motors, one to start and run for a short time after the other stops. In this system it is desired to have a second motor stared automatically when the first is stopped. The second motor is to run only for a given length of time. Such an application might be found where the second motor is needed to run a cooling fan or a pump.

To accomplish this, an off-delay timer (TR) is used. When the START button is pressed, it energizes both M1 and TR. The operation of TR closes its time-delay contact but the circuit to M2 is kept open by the opening of the instantaneous contact. As soon as the STOP button is pressed, both M1 and M2 will continue to run until TR times out and the time-delay contact opens.

PROGRAMMABLE TIMERS

With the advent of microprocessors and digitized equipment, many types of programmable timers have become available. One good example of some of the possibilities that timers offer is shown in the Watch-

L1 L2
Stop Start TR1 (2, 3)
1
TR1
2
All OL'S
3 TR1 M1

FIGURE 9–20 *Time-delay circuit.*

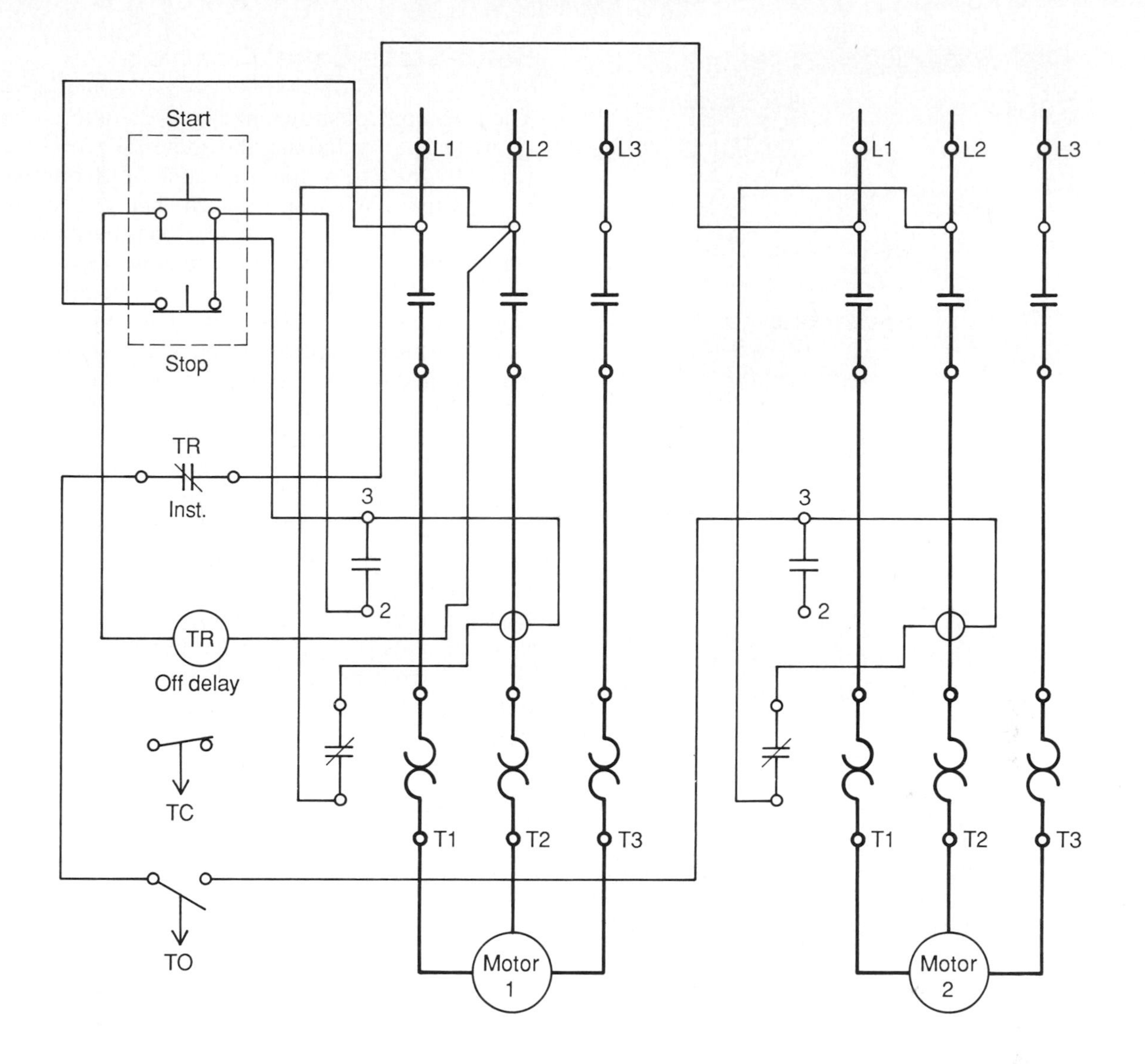

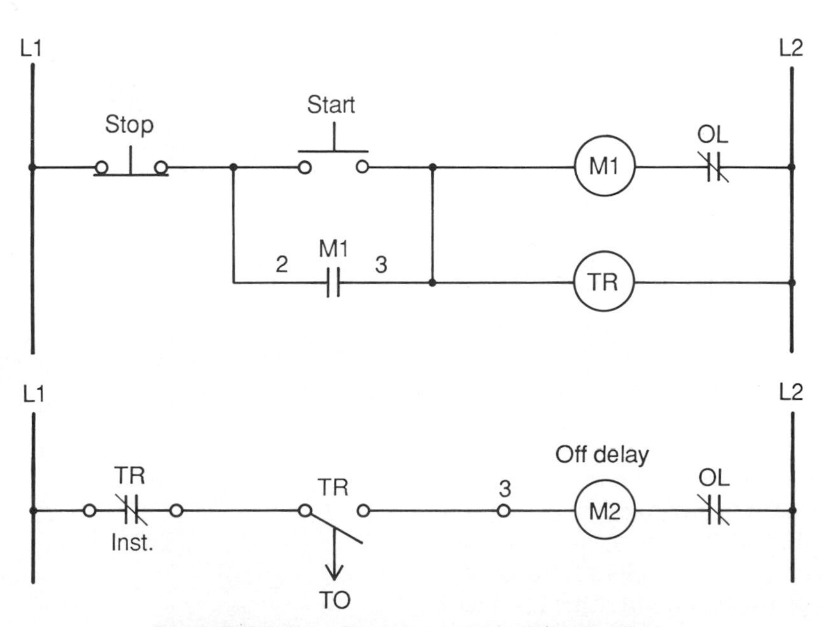

FIGURE 9–21 *Sequence control. (Allen-Bradley)*

FIGURE 9–22 *Watchdog Time Commander programmable timer. (Square D)*

dog Time Commander (Fig. 9–22). It is a precision, wall-mounted, 365-day timer that can be programmed to operate to an exact second. The timer consists of a clock and an enclosure with a four-circuit capacity. Another model has 16-circuit capability. Relay modules may be added to both models. Each relay module controls four electrical circuits.

The timer accepts 60 operational programs, each capable of controlling any or all circuits. Up to 20 holidays may be programmed, and each holiday can be a single day or an unlimited span of days. Battery backup retains programs, during power failures, for up to three years. Leap-year adjustment is automatic and the start and end of daylight savings time can be programmed. Commands may also be programmed relative to sunrise and sunset.

Some of the purposes of this type of timing are energy management, security, and a wide range of industrial and commercial applications. Lights, water heaters, fans, furnaces, and pumps may be turned on and off to best utilize energy dollars. Door locks, burglar alarms, and elevators may be cycled to ensure security during unoccupied hours. Industrial and commercial uses include sprinkler and irrigation systems, traffic control, schools, and industrial bells for life testing. One-second resolution provides precise, accurate timed control. Programming instructions come with the unit. The detailed instructions manual ensures that you are able to program the unit for its intended purposes.

SENSORS

Sensors are used for many purposes in industry. They detect the presence or absence of materials or products. They can be used to set off an alarm if one or a number of conditions are not met. Sensors are made to sense temperature, pressure, and in some instances, levels of materials in tanks or storage facilities.

Solid-State Level Controls

The level control shown in Fig. 9–23 detects the level of materials by having the material's presence detected by the two probes that stick out from the body of the sensor. When the material being checked is between the two probes, it changes the characteristics of a transistorized circuit and causes it to energize a relay. A number of approaches can be taken in electronics to accomplish the sensing action. The Hall effect can be utilized (discussed in Chapter 10), the eddy current method can be used, or the capacitance between the two probes can be changed by having the dielectric altered. The sensor shown in Fig. 9–23 has the ability to respond in 0.5 second to level changes. It can be used to detect either rising or falling liquid levels in storage vessels, mixing tanks, or pipelines. It can be used with any liquid compatible with the stainless steel housing. It is ideally suited for applications that have material buildup, foam, gas bubbles, or suspended solids. It is unaffected by agitation, wave action, or turbulence and can be installed directly without the need for measuring chambers or bypasses. It is designed to operate between 24 and 250 V ac at 50/60 Hz, in series with the relay, solenoid, valve, or annunciator it is to control. Power should never be applied

FIGURE 9–23 *(A) Solid-state level sensor; (B) note the shape and size of the oscillating fork. (Square D)*

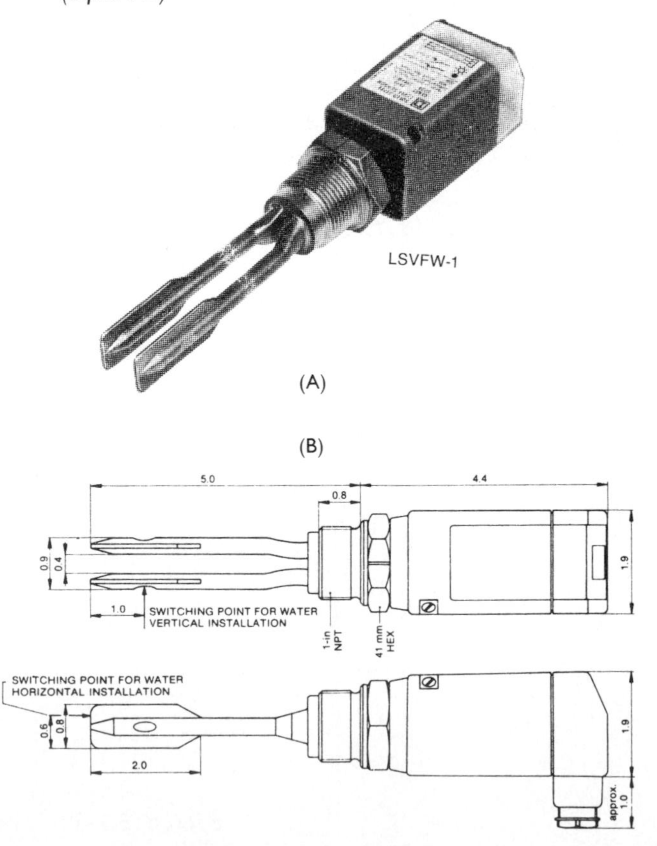

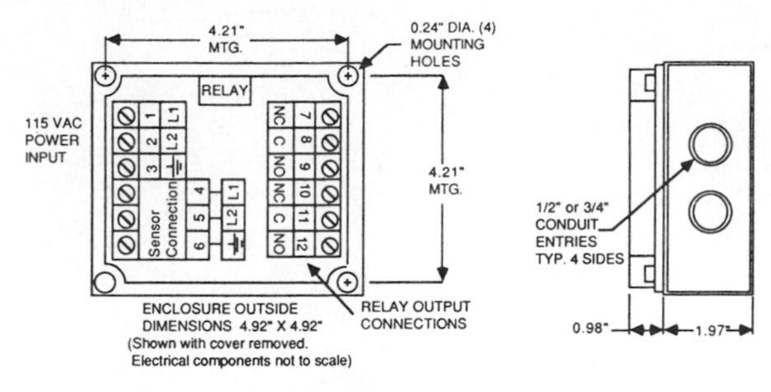

FIGURE 9–24 *Relay operated by the level sensors. (Square D)*

TABLE 9–2 DIELECTRIC CONSTANTS

Air	1.00006	Olive oil	3.11
Ammonia (liquid)	15.5	Paper	2–6
Asphalt	2.68	Paraffin	2.0–2.5
Beeswax	2.75–3.0	Polyethylene	2.3
Ceramics	80–1200	Porcelain	6–8
Glass, Pyrex	3.8–6.0	Rubber	2.8
Glass, Corning	9.5	Sulfur	4.0
Mica	7–9	Vacuum	1.0
Nylon	3.5	Water	80
Oil	2–5		

to the sensor without an external load connected in series. Figure 9–24 shows how the power input is connected and how the contacts are arranged at terminals 7 through 12. Even with the load switched off, a small current of less than 5 milliamperes still passes through the sensor and load to maintain power to the unit's sensing electronics.

Capacitance-Type Sensor

The capacitance-type level limit sensor is ideal for use in applications where control and measurement of powders, granulars, or peltized solids is required. The extremely sensitive sensing head detects materials with dielectric constants as low as 1.5.

A dielectric is the material (air and vacuum or other substances) that separates the two plates of a capacitor. This material varies in its influence on the capacitance of a capacitor. Each material has a constant when compared with a vacuum, which has a constant of 1. Table 9–2 shows the constants for a variety of substances.

The capacitance-type level limit control shown in Fig. 9–25 has the ability to be mounted in any number of positions. The field-selectable *fail-safe* mode is advantageous since it allows the user to apply the unit as either normally open (material absent) or normally closed (material present). The connections shown in Fig. 9–24, which also apply here for the sensor connections, are marked next to the 115-V ac power input. The internal relay has its output connections from terminals 7 to 12.

Temperature Sensing

The ability to sense the actual temperature or the ability to detect a difference between temperatures makes it possible to control the temperature of liquid baths, bearings, internal combustion engines, and large air compressors, to mention but a few devices. Figure 9–26 shows some of the temperature switches used for automatic control of temperature-maintenance equipment in industrial and general-duty applications. Temperature sensing may be done by a number of methods, including the thermocouple, thermistor, resistance temperature detector, and semiconductor temperature sensor.

Thermocouple. The thermocouple is the most widely used temperature-sensing device. It dates back to 1821, when Thomas Seebeck found that joining two

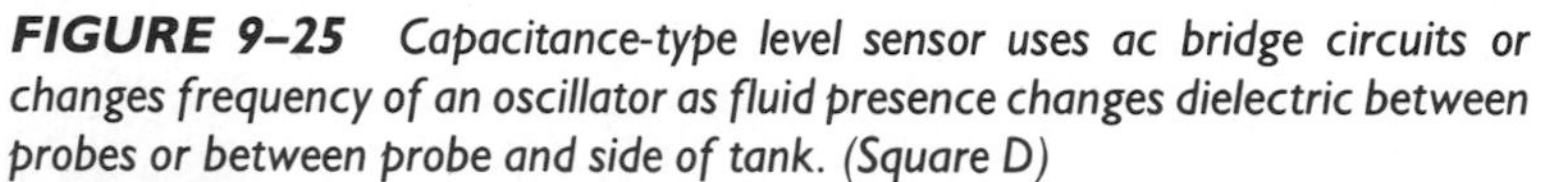

FIGURE 9–25 *Capacitance-type level sensor uses ac bridge circuits or changes frequency of an oscillator as fluid presence changes dielectric between probes or between probe and side of tank. (Square D)*

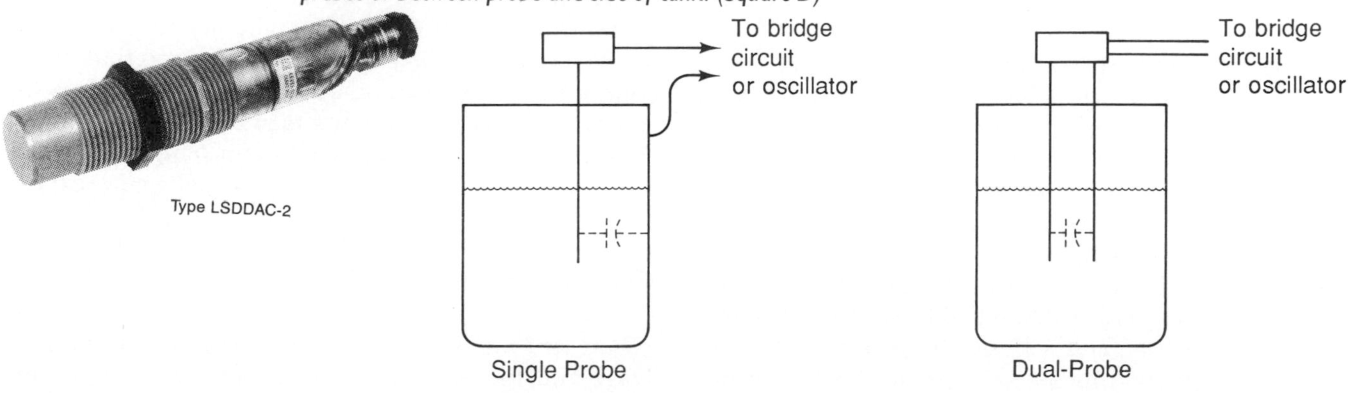

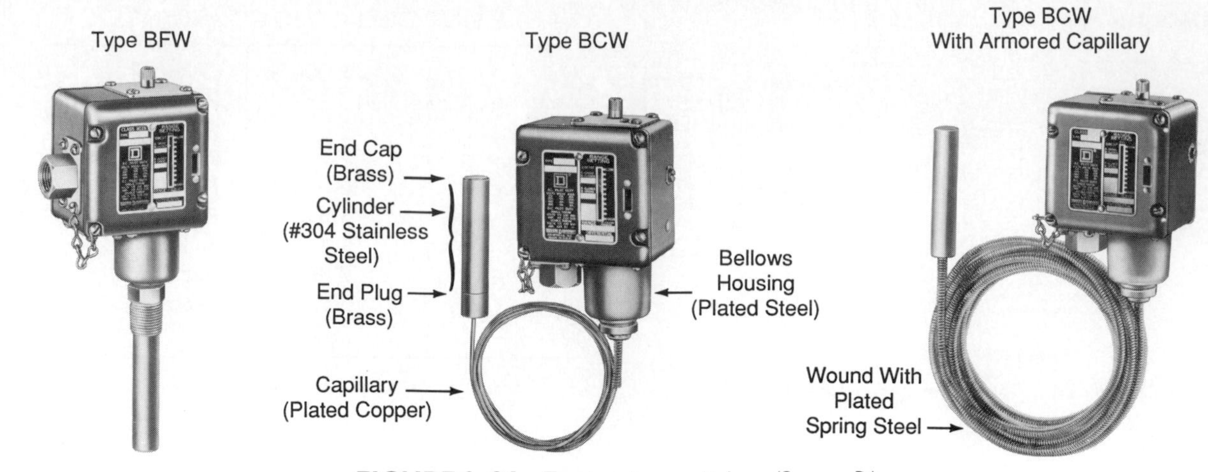

FIGURE 9–26 *Temperature switches. (Square D)*

wires made of different materials generate an EMF when heated at one end. The amount of heat is directly proportional to the amount of output EMF or voltage. This is a simple, rugged device that can be made inexpensively. It also has the advantage of being able to measure a 4500°F range of temperatures. It is, however, unstable and needs a reference junction to make sure that its output is useful. Figure 9–27 shows how the thermocouple works. Industry uses thermocouples for measuring the temperature of ovens and furnaces and of flowing liquids, and the core temperatures of nuclear reactors.

Thermistor. The thermistor is a device that produces a decrease in resistance when heated. As the temperature increases, the resistance decreases. Thermistors come in many sizes and a multitude of shapes (Fig. 9–28). This is a most sensitive temperature device, and its resistance varies from 0.5 Ω to 80 MΩ. It does have disadvantages, however, including nonlinear response and limited temperature range. Thermistors are used where it is necessary to detect temperature changes of 1°C in chemical processes. They can also be used as a liquid or fluid-flow control device. They detect the presence of a liquid through the temperature change in the material. Once immersed in the liquid, a thermistor can determine the temperature of the material and serve to energize the proper circuitry to cause the level to be corrected.

Resistance Temperature Devices. The thermistor changes resistance when heated in an inverse direction, as is normal for most metals or materials. This means that gold, silver, iron, or pure elements such as these increase in resistance as they are heated. This particular characteristic can also be used in designing devices to control temperature.

Some of the best known devices are the bridge circuit (shown in Fig. 9–29), metal film resistance temperature detectors, and helical resistance temperature devices. The metal film detector is made by depositing

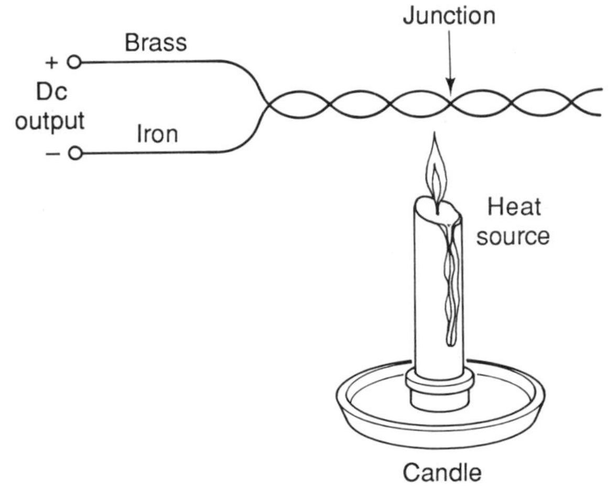

FIGURE 9–27 *Thermocouple produces the Seebeck effect.*

a thin film of platinum on a flat ceramic substrate and etching it with a laser. It is then trimmed and sealed. This makes a very sensitive device for detecting temperature changes. Disadvantages include the requirement of an amplifier because of the small temperature changes and the fact that response time is rather slow.

Semiconductor Temperature Sensors. Silicon and germanium can both be used for temperature sensing. Germanium-doped crystals can be used to detect temperatures near absolute zero. The negative temperature coefficient of germanium is similar to that of the thermistor. However, silicon has a positive temperature coefficient. It is also limited in its range of sensing from −67 to 275°F. Semiconductor temperature sensors have good linearity and are small and inexpensive.

Pressure Sensors

There are many ways to sense pressure. The strain gage and the piezoelectrical are the two most com-

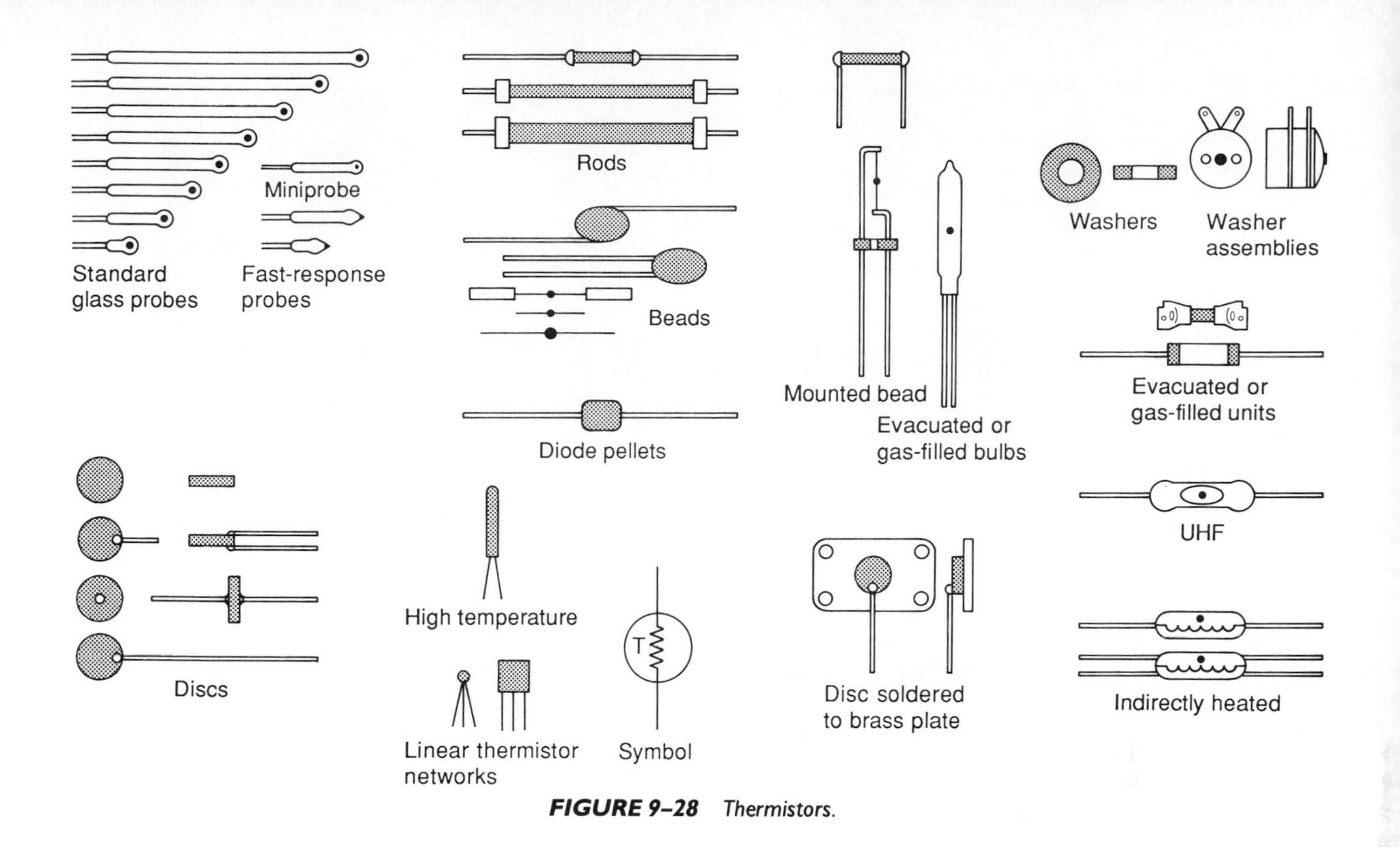

FIGURE 9–28 *Thermistors.*

FIGURE 9–29 *(A) Helical and (B) metal film resistance temperature devices with (C) a bridge circuit.*

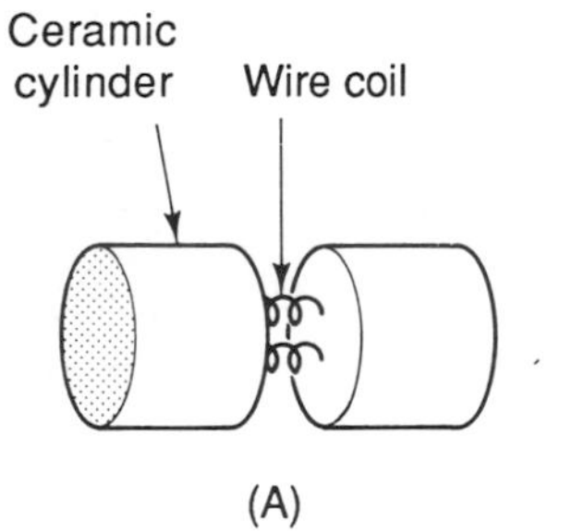

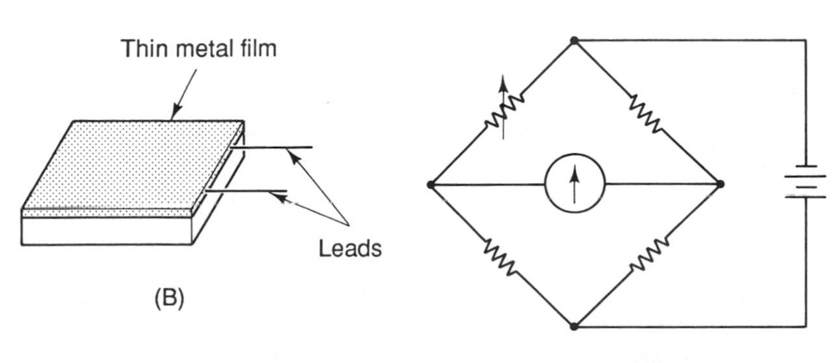

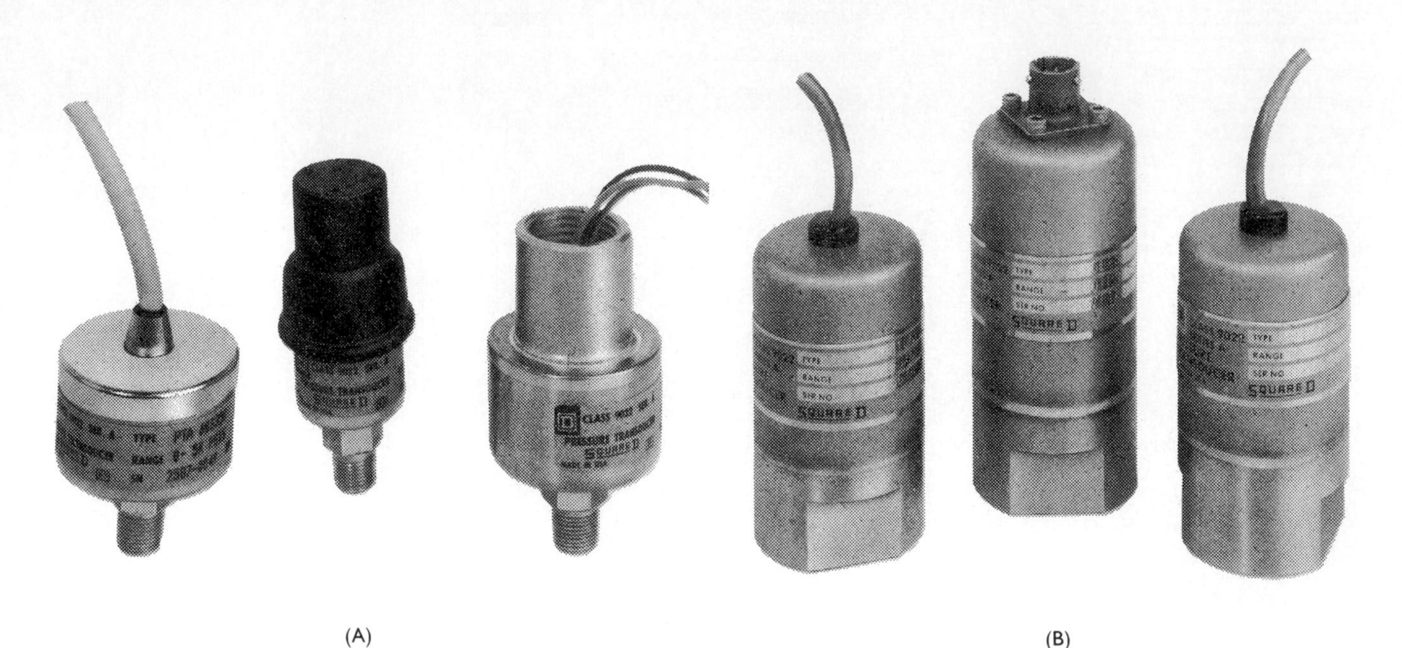

FIGURE 9–30 *Pressure transducers: (A) PTA; (B) PTB. (Square D)*

monly used to produce an electrical output that can be utilized to control a process. Different manufacturers use different methods. Each appears to have a specialty that has worked for years, and it is improved over the years to make use of the latest technology in the electronics field.

Strain Gage Transducers. Figure 9–30 shows two types of pressure transducers that fulfill the requirements of many industrial users. They are reliable, accurate, and provide a continuous (analog) pressure input in process and factory automation. These strain gage–based sensors measure strain or tension. The gages are attached to one side of a pressure-sensing diaphragm, then pressure is applied to the diaphragm, producing a minute deflection that induces strain to the gages, thus changing their resistances. By applying a constant voltage across the strain gages, the change in resistance is measured. The resulting outputs are conditioned to user needs through amplifiers and other circuitry within the transducer (Fig. 9–31).

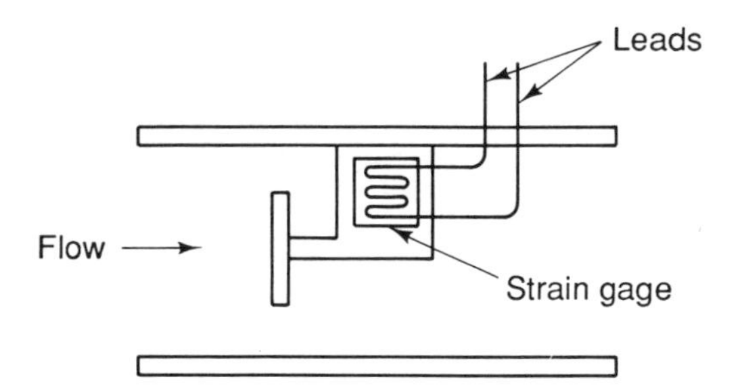

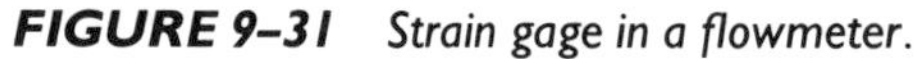

FIGURE 9–31 *Strain gage in a flowmeter.*

Piezoresistive Transducers. This type of transducer or sensor is used to sense ranges from 10 to 5000 psi. A strain sensor is incorporated in the semiconductor circuitry.

Pressure sensors are used in pneumatic systems; hydraulic systems; irrigation systems; lubrication systems; heating, air-conditioning, and ventilation systems; energy management systems; robotics; automated process equipment; plant utilities; and machine tools. Figure 9–32 shows how the transducers are wired into the circuitry.

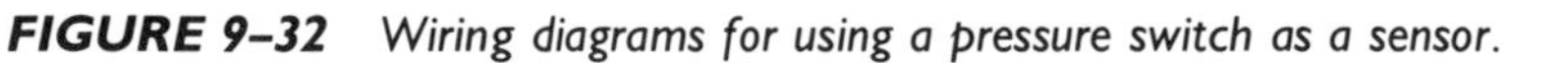

FIGURE 9–32 *Wiring diagrams for using a pressure switch as a sensor. (Square D)*

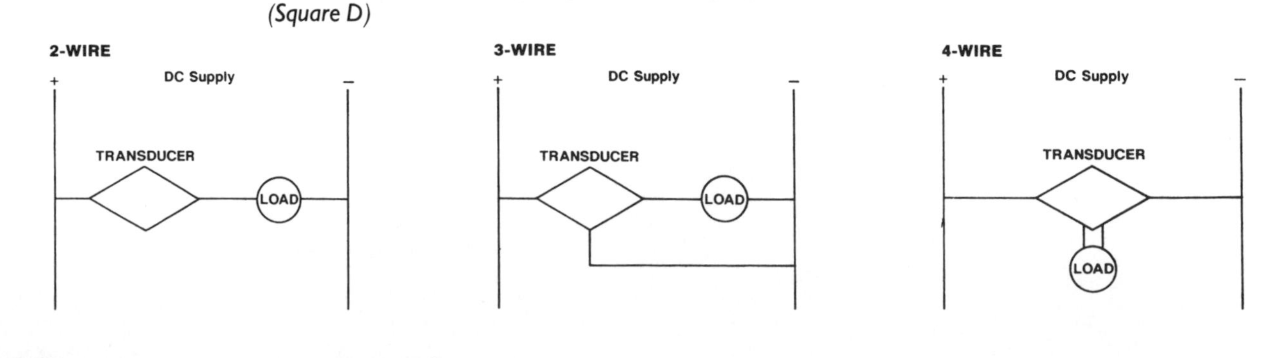

QUESTIONS

1. Name the three categories of timers.
2. Explain how the dashpot timer works.
3. What is the advantage of the pneumatic-type relay over the dashpot type?
4. Explain the terms *on-delay, off-delay,* and *times-out.*
5. Where is the general-purpose timing relay used?
6. How is the time delay set on electronic timers?
7. What are the four modes for solid-state timer/counters?
8. What are thumbwheel switches used for?
9. How are DIP switches set?
10. Where are pneumatic timing relays used?
11. Draw the symbols for a timer's on-delay mode and off-delay mode.
12. What is a dielectric? Where is the term useful?
13. Name four methods of temperature sensing.
14. How does the thermistor operate?
15. What are the two most commonly used methods used to detect pressure changes?
16. What does *DIP* mean?
17. Which timer is becoming the standard in industry?
18. What is an *RC* network?
19. What does *CR* stand for?
20. How is the time period adjusted on a general-purpose timing relay?

REVIEW PROBLEMS

Timing has been done in electronic circuits for some time, utilizing the charge and discharge rate of a capacitor. By utilizing the time constant in a resistor–capacitor combination it is possible to time various operations. Inasmuch as many timers are now electronically based, it may be good to review the basic principles of operation of the $T = R \times C$ circuit.

1. A capacitor of 10 microfarads (μF) and a resistor of 1.5 MΩ are connected in series. What is the time constant of the circuit?
2. If a 0.5-μF capacitor and a 2-MΩ resistor are in series and are connected across a source of 120 V dc, how long after energizing the circuit will it take for the voltage across the capacitor to reach 75.84 V?
3. If a 5.0-μF capacitor and a 2-MΩ resistor in series are connected across a 250-V dc source, how long after energizing the circuit will it take for the voltage across the capacitor to reach 248 V?
4. What is the time constant of a 10-μF capacitor in series with a 1-MΩ resistor?
5. What is the time constant of a 9-μF capacitor in series with a 10-MΩ resistor?
6. What is the second time constant of a 10-μF capacitor and a 2-MΩ resistor connected in series?
7. What is the third time constant of a 10-μF capacitor and a 2-MΩ resistor connected in series?
8. What is the fourth time constant of a 10-μF capacitor and a 2-MΩ resistor connected in series?
9. What is the fifth time constant of a 10-μF capacitor and a 2-MΩ resistor connected in series?
10. How many time constants does it take for a capacitor to charge to its full value?

Sensors and Sensing

Objectives

After studying this chapter, you will be able to:

1. List the classes of sensors.
2. Describe the difference between presence and noncontact sensing.
3. Explain how limit switches are used as sensors.
4. Explain how speed sensing is accomplished.
5. Describe antiplugging.
6. Explain the operation of pressure controls.
7. Explain the operation of temperature sensors.
8. Describe the function of wells and packing glands.
9. Understand how float switches are used as sensors.
10. Understand how optical encoders operate.
11. Define BCD and LED.
12. Define multiplexing.
13. Explain the operation of various types of proximity switches.
14. Describe the functioning of photoelectric switches.
15. List the advantages of LEDs as a light source.
16. Describe the operation of RF identification systems.
17. Describe the operation of scanners.
18. Explain how the vision system operates.

Sensors are the foundation of the control system. They must provide usable input to the high-level controller. They must gather data and communicate the data, on a real-time basis, to the appropriate level within the control system. A wide range of sensing devices are available from many different technologies. Sensors include a broad range of technologies and physical configurations. They range from a simple electromechanical limit switch, which provides a single bit of information to the control system, to radio-frequency-activated tags storing up to 2K bytes of manufacturing information.

CLASSES OF SENSORS

Sensors may be classified as either contact or noncontact. They may be further classified as internal or external and as passive or active.

Contact Sensors

A limit switch is a *contact sensor* that permits a system to sense whether an object is present or missing. If the object makes contact with the limit switch, the system knows that the object is near enough to begin the next operation. If the switch is not closed, it means that the object is missing and the system must react accordingly. That usually generates what is referred to as an *alarm condition.*

Force, pressure, temperature, and tactile sensors all respond to contact. They all send their signals to another device or to a central location for processing.

Noncontact Sensors

Pressure changes, temperature changes, and electromagnetic changes can all be sensed by contact methods. They usually react to a change in a magnetic field or a light pattern. If an electromagnetic field is disturbed, it is sensed and fed to the controller. The same is true of the disturbance of a light beam. Changes in the light beam—its intensity or whether or not it is present—are sent to an electronic circuit or controller for processing and then sent back to a device so that it can react according to the disturbance.

PRESENCE SENSING

Presence sensors indicate whether a part or piece is in position. These are used to indicate an on-off condition. They are status devices that communicate a *single bit* of information. Limit switches, proximity switches, and photoelectric cells fall into this category. These devices are becoming smaller and faster, and have greater reliability, with increasing emphasis on noncontact sensing.

LIMIT SWITCHES

Perhaps the most commonly used type of sensing device is the limit switch. It has been around for many years and has been used in almost all electrically operated devices. This wide usage translates into many varied types of devices. Some are designed for general purposes and others are designed for specific operations. For instance, they may be designed for corrosion resistance, with sealed contacts that are oil-tight, or they may be the plug-in style or non-plug-in type. This type of switch has been designed as a rating cam limit type with programming capability.

Various types of sensing devices are attached to the limit switch (Fig. 10–1). The device may be a lever type with a spring return, a lever type with a maintained contact, a low-operating-torque type with a return spring, a push type with spring return, a wobble stick, or a cat whisker for making contact with the object being tested, counted, stopped, or started. Limit switches can also be designed for programmable operation. Figure 10–2 is a solid-state control device used with machinery and equipment having a repetitive operation cycle where motion can be correlated to shaft rotation. Its limits are controlled by three screwdriver locations (Fig. 10–3). Chapter 5 provides many examples of limit switches.

SPEED SWITCHES

Speed sensing can be used to sequence conveyors where it is necessary for one conveyor to be running at nearly full speed before a second conveyor is started. The switch can also be used to indicate in which direction material on a conveyor is moving from the rotation of a suitable driven shaft (Fig. 10–4). This switch is used in conjunction with automatic starters arranged for reversing or plugging duty, to provide plugging or antiplugging of squirrel-cage motors. The pilot device can be used as a speed-sending switch or indicate direction of rotation from the driven shaft.

Figure 10–5 shows a speed switch used in a plugging operation. A switch with normally open contacts is used. It is designed to interrupt reverse braking power automatically as the motor approaches zero speed. The speed at which the contacts operate can be adjusted so as to avoid coasting or reverse rotation of the motor. A speed switch can be wired for plugging in either or both directions.

Antiplugging can also be accomplished by using this switch. Then a switch with normally closed contacts is used. It is designed to keep the reverse contacts open until the machine being driven has slowed to a predetermined safe speed. At this speed the contacts are designed to close, permitting reversing or breaking by a designated method. The switch can be used for antiplugging in either or both directions.

Operation of the Speed Switch

When the shaft of the switch is rotated, a magnet induction linkage operates a contact. One contact is provided for forward operation and one for reverse. In plugging circuits, the forward or reverse contact is closed (depending on the direction of rotation) at any speed above the point at which the contact is set to operate. However, as the shaft speed is reduced, the electromagnetic torque holding the contact closed is also reduced, until a point is reached where the contact returns to the normally open position.

The contacts can be adjusted to determine the speed at which they will operate. This is done by altering two external screws: one for each set of contacts. After the operating temperature has been reached, the screw is turned to adjust the point at which the con-

ROLLER LEVER			
Type	**Material**	**Dia-meter**	**Width**
Non-Adjustable 3/4″ Radius	Nylon	3/4″	9/32″
	Metal	3/4″	17/64″
Non-Adjustable 1 1/2″ Radius	Nylon	3/4″	9/32″
	Nylon	3/4″	1″
	Steel	3/4″	1/4″
	Ball Bearing	3/4″	15/64″
	Beryllium Copper (Non-Sparking)	3/4″	9/32″
Non-Adjustable Roller on Back 1 1/2″ Radius	Nylon	3/4″	9/32″
	Nylon	3/4″	1″
	Nylon	1 1/2″	9/32″
	Steel	3/4″	1/4″
	Steel	3/4″	3/4″
Adjustable 1 3/16″–3″ Radius	Nylon	3/4″	9/32″
	Nylon	3/4″	1″
	Nylon	1 1/2″	9/32″
	Steel	3/4″	1/4″
	Ball Bearing	3/4″	15/64″
Fork Lever 1 1/2″ Radius	Nylon L.H. Roller on Front R.H. Roller on Back	3/4″	9/32″
	Steel L.H. Roller on Front R.H. Roller on Back	3/4″	1/4″
	Nylon Both Rollers on Front	3/4″	9/32″
	Nylon Both Rollers on Front	3/4″	1″
	Steel Both Rollers on Front	3/4″	1/4″
	Nylon L.H. Roller on Back R.H. Roller on Front	3/4″	9/32″
Micrometer Adjustment 1 1/2″ Radius	Nylon R.H. Adjustment	3/4″	9/32″
	Steel R.H. Adjustment	3/4″	1/4″
	Ball Bearing R.H. Adjustment	3/4″	15/64″
	Nylon L.H. Adjustment	3/4″	9/32″
	Steel L.H. Adjustment	3/4″	1/4″
	Ball Bearing L.H. Adjustment	3/4″	15/64″
	Nylon R.H. Adjustment	3/4″	1″
One-Way 1 1/2″ Radius	Nylon	3/4″	9/32″
	Steel	3/4″	1/4″
	Ball Bearing	3/4″	15/64″

ROLLER LEVER			
Type	**Material**	**Dia-meter**	**Width**
Non-Adjustable Offset 1 7/16″ Radius	Nylon	3/4″	9/32″
	Steel	3/4″	1/4″

ROD LEVER		
Type	**Material**	**Diameter**
	Stainless Steel Rod 5″ Long	1/8″
	Stainless Steel Rod 8 1/2″ Long	1/8″
	Stainless Steel Rod 11 1/2″ Long	1/8″
		5/64″
	Nylon Rod 12″ Long	1/4″
	Stainless Steel Rod 5″ Long	1/16″
	Stainless Steel Rod 5″ Long	1/16″
	Nylatron Looped Rod 6″ Long 2″ Wide Loop	3/16″

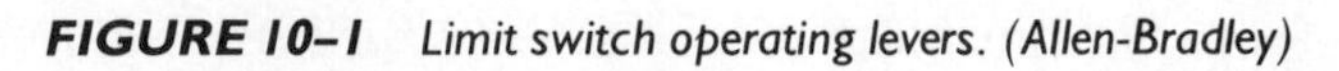

FIGURE 10–1 *Limit switch operating levers. (Allen-Bradley)*

FIGURE 10–2 *Programmable limit switch. (Allen-Bradley)*

FIGURE 10–3 *Rotating cam limit switch adjustment screws. (Allen-Bradley)*

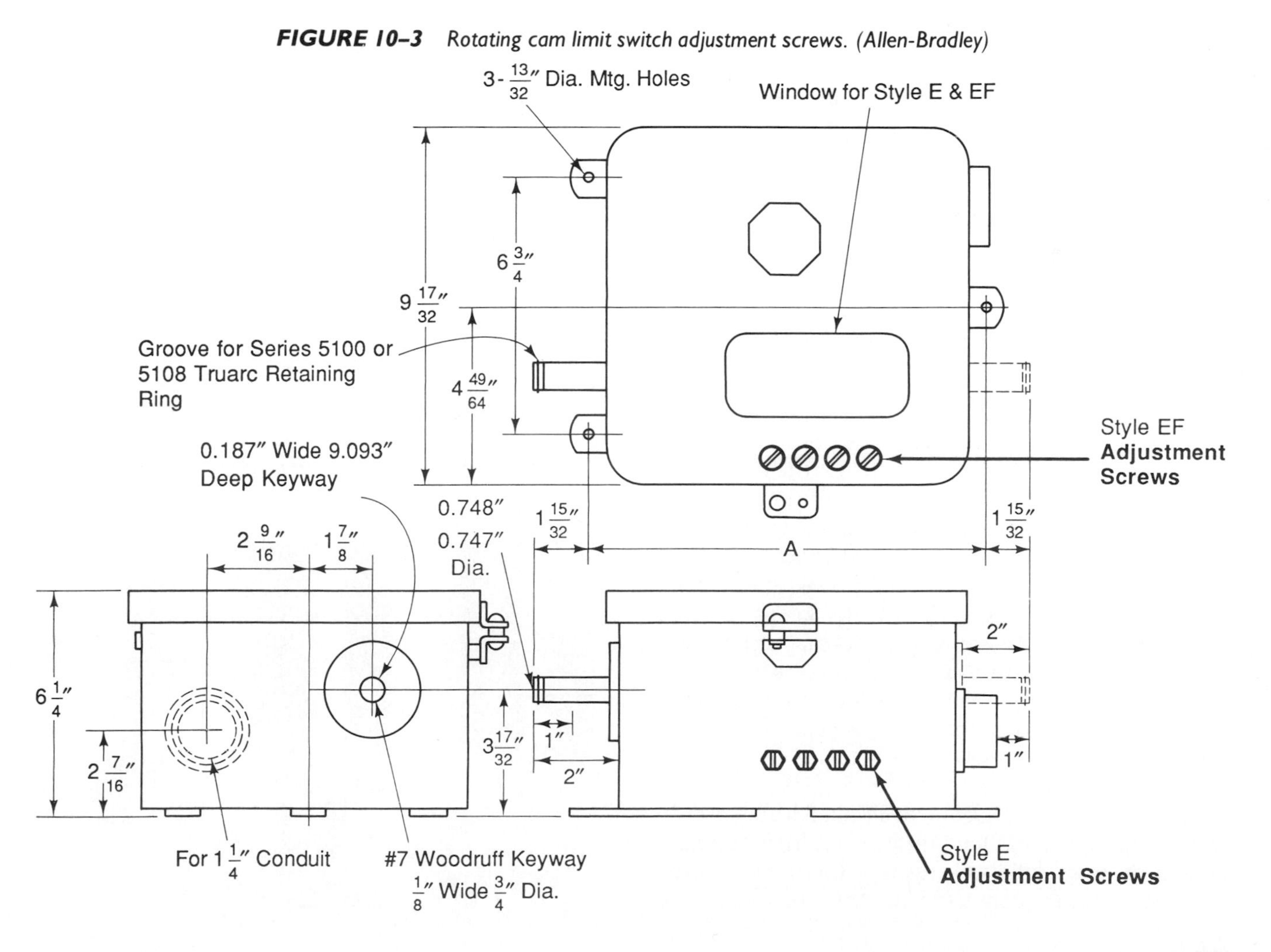

FIGURE 10–4 *Speed switches: (A) speed switch with lockout solenoid; (B) speed switch less lockout solenoid; (C) speed switches with various mountings. (Allen-Bradley)*

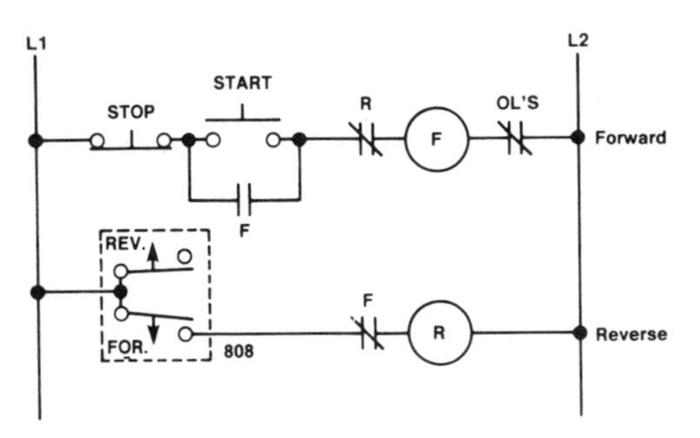

FIGURE 10–5 *Typical plugging circuit. (Allen-Bradley)*

tacts are to operate. Changes in inertia of moving equipment may require readjustment of the set points.

In some cases an accidental turn of the shaft may close the switch contacts and start the motor. To prevent this, the switch can be equipped with a lockout solenoid. The solenoid mechanically keeps the contacts from operating unless the lockout coil is energized. This is a factory-mounted feature or can be field mounted from a kit. If a lockout solenoid is used, a slight change in the wiring is necessitated. Figure 10–6 shows which terminals are used for the lockout (LO) solenoid.

In Figure 10–5 the control circuit for forward-direction plugging (with lockout protection) is optional. Operation is as follows—pushing the START button closes the forward contactor and the motor runs FORWARD. The normally closed contact (F) opens the circuit to the reverse contactor. The FORWARD contact on the speed switch *closes.*

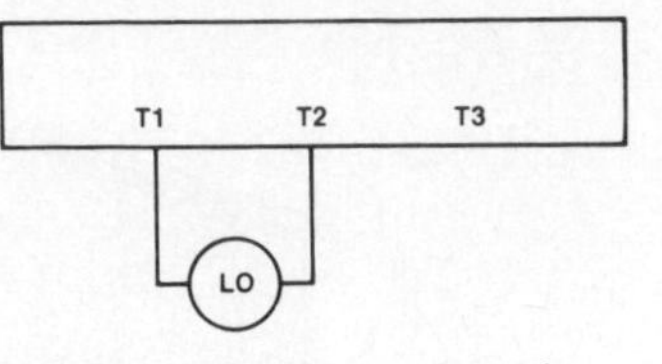

FIGURE 10–6 *Lockout solenoid connections. (Allen-Bradley)*

Pushing the STOP button drops out the forward contactor. The reverse contactor is energized and the motor is plugged. The motor speed decreases to the preset speed setting of the switch, at which point the speed switch contact *opens* and drops out the REVERSE contactor.

PRESSURE CONTROLS

Pressure controls operate on the principle of responding to changes in pneumatic (air or gas) or hydraulic (water or oil) pressure applied to a bellows or piston. This force is opposed by a main spring. Varying the force on the main spring (by turning the range adjustment screw) allows setting the contacts to trip at the upper pressure setting. Turning the differential adjusting screw (when provided) varies the force on a secondary spring and allows setting the lower pressure setting, where the contacts reset to their static state. Many ranges and differentials can be achieved by using bellows or pistons of different sizes to meet various requirements (Fig. 10–7).

Copper alloy bellows are used with air, water, oil, noncorrosive liquids, vapors, or gases in a series

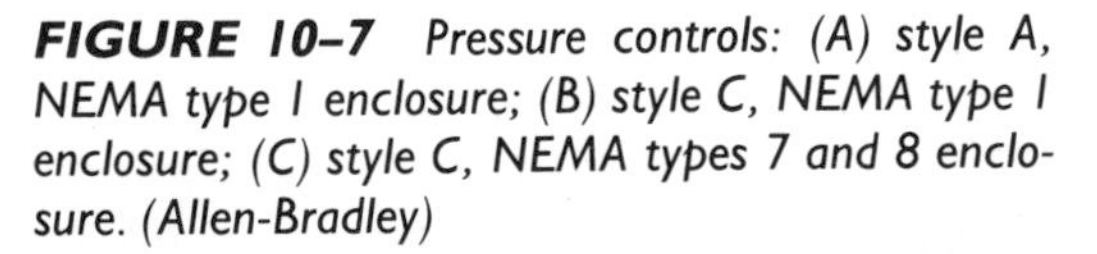

FIGURE 10–7 *Pressure controls: (A) style A, NEMA type 1 enclosure; (B) style C, NEMA type 1 enclosure; (C) style C, NEMA types 7 and 8 enclosure. (Allen-Bradley)*

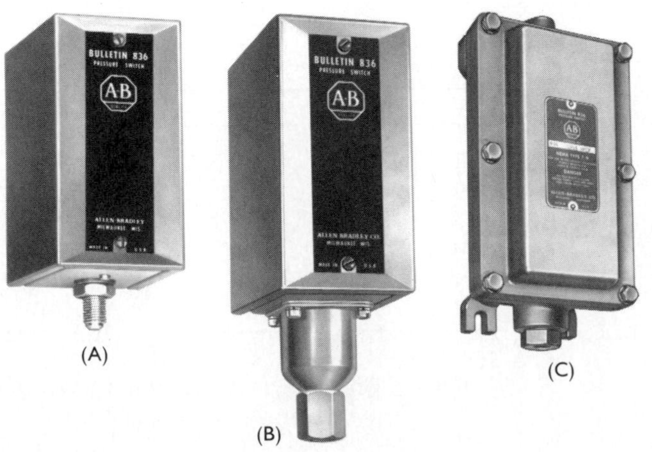

FIGURE 10–8 Contact block replacement kit for pressure controls. (Allen-Bradley)

of pressure ranges from 30 in. vacuum (Hg) to 900 psi. Stainless steel bellows are used with many of the more corrosive liquids or gases at pressures to 375 psi.

Contact blocks are single-pole, double-throw and can be wired to open or close on increasing or decreasing pressures. Contact blocks can be obtained as a replacement kit (Fig. 10–8).

Refrigeration-type controls may have low pressures ranging from 20 in. Hg vacuum to 120 psi. High-pressure cutout controls have ranges from 100 to 500 psi. They are available with or without maximum limited range stops to meet high-pressure safety adjustment requirements. Figure 10–9 shows the trip and reset pressure settings with minimum adjustment and maximum adjustment as well as differentials. Figure

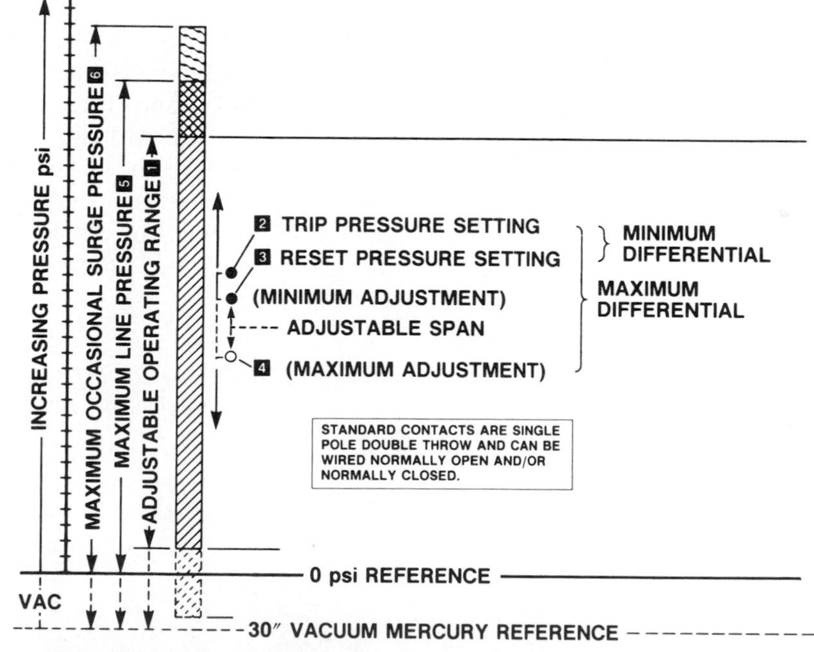

FIGURE 10–9 *Tripping points for pressure controls: 1, The total span within which the contacts can be adjusted to* TRIP *(2) and* RESET *(3) when adjusted to minimum difference. 2,* TRIP*—the higher pressure (temperature) setting; 3, the contacts change state.* RESET*—the lower pressure (temperature) setting 4, the contacts return to their normal state. [Note: Differential is the difference between* TRIP *(2) and* RESET *(3) (minimum adjusted setting) to (4) (maximum adjusted setting).] 5, The maximum sustained pressure that can be applied to the bellows without permanent damage. The control should not be cycled at this pressure. Note: Does not apply to piston-type controls. 6, A transient(s) [pulse(s)] that can occur in a system prior to reaching a steady state condition, expressed in milliseconds. Complex electronic instrumentation is required to measure the varying amplitude, frequency, and duration of this waveform. Frequent occurrence of extreme surge pressures could reduce bellows life. Surge pressure within published value generated during startup or shutdown of a machine or system, not exceeding eight times per day, are negligible. 7, Vac.—vacuum (negative pressure) inches of mercury. 8, Copper alloy bellows may be used on water, air, and other liquids or gases not corrosive to this alloy. Type 316 stainless steel bellows are available for the more corrosive liquids or gases. (Allen-Bradley)*

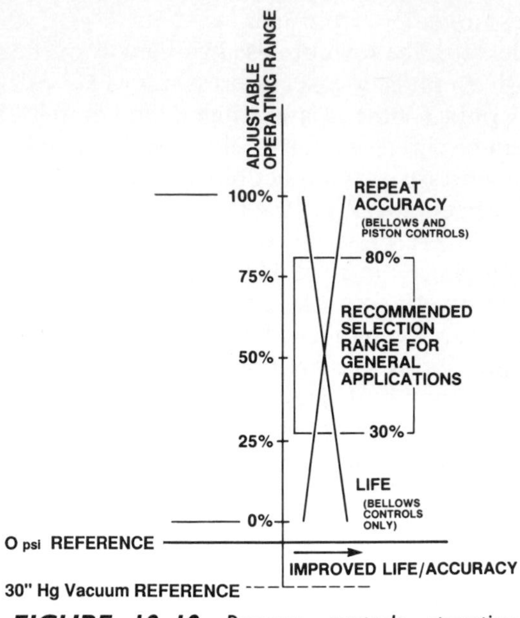

FIGURE 10–10 *Pressure controls operating points. (Allen-Bradley)*

10–10 shows the selection range for general applications. Bellows are used in low-pressure ranges up to 650 psi. Piston assemblies are used in high-pressure ranges up to 5000 psi. A diaphragm is used on the vacuum controls.

TEMPERATURE CONTROLS

Temperature controls may have direct vertical immersion bulbs for sensing changes in temperature or they may have remote capillary and bulb (Fig. 10–11). Temperature controls are similar to pressure controls. They differ in that a closed chemically filled bellows system is used. The pressure in the system changes in proportion to the temperature of the bulb. The temperature response medium in the system is a liquid whose vapor pressure increases as the temperature of the bulb rises. As the temperature of the bulbs falls, the vapor pressure decreases. The pressure change is transmitted to the bellows through a capillary tube operating the control at a predetermined setting. Temperature controls using this accurate and long-life vapor-pressure method of sensing temperatures are available in a series of ranges from −150 to +570°F.

Armored capillary is available for added protection of all bulb and capillary-type temperature controls. Copper bulb and capillary, bronze armor, and controls with stainless steel bulb and capillary are available for various environments.

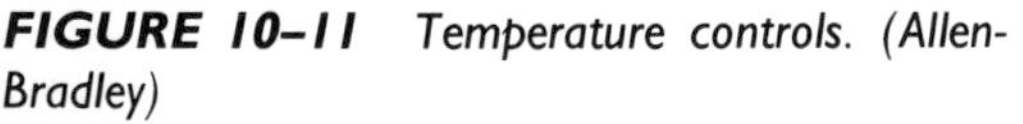

FIGURE 10–11 *Temperature controls. (Allen-Bradley)*

Wells and Packing Glands

Wells are used to mount and protect as well as permit removal of the bulb when necessary without discharging the system. Glands form seals at any desired position along the standard capillary. Armored capillary wells and glands include a setscrew to hold the armor so as not to expose the capillary. Wells are also available with a retaining nut to secure the bulb (Fig. 10–12). Figure 10–13 shows some contact blocks for these controls.

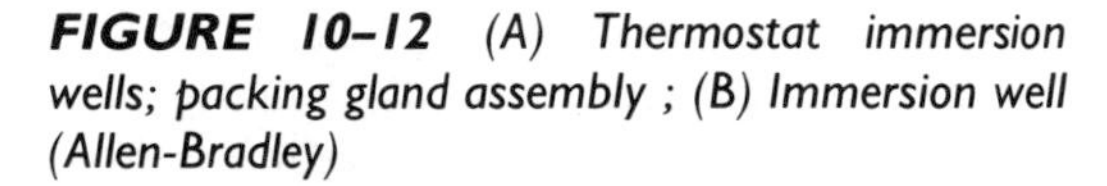

FIGURE 10–12 *(A) Thermostat immersion wells; packing gland assembly ; (B) Immersion well (Allen-Bradley)*

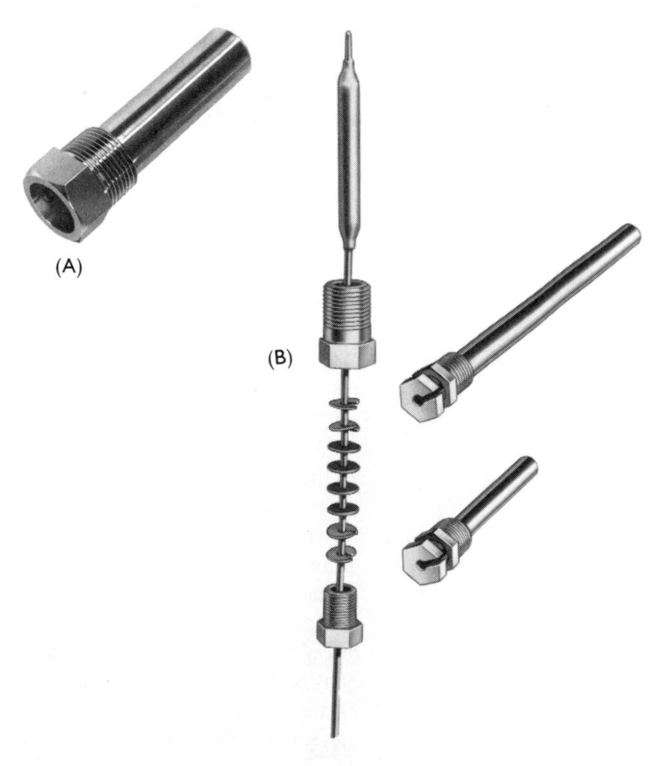

CONTACT BLOCKS

Symbol: Pressure Controls	Symbol: Temperature Controls	Description	Rating
		AUTOMATIC OPERATION	
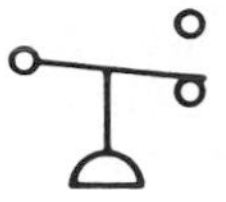	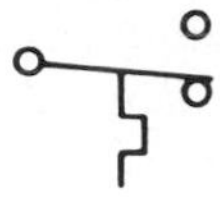	Single pole double throw — automatically opens or closes on rise or fall.	Non-inductive 5A, 240V 3A, 600V Control Circuit Rating AC-125VA, 24 to 600V DC-57.5VA, 115 to 230V
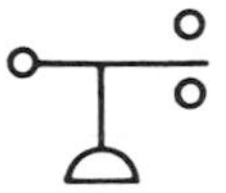	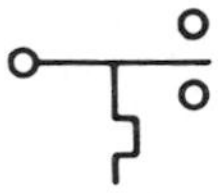	Single pole double throw — slow acting contact with no snap action. Contacts close on rise and close on fall with an open circuit between contact closures.	Control Circuit Rating AC-125VA, 24 to 250V
		Single pole single throw, normally open — closes on rise.	1 H.P., 230V AC .5 H.P., 115V AC Control Circuit Rating AC-125VA, 24-110V AC-345VA, 110-600V DC-57.5VA, 110-250V
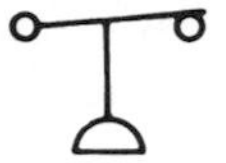	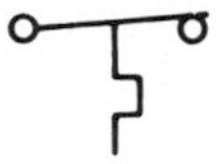	Single pole single throw, normally closed — opens on rise.	
		Single pole single throw, normally open — closes on rise.	1 H.P., 115V AC 1.5 H.P., 230V AC Control Circuit Rating AC-600VA, 110-600V DC-57.5VA, 110-250V
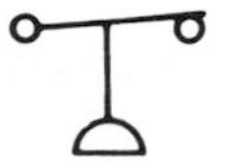	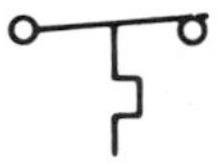	Single pole single throw, normally closed — opens on rise.	
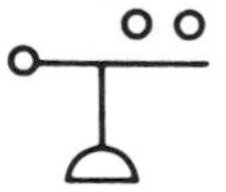	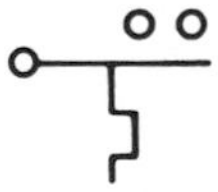	Two circuit, single pole single throw, normally open — a common terminal is connected to 2 separate contacts which close on rise.	Non-inductive 5A, 240V 3A, 600V Control Circuit Rating AC-125VA, 24 to 600V DC-57.5VA, 115 to 230V
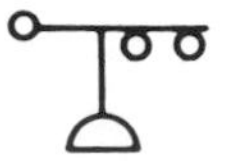		Two circuit, single pole single throw, normally closed — a common terminal is connected to 2 separate contacts which open on rise.	
		MANUAL RESET	
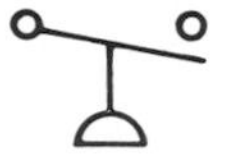	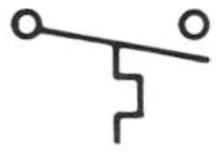	Single pole single throw, normally open — contacts open at a predetermined setting on fall and remain open until system is restored to normal run conditions at which time contacts can be manually reset.	Non-inductive 5A, 240V 3A, 600V Control Circuit Rating AC-125VA, 24 to 600V DC-57.5VA, 115 to 230V
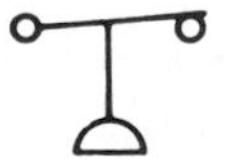	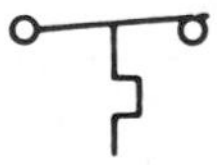	Single pole single throw, normally closed — contacts open on rise and remain open until system is restored to normal run conditions at which time contacts can be manually reset.	
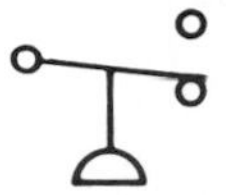	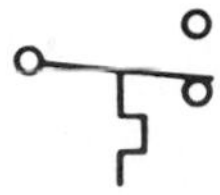	Single pole double throw, one contact normally closed — contact opens on rise and remains open until system is restored to normal run condition at which time contact can be manually reset. A second contact closes when the first contact opens.	

FIGURE 10–13 *Pressure and temperature controls. (Allen-Bradley)*

FLOAT SWITCHES

Float switches are also sensing devices. They provide automatic control for motors that pump liquids from a sump or tank. They have both motor and pilot duty ratings (Fig. 10–14).

Float switches have a snap-action mechanism for quick-make and quick-break contact operation. This feature provides high snap-through forces once the mechanism has traveled the required distance. This type of switch is available in copper, brass, and stainless steel to allow for wall or floor mounting to accommodate different tank or sump depths. Figure 10–15 shows various types of float switch arrangements.

FIGURE 10–14 *Float switches: (A) style A, wall or floor mount for tank or sump, NEMA type 4; (B) style C and (C) style B for one- or three-phase or dc; (D) style D and DS. (Allen-Bradley)*

(A)

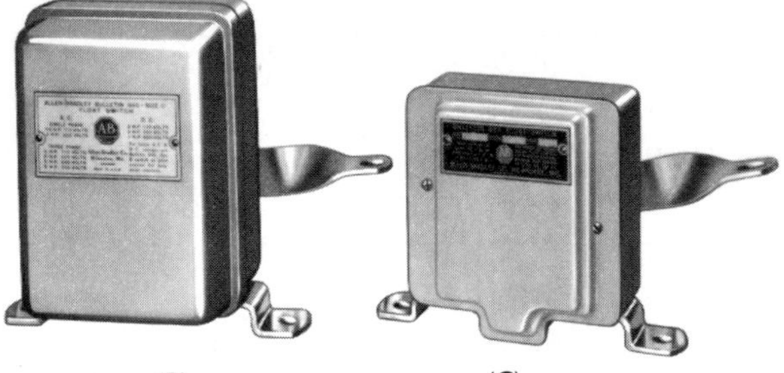

(B) (C)

(D)

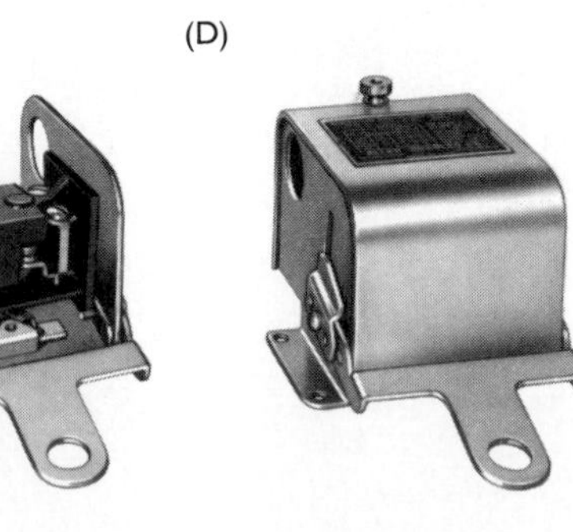

ENCODERS

Industrial optical encoders are designed to convert mechanical shaft rotation to an accurate electrical output in binary-coded decimal (BCD) or optional Gray code format. *Gray code* is defined as sequential numbers by binary values in which only one value changes at a time. This type of encoder (Fig. 10–16) is designed for use in industrial environments. It is a noncontacting optical design that uses high-speed, low-torque operation. Long life and good reliability are a result of using electronic components. They include a light-emitting diode as the source of light. A metal code disk is used with the single-LED light source, utilizing a fiber-optic light guide. Operating speed for BCD is 800 rpm and 2000 rpm for Gray code.

Many digital circuits, including some combinational logic circuits, are designed to handle BCD data. Binary-coded decimal is a code that uses a *4-bit binary number* to represent each digit in the decimal system. This is a conventional code for taking decimal information from a device such as a calculator keyboard and converting it to binary information for processing by digital circuits. It is equally useful for converting the binary output from digital circuits to decimal information for displaying on an output device such as a seven-segment light-emitting diode (LED) or liquid-crystal display (LCD) (Table 10–1). This information generated by the encoder can be fed to circuits that are designed to utilize the information and present it as a speed in rpm, and it can also be utilized to control the speed of rotation by adjusting the motor controller output.

Optical Programmable Controller Encoders

Encoders are also designed for interface with programmable controllers. The encoder contains all the

TABLE 10–1 BCD CODE TO DECIMAL DIGIT

BCD Code	*Decimal Digit*
0000	0
0001	1
0010	2
0011	3
0100	4
0101	5
0110	6
0111	7
1000	8
1001	9

DESCRIPTION	WALL MOUNTED	FLOOR MOUNTED
Single Arm Lever The float is fixed to one end of the rod. Adjustable stop collars at the top of the rod operate the switch. The rod may need guides to stabilize vertical float movement. **Maximum Rod Lengths** **Style / Copper/Brass / St. Steel** A / 9-ft. / 6-ft. B / 18-ft. / 6-ft. C / 18-ft. / 6-ft. D, DS / 9-ft. / 6-ft.	 *Includes two 3-foot lengths of ⅜-inch tubing with couplings, stop collars and float.*	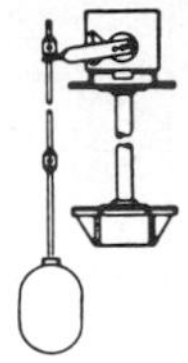 *Includes floor mounting bracket, 20-inch length of 1-inch pipe, mounting accessories, two 3-foot lengths of ⅜-inch tubing with couplings, stop collars and float.*
Double Arm Lever The double arm has a counterweight to offset weight of rod and float. The float moves up and down between stops on the rod so that large changes in liquid level move the rod only a short distance. Top of rod is fixed to switch lever. The rod may need guides to stabilize vertical float movement. **Maximum Rod Lengths** **Style / Copper/Brass / St. Steel** A / 33-ft. / 6-ft. B / 33-ft. / — C / 33-ft. / —	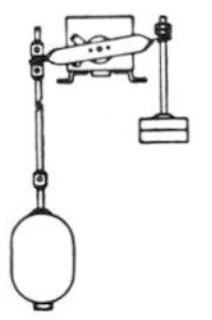 *Includes double arm lever for float switch, counterweight, two 3-foot lengths of ⅜-inch tubing with couplings, stop collars and float.*	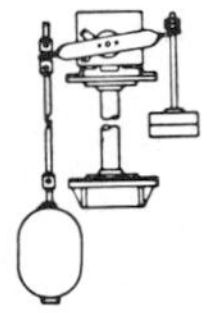 *Includes double operating arm for float switch, floor mounting bracket, 20-inch length of 1-inch pipe, mounting accessories, two 3-foot lengths of ⅜-inch tubing with couplings, stop collars and float.*
Double Parallel Arms Used with unguided rods up to 33-ft. long. The parallel levers keep the rod vertical and help limit sideways movement. The float moves up and down between stops on the rod so that large changes in liquid level move the rod only a short distance. Top of rod is fixed to switch lever. **Maximum Rod Lengths** **Style / Copper/Brass / St. Steel** A / 33-ft. / 6-ft. B / 33-ft. / — C / 33-ft. / —	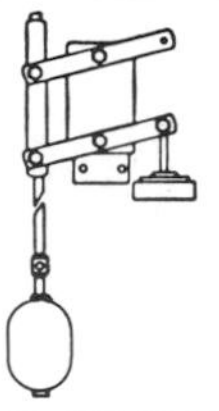 *Includes two double arms for float switch, counterweight, two 3-foot lengths of ⅜-inch tubing with couplings, stop collars and float.*	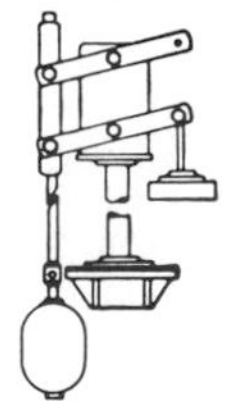 *Includes two double arm levers for float switch counterweight floor mounting bracket, 20-inch length of 1-inch pipe, mounting accessories, two 3-foot lengths of ⅜-inch tubing, stop collars and float.*

FIGURE 10–15 *Automatic float switches. (Allen-Bradley)*

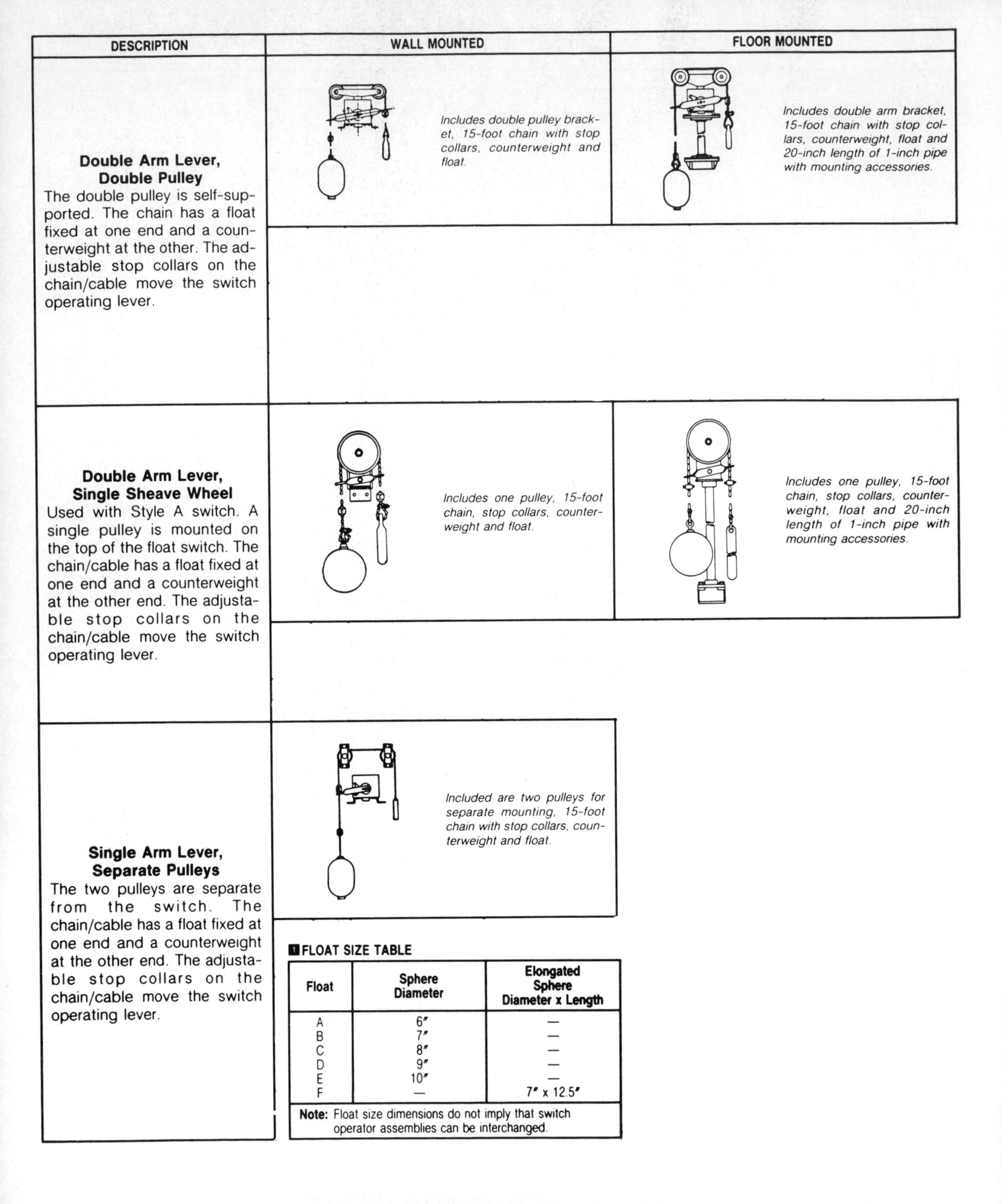

DESCRIPTION	WALL MOUNTED	FLOOR MOUNTED
Double Arm Lever, Double Pulley The double pulley is self-supported. The chain has a float fixed at one end and a counterweight at the other. The adjustable stop collars on the chain/cable move the switch operating lever.	*Includes double pulley bracket, 15-foot chain with stop collars, counterweight and float.*	*Includes double arm bracket, 15-foot chain with stop collars, counterweight, float and 20-inch length of 1-inch pipe with mounting accessories.*
Double Arm Lever, Single Sheave Wheel Used with Style A switch. A single pulley is mounted on the top of the float switch. The chain/cable has a float fixed at one end and a counterweight at the other end. The adjustable stop collars on the chain/cable move the switch operating lever.	*Includes one pulley, 15-foot chain, stop collars, counterweight and float.*	*Includes one pulley, 15-foot chain, stop collars, counterweight, float and 20-inch length of 1-inch pipe with mounting accessories.*
Single Arm Lever, Separate Pulleys The two pulleys are separate from the switch. The chain/cable has a float fixed at one end and a counterweight at the other end. The adjustable stop collars on the chain/cable move the switch operating lever.	*Included are two pulleys for separate mounting, 15-foot chain with stop collars, counterweight and float.*	

FLOAT SIZE TABLE

Float	Sphere Diameter	Elongated Sphere Diameter x Length
A	6″	—
B	7″	—
C	8″	—
D	9″	—
E	10″	—
F	—	7″ x 12.5″

Note: Float size dimensions do not imply that switch operator assemblies can be interchanged.

FIGURE 10–15 *continued*

FIGURE 10–16 *Optical single-turn absolute encoder. (Allen-Bradley)*

electronics necessary to provide a latched output on command from a programmable controller. A data-ready output signals the programmable controller when the encoder data are latched. The encoders are capable of being multiplexed, allowing one programmable controller with one set of input cards/modules to accept data from many encoders. The multiplex and latch circuits operate independently.

Multiplexing

Placing a logic zero (0) on the multiplex control line will force all outputs to high-impedance state, allowing the programmable controller to scan multiple encoders using one set of input cards/modules. A multiplexer is a type of combinational digital circuit that is readily available in an IC package. Multiplexing allows a single conductor (bus line) to carry signals alternately from a variety of signal sources. Multiplexers are also called data selectors. They operate like an electrically controlled rotary switch. The output line is like a pole of a rotary switch, and the input lines are like the positions of a rotary switch. Data selectors are available with as many as 16 input lines.

PROXIMITY SWITCHES

Proximity switches are designed for industrial environments in places where it is required to sense the presence of metal objects (ferrous and nonferrous) without touching them. These are self-contained two-wire devices used in 120-V ac control circuits (Fig. 10–17).

Operation

Proximity switches are designed for general-purpose use, with a high output rating of 1 A maximum continuous and 10 A maximum inrush. They are capable of energizing such external loads as relays, contactors, motor starters, and solenoids. Switches with programmable normally open/normally closed output have

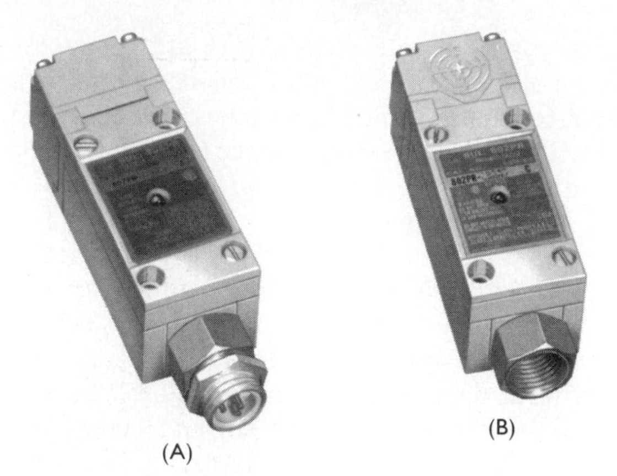

FIGURE 10–17 *Self-contained proximity switches: (A) top sensing; (B) front sensing. (Allen-Bradley)*

two LED indicators: a power LED indicator that is ON when power is applied to the device, and an output LED indicator that is ON when the switch output is conducting current.

Solid-State Switches

Switches for solid-state applications interface directly with programmable controllers, hard-wired solid-state logic, and similar high-impedance loads. In addition, they have circuit protection against overload and short-circuit conditions. In switches with a normally open fixed output, the LED indicator is ON when a metal object is within the sensing field and the load is energized.

Self-contained proximity switches include both front-sensing and top-sensing types. The front-sensing version can be arranged to sense in any of four directions: front, rear, or either side. The switches come with a conduit coupler, threaded conduit opening, prewired receptacle, or prewired cable base.

Inductive Cylindrical Switch

Inductive cylindrical switches are self-contained and solid state (Fig. 10–18). They are also designed for in-

FIGURE 10–18 *Inductive cylindrical proximity switches. (Allen-Bradley)*

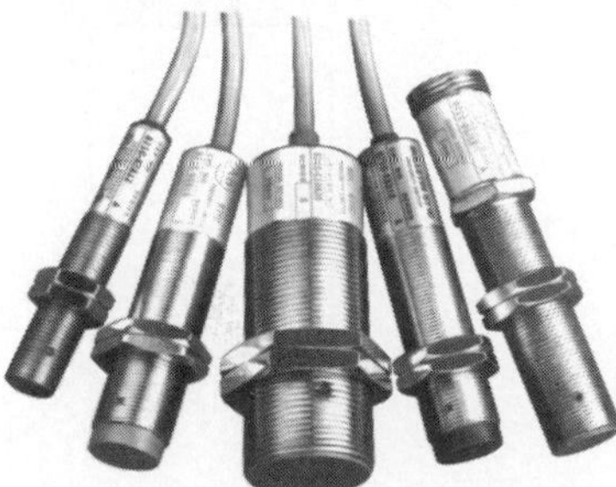

dustrial environments where it is required to sense the presence of metal objects, both ferrous and nonferrous, without touching them. The switches are equipped with a red LED to indicate the presence (NO output configuration) or absence (NC output configuration) of a target in the sensing field.

This type of switch interfaces with programmable controller input modules without the use of an external loading resistor. Switches can also be used to energize relays, contactors, and motor starters. The switch housing is made of nickel-plated brass. The electronic circuitry is potted for protection against shock, vibration, and contaminants.

Extended-Sensing-Range Inductive Proximity Switch

The extended-sensing-range switch is also self-contained and solid state (Fig. 10–19). It is designed for industrial use such as materials handling (transfer lines, roller-belt conveyors, and the like). The position of objects traveling these lines cannot always be accurately controlled, therefore causing the need to sense an extended distance. This switch interfaces with programmable controllers without the use of loading resistors.

FIGURE 10–19 *Extended range inductive proximity switch. (Allen-Bradley)*

Compact Inductive Proximity Switch

The compact inductive proximity switch is square, is PC compatible, and has a 15-mm sensing range (Fig. 10–20). The two-wire device energizes and deenergizes an external load when metal targets are sensed.

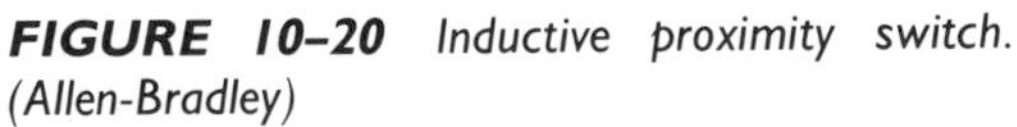

FIGURE 10–20 *Inductive proximity switch. (Allen-Bradley)*

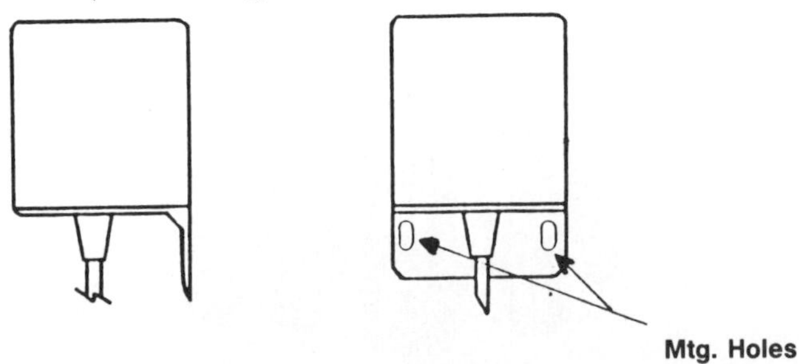

Self-Contained Low Profile Proximity Switch

The self-contained low profile proximity switch will operate with PCs and sense the presence of metal objects (Fig. 10–21). There are two types: weld field and general purpose. The weld-field type has a built-in tolerance to electromagnetic fields generated by resistance welding equipment and can be mounted within 1 in. of a 20,000-A current-carrying bus.

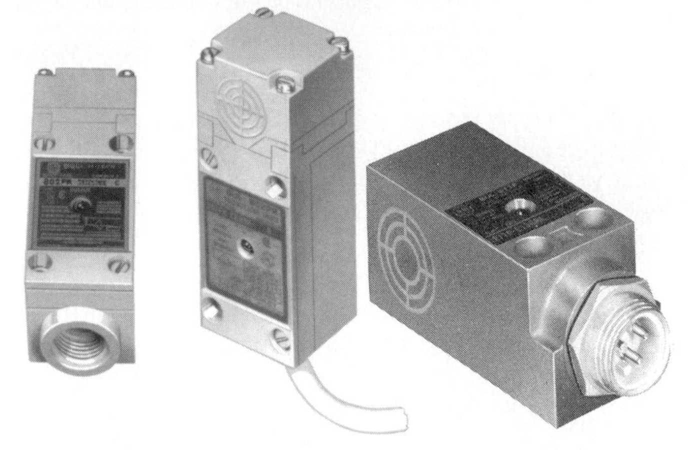

FIGURE 10–21 *Self-contained proximity switches. (Allen-Bradley)*

PHOTOELECTRIC SWITCHES

The photoelectric industrial switch is self-contained and has solid-state electronics as part of the package (Fig. 10–22). The industrial-type photoelectric switch is capable of sensing the presence of a wide variety of objects without making contact.

The switch shown in Fig. 10–22 uses a modulated infrared LED light source-detection system. It is modulated or pulsed in the near-infrared portion of the electromagnetic spectrum. This switch is highly tolerant of ambient light, including sunlight.

Operation

This switch is a retroflective device that operates loads between 10 and 30 V dc. It can be wired in *current sink* or *current source* mode, for energizing a variety of loads. The operating range is 18 feet for a 3-in. reflector. Other types use the through-beam photoelectric system. The through-beam pair of 120-V ac devices is especially well suited for demanding applications that require a maximum range of 25 feet. These switches can be used to operate relays, contactors, and motor starters and interface directly with most programmable controllers. The output condition can be programmed between *light* and *dark* operation by means

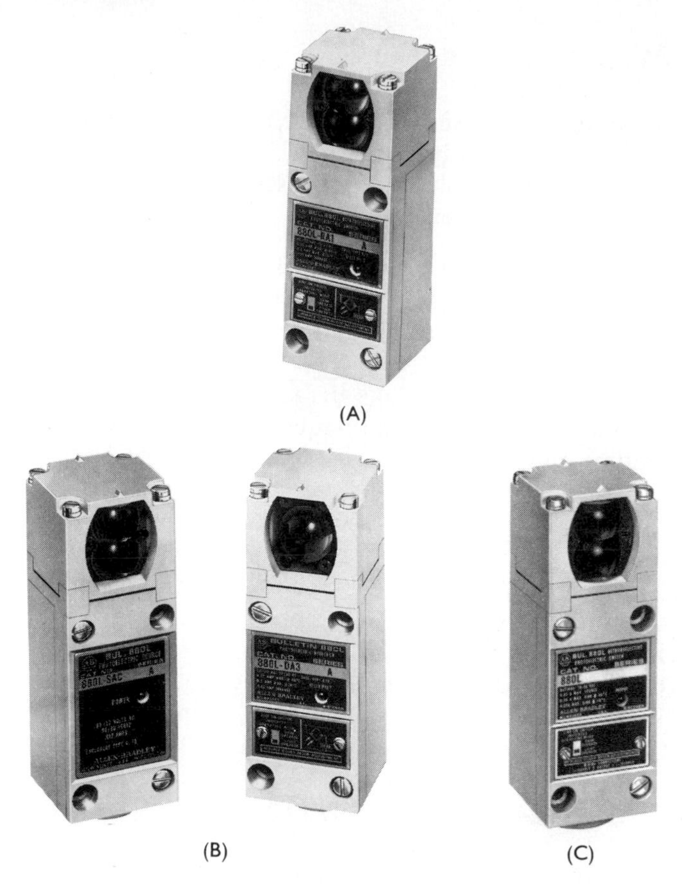

FIGURE 10–22 *Self-contained photoelectric switches: (A) type RA uses a modulated infrared LED light source and detector; (B) type SAC source (left) and type DA detector (right); (C) type RL operates loads between 10 and 30 V dc. (Allen-Bradley)*

of a recessed rocker switch. A rotatable sensing head can be arranged to sense in any of four directions.

Features

Reliable performance is the needed feature on any sensing operation or system. One means of ensuring proper operation of a photoelectric switch is to use a stability indicator. A green stability LED is provided

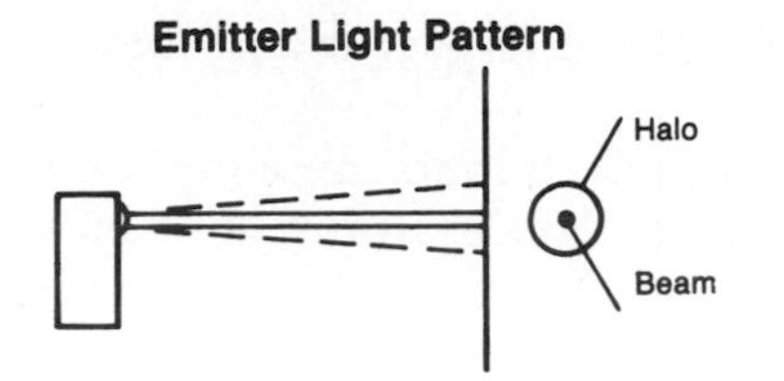

FIGURE 10–23 *Emitter (LED) light pattern. (Square D)*

on some switches to indicate reliable system performance. An unstable indication from the LED means that the light intensity level at the receiver is staying within +20% of the switch trip point, resulting in marginal switch operation (Fig. 10–23). Common situations causing such an indication include:

1. Sensitivity adjustment set too low
2. Switch misalignment, particularly when the emitter halo rather than the emitter beam is detected (Fig. 10–23)
3. Target or reflector set too far from the switch
4. Dust or other residue buildup on the lens or reflector
5. Target not large enough to block the light beam completely
6. Insufficient contrast between the target and background (in diffusing applications)

This setup and use of a photoelectric switch system without a stability indicator can result in premature system failure. Failure modes can include erratic tripping or no switching response.

Figure 10–24 shows selectable light-dark operation of the photoelectric switch. Polarized lenses are found only on retroreflective devices; a photoelectric switch with polarized lenses detects only light signals bounced off a retroreflector. The switch is immune to any undesirable light reflection from targets such as metals, foils, mirrors, tiles, shrink wrap, and shiny plastics.

FIGURE 10–24 *Selectable light/dark operation. (Square D)*

LIGHT/DARK OPERATE DEFINED

	Thru Beam Types	Retroreflective Types	Diffuse Types
DARK OPERATE — output energized when light is NOT incident on the receiver	Target	Target	
LIGHT OPERATE — output energized when light is incident on the receiver			Target

Photoelectric Light Sources

Generally, switch emitters are either an incandescent light or an LED light source. LED light sources have many advantages compared to incandescent. LEDs have:

1. Longer life
2. Virtually no heat dissipation
3. Resistance to shock and vibration
4. Longer sensing ranges
5. Cost-effectiveness
6. The ability to be pulsed at high frequencies

Pulsing an LED beam creates a stronger light beam capable of longer sensing ranges. By modulating the pulse frequency, emitters can be keyed to the light receiver. This helps eliminate ambient-light interference problems.

LEDs are available in four basic color bands: infrared, red, green, and yellow. All but yellow are used as photoelectric switch emitters. Of the three color bands used, infrared LED emitters are found in a large majority of photoelectric switches. Infrared light does the best job of penetrating dust, fog, and other airborne particles that may interfere with a photoelectric beam.

Red LED emitters can be found in fiber-optic units using plastic fiber cable and in retroreflective units with polarized lenses. Mark sensors use either a red or green LED emitter.

Mark Sensors

Selecting a mark sensor with the proper LED emitter color depends on both the mark color and its background color. Table 10–2 is a useful guide for determining if a mark sensor will:

1. Require a green or red LED emitter
2. Function stably in the light-operate or dark-operate mode

Mark sensors detect (diffuse setting) color marks on most materials, both opaque and transparent (including film). They have either green or red emitter LEDs and can detect a wide variety of color marks on different backgrounds. Special optics and adjustable lensing allow precise setting of the switch focal point. This feature is useful in pinpoint detection (narrow visibility types) or for setting a precise limit to a detecting distance (wide-visibility types). Special applications for this device include detecting objects as shown in Fig. 10–25. A miniature speciality focusable diffuse switch is shown in Fig. 10–26.

Clear Material Detection

The retroreflective type photoelectric switch is ideal for transparent bottle detection. A special optical system permits stable detection of clear materials not previously possible with conventional retroflective photoelectric switches. The 1-meter sensing range of this switch makes the system flexible compared to stan-

TABLE 10–2 MARK SENSOR GUIDE[a]

Background Color	*Emitter LED Color*	*Color of Mark*						
		White	*Black*	*Transparent*	*Red*	*Yellow*	*Green*	*Blue*
White	Green	—	•	•	•	—	▲	•
	Red	—	ù	•	—	-	•	•
Black	Green	■	—	-	—	■	■	—
	Red	■	—	-	■	■	—	-
Transparent	Green	■	—	-	—	■	■	—
	Red	■	—	-	■	■	—	-
Red	Green	■	—	-	—	◆	■	•
	Red	—	ù	•	—	-	•	•
Yellow	Green	—	ù	•	▲	—	-	•
	Red	—	ù	•	—	-	▲	•
Green	Green	◆	•	•	•	—	-	▲
	Red	■	—	-	■	◆	—	-
Blue	Green	■	—	-	■	■	◆	—
	Red	■	—	-	■	■	—	-

[a]In the dark-operate mode;
•means detects the mark stably
▲means may detect the mark.
In the light-operate mode;
■ means detects the mark stably and
◆means may detect the mark.

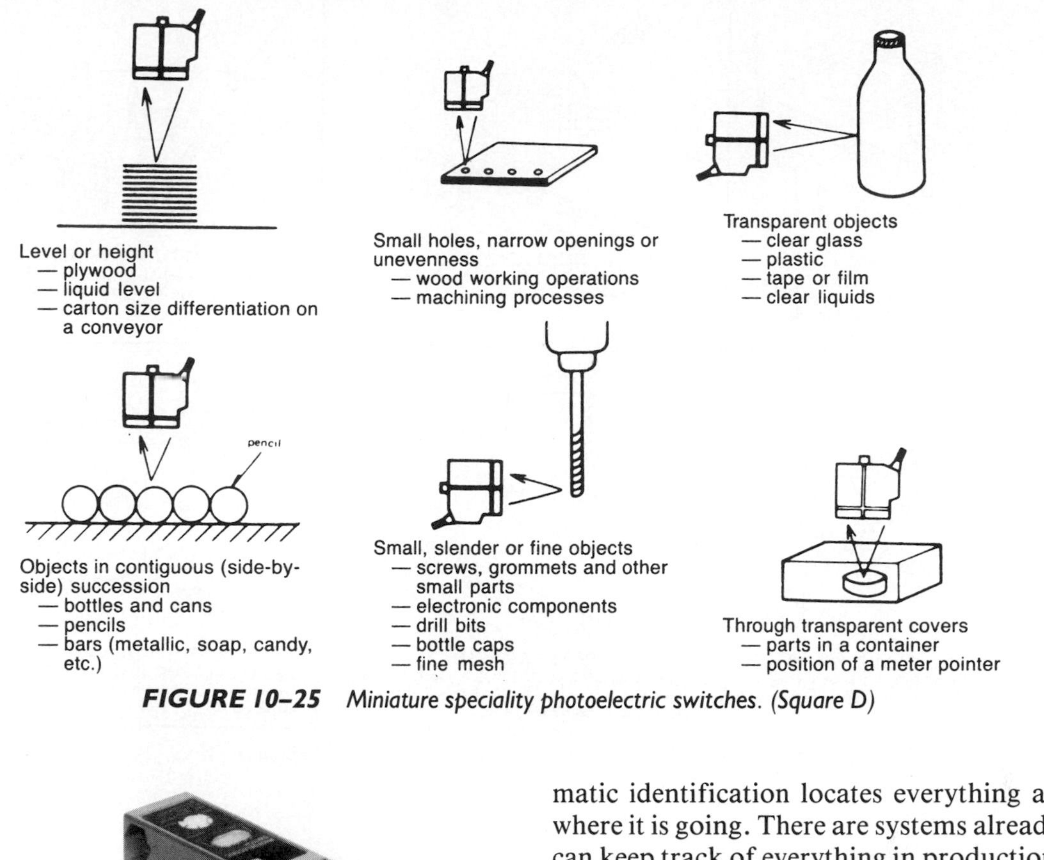

FIGURE 10–25 *Miniature speciality photoelectric switches. (Square D)*

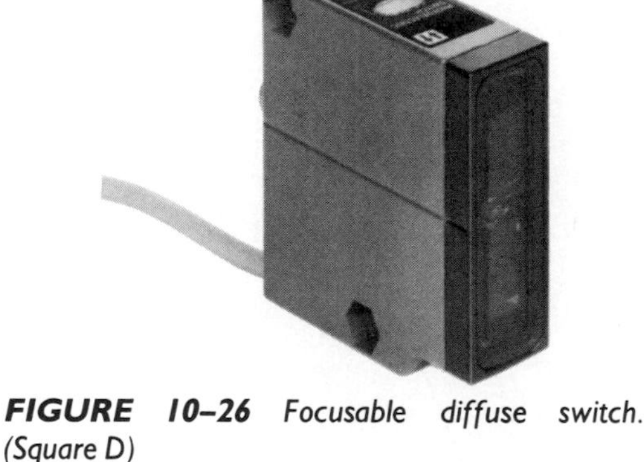

FIGURE 10–26 *Focusable diffuse switch. (Square D)*

dard diffuse photoelectric switches normally used in these applications.

Switch Applications

Figure 10–27 shows the typical photoelectric switch applications. Note the diffuse mode, specular mode, through-beam mode, retroreflective mode, and diffuse mode. Sensing ranges are based on clean lenses and a clean air environment. Air contaminants and dirty lenses affect switch performance.

AUTOMATIC IDENTIFICATION

Automatic identification keeps a control system or operation under scrutiny. It supplies the control system and data bases with information they need. Automatic identification locates everything and tells you where it is going. There are systems already tested that can keep track of everything in production. They consist of a radio-frequency (RF) identification system, bar code readers and decoders, and multiplexing.

All of these involve hardware, software, and networking communications. They require sensors made for special systems applications. Identification devices can provide a few characters of data in the case of a bar code system, or up to 2K bytes of data when a radio-frequency system is used.

RADIO-FREQUENCY IDENTIFICATION SYSTEMS

Figure 10–28 shows the Allen-Bradley radio-frequency (RF) system as it is set up for operation. The system keeps track of parts and follows them through the manufacturing process, without the need for a paper chase or factoring in an operator error. Data can be transmitted or received by an RF tag whenever they pass an antenna.

The system is based on radio waves that can penetrate plastic, wood, and various types of factory contaminants. It can be installed and operated in a wide range of industrial environments. It is dependable in transferring its information to a control system (Fig. 10–29). Note how the system consists of three main pieces of hardware: the antenna, RF tags, and the universal identification interface (UII).

Using this method it is possible to use a single

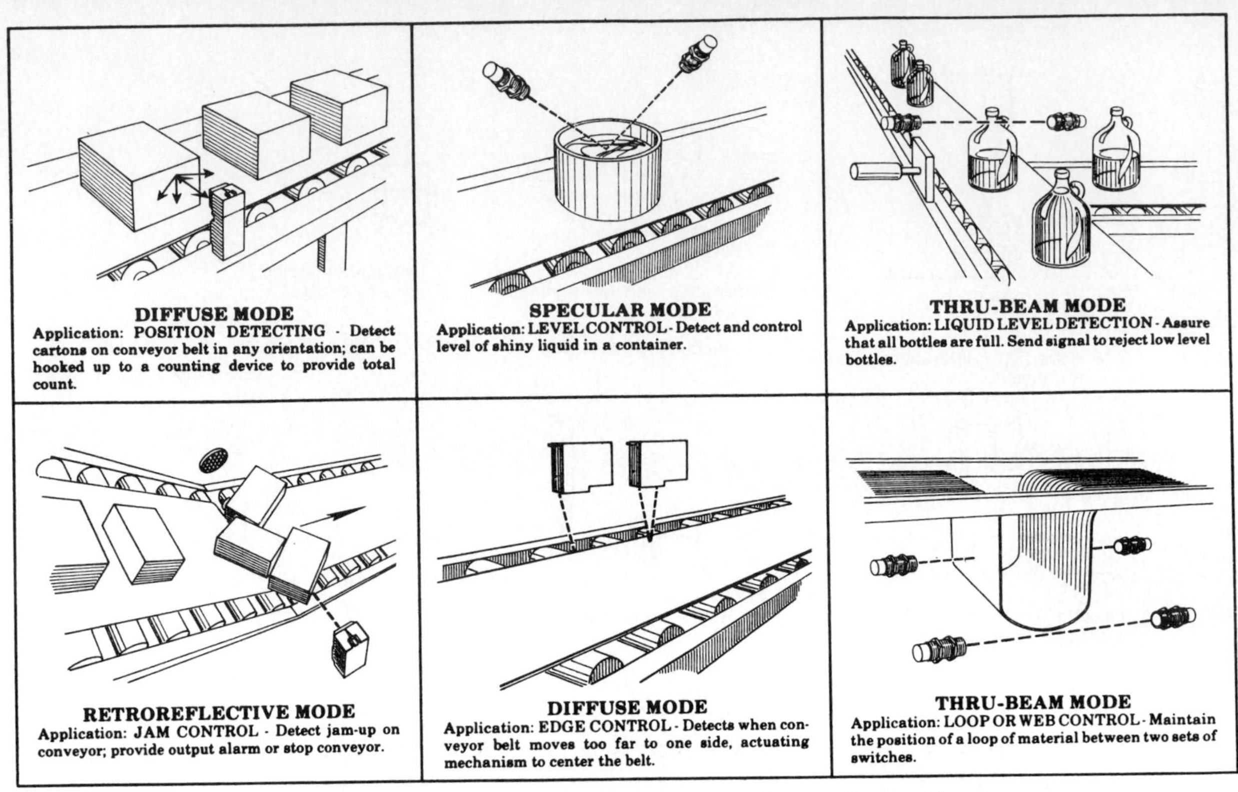

FIGURE 10–27 *Typical photoelectric switch uses. (Square D)*

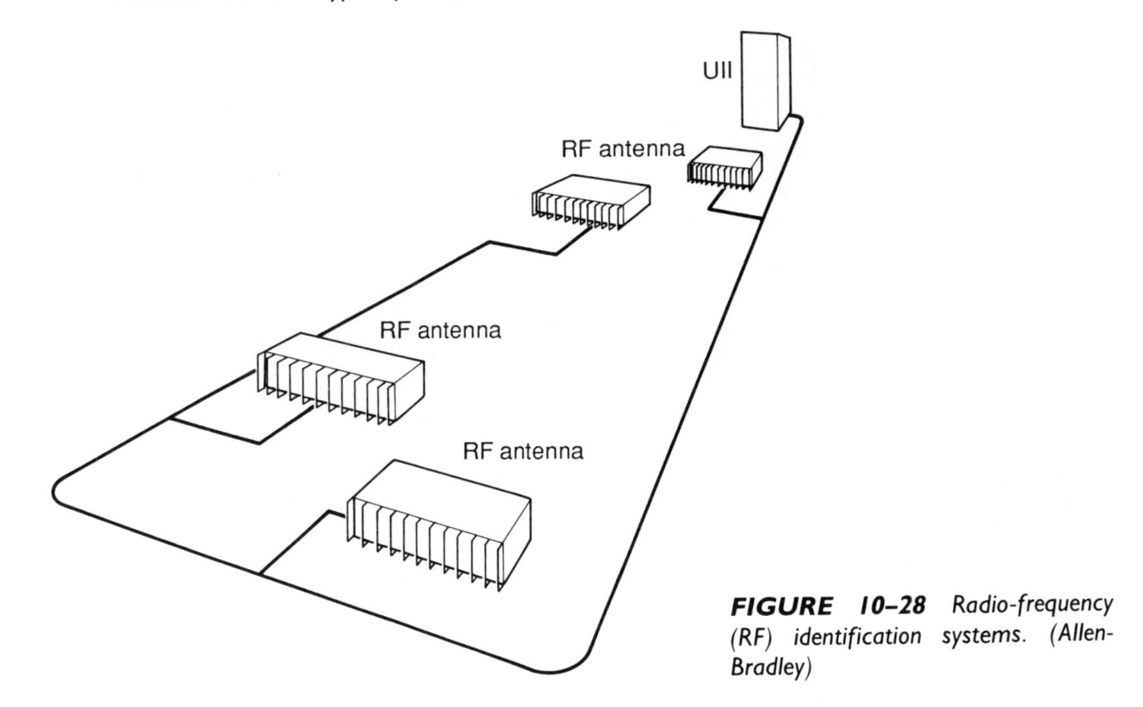

FIGURE 10–28 *Radio-frequency (RF) identification systems. (Allen-Bradley)*

method of product identification in all phases of production and processing. Data can be stored on a read/write RF tag and can be added to, altered, or deleted whenever the tag passes an antenna.

The Antenna

The antenna contains both the transmit and receive antennas (Fig. 10–30A). It decodes the data stored in the tags and transmits this information to the UII or the host, depending on the control architecture utilized. Housed in a die-cast aluminum enclosure, the unit can be installed on the shop floor or in a wide range of industrial environments.

The RF Tag

There are two RF tags available: the read/write and the programmable (Fig. 10–30B). The read/write tag is a small battery-powered tag that can store up to 2K

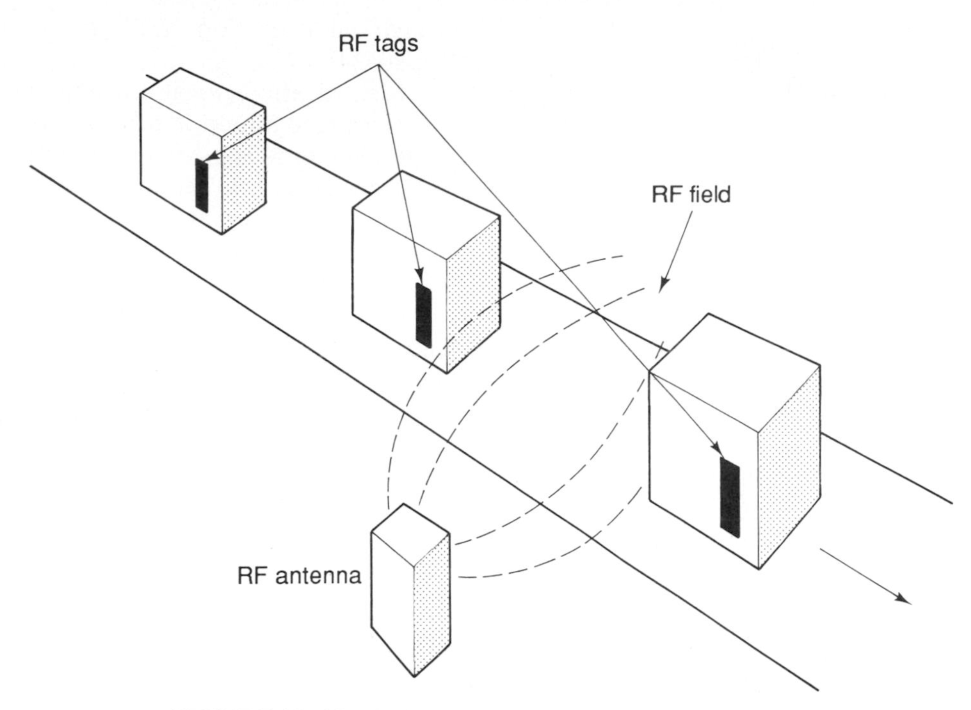

FIGURE 10–29 *Radio-frequency identification. (Allen-Bradley)*

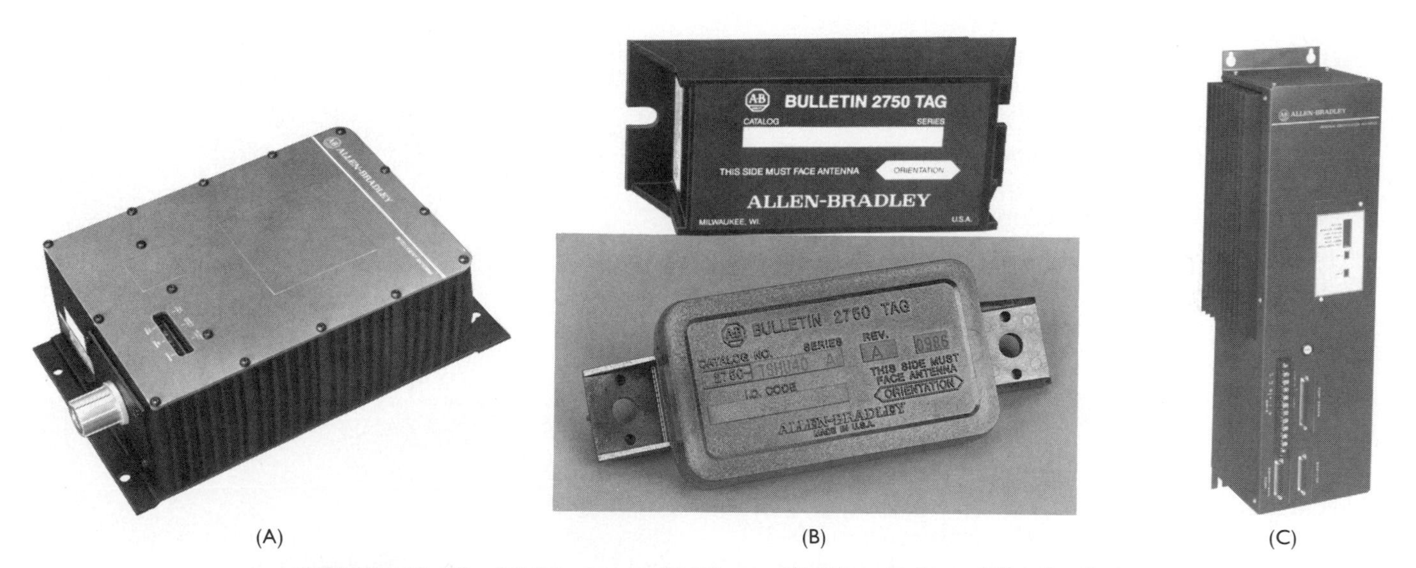

FIGURE 10–30 *(A) RF antenna; (B) RF tags; (C) UII multiplexer. (Allen-Bradley)*

bytes of data and is reusable. The tag is activated when it enters the antenna's RF field. At that time the information can be read from or written into the tag's memory (Fig. 10–30). The programmable tag is small and rugged. It is designed for industrial applications and is also reusable. The tag can be exposed to temperatures as high as 180°C for up to 1 hour, without damaging data stored in the tag. Up to 40 bytes of data can be programmed into the tag whenever the tag enters an antenna's RF field.

BAR CODE READERS AND DECODERS

Bar coder readers offer fast entry of data, with speed and accuracy not humanly possible through keyboard entry (Fig. 10–31 shows a typical bar code). Scanning devices can be stationary or portable. They can be automatic or manually operated. Allen-Bradley has a variety of scanners to meet industrial needs.

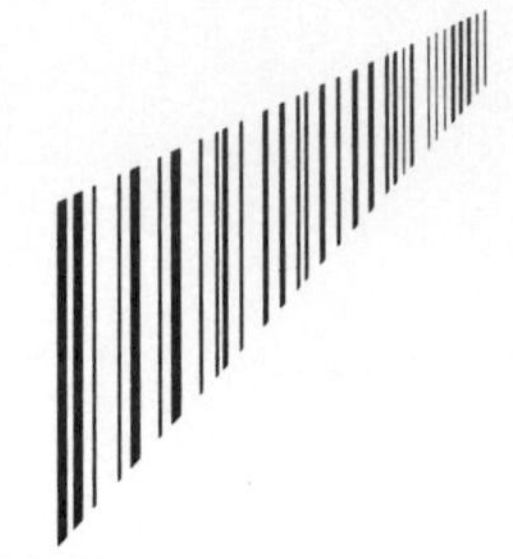

FIGURE 10–31 Bar code. (Allen-Bradley)

Hand-Held Scanners

This reader (Fig. 10–32A) accepts input from a variety of bar code scanners, offers high-performance decoding, and full autodiscrimination of most popular symbologies. Its RS-232 port makes it compatible with most computers. The reader works with either a wand or a moving-beam laser scanner. The *laser scanner* (Fig. 10–32B) is a noncontact device that can operate in harsh environments. It can be dropped on concrete and still survive to work again. The scanning rate is 36 scans per second. Labels can be decoded within 1 to 23 in. from the scanner, depending on the density of the bar code symbol.

Moving-Beam Scanners

This type of scanner (Fig. 10–32C) operates at 175 scans per second and can decode symbols up to 20 in. away. The scanner head is separate from the decoder console, so it can be installed in areas where space is a problem. The decoder automatically decodes Code 39, interleaved 2-of-5, UPC/EAN, and Codabar symbols.

MULTIPLEXING

The universal identification interface (UII) is a new idea in automatic identification control. With one UII you can multiplex up to eight sensors (Fig. 10–33). This type of sensor utilization makes it possible for programmable controllers and robots to function as intended.

VISION SYSTEM

Sensing involves the five senses: touch, feel, smell, taste, and seeing. The vision system makes use of the size and shape of objects for its operation. Of course, the ideal inspection rate is 100% at production-line speeds. With a vision system it is possible to inspect, visually, all products coming off a line. In fact, it is also possible to inspect each product as it passes through an operation heading toward completion (Fig. 10–34).

Identification, location, and sorting of objects by pattern recognition is possible with the utilization of video imagery and computer technology. The system offers a means of reducing product waste, inspection errors, and human fatigue. This inspection system can operate 24 hours a day and 7 days a week without endangering product integrity.

(A)

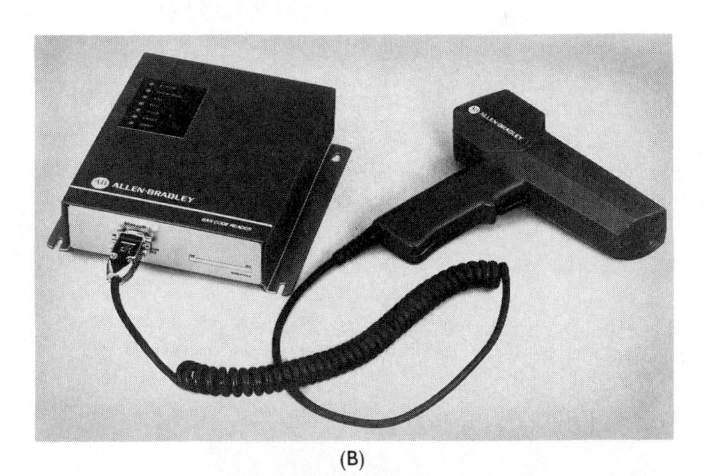

(B)

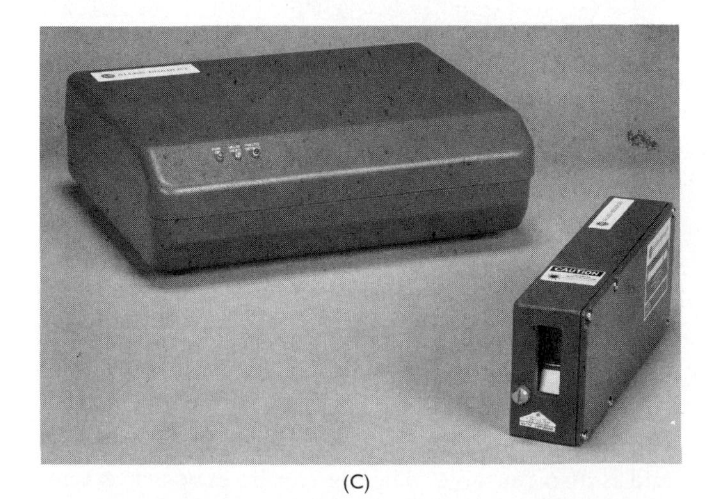

(C)

FIGURE 10–32 (A) Hand-held reader; (B) Hand-held laser scaner; (C) decoder console and scanner head. (Allen-Bradley)

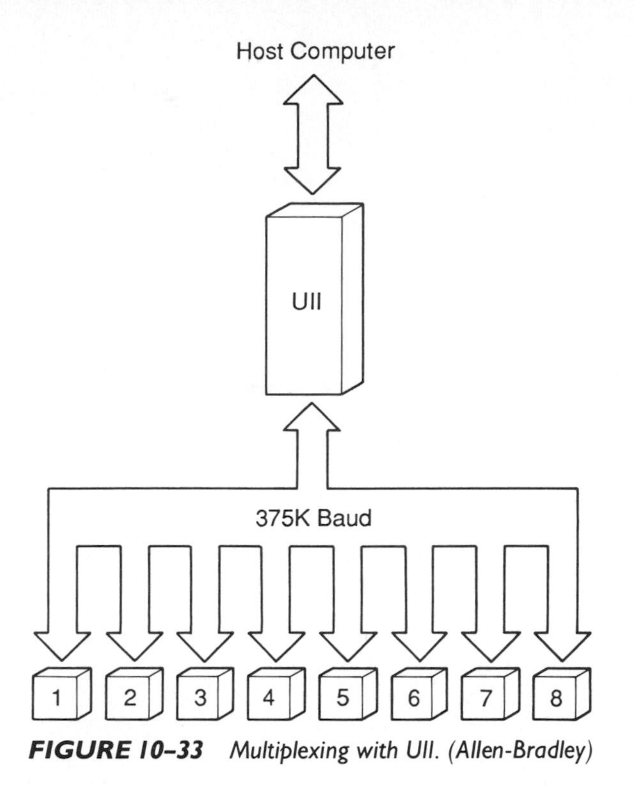

FIGURE 10–33 *Multiplexing with UII. (Allen-Bradley)*

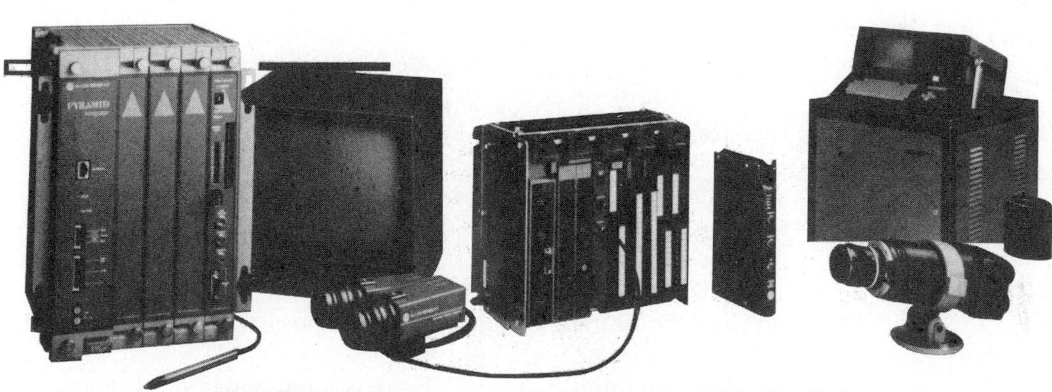

FIGURE 10–34 *Programmable vision system. (Allen-Bradley)*

QUESTIONS

1. How are sensors classified?
2. What is another name for a limit switch?
3. What type of switches respond to contact?
4. What type of switches are used for presence sensing?
5. Other than speed, what can a speed switch sense?
6. How do you adjust pressure controls for a given point?
7. Why are wells and packing glands needed?
8. What is an optical encoder?
9. What does *BCD* stand for?
10. What is multiplexing?
11. Where are proximity switches used?
12. What is the most needed feature of a sensing system?
13. In what colors are LEDs available?
14. What are the three pieces of hardware that make up the RF identification system?
15. What does *UII* stand for?
16. How are scanners used to sense and input information into a system?
17. What type of switch is used for antiplugging?
18. What does *LO* stand for?
19. What is the Gray code?
20. What does placing a logic ''0'' on the multiplex control line do?

11 Solenoids and Valves

Objectives

After studying this chapter, you will be able to:

1. Explain how a solenoid provides mechanical motion.
2. Identify various types of industrial solenoids.
3. List applications for solenoid devices.
4. Select a solenoid according to its class.
5. Describe how to service solenoid coils.
6. Explain how solenoid valves work in circuits.

A solenoid where the length is greater than the diameter is one of the most common types of coil construction used in electricity and electronics. The field intensity is the highest at the center in an iron-core solenoid. At the ends of the air-core coil, the field strength falls to a lower value.

A solenoid that is long compared to the diameter has a field intensity at the ends approximately one-half of that at the center. If the solenoid has a ferromagnetic core, the magnetic lines pass uniformly through the core.

Mechanical motion can be produced by the action of a solenoid or it can generate a voltage that is a result of some mechanical movement. The term *solenoid* has commonly come to mean a coil of wire with a moving iron core that can center itself lengthwise within the coil when current is applied to the coil. Then if a ferromagnetic core is properly suspended and under suitable tension, it can be moved in and out of a solenoid coil form with the application of coil current. This is the operating basis of some relays and a number of other electromechanical devices. If an outside force is used to move the ferromagnetic core physically, it is possible to induce a voltage in the solenoid coil.

There is a tendency in a solenoid for the core to move so that it encloses a maximum number of magnetic lines of force. Each line of force has the shortest possible length (Fig. 11-1). In the illustration the core is outside the coil. Because it is a ferromagnetic mate-

FIGURE 11-1 *Solenoid pulls core into the coil. Sucking effect of a coil.*

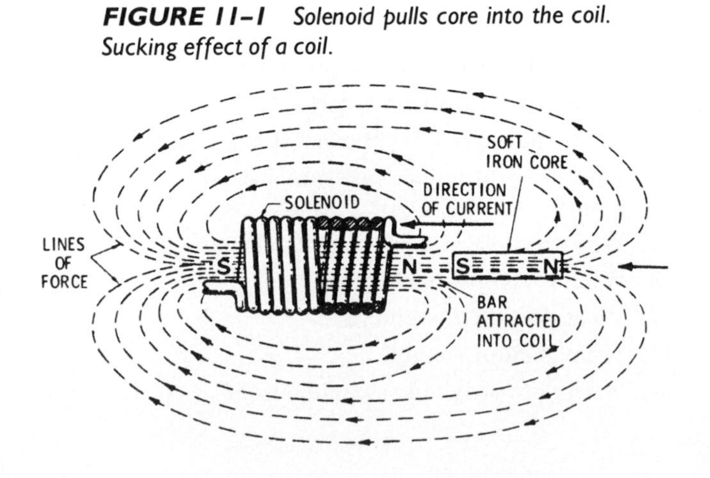

rial, the coil presents a low-reluctance path to the magnetic lines of force at the north end of the coil. These lines of force concentrate the soft-iron core and then complete their paths back to the south pole of the electromagnet.

Electromagnetic lines of force that pass through the core magnetize it. That means that the induced magnetic field in the core has a south pole near the coil's north pole. Inasmuch as unlike poles attract, the core is attracted toward the hole in the solenoid coil. This attraction tends to pull the core into the coil. As the iron core is pulled into the coil, the magnetic field becomes increasingly shorter and the magnetic lines of force travel the shortest possible distance when the core centers itself in the coil.

By attaching a spring to the core, it is possible to have the core return to its outside position once the power is interrupted to the coil. When the power is then turned on again, it pulls the core back into the coil. It is this type of movement that is utilized in the construction of industrial solenoids that operate switch contacts in relays and motor starters and valves in gas, air, and liquid lines of various types.

INDUSTRIAL SOLENOIDS

Tubular Solenoids

There are various uses for solenoids. Figure 11-2 shows tubular solenoids. Notice the type, voltage rating, coil resistance, and the minimum and maximum lifts and strokes. Some are pull types and others are push types. They are also specified as to intermittent and continuous duty.

Frame Solenoids

The frame-mounted types of solenoids (Fig. 11-3) are available in intermittent and continuous duty as well as usable on either ac or dc. Types 11 and 28 can operate on ac/dc. The other types are identified as to whether they operate best on ac or dc.

APPLICATIONS

Solenoids are devices that turn electricity, gas, oil, or water on and off. Solenoids can be used, for example, to turn the cold water on, and the hot water off, to get the proper mix of warm water in a washing machine. To control the hot water solenoid, a thermostat is inserted in the circuit.

Figure 11-4 shows a solenoid for controlling natural gas flow in a hot-air furnace. Note how the coil is wound around the plunger. The plunger is the core of the solenoid. It has a tendency to be sucked into the coil whenever the coil is energized by current flowing through it. The electromagnetic effect causes the plunger to be attracted upward into the coil area. When the plunger is moved upward by the pull of the electromagnet, the soft disk (10) is also pulled upward, allowing gas to flow through the valve. This basic technique is used to control water, oil, gasoline, or any other liquid or gas.

The starter solenoid on an automobile uses a similar procedure except that the plunger has electrical contacts on the end that complete the circuit from the battery to the starter. The solenoid uses low voltage (12 V) and low current to energize the coil. The coil in turn sucks the plunger upward. The plunger, with a heavy-duty copper washer attached, then touches heavy-duty contacts that are designed to handle the 300 A needed to start a cold engine. In this way, low voltage and low current are used, from a remote location, to control low voltage and *high current*.

Solenoids as Electromagnets

An electromagnet is composed of a coil of wire wound around a core of soft iron. A solenoid is an electromagnet. When current flows through the coil, the core becomes magnetized.

The magnetized core can be used to attract an armature and act as a magnetic circuit breaker (Fig. 11-5). Note how the magnetic circuit breaker is connected in series with both the load circuit to be protected and with the switch contact points. When excessive current flows in the circuit, a strong magnetic field in the electromagnet causes the armature to be attracted to the core. A spring attached to the armature causes the switch contacts to open and break the circuit. The circuit breaker must be reset by hand to allow the circuit to operate properly again. If the overload is still present, the circuit breaker will "trip" again. It will continue to do so until the cause of the short circuit or overload is found and corrected.

Solenoid Coils

The coil is the most important part of the solenoid, inasmuch as the valve or switch contacts that it operates cannot work unless the coil is capable of being energized. There are at least three types of coils you should be aware of in solenoids used in air conditioning, refrigeration, and heating circuits. For various applications they are divided into classes, as outlined in Table 11-1 (see Fig. 11-6).

Servicing Coils

Coils can be replaced when they malfunction. Excessive heat causes coil malfunction. Make sure that the valve is not heated to a temperature above the coil rat-

T DENOTES PULL TYPE; TP DENOTES PUSH TYPE; LT DENOTES LONG LIFE PULL TYPE.

T4x7
LT4x7

TP4x7

T4x12

TP4x12

T4x16
LT4x16

TP4x16

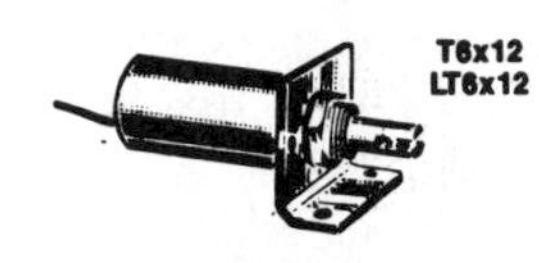

T8x9
LT8x9

T8x16
LT8x16

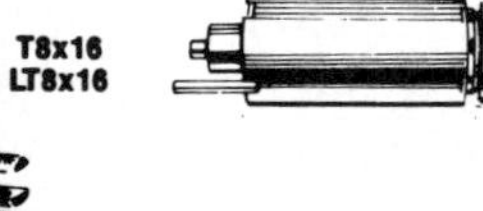

TP12x13

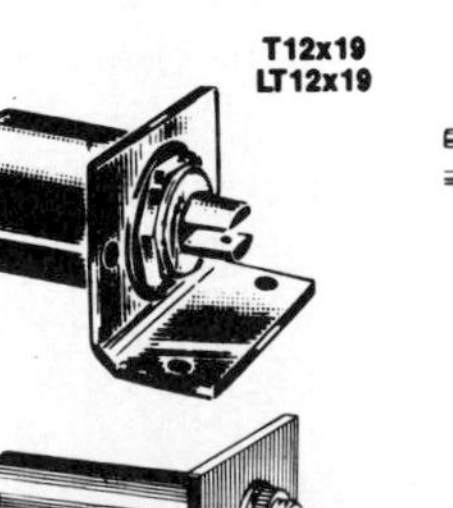

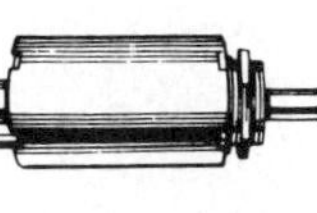

T, TP AND LT SERIES SOLENOIDS

T and **TP series** are UL recognized.

Duty: I=intermittent. C=Continuous.

Type	Duty-Volt.	Coil Resis. Ohms	Lifts and Strokes			
			Min.		Max.	
			Oz.	@in	Oz	@in
T3.5X9	C-12D	60.2	4½	1/32	.3	5/16
	I-12D	31.1	7	1/32	.7	1/2
	I-24D	122	7	1/32	.7	1/2
	C-24D	254	4½	1/32	.3	5/16
LT3.5X9	C-12D	52.4	2.5	1/32	.2	3/8
	C-24D	221	2.5	1/32	.2	3/8
TP3.5X9	I-24D	122	6	1/32	.5	1/2
	C-24D	254	4	1/32	.3	5/16
T4X7	I-24D	131	7	1/32	1	1/4
	C-24D	270	5	1/32	.5	1/4
LT4X7	C-12D	63.3	2.5	1/32	1	.15
	C-24D	264	2.5	1/32	1	.15
TP4X7	I-24D	131	6.5	1/32	.75	.15
	C-24D	270	3.5	1/32	.50	1.5
T4X16	C-12D	45.1	4.5	1/16	1.2	1/2
	I-12D	17.7	7	1/16	2.5	1
	C-24D	173	4.5	1/16	1.2	1/2
	I-24D	72.7	7	1/16	2.5	1
LT4X16	C-12D	42.5	4	1/16	2	1/2
	I-12D	14	8	1/16	2.5	1/2
	C-24D	168	4	1/16	2	1/2
	I-24D	69.1	8	1/16	2.5	1/2
TP4X16	C-24D	173	3.8	1/16	1	1/2
	I-24D	72.7	5.5	1/16	2	1/2
T4X12	C-24D	195	6	1/32	1.5	.15
	I-24D	96.7	9	1/32	2	.15
LT4X12	C-12D	49.3	7.5	1/32	.5	1/2
	C-24D	19.2	7.5	1/32	.5	1/2
TP4X12	I-24D	96.7	7	1/32	1.5	.15
	C-24D	195	5	1/32	1.0	.15
T6X12	C-12D	31.7	15	1/16	1	3/4
	I-12D	12.1	23	1/16	3	3/4
	C-24D	121	15	1/16	1	3/4
	I-24D	60.6	23	1/16	3	3/4
LT6X12	C-12D	35	18	1/16	1	1/8
	I-12D	13.8	35	1/16	2.5	1/2
	C-24D	138	18	1/16	1	1/2
TP6X12	C-24D	121	12	1/16	.8	3/4
	I-24D	60.6	18	1/16	2.5	3/4
T8X9	C-24D	135	18	1/16	2.5	7/16
	I-24D	44	45	1/16	5	7/16
LT8X9	C-24D	109	18	1/16	1	1/2
	I-24D	44.6	35	1/16	3	1/2
TP8X9	I-24D	44	36	1/16	4	7/16
	C-24D	135	14	1/16	2	7/16
T8X16	I-12D	9.3	58	1/16	5	.60
	C-12D	28.3	30	1/16	2.5	.60
	I-24D	36.1	58	1/16	5	.60
	C-24D	110	30	1/16	2.5	.60
LT8X16	I-12D	6.2	67	1/16	12	1/2
	C-12D	19.3	33	1/16	2.5	1/2
	I-24D	29.7	62	1/16	10	1/2
	C-24D	77.2	33	1/16	2.5	1/2
T12X13	I-24D	28.4	100	1/8	15	.60
	C-24D	90.4	40	1/8	5	.60
LT12X13	I-12D	5.9	110	1/8	10	.5
	C-12D	18.3	35	1/8	1	.5
	I-24D	22	110	1/8	10	.5
	C-24D	71.2	35	1/8	1	.5
TP12X13	I-24D	28.4	80	1/8	12	.60
	C-24D	90.4	32	1/8	4	.60
T12X19	I-24D	22.1	130	1/8	40	3/4
	C-24D	68	70	1/8	10	3/4
LT12X19	I-12D	4.66	130	1/8	15	3/4
	C-12D	14.8	70	1/8	5	3/4
	I-24D	18.6	130	1/8	15	3/4
	C-24D	71.8	70	1/8	3	3/4
TP12X19	I-24D	22.1	100	1/8	30	3/4
	C-24D	68	55	1/8	8	3/4

FIGURE 11–2 *Tubular solenoids.*

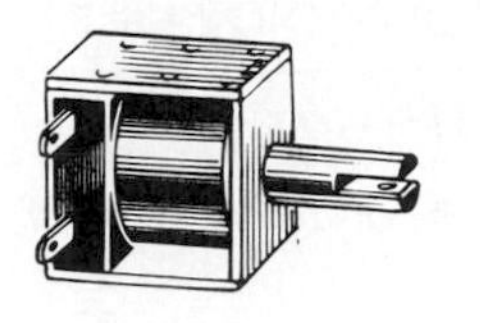

Type 2

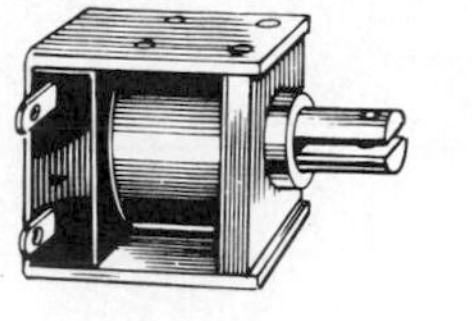

Type 2HD

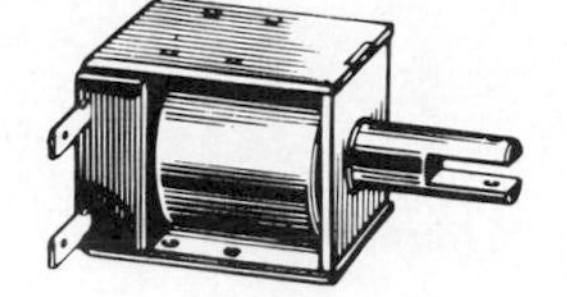

Type 4

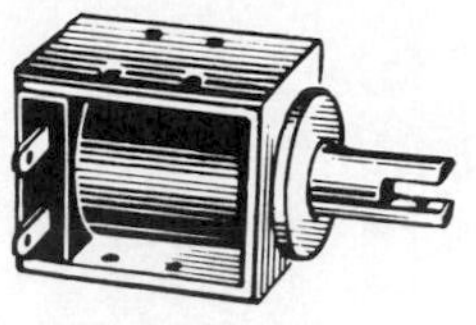

Type 4HD

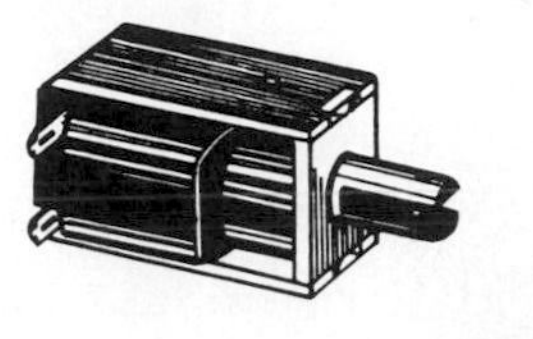

Type 11

Type 11HD

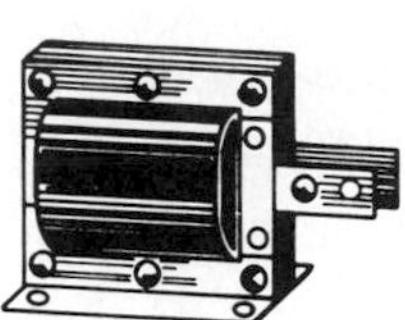

Type 12

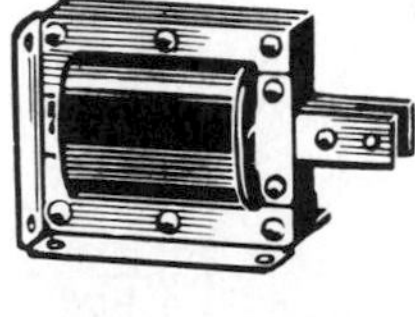

Type 14

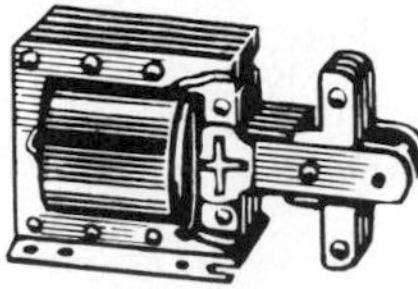

Type 16

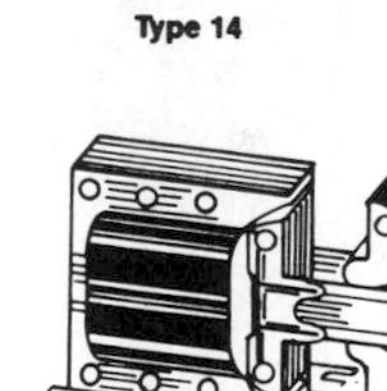

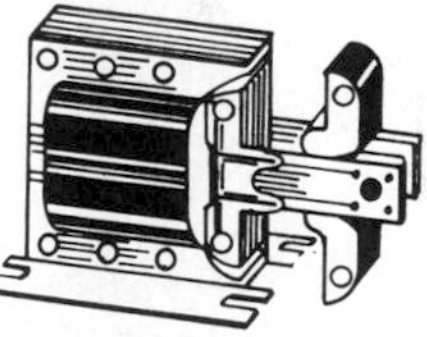

Type 18

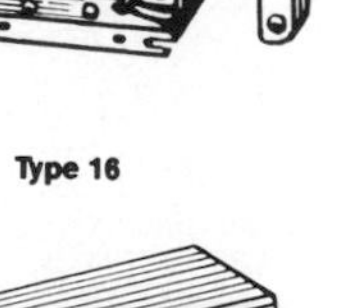

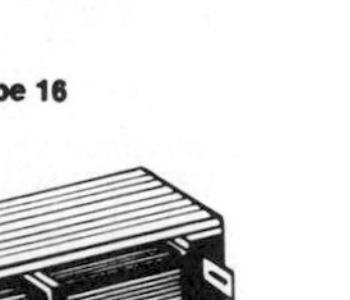

Type 22

DC VOLTAGE MODEL SOLENOIDS

Type	Duty	Volts	Ohms	Oz. @ Inch of Stroke	
				Minimum	Maximum
2HD	Intermittent	24	22.6	96@⅛″	15@½″
2HD	Continuous	24	71	48@⅛″	5@½″
4	Intermittent	24	15.8	100@⅛″	20@1″
4	Continuous	24	61.3	60@⅛″	7@⅘″
4	Intermittent	110	296	100@⅛″	20@1″
4	Continuous	110	1215	60@⅛″	7@⅘″
4HD	Intermittent	24	18.9	130@⅛″	25@¾″
4HD	Continuous	24	57.5	80@⅛″	5@¾″
4HD	Intermittent	110	354	130@⅛″	25@¾″
4HD	Continuous	110	1140	80@⅛″	5@¾″
11	Intermittent	6	1.88	45@⅛″	10@½″
11	Continuous	6	4.69	30@⅛″	4@½″
11	Intermittent	24	29.1	45@⅛″	10@½″
11	Continuous	24	93.1	30@⅛″	4@½″
11HD	Intermittent	24	29.3	70@⅛″	5@¾″
11HD	Continuous	24	76.3	30@⅛″	2@¾″
11P	Continuous	24	93.1	24@⅛″	3.2@½″
22	Intermittent	6	5.8	17@1/16″	2@0.3″
22	Continuous	6	11.5	11@1/16″	1@0.3″
22	Intermittent	24	93.2	17@1/16″	2@0.3″
22	Continuous	24	182	11@1/16″	1@0.3″
28	Intermittent	6	3.03	40@1/16″	3@½″
28	Continuous	6	7.5	23@1/16″	2@½″
28	Intermittent	12	11.9	40@1/16″	3@½″
28	Continuous	12	29.8	23@1/16″	2@½″
28	Intermittent	24	47.4	40@1/16″	3@½″
28	Continuous	24	116	23@1/16″	2@½″

AC VOLTAGE MODEL SOLENOIDS

2	Intermittent	120	60	45@⅛″	11@⅞″
2	Continuous	120	166	14@⅛″	3@⅞″
2HD	Intermittent	120	36	70@⅛″	16@¾″
2HD	Continuous	120	113	25@⅛″	6@¾″
4	Intermittent	120	37	36@⅛″	26@1″
4	Continuous	120	133	8@⅛″	7@1″
11	Intermittent	120	85	21@⅛″	11@¾″
11	Continuous	120	200	12@⅛″	6@¾″
11HD	Continuous	120	165	12@⅛″	3½@1″
11P	Continuous	120	200	9.6@⅛″	4.8@¾″
12	Intermittent	120	100	48@⅛″	9@⅞″
12	Continuous	120	150	28@⅛″	6@⅞″
14	Intermittent	120	11	108@⅛″	56@1½″
14	Continuous	120	18	75@⅛″	40@1½″
16	Intermittent	120	41	110@⅛″	28@¾″
16	Continuous	120	85	63@⅛″	15@¾″
16	Continuous	240	350	63@⅛″	15@¾″
16P	Intermittent	120	41	88@⅛″	22.5@¾″
16P	Continuous	120	85	50.5@⅛″	12@¾″
18	Intermittent	120	8.8	350@⅛″	208@⅞″
18	Continuous	120	19.7	152@⅛″	100@⅞″
18	Intermittent	240	45	350@⅛″	208@⅞″
18	Continuous	240	78	152@⅛″	100@⅞″
18P	Intermittent	120	8.8	315@⅛″	187@⅞″
18P	Continuous	120	19.7	137@⅛″	90@⅞″
24	Continuous	120	500	10@1/16″	2@⅝″
28	Continuous	24	17.4	24@1/16″	5@⅝″
28	Continuous	120	400	24@1/16″	5@½″

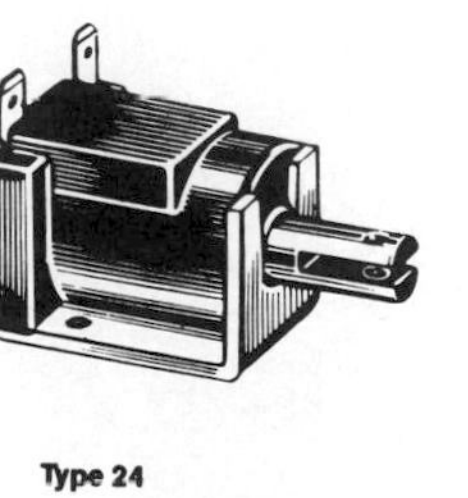

Type 24

Type 28

- **All Models are UL Recognized**

These intermittent and continuous duty solenoids are available in AC and DC versions, and in three constructions: box frame, U-frame and laminated. **Types 2, 2HD, 4, 4HD, 11, 11HD, 11P, 22** and **28** are box frame. **Types 12, 14, 16, 16P, 18** and **18P** are laminated. **Type 24** is U-frame. Suffix **P** indicates a push type model. Suffix **HD** indicates a heavy duty model. All box frame models have quick connect terminals.

FIGURE 11–3 *Frame solenoids.*

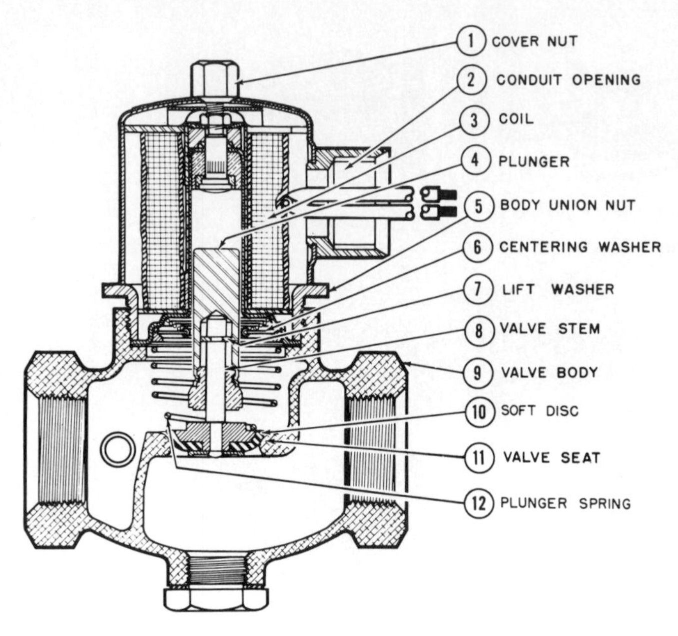

FIGURE 11–4 *Solenoid for controlling natural gas flow to a hot-air furnace. (Honeywell)*

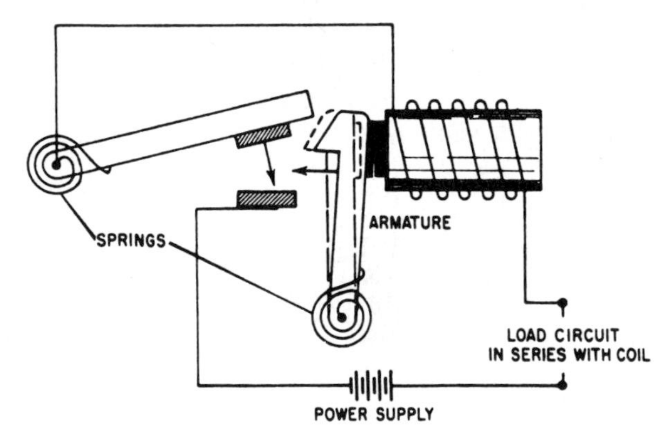

FIGURE 11–5 *Magnetic circuit breaker.*

FIGURE 11–6 *Solenoid coils.*

TABLE 11–1 CLASSES OF SOLENOID COILS

Class	*Application*
A	Moisture-resistant coil for normal use of gas or fluid up to 175°F.
B	Ambient and fluid temperature up to 200°F.
H	Temperatures up to 365°F, high steam pressure, rapid valve cycling, high voltage, fungus-proof.
BW	Same as coil B, and waterproof, fungus-proof, plastic-encapsulated for temperatures up to 200°F.
W	Same as coil A, and waterproof, fungus-proof, plastic-encapsulated for temperatures up to 175°F.

FIGURE 11–7 *Exploded view of balanced diaphragm valve.*

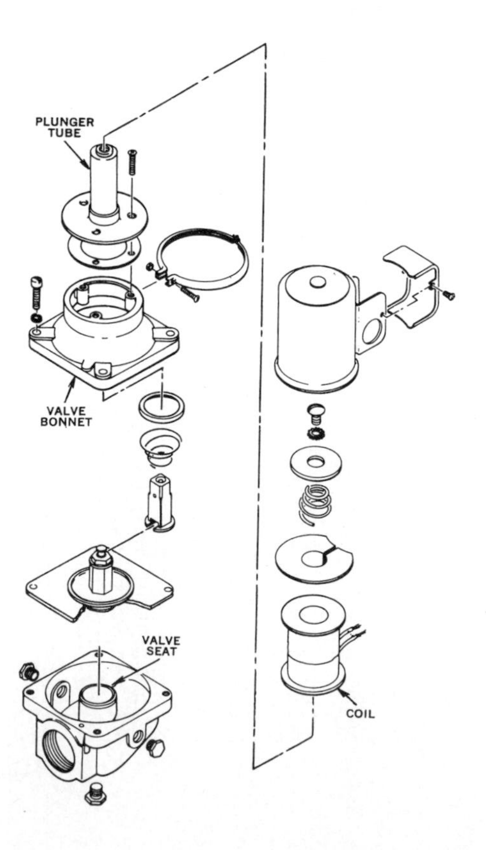

ing. When replacing a coil, reassemble the solenoid correctly. A missing part or improper reassembly causes excessive coil heat. See the exploded view in Fig. 11–7.

Applied voltage must be at the coil-rated frequency and voltage. A damaged plunger tube or tube sleeve causes heat and can prevent the solenoid from operating. For applications requiring greater resistance or different electrical requirements, use the proper coil in the solenoids. Do not change from ac to dc or dc to ac without changing the entire solenoid assembly (coil, plunger, plunger tube, and base fitting).

When replacing a coil, first be sure to turn off the electric power to the solenoid. It will not be necessary in most instances to remove the valve from the pipeline. Disconnect the coil leads. Disassemble the solenoid carefully and reassemble in reverse order. Failure to reassemble the solenoid properly can cause coil burn out.

Surge suppressors are available to protect the coil from unusual line surges. Figure 11–8 shows how the coil leads can be connected to allow for 120- or 240-V operation. These are referred to as *dual-voltage coils.*

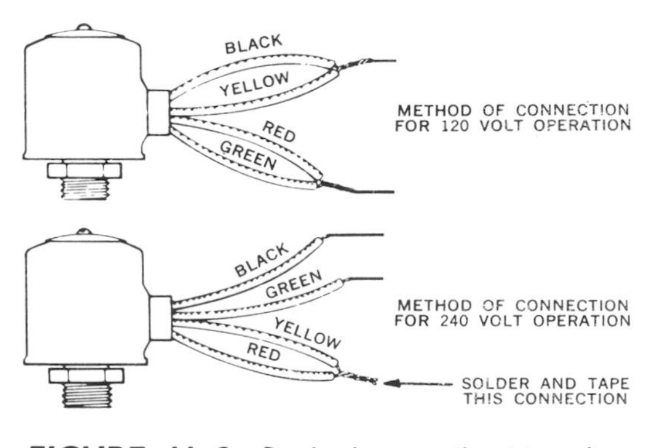

FIGURE 11–8 *Dual-voltage coil wiring diagrams.*

The valve shown in Fig. 11–7 is a series-balanced diaphragm solenoid valve that provides on–off control for domestic and industrial furnaces, boilers, conversion burners, and similar units using thermostats, limit controls, or similar control devices. The valve uses a balanced diaphragm for high operating pressure with low electrical power consumption. It is suitable for use with all gases and comes in a variety of sizes, capacities, and pressures.

Presence of a low, barely audible hum is normal when the coil is energized. If the valve develops a buzzing or chattering noise, check for proper voltage. Thoroughly clean the plunger and the interior of the plunger tube. Make sure that the plunger tube and solenoid assembly are tight (Fig. 11–9).

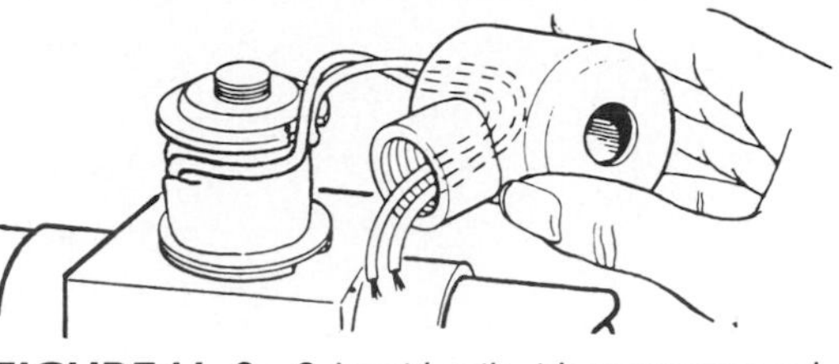

FIGURE 11–9 *Solenoid coil with cover removed.*

SOLENOID VALVES IN CIRCUITS

Solenoid valves are used on multiple installations in refrigeration systems. They are electrically operated as shown in Fig. 11–10. When connected as shown in the illustration, the valve remains open when current is supplied to it. It closes when the current is turned off. In general, solenoid valves are used to control the liquid refrigerant flow into the expansion valve or the refrigerant gas flow from the evaporator when it or the fixture it is controlling reaches the desired temperature. The most common application of the solenoid valve is in the liquid line and operates with a thermostat. With this hookup, the thermostat is set for the desired temperature in the fixture. When this temperature is reached, the thermostat opens the electrical circuit and shuts off the current to the valve. The solenoid valve closes and shuts off the refrigerant supply to the expansion valve. The condensing unit operation is controlled by the low-pressure switch. In other applications, where the evaporator is in operation for only a few hours each day, a manually operated snap switch is used to open and close the solenoid valve.

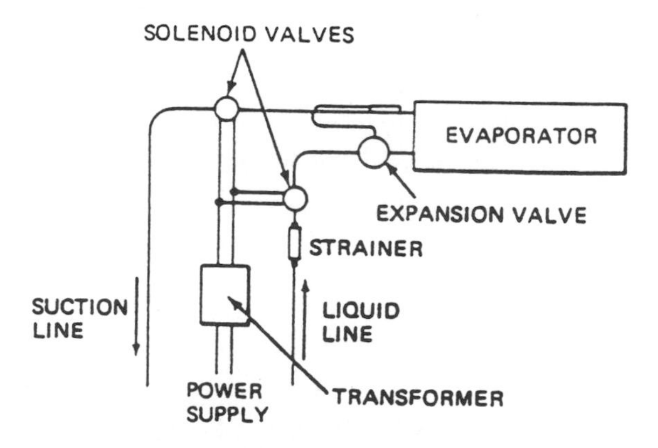

FIGURE 11–10 *Solenoid valves connected in the suction and liquid evaporator lines of a refrigeration system.*

Refrigeration Valve

The solenoid valve shown in Fig. 11–11 is operated with a normally closed status. A direct-acting metal

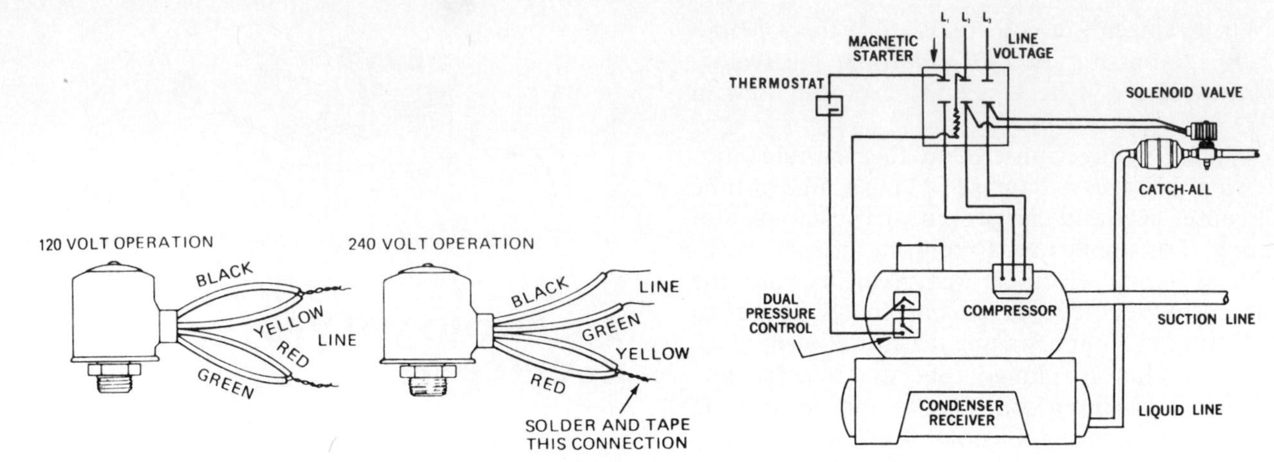

FIGURE 11–11 *Schematic of a refrigeration installation.*

ball and seat assure tight closing. The two-wire, class W coil is supplied standard for long life on low-temperature service or sweating conditions. Current failure or interruption causes the valve to fail-safe in the closed position. Explosion-proof models are available for use in hazardous areas.

This solenoid valve is usable with all refrigerants except ammonia. It can also be used for air, oil, water, detergents, butane or propane gas, and other noncorrosive liquids or gases.

A variety of temperature control installations can be accomplished with these valves. Such installations include bypass, defrost, suction line, hot gas service, humidity control, alcohols, unloading, reverse cycle, chilled water, cooling tower, brine, and liquid-line stop installations and ice makers.

The valves are held in the normally closed position by the weight of the plunger assembly and fluid pressure on top of the valve ball. The valve is opened by energizing the coil. This magnetically lifts the plunger and allows full flow by the valve ball. Deenergizing the coil permits the plunger and valve ball to return to the closed position.

QUESTIONS

1. Define *solenoid.*
2. What is meant by the term *sucking effect* of a solenoid?
3. What does the armature of an electromagnet do?
4. What is the most important part of the solenoid?
5. List the five classes of solenoids.
6. What are dual-voltage coils? How are they wired?
7. Where are series-balanced diaphragm solenoid valves used?
8. What does a barely audible hum indicate when a coil is energized?
9. Where are solenoid valves used?
10. How are valves made fail-safe?

12

Motor Starting Methods

Objectives

After studying this chapter, you will be able to:

1. Describe how a split-phase motor can be reversed.
2. List uses for the split-phase motor.
3. Identify the repulsion-induction motor.
4. Identify the capacitor-start motor.
5. Explain how to reverse the direction of rotation of a capacitor-start motor.
6. Describe the best use for the permanent-split capacitor motor.
7. Explain how a shaded-pole motor operates.
8. List various motor starter methods.
9. Explain how primary resistor starting operates.
10. Explain how part-winding starting is accomplished.
11. List the advantages and disadvantages of part-winding starters.
12. Describe how wye–delta and star–delta starters work.
13. List the advantages and disadvantages of wye–delta and star–delta starters.
14. Describe the operation of consequence pole motor controllers.
15. Identify the least expensive type of starter.
16. List basic characteristics of five types of starting methods.
17. Select a starter for a desired characteristic.

Electric motors are designed to deliver their best overall performance when operated at the design voltage shown on the nameplate. However, this voltage is often not maintained. Instead, it varies between minimum and maximum limits over what is termed *voltage spread.* The voltage spread is usually due to the wiring and transformers of the electrical distribution system and varies in proportion to motor or load currents.

In most modern plants using load-center power distribution systems, variations in voltage normally will be within recommended limits of 110–220, 220–240, 440–480, and 550–600 for single-phase and three-phase squirrel-cage and synchronous motors. However, there are older plants throughout the country with large low-voltage systems. Long low-voltage feeders often cause voltage drops that result in below-standard voltages at the motor terminals, especially during motor starting, when currents may be up to six times normal full-load. Table 12–1 shows the effect of voltage variations on the performance of polyphase induction motors.

Single-phase and polyphase motors call for different approaches or methods to start them under various conditions of operation. Most single-phase motors are started by the turning on of an on–off switch or a magnetic starter.

TABLE 12-1 VOLTAGE VARIATIONS

Rated Voltage	*Lower Limit*	*Upper Limit*
220	210	240
440	420	480
550	525	600
2300	2250	2480
4000	3920	4320
4600	4500	5000
6600	6470	7130

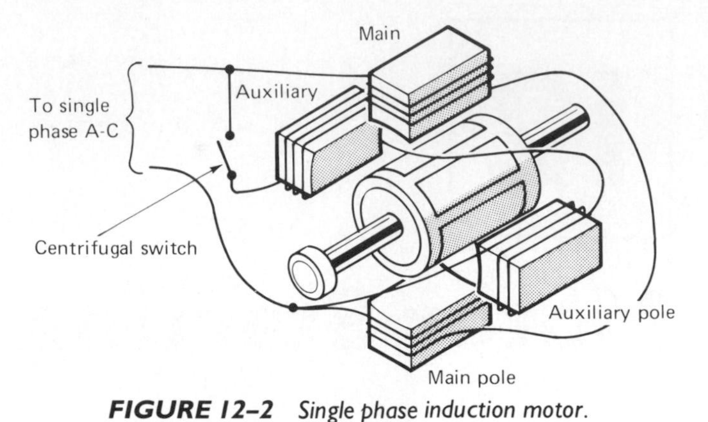

FIGURE 12-2 *Single phase induction motor.*

STARTING THE MOTOR

One of the most important parts of the electric motor is the start mechanism. A special type is needed for use with single-phase motors. A centrifugal switch is used to take a start winding out of the circuit once the motor has come up to within 75% of its run speed. The split-phase, capacitor-start, and other variations of these types all need the start mechanism to get them running.

The stator of a split-phase motor has two types of coils, one called the run winding and other the start winding. The *run winding* is made by winding the enamel-coated copper wire through the slots in the stator punchings. The *start winding* is made in the same way except that the wire is smaller. Coils that form the start windings are positioned in pairs in the stator directly opposite each other and between the run windings. When you look at the end of the stator, you see alternating run windings and start windings (Fig. 12-1).

The run windings are all connected together, so the electrical current must pass through one coil completely before it enters the next coil, and so on through all the run windings in the stator. The start windings are connected together in the same way and the current must pass through each in turn (Fig. 12-2).

The two wires from the run windings in the stator are connected to terminals on an insulated terminal block in one end bell where the power cord is attached to the same terminals. One wire from the start winding is tied to one of these terminals also. However, the other wire from the start winding is connected to the stationary switch mounted in the end bell. Another wire then connects this switch to the opposite terminal on the insulated block. The stationary switch does not revolve but is placed so that the weights in the rotating portion of the switch, located on the rotor shaft, will move outward when the motor is up to speed and open the switch to stop electrical current from passing through the start winding.

The motor then runs only on the main winding until such times as it is shut off. Then, as the rotor decreases in speed, the weights on the rotating switch again move inward to close the stationary switch and engage the start winding for the next time it is started.

Reversibility. The direction of rotation of the split-phase motor can be changed by reversing the start winding leads.

Uses. This type of motor is used for fans, furnace blowers, oil burners, office appliances, and unit heaters.

FIGURE 12-1 *Split-phase motor windings. (Bodine Electric Co.)*

REPULSION-INDUCTION MOTOR

The repulsion-induction motor starts on one principle of operation and, when almost up to speed, changes over to another type of operation. Very high twisting forces are produced during starting by the repulsion between the magnetic pole in the armature and the same kind of pole in the adjacent stator field winding. The repulsing force is controlled and changed so that the armature rotational speed increases rapidly, and if not stopped, would continue to increase beyond a practical operating speed. It is prevented by a speed-actuated mechanical switch that causes the armature to act as a rotor that is electrically the same as the rotor in single-phase induction motors. That is why the motor is called a repulsion-induction motor.

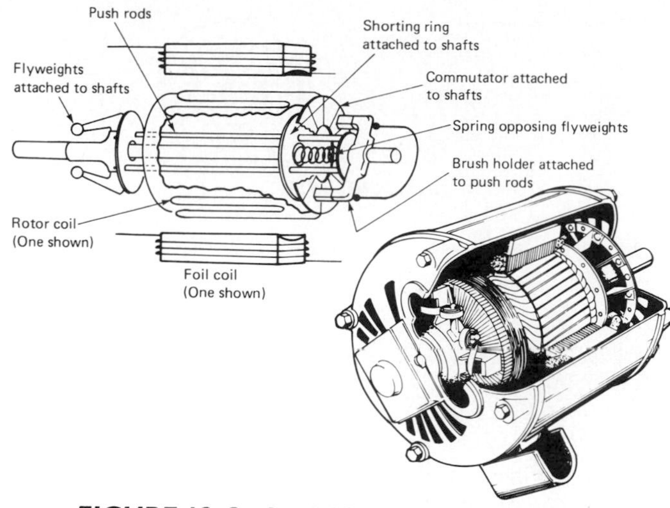

FIGURE 12–3 *Brush-lifting, repulsion-start, induction-run, single-phase motor.*

The stator of this motor is constructed very much like that of a split-phase or capacitor-start motor, but there are only run or field windings mounted inside. End bells keep the armature and shaft in position and hold the shaft bearings.

The armature consists of many separate coils of wire connected to segments of the commutator. Mounted on the other end of the armature are governor weights which move pushrods that pass through the armature core. These rods push against a short-circuiting ring mounted on the shaft on the commutator end of the armature. Brush holders and brushes are mounted in the commutator end bell, and the brushes, connected by a heavy wire, press against segments on opposite sides of the commutator (Fig. 12–3).

When the motor is stopped, the action of the governor weights keeps the short-circuiting ring from touching the commutator. When the power is turned on and current flows through the stator field windings, a current is induced in the armature coils. The two brushes connected together form an electromagnetic coil that produces a north and south pole in the armature, positioned so that the north pole in the armature is next to a north pole in the stator field windings. Since like poles try to move apart, the repulsion produced in this case can be satisfied in only one way, by the armature turning and moving the armature coil away from the field windings.

The armature turns faster and faster, accelerating until it reaches what is approximately 80% of the run speed. At this speed the governor weights fly outward and allow the pushrods to move. The pushrods, which are parallel to the armature shaft, have been holding the short-circuiting ring away from the commutator. Now that the governor has reached its designed speed, the rods can move together electrically in the same manner that the cast aluminum disks did in the cage of the induction motor rotor. This means that the motor runs as an induction motor.

Uses. The repulsion-induction motor can start very heavy, hard-to-turn loads without drawing too much current. They are made from $\frac{1}{2}$ to 20 hp. This type of motor is used for such applications as large air compressors, refrigeration equipment, large hoists, and are particularly useful in locations where low line voltage is a problem. This type of motor is no longer used in the refrigeration industry. Some older operating units may be found with this type of motor still in use.

CAPACITOR-START MOTOR

The capacitor motor is slightly different from a split-phase motor. A capacitor is placed in the path of the electrical current in the start winding (Fig. 12–4). Except for the capacitor, which is an electrical component that slows any rapid change in current, the two motors are the same electrically. A capacitor motor can usually be recognized by the capacitor can or housing that is mounted on the stator (Fig. 12–5).

By adding the capacitor to the start winding, it increases the effect of the two-phase field described in connection with the split-phase motor. The capacitor means that the motor can produce a much greater twisting force when it is started. It also reduces the amount of electrical current required during starting to about 1.5 times the current required after the motor is up to speed. Split-phase motors require three or four times the current in starting that they do in running.

Reversibility. An induction motor will not always

FIGURE 12–4 *Single-phase diagram for the AH air conditioner and heat pump compressor.*

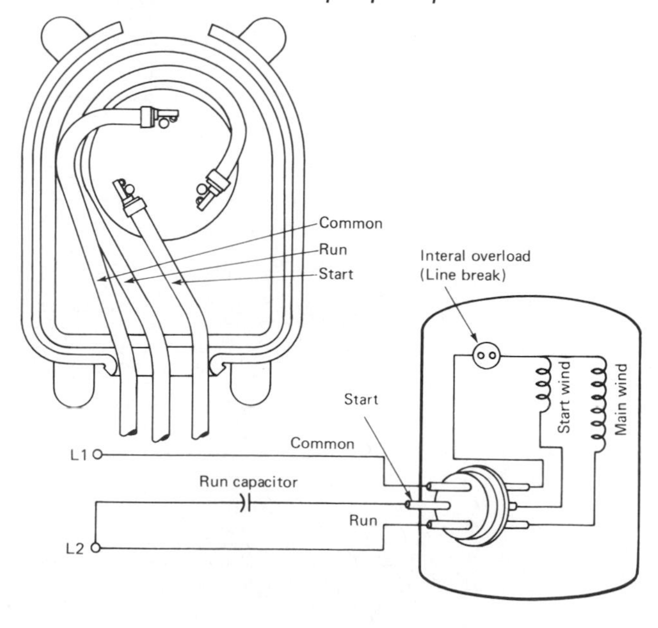

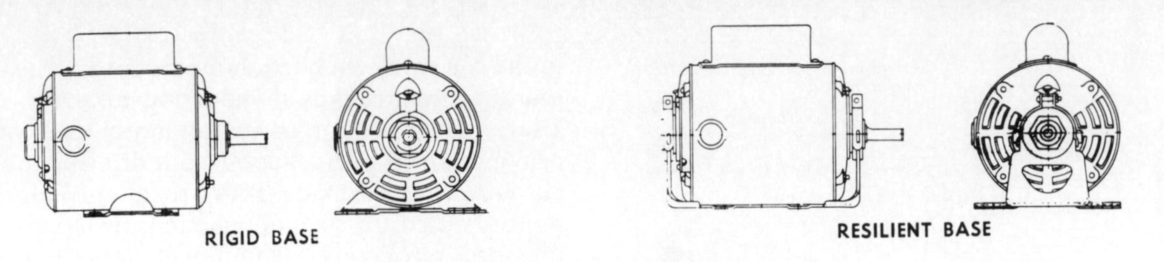

FIGURE 12–5 *Capacitor-start motor.*

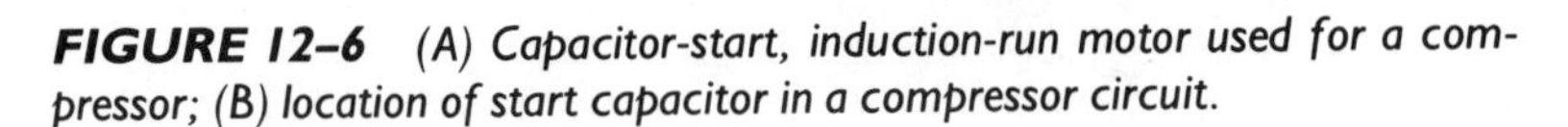

FIGURE 12–6 *(A) Capacitor-start, induction-run motor used for a compressor; (B) location of start capacitor in a compressor circuit.*

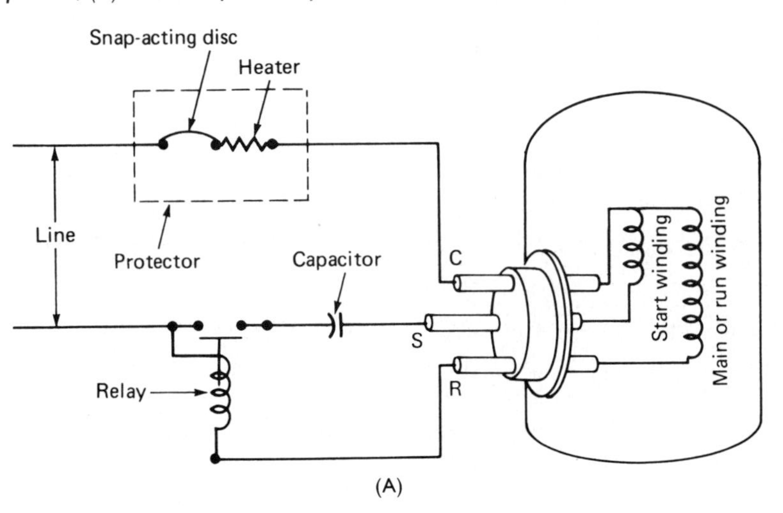

(A)

(B)

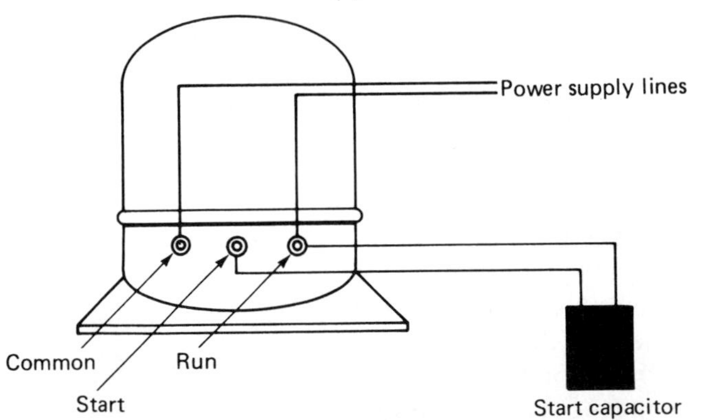

Start capacitor* sizes	
Compressor 1/8 hp	Capacitor size is 95 to 200 μF
Compressor 1/6 hp	Capacitor size is 95 to 200 μF
Compressor 1/4 hp	Capacitor size is 200 to 300 μF
Compressor 1/3 hp	Capacitor size is 250 to 350 μF
Compressor 1/2 hp	Capacitor size is 300 to 400 μF
Compressor 3/4 hp	Capacitor size is 300 to 400 μF

Black case (Bakelite)

reverse while running. It may continue to run in the same direction but at a reduced efficiency. An inertia-type load is difficult to reverse. Most motors that are classified as reversible while running will reverse with a noninertial-type load. They may not reverse if they are under no-load conditions or have a light load or an inertial load.

One of the problems related to the reversing of a motor while it is still running is the damage done to the transmission system connected to the load. In some cases it is possible to damage a load. One of the ways to avoid this is to make sure that the right motor is connected to a load.

Reversing (while standing still) the capacitor-start motor can be done by reversing its start winding connections. This is usually the only time that will work on a motor. The available replacement motor may not be rotating in the direction desired, so the electrician will have to locate the start winding terminals and reverse them in order to have the motor start in the desired direction.

Figure 12–6A shows a capacitor-start, induction-run motor used in a compressor. This type uses a relay to place the capacitor in and out of the circuit. More details regarding this type of relay will be given later. Figure 12–6B shows how the capacitor is located outside the compressor.

Uses. Capacitor motors are available in sizes from $\frac{1}{6}$ to 20 hp. They are used for fairly hard-starting loads that can be brought up to run speed in under 3 seconds. They may be used in industrial machine tools, pumps, air conditioners, air compressors, conveyors, and hoists.

PERMANENT SPLIT-CAPACITOR MOTOR

The permanent split-capacitor (PSC) motor is used in compressors for air-conditioning and refrigeration units. It has an advantage over the capacitor-start motor inasmuch as it does not need the centrifugal switch and its associated problems.

The PSC motor has a run capacitor in series with the start winding. Both run capacitor and start winding remain in the circuit during start and after the motor is up to speed. Motor torque is sufficient for capillary and other self-equalizing systems. No start capacitor or relay is necessary. The PSC motor is basically an air-conditioner compressor motor. It is very common through 3 hp. It is also available in the 4- and 5-hp sizes (Fig. 12–7).

SHADED-POLE MOTOR

The shaded-pole induction motor is a single-phase motor. It uses a unique method to start the rotor turning. The effect of a moving magnetic field is produced by constructing the stator in a special way (Fig. 12–8).

Portions of the pole-piece surfaces are surrounded by a copper strap called a shading coil. The

FIGURE 12–7 *Permanent split-capacitor motor schematic.*

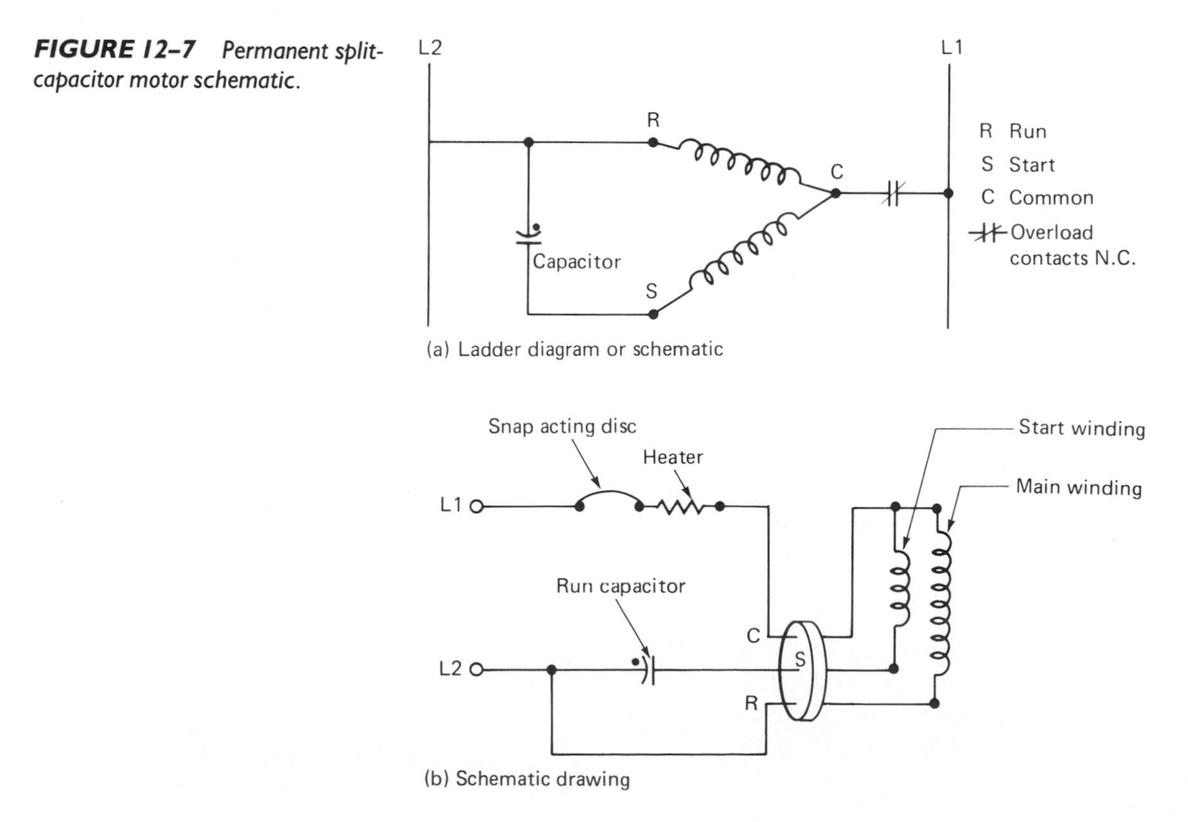

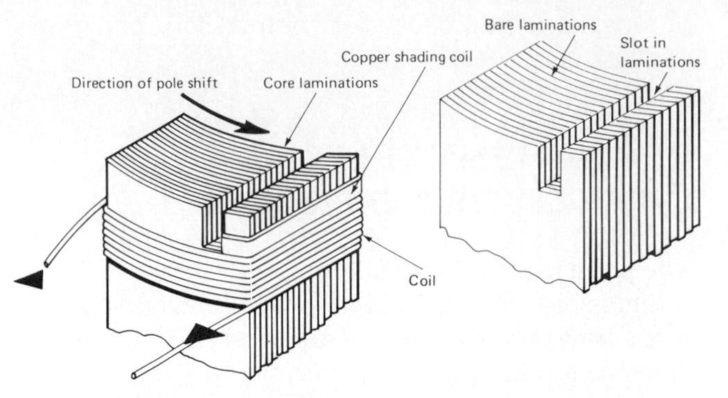

FIGURE 12–8 *Shading of the poles of a shaded-pole motor.*

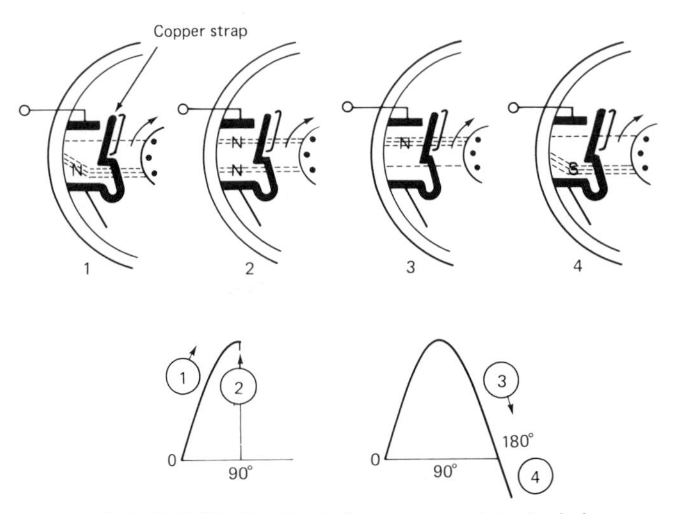

FIGURE 12–9 *Shaded poles as used in shaded-pole ac motors.*

strap causes the field to move back and forth across the face of the pole piece. In Fig. 12–9 the numbered sequence and points on the magnetization curve are shown. As the alternating stator field starts increasing from zero (1), the lines of force expand across the face of the pole piece and cut through the strap. A voltage is induced in the strap. The current that results generates a field that opposes the cutting action (and decreases the strength) of the main field. This action causes certain actions: As the field increases from zero to a maximum of 90°, a large portion of the magnetic lines of force are concentrated in the unshaded portion of the pole (1). At 90° the field reaches its maximum value. Since the lines of force have stopped expanding, no EMF is induced in the strap, and no opposite magnetic field is generated. As a result, the main field is uniformly distributed across the poles as shown in (2).

From 90 to 180° the main field starts decreasing or collapsing inward. The field generated in the strap opposes the collapsing field. The effect is to concentrate the lines of force in the shaded portion of the poles as shown in (3).

Note that from 0 to 180°, the main field has shifted across the pole face from the unshaded to the shaded portion. From 180 to 360°, the main field goes through the same change as it did from 0 to 180°. However, it is now in the opposite direction (4). The direction of the field does not affect the way the shaded pole works. The motion of the field is the same during the second half-hertz as it was during the first half-hertz.

The motion of the field back and forth between shaded and unshaded portions produces a weak torque. This torque is used to start the motor. Because of the weak starting torque, shaded-pole motors are built in only small sizes. They drive such devices as fans, clocks, and blowers.

Reversibility. Shaded-pole motors can be reversed mechanically. Turn the stator housing and shaded poles end for end. These motors are available from $\frac{1}{250}$ to $\frac{1}{2}$ hp.

Uses. As mentioned previously, this type of motor is used as a fan motor in refrigerators and freezers. They can also be used as fan motors in some types of air-conditioning equipment where the demand is not too great. They can also be used as part of the timing devices used for defrost timers and other sequenced operations.

The fan and motor assembly are located behind the provisions compartment in a refrigerator, directly above the evaporator in the freezer compartment. The suction fan pulls air through the evaporator and blows it through the provisions compartment air duct and freezer compartment fan grille (Fig. 12–10). This is a shaded-pole motor with a molded plastic fan blade. For maximum air circulation the location of the fan on the motor shaft is most important. Mounting the

FIGURE 12–10 *Fan, motor, and bracket assembly.*

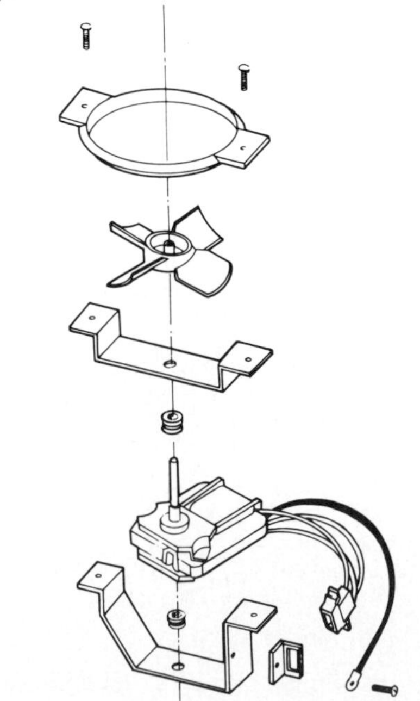

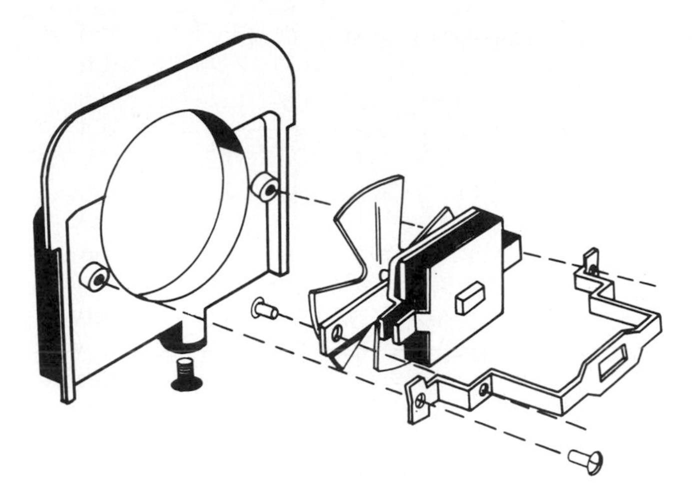

FIGURE 12–11 *Fan and fan motor bracket assembly.*

FIGURE 12–12 *Single-phase starting switch and governor mechanism.*

fan blade too far back or too far forward on the motor shaft, in relation to the evaporator cover, will result in improper air circulation. The freezer compartment fan must be positioned with the lead edge of the fan $\frac{1}{4}$ in. in front of the evaporator cover.

The fan assembly shown in Fig. 12–11 is used on the top-freezer, no-frost, fiberglass-insulated model refrigerators. The freezer fan and motor assembly is located in the divider partition directly under the freezer air duct.

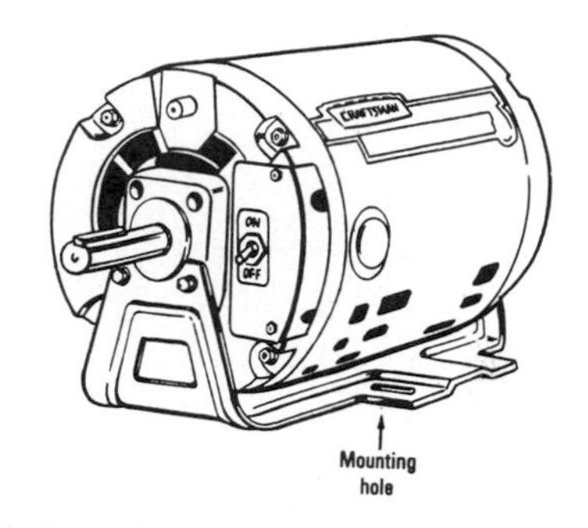

FIGURE 12–13 *Single-phase, split-phase motor.*

SPLIT-PHASE MOTOR

Instead of rotating, the field of a single-phase motor merely pulsates. No rotation of the rotor takes place. A single-phase pulsating field may be visualized as two rotating fields revolving at the same speed but in opposite directions. It therefore follows that the rotor will revolve in either direction at nearly synchronous speed—if it is given an initial impetus in either one direction or the other. The exact value of this initial rotational velocity varies widely with different machines. A velocity higher than 15% of the synchronous speed is usually sufficient to cause the rotor to accelerate to the rated or running speed. A single-phase motor can be made self-starting if means can be provided to give the effect of a rotating field.

To get the split-phase motor running, a run winding and a start winding are incorporated into the stator of the motor. Figure 12–12 shows the split-phase motor with the end cap removed so that you can see the starting switch and governor mechanism.

This type of motor is difficult to use with air-conditioning and refrigeration equipment inasmuch as it has very little starting torque and will not be able to start a compressor since it presents a load to the motor immediately upon starting. This type of motor, however, is very useful in heating equipment (Fig. 12–13).

POLYPHASE MOTOR STARTERS

The simple manual starter works for single-phase motors and also, in some instances, for polyphase motors. Most of the polyphase manual starters consisting of an on–off switching arrangement are designed for motors of 1 hp or less. Figure 12–14 shows the mag-

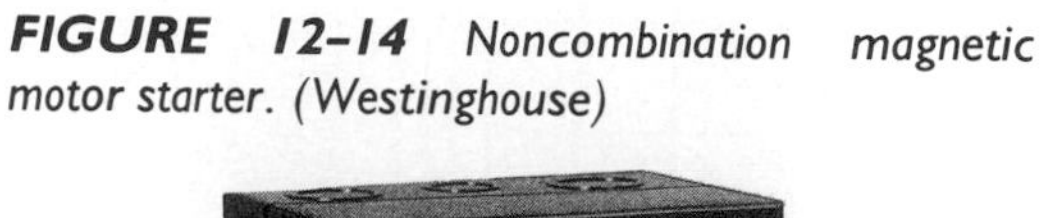

FIGURE 12–14 *Noncombination magnetic motor starter. (Westinghouse)*

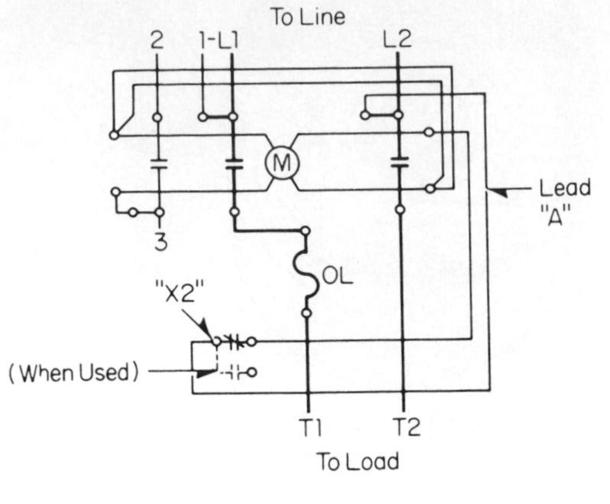

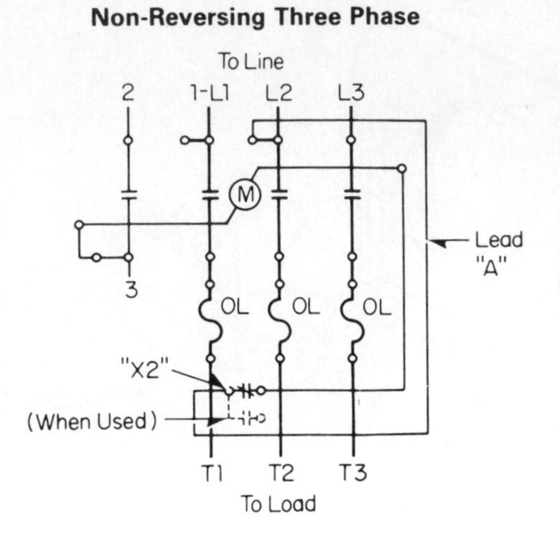

(A)

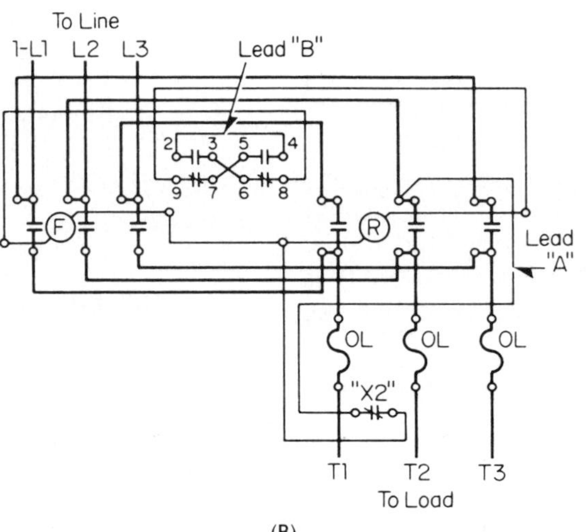

(B)

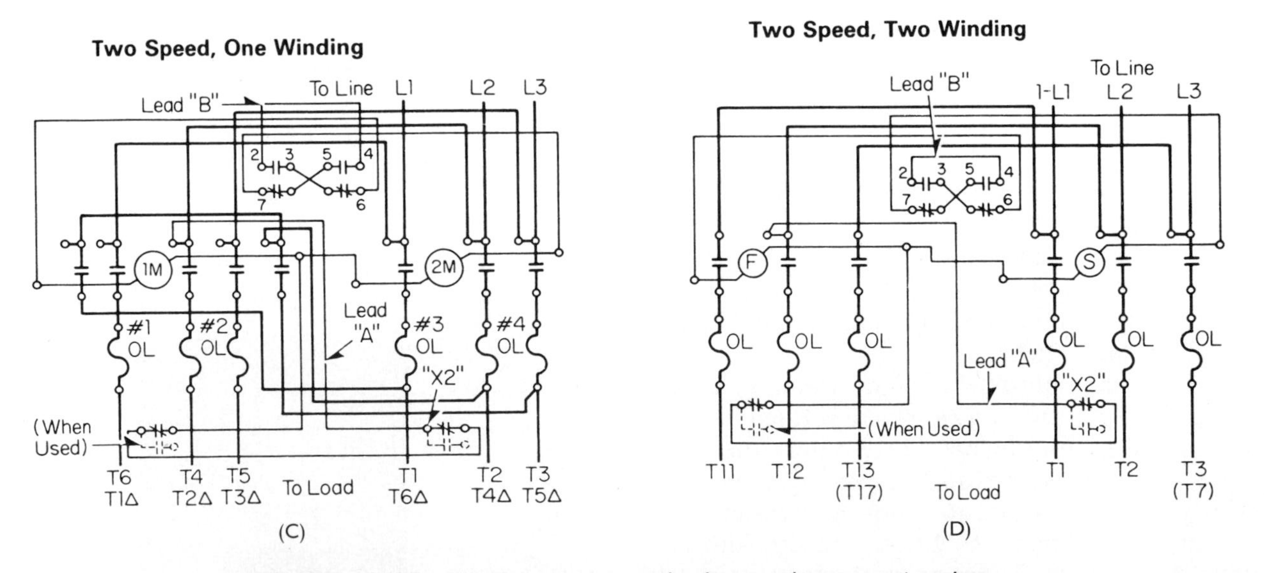

FIGURE 12–15 *(A) Nonreversing single-phase and nonreversing three-phase wiring diagrams; (B) reversing three-phase wiring diagram; (C) two-speed, one-winding, three-phase delta; (D) two-speed, two-winding, three-phase starter diagram. (Westinghouse)*

netic motor starter designed for across-the-line control of squirrel-cage motors or as primary control for wound-rotor motors. They are available for nonreversing, reversing, and two-speed applications. The drawing in Fig. 12–15A shows the difference between the single- and three-phase nonreversing type of starter. Figure 12–15B shows the reversing drawing; Fig. 12–15C is the two-speed, one winding starter and Fig. 12–15D is the two-speed, two-winding starter for motors up to 100 hp.

During across-the-line starting, motor input current is five to eight times normal full-load current. This can cause an excessive temporary voltage drop on power lines that causes lights to flicker or may even interrupt the service.

To control these temporary voltage drops, power companies have restrictions such as:

1. A specific maximum starting current (or kVA)
2. A specific limit on kVA/hp
3. A maximum horsepower motor size which can be started across-the-line
4. A specific maximum line current that can be drawn in steps (increment starting)

The specified restrictions vary considerably between power companies, even within one company's service area. It is wise to check local power company restrictions before making a larger motor installation.

REDUCED-VOLTAGE STARTING METHODS

Reduced-voltage starters operate such that input current and consequently torque are reduced during starting. Table 12–2 briefly describes the various methods of starting and gives features and limitations of each.

When motors are started at reduced voltage, the current at the motor terminals is reduced in direct proportion to the voltage reduction, while the torque is reduced as the *square* of the voltage reduction. For example, if the "typical" motor were started at 65% of line voltage, the starting current would be 42% and the torque would be 42% of full-voltage values. Thus reduced-voltage starting provides an effective means of reducing both current and torque (Fig. 12–19).

PRIMARY RESISTOR STARTING

In primary resistor starting, a resistor is connected in each motor line (in one line only in single-phase starters) to produce a voltage drop due to the motor starting current. A timing relay shorts out the resistors after the motor has accelerated. Thus the motor is started at reduced voltage but operates at line voltage.

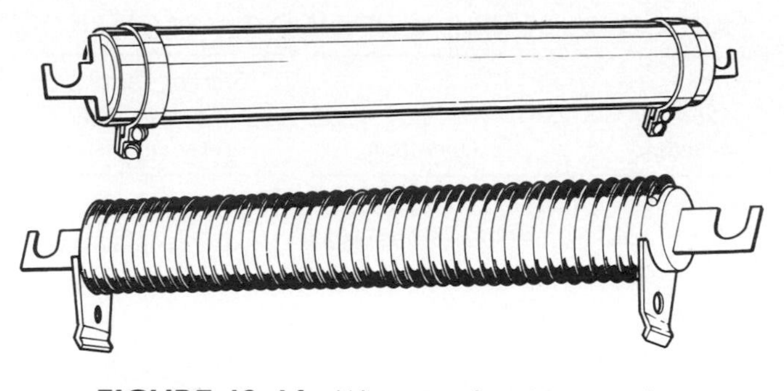

FIGURE 12–16 *Wire-wound resistors used in primary resistor starter circuits. (Westinghouse)*

Figure 12–16 shows two types of motor starter resistors. The resistance element will retain its mechanical and electrical properties both during and after repeated heating and cooling. All metal parts are either plated with or fabricated of corrosion-resistant material for overall corrosion protection. Under certain conditions operating temperatures may reach 600°C and not change the resistance value. These are 11, 14, 17, and 20 in. long and come in wattage ratings of 450 to 1320. Table 12–3 shows the resistance ranges and other factors. Note the current-handling ability of the resistors.

Primary resistor starters are sometimes known as *cushion* starters. The main reason for the name is their ability to produce a smooth, cushioned acceleration with closed transition. However, this method is not as efficient as other methods of reduced-voltage starting, but it is ideally suited for applications such as conveyors, textile machines, or other delicate machinery where reduction of starting torque is of prime consideration.

Operation

Figure 12–17 is the reduced-voltage magnetic starter that uses resistors to operate a three-phase motor properly at start. Closing the START button or other pilot device energizes the start contactor (S) shown in Fig. 12–18. This connects the motor in series with the starting resistors for a reduced-voltage start. The contactor (S) is now sealed in through its interlock (Sa). Timing relay (TR) is energized, and after a preset time interval its contacts (TR_{TC}) close. This energizes the run contactor, RUN, which seals through its interlock (RUN_a). The contacts (RUN) close, bypassing the starting resistors, and the motor will now running at full voltage. The contactor (S) and timing relay (TR) are deenergized when the interlock (RUN_b) opens.

An overload, which opens the STOP button or other pilot device, deenergizes the (RUN) contactor. This removes the motor from the line.

TABLE 12–2 STARTING METHOD CHARACTERISTICS[a]

Starting Method	*Operation*	*Starting Current (% of locked rotor current)*	*Starting Torque (% of locked rotor torque)*	*Open or Closed Transition*	*Basic Characteristics*	
					Advantages	*Limitations*
Across-the-line	Initially connects motor directly to power lines.	100%	100%	None	1. Lowest cost. 2. Highest starting torque. 3. Used with any standard motor. 4. Least maintenance.	1. High starting current. 2. High starting torque. 3. May shock driven machine
Primary resistance reduced voltage	Inserts resistance units in series with motor during first step(s).	50–80%	25–64%	Closed	1. Smoothest starting. 2. Least shock to driven machine. 3. Most flexible in application. 4. Used with any standard motor.	1. High power loss because of heating resistors 2. Heat must be dissipated. 3. Low torque per ampere input. 4. Highest cost.
Autotransformer reduced voltage	Uses autotransformer to reduce voltage applied to motor. **Tap** 50% 65% 80%	25% 42% 64%	25% 42% 64%	Closed	1. Best for hard to start loads. 2. Adjustable starting torque. 3. Used with any standard motor. 4. Less strain on motor.	1. May shock driven machine 2. High cost.
Wye–Delta	Starts motor with windings wye connected, then reconnects them in delta connection for running.	33%	33%	Open or closed	1. Medium cost. 2. Low starting current. 3. Low starting torque. 4. Less strain on motor.	1. Low starting torque. 2. Requires delta-wound motor.
Part Winding	Starts motor with only part of windings connected, then adds remainder for running.	70–80%	50–60% Minimum pull-up torque 35% of full-load torque.	Closed	1. Low cost. 2. Popular method for medium starting torque applications. 3. Low maintenance.	1. Not good for frequent starts 2. May require special wound motor. 3. Low pull-up torque. 4. May not come up to speed on first step when started with load applied.

NOTE: The reduced starting torque (LRT) indicated in this table for the various reduced starting methods can prevent starting high-inertia loads and must be considered when sizing motors and choosing starters.

TABLE 12–3 RESISTOR RANGES AND PROPERTIES

	Low R-High Current			*High R-Low Current*		
Unit Length (in.)	*Resistance Range (Ω)*	*Current Range (A)*	*Heat Dissipation (Watts per unit[a])*	*Resistance Range (Ω)*	*Current Range (A)*	*Heat Dissipation (Watts per unit[a])*
11	0.051–4.3	11–104	450–630	4.0–2000	0.46–10.3	426
14	0.069–5.7	11–104	620–820	5.0–2500	0.48–10.8	575
17	0.085–7.1	11–104	770–1080	5.0–2500	0.53–12.0	700
20	0.10–8.6	11–104	900–1320	6.4–4000	0.47–11.8	900

[a]Approximate only.

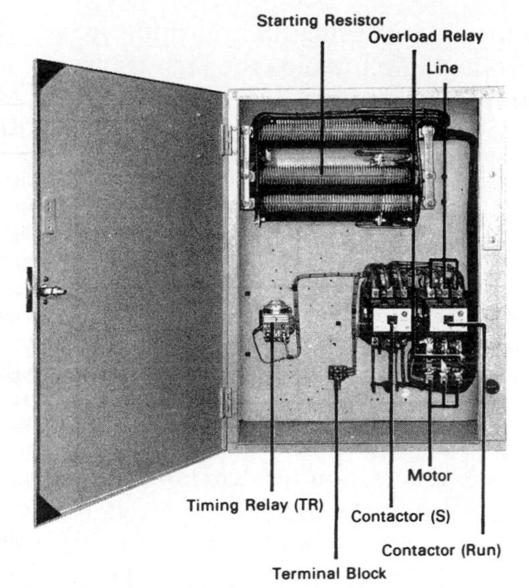

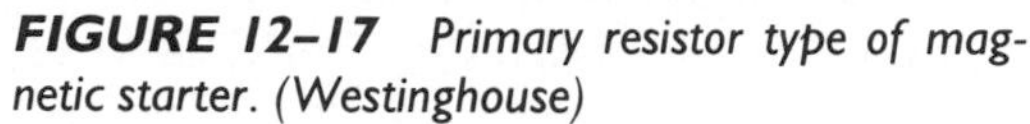

FIGURE 12–17 *Primary resistor type of magnetic starter. (Westinghouse)*

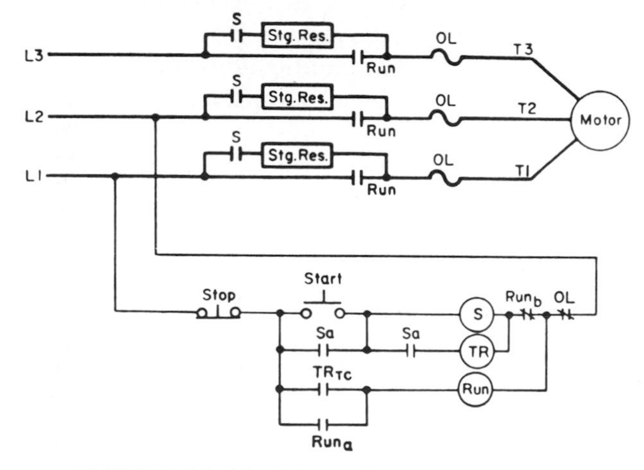

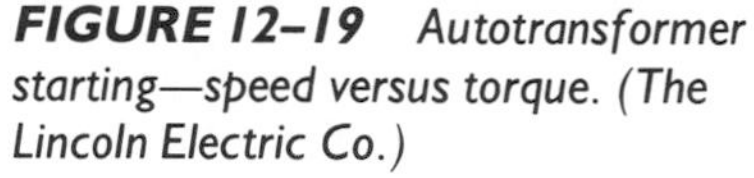

FIGURE 12–18 *Wiring diagram for a primary resistor type of starter. (Westinghouse)*

Primary resistor starters provide extremely smooth starting due to the increasing voltage across the motor terminals as the motor accelerates. Since motor current decreases with increasing speed, the voltage drop across the resistor decreases as the motor accelerates—and the motor terminal voltage increases. Thus if a resistor is shorted out as the motor reaches maximum speed, there is little or no increase in current or torque.

AUTOTRANSFORMER STARTING

Autotransformer starters provide reduced-voltage starting at the motor terminals through the use of a tapped, three-phase autotransformer. Upon initiation of the controller pilot device, a two- and a three-pole contactor close to connect the motor to the preselected autotransformer taps. A timing relay causes the transfer of the motor from the reduced-voltage start to line-voltage operation without disconnecting the motor from the power source. This is known as *closed transition starting*.

Taps on the autotransformer provide selection of 50%, 65%, or 80% of line voltage as a starting voltage. Starting torque will be 25%, 42%, or 64%, respectively, of line-voltage values. However, because of transformer action, the controller line current will be less than motor current, being 25%, 42%, or 64% of full-voltage values. This autotransformer starting may be used to provide maximum torque available with minimum line current, together with taps to permit both of these factors to be varied. Figure 12–19 shows torque and voltage tap points.

Manual autotransformer starters are used to start squirrel-cage polyphase motors when the characteristics of the driven load or power company limita-

FIGURE 12–19 *Autotransformer starting—speed versus torque. (The Lincoln Electric Co.)*

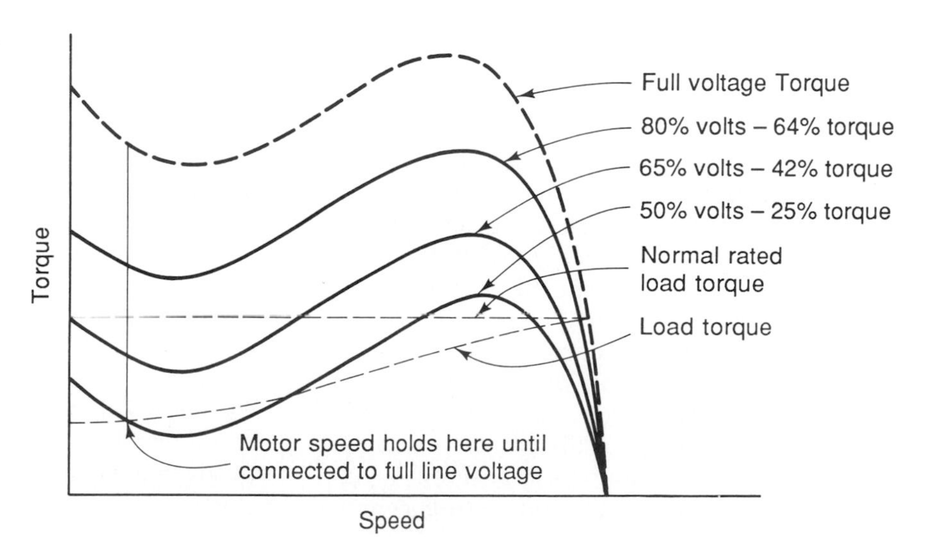

(A)

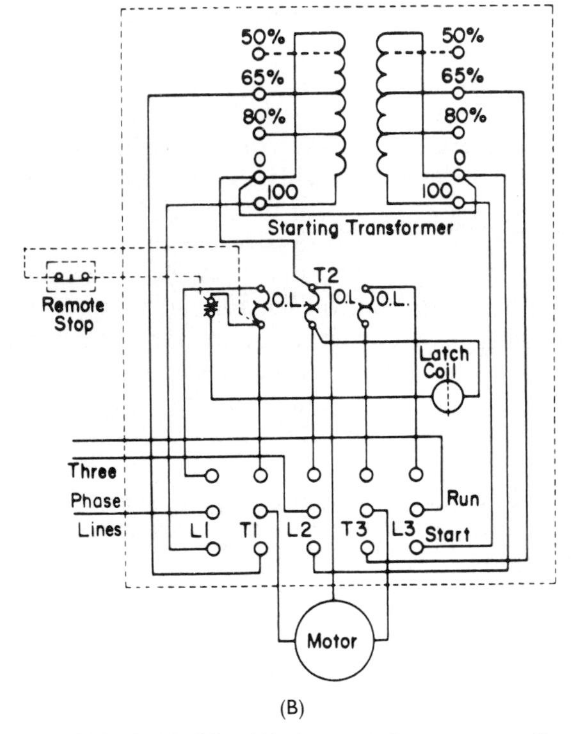

(B)

FIGURE 12–20 *(A) Autotransformer type of magnetic starter; (B) corresponding wiring diagram. (Allen-Bradley)*

tions require starting at reduced voltage (Fig. 12–20). NEMA (National Electrical Manufacturers Association) permits one start every 4 minutes, for a total of four starts followed by a rest period (2 hours). Each starting period is not to exceed 15 seconds. Figure 12–21 shows a autotransformer type of starter. Note the location of the taps on the starting transformer.

The autotransformer provides the highest starting torque per ampere of line current. Thus it is an effective means of motor starting for applications where the inrush current must be reduced with a minimum sacrifice of starting torque. This type of starter arrangement features closed-circuit transition, an arrangement that maintains a continuous power connection to the motor during the transition from reduced to full voltage. This avoids the high transient switching currents characteristic of starters using open-circuit transition. It provides smoother acceleration as well.

Operation

Operating an external START button or pilot device closes the neutral and start contactors, applying reduced voltage to the motor through the autotransformer. After a preset interval, the timer contacts drop out the neutral contactor, breaking the autotransformer connection but leaving part of the windings connected to the motor as a series reactor. The RUN contactor then closes to short out this reactance and apply full voltage to the motor. Transition from reduced to full voltage is accomplished without opening the motor circuit.

For starters rated up to 200 hp you should allow a 15-second operation out of every 4 minutes for 1 hour followed by a rest period of 2 hours. For starters rated above 200 hp, you should allow three 30-second operations separated by 30-second intervals followed by a rest period of 1 hour. The major disadvantages of this type of starter are its expense for lower horsepower ratings and its low power factor.

PART-WINDING STARTING

Part-winding motors have two sets of identical windings—intended to be operated in parallel—which can be energized in sequence to provide reduced starting current and reduced starting torque. Most (but not all) dual-voltage 230/460-V motors are suitable for part-winding starting at 230 V.

When one winding of a part-winding motor is energized, the torque produced is about 50% of "both winding" torque, and line current is 60 to 70% (depending on motor design) of comparable line voltage values. Thus, although part-winding starting is not truly a reduced voltage means, it is usually also classified as such because of its reduced current and torque.

When a dual-voltage delta-connected motor is operated at 230 V from a part-winding starter having a three-pole start and a three-pole run contactor, an unequal current division occurs during normal operation resulting in overloading of the starting contactor. To overcome this defect, some part-winding starters use a four-pole starting contactor and a two-pole run contactor. This arrangement eliminates the unequal current division obtained with a delta-wound motor, and it enables wye-connected part-winding motors to be given either a one-half or two-thirds part-winding start.

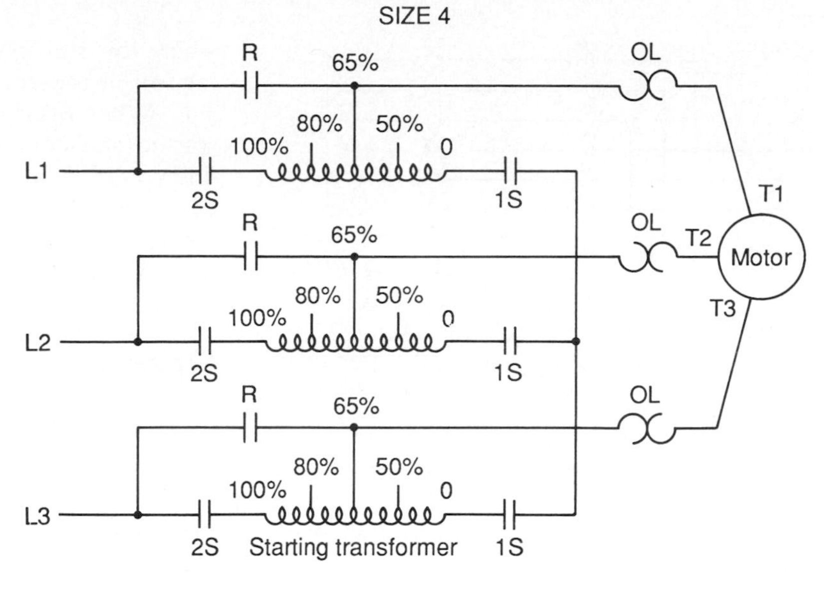

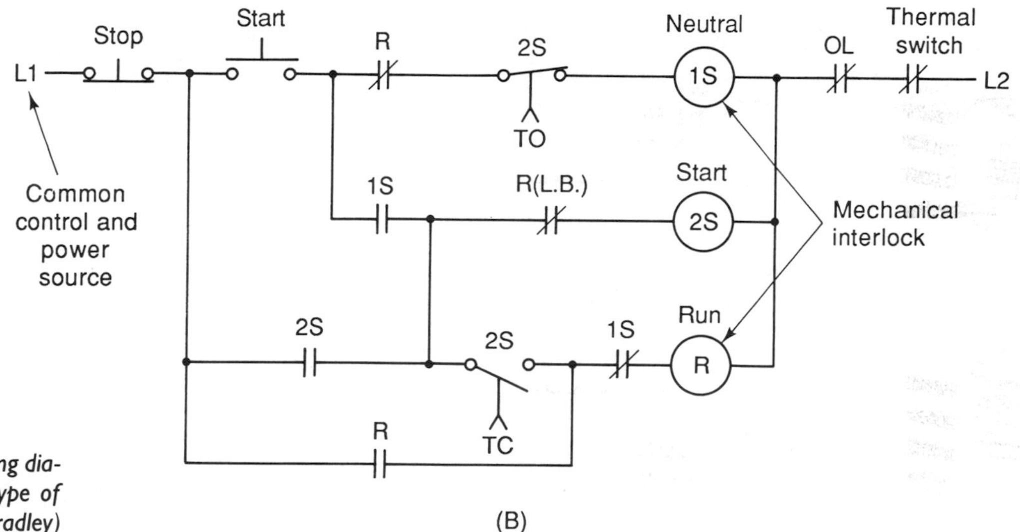

FIGURE 12–21 *Typical wiring diagram for an autotransformer type of reduced-voltage starter. (Allen-Bradley)*

The class 8640 starters have a start contactor, a timing relay, a run contactor, and necessary overload relays. Closing the pilot device contact causes the start contactor to close to connect the start winding and to initiate the time cycle. After expiration of the preset timing, the run contactor closes to connect the balance of the motor windings. A time setting of 1 second is recommended. Most motor manufacturers do not permit energization of the start winding alone for longer than 3 seconds. Part-winding starters provide closed transition starting.

Operation

The part-winding type of starter is shown in Fig. 12–22. The parts are located for ease in understanding the operation. By taking a look at the schematic in Fig. 12–23 you can see how the starter operates. Closing the START button or other pilot device energizes

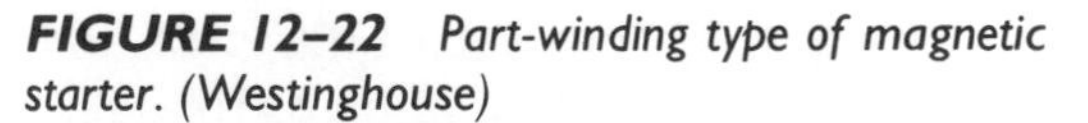

FIGURE 12–22 *Part-winding type of magnetic starter. (Westinghouse)*

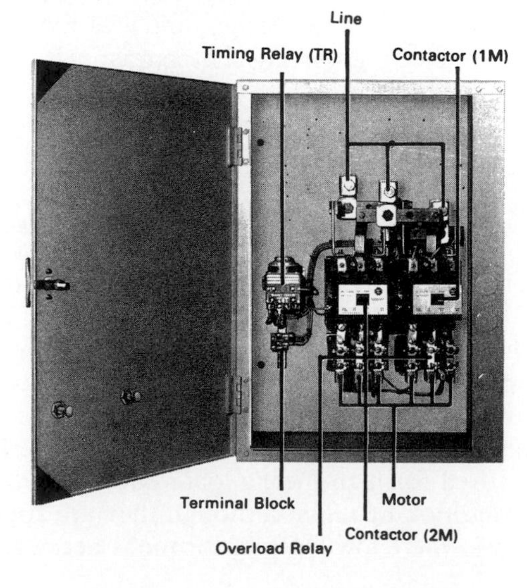

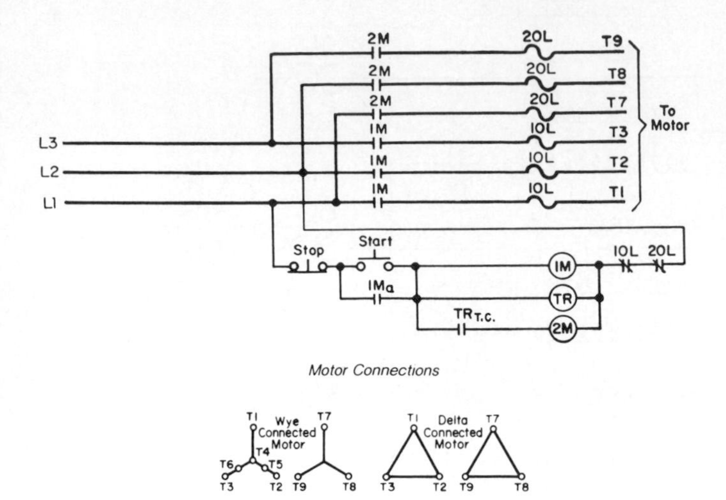

FIGURE 12-23 *Typical wiring diargram for part-winding type of starter. (Westinghouse)*

the start contactor (1M) that seals in through its interlock ($1M_a$) and energizes the timer (TR). The (1M) contacts connect the first half-winding of the motor across the line. After a preset time interval, timer (TR_{TC}) contacts close the energizing contactor (2M). The (2M) contacts connect the second half-winding of the motor across-the-line.

An overload, which opens the STOP button or other pilot device, deenergizes contacts 1M, 2M, and timer TR, removing the motor from the line. The three-pole contactor (1M) connects only the first half-winding of the motor for reduced inrush current on starting. A three-pole contactor (2M) connects the second half-winding of the motor for running.

Advantages and Disadvantages

Part-winding starters are the least expensive reduced-voltage controller. They use closed transition starting and are small in size. The disadvantages are that they are unsuitable for long acceleration or frequent starting, require special motor design, and that there is no flexibility in selecting starting characteristics.

WYE–DELTA OR STAR–DELTA STARTERS

Wye–delta or star–delta starters are used with delta-wound squirrel-cage motors that have all leads brought out to facilitate a wye connection for reduced-voltage starting. This starting method is particularly suitable for applications involving long accelerating times or frequent starts. Wye–delta starters are typically used for high-inertia loads such as centrifugal air-conditioning units, although they are applicable in cases where low starting torque is necessary or where low starting current is necessary and low starting torque is permissible.

When 6- or 12-lead delta-connected motors are started star-connected, approximately 58% of full-line voltage is applied to each winding and the motor develops 33% of full-voltage starting torque and draws 33% of normal locked-rotor current from the line. When the motor has accelerated, it is reconnected for normal delta operation.

Operation

Operating an external START button energizes the motor in the wye connection (Fig. 12–24). This applies approximately 58% of full-line voltage to the windings. At this reduced voltage, the motor will develop about 33% of its full-voltage starting torque and will

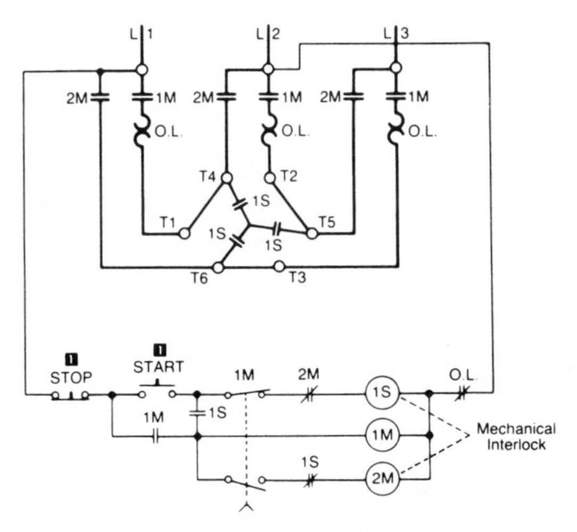

FIGURE 12–24 *Typical wiring diagram for wye–delta starter, open-circuit transistion. (Allen-Bradley)*

FIGURE 12–25 *Typical wiring diagram for wye–delta starter, closed-circuit transistion. (Allen-Bradley)*

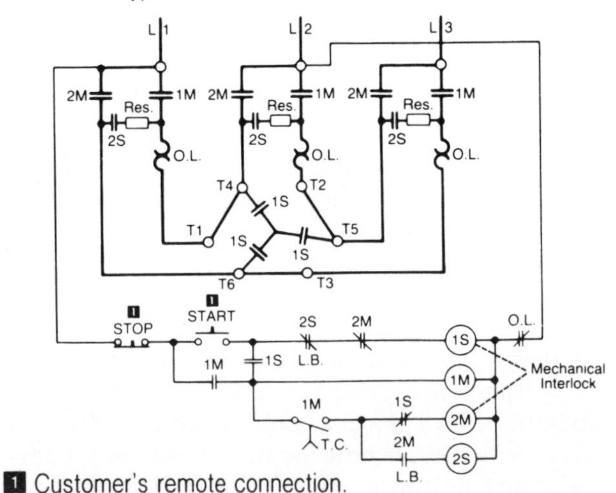

■ Customer's remote connection.

draw about 33% of its normal locked-rotor current. After an adjustable time interval, the motor is automatically connected in delta, applying full line voltage to the windings. In starters with open-circuit transition the motor is momentarily disconnected from the line during the transition from the wye to delta. With closed transition (Fig. 12–25) the motor remains connected to the line through the resistors. This avoids the current surges associated with open-circuit transition.

Advantages and Disadvantages

The advantages are moderate cost and its suitability for high-inertial, long-acceleration loads. It does have torque efficiency. However, the disadvantages are that it requires special motor design, starting torque is low, and it is inherently open transition—closed transition is available at added cost. There is no flexibility in selecting starting characteristics.

Star–Delta (Wye–Delta) Connections

There is the 12-lead motor wound for Y–Δ starting operation on either low voltage or a higher voltage (Fig. 12–26). There is also a six-lead single-voltage motor suitable for Y–Δ starting. Figure 12–26B shows the connection to the lines for the six-lead motor. Keep in mind that overload relay protection is required by the *National Electrical Code®*. The size of the protection is determined by the manufacturer of the motor (Table 12–4).

MULTISPEED STARTERS

Multispeed starters are designed for the automatic control of two-speed squirrel-cage motors of either the consequent pole or separate winding types. These starters are available for constant-horsepower, constant-torque, or variable-torque three-phase motors. Multispeed motor starters are commonly used on machine tools, fans, blowers, refrigeration compressors, and many other types of equipment.

Low-Speed Compelling Relay

When added to a standard starter, the low-speed compelling relay compels the operator always to start the motor in low speed before switching to a higher speed. This is a safety feature where damage to equipment may result when the motor is started at high speed (Fig. 12–27).

Automatic Sequence Accelerating Relay

The automatic sequence accelerating relay will control the sequence of acceleration from low speed up to high speed.

Automatic Sequence Decelerating Relay

The automatic sequence decelerating relay is used with large-inertia loads. The braking effect caused by a

FIGURE 12–26 *Star–delta connections. (The Lincoln Electric Co.)*

STAR-DELTA (YΔ) CONNECTIONS

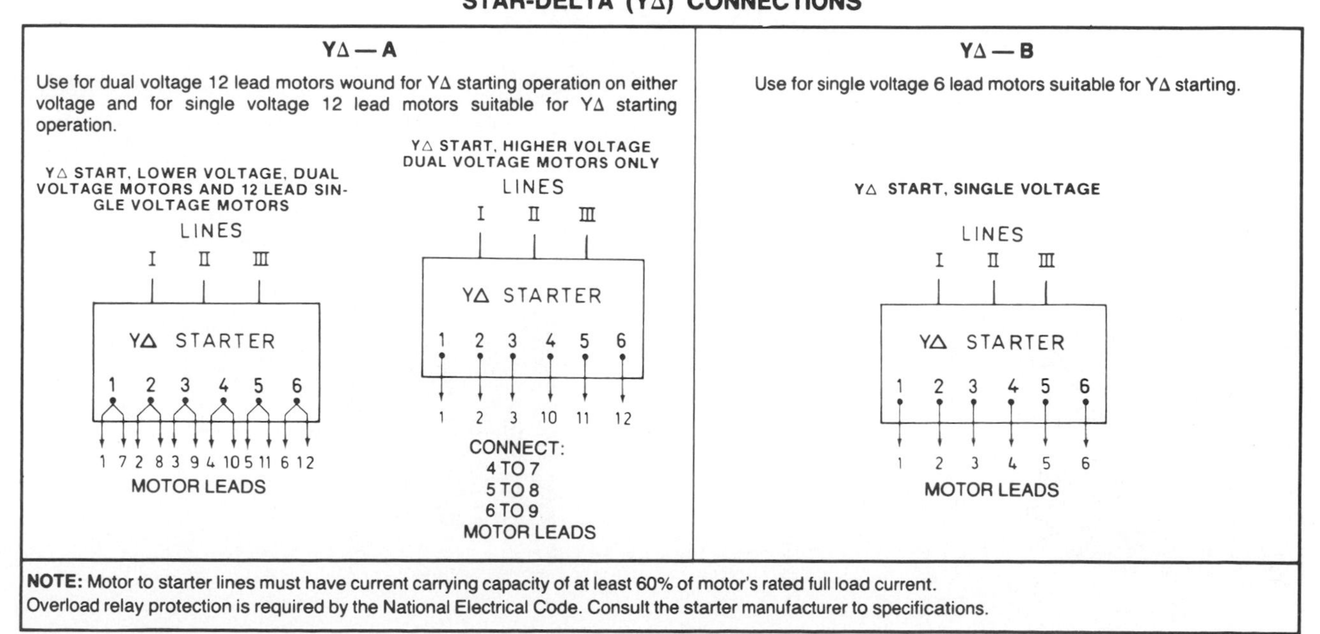

TABLE 12–4 SELECTION OF A CONTROLLER BEST SUITED FOR A PARTICULAR CHARACTERISTIC

Characteristic Wanted	*Type of Starter to Use (Listed in Order of Desirability)*	*Comments*
Smooth acceleration	1. Solid state (class 8660) 2. Primary resistor (class 8647) 3. Wye–delta (class 8630) 4. Autotransformer (class 8606) 5. Part-winding (class 8640)	Little choice between 3 and 4.
Minimum line current	1. Autotransformer (class 8606) 2. Solid state (class 8660) 3. Wye–delta (class 8630) 4. Part winding (class 8640) 5. Primary resistor (class 8647)	
High starting torque	1. Autotransformer (class 8606) 2. Solid state (class 8660) 3. Primary resistor (class 8647) 4. Part winding (class 8640) 5. Wye–delta (class 8630)	
High torque efficiency (torque vs. line current)	1. Autotransformer (class 8606) 2. Wye–delta (class 8630) 3. Part winding (class 8640) 4. Solid state (class 8660) 5. Primary resistor (class 8647)	Little choice between 3, 4, and 5.
Suitability for long acceleration	1. Wye–delta (class 8630) 2. Autotransformer (class 8606) 3. Solid state (class 8660) 4. Primary resistor (class 8647)	For acceleration time greater than 5 seconds, primary resistor requires non-standard resistors. Part-winding controllers are unsuitable for acceleration time greater than 2 seconds.
Suitability for frequent starting	1. Wye–delta (class 8630) 2. Solid state (class 8660) 3. Primary resistor (class 8647) 4. Autotransformer (class 8606)	Part-winding is unsuitable for frequent starts.
Flexibility in selecting starting characteristics	1. Solid state (class 8660) 2. Autotransformer (class 8606) 3. Primary resistor (class 8647)	For primary resistor, resistor change required to change starting characteristics. Starting characteristics cannot be changed for wye–delta or part-winding controllers.

Source: Courtesy of Square D.

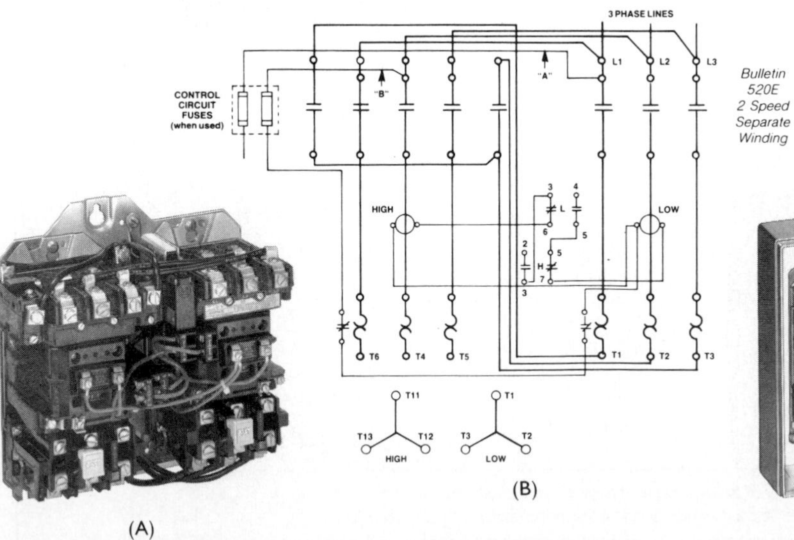

(A)

(B)

(C)

FIGURE 12–27 *(A) Multispeed starter and two-speed consequent pole starter without enclosure; (B) typical wiring diagram for two-speed separate winding motor starter; (C) general-purpose enclosure with cover removed. (Allen-Bradley)*

sudden change from high to low speed may cause damage to the motor or to the driven machine. To avoid this danger, the operation should give the motor sufficient time to slow down by pushing the STOP button and then waiting a short interval before pushing the button for a lower speed.

To help provide correct operation, multispeed starters can be equipped with an automatic sequence decelerating relay for each lower-speed step. This relay automatically interposes a time delay between the speed steps and makes it unnecessary to press the STOP button when switching to a lower speed.

CONSEQUENT-POLE MOTOR CONTROLLER

By increasing the number of poles a motor has it is possible to change its speed. By increasing the number of poles, the speed of the motor is decreased. Inasmuch as a motor is wound and mounted rather permanently on a frame, it is not easily possible to take out or put in poles or the associated windings. Therefore, an electrical means must be found if the speed of the motor is to be changed by using the number of poles method to do so. One method of doing this is the consequent-pole arrangement. This method can be used for two-speed, one-winding motors or four-speed, two-winding motors.

The reversal of some of the currents in the windings has the same effect as physically increasing or decreasing the number of poles. Three-phase motors are wound, in some cases, with six leads brought out for connection purposes. It is possible to connect the windings, using combinations of the terminals for connection purposes, either in series delta or in parallel wye (Fig. 12–28). By tapping the windings it is possible to send current in two different directions, effectively creating more poles and decreasing the speed of the motor. The number of poles is doubled by reversing through half a phase. Two speeds are obtained by producing twice as many consequent poles for low-speed operation as for high speed.

FIGURE 12–28 *Connections made by the consequent-pole starter for constant torque or variable torque. (Allen-Bradley)*

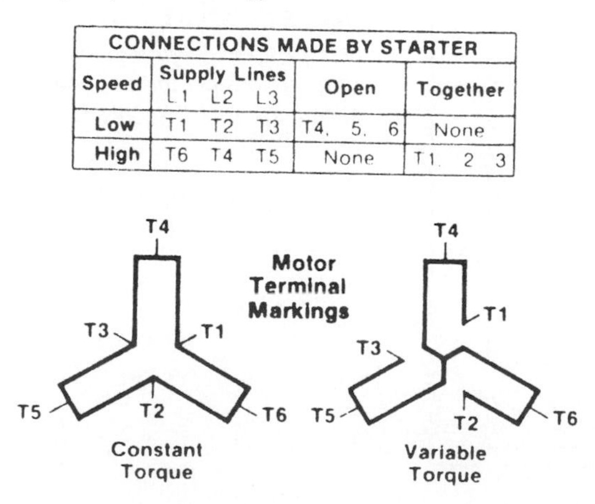

CONNECTIONS MADE BY STARTER			
Speed	Supply Lines L1 L2 L3	Open	Together
Low	T1 T2 T3	T4, 5, 6	None
High	T6 T4 T5	None	T1, 2, 3

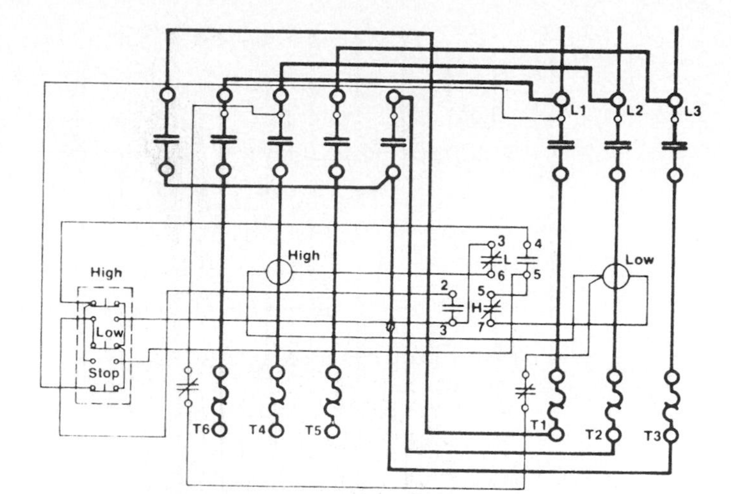

FIGURE 12–29 *Wiring diagram for a two-speed, consequent-pole, constant-horsepower motor, NEMA size 0–4. (Square D)*

Figure 12–29 shows how the controller is wired to produce consequent poles for constant torque or variable torque. The wiring diagram and the line drawing (Fig. 12–30) illustrate connections for the following method of operation: The motor can be started in either HIGH or LOW speed. The change from LOW to HIGH or from HIGH to LOW can be made without first pressing the STOP button. Figure 12–31 shows pilot devices with connections that can be made to obtain different sequences and methods of operation. The series delta arrangement produces high speed. It also produces the same horsepower rating at high and low speeds.

The torque rating is the same for both speeds if the winding is such that the series delta connection gives the low speed and the parallel wye connection gives the high speed. Consequent-pole motors that have a single winding for two speeds have the extra tap at the midpoint of the winding. This permits the various connection possibilities. However, the speed range is limited to a 1:2 ratio of or 600/1200 or 900/1800 rpm.

Figure 12–32 shows the motor terminal markings and connections for a constant-horsepower delta. The wiring diagram (Fig. 12–33) and the line drawing (Fig. 12–34) illustrate connections for the following method of operation: Motor can be started in either HIGH or LOW speed. The change from LOW to HIGH can be made without first pressing the STOP button. When changing from HIGH to LOW, the STOP button must be pressed between speeds. The pilot devices shown in Figure 12–35 show the other connections that can be

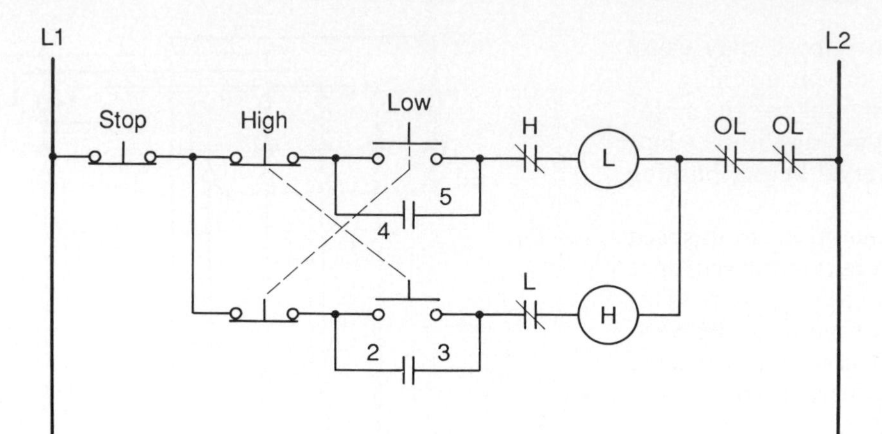

FIGURE 12-30 Line diagram for a two-speed motor. (Allen-Bradley)

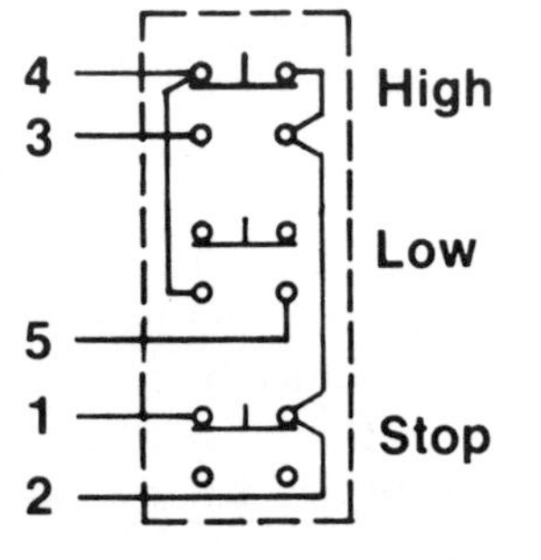

Connections above allow speed change from "LOW" to "HIGH" only without using "STOP." Start in either speed.

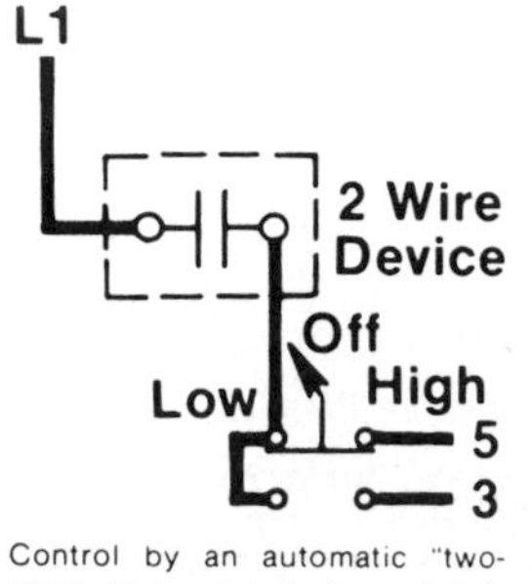

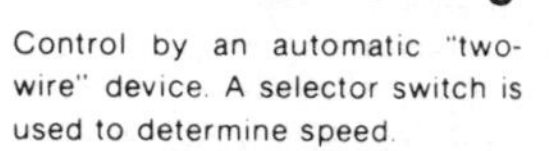

Control by an automatic "two-wire" device. A selector switch is used to determine speed.

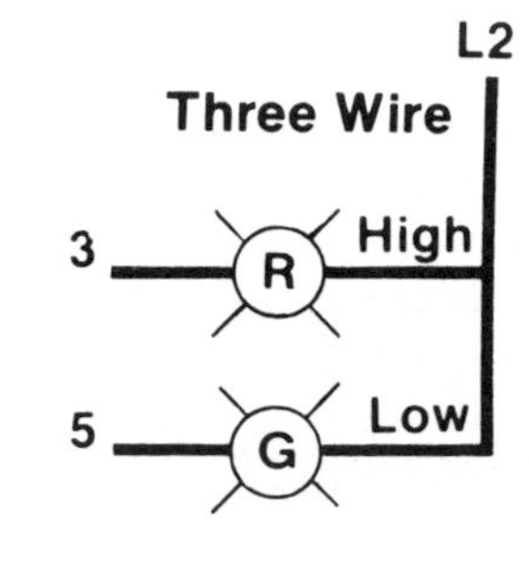

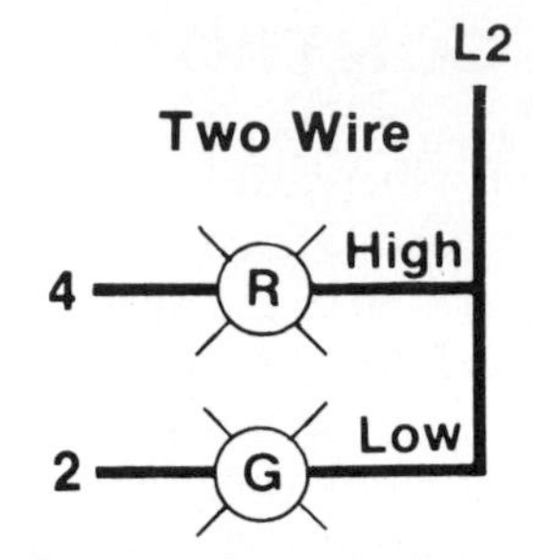

Connections for speed-indicating pilot lights. Can be added to any of the control schemes.

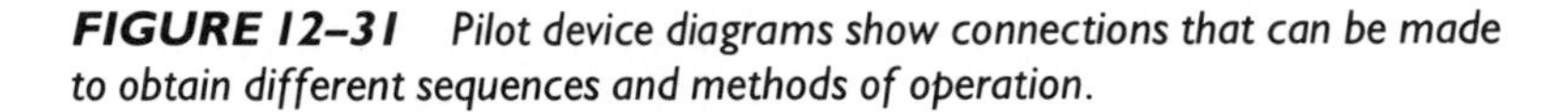

FIGURE 12-31 Pilot device diagrams show connections that can be made to obtain different sequences and methods of operation.

CONNECTIONS MADE BY STARTER			
Speed	Supply Lines L1 L2 L3	Open	Together
Low	T1 T2 T3	None	T4, 5, 6
High	T6 T4 T5	T1, 2, 3	None

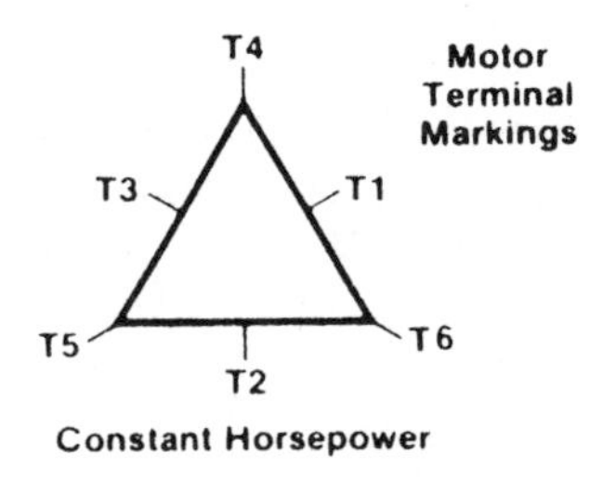

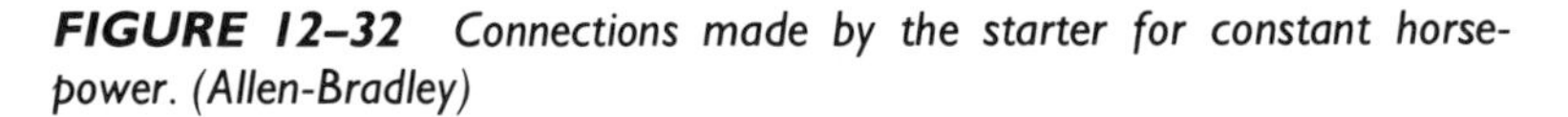

FIGURE 12-32 Connections made by the starter for constant horsepower. (Allen-Bradley)

FIGURE 12-33 Wiring diagram for two-speed, consequent-pole, constant- or variable-torque motor starter, NEMA size 0–4. (Square D)

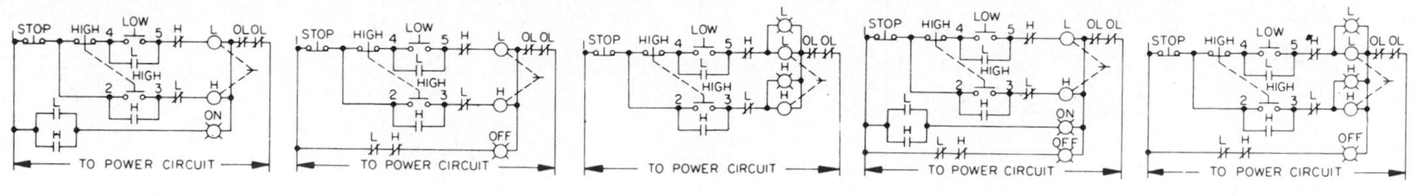

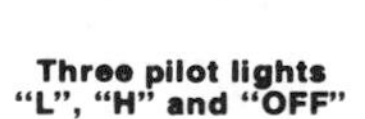

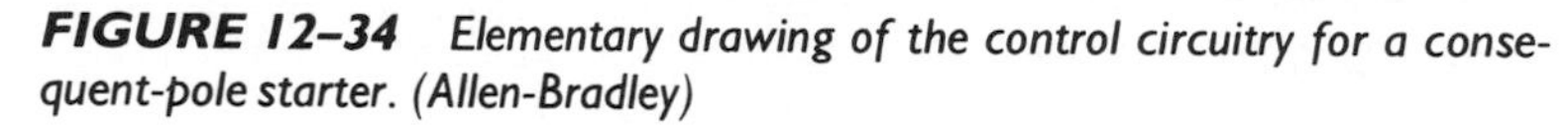

FIGURE 12–34 *Elementary drawing of the control circuitry for a consequent-pole starter. (Allen-Bradley)*

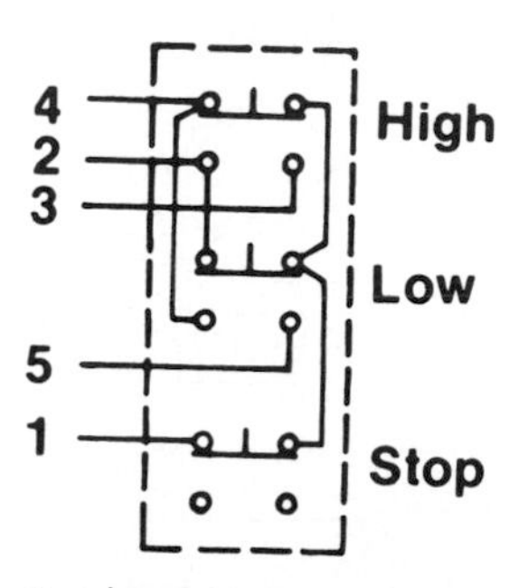

Push button connections to allow starting in either speed and changing from one speed to another without first pressing the "STOP" button.

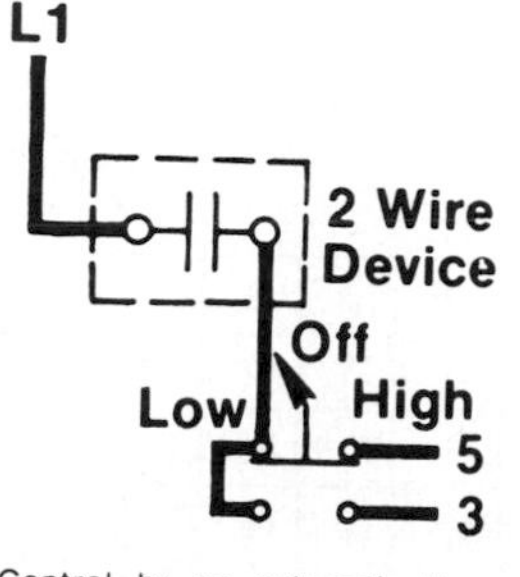

Control by an automatic "two-wire" device. A selector switch is used to determine speed.

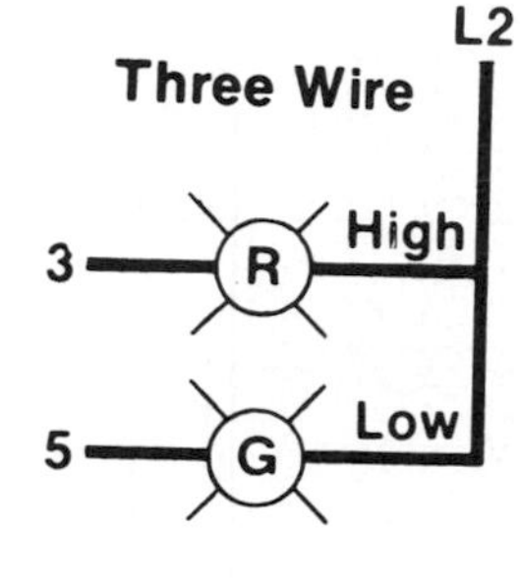

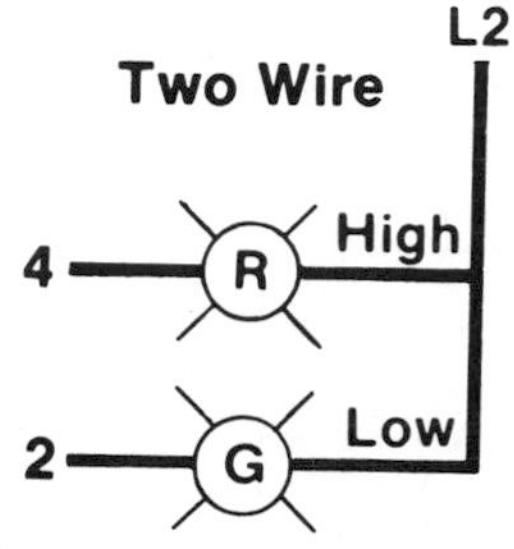

Connections for speed-indicating pilot lights. Can be added to any of the control schemes.

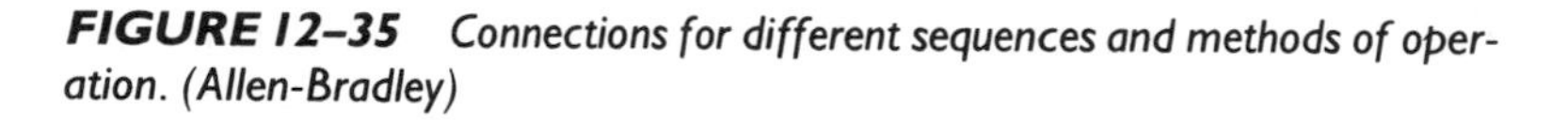

FIGURE 12–35 *Connections for different sequences and methods of operation. (Allen-Bradley)*

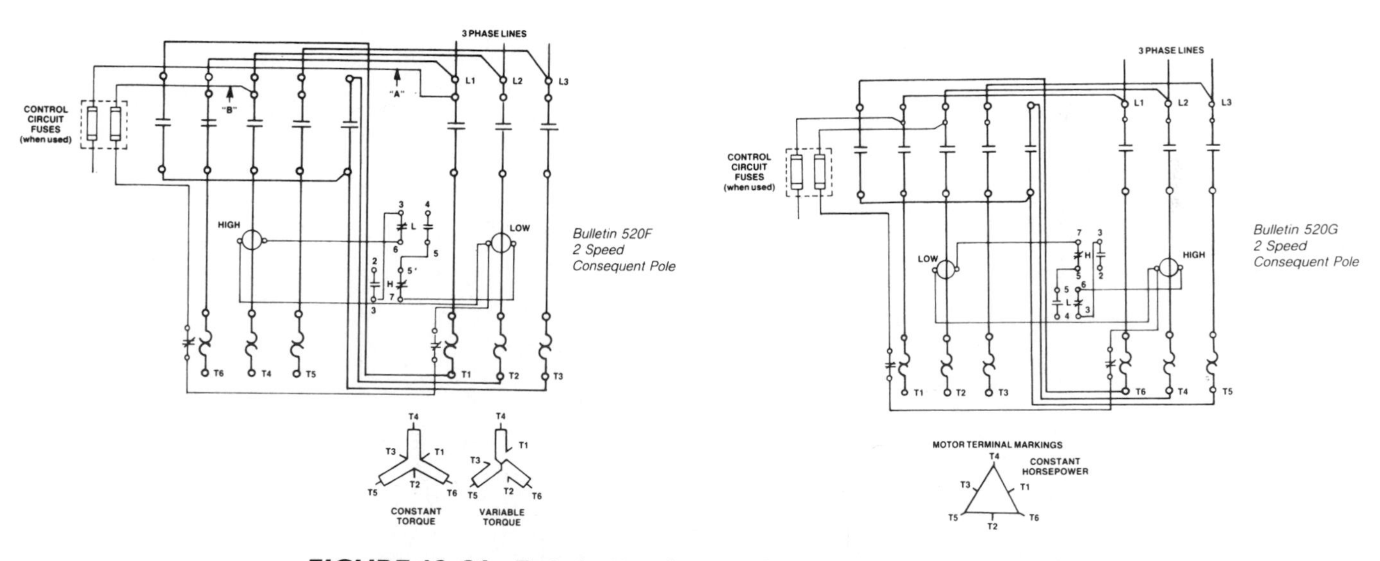

FIGURE 12–36 *Typical wiring diagrams for two-speed consequent-pole starter. (Allen-Bradley)*

made to obtain different sequences and methods of operation.

Four-speed, two-winding consequent-pole motor controllers can be used on squirrel-cage motors that have two reconnectable windings and two speeds for each winding. This type of motor does need a special type of starting sequence. This means that it must use the properties of the compelling relay, accelerating relay, and decelerating relay to operate correctly.

Figure 12–36 shows the two-speed consequent-pole starter with variable-torque and constant-torque connections. Figure 12–37 shows how the four-speed, two-winding controller is connected for the possible arrangements using this type of motor.

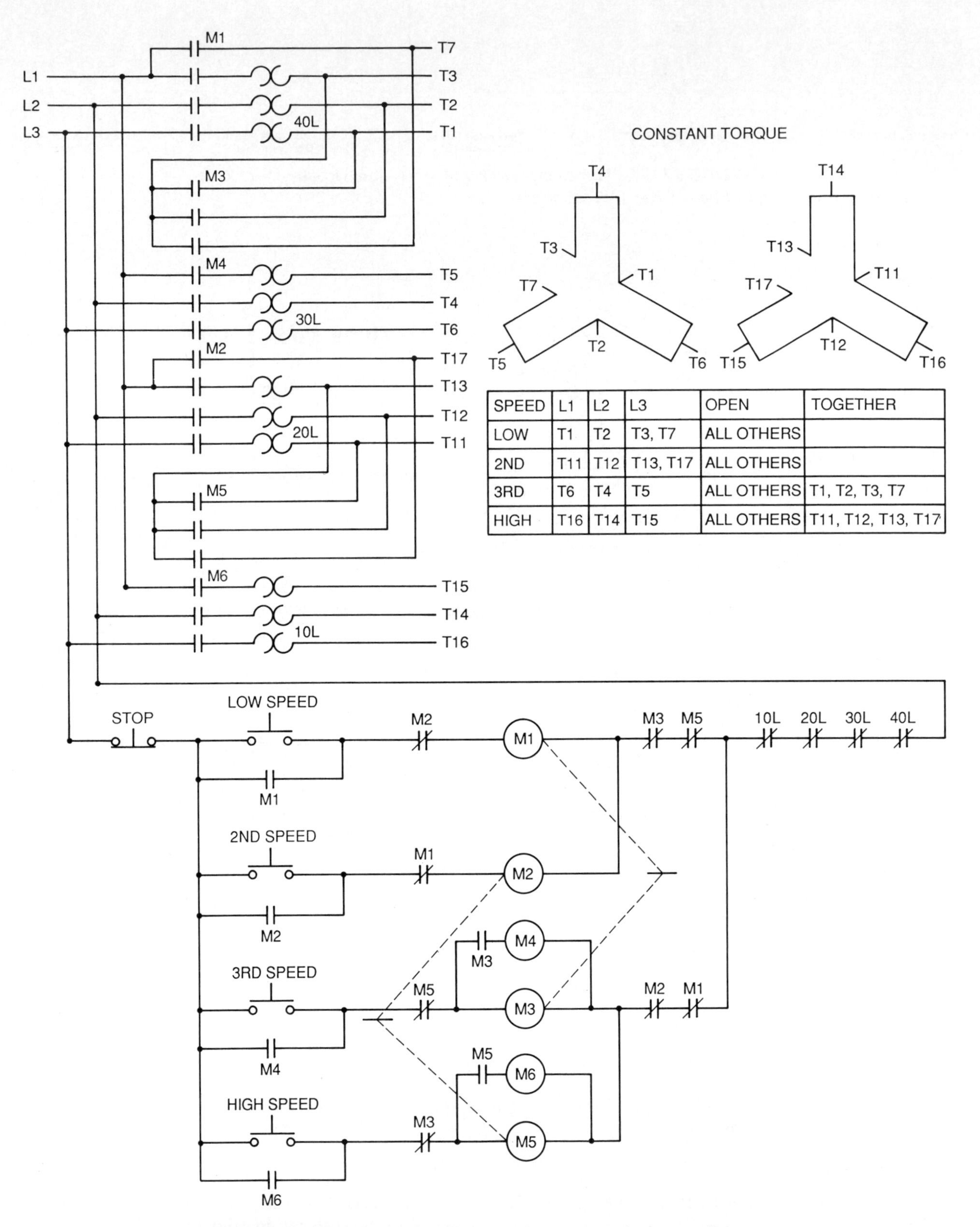

SPEED	L1	L2	L3	OPEN	TOGETHER
LOW	T1	T2	T3, T7	ALL OTHERS	
2ND	T11	T12	T13, T17	ALL OTHERS	
3RD	T6	T4	T5	ALL OTHERS	T1, T2, T3, T7
HIGH	T16	T14	T15	ALL OTHERS	T11, T12, T13, T17

FIGURE 12–37 *Elementary diagram of a four-speed, two-winding controller and the possible arrangements for motor connections. (Allen-Bradley)*

CONSTANT HORSEPOWER

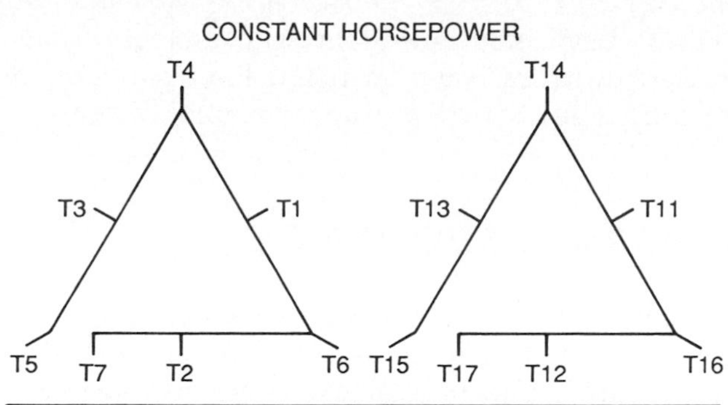

SPEED	L1	L2	L3	OPEN	TOGETHER
LOW	T1	T2	T3	ALL OTHERS	T4, T5, T6, T7
2ND	T6	T4	T5, T7	ALL OTHERS	
3RD	T11	T12	T13	ALL OTHERS	T14, T15, T16, T17
HIGH	T16	T14	T15, T17	ALL OTHERS	

(A)

CONSTANT HORSEPOWER

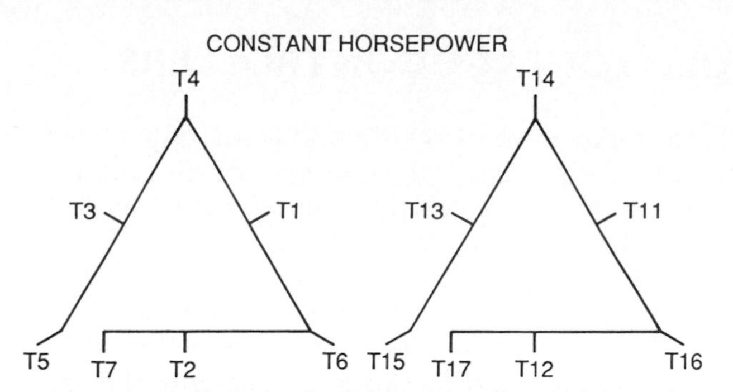

SPEED	L1	L2	L3	OPEN	TOGETHER
LOW	T1	T2	T3	ALL OTHERS	T4, T5, T6, T7
2ND	T11	T12	T13	ALL OTHERS	T14, T15, T16, T17
3RD	T6	T4	T5, T7	ALL OTHERS	
HIGH	T16	T14	T15, T17	ALL OTHERS	

(B)

CONSTANT TORQUE

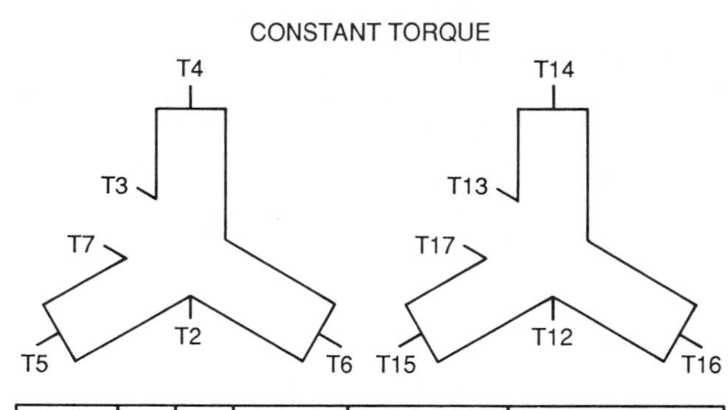

SPEED	L1	L2	L3	OPEN	TOGETHER
LOW	T1	T2	T3, T7	ALL OTHERS	
2ND	T6	T4	T5	ALL OTHERS	T1, T2, T3, T7
3RD	T11	T12	T13, T17	ALL OTHERS	
HIGH	T16	T14	T15	ALL OTHERS	T11, T12, T13, T17

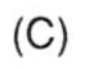

CONSTANT TORQUE

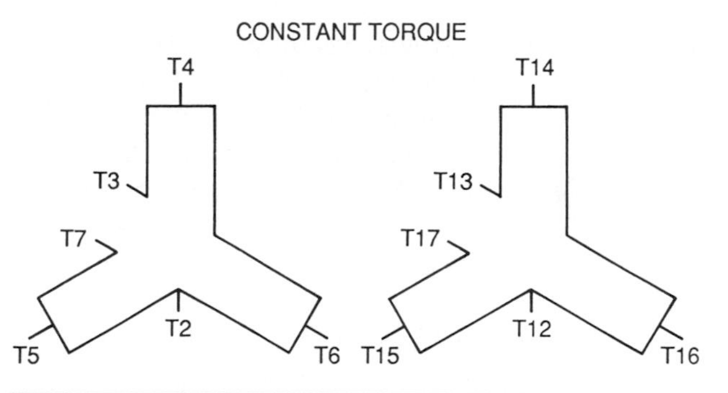

SPEED	L1	L2	L3	OPEN	TOGETHER
LOW	T1	T2	T3, T7	ALL OTHERS	
2ND	T11	T12	T13, T17	ALL OTHERS	
3RD	T6	T4	T5	ALL OTHERS	T1, T2, T3, T7
HIGH	T16	T14	T15	ALL OTHERS	T11, T12, T13, T17

(D)

VARIABLE TORQUE

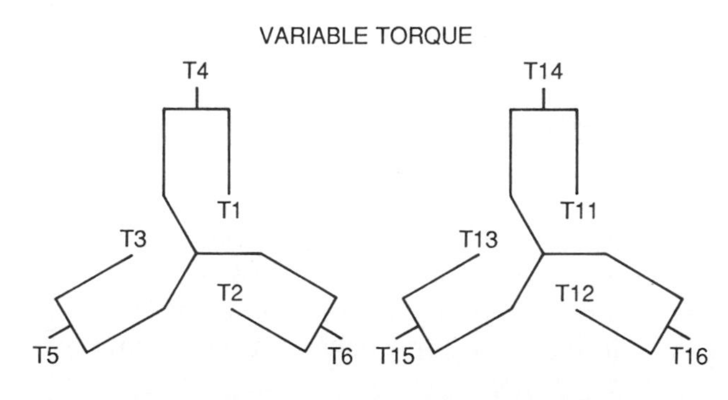

SPEED	L1	L2	L3	OPEN	TOGETHER
LOW	T1	T2	T3	ALL OTHERS	
2ND	T6	T4	T5	ALL OTHERS	T1, T2, T3
3RD	T11	T12	T13	ALL OTHERS	
HIGH	T16	T14	T15	ALL OTHERS	T11, T12, T13

(E)

VARIABLE TORQUE

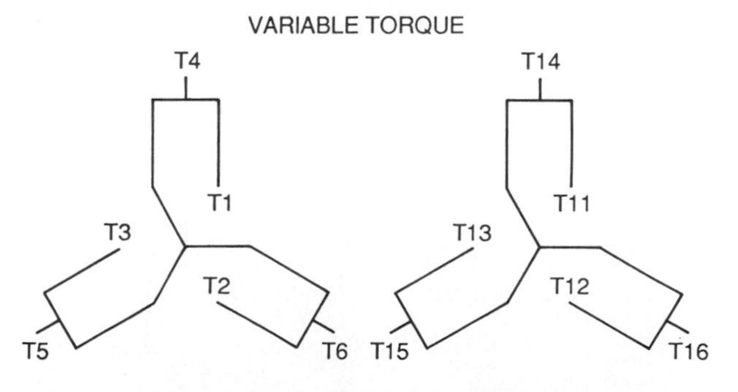

SPEED	L1	L2	L3	OPEN	TOGETHER
LOW	T1	T2	T3	ALL OTHERS	
2ND	T11	T12	T13	ALL OTHERS	
3RD	T6	T4	T5	ALL OTHERS	T1, T2, T3
HIGH	T16	T14	T15	ALL OTHERS	T11, T12, T13

(F)

FIGURE 12–37 *continued*

FULL-VOLTAGE CONTROLLERS

The least expensive of the starters is the full-voltage type. There is no limit to the horsepower, size, voltage rating, or type of motor that can be started on full voltage when the power is available.

Full-voltage starters are always the first choice *when the power system can supply initial inrush current,* and the motor and the driven machine can withstand the sudden starting shock. Examples of this are machines that start unloaded, as well as those that require little torque; or machines may be equipped with some form of unloading device to reduce starting torque, as in the use of an unloader valve in a compressor. A clutch may be inserted between a machine and motor so that the motor may be started unloaded. When the motor is up to speed the clutch is engaged. Clutches are sometimes used on large machines so that maximum horsepower can be exerted during breakaway without serious power system disturbance. Use of clutches also permits using motors with lower torque and locked-rotor currents. In most instances, up-to-date installations use *solid-state motor controllers* to better advantage. Many of the older types of starters are still in use and will continue to provide good service for many more years. As they deteriorate, they are usually replaced by a solid-state type of starter so that the clutch arrangements are unnecessary.

Figure 12–38 shows the general-purpose enclosure for a full-voltage starter. This type of starter is designed for full-voltage starting of polyphase squirrel-cage motors and primary control of slip-ring motors. This type of starter may be operated by remote control with pushbuttons, float switches, thermostats, pressure switches, snap switches, limit switches, or any other suitable two- or three-wire pilot device.

FIGURE 12–38 *Full-voltage starters (NEMA), open type, without enclosure: (A) size 3; (B) size 5. (Allen-Bradley)*

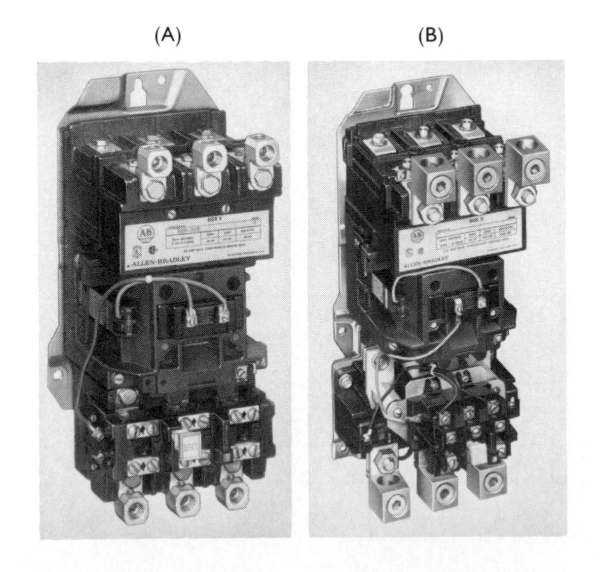

STARTING SEQUENCE

If full-voltage starting produces excessive current demands on the distribution system, motors should be started individually or in blocks of permissible size by using some method of time delay, such as motor driver, pneumatic, or mercury plunger timing relays. When large and small motors are to be started on a common power system, best results are obtained by starting the largest sizes first. This gives larger motors the advantage of full-line capacity. If synchronous motors are on the system with other types of ac motors, the synchronous units should always be started first since they provide voltage stability for starting the induction motors.

PROTECTION AGAINST LOW VOLTAGE

Low-voltage protection is needed while the motors are running even though systematic starting permits all motors to be started without excessive line voltage drop. When three-wire control circuits are used, a severe dip in line voltage or a momentary complete outage breaks the control-sealing circuits, and the controller drops out and stops the motor. This provides low-voltage protection and prevents simultaneous acceleration of all motors to full speed after being slowed down by a voltage dip. However, all motors are disconnected from the line during the voltage dip, and each must be restarted.

TIME-DELAY PROTECTION

It is possible to wire the circuitry so that a time-delay undervoltage arrangement can be used. This permits dropout of the controllers on low-voltage dips but allows restarting automatically if normal voltage is restored within a preset time delay. The usual time delay is 2 seconds or less.

Time-delay undervoltage protection on controllers will prevent some complete shutdowns but should be applied with caution. If used on all motor controllers, restoration of voltage within the time-delay setting after a voltage dip causes each motor to attempt to accelerate simultaneously, thus producing excessive currents that may operate backup protection and starter overload devices and disconnect the motors.

Pilot devices such as pressure, float, or temperature switches automatically start and stop motors as the demand arises. On severe voltage dips or voltage failure, motor controllers drop open even though the demand switch is closed. Upon restoration of full voltage all units attempt to restart at the same time. This operating hazard can be overcome by adding a time delay in the starting circuit of each motor and timing the demand for starting at slightly different intervals. Time delays of various units can then be staggered so that at the restoration of voltage only one unit at a time will be started.

QUESTIONS

1. What is voltage spread?
2. What is the purpose of a centrifugal switch on a single-phase motor?
3. How can direction of rotation be reversed on a split-phase motor?
4. What type of motor uses pushrods and a wound armature?
5. Where are capacitor-start motors used?
6. How are capacitor-start motors reversed when standing still?
7. What advantage does the permanent-split capacitor motor have?
8. What are shaded-pole motors most likely to be used for?
9. What is needed to get a split-phase motor to run?
10. How much current does an across-the-line motor draw when it starts?
11. What is the advantage of reduced-voltage motor starting?
12. What is another name for primary resistor starters?
13. What is the major disadvantage of the autotransformer starter?
14. What type of starting does part-winding starters provide?
15. What is the least expensive method of motor starting?
16. Where are wye–delta starters typically used?
17. Why are wye–delta starters used with delta–wound squirrel-cage motors?
18. Why are compelling relays needed?
19. What happens to motor speed when more poles are added?
20. How do consequence pole motors obtain two speeds?

13

Solid-State Reduced-Voltage Starters

Objectives

After studying this chapter, you will be able to:

1. Define thyristor operation.
2. Explain what gating does.
3. Explain solid-state stepless acceleration.
4. Describe the operation of a diac in a control circuit.
5. Describe the operation of a triac in a control circuit.
6. Explain how surge supressors are installed on magnetic devices.
7. Describe lightning surge protection.

The electromechanical devices used for years are still reliable and working in many installations. They are used to provide sequencing and interlocking tasks. They are simple in construction, flexible in use, and have many contact combinations. They can also handle large currents and break the circuit as required.

Solid-state devices have no moving parts and no contacts to clean, replace, or adjust. They use transistors, triacs, diacs, and SCRs to do the switching. These logic elements can perform the same functions in a solid-state system as relays do in the electromechanical systems (Fig. 13–1).

The solid-state control device has many advantages that make it desirable for the various environments in which it has to operate. It has no contacts to become dirty or malfunction when needed to control a critical sequence of operations. The solid-state control devices are more reliable than electromechanical devices. They come in sealed-in modules that can be plugged into a rack and replaced as a unit if anything goes wrong with the circuitry.

FIGURE 13–1 *Solid-state reduced-voltage controller. (Square D)*

REDUCED-VOLTAGE STARTING

Reduced-voltage starting can be accomplished in a number of ways. However, in solid-state circuitry it is somewhat simpler than described previously. The exact details of the circuit functions are somewhat more complex than those of the electromechanical system; however, a complete understanding of solid-state physics and/or electronics is not necessary in order to grasp the workings of the simple devices utilized to perform the operations of solid-state switching and control.

SILICON-CONTROLLED RECTIFIERS

The silicon-controlled rectifier (SCR) is the device used most often to control electric motors. The proper name for an SCR is *thyristor.* However, popular use of the term SCR has made it part of the literature and accepted by everyone working in the field. It is a specialized type of semiconductor used for control of electrical circuits.

An SCR conducts current in a forward direction only. The symbol for an SCR is shown in Fig. 13-2. Current flows through an SCR from the cathode (C) to the anode (A). The illustration indicates the SCR also has a gate (G).

The function of the SCR is shown in the circuit diagram in Fig. 13-3. The most typical use of an SCR is for a controlled circuit. Examples include a light dimmer or a speed control for a motor. This type of circuit is illustrated in Fig. 13-3. The resistor in this circuit, R_1, is a rheostat, or adjustable resistor. This is used to control the amount of voltage delivered to the gate of the SCR. The more voltage delivered, the greater the flow. Thus, adjusting the rheostat can serve to control the circuit. If the circuit illuminates a lamp, lowering the voltage to the rheostat dims the bulb. If the load is a motor, its speed is slowed. Figures 13-4, 13-5, and 13-6 show what typical SCRs look like with their leads identified according to cathode, gate, and anode connections.

One of the main reasons for using semiconductor devices for motor control is the device's ability to start a motor under reduced-voltage conditions and thus allow the motor to accelerate to full speed at a lower torque level. By reducing the high current inrush the mechanical shock to the driven equipment is reduced.

A reduced-voltage solid-state motor starter uses SCRs for power control. Inasmuch as an SCR conducts in the direction of the arrow in the symbol, it

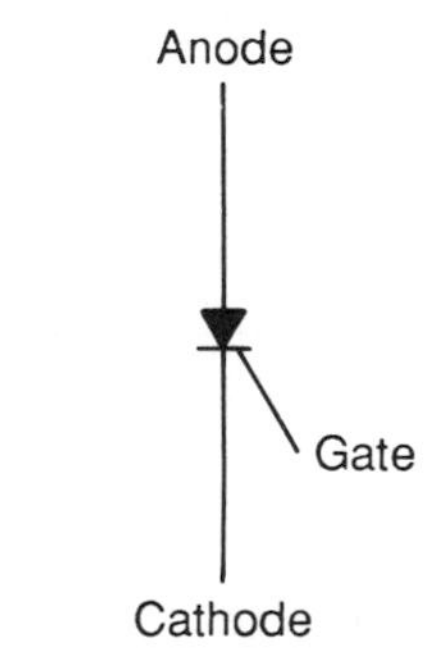

FIGURE 13-2 *Symbol for SCR.*

FIGURE 13-3 *Schematic of SCR-controlled circuit.*

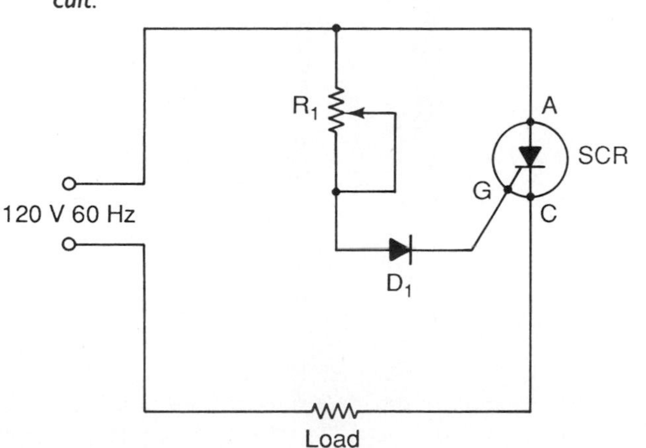

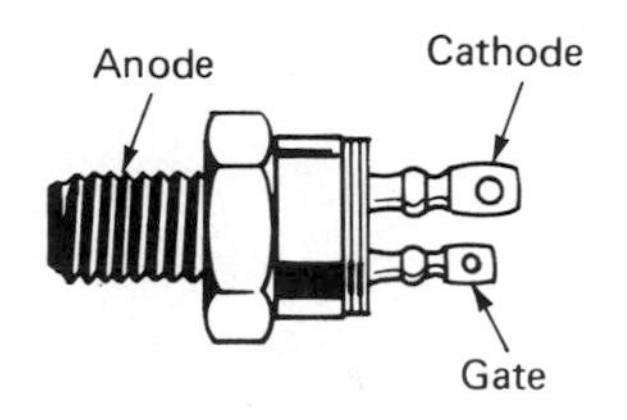

FIGURE 13-4 *Drawing of a typical SCR.*

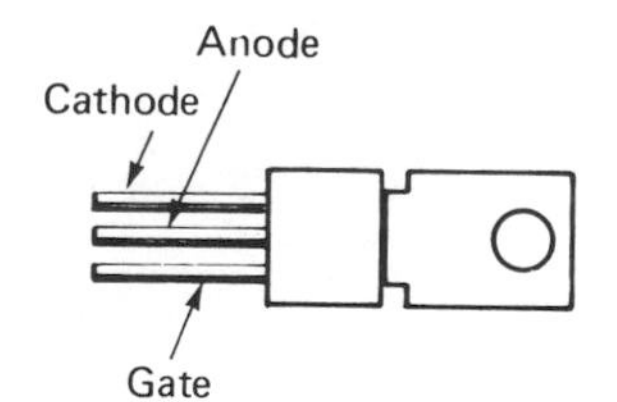

FIGURE 13-5 *Drawing of typical SCR.*

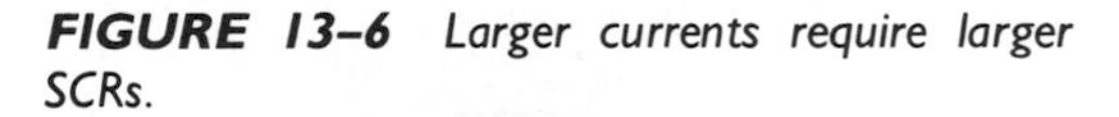

FIGURE 13-6 *Larger currents require larger SCRs.*

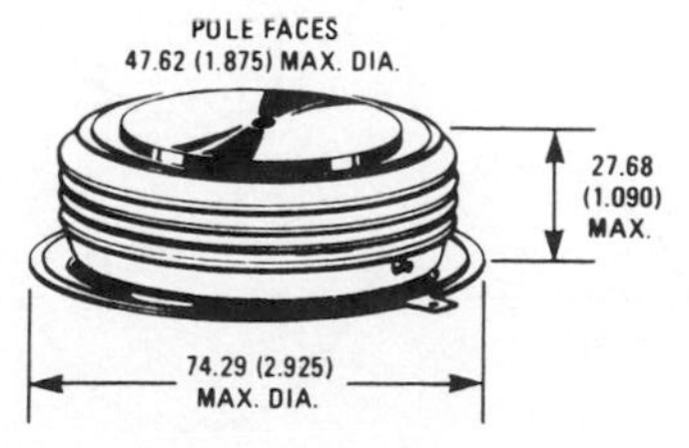

means that current flows only one way in an SCR. To use an SCR to its advantage on ac it is necessary to use two of them in reverse parallel (Fig. 13–7). SCRs have to be turned on in order to conduct current through them; that is, they need a gate pulse to turn them on. Once an SCR is turned on or gated, it does not stop forward current flow. Full wave control uses two SCRs in *each phase*. Three-phase operation must utilize six diodes, connected as shown in Fig. 13–8.

The current through an SCR can be controlled by *gating* the SCR at different times within the half-cycle. This also controls the acceleration time of the motor. If the gate pulse is applied early in the half-cycle, the output is high. If the gate pulse is applied late in the half-cycle, only a small part of the waveform is passed through and the output is low. So by controlling the SCR's output voltage the motor acceleration characteristics can be controlled (Fig. 13–9).

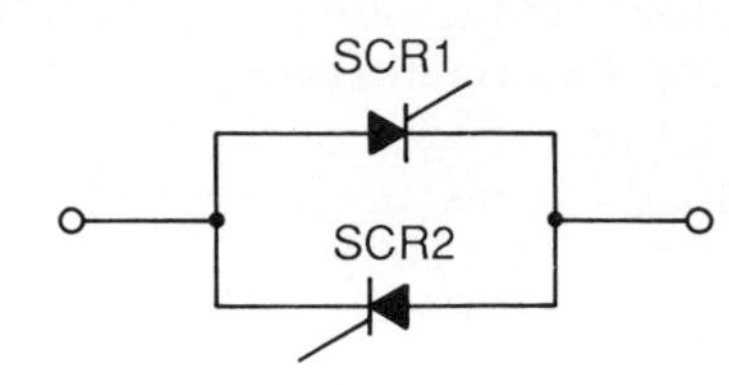

FIGURE 13–7 *Parallel SCRs for one phase.*

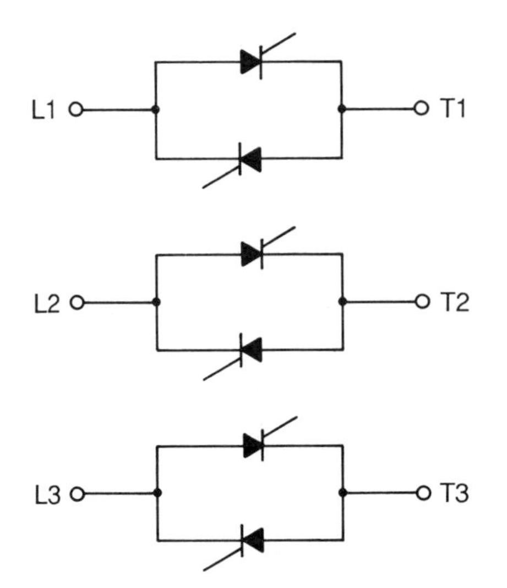

FIGURE 13–8 *Three-phase SCR arrangement.*

FIGURE 13–9 *Outputs from differently gated SCRs.*

25% output
(Late gated)
50% output
(Half gated)
75% output
(Early gated)
Full voltage
(Fully gated)

SOLID-STATE STEPLESS ACCELERATION

The class 8660 solid-state reduced-voltage controller provides smooth, stepless acceleration of a three-phase induction motor. The controller offers several standard and option features to control, monitor, and protect the motor during the start and run modes of operation. Modular construction of the controller adds flexibility and ease of maintenance (Fig. 13–10). Soft start is accomplished by gradually turning on six silicon-controlled rectifiers. Two SCRs are connected in a back-to-back or reverse-parallel arrangement and mounted on a heat sink to make up a power pole. The power pole also contains a printed circuit board and a thermal sensor.

Firing of the SCRs is controlled by the modules on the logic rack. These modules also check for correct startup and running conditions and provide a visual indication of controller status through the use of light-emitting diodes (LEDs). Each module has a spe-

FIGURE 13–10 *Power pole. (Square D)*

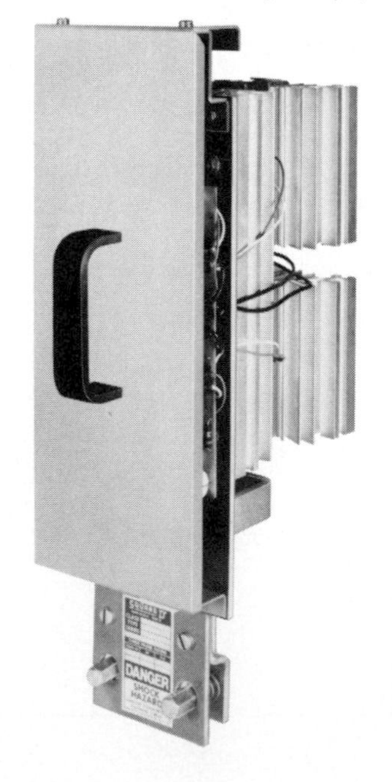

FIGURE 13–11 *Logic module rack. (Square D)*

cific location and function. Figure 13–11 shows a logic module rack.

LOGIC RACK

The logic rack is located on the lower part of the controller and has sockets for eight plug-in modules (Fig. 13–12). Each module has a specific location and performs a specific function in the operation of the controller. The module in the first position is internal to the controller and provides wiring connections between the power pole and the logic modules. The modules in positions 2 through 8 control the firing of the SCRs, check for correct startup and running conditions, and provide a visual indication of the controller status through the use of LEDs. The B-2 module goes in the second position, one of the B-3 modules goes in the third position, and so on. The specific module functions are described below.

FIGURE 13–12 *Module position in logic rack. (Square D)*

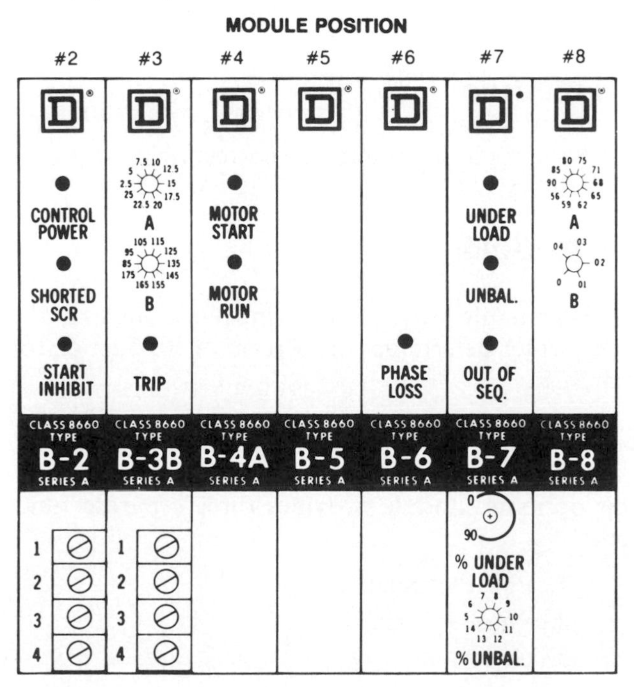

B-2 Module

This module provides logic voltages and checks for correct starting conditions. The control can be started if the control power LED is ON and the start inhibit LED is OFF.

B-3 Module

A three-phase, temperature-compensated, solid-state overload relay is supplied as an integral part of the controller. It provides class 10, inverse-time trip characteristics that protect against harmful motor overloads. There is a different B-3 module for each of the four controller current ratings of 200, 320, 500, and 720 A. Motor full-load current settings are adjustable by the use of potentiometers on the B-3 module. An overload condition will automatically deenergize the controller, close the alarm contact, and light the TRIP and START INHIBIT LEDs. An overload test feature on the logic rack assembly provides a check for operation of the solid-state overload circuitry. Overload trip time is a function of the current limit setting. The lower the current limit setting, the longer the trip time. Trip times for three current limit settings are shown in Table 13–1. Longer trip times for high-inertial loads can be provided on the other types of controllers. Form Z72 provides class 30 inverse-time trip characteristics by using a special B-3 module and power poles with higher current ratings. Trip times for class 30 overloads are shown in Table 13–1.

B-4 Module

The starting method that is used is determined by the B-4 module. Current limit starting is standard. The current limit setting is adjustable by the use of a potentiometer on the logic rack assembly. Optional starting methods are available. A description of each of the starting methods follows.

Current Limit (B-4A Module). The current-limit feature will limit the motor current to a preset level at all times during start and run conditions. Current limit

TABLE 13–1 OVERLOAD TRIP TIMES

Current Limit (% of MFLC)	*Trip Time (seconds)*	
	Standard Class 10	*Form Z72 Class 30*
150	90	250
300	30	90
425	5	40

is adjustable between 150 and 425% of motor full-load current by way of a potentiometer located on the logic rack. If a shorting contactor is used, this feature will be present only in the start condition (Fig. 13–13).

Linear Timed Acceleration (B-4B) Tachometer Feedback. This option allows the motor speed to be increased linearly with the time until the motor reaches full speed (Fig. 13–14). Start time is adjustable from 3 to 30 seconds and does not fluctuate with motor loading. This method gives the smoothest acceleration but requires a tachometer input. Motor current is limited to the current-limit setting.

Voltage Ramp (B-4C Module). This option allows the applied motor voltage to increase linearly from 0 to 100% over an adjustable period of 3 to 30 seconds. The motor current is limited to the current-limit setting. This method provides acceleration that is approximately linear from zero to full speed but does not require a tachometer. The actual acceleration time depends on the motor and load (Fig. 13–15).

Current Ramp (B-4D Module). This option supplies a breakaway current to the motor at start and then linearly ramps the current up to the current-limit setting. Breakaway current is adjustable from 0 to 150% of motor full-load current. Ramp time is adjustable from 0 to 7 seconds. This method provides the greatest control of starting current (Fig. 13–16).

Accel/Decel (B-4E Module). This option provides both a soft start and a soft stop. Starting characteristics are identical to the voltage ramp start (B-4C). This option also allows the applied motor voltage to decrease linearly from 50% to 0% over an adjustable period of 3 to 30 seconds to provide a soft stop. Provisions for an emergency stop are included (Fig. 13–17).

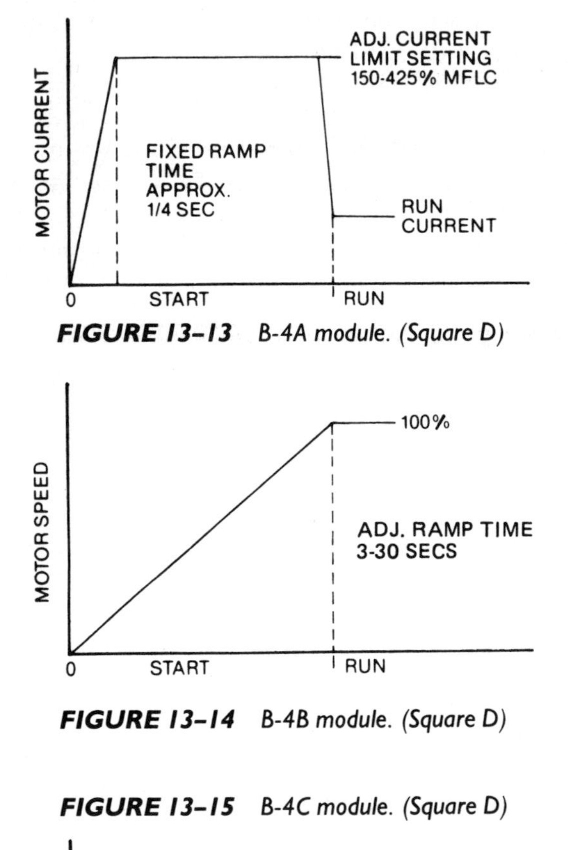

FIGURE 13–13 *B-4A module. (Square D)*

FIGURE 13–14 *B-4B module. (Square D)*

FIGURE 13–15 *B-4C module. (Square D)*

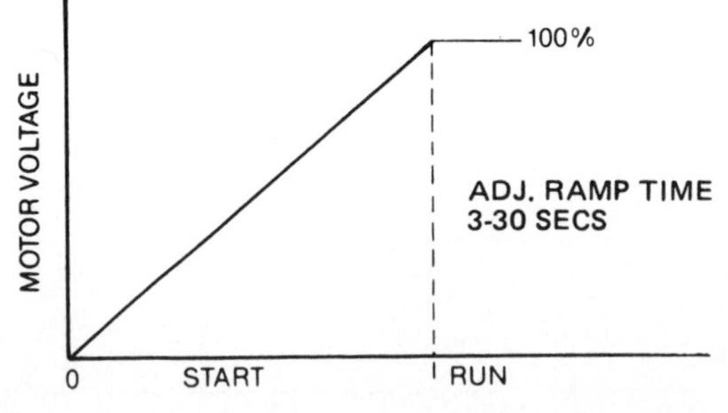

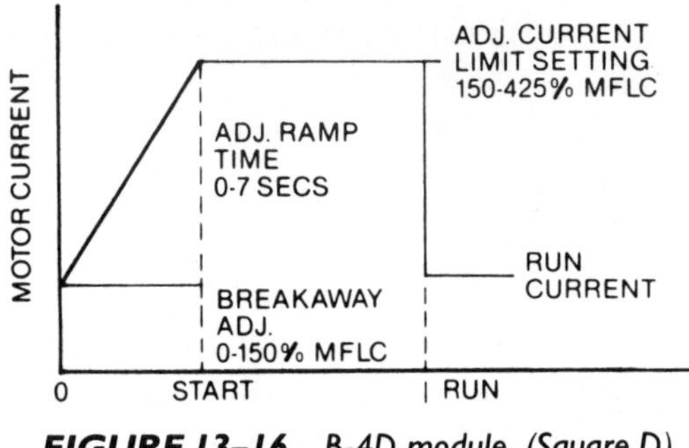

FIGURE 13–16 *B-4D module. (Square D)*

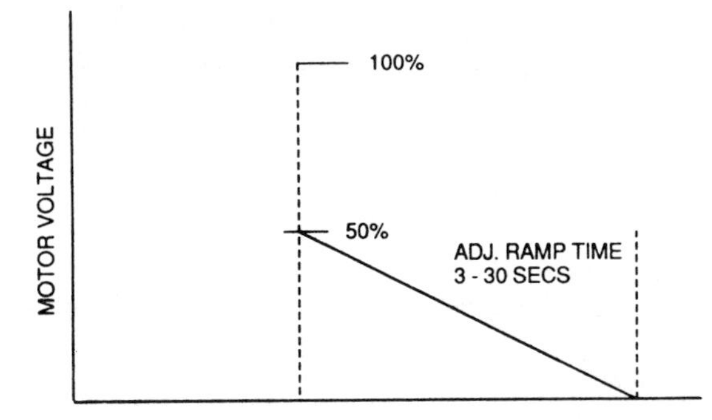

FIGURE 13–17 *B-4E module. (Square D)*

B-5 Module

The B-5 module determines the correct firing sequence of the SCRs.

B-6 Module

The B-6 module provides the firing phase angles of the SCRs, which determines the percent of conduction for each SCR.

B-7 Voltage Monitor Module

This optional module provides three separate functions:

1. Phase unbalance
2. Phase reversal
3. Underload

If any one of these occurs, the controller will shut off and the appropriate LEDs will be lighted.

The *phase unbalance function* is activated whenever three-phase power is present at the controller line terminals but is disabled during starting. A fault condition occurs when voltage unbalance is greater than the unbalance setting. The voltage unbalance setting is adjustable from 5% to 14% as defined by NEMA standards.

The *phase-reversal function* is activated whenever three-phase power is present at the controller line terminals. A fault condition occurs if the three phases are not in correct sequence. Without the B-7 module, the controller is phase insensitive and will operate with any phase sequence.

The *underload function* is activated after the motor is "up to speed." A fault condition occurs when the motor drops below the underload setting, which is adjustable from 0 to 90% of motor full-load current. This can be disabled by adjusting the setting to zero.

B-8 Energy-Saving Module

The energy-saving module will automatically adjust the voltage to the motor when load fluctuations occur. The motor will maintain full speed and required torque but draw less kVA when the load decreases. If the load increases, the module will respond by increasing the kVA so that the motor and load do not slow down in speed. This feature cannot be used on controllers with shorting contactors.

SHORTED SCR SWITCH

If an SCR shorts, the short is detected and the shorted SCR switch will flip to the YES position. This switch will also trip the shunt trip circuit breaker (if used) ahead of the controller. If there is an open circuit between the controller and the motor, the shorted SCR circuitry will trip. This can occur if there is an open disconnect switch between the controller and motor. Isolation contactors should be placed ahead of the controller. A motor load must be connected to the controller to prevent nuisance tripping of the shorted SCR circuitry.

ELEMENTARY WIRING DIAGRAMS FOR SOLID STATE

The solid-state reduced-voltage controller with an isolation contactor is shown in Fig. 13–18. Keep in mind that the M, SR2, OT, ALARM, SHORTED SCR, and UP-TO-SPEED relays are mounted on the controller and wired internally. Figure 13–19 shows the solid-state reduced-voltage controller with a shorting contactor and Fig. 13–20 shows the controller with a shorting contactor and an isolation contactor.

DIAC

The diac is basically a two-terminal device. It has a parallel-inverse combination of semiconductor layers.

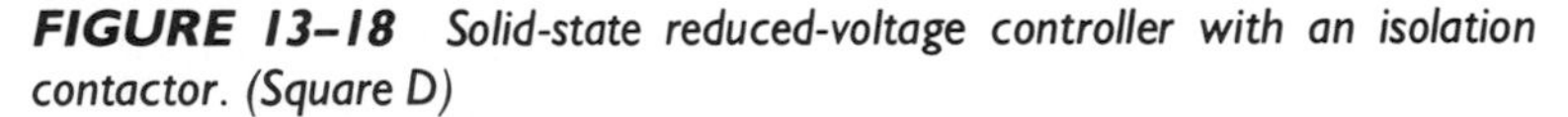

FIGURE 13–18 *Solid-state reduced-voltage controller with an isolation contactor. (Square D)*

NOTES:

1. M, SR2, OT, ALARM, SHORTED SCR, AND UP-TO-SPEED RELAYS ARE MOUNTED ON THE CONTROLLER AND WIRED INTERNALLY.
2. M DENOTES THE COIL FUNCTION OF THE SOLID STATE REDUCED VOLTAGE CONTROLLER.
3. THE SR2 RELAY CONTROLS THE START AND STOP SEQUENCE, AND ALSO HAS CONTACTS THAT MAY BE USED AS ELECTRICAL INTERLOCKS.
4. OT IS AN OVER TEMPERATURE SWITCH THAT OPENS WHEN THAT CONDITION EXISTS.
5. OL IS THE OVERLOAD RELAY CONTACT. IT OPENS WHEN: AN OVERLOAD IS DETECTED; L1, L2 OR L3 VOLTAGE IS NOT PRESENT; OR THE 120V CONTROL VOLTAGE IS MISSING.
6. THE ALARM CONTACT CLOSES WHEN AN OVERLOAD IS DETECTED.
7. THE SHORTED SCR CONTACT CLOSES WHEN THAT CONDITION EXISTS. IT IS USED WITH A CIRCUIT BREAKER OR DISCONNECTING SWITCH WITH A SHUNT TRIP COIL.
8. THE UP-TO-SPEED CONTACT CLOSES WHEN THE SCR'S ARE IN FULL CONDUCTION. IT IS USED WITH A SHORTING CONTACTOR.

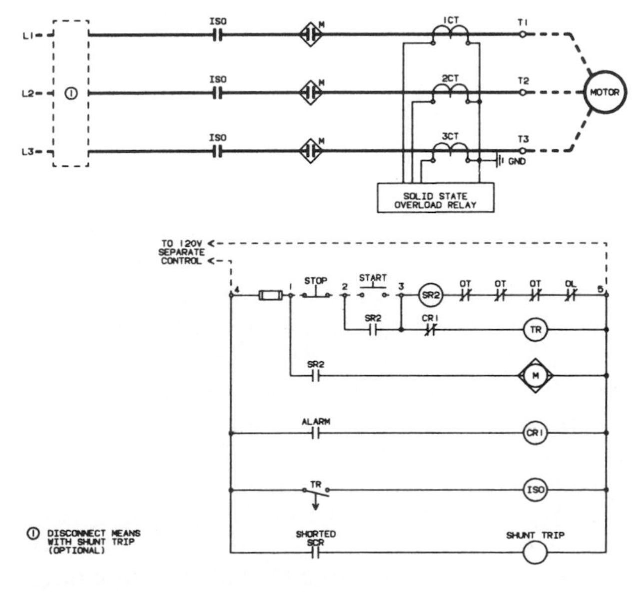

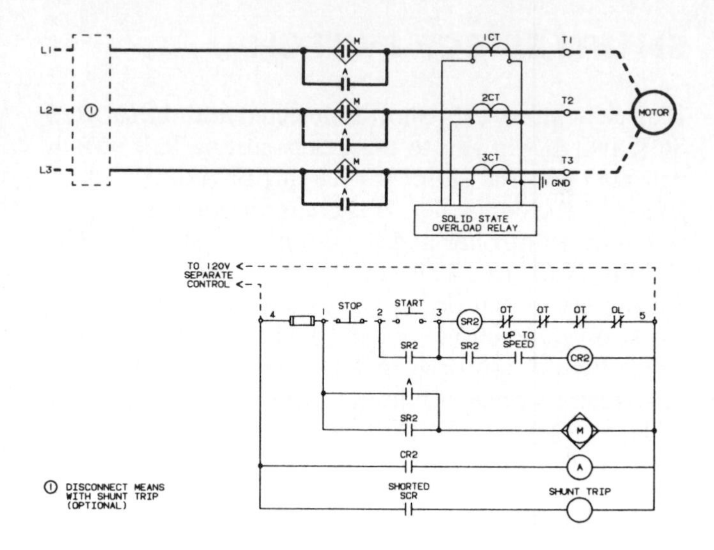

FIGURE 13–19 *Solid-state reduced-voltage controller with a shorting contactor. (Square D)*

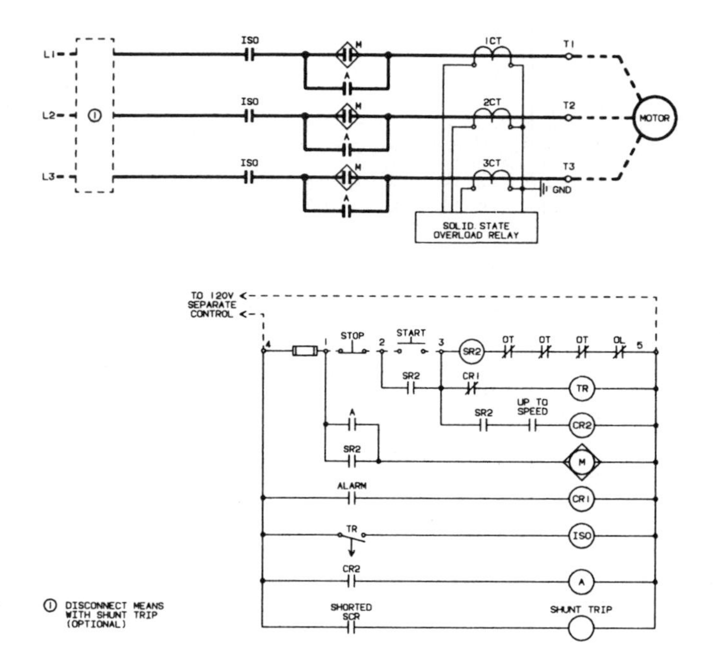

FIGURE 13–20 *Solid-state reduced-voltage controller with a shorting contactor and an isolation contactor. (Square D)*

This combination of layers permits the triggering of the device in either direction (Fig. 13–21). As you remember, the SCR allowed triggering in only one direction. Thus the diac has the ability to conduct in both directions when an ac signal voltage is applied across its terminals. There are a number of applications for such a device. One of them is in the control of ac electric motors. They may also be used in proximity detectors.

Note in the symbol that the diac does not have a gate or control element. It can be used as a bidirec-

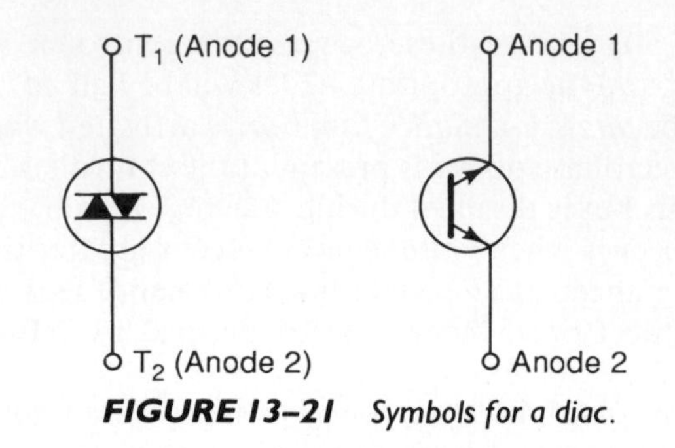

FIGURE 13–21 *Symbols for a diac.*

tional trigger diode (Fig. 13–23B). Current can flow either way when enough voltage is supplied for breakover. Typically, the firing potential is about 30 V in either direction. The diac is in its OFF state until the voltage across terminals T1 and T2 exceeds the breakover voltage. In power control circuits a diac can be used for more effective control of the *turn-on point* for the gate electrode of either a triac or an SCR.

TRIAC

The triac is basically a diac with a gate terminal. The gate terminal controls the *turn-on* conditions of this bilateral device. The gate current can control the action of the device in either direction. This is similar to that of the SCR. However, the characteristics of the triac are somewhat different from those of the diac. Figure 13–22 shows the symbol and the location of the gate terminal.

By placing the triac in a circuit it is possible to indicate how it works (Fig. 13–23). In this arrangement the switch is used to select various conditions for the triac. The load can be either a light bulb or an ac motor. When the switch is in position 1 there is no gate connection. The triac does not conduct. The motor does not run. There is no trigger voltage applied to the gate. In position 2 a diode is placed in the circuit and with its polarity so arranged to allow a trigger voltage applied to the gate on the positive pulse of the ac applied to the circuit. The triac conducts, but only dur-

FIGURE 13–22 *Symbol for a triac.*

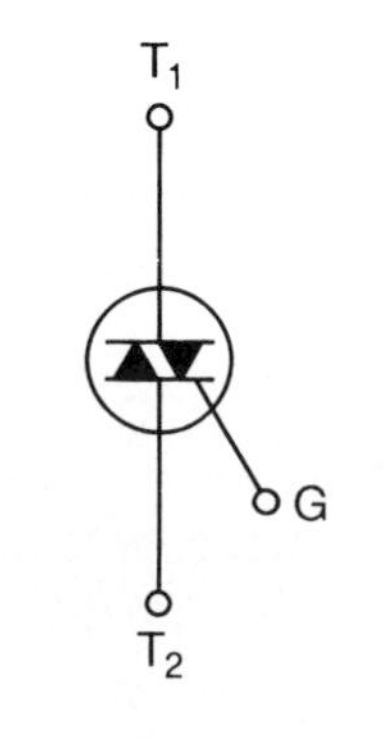

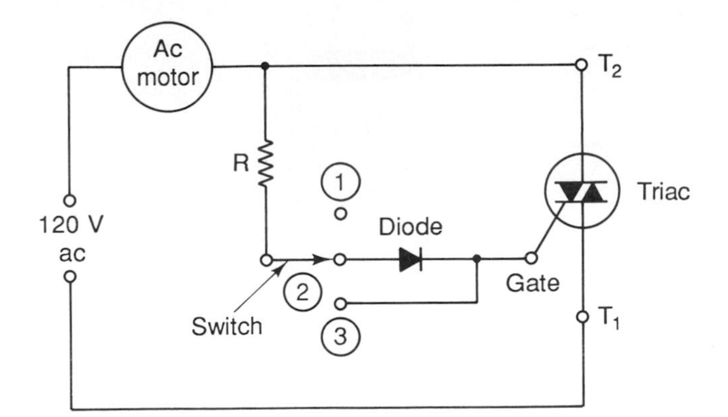

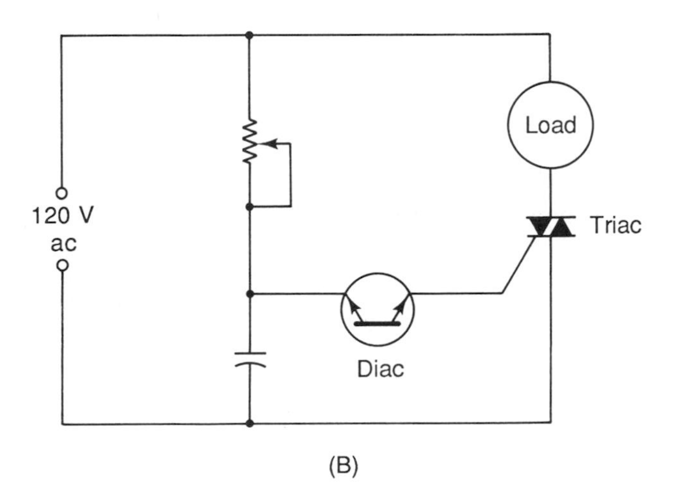

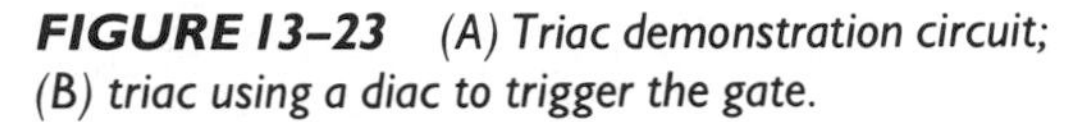

FIGURE 13–23 *(A) Triac demonstration circuit; (B) triac using a diac to trigger the gate.*

ing on one-half of the ac sine wave. This means that only about one-half of the normal current is applied to the motor. This is the same arrangement as with an SCR. An ac motor may have a problem with this type of pulsating dc voltage. When the switch is moved to position 3, the full ac sine-wave voltage is applied to the gate, with, of course, a reduction in value caused by the resistor R. Now that both halves of the ac sine wave are applied to the gate, the triac conducts full time and the full value of ac is applied to the ac motor. The motor then runs at full speed. R can be made a variable type and its value would then control the amount of ac current that passes through the triac and to the motor.

Another arrangement for the triac is shown in Fig. 13–23B. Here a diac is used to trigger the triac. The trigger voltage is controlled by the variable resistor. This allows for better regulation of the motor.

Triacs are packaged in the same types of cases as SCRs, so it is difficult or impossible to tell by a visual inspection which type is in the package. The numbers on the package indicate whether it is an SCR or a triac. There are triacs available today that can handle in excess of 10-kW loads.

LIGHT-EMITTING DIODES

Light-emitting diodes (LEDs) are used as indicator lights on the module panels for solid-state controllers. They are small, give off enough light for the purpose, and draw very little current. They are available in red, green, and amber (Fig. 13–24).

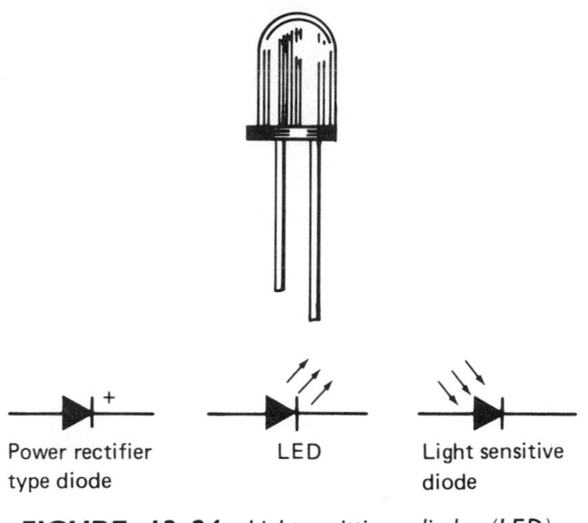

FIGURE 13–24 *Light-emitting diode (LED); symbol for LED.*

LEDs are made of gallium–arsenide junctions, a semiconductor material. Creation of electron-hole pairs is a reversible process. Energy is released when an electron recombines with a hole. In gallium–arsenide, an electron drops directly into a hole and a photon of energy is emitted. The gallium–arsenide junctions provide the best conditions for the generation of radiation in the visible range. Some are made for infrared radiation.

LEDs are used as indicator lamps. In most instances, they must be used in series with a resistor. They are also used as logic indicators for computer circuits. When reverse biased, the LED is nonconducting. This means that you have to have the proper polarity connections to the cathode and anode in order for it to glow. It is capable of conducting current when it is forward biased. It emits light when conducting with a forward bias current. An LED usually operates on 1 to 3 V. Excessive current will destroy an LED, and this calls for a series resistor in most circuits.

USING SOLID-STATE CONTROL AND ELECTROMAGNETIC DEVICES

When solid-state controls are utilized in circuits that have electromagnetic devices, there are problems with the "dirty" power source. The buildup and collapse of a magnetic field whenever a coil of wire or inductor is energized and deenergized produces spikes and other types of electrical *noise*. These spikes can cause problems with solid-state devices since they are susceptible to voltage surges and spikes that are commonplace with the energizing of relay coils and the turning on and off of electric motors.

SURGE SUPPRESSORS

Surge suppressors are installed on magnetic device coils, such as relays, contactors, and motor starters. A voltage-surge suppressor may have its leads connected to the coil terminals. The purpose of the suppressor is to limit voltage noise and overvoltage spikes produced by the starter coil when the coil circuit is opened.

The surge suppressor shown in Fig. 13-25 is made to be easily mountable directly across the coil terminals of contactors and starters with 120 and 240 V ac coils. The purpose of the suppressor is to limit voltage transients for applications requiring interface with solid-state components. One suppressor is required for each coil.

Figure 13-26 shows two types of surge suppressors used to reduce the high transient voltages generated when the coil circuit is opened. These suppressors are used with relay coils and other electromechanical devices.

Figure 13-27 is a surge suppressor used to protect solid-state devices against electrical transients that can result whenever electromechanical devices are op-

FIGURE 13-25 *Surge suppressor for mounting across coil terminals. (Allen-Bradley)*

FIGURE 13-26 *Surge suppressor: (A) for mounting under a relay; (B) for mounting on coil terminals. (Allen-Bradley)*

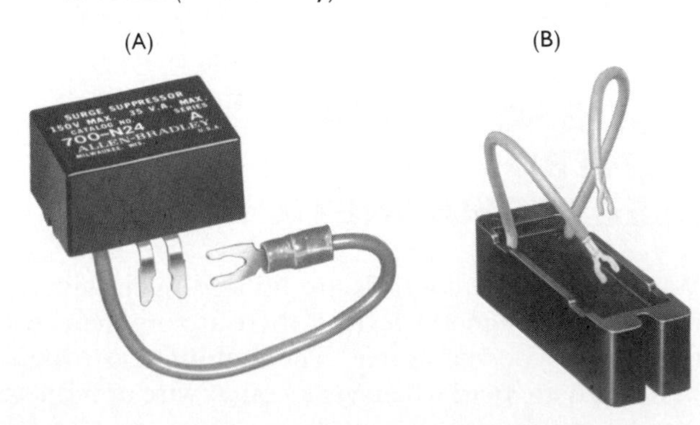

FIGURE 13-27 *Resistor–capacitor combination surge suppressor. (Allen-Bradley)*

erated. Suppressors are for use with relays, timers, ac contactors, and starters. This suppressor consists of a resistor–capacitor combination sealed in epoxy.

LIGHTNING PROTECTION

Lightning can also present some high-voltage surges. Secondary surge arrestors can be installed to prevent problems associated with lightning (Fig. 13-28). The arrestor shown in Fig. 13-28A is for single-phase, two- or three-wire grounded service. Two of them may be installed to provide protection on 208Y/120 V ac three-phase four-wire services. This suppressor will handle 1500 A at 940 V, 5000 A at 1600 V, 10,000 A at 2200 V, and 20,000 A at 3250 V.

A suppressor for use on 650-V ac phase-to-ground maximum is shown in Fig. 13-28B. It is used for three- or four-wire grounded service such as single-phase three-wire, three-phase three-wire, or three-phase four-wire systems. This suppressor will handle 1500 A at 2200 V, 5000 A at 2900 V, and 10,000 A at 3400 V, and 20,000 A at 4000 V. These are maximum discharge voltages that appear across the arrestor during the passage of the discharge current. Discharge current is the current at the arrester during sparkover.

FIGURE 13-28 *(A) Secondary surge arrestors used in lightning protection for electrical systems; 175-V ac phase-to-ground maximum. (B) Secondary surge arrestors for lightning protection used for electrical systems; 650-V ac phase-to-ground maximum. (Square D)*

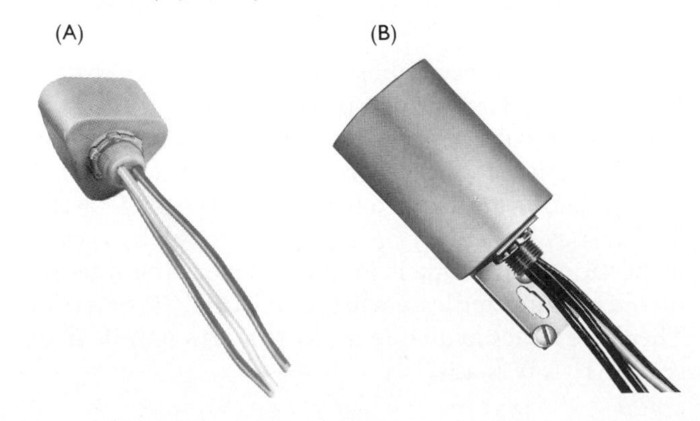

QUESTIONS

1. What does *SCR* stand for?
2. What is another name for the SCR?
3. What does *phase unbalance detection* mean?
4. What does *phase reversal* mean?
5. How are LEDs used on solid-state devices?
6. What is a ''dirty'' power source?
7. What is the purpose of a surge suppressor?
8. How are solid-state devices protected from lightning?
9. What is a diac used for?
10. What is a triac?

Speed Control and Monitoring

Objectives

After studying this chapter, you will be able to:

1. Describe how the speed of an electric motor is controlled.
2. Explain the difference between a synchronous motor and an induction motor.
3. Describe how field excitation of a synchronous motor is obtained.
4. Explain how the speed of a synchronous motor is determined.
5. Describe an amortisseur.
6. Explain various starting methods for synchronous motors.
7. Describe Korndorfer starting.
8. Explain how speed is regulated by resistance.
9. List types of speed control for wound-rotor motors.
10. Explain how secondary resistances are used for wound-rotor induction motor control.
11. List the reason for using solid-state adjustable-speed controllers.
12. Describe how frequency changing is used to change motor speed.

Controlling the speed of an electric motor is possible in some instances. However, there is always some price to pay for speed control. A motor is usually designed to operate at a given speed, and any deviation from that speed causes it to have less starting torque or to run hot.

Industrial processes call for motors that can be varied in their speed. This means that methods must be designed to handle the variable-speed requirement. The first place to look is at the characteristics of the motors themselves as to what they will or will not tolerate in terms of speed variations.

SQUIRREL-CAGE MOTORS

The speed of a squirrel-cage induction motor (Fig. 14–1) is nearly constant under normal load and voltage conditions but is dependent on the number of poles and the frequency of the ac source. This type of motor slows down, however, when loaded an amount that is just sufficient to produce the increased current needed to meet the required torque.

The difference in speed for any given load between synchronous and load speed is called the *slip* of

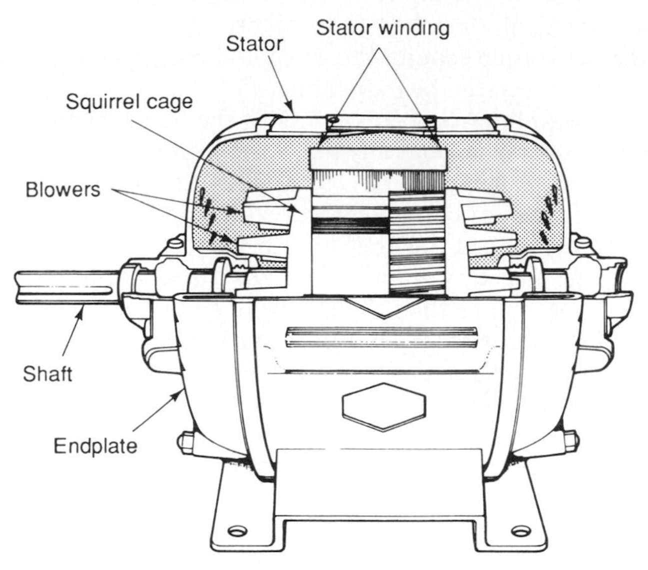

FIGURE 14–1 *Squirrel-cage electric motor.*

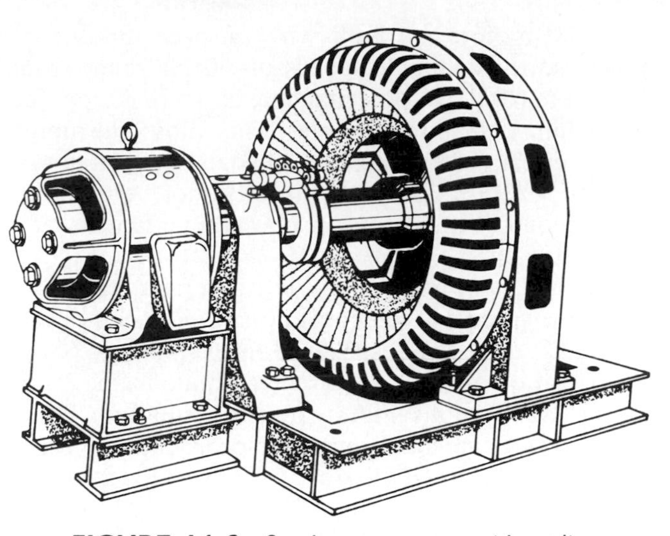

FIGURE 14–2 *Synchronous motor with a directly connected exciter.*

the motor. Slip is usually expressed as a percentage of the synchronous speed. Synchronous speed equals the speed of the rotating field. Since the amount of slip is dependent on the load, the greater the load, the greater the slip will be, that is, the slower the motor will run. This slowing of the motor, however, is very slight, even at full load, and amounts to from 1 to 4% of synchronous speed. Thus the squirrel-cage type is considered a constant-speed motor.

This type of motor is not suitable for industrial applications where a great amount of speed regulation is required, because the speed can be controlled only by a change in frequency, number of poles, or slip. Speed control by changing the frequency is becoming very popular. The number of poles is sometimes changed either by using two or more distinct windings or by reconnecting the same winding for a different number of poles.

SYNCHRONOUS MOTORS

A synchronous motor is by definition one that is in unison or in step with the phase of the alternating current that operates it. This condition is only approximated in practice because there is always a slight phase difference. Any single-phase or polyphase alternator will operate as a synchronous motor when supplied with current at the same potential, frequency, and wave shape that it produces as an alternator, the essential condition in the case of an alternator being that it must be speeded up to synchronism before being put into the circuit.

A synchronous motor may have either a revolving armature or a revolving field. Most synchronous motors are of the revolving field type. The stationary armature is attached to the stator frame, while the field magnets are attached to a frame that revolves with the shaft.

The field coils are excited by direct currents, either from a small dc generator (usually mounted on the same shaft as the motor and called an *exciter*) or from some other source. Figure 14–2 shows a directly connected exciter.

In most industrial applications today, solid-state electronics provide the dc needed for operation of this type of motor. Solid-state electronics also provides the change in frequency needed to control the speed of the motor.

Excitation

Field excitation for a synchronous motor is obtained from a separate exciter set driven by an induction motor, from a direct-connected or belted exciter, or from a constant dc voltage supply such as a station bus. Standard excitation voltage is either 125 or 250 V, but the motor field winding is designed for an excitation voltage approximately 10% below this, to allow for voltage drop in the line.

Speed

The speed of a synchronous motor is determined by the frequency of the supply current and the number of poles of the motor. This means that the operating speed is constant for a given frequency and number of poles. The equation for the determination of motor speed is

$$\text{rpm} = \frac{\text{frequency} \times 120}{P}$$

where P is the number of poles of the motor.

All motors are built with an even number of poles, so the available speeds on 60 Hz range from 3600 rpm for a two-pole machine down to 80 rpm for a machine containing 90 poles. This allows the motor to be directly connected to its load, even at lower speeds, where induction motors cannot be used advantageously because of low operating efficiency and power factor.

Some motors are required to operate at more than one speed but are constant-speed machines at a particular operating speed. For example, when a speed ratio of 2:1 is required, a single-frame, two-speed synchronous motor may be suitable. Four-speed motors are used when two speeds that are not in the ratio of 2:1 are desired.

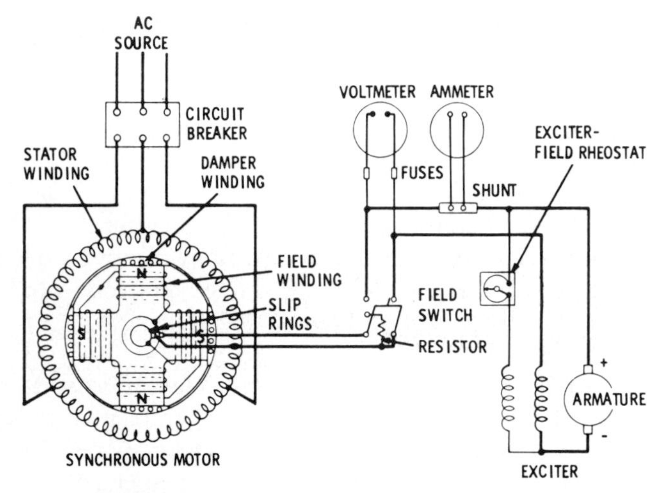

FIGURE 14–3 *Synchronous motor and exciter with rheostat, switch, and meters.*

The single-frame, two-speed motor is usually of the salient-pole type of construction, with the number of poles corresponding to the low speed. High speed is obtained by regrouping the poles so as to obtain two adjacent poles of the same polarity, followed by two poles of opposite polarity. This gives the effect of reducing the number of poles on the rotor by one-half for high-speed operation. Corresponding changes in the stator connections are also made. This switching is usually accomplished automatically by means of magnetic starters, by manually operated pole-changing equipment. Figure 14–3 shows a synchronous motor and exciter with the exciter-field rheostat, field switch, and exciter-field meters.

Starting

To make a synchronous motor self-starting, a squirrel-cage winding is usually placed on the rotor. After the motor reaches a speed slightly below synchronous, the rotor is energized. When synchronous motors are started, their dc fields are not excited until the rotor has practically reached full synchronous speed. The starting torque required to bring the rotor up to this speed is produced by induction.

In addition to a dc winding on the field, synchronous motors are generally provided with a damper or *amortisseur* winding. It consists of short-circuited bars of brass or copper embedded in slots in the pole faces and joined together at either end by means of end rings. This winding, usually termed a *squirrel-cage winding,* enables the motor to obtain sufficient starting torque for the motor to start under load.

The starting torque necessary to bring the motor up to synchronous speed is termed the *pull-in torque.* The maximum torque that the motor will develop without pulling out of step is termed the *pull-out torque.*

When the stator winding in the synchronous motor is being excited by the ac line connection, it immediately sets up a rotating magnetic field. The rotating flux of this field cuts across the damper winding of the rotor and induces secondary currents in the bars of this winding. The reaction between the flux of these secondary currents and that of the rotating stator field produces the torque necessary to start the rotor and to bring it up to speed.

When the rotor has been brought up to nearly synchronous speed (as an induction motor because of the damper winding), the dc field poles are excited and the strong flux of these poles causes them to be drawn into step or full synchronous speed with the poles of the rotating magnetic field of the stator. During normal operation, the rotor continues to revolve at synchronous speed as if the dc poles were locked to the poles of the rotating magnetic field of the stator. Because a synchronous motor has no slip after the rotor is brought up to full speed, no secondary currents are induced in the bars of the damped windings during normal operation.

Starting Methods

In starting a synchronous motor as an induction motor, the voltage impressed on the motor should be reduced in starting and while coming up to speed. This reduced starting voltage is usually obtained from a starting compensator (autotransformer) similar to that used in the starting of an induction motor.

Autotransformer Starting. With the autotransformer method of starting, the usual practice is to close the starting contactor first. This connects the stator to the reduced voltage. When a speed near synchronization has been reached, the starting contactor is opened and the running contactor is closed, thus connecting the motor to the full-line voltage. After synchronous speed has been reached, the field switch is closed through a moderate amount of resistance.

The field current is now adjusted in order to make the motor operate at the desired power factor.

Full-Line Voltage Starting. If the motor is to operate at high starting torque, it is common practice to use a full-line starting voltage in connection with a time-delay overcurrent relay. The relay will operate before the surge of starting current can damage the motor windings. Figure 14–4 shows the typical automatic across-the-line synchronous motor starter.

Across-the-line starting and reduced-voltage starting have already been discussed. However, the reduced-voltage diagram is shown in Fig. 14–5. Note how the autotransformers are switched into the circuit. This method of reduced-voltage starting uses

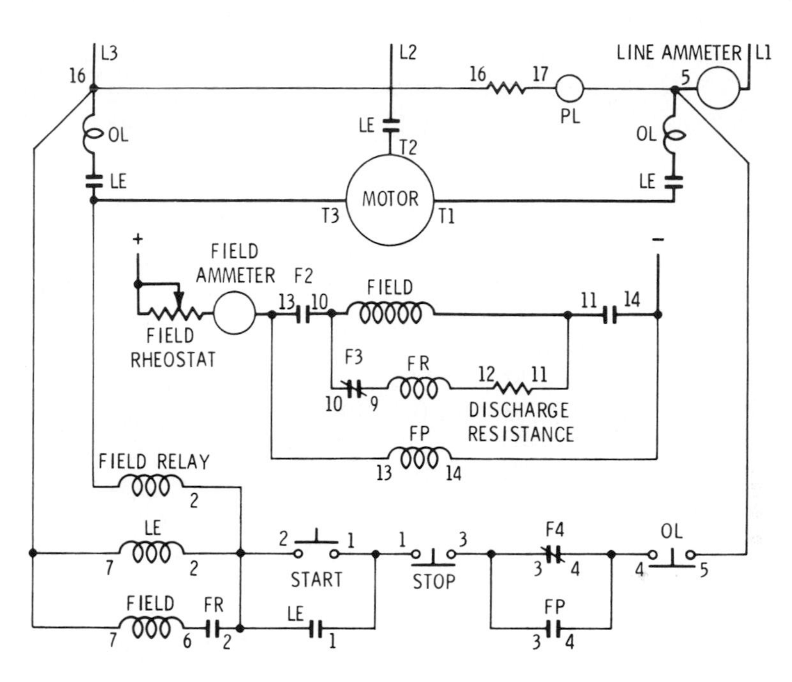

FIGURE 14–4 Elementary diagram of an across-the-line motor starter.

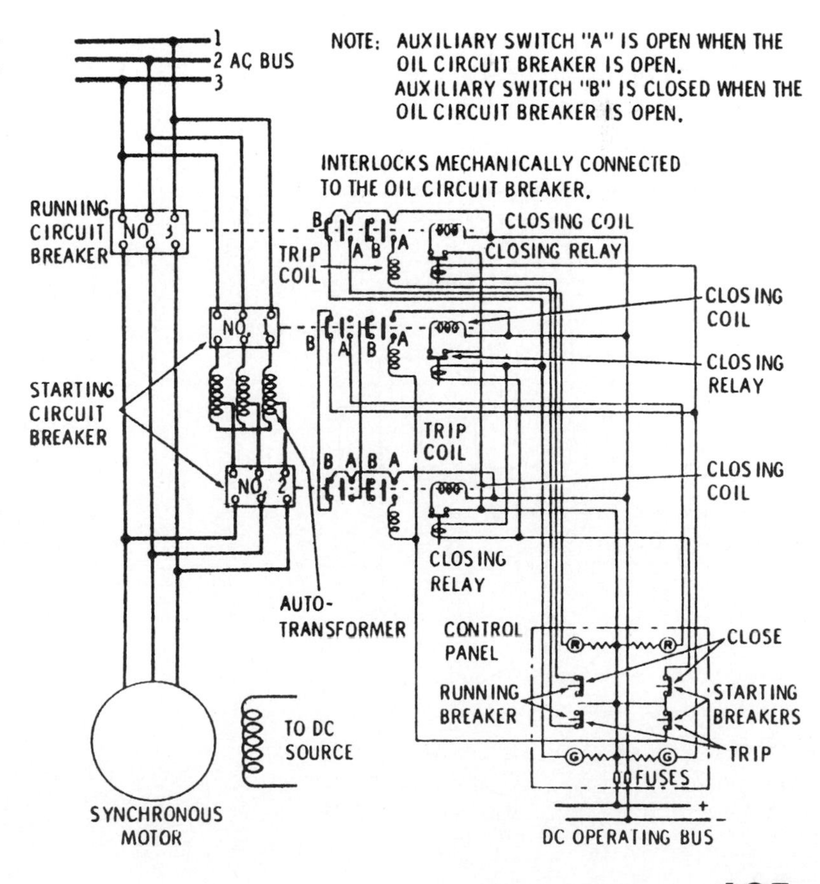

FIGURE 14–5 Diagram of a reduced-voltage starter for a synchronous motor.

three switches or contactors. The two starting switches or contactors are connected to each side of the autotransformer or compensator and are both closed at the same time. After the motor has attained nearly synchronous speed, the starting switches are opened and the running switch is closed. This sounds great in theory but is not always successful. The motor may not become synchronized when it is supposed to, so you may have to start over again and try to get it to synchronous speed or use an automatic method.

Reactance Starting. Reactance starting is similar to the reduced-voltage starting methods, except that the first step is obtained by reactance in series with the motor armature instead of autotransformers. In the reactance method of starting, more current is required from the line for the same torque on the first step than when compensators are used. It has one advantage. No circuit opening is required when the motor is transferred to running voltage. The transfer is accomplished by short-circuiting the reactance.

Resistance Starting. A typical circuit using the resistance method of reduced-voltage starting is shown in Fig. 14–6. Switch 1 is closed first. This connects the motor to the line through the entire resistance. Switches 2, 3, and 4 are then closed, with a time interval between each closing. Each switch, in turn, short-circuits a part of the resistance. This method of starting is sometimes used when power company rulings require several progressive steps of starting current.

FIGURE 14–6 *Schematic diagtram of a resistance-type synchronous motor starter.*

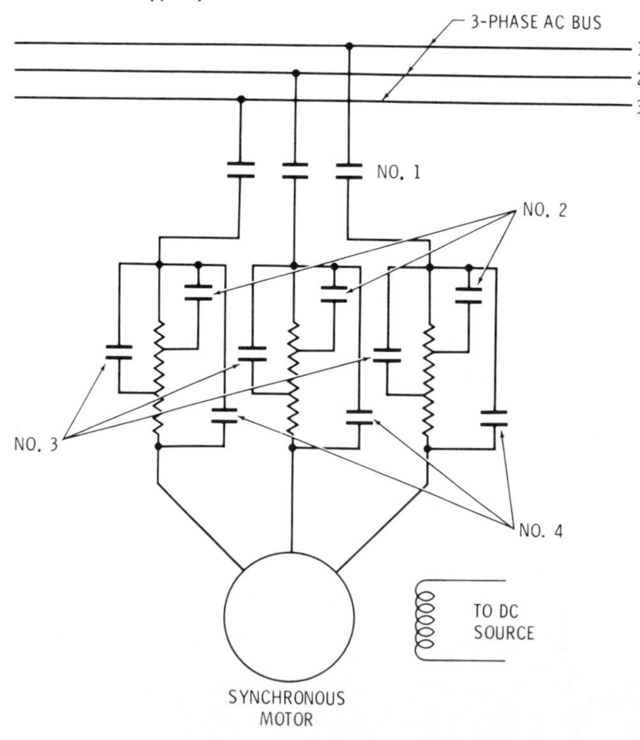

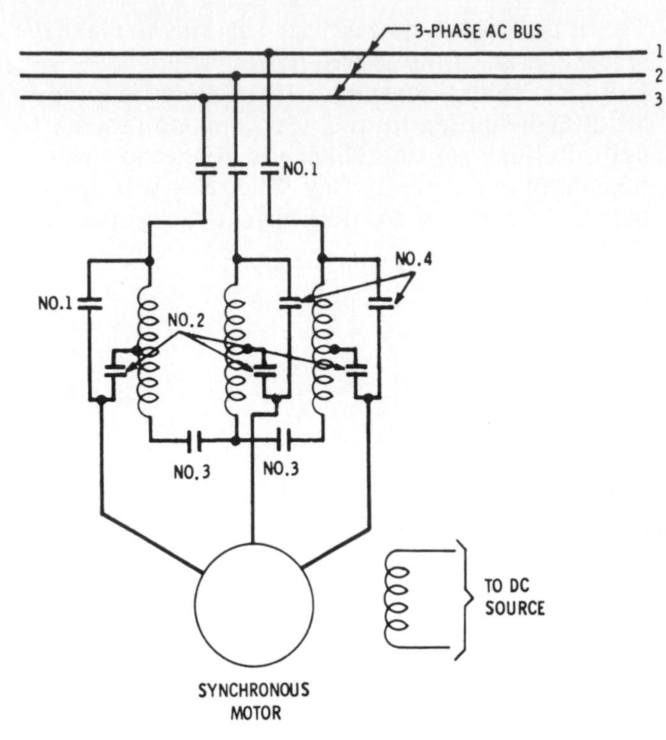

FIGURE 14–7 *Schematic diagram of a Korndorfer-type synchronous motor starter.*

Korndorfer Starting. The reactance and resistance methods are similar to starting the motor using the *Korndorfer method.* It permits the motor to be started without opening the motor circuit. The motor is first connected through suitable taps of a compensator, and then started by connecting the compensator to the line. Full voltage is connected by first opening the neutral of the starting compensator. This allows the motor to run with part of the compensator winding in series with the motor. Then the entire compensator winding is short circuited (Fig. 14–7).

Switch 1 is closed first. This connects one of the compensator windings to the line. Then switch 2 is closed, completing the motor circuit at reduced voltage. As the motor increases its speed, a timing relay, operated by switch 2, opens the circuit of 3. This, in turn, opens the transformer neutral. Switch 4 is closed next. This connects the motor to full-line voltage by shorting the compensator sections. By opening switch 2, the reduced-voltage taps of the compensator are disconnected and the permanent running connection to the motor is completed.

Other Methods of Starting. An auxiliary prime mover, usually an induction motor, may be used as a starter. This method of starting is applied to the motors that have no squirrel-cage winding, or it is used with alternators converted to motor use. This type of motor cannot start under load.

Uses for Synchronous Motors

Synchronous motors may be used for power factor correction; for constant-speed, constant-load drives; and for voltage regulation. Because of the higher efficiency possible with synchronous motors, they can be used advantageously on most loads where constant speed is required. Typical applications are compressors, fans, blowers, line shafts, centrifugal pumps, rubber and paper mills, and to drive dc generators.

WOUND-ROTOR MOTORS

The wound-rotor motor differs from the squirrel-cage type. It has wire-coil windings in its rotor instead of a series of conducting bars in the rotor. Inserting external resistance in the motor circuit when starting will develop a high torque with a comparatively low starting current. As the motor comes up to speed, the resistance is gradually removed until, at full speed, the rotor is short-circuited. Speed can be regulated, within limits, by varying the amount of resistance in the rotor circuit (Fig. 14–8).

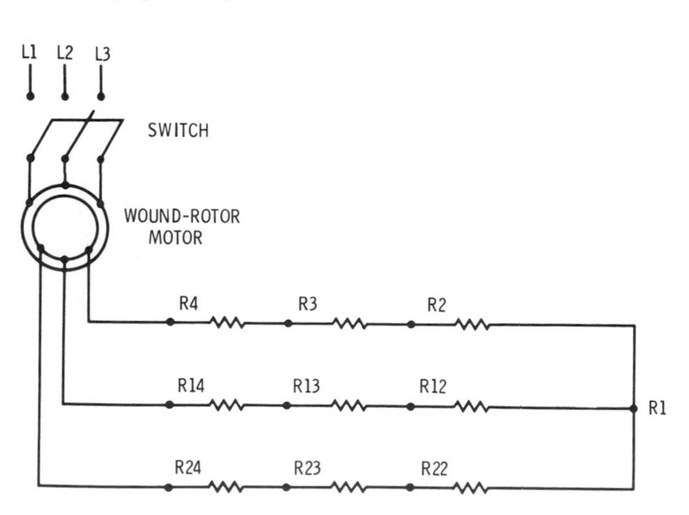

FIGURE 14–8 *Wiring diagram with resistor connections.*

Speed Regulation by Resistance

Resistors can be used to regulate the speed if they are of the proper size to prevent overheating from constant use. The resistors used in starting are used only for a short time, but those used for continuous motor speed reduction are in use for longer periods of time. This means that a resistor must be selected for its intended purpose.

Dc motors produce the most effective variable-speed outputs. However, wound-rotor motors, because of their adjustable rotor resistance, are one of the few means of speed control available for ac motors.

Wound-rotor motors are just that—they have a wound rotor. They are insulated coils of wire that are not permanently short circuited, as in the squirrel-cage motor, but are connected in regular succession to form a definite polar area having the same number of poles as the stator. The ends of these rotor windings are brought out to collector rings, usually referred to as *slip rings.*

Currents induced in the rotor are carried by means of slip rings (and carbon brushes riding on the slip rings) to an externally mounted resistance (Fig. 14–9). These resistances can then be regulated or changed according to the needs of the start sequence. By changing the resistance in the rotor circuit it is possible to change the speed of the motor. Once it has come up to synchronous speed the resistors are then short circuited and the motor runs with characteristics similar to a squirrel-cage type.

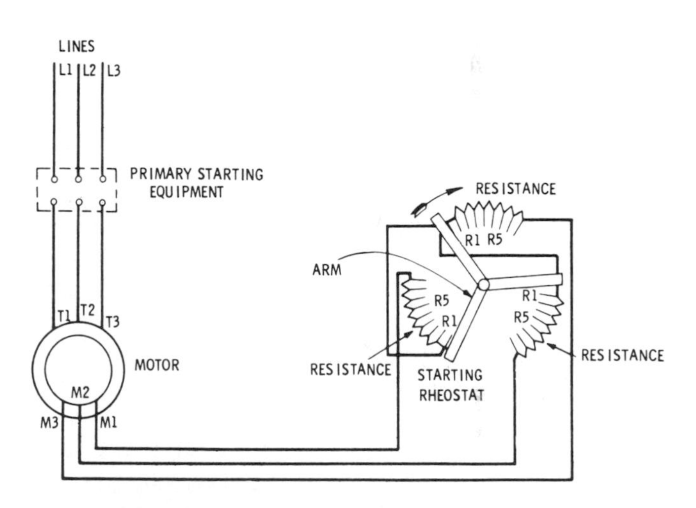

FIGURE 14–9 *Starter–controller for a wound-rotor induction motor.*

However, some resistance can be left in the circuit to aid in speed control, that is, of course, if the size (wattage rating) of the resistors is such as to withstand the constant current flow through them. By placing a high resistance in the rotor circuit, it is possible to start the motor and have it produce high starting torque with low starting current.

Types of Speed Control

The wound-rotor motor can be used where the speed range is small, where the speeds desired do not coincide with a synchronous speed of the line frequency, and where the speed must be gradually or frequently changed from one value to another. This includes compressors, pulverizers, stokers, and conveyors.

A smooth, no-jerk start can be obtained by using the wound-rotor motor. It is simply a matter of supplying the right control equipment.

Multiswitch Starters

Figure 14–10 shows how a typical multiswitch starter for a wound rotor is wired into the rotor circuit. This type of starter is used in the secondary circuits of large wound-rotor induction motors up to 2000 hp with rotor currents up to 1000 A. Contact levers are of the double-pole type and are mechanically arranged in such a manner that they must be closed in a predetermined sequence, and only one at a time. Since the switches are designed for hand-over-hand operation, a desirable time element is introduced that prevents too-rapid acceleration of the motor. When the final switch has been closed, it is held in place by a magnetic coil, and because of the mechanical interlocking feature, all other switches remain closed. This type of starter is just that—a starter, it is not useful as a speed regulator.

Drum Controllers

Drum controllers can be used for starting and for speed control of the wound-rotor motor (Fig. 14–11). Drum controllers are made to handle both stator and rotor circuits. The cylinder mounting the contact segments are built in two insulated sections. When they are built to handle the rotor circuit, only the stator circuit is controlled by a circuit breaker or line starter. In addition to starting and regulating, speed-regulation drum collectors are commonly used for speed-reversing duty as well.

Motor-driven controllers are used in certain drives requiring close automatic speed regulation such as in large air-conditioning plants, blowers, stokers, and similar applications. Some of these installations have been in use for a number of years and are gradually being replaced by more modern motor control methods.

Magnetic Starters

Magnetic starters are built to regulate motor speed, start the motor, and to set the speed of the motor. They consist of a magnetic contactor for connecting the stator circuit to the line, and one or more accelerating contactors to commutate the resistance in the rotor circuit. The number of secondary accelerating contactors varies with the rating, a sufficient number being used to assure smooth acceleration and to keep the inrush current within practical limits. The operation of the accelerating contactors is controlled by a timing device, which provides definite time acceleration. For high-voltage service, the primary contactor is usually of the oil-immersed type. The diagram of a typical magnetic starter for use with a wound-rotor induction motor is shown in Fig. 14–12.

Resistors

Generally, the secondary resistors for wound-rotor induction motors are designed for star connection. Resistors for most manual controllers may be connected with all three secondary phases closed or with one secondary phase open on the first point of the controller. Resistors for magnetic controllers are connected with all three phases closed in the secondary on the first point. The torque obtained with a resistor of a given class number varies with the connection used on the first point of the controller.

Keep in mind that wound-rotor motors can be

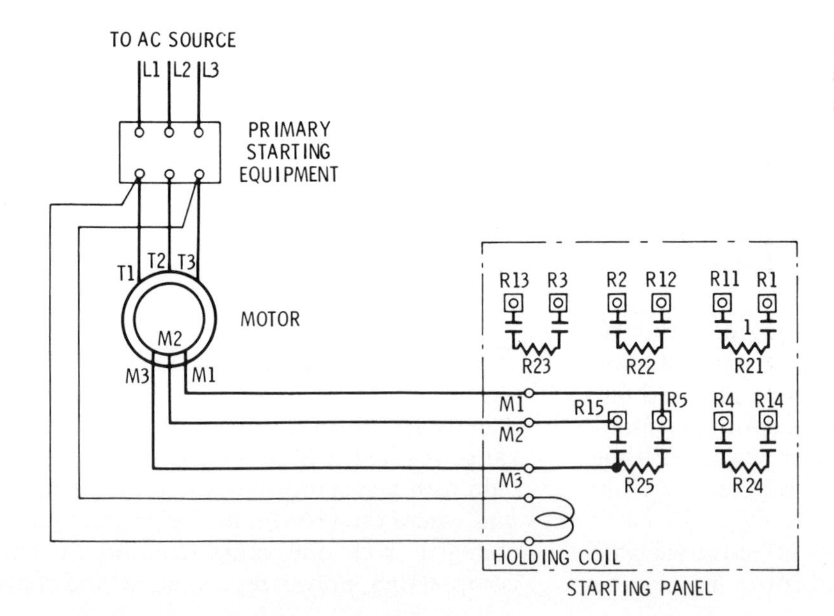

FIGURE 14–10 *Diagram of a typical multiswitch starter for a wound-rotor motor.*

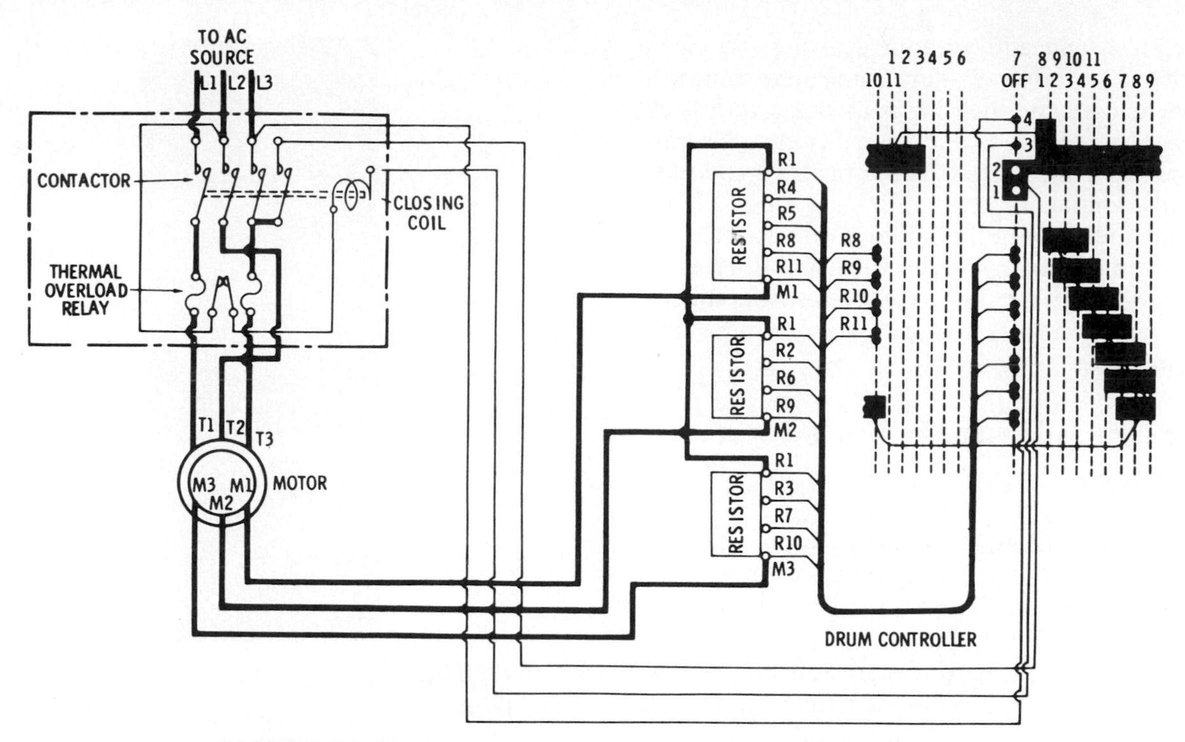

FIGURE 14–11 *Nonreversing drum controller for a wound-rotor motor with a three-phase secondary.*

FIGURE 14–12 *Magnetic starter contactor for use with a wound-rotor motor.*

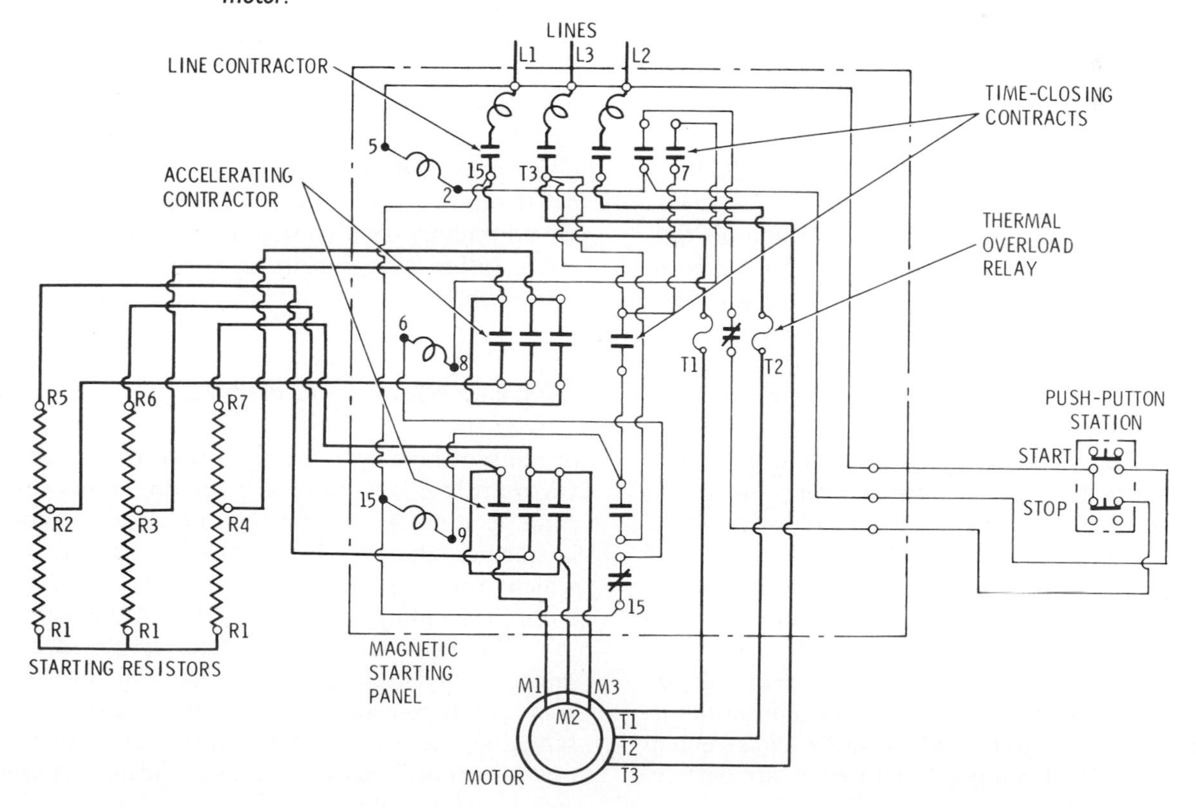

started with a load and without drawing too much current. They do, however, have some disadvantages. They can be used for such loads as those with back pressures set up by fluids and gases, as in reciprocating pumps and compressors. They are also used in elevators and cranes.

Disadvantages are the initial cost since they do require a wound rotor. The slip rings and brushes do need maintenance from time to time. Resistors and the switching arrangements require periodic inspection and maintenance.

Solid-State Adjustable-Speed Controllers

Solid-state adjustable-speed controllers are available to produce smooth starts and energy savings. They are, in most instances, maintenance free and are easy to operate and make it simple to train new operators.

The reason for using solid-state adjustable-speed controllers is because they provide stepless, smooth adjustable-speed control of the ac wound-rotor motor. This means the elimination of resistors, liquid rheostats, and reactors as well as magnetic clutches. They all consume energy that is not the case with the solid-state controller. The solid-state circuitry is used to provide the excitation to the rotor. By controlling the rotor current it is possible to control the motor speed and thereby its torque.

FREQUENCY SPEED CONTROL

Solid-state ac motor control is accomplished by changing the frequency of the power source. Westinghouse's ACCUTROL line is an adjustable-speed ac drive package in ratings from 1 to 5 hp at 230 V and 3 to 250 hp at 460 V, three-phase, 60 Hz (Fig. 14–13).

Motor speed is adjusted by controlling the output voltage and frequency of the unit. This is accomplished by rectifying the incoming ac supply voltage and changing it to dc. The dc voltage is *inverted* by a three-phase inverter section to an adjustable frequency output whose voltage is adjusted proportionately to the frequency to provide constant volts per hertz excitation to the motor terminals up to 60 Hz. Above 60 Hz, the voltage may remain constant at rated volts. In this way energy-efficient low-loss speed control is obtained in the range 2 to 120 Hz.

This type of speed control does have advantages over dc machines inasmuch as the dc motors are hard to maintain and have problems in environments that are wet, corrosive, or explosive. These controls are found in food-packing plants, dairies, chemical plants, sand and gravel plants, paper mills, and cement plants. Centrifugal pumps and blowers are particularly suited for use with this type of control, as considerable reduction in energy consumption can be achieved by varying the speed to control the flow of gas or fluids instead of using throttling devices such as valves, dampers, or fluid recirculators.

FIGURE 14–13 *Soliid-state ac motor control. (Westinghouse)*

MULTISPEED STARTERS

Multispeed starters are designed for the automatic control of two-speed squirrel-cage motors of either the consequent pole or separate winding types (Fig. 14–14). These starters are available for constant-horsepower, constant-torque, or variable-torque three-phase motors (Fig. 14–15). Multispeed motor starters are commonly used on machine tools, fans, flowers, refrigeration compressors, and many other types of equipment.

SPEED MONITORING

Speed sensing a switch can be used to sequence conveyors where it is necessary for one conveyor to be running at nearly full speed before a second conveyor is started. The switch can also be used to indicate which direction material on a conveyor is moving from the rotation of a suitable driven shaft.

The electronic speed switch is a rugged, self-contained rotary shaft speed detector (Fig. 14–16). If the shaft speed exceeds or falls below an adjustable, preset value, the speed switch detects the change and actuates external relays, audible alarms, or warning

FIGURE 14–14 Open-type, two-speed separate winding multispeed starter. (Square D)

FIGURE 14–15 Starter diagram for a consequent-pole motor with constant torque or variable torque. (Allen-Bradley)

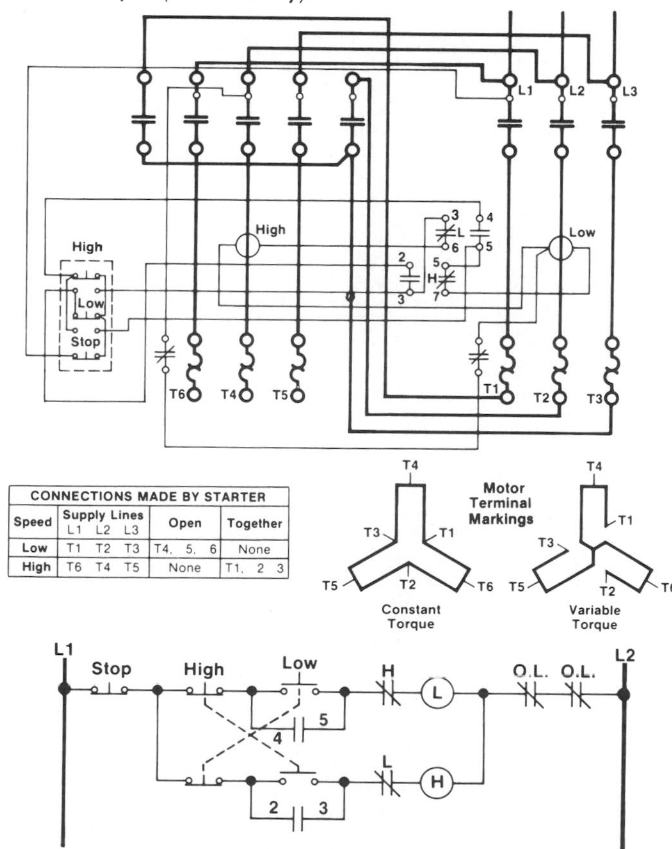

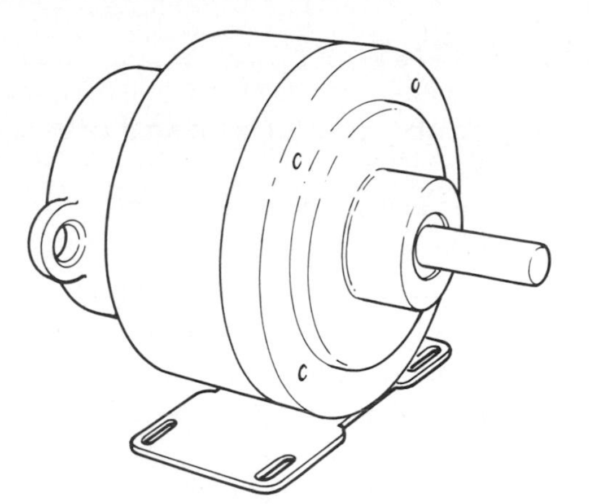

FIGURE 14–16 Electronic speed switch. (Reliance)

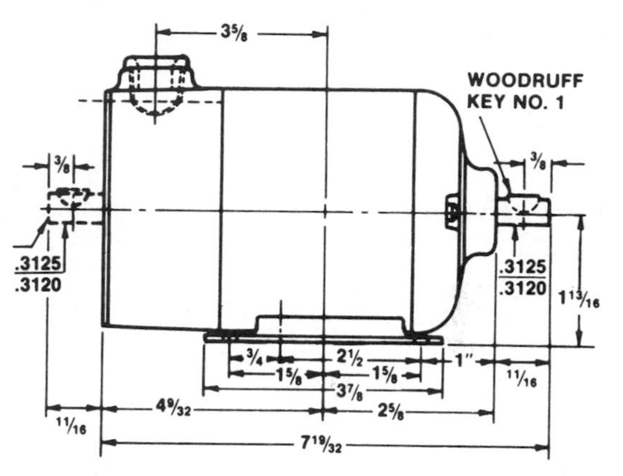

FIGURE 14–17 Tachometer generator. (Reliance)

lamps. Output power is switched by a triac solid-state switch. This model is available in the pictured foot-mounted model or the flange-mounted model. Table 14–1 shows the available speed ranges for the speed switch.

A *tachometer* generator allows accurate monitoring of machine operating speeds. When this is tied into a closed-loop speed regulator, the tachometer generator can be used to control the machine speed (Fig. 14–17).

TABLE 14–1 SPEED RANGES[a]

Range Dial Setting	*Speed Range* 5–5000 rpm (Standard)	0.7–700 rpm (Option D)
1	5–15	0.7–2
2	15–50	2–7
3	50–100	7–20
4	150–500	20–70
5	500–1500	70–100
6	1500–5000	200–700

[a]Speed switch range is field-adjustable by dial setting to the range limits shown. After the desired general speed range is selected, specific speed is set by turning an adjustable potentiometer.

QUESTIONS

1. How can the speed of a squirrel-cage motor be changed?
2. What is an exciter? Where is it useful?
3. What is the rpm range of motors?
4. What is a damper winding on a motor?
5. What is the Korndorfer method of starting a motor?
6. What is another function of a synchronous motor?
7. What type of motors produce the most effective variable-speed outputs?
8. What types of motors use multiswitch starters?
9. What two purposes do drum controllers serve on wound-rotor motors?
10. What are the disadvantages of wound-rotor motors?
11. What is the main reason for using solid-state adjustable-speed controllers?
12. What does an inverter do?
13. How is a tachometer generator useful in regard to machine speeds?
14. Define a synchronous motor.
15. How is field excitation for a synchronous motor obtained?
16. What determines the speed of a synchronous motor?
17. When are four-speed motors used?
18. How are synchronous motors made self-starting?
19. What is another name for a damper winding?
20. What is another name for a starting compensator?

15

Motor Control and Protection

Objectives

After studying this chapter, you will be able to:

1. Describe manual starter operation.
2. Describe the "soft start" method of motor control.
3. Explain how sequence control is accomplished.
4. Explain how automatic sequence control is accomplished.
5. Describe how jogging is accomplished.
6. List advantages and disadvantages of plugging a motor.
7. Define antiplugging.
8. Describe electronic motor braking.
9. List the disadvantages of electronic motor braking.
10. Describe how mechanical braking works.
11. Understand how thruster brakes are used.
12. List the advantages of hydraulic brakes and magnetic brakes.
13. Explain what causes chattering brakes.
14. Describe overload protection for a motor.
15. List the advantages of a line-voltage monitor.
16. Explain how programmable motor protection works.
17. Explain how a remote temperature detector module works.

Controlling a motor can be a simple task or it can be complicated, depending on the needs of the machine that is powered by the motor. Motors are incorporated into many machines. This makes it necessary for the motor to be controlled according to the needs of the machine it powers. Then, of course, as the motor is used, it is also abused. This means that it must be protected if it is to run properly and for long periods of time as needed for any given purpose.

Motors must be started, stopped, reversed, and the speed controlled. They are also jogged, plugged, and in some instance stopped rather quickly. All these operations require equipment and circuitry to accomplish the job correctly without damage to the motor. This chapter covers these operations and the equipment needed to cause proper operation of an electric motor under varying load conditions.

MANUAL STARTERS

Manual starters (Fig. 15-1) provide full-line voltage starting, reliable thermal overload protection, as well as undervoltage protection. Typical applications are on woodworking machinery, metal sawing machines, and many other machine tools where undervoltage protection is needed to meet safety standards. By removing jumper A, a remote emergency stop operator

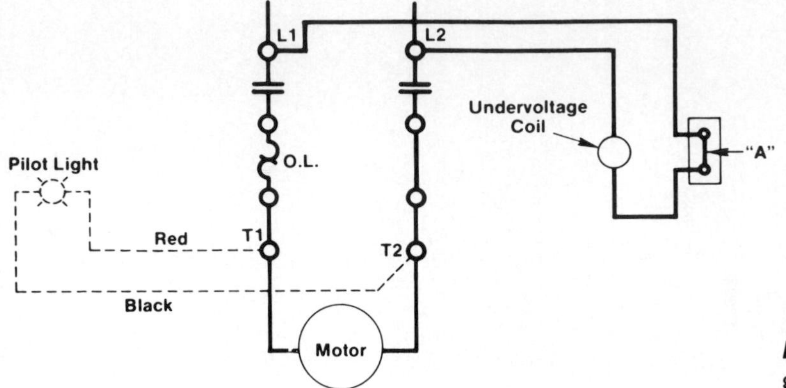

FIGURE 15–1 *Manual starter, single-phase. (Allen-Bradley)*

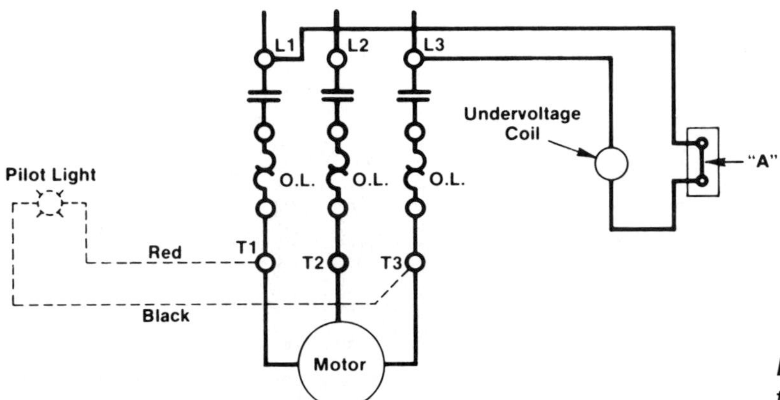

FIGURE 15–2 *Manual starter, three-phase. (Allen-Bradley)*

FIGURE 15–3 *Three-phase starters. (Allen-Bradley)*

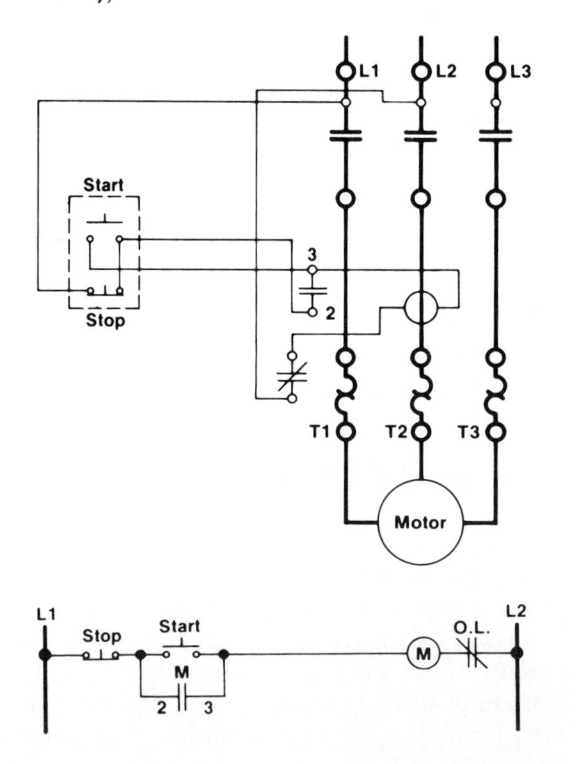

may be wired to the vacated terminals. Note how the pilot light is wired in the circuit so that the light is on when the motor is energized. Three-phase operation using the same type of starter is shown in Fig. 15–2. The pushbutton wiring is shown in Fig. 15–3. Note how the wiring diagram and the elementary diagram differ but contain the same information. The elementary diagram shows how the pilot device is energized. The M shown in the diagram represents the contactor coil that will close the contacts in the three-phase lines to the motor when energized. A complete circuit for energizing the coil is provided by depressing the START button.

A variation on the standard start–stop starter is shown in Fig. 15–4, where the three-phase starter is used with a single-phase motor. Figure 15–5 is a variation that is used with a three-phase motor. It provides more than one start–stop station. The elementary diagram shows how the start buttons are in parallel with the contacts of the starter M coil. This is a useful arrangement when a motor must be started and stopped from any of several widely separated locations. Using this particular circuitry, it is also possible to use only one start–stop station and have several STOP buttons

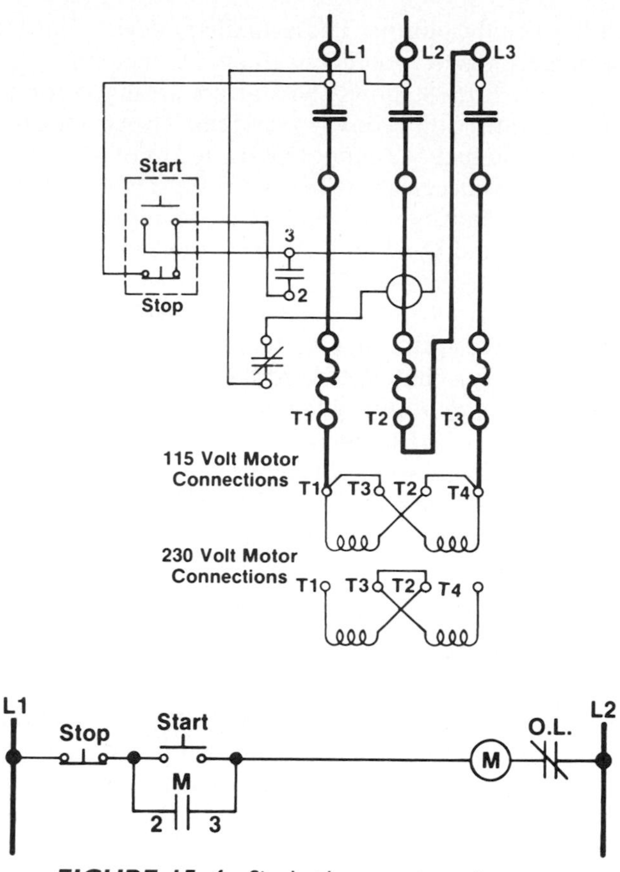

FIGURE 15–4 *Single-phase motor using standard three-phase starter. (Allen-Bradley)*

at different locations to serve as emergency stops. Standard-duty start–stop stations are provided with connections A shown in Fig. 15–5. This connection must be removed from all but one of the start–stop stations used.

Low-voltage protection is a method of protecting an operator from injury from automatic restart of a machine upon resumption of voltage after a power failure. This is normally accomplished with a magnetic starter with three-wire control. This protection can be had with standard manual starters (Fig. 15–6). The low-voltage feature is accomplished by a continuous-duty solenoid assembly built into the overload relay mechanism. When a power failure occurs or the line voltage is disconnected, the solenoid will deenergize and mechanically open the starter contacts. When power is restored, the starter must be manually reset before the contacts can be closed and normal operation resumed.

Low-voltage protection is required by OSHA 1910.213b3 and 1910.217b8iii on certain woodworking machines and all mechanical power presses. NFPA 79 Section 130-21 requires it on certain metalworking machines. Some local safety regulations have extended it to other applications, such as mixers, conveyors, or wherever operator safety could be in jeopardy.

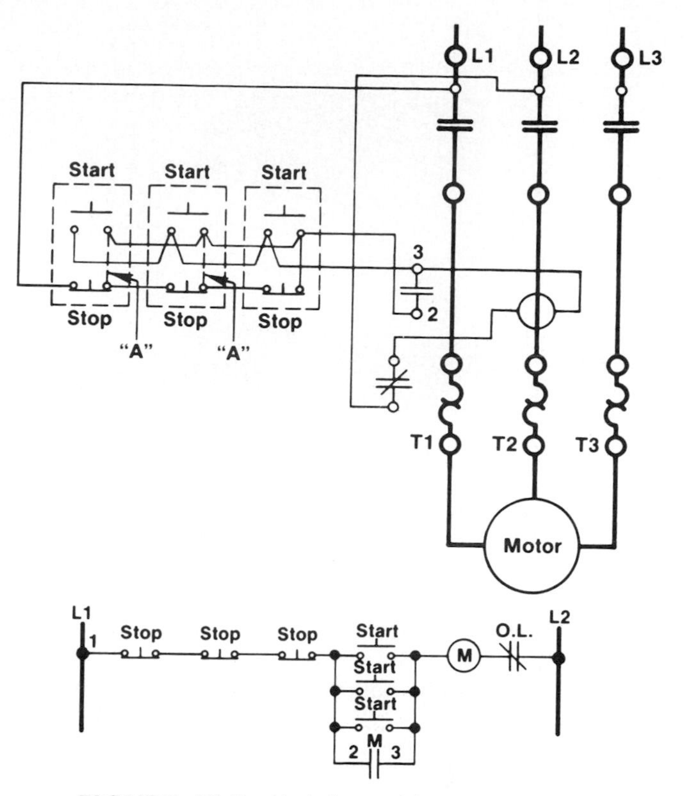

FIGURE 15–5 *Variations with start–stop stations. (Allen-Bradley)*

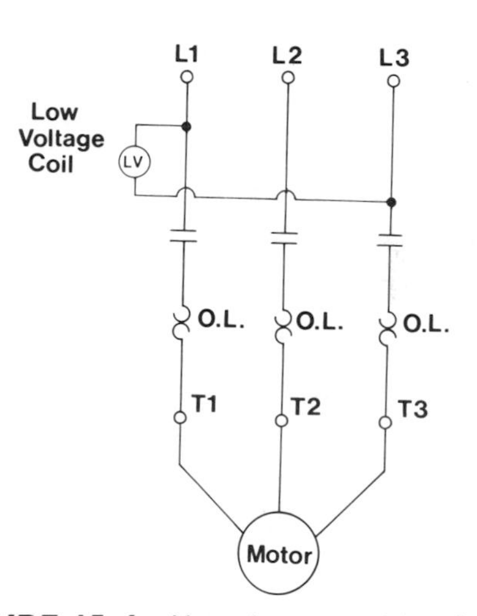

FIGURE 15–6 *Manual starter wiring diagram with low-voltage coil. (Square D)*

SOLID-STATE MOTOR CONTROLLER

It is possible to obtain the advantages of solid-state electronics in motor controllers. In fact, an entire chapter is dedicated to the programmable controller and its advantages and uses. In this chapter, however, the smaller microcomputer-controlled starters for

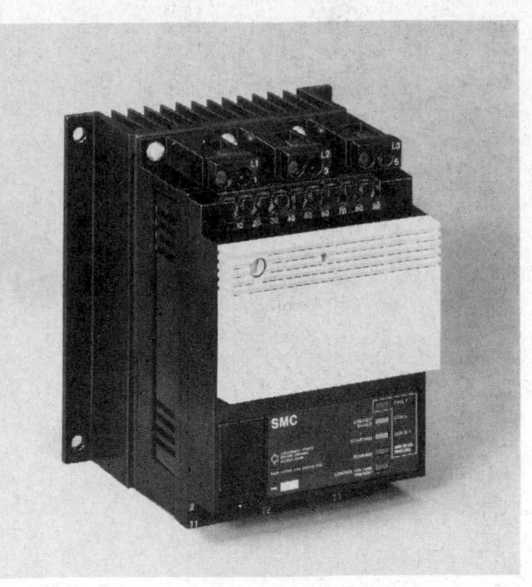

FIGURE 15–7 SMC (Smart Motor Controller), microprocessor-controlled starting. (Allen-Bradley)

standard squirrel-cage induction motors are highlighted.

Allen-Bradley makes the SMC (Smart Motor Controller), which has a soft start, current limiting, and full-voltage starting. Three modes are possible with the same controller: the soft start, current limit, and full voltage (Fig. 15–7).

Soft Start. This method has the most general application. The motor voltage gradually increases during the acceleration ramp period. The ramp period can be adjusted from 2 to 30 seconds. Then the user sets it for the best starting performance over the required load range.

Current Limit. This starting mode is used when it is necessary to limit the maximum starting current. The current limit is adjusted according to the starting current restriction. This can be adjusted for 200 to 450% of full-load amperes.

Full Voltage. For applications requiring a full-load start, the acceleration ramp time is set to a minimum of 0.25 second. This, in effect, allows the controller to start the load across-the-line.

Solid-state electronics have been utilized successfully in the production of motor controls. In time, they will probably replace the electromechanical devices. However, the cost of the electronic devices will continue to decrease with better circuitry and more competition. Older machines will be retrofitted with the newer electronics. Some packages are already available to be attached to existing setups.

SEQUENCE CONTROL

In some applications it is necessary to make sure that one starter is not operational until the other has been energized. This type of use is found in equipment that may have the need for high-pressure lubrication or with hydraulic pumps. These auxiliary devices have to be operational before the machine is turned on.

Figure 15–8 shows the starters arranged for sequence control of a conveyor system. The two starters are wired so that M2 cannot be started until M1 is running. This is necessary if M1 is driving a conveyor fed by another conveyor driven by M2. Material from M2 conveyor would pile up if the M1 conveyor could not move and carry it away.

If a series of conveyors is involved, the control circuits of the additional starters can be interlocked in the same way. That is, M3 would be connected to M2 in the same *step* arrangement that M2 is now connected to M1, and so on.

The M1 button STOP button or an overload on M1 will stop both conveyors. The M2 STOP button or an overload on M2 will stop only M2.

When Starting Any One Requires Another

Several motors can be run independently of each other with some of the starters actuated by two-wire and some by three-wire pilot devices. Whenever any one of these motors is running, a pump or fan motor must also run (Fig. 15–9).

A master start–stop pushbutton station with a control relay is used to shut down the entire system in an emergency. Control relay (CR) provides three-wire control for M1, which is controlled by a two-wire control device such as a pressure switch. Motors M2 and M3 are controlled by start–stop pushbutton stations.

Auxiliary contacts on M1, M2, and M3 control M4. These auxiliary contacts are all wired in parallel so that any one of them may start M4. On some starters auxiliary contacts have been added to M2 and M3 for this purpose. The standard *hold-in contact* on M1 may be used as an auxiliary if wire Y is removed. Hold-in contacts are not required when a two-wire control device is used.

When this system is used, the phase connections on all of the starters must be the same. That is, L1 of each starter must be connected to the same incoming phase line; L2 and L3 of each starter must be phased out similarly.

Automatic Sequence Control

Having automatic sequence control is also possible with the arrangement shown in Fig. 15–10. In this system it is desired to have a second motor start automatically when the first one is stopped. The second motor is to run only for a given length of time. A good application of this might be found where the second motor is needed to run a cooling fan or a pump after the first motor has stopped.

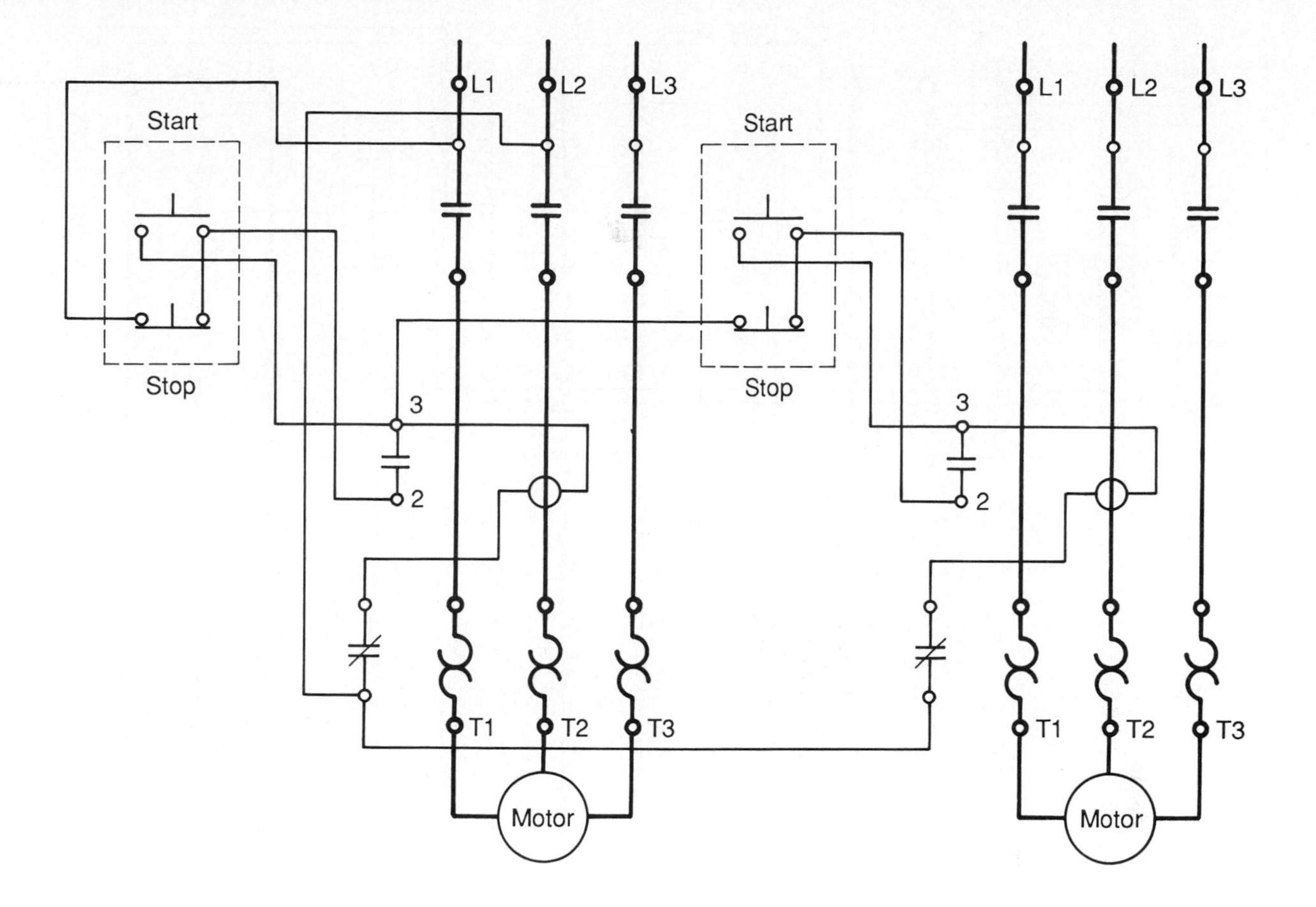

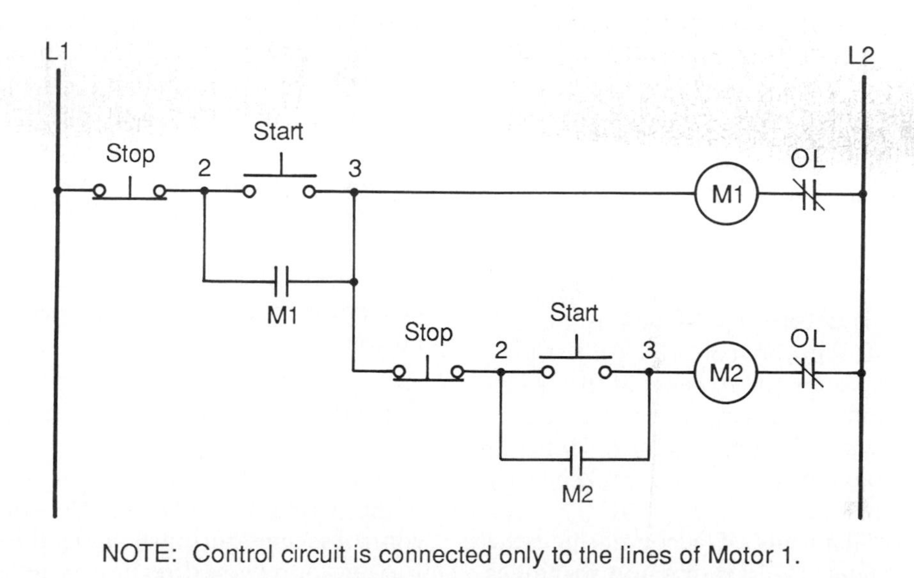

NOTE: Control circuit is connected only to the lines of Motor 1.

FIGURE 15–8 *Sequence control diagrams. (Allen-Bradley)*

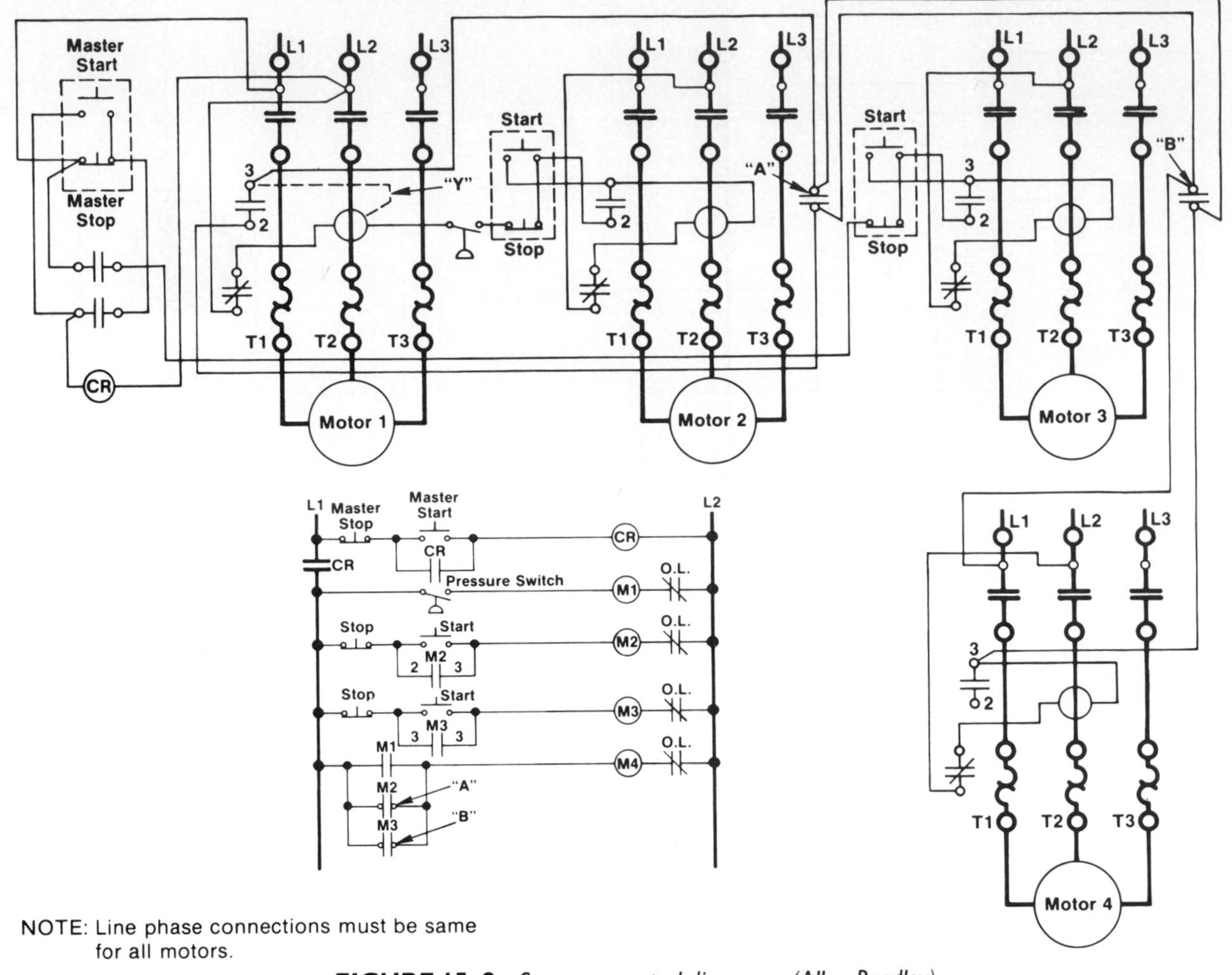

FIGURE 15–9 *Sequence control diagrams. (Allen-Bradley)*

To accomplish this, an off-delay timer (TR) is used. When the START button is pressed, it energizes both M1 and TR. This operation of TR closes its time-delay contact, but the circuit to M2 is kept open by the opening of the instantaneous contact. As soon as the STOP button is pressed, both M1 and TR are dropped out. This closes the instantaneous contact on TR and starts M2. M2 will continue to run until TR times out and the time-delay contact opens.

JOGGING

Jogging, or inching, is defined by NEMA as the momentary operation of a motor from rest for the purpose of accomplishing small movements of the driven machine. One method of jogging is shown in Fig. 15–11. The selector switch disconnects the holding circuit interlock and jogging may be accomplished by pressing the START button.

There are several means of accomplishing the jogging operation. Figure 15–12 shows how jogging is done using a control relay. Pressing the START button energizes the control relay that in turn energizes the starter coil. The normally open starter interlock and relay contact then form a holding circuit around the START button. However, pressing the JOG button energizes the starter coil independent of the control relay and no holding circuit forms. Then jogging can be obtained simply by pushing the JOG button and releasing it independent of the START button.

Jogging can also be accomplished by using a *selector* pushbutton. The use of a selector pushbutton to obtain jogging is shown in Fig. 15–13. In the RUN position the selector pushbutton gives normal three-wire control. In the JOG position, the holding circuit is broken and the jogging is accomplished by depressing the button.

Forward or Reverse Jogging

Jogging in the forward or reverse direction is possible if the wiring shown in Fig. 15–14 is followed. This control scheme permits jogging the motor either in the forward or reverse direction, whether the motor is at a standstill or is rotating in either direction. Pressing the

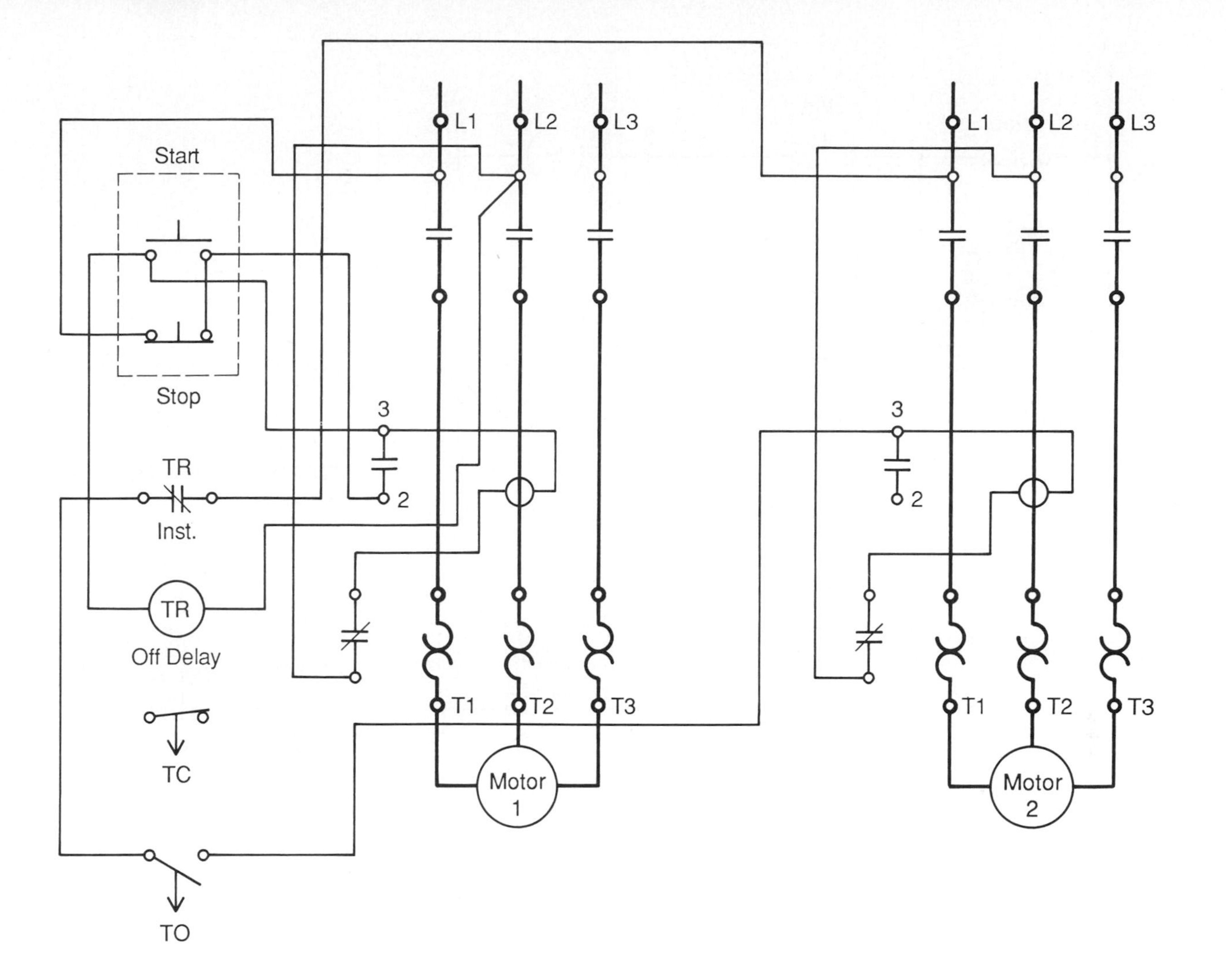

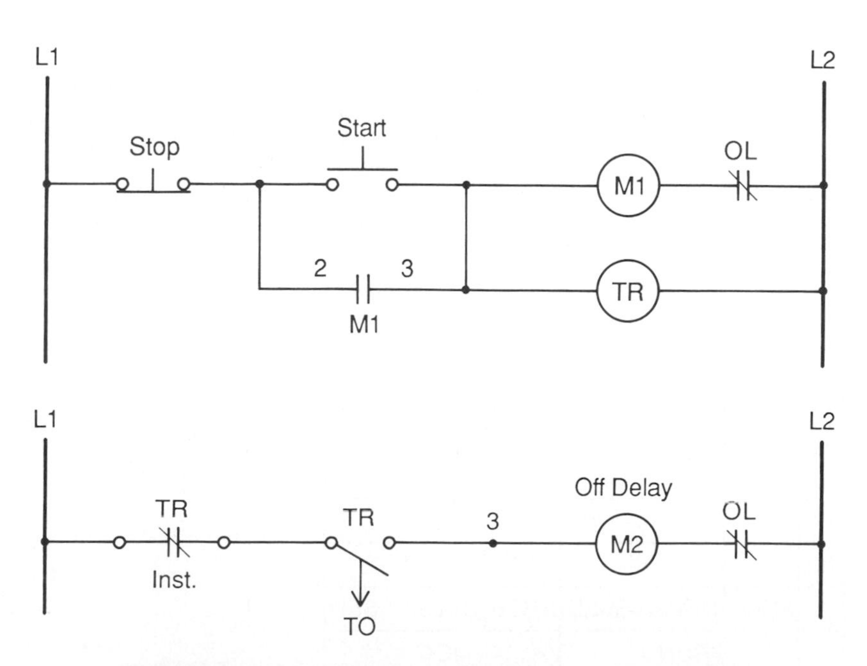

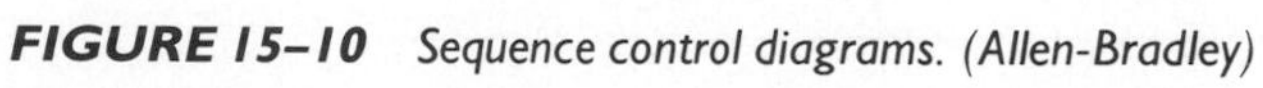
FIGURE 15–10 *Sequence control diagrams. (Allen-Bradley)*

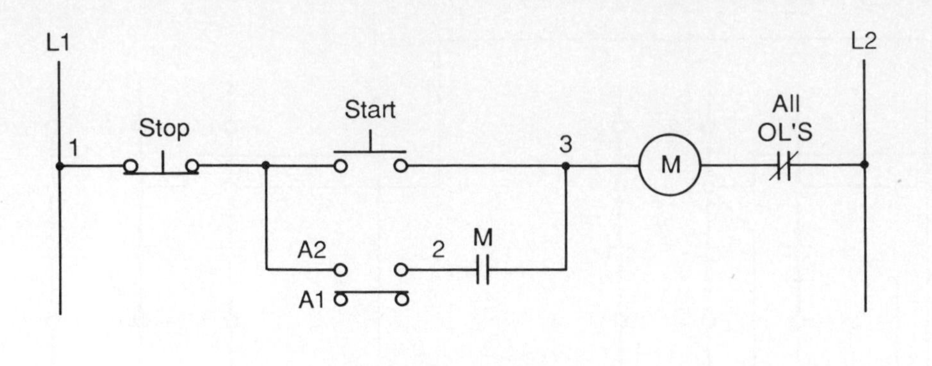

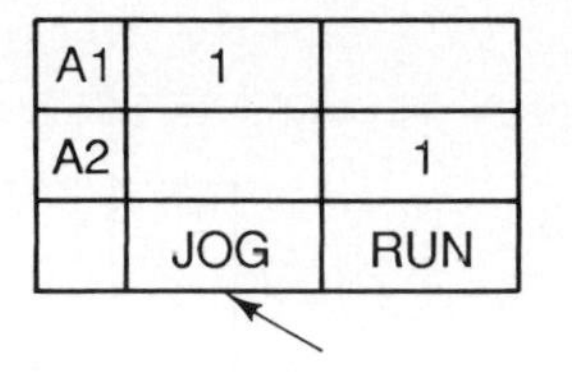

A1	1	
A2		1
	JOG	RUN

FIGURE 15–11 *Jogging with a selector switch.*

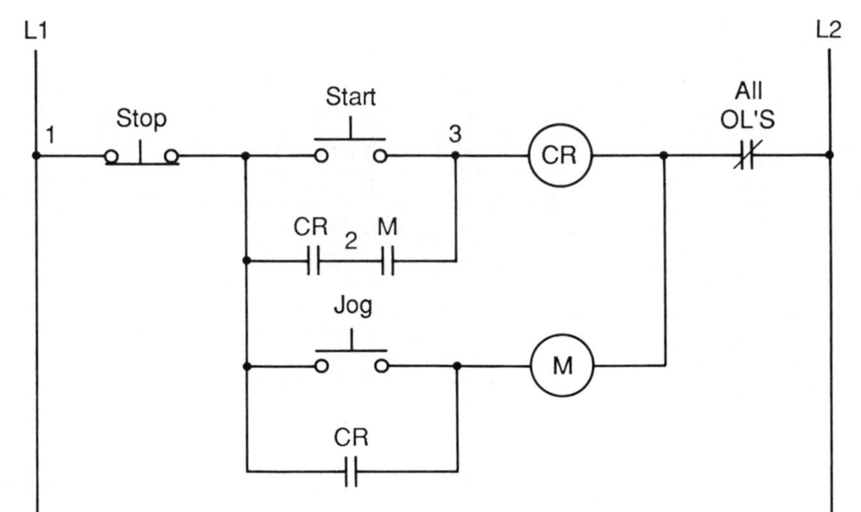

FIGURE 15–12 *Jogging with a control relay.*

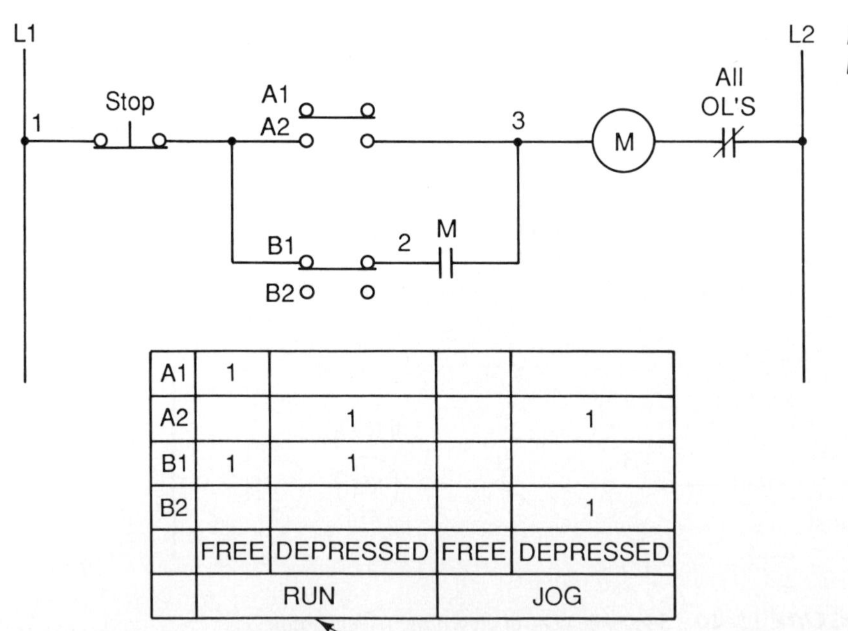

A1	1			
A2		1		1
B1	1	1		
B2				1
	FREE	DEPRESSED	FREE	DEPRESSED
	RUN		JOG	

FIGURE 15–13 *Jogging using a selector switch pushbutton.*

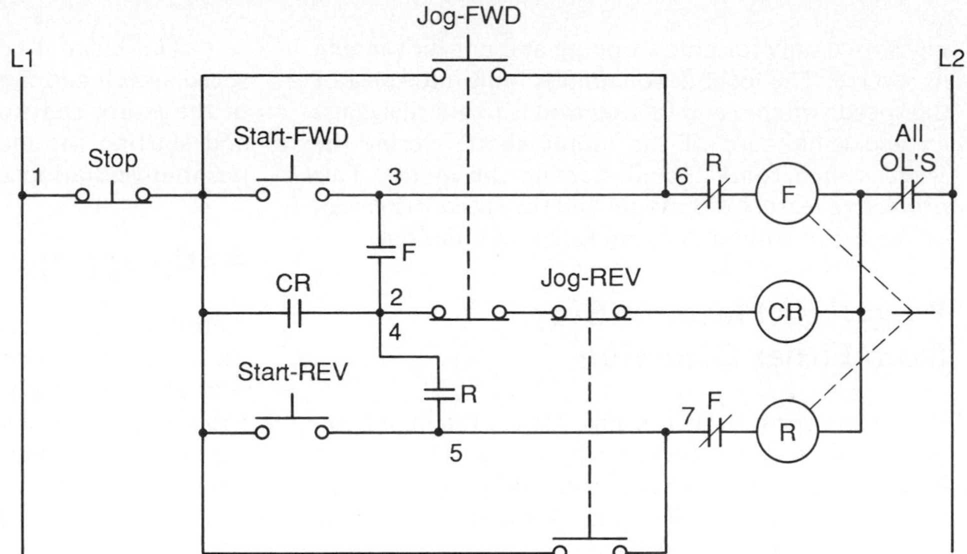

FIGURE 15–14 Jogging using a control relay for reversing starter.

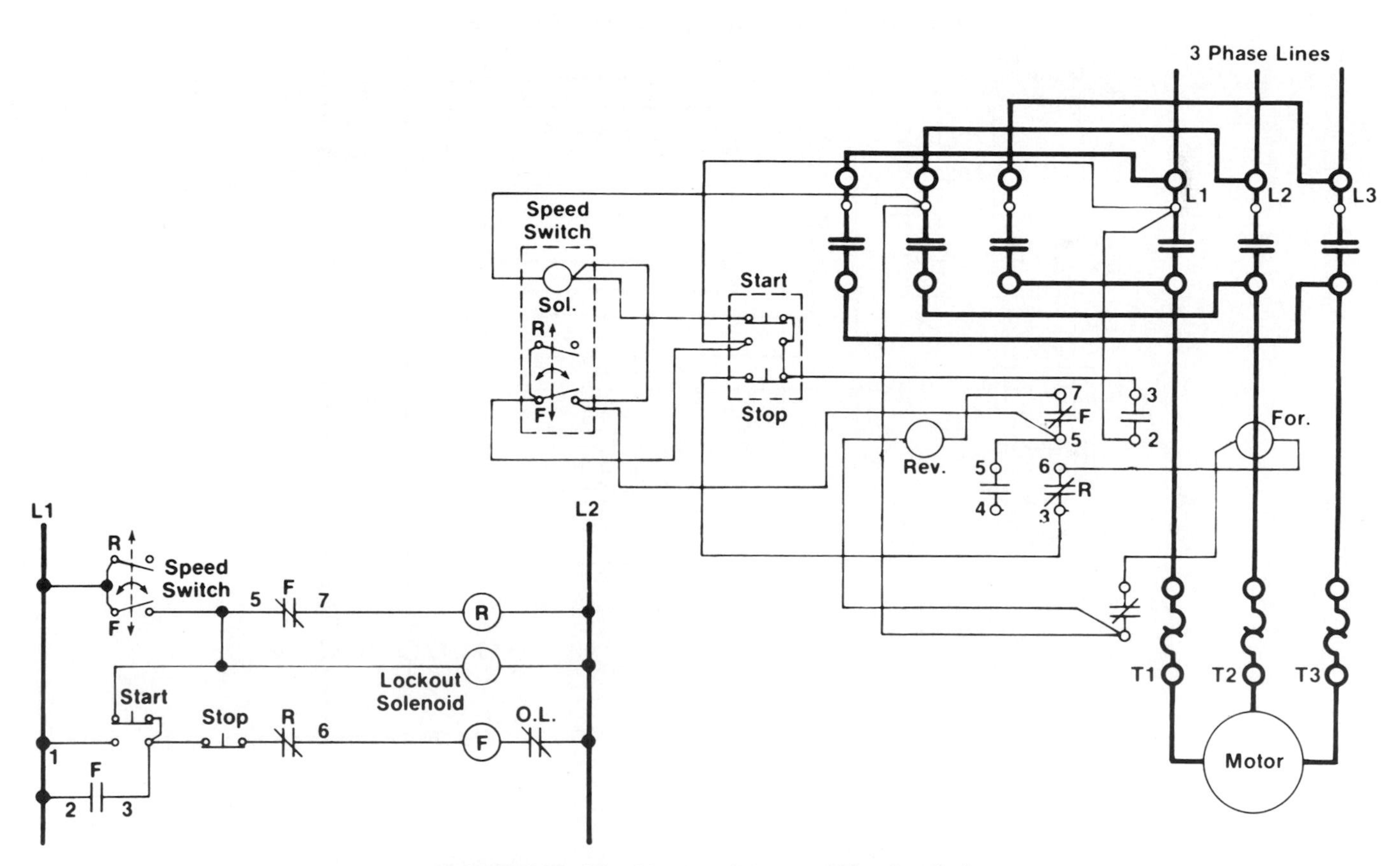

FIGURE 15–15 *Plugging diagrams. (Allen-Bradley)*

START-FORWARD or START-REVERSE buttons energizes the corresponding starter coil, which in turn closes the circuit to the control relay. The relay picks up and completes the holding circuit around the START button. As long as the relay is energized, either the forward or reverse contactor will remain energized. Pressing *either* JOG button will deenergize the relay, releasing the closed contactor. Further pressing of the JOG button permits jogging in the desired direction.

PLUGGING

Plugging is defined by the NEMA as a system of *braking* in which the motor connections are reversed so that the motor develops a countertorque. Thus it exerts a retarding force. In the scheme shown in Fig. 15-15 the motor is run in one direction only and must come to a complete stop when the STOP button is pressed. The reverse contactor of the reversing switching is used only for plug stopping and not for running in reverse. The lockout solenoid is built into some of the speed switches and its function is to guard against an accidental turn of the motor shaft, closing the speed switch contacts and starting the motor. This protective feature is optional and the speed switch can be furnished without lockout solenoid if desired.

Plugging a Motor to Stop from Either Direction

With the system shown in Fig. 15–16, the motor can be started in either direction by pressing the proper button. Pressing the STOP button will plug the motor to stop from either direction. A standard reversing switch is used for this purpose.

The lockout solenoid is a built-in part of the speed switch and it guards against an accidental turn of the motor shaft closing the speed switch contacts and starting the motor. The control relay and the pushbutton station are standard parts.

Antiplugging

Antiplugging protection is defined by the NEMA as the effect of a device that operates to prevent application of countertorque by the motor until the motor speed has been reduced to an acceptable value. With the motor operating in one direction, as shown in Fig. 15–17, a contact on the antiplugging switch opens the control circuit of the contactor used for the opposite direction. This contact will not close until the motor

FIGURE 15–16 *Plugging diagrams. (Allen-Bradley)*

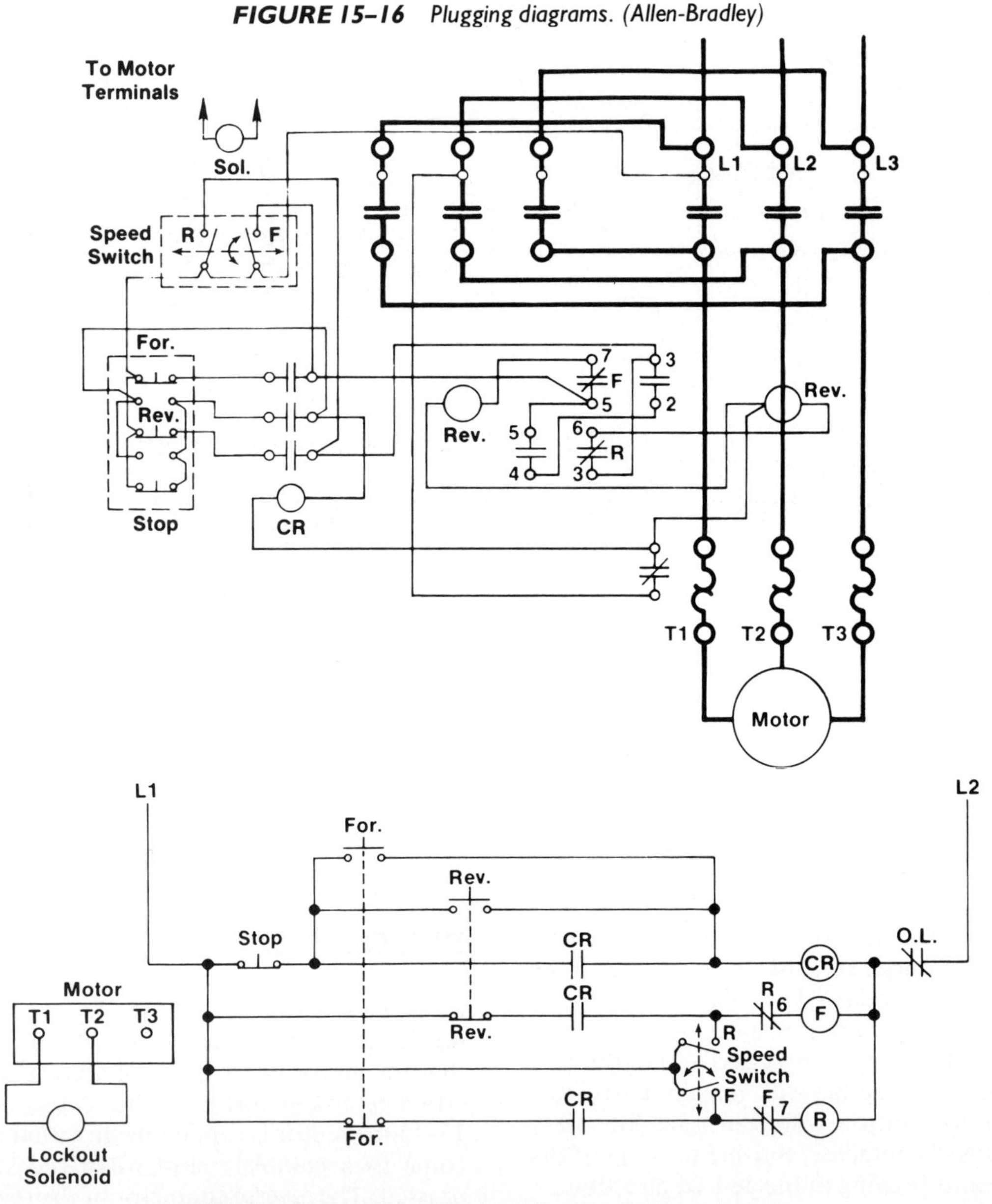

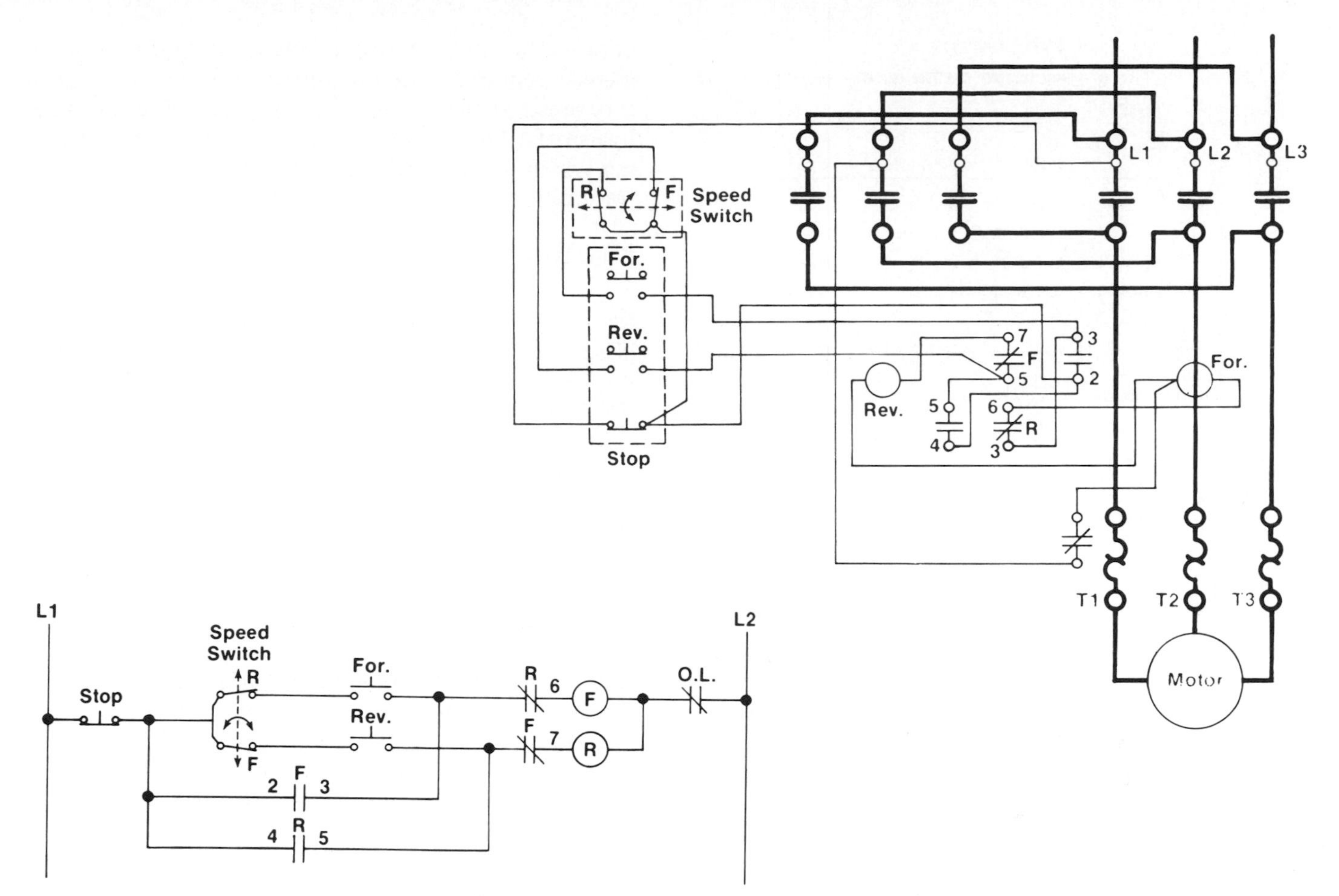

FIGURE 15–17 *Antiplugging diagrams. (Allen-Bradley)*

has slowed down, after which the other contactor can be energized. In this schematic the motor can be reversed, but it must not be plugged.

BRAKING

Electric motors can be brought to a stop or braked both electrically and mechanically. In some instances it is necessary to use a combination of both. This usually happens when the motor is connected to a load that is not easily stopped or cannot be disconnected easily.

Electronic Motor Brake

The electronic motor brake made by Square D provides a simple, effective means of braking an ac squirrel-cage motor (Fig. 15–18). It can be used for woodworking machines such as saws and sanders, and for machine tools such as lathes and drills, as well as for conveyor systems, textile machinery, and centrifuges. Heating, venting, and air-conditioning fans and many other machines in varied industries may also use this type of braking.

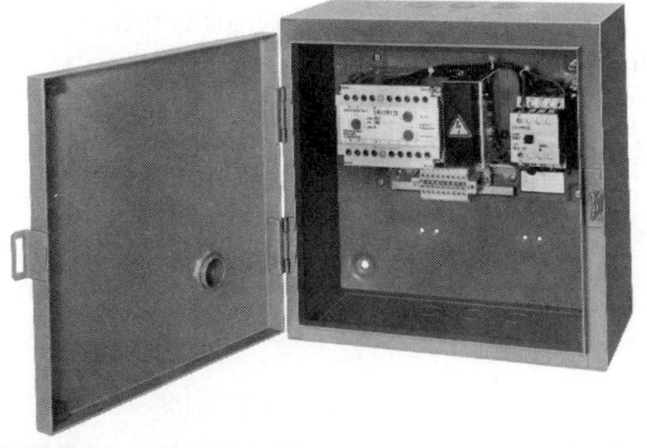

FIGURE 15–18 *Electronic motor brake. (Square D)*

The major advantages of the electronic methods versus the mechanical brake system are:

1. No friction, wear, or maintenance
2. Adjustable soft-stop capability
3. No mechanical connection to the motor shaft
4. Multimotor braking capability
5. Easily wired to a new or existing machinery
6. Unaffected by hostile motor environment

Electronic braking is commonly known as dynamic braking. Dynamic braking of an ac induction

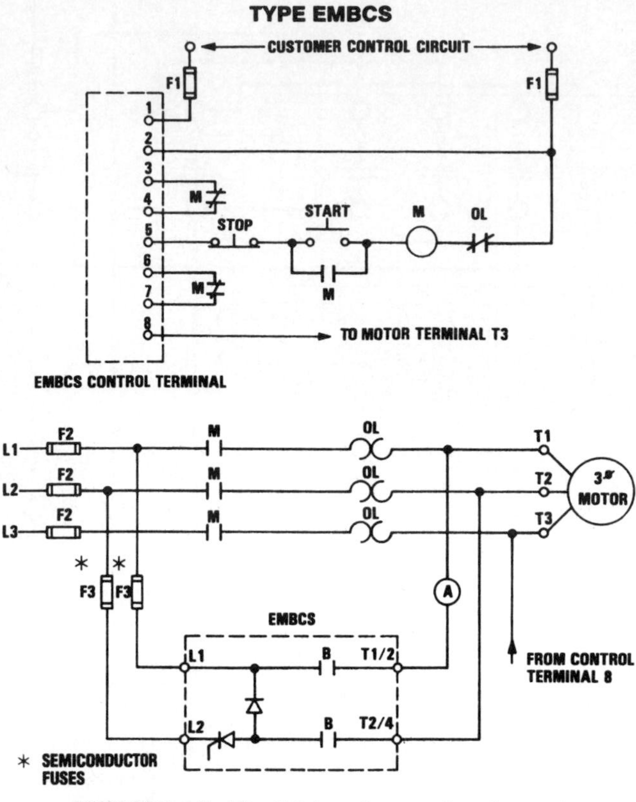

FIGURE 15–19 *Wiring diagram for electronic motor brake. (Square D)*

motor is generally accomplished by exciting its stator windings with dc current. The amount of braking torque is directly proportional to the dc current passing through the stator windings of the motor (Fig. 15–19).

Dynamic braking of a motor may cause threaded fasteners connected to the motor shaft to loosen, due to the reverse torque applied. Use positive-locking fasteners or fastening compound to prevent such loosening.

Note: Electronic motor brakes will not stop the motor if power is lost or disconnected.

This type of electronic motor brake can be used to stop a load and signal a mechanical brake system to hold it. In addition, the brake will interface with either jogging, reversing, multispeed, or reduced-voltage motor starter applications.

The electronic motor brake is designed such that the braking contactorcloses before the thyristor (SCR) switches the braking current on. The contactor will not open until after the braking current has been switched off. This allows the braking contactor to be rated for current-carrying capacity only and not for the higher make-and-break duty.

An additional circuit detects when the motor has come to a halt, switches off the braking current, and permits the motor to restart. No braking time adjustment is required. The maximum braking time is factory preset at 10 seconds. Braking torque is adjustable by use of a single potentiometer. This is an ideal braking system for jobs where there is a variable load and for multispeed three-phase motors.

Mechanical Braking

Electric motors can also be stopped when necessary by using a mechanical means. These are similar to what is used with automobiles. Some of them rely on an electric current to energize the solenoid to cause the brake shoes to tighten around the motor shaft and stop it.

Reliance makes an electromechanical brake for motors up to 10 hp and 3600 rpm (Fig. 15–20). It has friction disks that are self-resetting with a manual release lever. The magnet coils are encapsulated to protect them from dirt and moisture. Antirattle springs are incorporated to reduce vibration and noise.

In some instances it is necessary to have a mechanical brake since dynamic braking is not sufficient to stop the motor rotation completely after power is removed. These brakes may be actuated whenever power is removed from the motor circuit. An electromagnet holds the brake shoes away from the motor shaft whenever the motor is energized. Once power has been removed the brake is automatically applied by spring action (Fig. 15–21). This type of braking is very useful in elevators and similar installations.

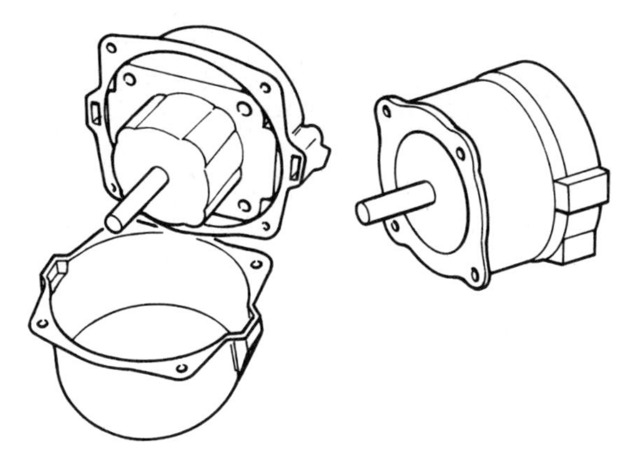

FIGURE 15–20 *Electromechanical brakes. (Reliance)*

Thruster Brakes

Thruster brakes are used with ac or dc motors and provide a smoothly applied fixed torque for hold or for stopping (Fig. 15–22). They are used on crane travel drives, lift bridges, conveyors, and similar applications to reduce load sway, and affect loading to motors and the mechanical system. These brakes are released by a thruster mechanism. This self-contained

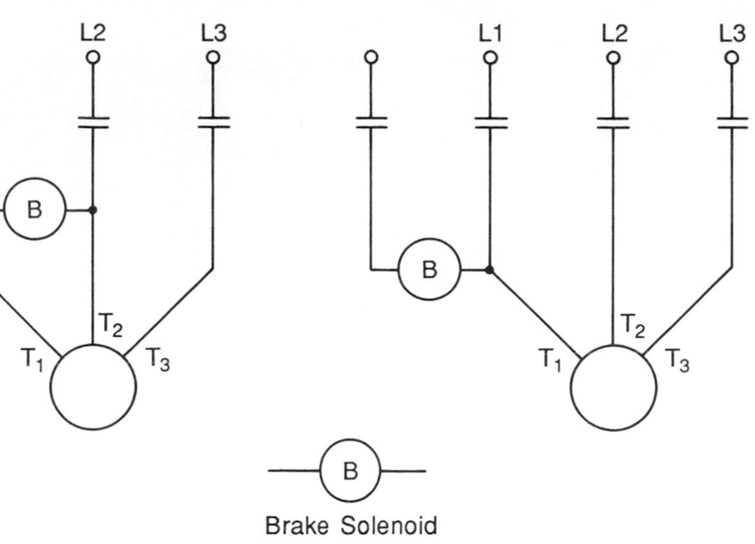

FIGURE 15–21 *Ac brake coil hook ups for across-the-line starting.*

FIGURE 15–22 *Thruster brakes. (Square D)*

mechanism contains an ac squirrel-cage motor and hydraulic pump. When deenergized the brake sets smoothly as the pumping action ceases.

Magnetic Brakes

Brakes are selected by the amount of torque required for the particular application. Generally, the full-load torque of the motor is used as a basis for determining the brake torque required. This can be calculated by using the following formula for both ac and dc motors:

$$\text{torque} = \frac{\text{rated hp} \times 5252}{\text{rated rpm}}$$

Depending on the characteristics of the drive, the braking torque required may be more or less than the full-load torque of the motor. In addition to being selected to meet the torque requirements of the particular application, the magnetic brake used for stopping must be selected to prevent overheating of the brake wheel when operated on the anticipated duty cycle.

Hydraulic Brakes

Hydraulic brakes are used with ac or dc motors to provide an operator-controlled infinitely adjustable torque for slowing and stopping. These are used on crane travel drives, mill machines, conveyors, and similar jobs. They are spring released, hydraulically applied shoe-type friction brakes designed to meet AISE (American Iron and Steel Engineers) standards for mounting. The standard brake includes corrosion-resistant hardware and grease fittings (Fig. 15–23). Figure 15–24 shows the typical piping diagram for one brake.

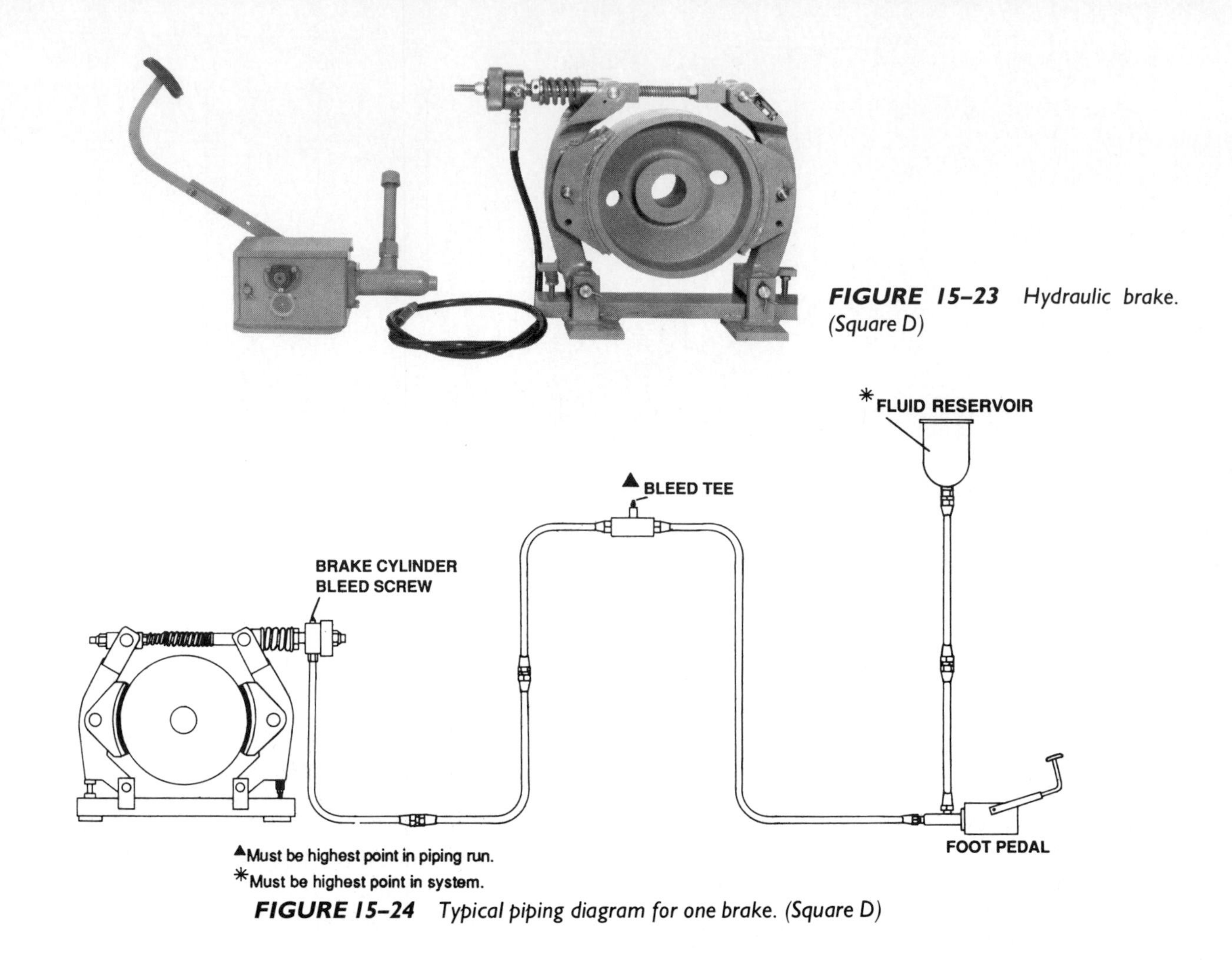

FIGURE 15–23 *Hydraulic brake. (Square D)*

FIGURE 15–24 *Typical piping diagram for one brake. (Square D)*

MOTOR PROTECTION

Motor protection can take many forms, inasmuch as various types of motors are used to drive many types of machines. One of the most commonly used motor applications is to drive pumps. Surge protection and backspin are both present in pumps and must be considered in the circuitry design of the starters.

Surge Protection and Backspin

Surge protection and "backspin" are two of the factors to be considered when protecting motor used to power pumps. Surge protection is often necessary when the pump is turned off and the long column of water is stopped by a check valve. The force of the sudden stop may cause surges that operate the pressure switch contacts, subjecting the starter to *chattering*.

Figure 15–25 shows how the system provides protection on both starting and stopping. Backspin is included automatically. Two timing relays are used here, one to provide surge protection on starting and one to provide surge protection on stopping and backspin protection. TR1 is an *on-delay timer* used for

FIGURE 15–25 *Pump operation with surge protection on starting and stopping. (Allen-Bradley)*

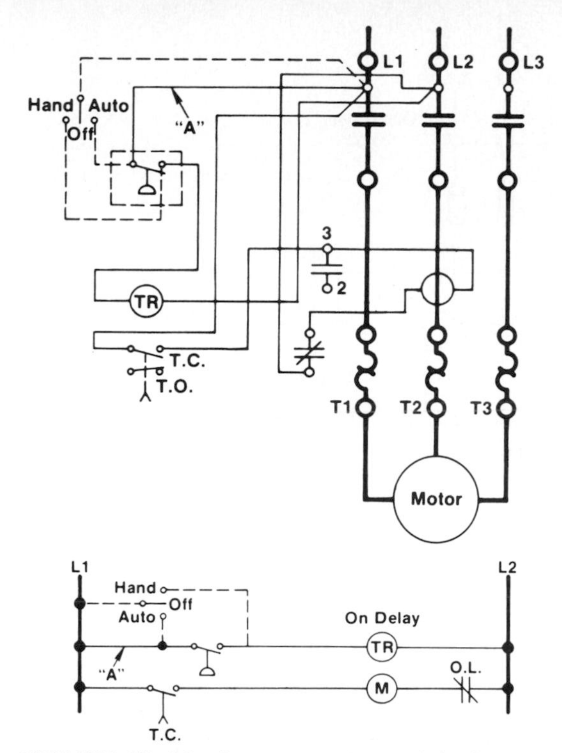

FIGURE 15–26 *Pump operation with backspin protection and surge protection. (Allen-Bradley)*

surge protection on starting. When the pressure switch contact closes, relay CR, the starter and two timers are energized. The instantaneous contact on TR1 closes, bypassing the pressure switch contact and preventing the pump motor starter from dropping out even though starting surges open the pressure contact. After the timing period, the time-delay contact TR1 opens the bypass and PS can then stop the pump at the proper pressure. TR2 is an *off-delay timer* for surge protection on stopping and backspin protection. Once turned off the system cannot be operated again until timer TR2 has timed out and its normally closed contact is closed.

Another system that provides backspin protection and surge protection on stopping is shown in Fig. 15–26. It also has time delay between pressure switch closing and motor starting. The pressure switch energizes the timer (TR), but the motor cannot start until the time-delay contact has closed. The timer can thus be set for a time long enough to allow all surges and backspin to stop.

The dashed lines show how a selector switch can be added to bypass the pressure switch if necessary. This is often used for motor testing purposes. It does not eliminate the time delay, however. If the selector switch is added, wire A must be removed.

Overload Protection

The overload relay is a manual reset, eutectic alloy, thermal-type overload device (Fig. 15–27). When co-

FIGURE 15–27 *Overload relay, eutectic alloy, thermal type. (Allen-Bradley)*

ordinated with the proper short-circuit protection, the overload relay is intended to protect the motor, motor controller, and power wiring against overheating due to excessive overcurrents.

The ratchet stud assembly is heated by current flowing through the heater element. Relay operation occurs when the temperature of the ratchet and stud reaches the melting point of the eutectic alloy, freeing the ratchet stud and opening the NC contact.

To reset the overload relay contact, it is necessary to press and release the reset operator after the eutectic alloy has solidified. Approximately 2 minutes are required for the alloy to solidify.

Automatic Reset Overload Relay

An indirectly heated, automatic reset noncompensated thermal relay is shown in Fig. 15–28. The temperature-sensitive unit of this relay is a bimetallic U-shaped strip that is mechanically coupled to a precision snap-action switch. The bimetallic strip is heated by the motor current that flows through a heater element affixed in close proximity. As the bimetal is heated, a deflection is produced in it by the different rates of expansion of the two metals. A sustained current, greater than the rating of the heater element, will develop sufficient deflection to open the contact of the snap switch.

FIGURE 15–28 *Overload relay-automatic reset. (Allen-Bradley)*

Inverse Time Current Relay

A magnetically operated inverse time current overload relay that can be used in the protection of ac or dc motors is shown in Fig. 15–29. Both the tripping current and tripping time are easily adjustable. The relay is usually supplied with normally closed contacts and automatic reset. To prevent relay damage, current through the relay coil should be interrupted after the relay trips.

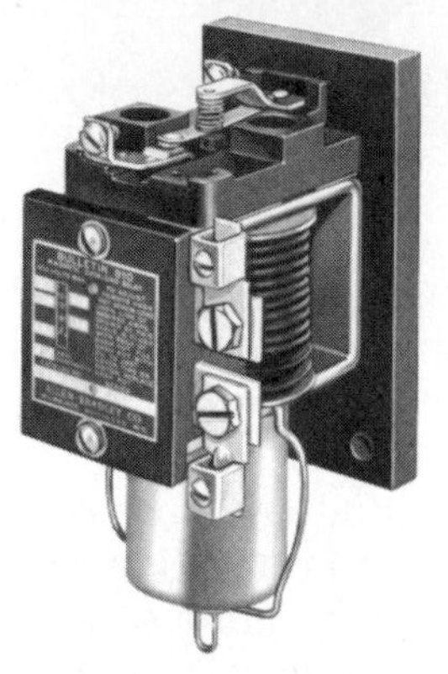

FIGURE 15–29 *Inverse time current relay. (Allen-Bradley)*

Solid-State Line Voltage and Current Monitor Relays

The line voltage monitor and line current monitor are solid-state devices designed for use in three-phase systems to protect motors and other loads against abnormal voltage/current conditions. In general, the line voltage monitor is applied where prestart protection and line-side protection are important, whereas the line current monitor is applied where line- *and* load-side protection is important (Fig. 15–30).

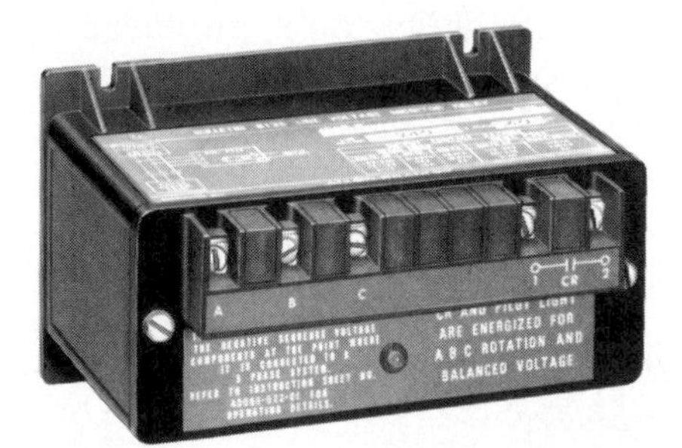

FIGURE 15–30 *Line-voltage monitor. (Allen-Bradley)*

Line Voltage Monitor. The line voltage monitor detects phase failure, voltage imbalance, phase reversal, and undervoltage (Fig. 15–31A). It provides prestart and running protection on the line side of the point of connection and connects directly to three-phase lines with a 600-V maximum. It can be used with potential

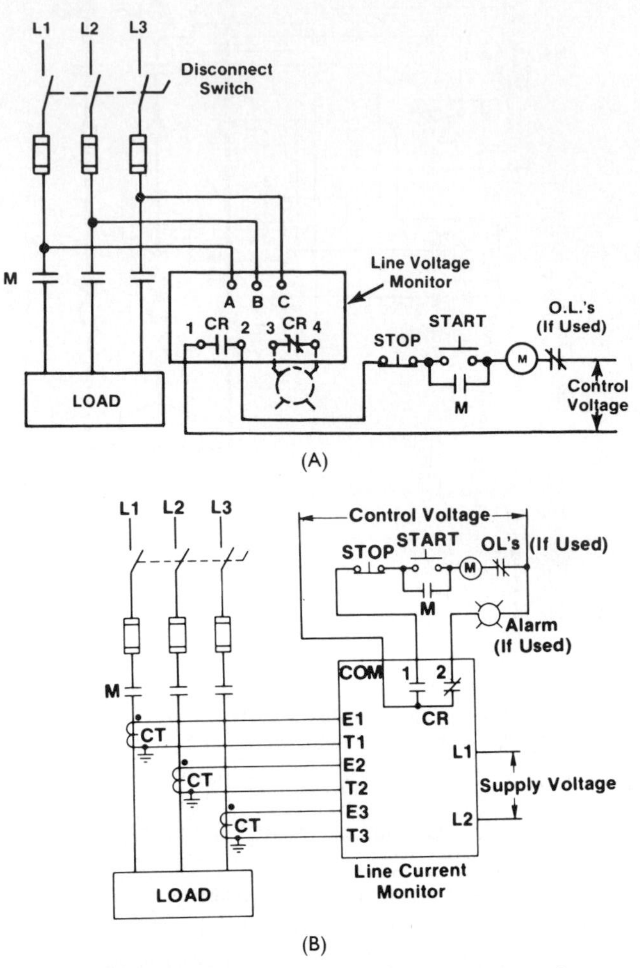

FIGURE 15–31 *(A) Line-voltage monitor in the circuit; (B) line-current monitor in the circuit. (Allen-Bradley)*

transformers where line voltage exceeds 600 V. It has automatic reset and an LED indicates normal voltage conditions and energized output relay CR.

Line Current Monitor. The line current monitor detects phase failure, current imbalance, and phase reversal. It also provides running protection on the line and load side of the point of connection when used in a single motor branch circuit. Inputs connect to standard current transformers (5 A secondary). An LED indicates that supply voltage is present. The manual reset device also has an LED to indicate normal current conditions and an energized output relay CR (Fig. 15-31B). Current imbalance protection is effective during the motor running period only. Phase failure and reversal protection is provided during both starting and running periods.

Programmable Motor Protection

Allen-Bradley's programmable motor protector combines sophisticated, comprehensive, and coordinated motor protection into a modular system (Fig. 15–32).

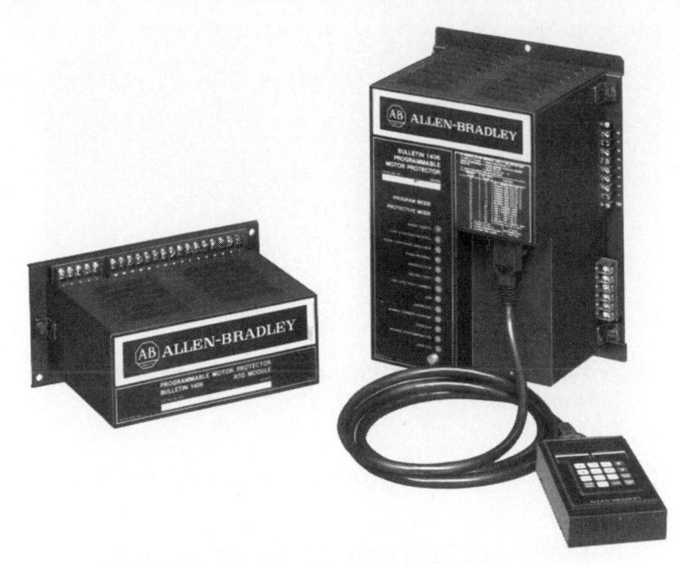

FIGURE 15–32 *Programmable motor protector: (A) remote RTD module; (B) programmer monitor. (Allen-Bradley)*

It is intended for protection of large expensive motors that are often critical to a system or process. Typically, these motors are medium-voltage (2300 to 7200 V) or large low-voltage (200 hp or larger) motors.

By processing incoming data and looking for trends in various motor parameters, the device can provide a high degree of coordinated motor protection. It annunciates an abnormal condition and provides an output that trips the motor off-line. Alarm and trip contacts can also be used to initiate an orderly shutdown of the process.

Protective Module

The *protective module* is the main module in the system. It has inputs, an annunciator panel, and outputs. This module receives inputs from external potential transformers (for three-phase line voltage) and current transformers (for three-phase line current). These inputs represent a portion of real-time motor data coming into the device for processing (Fig. 15–33). The remaining inputs to the protective module consist of the outputs from the programmer/monitor and the remote resistance temperature detector (RTD) module.

The protective module examines incoming motor data and compares them to user-preprogrammed limits. The module then determines if the motor should be taken off-line. It provides alarm and trip signals, and visually indicates the abnormality.

FIGURE 15–33 *System configuration with communications option. (Allen-Bradley)*

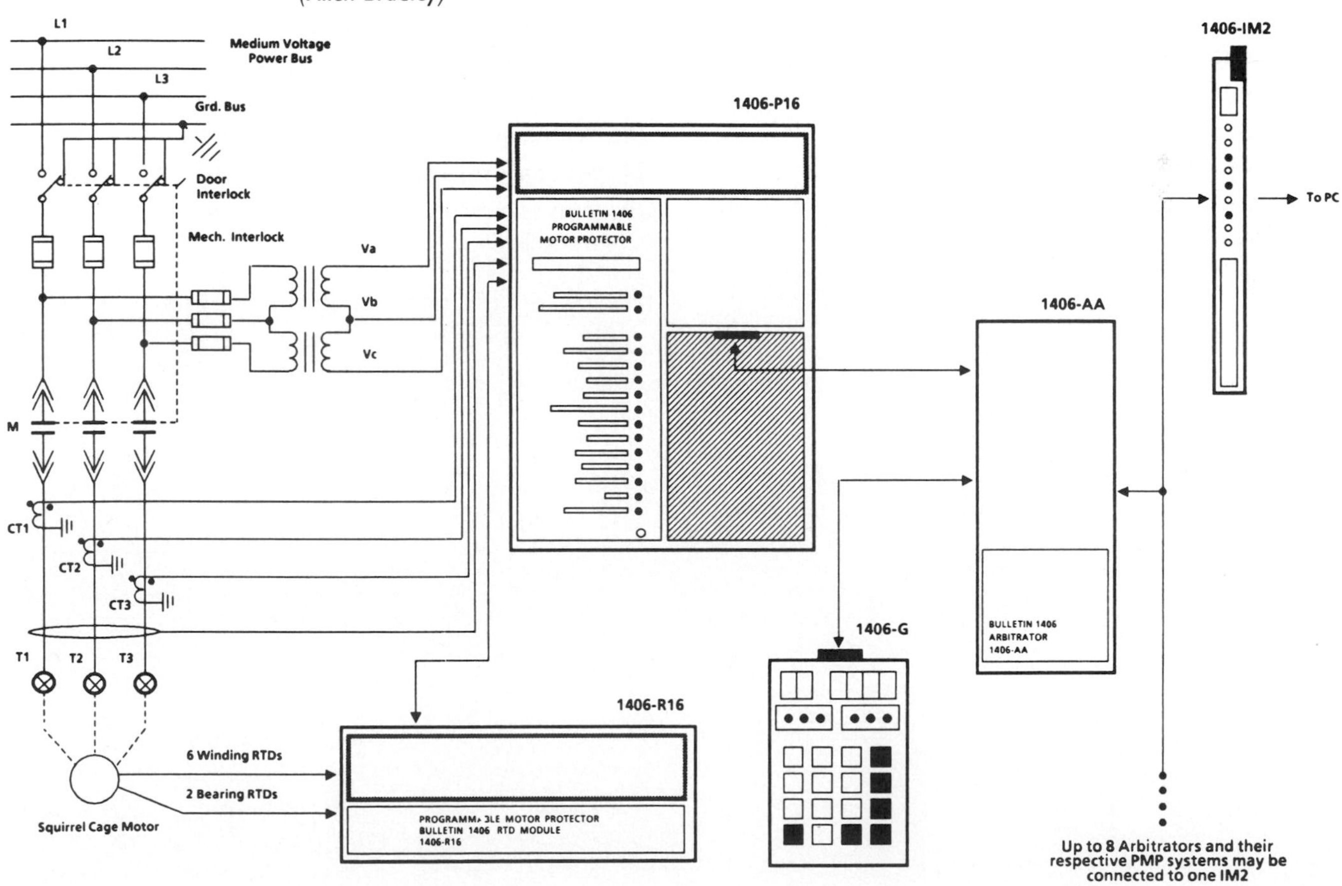

Remote Temperature Detector Module

The remote RTD module serves as a remote temperature gathering panel. It is a microprocessor-based device that performs temperature data acquisition and coordination on demand from the protective module. The temperature data are used by the protective module to help construct a thermal model of the motor copper and iron.

This remote RTD module scans eight RTDs (two bearing and six winding). It takes an analog signal, digitizes it, performs a linearization function (to compensate for nonlinear RTD characteristic), scales it, and communicates the value to the protective module on demand. For motors not having RTDs embedded in the stator windings and bearings, motor protection can still be provided using just the protective module.

An optional instrumentation card allows the PMP to provide a variety of metering functions (Fig. 15-34). The following information can be provided:

FIGURE 15-34 *Instrumentation card for PMP. (Allen-Bradley)*

1. Elapsed time running: hours
2. Total energy consumption: MWh
3. Power factor: lead/lag
4. Power: kilowatts

QUESTIONS

1. What do manual starters provided provide for a motor?
2. What is meant by low-voltage protection?
3. How is low-voltage protection provided?
4. What is meant by *soft start*?
5. What is meant by *current limit*?
6. What is jogging? How is it done?
7. What is plugging? How is it done?
8. How are motors braked or stopped?
9. What are six advantages of the electronic braking method?
10. What is another name for electronic braking?
11. What is backspin protection?
12. What causes chattering in a starter?
13. When are line monitors used?
14. What does a line voltage monitor do?
15. What is meant by *annunciates*?
16. Where are manual starters used?
17. What does the symbol Ⓜ represent?
18. Who requires low-voltage protection?
19. What are now replacing electromechanical devices?
20. Why is a master start–stop button used with a control relay?

16

Three-Phase Controllers

Objectives

After studying this chapter, you will be able to:

1. List factors to be considered in the operation and control of any motor.
2. List the types of three-phase starters.
3. Draw a ladder diagram for the operation of start–stop–jog on a three-phase motor.
4. Describe how to reverse a three-phase motor.
5. List the advantages of solid-state controllers for three-phase motors.

The squirrel-cage three-phase motor is highly reliable and efficient at essentially constant speed and requires little or no maintenance. Depending on construction, it may be classified as normal torque, normal starting current; normal torque, low starting current; high torque, low starting current; high slip; low starting torque, normal starting current; low torque, low starting current.

The three-phase squirrel-cage motor can be used in many different locations and for various applications, including rotary compressors, machine tools, large fans, light conveyors, milling machines, agitators, elevators, hoists, punch presses, centrifugal pumps, and blowers. It is made in the range $\frac{1}{2}$ to 400 hp.

The wound-rotor motor is used where limited speed control and speed adjustments under fluctuating load are required. This type of motor is made in the range $\frac{1}{2}$ up to several thousand horsepower. It can be found driving conveyors, fans, lift bridges, cranes, hoists, and metal rolling mills.

The synchronous motor is made in the range of 20 to several thousand horsepower. It is used for power factor correction and for exact slow-speed drives and maximum efficiency on continuous loads above 75 hp.

These motors can be started with comparative ease with the equipment available today. However, there are some uses that require special consideration. A controller is something more than just a starter. A starter causes the motor to start and stop. The controller not only starts the motor, but it also has the ability (according to its intended design) to reverse the motor and to vary its speed and torque as needed. It is this facet of motor control that requires special consideration and slightly different use of electrical circuitry and electromechanical devices. More solid state devices such as programmable controllers are being introduced into the field of control and will require a better grasp of electronics to be fully understood.

The following factors are to be considered in the operation and control of any motor. These conditions can cause the motor to be damaged or destroyed.

1. *Low voltage.* The motor tried to do the work, but the low voltage causes excessive currents that overheat the motor.
2. *Heavy-duty load cycles.* Constant starting, stopping, jogging, inching, or plugging overloads the motor.
3. *Excessive loads.* The load is too big for the motor.
4. *Locked rotor.* Motor jams or cannot get running.
5. *High-inertia load.* Motor takes abnormally long time to accelerate.
6. *High ambient temperature.* The surrounding temperature is high, which heats the motor. Thus the motor may not be able to handle as big a load safely as in a lower ambient temperature.
7. *Single phasing.* One of the power lines fails. The motor will draw higher line current. Two overload relays with properly sized elements will protect the motor except in rare cases such as when the motor is fed by an ungrounded wye-delta supply transformer.
8. *Poor ventilation.* Motor cannot get cool air to take away some of the heat (cannot be protected by OL relays).

STARTERS

Simplest Type

The simplest type of three-phase starter is shown in Fig. 16–1. The wiring diagram shown in Fig. 16–2 makes it easy to visualize the wiring of the starter. The armature and crossbar or the overload reset mechanism is not shown in the wiring diagram since these parts need not be considered from the wiring standpoint. Note the location of L1, L2, and L3 at the top of the starter and T1, T2, and T3 at the bottom. As can be seen here, overload protection is built into the device.

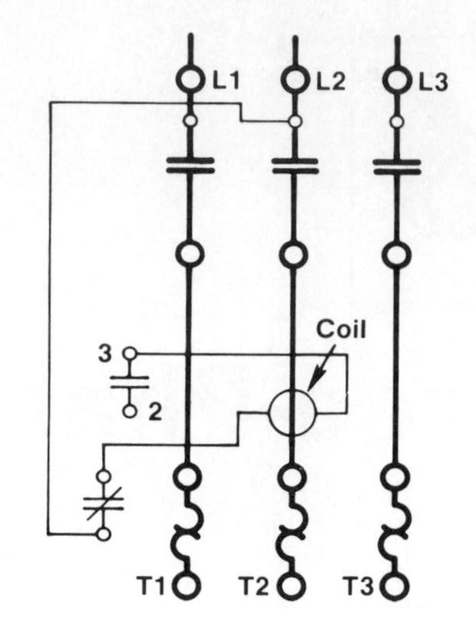

FIGURE 16–2 *Diagram for starter. (Allen-Bradley)*

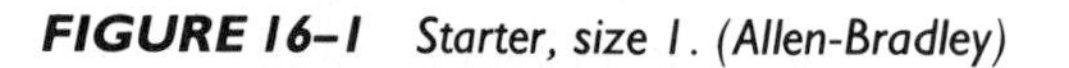

FIGURE 16–1 *Starter, size 1. (Allen-Bradley)*

Full-Voltage Reversing Starter

Full-voltage starting with the capability of reversing the motor calls for a little more complex wiring (Fig. 16–3). Remember: To reverse the direction of rotation of a three-phase motor it is necessary to change two of the incoming lines. Note in Fig. 16–4 that L1 and L3 are reversed when the contacts on the left are closed. It also means that the contacts in L1, L2, and L3 are opened when the reversing contacts close. L2 is not switched, so look at the diagram closely to see how the reversing contacts are directly across the L2 contacts, indicating that there is no change in this line. Also notice how the starters are electrically interlocked to avoid both contactors being closed simultaneously. The contacts are also mechanically interlocked so that both cannot be closed at the same time. Normally, L1 and L2 are switched to reverse direction.

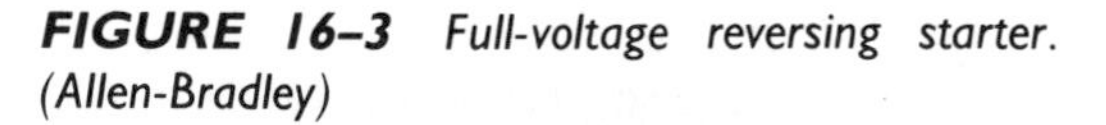

FIGURE 16–3 *Full-voltage reversing starter. (Allen-Bradley)*

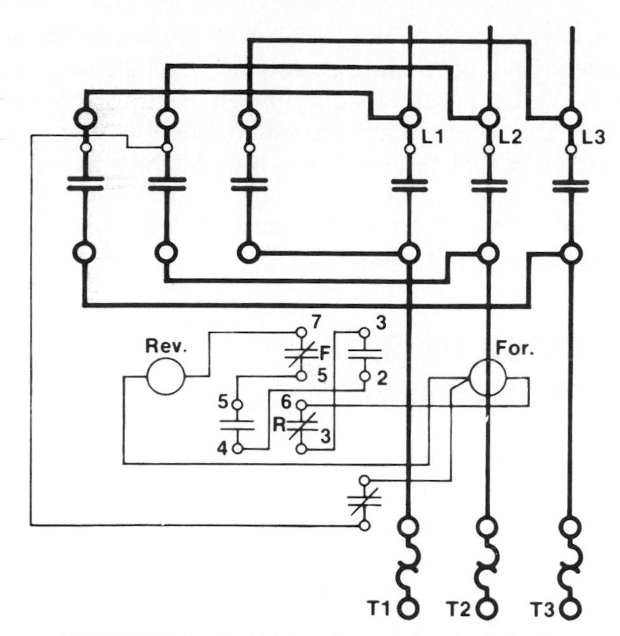

FIGURE 16–4 *Wiring diagram for the full-voltage reversing starter. (Allen-Bradley)*

Forward–Reverse–Stop

Standard wiring with the FORWARD-REVERSE-STOP pushbutton station is shown in Fig. 16–5. The STOP button must be depressed before changing direction.

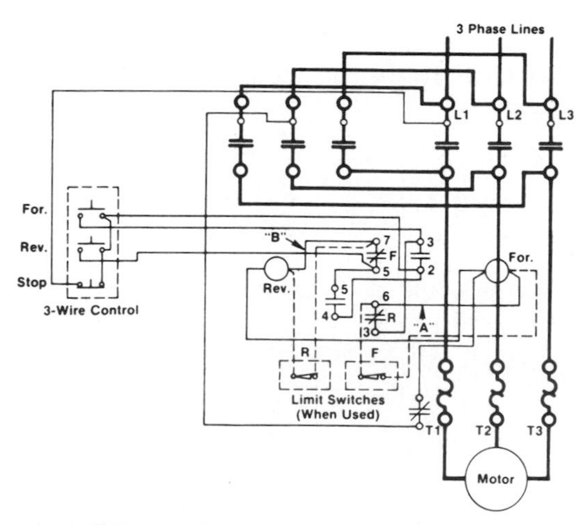

FIGURE 16–5 FORWARD–REVERSE–STOP *three phase starter. (Allen-Bradley)*

A mechanical interlock and electrical interlocks are supplied as standard on all reversing starters. Limit switches can be added to stop the motor at a certain point in either direction. Connections A and B must be removed when the limit switches are used.

Start–Stop–Jog

The purpose of jogging is to have the motor operate only as long as the JOG button is held down (Fig.

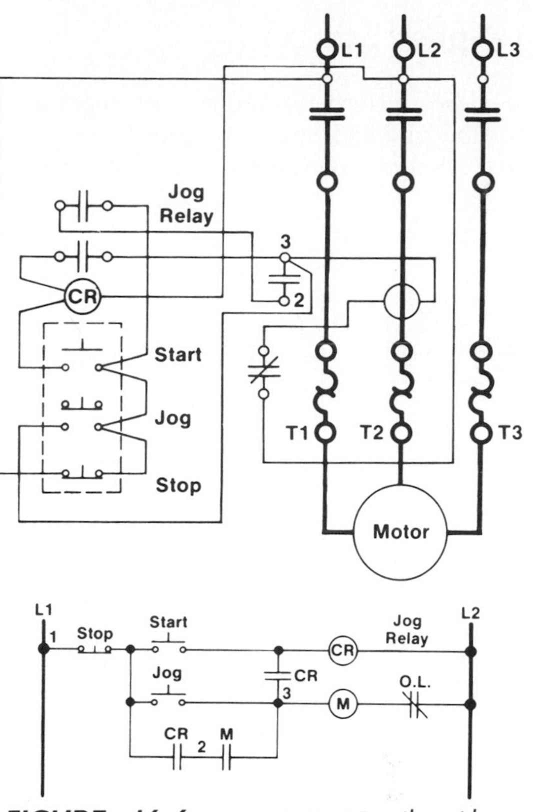

FIGURE 16–6 START–STOP–JOG *three-phase starter. (Allen-Bradley)*

16–6). The starter must not *lock-in* during jogging. That is why the jog relay (CR) is used.

Pushing the START button operates the jog relay. This causes the starter to lock in through one of the relay contacts. When the JOG button is pressed, the starter operates, but this time the relay is not energized and thus the starter will not lock in.

Reversing Starter

Figure 16–7 shows how a manually operated reversing starter operates. Note the absence of a low-voltage control circuitry. This means that the contacts are closed, either to start or to reverse by manually (by hand) setting the contacts. L1 is changed with L3 and L2 just shorted in the reverse positions with the contacts shorting or paralleling the start contacts.

Figure 16–7 *Manual reversing starter. (Allen-Bradley)*

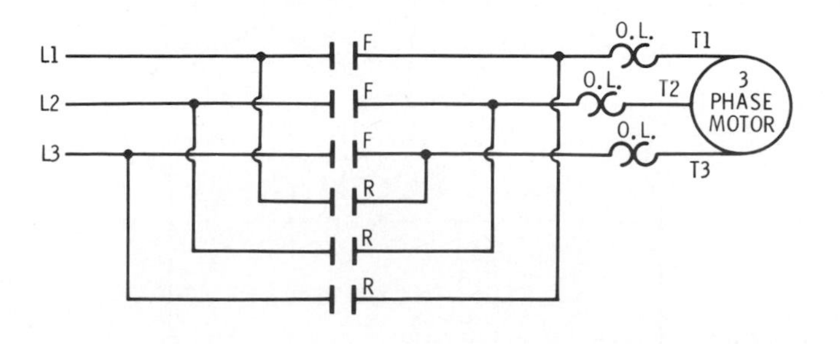

Two-Speed Starter

A two-speed starter for a motor with separate windings for low and high speeds resembles that shown in Fig. 16–8. As the situation becomes more demanding, it is necessary to wire up motors to do various things under different conditions. A typical connection for a two-speed separate winding motor is shown in Fig. 16–9. The motor can be started in either HIGH or LOW-speed. The change from low to high can be made without first pressing the STOP button. However, when changing from high to low, the STOP button must be pressed between speeds.

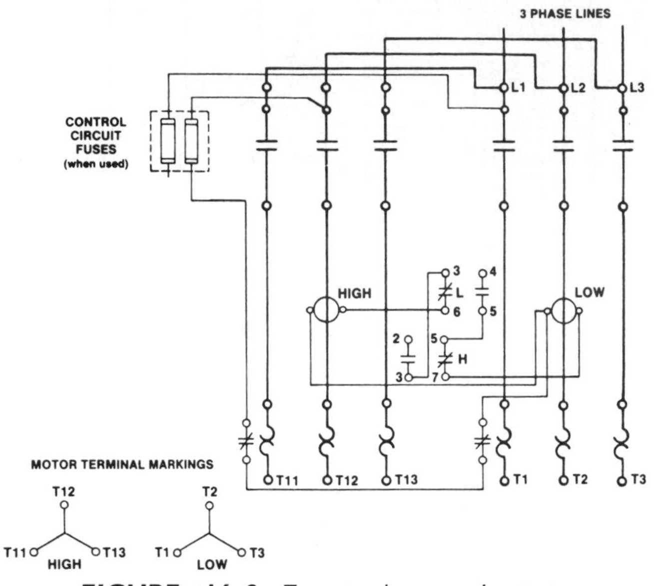

FIGURE 16–8 *Two-speed manual starter. (Allen-Bradley)*

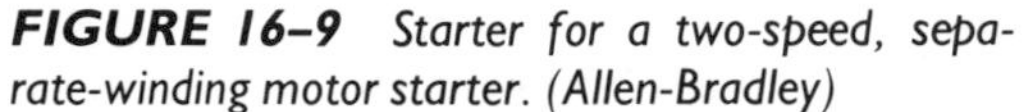

FIGURE 16–9 *Starter for a two-speed, separate-winding motor starter. (Allen-Bradley)*

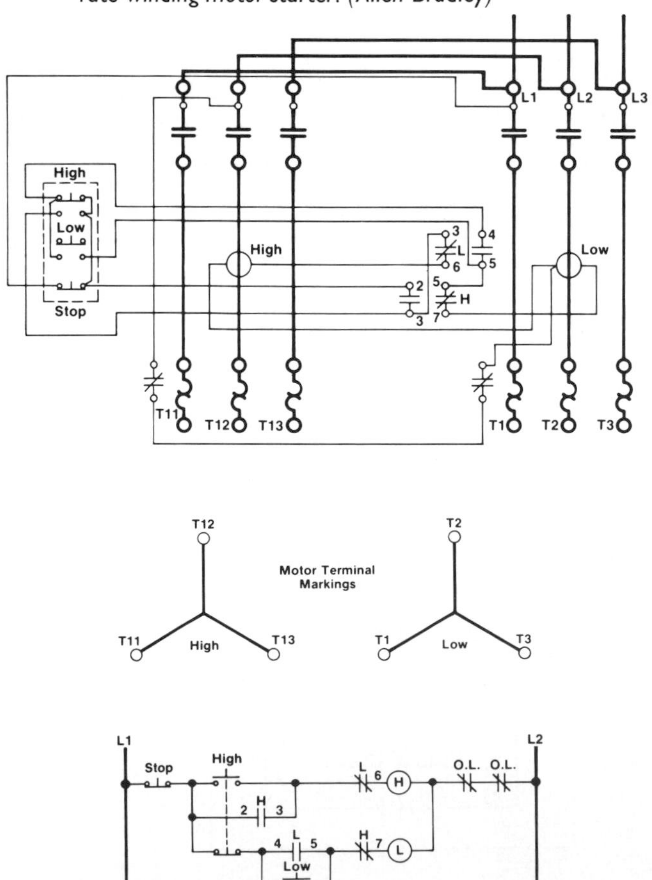

DUPLEX MOTOR CONTROLLERS

The controller operates first one motor and then the other on each successive closing of pilot device A. (Fig. 16–10). When pilot device B closes, both motors are energized. Typical applications include pump motors, where a second pump is required for peak demand periods. For this application, both pilot devices may be float or pressure switches and B is set to operate after A and only if both pumps are required.

Both pilot devices must be two-pole, but B can be omitted if only alternation is required and both motors are never required to run simultaneously. If one motor is running and its disconnect switch is opened, an overload relay trips, or the starter is deenergized for any reason, the other motor will automatically be started.

FIGURE 16–10 *AC duplex motor controllers. (Square D)*

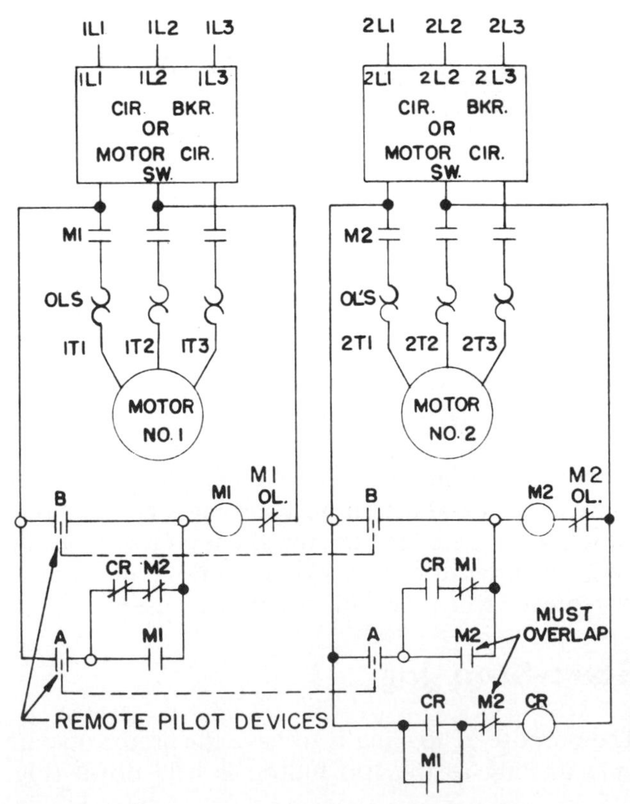

MEDIUM-VOLTAGE CONTROLLERS

Medium voltage usually refers to the range from 2200 to 7200 V. This higher voltage range calls for protection for those who work around the equipment. Larger cabinets are needed to contain all the devices utilized in the control of these higher-horsepower motors. Figure 16–11 shows the housing for two of the controllers. Part A is used for squirrel-cage reduced-voltage, nonreversing reactor controller, and part B is used for synchronous reduced-voltage, nonreversing autotransformer and reactor controllers.

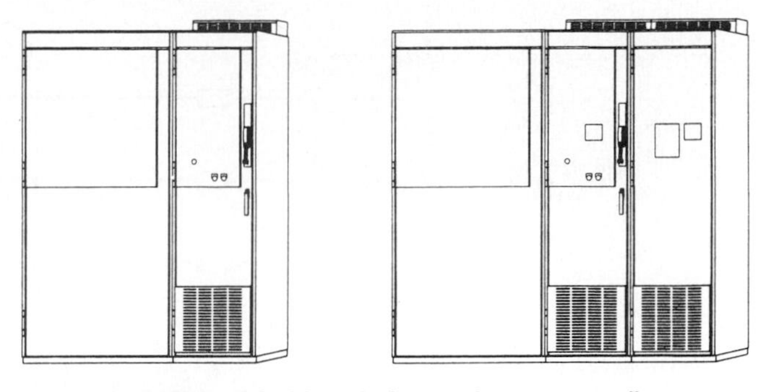

FIGURE 16–11 *Medium-voltage controllers. (Square D)*

Full-voltage controllers are used when full starting torque and resulting inrush current are not objectionable. One-high construction provides complete isolation for each controller and permits space for adding optional power and control devices. Figure 16–12 shows how the squirrel-cage, full-voltage, nonreversing motor is controlled. Note the three fuses used for current limiting and the three current transformers plus the bridge circuit made up of diodes.

Figure 16–13 shows a feeder disconnect circuit. Feeder disconnect controllers use mechanically held contactors that remain closed on loss of power. They are opened by use of a manual-trip pushbutton or an optional electrical solenoid release. Feeder disconnect controllers are available with either vacuum or air brake contactors of the bolted or draw-out design. These controllers are used frequently to disconnect transformers and in transfer schemes in place of metal-clad circuit breakers or disconnect switches.

A full-voltage, squirrel-cage reversing controller is shown in Fig. 16–14. These controllers are used to control motors being operated in forward and reverse directions, where full starting torque and resulting inrush current are not objectionable to the motor. Reversing controllers are available with either vacuum or air-break contactors.

The squirrel-cage, reduced-voltage motor using

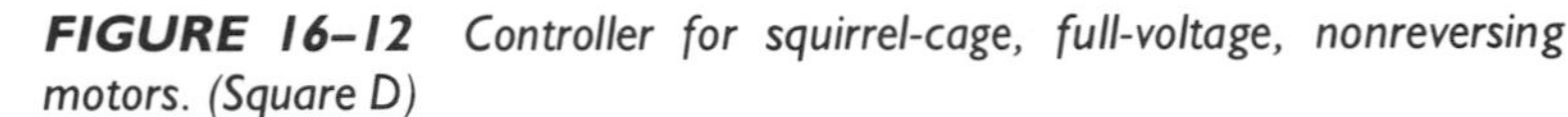

FIGURE 16–12 *Controller for squirrel-cage, full-voltage, nonreversing motors. (Square D)*

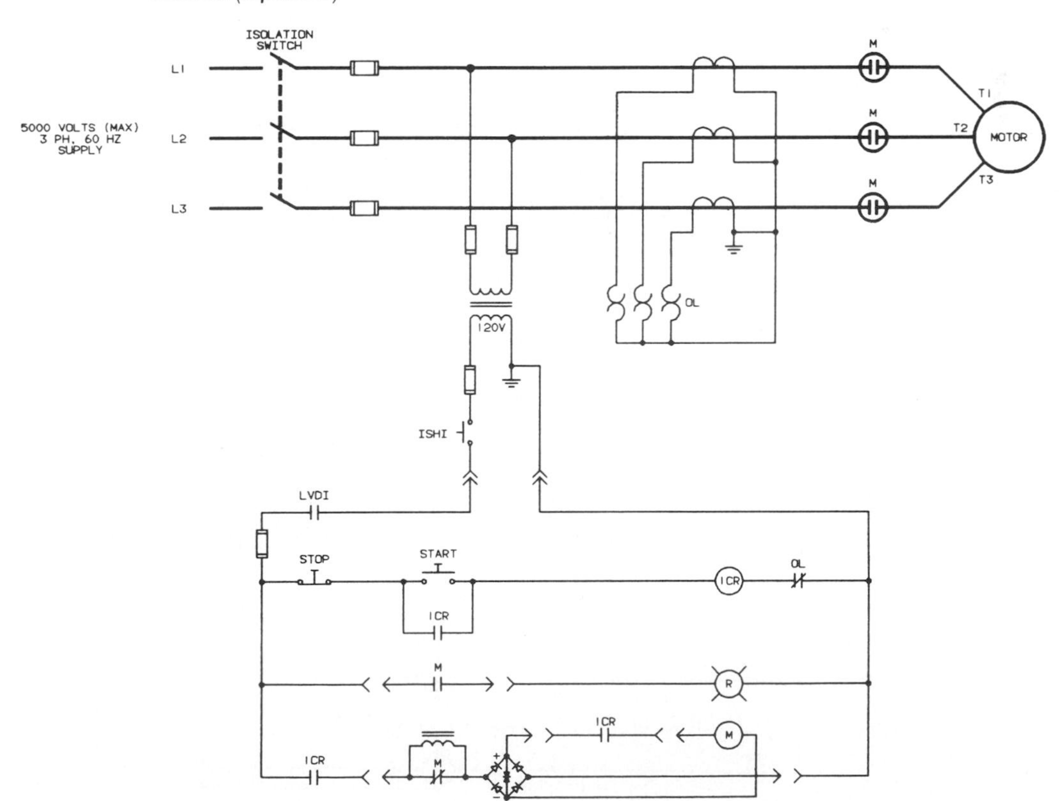

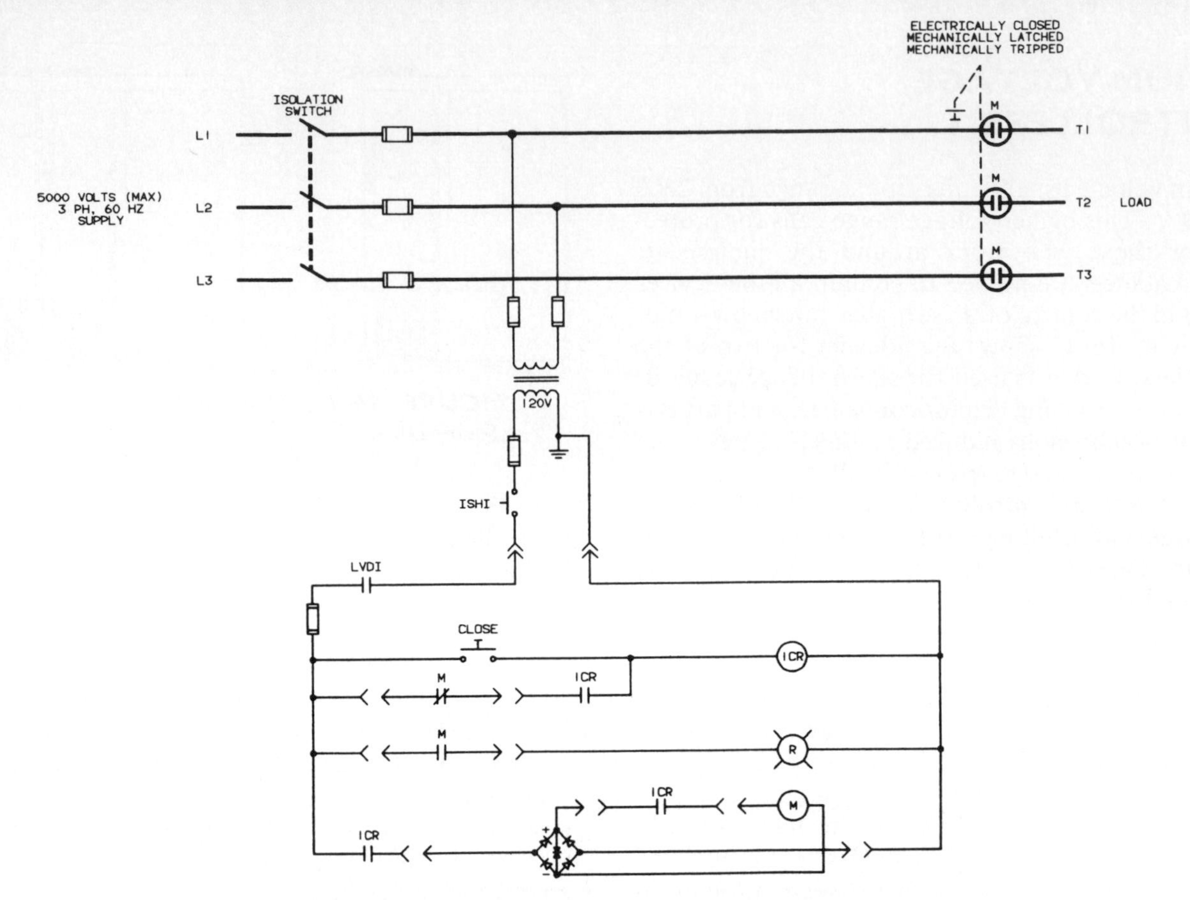

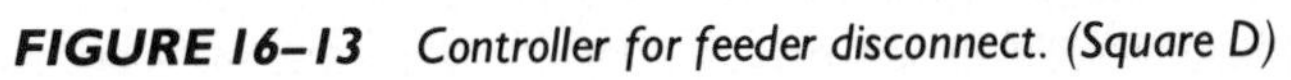

FIGURE 16–13 *Controller for feeder disconnect. (Square D)*

FIGURE 16–14 *Controller for squirrel-cage, full-voltage, reversing motors. (Square D)*

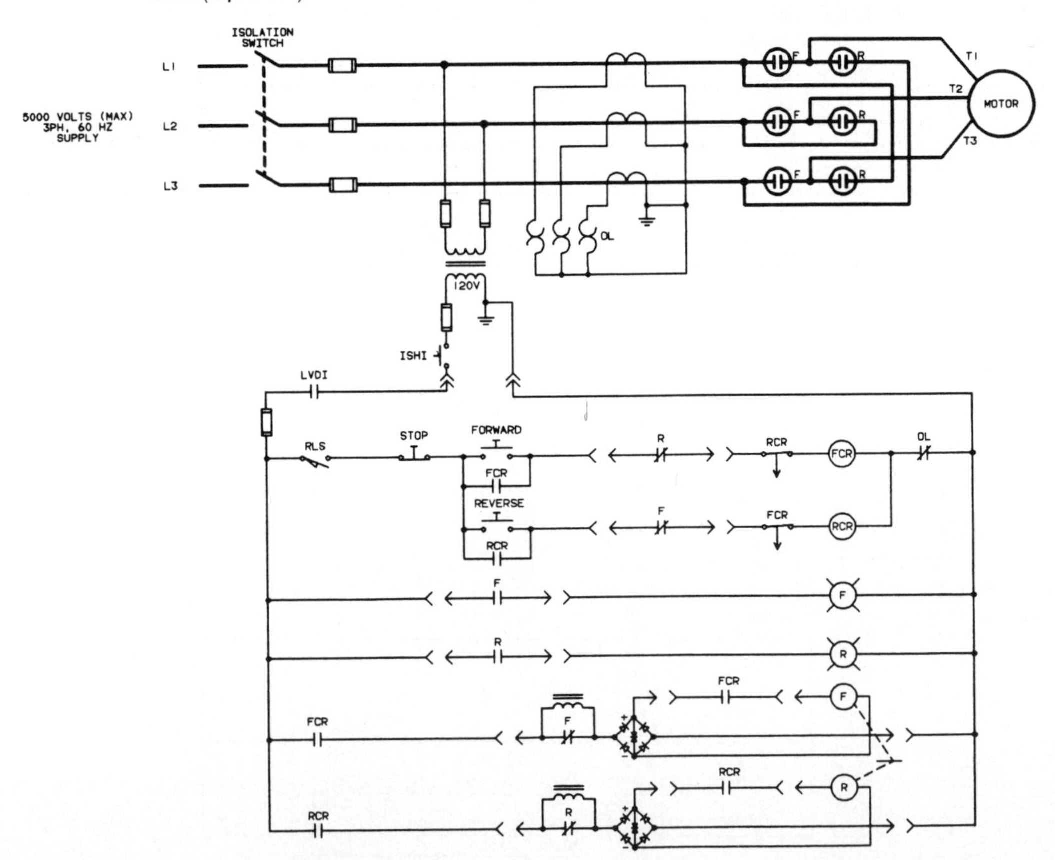

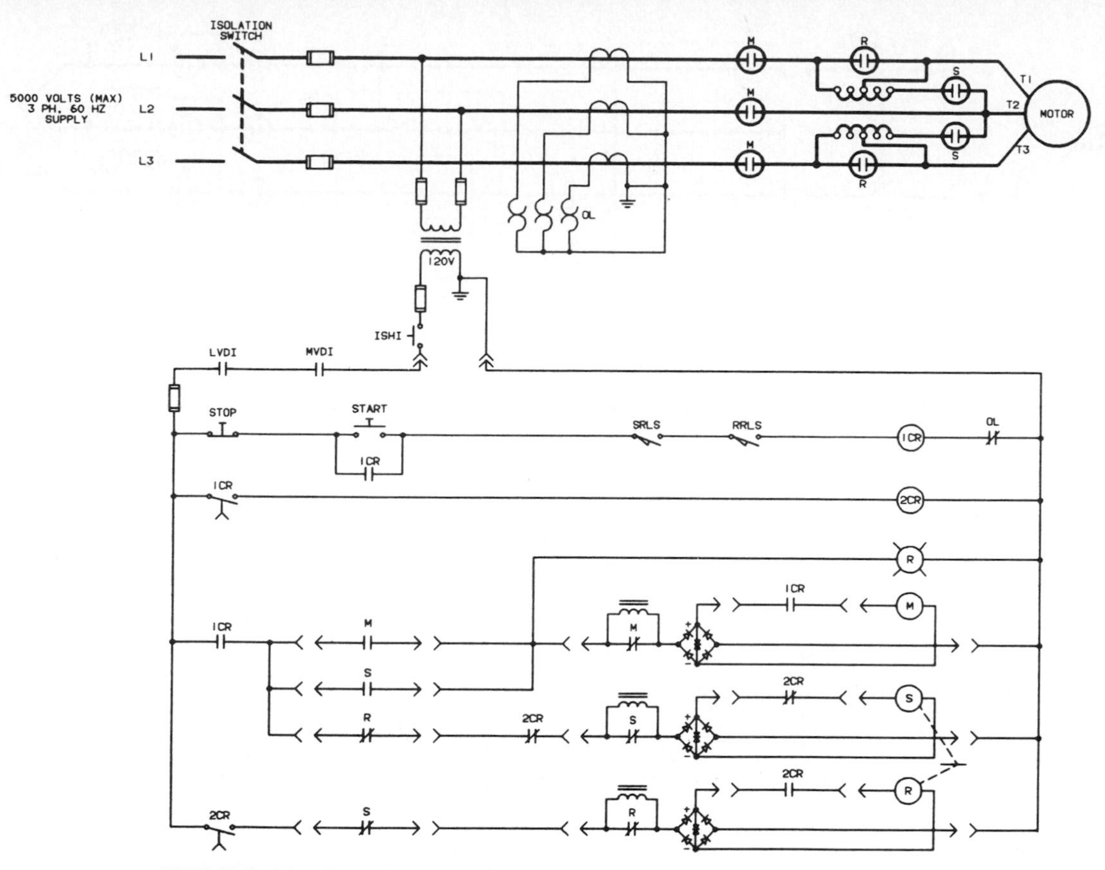

FIGURE 16–15 *Controller for squirrel-cage, reduced voltage, autotransformer motors. (Square D)*

an autotransformer is shown in its controller configuration in Fig. 16–15. These controllers provide maximum torque with a minimum of line current while providing taps to permit torque and line current to be varied. Vacuum or air-break contactors are available.

The reduced-voltage, primary reactor, squirrel-cage motor controller permits the starting of motors without the high inrush currents and voltage variations associated with full-voltage starting (Fig. 16–16).

Synchronous, full-voltage, nonreversing motor controllers are used with motors where constant speed and plant power factor correction are desired. Some typical industrial applications are pulp and paper mills, lumber mills, rubber mills, metal rolling mills, gas compressors, centrifugal fans, blowers, generators, crushers, and grinders (Fig. 16–17).

The brushless synchronous, full-voltage motor controller is shown in Fig. 16–18. It is used for synchronous motors required in explosive atmospheres. Brushless motors have the same advantages as regular synchronous motors, and since brushes are not used, less maintenance is required.

The controller shown has brushless field control with an incomplete sequence relay, dc power supply for the exciter field with a thyrite protector, a powerstat for field adjustment, loss of excitation protection, and pull-out protection.

The synchronous, reduced-voltage, autotransformer controller shown in Fig. 16–19 provides maximum torque with a minimum of line current while providing taps to permit torque and line current to be varied. Further, these controllers are used where constant speed and plant power factor correction are desired.

The primary reactor synchronous motor controllers are used for drives where maximum efficiencies are required and when full starting torque and resulting inrush current are objectionable to the system. Further, these controllers are used where constant speed and plant power factor correction are desired (Fig. 16–20).

SOLID-STATE MOTOR CONTROLLER

The Bulletin 2050 controller made by Allen-Bradley provides controlled current (torque) starting of squirrel-cage motors from 30 to 1200 hp rated at 208 through 575 V ac at 50/60 Hz. These controllers are used on applications such as conveyors, pumps, compressors, and various other loads where minimum shock starting and smooth stepless acceleration is re-

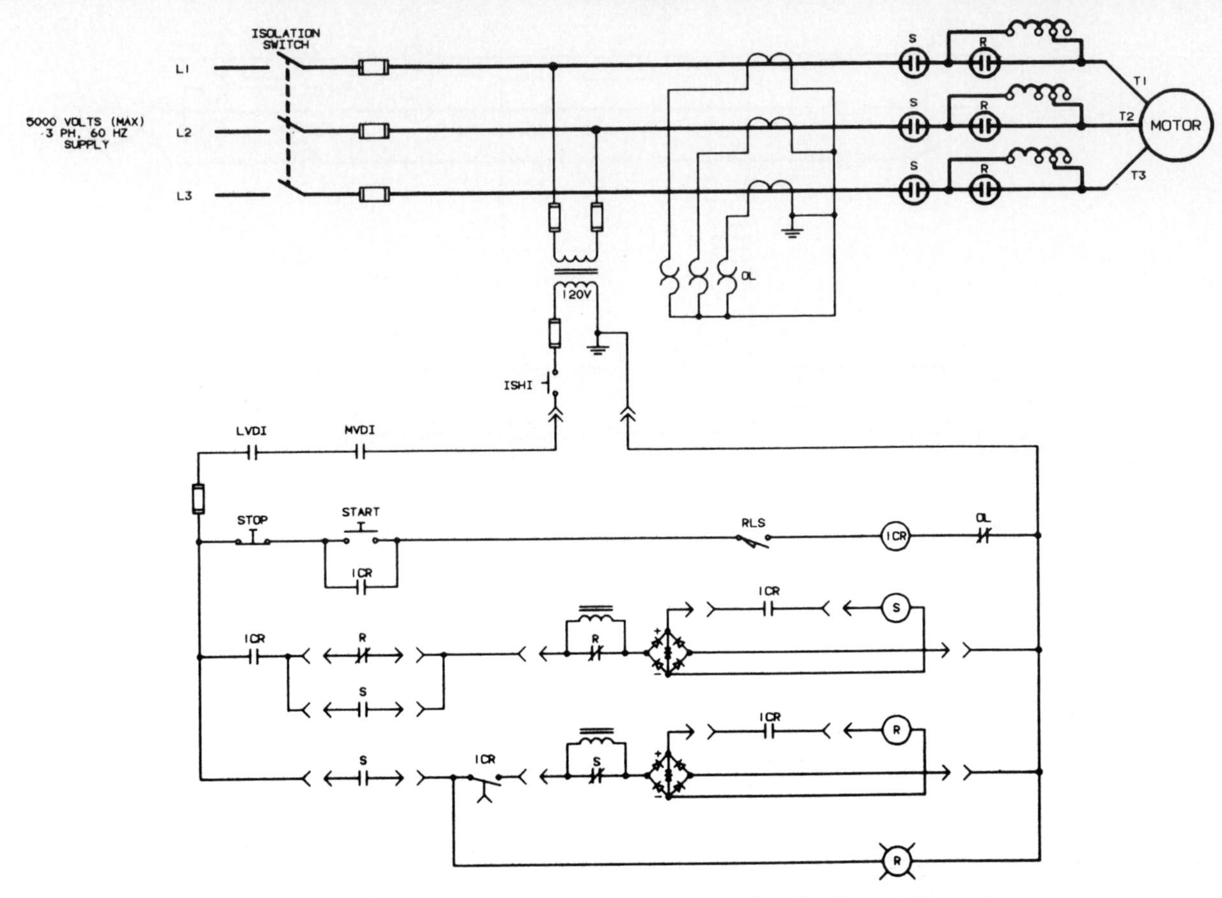

FIGURE 16–16 *Controller for squirrel-cage, reduced-voltage, primary reactor motors. (Square D)*

FIGURE 16–17 *Controller for synchronous, full-voltage, nonreversing motors. (Square D)*

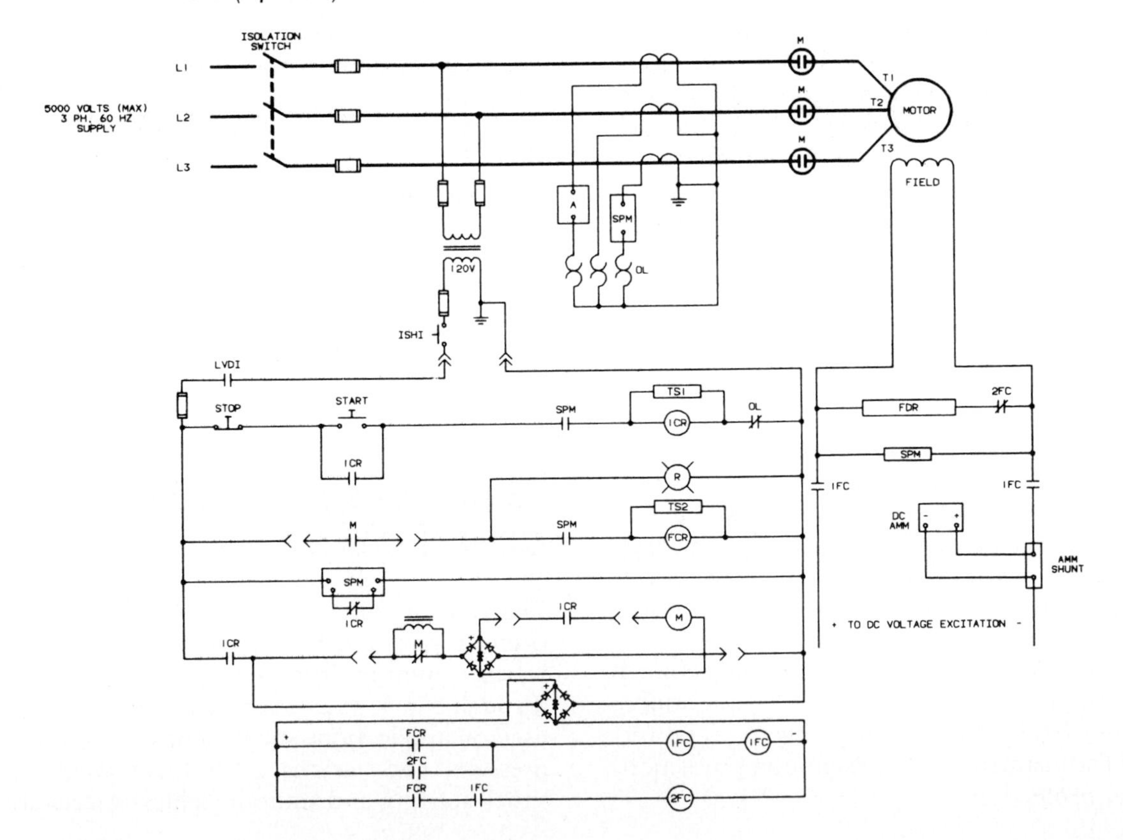

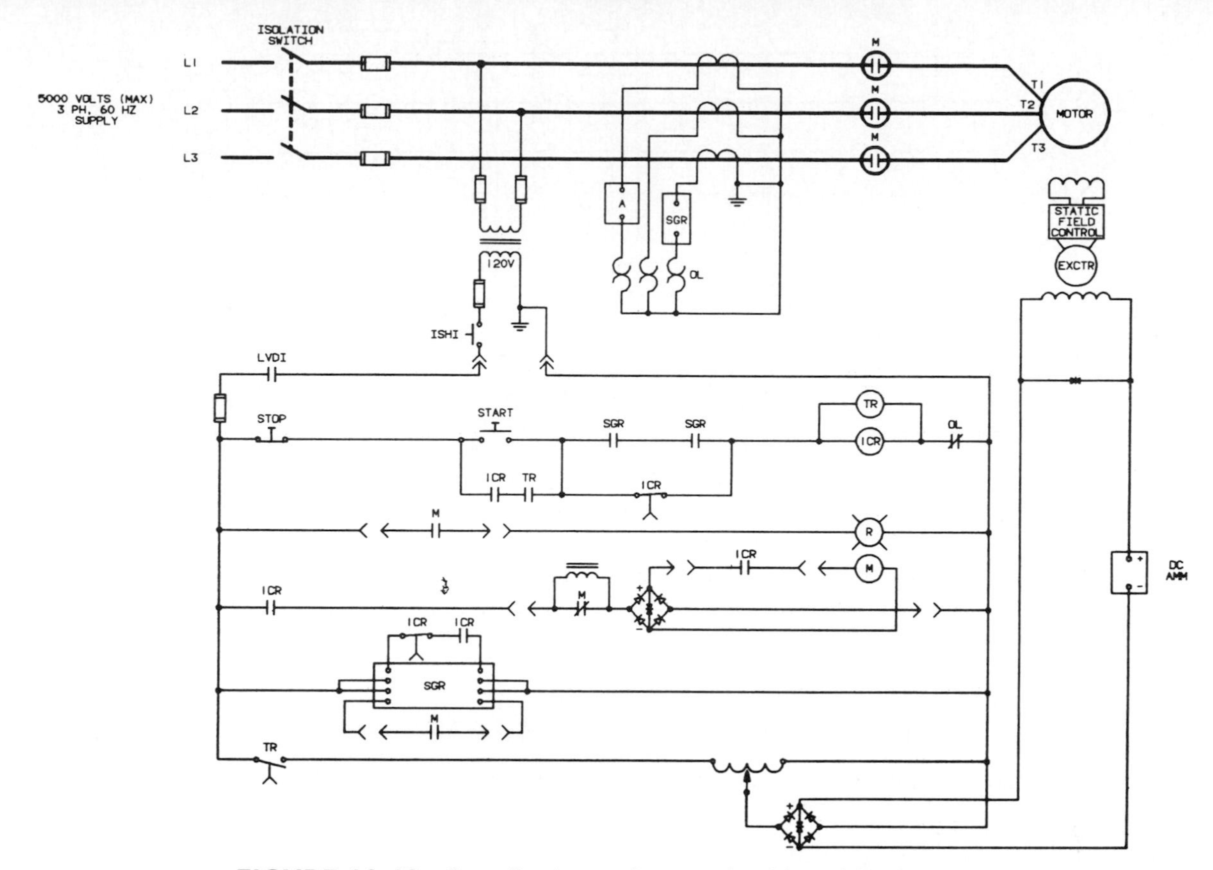

FIGURE 16–18 *Controller for synchronous, brushless, full-voltage motors. (Square D)*

FIGURE 16–19 *Controller for synchronus, reduced-voltage, autotransformer motors. (Square D)*

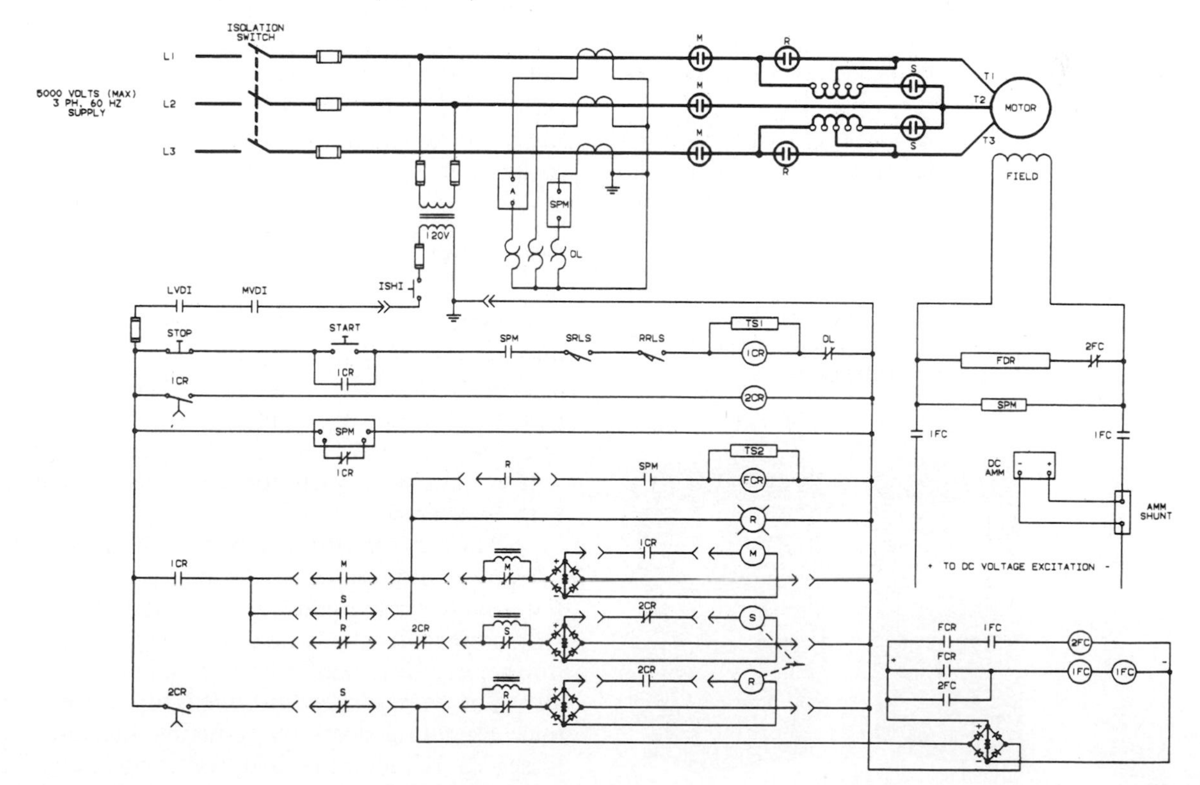

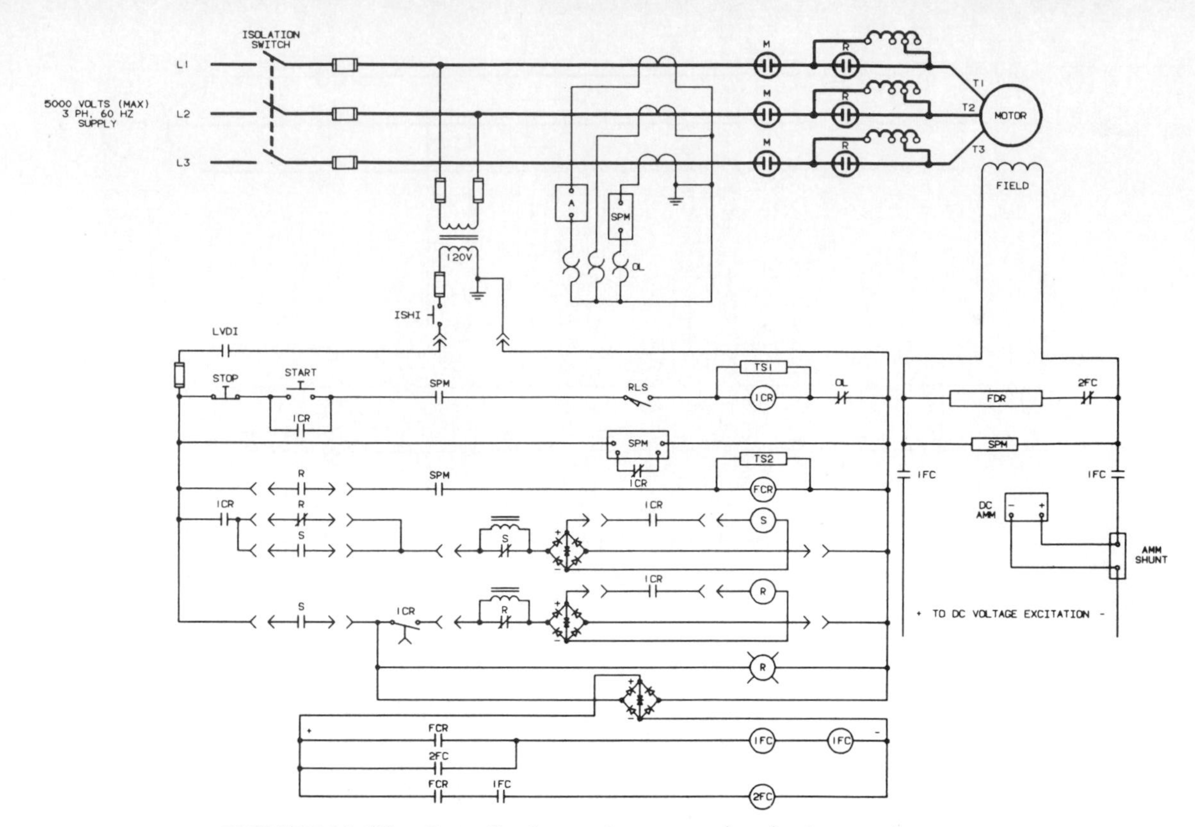

FIGURE 16–20 *Controller for synchronous, reduced-voltage, primary reactor motors. (Square D)*

quired (Fig. 16–21). Three acceleration modes are provided: current ramp, constant current, and linear timed.

Current Ramp. This method has the most general application. During acceleration, a low initial current is gradually increased to a limiting-start current value. Smooth starting and optimum performance are achieved by adjusting the *acceleration ramp* time (rate of current increase). In most cases no other adjustments are required.

Constant Current. This is a variation of the *current ramp* method, used where the principal requirement is to limit the starting current. With the acceleration ramp time set at minimum, the start current limit is adjusted in accordance with starting current restrictions.

Linear Timed Acceleration. For applications requiring a controlled acceleration time or linear rate of speed increase, the controller can be set up for *linear timed acceleration.* A tachometer is installed on the motor and provides a feedback signal that is used by the controller to increase the motor speed linearly with time, according to the acceleration ramp potentiometer setting.

FIGURE 16–21 *Solid-state motor controller: (A) separate 200-hp power module and logic module with mounting plate; (B) 200-hp logic module mounted on power module. (Allen-Bradley)*

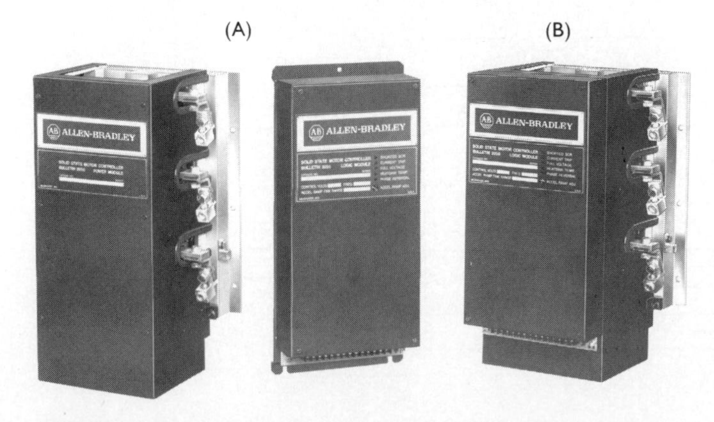

Modules

The controller consists of modules. The *power module* contains three power pole assemblies, each having back-to-back SCRs with single-bolt clamping. The power poles share a single heat sink/base and cooling fan. See Fig. 16–22 for a block diagram of the combination controller with some of the options marked with an asterisk.

The *logic module* can be mounted on the front of the power module or it can be mounted separately. External features include diagnostic LEDs, a terminal strip connecting the control devices, and the acceleration ramp adjustment potentiometer.

The *energy-saver module* is an option. It can reduce operating costs by reducing the motor power losses. It is recommended for certain applications,

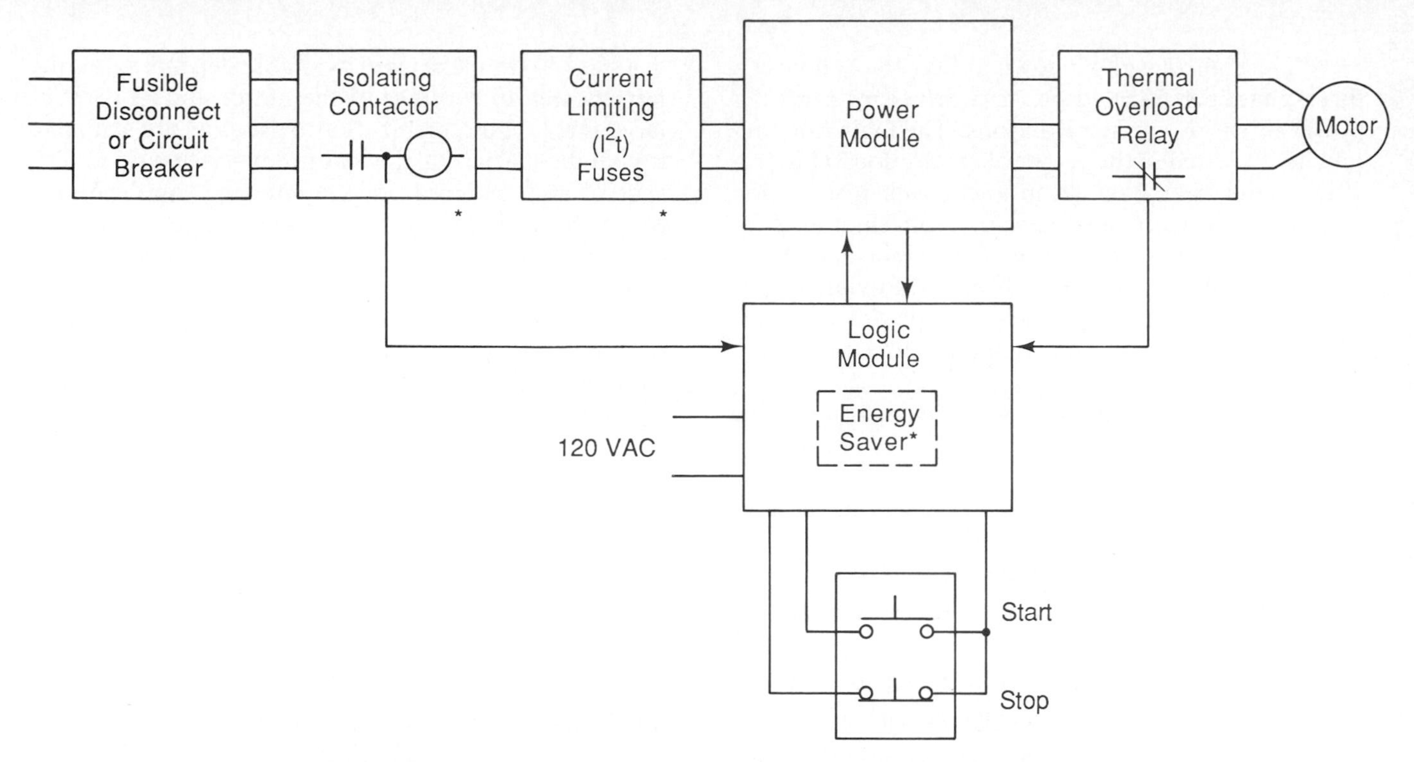

FIGURE 16–22 *Block diagram of combination controller with options designated with asterisk. (Allen-Bradley)*

such as where motors run unloaded for long periods. It can be installed within the logic module.

Solid-State Advantages

There are a few advantages claimed for solid-state controllers as opposed to electromechanical types. The main advantage is maintenance. Inasmuch as there are no moving parts such as contacts, there is little call for cleaning and adjusting them. It is possible to incorporate voltage regulation, transient suppression, and snubber circuits for protection against changes in temperature and changes in voltage. It is also possible to incorporate into the circuitry shorted SCR and open-phase protection. Phase-reversal detection is possible as well as protection against startup in an incorrect sequence. Overcurrent protection can also be designed for protection in various operating modes.

MOTOR CONTROL CENTERS

As technology becomes more complex, integrated control equipment is becoming more in demand. There is a decided advantage to integrating all control and power requirements into one centralized package. It can be preengineered, prewired, and fully tested before it is delivered, thereby saving time not only in installation but in testing.

Motor control centers are used in a wide variety of industrial and commercial applications, such as pulp and paper mills, sawmills, building products, food processing, can plants, wastewater treatment plants, coal and bulk handling, chemical plants, and oil and gas production, to name but a few. In other words, it can be used wherever three-phase motors are used (Fig. 16–23).

FIGURE 16–23 *Motor control center. (Allen-Bradley)*

The control center shown in Fig. 16–23 houses a three-phase 600-V-rated bus network that distributes power to various vertical sections. The main horizontal bus is located in the center of the section. This provides better heat dissipation and power distribution and makes the main bus accessible from floor level for easier and safer maintenance. The center-fed 300-A-rated vertical bus supplies power to individual units above and below the horizontal bus for effective 600-A capacity, providing virtually unrestricted unit arrangement.

Different sizes and ratings of units with varying degrees of complexity are available depending on the horsepower and voltage of the motor and its particular control requirements. Standard units are available to handle basic starting–stopping–reversing, multispeed, and reduced-voltage starting applications. Motor control centers are not only used to house basic control devices; they are used to package solid-state technology.

Programmable controllers, adjustable-frequency drives, solid-state reduced-voltage starters, and solid-state protective devices represent the kind of technology that is integrated into today's motor control centers.

QUESTIONS

1. Where are wound-rotor motors used?
2. What is the range of available horsepower for synchronous motors?
3. Why shouldn't a motor lock-in during jogging?
4. When are full-voltage controllers used?
5. Where are primary reactor synchronous motor controllers used?
6. What is an acceleration ramp?
7. When is an energy saver recommended?
8. What are motor control centers used for?
9. What is the simplest type of three-phase starter?
10. What wires are switched in order to reverse a motor?

17

Drives

Objectives

After studying this chapter, you will be able to:

1. Describe adjustable-frequency ac drives.
2. Identify three major types of inverters.
3. Figure rpm when frequency and number of poles are known.
4. Describe a variable-voltage inverter.
5. Explain how pulse-width-modulated inverters work.
6. Describe the operation of eddy current drives.
7. Explain how open-loop controls operate.
8. Explain how closed-loop controls operate.
9. List the advantages and disadvantages of dc drives.
10. Describe solid-state digital ac drive advantages.

Industry is constantly striving to find controls that will increase productivity and reduce energy costs. This striving has about ceased, inasmuch as electronics has taken over and produced the desired results. However, many electromechanical devices are still in use and will be for many years. It is both of these worlds—electromechanical and electronic—that we discuss in this chapter.

The ac induction motor is the mainstay for energy conversion in the United States. It is found in industry, in commerce, and in the home. Our lifestyle would be unimaginable without it. It is the major converter of electrical energy into another usable form. For this purpose, about two-thirds of the electrical energy produced is fed to motors.

FANS, BLOWERS, AND PUMPS

Fans, blowers, and pumps consume much of the electrical energy to operate the ac motors that power them. It has been estimated that approximately 50% of the motors in use today are attached to these types of loads. Fans, blowers, and pumps are particularly attractive to look at for energy savings. Several other methods of control for fans and pumps have been advanced recently that show energy savings over traditional methods.

Fans are designed for their maximum load. However, they do not always operate at maximum capacity. This means that there is a possibility of energy savings if the speed of the motor can be changed according to the load demand. In most instances the outlet dampers for fans and throttling valves for pumps were used to control their outputs. However, neither of these controls improved the efficiency of the pump or fan. What is needed is a control method to adapt fans and pumps to varying demands that do not decrease the efficiency of the system as much. Newer methods include direct variable-speed control of the fan or pump. This method produces a more efficient means of flow control than do the existing methods.

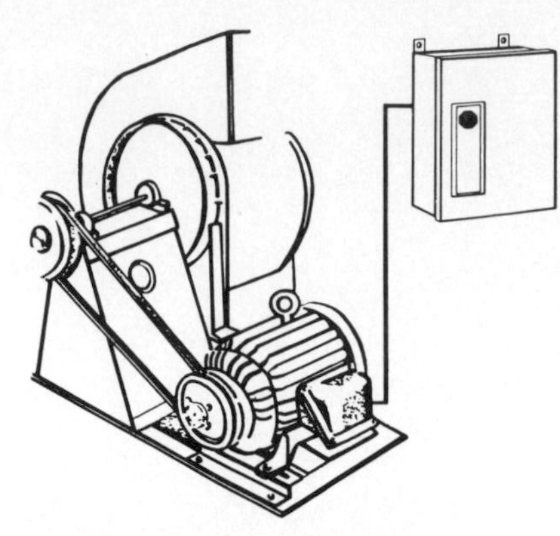

FIGURE 17–1 Variable-speed drives change fan rpm. (Allen-Bradley)

In addition, adjustable-frequency drives offer a distinct advantage over other forms of variable-speed control. As can be seen in Fig. 17–1, by changing the speed or actual rpm of the fan, the performance of the fan changes, producing a different airflow.

ADJUSTABLE-SPEED DRIVES

There are several types of adjustable-speed drives that can be used on fans. These include variable-pitch belt drives, eddy current drives, dc drives, and adjustable-frequency drives.

ADJUSTABLE-FREQUENCY AC DRIVES

Adjustable-frequency drives are commonly called *inverters.* They are available in a range of horsepowers from fractional to 1000. They are designed to operate standard induction motors. This allows them to be added easily to an existing system. The inverters are often sold separately because the motor may already be in place (Fig. 17–2).

The basic drive consists of the inverter itself, which converts the 60-Hz incoming power to a variable frequency and a variable voltage. The variable frequency is the actual requirement that will control the motor speed. Three major types of inverter designs are in use today: current source inverters (CSI), variable-voltage inverters (VVI), and pulse-width-modulated inverters (PWM).

FIGURE 17–2 Industrial ac drive. (Allen-Bradley)

Keep in mind that ac motor speeds are a function of *frequency* and the number of poles:

$$\text{rpm} = \frac{120 \times \text{frequency (Hz)}}{\text{number of poles}}$$

Synchronous motors will run at the synchronous speed as determined by the formula. Their speed will not change with load changes within the pull-out torque capacity of the motor. An induction motor's speed–torque characteristics are shown in Fig. 17–3.

Motor output requirements are dictated by the load. Friction-type loads such as conveyors require constant torque from the drive motor. Certain machine tool applications require constant horsepower. Most fans and blowers require variable torque. Generally, standard ac motors can produce constant torque throughout the speed range with a constant volts/Hz supply. The converter is adjusted to provide this supply (Fig. 17–4).

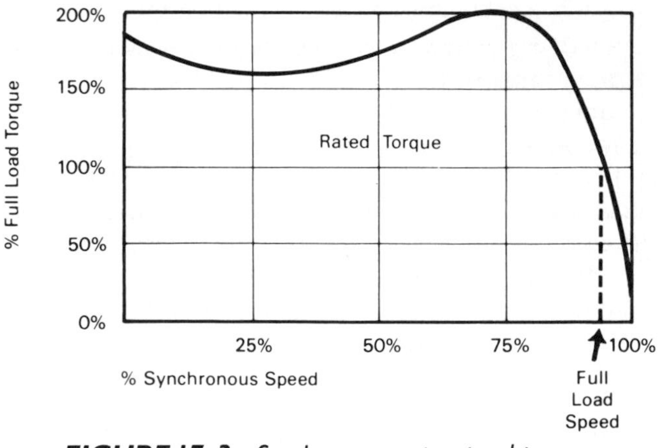

FIGURE 17–3 Synchronous motor speed-torque curve.

FIGURE 17–4 Ac motor operation curve.

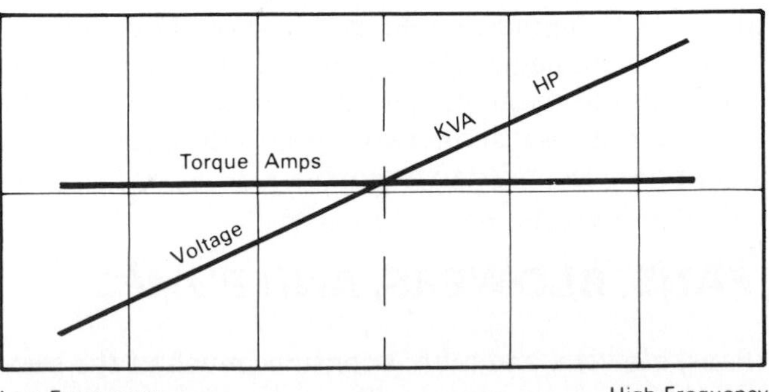

The converter and inverter are two different devices. The converter changes from ac to dc and the inverter changes dc to ac, whereas in actual practice the inverter changes the ac to dc and back to ac again at a different frequency. It may also be a reshaped waveform when changed back to ac again for frequency control purposes.

CSI INVERTERS

The CSI inverter controls the current output to the motor. The actual speed of the motor is sensed by the use of other circuits. This is then compared to the reference speed and an error is used to generate a demand for more or less current to the motor. The output switching devices, usually SCRs, are switched at the desired frequency to "steer" the current to the motor. Current source inverters are available in a wide range of horsepowers but most often are found in the range of 50 hp and above.

This type of inverter is used on a standard induction motor and is readily available, reliable, and easy to repair. If the inverter fails, the motor can be operated directly across the incoming line for continued operation. The inverter can adapt its operation to prevent overloads caused by accelerating the high-inertia loads found in some applications. The current control limits fault currents that will minimize damage on a major fault or overload condition. This inverter may require tach feedback for speed regulation. A tach generator must be added and is not a standard option for induction motors. If the tach feedback signal is lost during operation, the drive may run away to full speed. The inverter has to be matched to the motor's electrical characteristics. The inverter is sensitive to those characteristics, and improper operation may occur if the motor is replaced with a different type or size.

The inverter design requires the motor to be connected to operate at all. The inverter cannot be run or tested without the motor. The inverter uses a phase-controlled rectifier for current control. This method produces low power factor at low speeds.

The size of the major components usually causes these inverters to be the largest of the drives in overall size. All the power delivered to the system may go through a conversion within the inverter. Large power devices must be used to handle this.

VARIABLE-VOLTAGE INVERTERS

The variable-voltage inverter (VVI) controls the voltage and frequency to the motor to produce variable-speed operation. The distinguishing characteristic between this type of inverter and the PWM inverter is the scheme used to control the voltage. VVI inverters control the voltage in a separate section from the output section used for frequency generation. Usually, the voltage control is done using a phase-controlled input bridge rectifier circuit at the input of the inverter. The frequency control is accomplished by an output bridge circuit that switches the variable voltage to the motor at the desired frequency. These drives are available from fractional horsepower to about 500 hp.

The standard induction motor is readily available, reliable, and easy to repair and can be used with the VVI inverter. The inverter can achieve efficiencies of 90% at full speed and full load. If the inverter fails, the motor can still be operated directly across the incoming line for continued operation. The inverter can adapt its operation to prevent overloads caused by accelerating the high-inertia loads found in some applications.

Installation of this type of inverter is simple. Just three power leads to the motor are used. No tach feedback is required, and the drive can be located large distances from the motor being controlled. The drive can be tested and operated without requiring a motor to be connected. More than one motor can be operated from the same inverter. Also, the inverter is not sensitive to changing the combination of motors operated as long as the total load current does not exceed the inverter's rated current.

There are some drawbacks to the VVI inverter. The initial cost of the inverter system is high. The total power delivered to the motor must be converted by the inverter. This requires high-power components within the inverter. The inverter has a large portion of sophisticated circuits that require skilled technicians for service.

PULSE-WIDTH-MODULATED INVERTERS

These inverters accomplish both frequency and voltage control at the output section of the drive. The output voltage is always a constant amplitude and by chopping or pulse-width modulation, the average voltage is controlled (Fig. 17–5). These drives are available from 1 to 1000 hp.

Some of the features of this type of inverter are its use on induction motors. It has good efficiency, up to 90% at full speed and full load. A diode bridge rectifier is used to rectify the incoming power. This permits a good power factor throughout the full operating speed range of the inverter (Fig. 17–6).

Control and logic functions are integrated onto a single LSI chip. This chip contains approximately

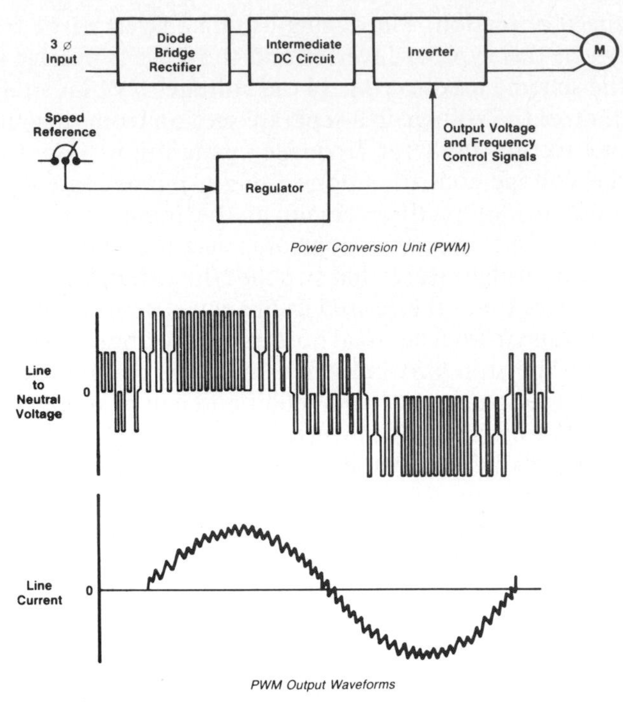

FIGURE 17–5 *Pulse-width-modulation (PWM) waveforms. (Allen-Bradley)*

6300 transistors to perform the complex logic and control function. The use of this chip actually simplifies drive construction by reducing the number of electrical connections needed, thus providing more reliability and quality.

If the inverter fails, the motor can be operated directly across the incoming line for continued operation. The inverter can adapt its operation to prevent overloads caused by accelerating the high inertial loads found in some applications. No tach is needed, so there are only three wires to connect. The control can be operated large distances from the motor.

This type of drive can also be tested without requiring a motor to be connected. More than one motor can be operated from the same inverter. Also, the inverter is not sensitive to changing the combination of motors operated as long as the total current is within the rated current limits of the inverter.

Some of the less desirable features of this type of inverter are its high cost initially and the fact that the total power delivered to the motor must be converted by the inverter. This requires high-power components within the inverter.

It does have some sophisticated circuits that require skilled technicians to repair or service. However, the use of large-scale-integrated (LSI) circuits and microprocessor circuits permit self-diagnostics that aids in troubleshooting. Printed circuit board substitution can be done by unskilled service workers.

EDDY CURRENT DRIVES

The eddy current drive consists of two distinct parts. One part, the mechanical unit, consists of the eddy current clutch and an induction motor. The motor runs at constant speed and provides a source of energy for the clutch. By controlling the excitation to the clutch, the amount of slip between the motor and the output shaft can be regulated and the output speed varied. If the excitation is high, the output speed increases toward the motor speed. If the excitation is lowered, the speed decreases toward zero speed. A given speed is maintained by balancing the excitation of the clutch to the load requirement.

Eddy currents are small currents created in the core when magnetic fields are changed. Eddy currents generate heat. They flow in the opposite direction to the current that induced them. The effect is to resist the flow of current in the core. Eddy currents can be minimized by using laminating techniques. Laminating is the building of an object through use of several layers of material. When this method is used, each

FIGURE 17–6 *PWM drive. (Allen-Bradley)*

lamination is varnished. Varnishing insulates the layers from each other, which increases resistance to eddy currents. Eddy currents in motors and transformers must be reduced as much as possible inasmuch as they represent power consumption for no work accomplished. By varying the excitation to the clutch in the eddy current drive it is possible to vary the eddy currents accordingly.

The clutch excitation is controlled by the eddy current controller. The controller uses high-gain amplifiers and a closed-loop speed control circuit to sense the need for clutch excitation. The mechanical unit has a tach generator mounted on the output shaft to provide speed feedback for the controller. The drive cannot regulate speed without this type of feedback.

Eddy current drives are available as integral units with the induction motor mounted or as just the eddy current coupling alone, which must be connected to an induction motor. There are also special designs available for use on vertical pumps. The drives are available from integral horsepower ratings all the way up to clutches large enough to handle several thousand horsepower.

Some of the advantages and good features of the eddy current drives are that the first costs are usually smaller than for adjustable-frequency drives, the controllers are much smaller than for other drives, and they need to handle 10% or less of the total power being delivered to the system. The control circuitry is less sophisticated and complex than that found in other systems. The eddy current coupling and the control respond well without overloading when operating high-inertia loads.

However, there are some disadvantages to this type of drive. Eddy current clutches are not common devices. On-site repair or even local repair may not be available. The special clutch does not permit the motor to operate the load directly. If the mechanical unit or the controller needs repair, the system is down.

The efficiency of the system decreases with speed. This is due to the output speed being controlled by slip within the clutch. Tach feedback is required to maintain speed control. If speed feedback is lost, the drive will go to full speed. Eddy current clutch brand names include Louis-Allis's Adjusto Speed and GE's Kinatrol.

VARIABLE-PITCH DRIVES

The variable-pitch drive is a method of speed control that uses the mechanical means of belts and variable-pitch sheaves or pulleys to change speed. The power source is a standard induction motor. Often these units are enclosed and have a gear reduction built in for reduced speed ranges. The horsepower range is generally limited from 5 to 50 hp, with not much available outside that range.

However, the cost is an important factor in this type of speed control. These systems are among the lowest-cost methods of achieving variable speed. The principle of operation is well known and easy to understand, and construction is simple.

The disadvantages are headed by remote control. This is not an inherent feature of the variable-pitch drive. This is because the drive uses mechanical means to vary the speed. Electrical control signals must be adapted to existing mechanical controls.

The stress of variable-speed operation on belts requires periodic checks and the replacement of belts from time to time. High-inertia loads may cause problems. This may require oversizing the drive or custom motors. Special shutdown and startup procedures may be required to prevent overloading the motor. Running at constant speed for extended periods of time may cause grooving in the sheaves. This degrades speed control and decreases belt life.

WOUND-ROTOR AC MOTOR DRIVES

Wound-rotor motor drives use a specially constructed motor to accomplish speed control. The motor rotor is constructed with windings that are brought out of the motor through slip rings on the motor shaft. These windings are connected to a controller that places variable resistors in series with the windings. The torque performance of the motor can be controlled using these variable resistors. Wound rotor motors are most common in the range of 300 hp and above.

There are a few advantages to this type of drive. The initial cost is moderate for the high-horsepower units. Not all the power need be controlled. That results in a moderate size and simple controller. The simple construction of the motor and control lends itself to maintenance without the need for a high level of training. High-inertia loads work well with this type of motor.

However, there are some disadvantages that should be taken into consideration for this type of drive. The motor has slip rings and is not readily available. Efficiency suffers at low speeds. The drive usually is limited to a speed range of 2:1. The speed regulation is also poor on the fan-type loads.

DIRECT-CURRENT DRIVES

The dc drive technology is the oldest form of electrical speed control. The drive system consists of a dc motor and a controller. The motor is constructed with armature and field windings. Both of these windings require dc excitation for motor operation. Usually, the field winding is excited with a constant level voltage from the controller.

REGULATED SPEED DRIVES

There are various speed control methods used on dc motors. One of the simplest is the *rheostat*. The rheostat has an effective speed control range of about 4:1 with poor regulation of motor speed against changes in load and torque and line voltage. This control method is very inefficient because of the power dissipated in the rheostat.

Another method, the variable transformer with rectifiers, can run dc motors over a wider speed range (up to 10:1) at an improved regulation compared to the rheostat control. It is also more efficient than the rheostat.

The SCR (thyristor) controls have half-wave operation and characteristics that are similar to the variable transformer control. However, SCR systems with full-wave rectification can achieve a 20:1 speed range when used with IR compensation techniques, that is, pseudo-closed-loop current sensing techniques.

If a tachometer is used for feedback, true closed-loop feedback control is possible with SCR circuits. A full-wave orthree-phase SCR control may achieve speed ranges of up to 100:1. Due to the pulsating nature of SCR control techniques, the speed stability of motors below 1 hp may not always be good in the lower speed range, due to the low moment of inertia of the motor and due to the load. However, in integral-horsepower machines, excellent speed range can be achieved.

Transistor controllers can handle closed-loop speed control with a speed range of over 1000:1 and with regulation better than 1 or 2% based on set speed. Transistor controls have their best operational advantage for motors up to 1 hp, with some servomotors up to 5 hp. For smaller motor sizes a continuous control method is used, but for motors above $\frac{1}{3}$ to $\frac{1}{2}$ hp a pulse-width-modulation (PWM) technique is the most efficient control method.

OPEN-LOOP CONTROLS (TRADITIONAL APPROACH)

Dc motor speed controls have been around a long time. Perhaps the oldest and most widely used control for small dc motors is the series resistor speed control, shown in Fig. 17-7. The variable resistance is inserted in series with the armature and field circuit. The control has good starting characteristics with large torque

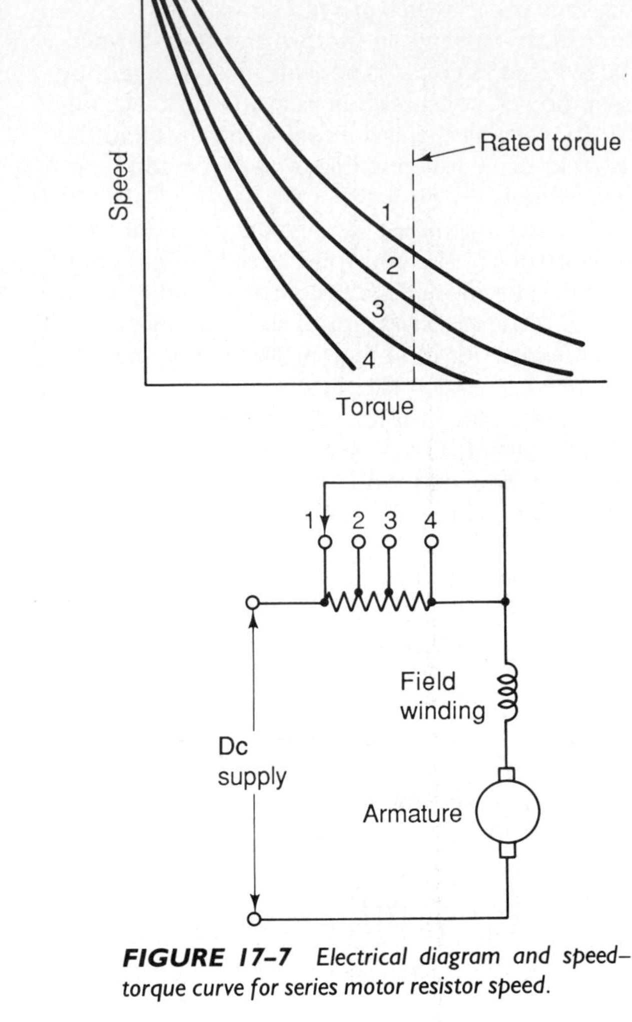

FIGURE 17-7 *Electrical diagram and speed–torque curve for series motor resistor speed.*

FIGURE 17-8 *Electrical diagram and speed–torque curve for shunt motor resistor speed.*

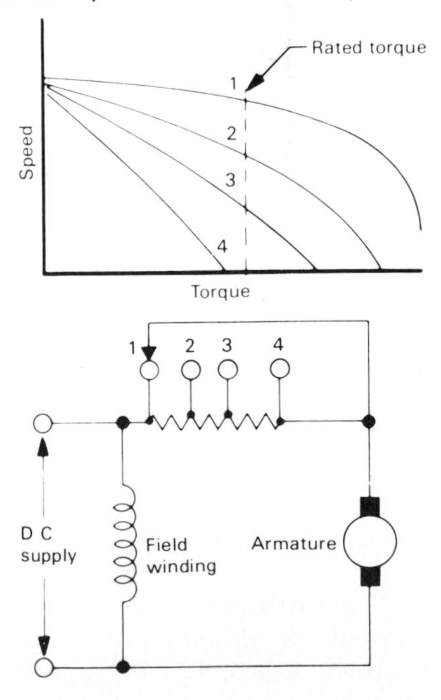

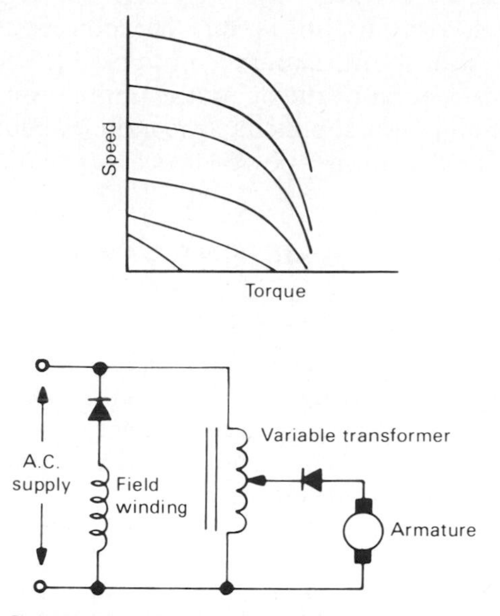

FIGURE 17–9 *Shunt motor with variable transformer to control armature voltage.*

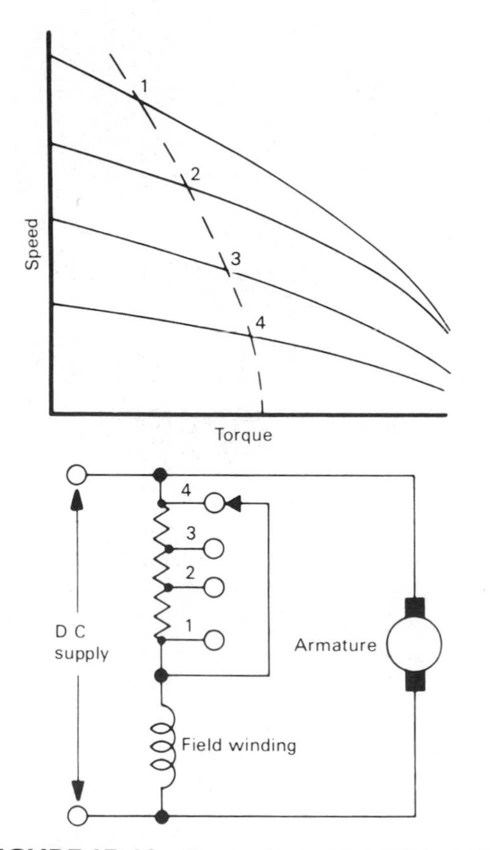

FIGURE 17–10 *Shunt motor with variable field resistor control.*

available at low speed but has a runaway speed tendency at small-load torque conditions. This makes the control useful only for control applications with somewhat fixed friction conditions.

The shunt motor variable resistor speed control is shown in Fig. 17–8. Here a series resistor is inserted in the armature circuit. The field winding is excited with a constant voltage. When more resistance is inserted in the armature circuit, speed regulation degenerates. This type of control works well for constant load torque rather than for a widely varying torque situation.

Figure 17–9 shows a control that uses a variable transformer to control the armature voltage of a shunt motor with constant field excitation. The resulting speed–torque characteristics are much improved over variable resistor control characteristics, with a more uniform speed regulation over a wider speed range.

Figure 17–10 shows how a shunt motor connected to a variable field resistor operates. The action of this control circuit is unique in that the motor speed is variable only above the speed it would have without the field resistor control. This has an undesirable effect in that the torque constant of the motor will decrease with increasing resistance insertion (field weakening). The net effect or result is that the armature current for a given torque will increase with higher speed. The motor can easily be overloaded. This control circuit is used only in unique cases where load conditions are both predictable and well controlled.

In some sophisticated open-loop control methods, such as the motor–generator armature control shown in Fig. 17–11, a constant-speed motor drives a generator with an adjustable control field voltage. The generator will produce an adjustable voltage that

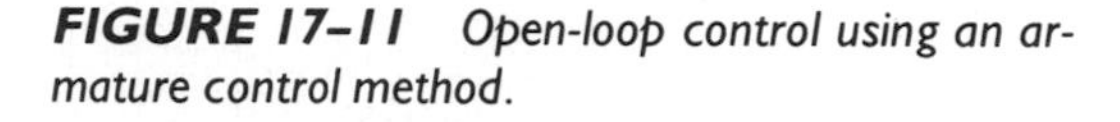

FIGURE 17–11 *Open-loop control using an armature control method.*

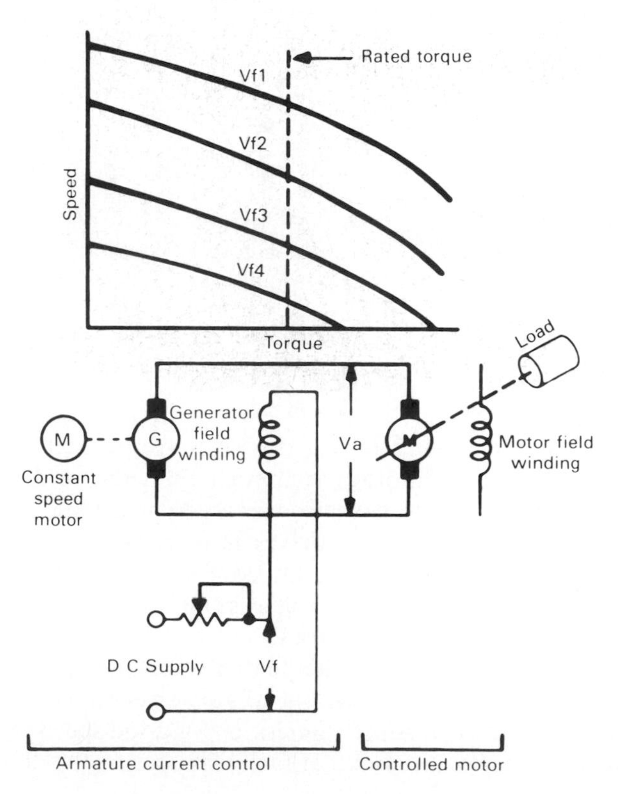

is delivered to the armature of the motor. The resulting torque–speed characteristics are improved over those shown in Fig. 17–8. This is because the regulation is essentially independent of the speed setting.

This results in superior motor speed control performance over any of the methods shown previously. However, due to the cost of the motor–generator set and associated field control, this method has not been practical for the small motor speed control applications most commonly used in home and industry. Therefore, the motor–generator control method has been confined primarily to industrial uses of large motor speed control, that is, for motors with more than 1 hp.

CLOSED-LOOP CONTROLS

Many of the open-loop controls are adequate, but the trend toward better speed regulation, such as in meeting servo systems requirements, has necessitated closed-loop control (Fig. 17–12). In the basic or elementary form, a closed-loop control consists of an actuator (the motor), a comparator, an amplifier, and a sensor (the generator). The more sophisticated controls use this basic circuitry and built on it to get motors to produce the torque and speed required when needed.

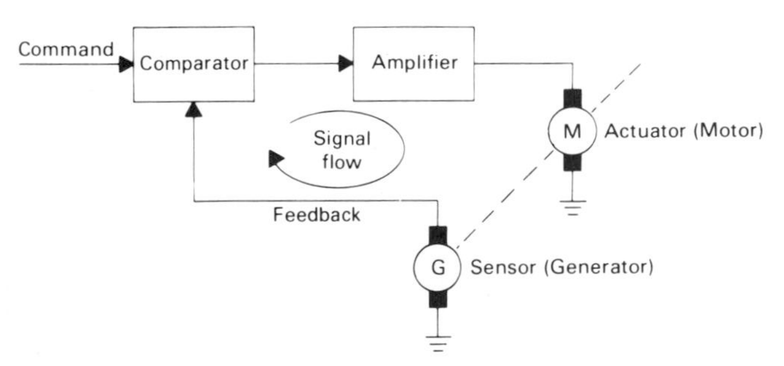

FIGURE 17–12 *Closed-loop speed control system.*

DC DRIVES: ADVANTAGES AND DISADVANTAGES

Among the advantages for dc drives are the fact that the dc drive technology is simpler than the ac drive technology. It has been in existence for a long time and is well known. Dc drives have good efficiency through the speed range. Dc controllers are smaller than adjustable-frequency drives, but the motors are larger than induction motors.

Disadvantages of the dc drives are the fact that the dc motor is not always available or is not considered a shelf item. A tach generator for good speed regulation is a requirement. If tach loss occurs, the drive may run away to full speed. The power factor decreases with speed. Bypass is not possible because of the construction of the dc motor. Full power conversion of all power supplied to the motor is required by the controller. Larger power devices are required.

DC DRIVES AND SYSTEMS

Custom-engineered drives and coordinated drive systems are available in ratings from $\frac{1}{4}$ to 1500 hp. A wide variety of packaging and functional modifications provide flexibility and performance levels for demanding applications as dictated by machine characteristics and process requirements.

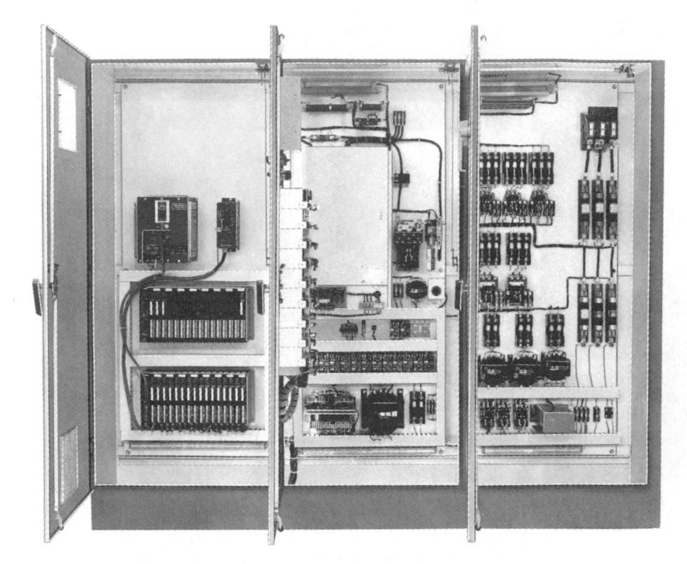

FIGURE 17–13 *Programmable logic controller for 75-hp motor. (Allen-Bradley)*

Figure 17–13 is a programmable logic controller with 75-hp Speedpak drive for a pull-through-slitter application. It is mounted in a standard enclosure. System-engineered drives serve complex manufacturing processes. In many cases they are integrated with programmable controllers and a variety of standard industrial control products to provide the user with a total control system.

SOLID-STATE DIGITAL AC DRIVES

The Bulletin 1352 is a versatile package capable of controlling the speed of a standard induction motor. It has the latest microprocessor and power semiconductor technology for controlling ac induction motors. The control panel is used to set up, operate, and troubleshoot the drive—all at the touch of a button. Take a look at Fig. 17-14. All parameters are entered in numerical format and stored in an electrically erasable programmable read-only memory chip (an EEPROM). There are no potentiometers to adjust and no other equipment required to calibrate the drive.

The control panel (Fig. 17-15) serves as a fully functional operator station, and a full complement of inputs and outputs are provided for handwiring to external devices or interfacing to other equipment. Faults are displayed on the control panel in an easy-to-understand digital format. In addition, all control logic values and drive input–output (I/O) points can be monitored to further enhance drive troubleshooting.

Dimensions in inches and (mm)

23 (589)

20.0 (508)

52.0 (1321)

54.75 (1391)

29.0 (737)

30-115 KVA Standard
NEMA Type 1 Enclosure

FIGURE 17–14 *Digital ac drive. (Allen-Bradley)*

FIGURE 17–15 *Control panel operation for the Bulletin 1352 digital ac drive. (Allen-Bradley)*

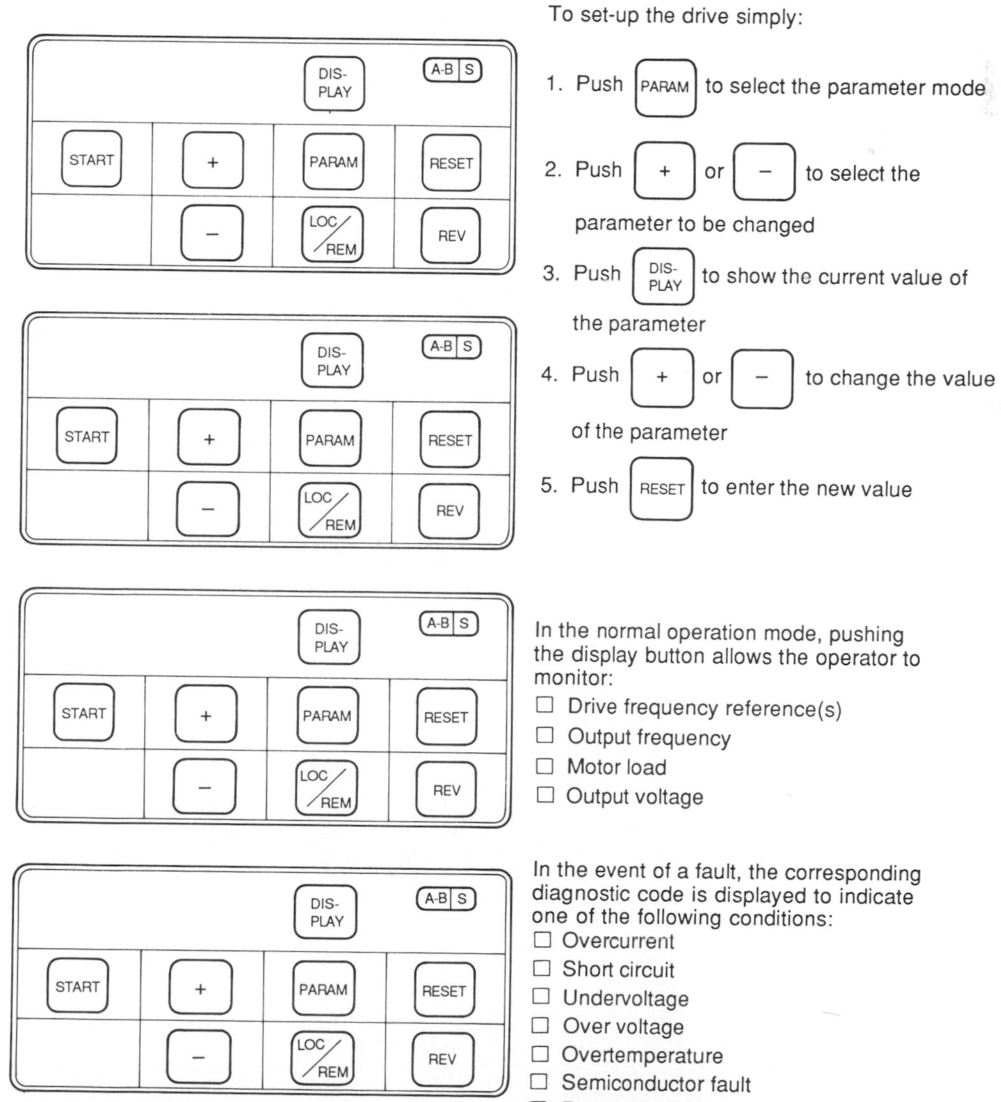

QUESTIONS

1. What are two types of adjustable-speed drives that can be used on fans?
2. What is another name for adjustable-frequency drives?
3. What are the three major types of inverter designs used today?
4. What dictates motor output requirements?
5. What is the range of horsepower on which CSI inverters operate best?
6. What does VVI mean?
7. What are some of the drawbacks to using VVIs?
8. What does PWM mean?
9. What are the two parts of an eddy current drive?
10. How are eddy current minimized?
11. What is the main advantage to an eddy current drive?
12. What is the common horsepower range of wound-rotor motors?
13. Why is rheostat control inefficient for dc motors?
14. What is the main disadvantage of open-loop controls?
15. What is used to make up a closed-loop control?
16. What is an EEPROM?
17. What is the purpose of a control panel for a digital ac drive?
18. List three advantages and three disadvantages of dc drives.
19. What is an SCR control?
20. List three advantages and three disadvantages for ac motor drives.

18

Transformers

Objectives

After studying this chapter, you will be able to:

1. Explain how a transformer operates.
2. Describe autotransformer operation.
3. List transformer losses.
4. Describe how transformers are made environmentally safe.
5. Define PCB.
6. Explain why three-phase transformers are chosen for motor control.
7. Define third harmonics.
8. Draw the schematic for a buck and boost transformer.
9. Define an askarel.
10. List three dry types of transformers.
11. Troubleshoot and maintain oil-filled transformers.
12. Locate proper current demands for various horsepower single-phase and three-phase motors.
13. Draw wye-to-wye connections.
14. Draw wye-to-delta connections.
15. Draw delta-to-delta connections.
16. Draw delta-to-wye-connections.

Transformers make it possible to utilize the high voltages generated at power plants. They can be used to step up voltages to allow electrical power to be transported from the generator site to the user. Transformers make it possible to do many things with electricity. This is a relatively silent device that is hidden in enclosures above and below ground. It has no moving parts (except ventilation fans in some cases) and is over 99% efficient. Transformers can range in size from tiny assemblies the size of a pea to behemoths weighing over 500 tons. However, the principles that govern the function of electrical transformers are the same no matter what the size or the use to which it is put.

A transformer functions with no physical connection between the input and output coils. The principle used is mutual inductance. Current flows into the transformer primary coil. This current creates a magnetic flux. The magnetic flux, in effect, couples the primary coil with the secondary coil. Voltage is induced in the secondary coil. The induced voltage may be varied by increasing or decreasing the magnetic field. The result of a transformer's operation is the induction of EMF in the secondary coil (Fig. 18–1).

IRON-CORE TRANSFORMER

Iron-core transformers use the mutual inductance principle to transfer power between primary and secondary windings. An iron core provides a low-reluc-

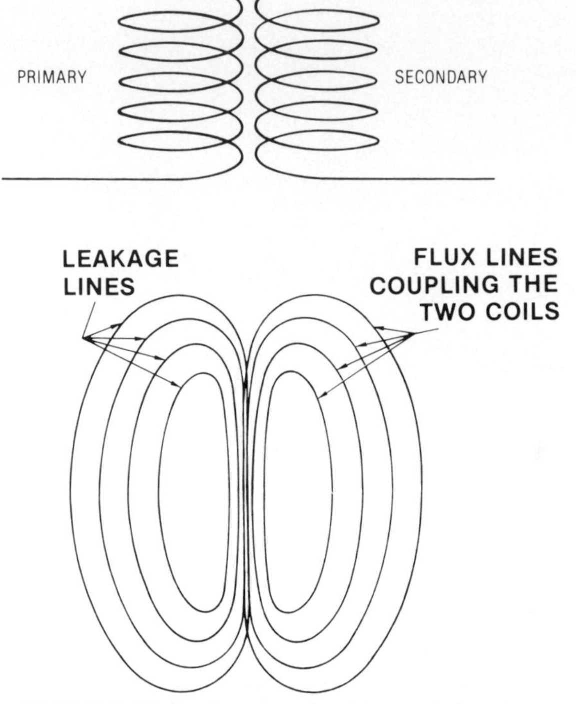

FIGURE 18–1 *Primary of coil is coupled to the secondary by the magnetic flux lines.*

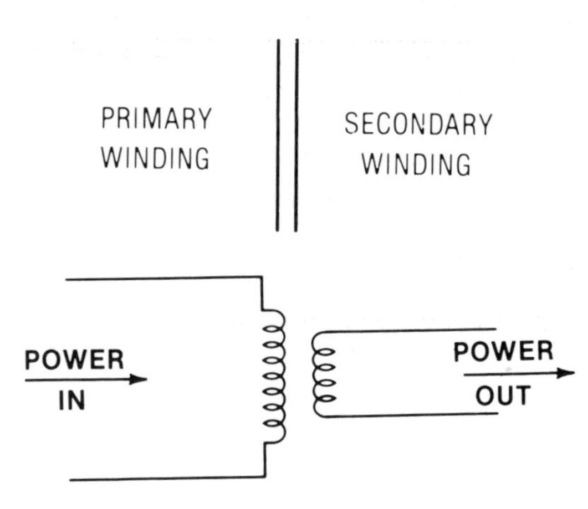

FIGURE 18–2 *Electric power is transferred from primary to secondary by the mutual coupling of the magnetic fields.*

tance path for the magnetic flux. The iron core is represented in symbols by two straight lines between the primary and the secondary coils.

Alternating current (ac) changes its magnetic field constantly so that a transformer can use this changing magnetic flux field to transfer ac power from primary to secondary (Fig. 18–2). Note that the two windings are not connected in this type of transformer. (Autotransformers will be examined later—they have only one winding.) The only means of transferring energy from the primary winding to the secondary winding is the magnetic field.

Turns Ratio

Transformers are either step-up or step-down. If the input voltage is higher than the output voltage, the device is a step-down transformer. In a step-up transformer the output voltage is higher than the input. The input and output voltage relationship depends on the *turns ratio,* which describes the relationship of windings of the primary and secondary coils. The first number in a given ratio is the secondary. The second number is the primary. The ratio of primary to secondary voltage is the same as the primary-to-secondary turns ratio:

$$\frac{E_s}{E_p} = \frac{N_s}{N_p}$$

We can also state that the power in the primary equals the power in the secondary, minus any losses.

A moderate overload causes the temperature of the transformer to be increased. Therefore, the current available is actually a function of the amount of heating permissible in the transformer itself.

AUTOTRANSFORMERS

Autotransformers provide a simple means of transforming voltage and impedance, but they do not provide the isolation between primary and secondary given by a regular transformer. The secondary load is really a part of the primary circuit whether the autotransformer is being used to step up or step down the voltage. A good example of autotransformer use in industry is the buck and boast transformer. It is used to increase or decrease the voltage as needed (Fig. 18–3).

FIGURE 18–3 *Primary and secondary are both part of a single winding on an autotransformer.*

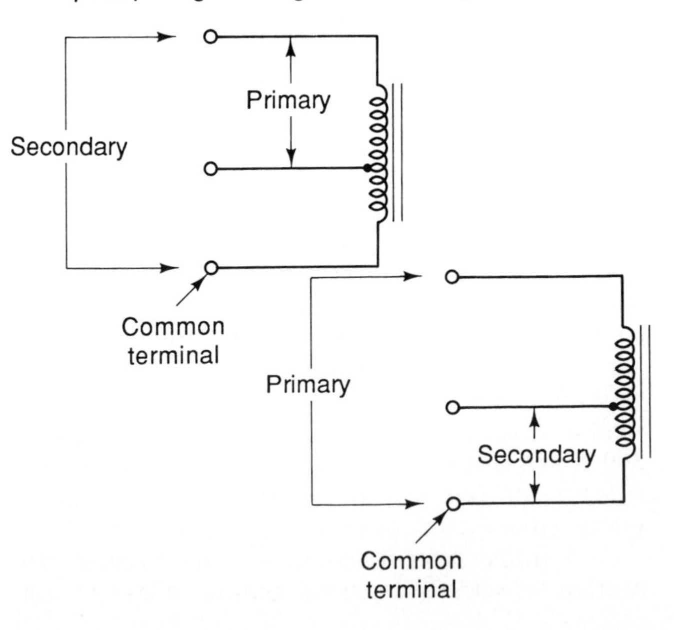

Autotransformers are restricted in their applications because it is not always possible to have one end grounded as required by the *National Electrical Code®* in some installations.

Turns Ratio

The voltage ratio for an autotransformer is directly related to the number of turns between the tap and the common terminal and over the entire winding. Voltages and impedances can be calculated for an autotransformer in the same way that they are calculated for a transformer with separate windings. The voltage is directly proportional to the number of turns in the winding. The impedance ratio is equal to the square of the turns ratio:

$$\frac{N_p}{N_s} = \sqrt{\frac{z_p}{z_s}} \quad \text{or} \quad \frac{z_p}{z_s} = \left(\frac{N_p}{N_s}\right)^2$$

TRANSFORMER LOSSES

Most transformers operate warm and some are quite hot when used for a period of time. This heat represents a power loss and means an efficiency of less than 100%.

Eddy Current Losses

The greatest loss comes from eddy currents. They are currents induced into the core material. Since the core material has low electrical resistance, there are many closed loops for high current. This current in the core material generates heat and is wasted power.

Eddy currents are not desired. They are produced by the magnetic flux that links the primary and secondary windings. The flux lines are necessary for the transfer of electrical energy from primary to secondary. However, they also induce voltages into the core material (Fig. 18–4A). Flux that is produced by the eddy currents opposes the desired flux that links the primary and secondary. This means that a greater power demand is made on the source, and the primary current is higher than the amount required to supply the secondary load.

Eddy current losses are reduced to a minimum by using a laminated core (18–4B). With a laminated core the eddy current paths in the core are parallel to the current in the windings. These thin layers of material create a high-resistance electrical path and reduce the circulation of eddy currents. At the same time, the iron laminations provide a low-reluctance magnetic path and the structure has no adverse effect on transformer action.

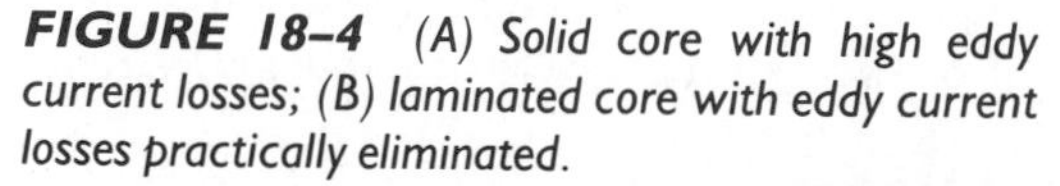
FIGURE 18–4 *(A) Solid core with high eddy current losses; (B) laminated core with eddy current losses practically eliminated.*

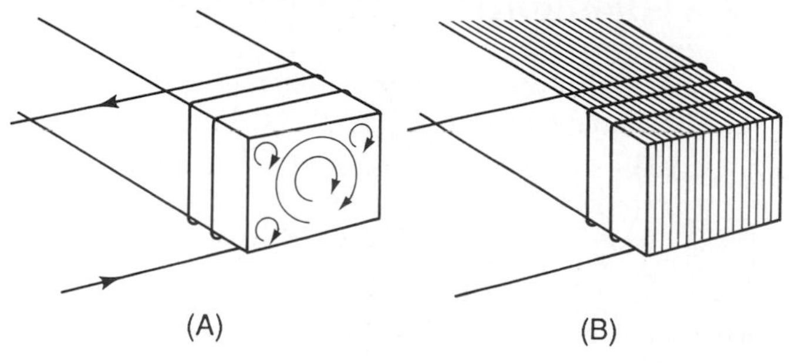

Hysteresis Losses

Hysteresis also plays a role in the losses presented by transformer operation. These losses are due to the properties of the iron core. Iron is slow to change polarity with changes in current and magnetic field polarity. The delay is known as *hysteresis,* or slowness to change properties (from north to south or south to north as the current changes polarity) (Fig. 18–5).

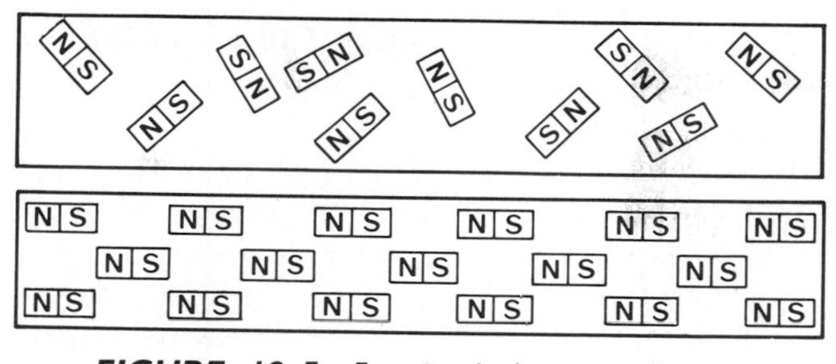

FIGURE 18–5 *Functional drawings showing hysteresis loss.*

Note how the north–south poles are aligned in the magnetized piece and scattered in the nonmagnetized piece in between the change from north to south pole. This changing from a north–south orientation to a south–north orientation consumes energy inasmuch as the metal structure is actually changed with the change in polarity. Hysteresis losses in transformers are minimized by using silicon steel inasmuch as it will change its polarity with a minimum of effort.

Copper Losses

Copper losses are due to the resistance of the copper wire in the primary and secondary coils. Large-size wire helps to minimize these losses.

CORE CONSTRUCTION

There are at least two usable core designs for power transformers (Fig. 18–6). The core form has the low-voltage and high-voltage windings at different loca-

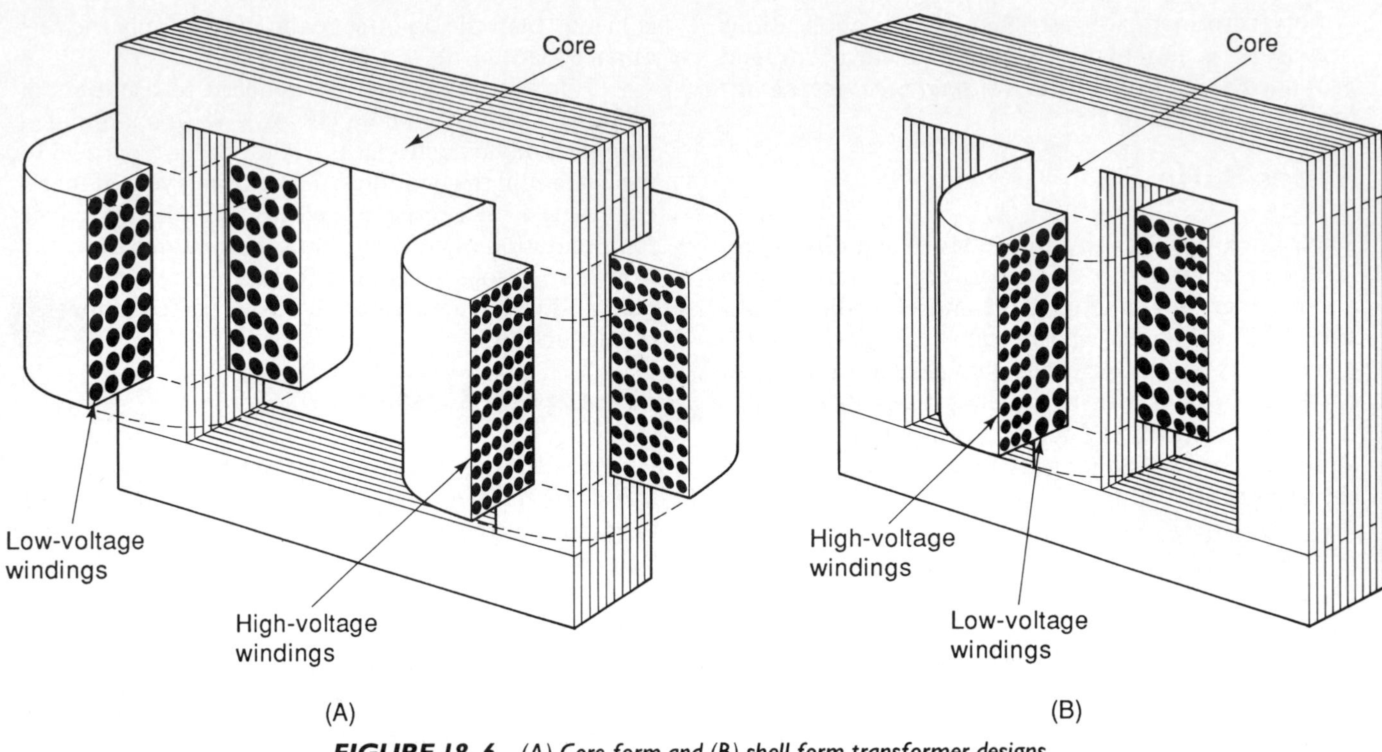

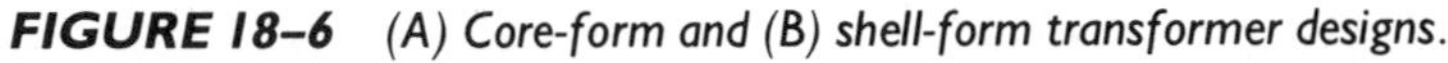
FIGURE 18–6 *(A) Core-form and (B) shell-form transformer designs.*

tions on the frame or core (18–6A). The shell form of construction makes a more efficient type of transformer inasmuch as the secondary is wound on top of the primary (18–6B). This allows for maximum flux to cut the secondary windings. The magnetic field is also concentrated by the middle leg and contained within the core by the other two legs of the core. The laminations are E-shaped and inserted within the coil first in one direction and then the other. Then I laminations are used to complete the flux path and the E–I combination makes for a complete path for the magnetic flux.

In a three-phase transformer the primary is wound and then the secondary on top of it, but there are three coil sets, one on each leg of the transformer core. Figure 18–7 shows how this is done with a cutaway view of a three-phase transformer. Note the tube coolers, where the hot oil moves up and then through the tubes to be cooled and to reenter the tank through the bottom.

TRANSFORMER OIL

In order to make a better transformer, the industry discovered in 1929 that polychlorinated biphenyls (PCBs) made a good insulating oil and also served to keep the coil from overheating since it carried the heat to an outside coiling arrangement (Fig. 18–7). However, PCBs have been determined to be toxic and to cause cancer in human beings and animals. The Environmental Protection Agency (EPA) keeps close tabs on transformers that utilize this cooling oil:

1. *PCB transformers:* all transformers that contain over 500 ppm (parts per million) of PCB. These cannot be rebuilt and must be labeled with EPA label.
2. *PCB-contaminated transformer:* all transformers not in category 1 must be assumed to be contaminated with between 50 and 500 ppm, unless proven otherwise by testing. They may be rebuilt, but PCB fluid must be burned in an EPA-approved incinerator.
3. *Non-PCB transformers:* have less than 50 ppm PCB.

The historical significance of this development in the transformer manufacturing process makes interesting reading, inasmuch as it also shows the development of the technology for the past 60 years.

EPA Highlights*

1929—Polychlorinated biphenyls (PCBs) invented by Swann Chemical Company.

1933—General Electric Company patented the application for use as a dielectric fluid in transformers and placed in service.

*Information courtesy of Square D Company.

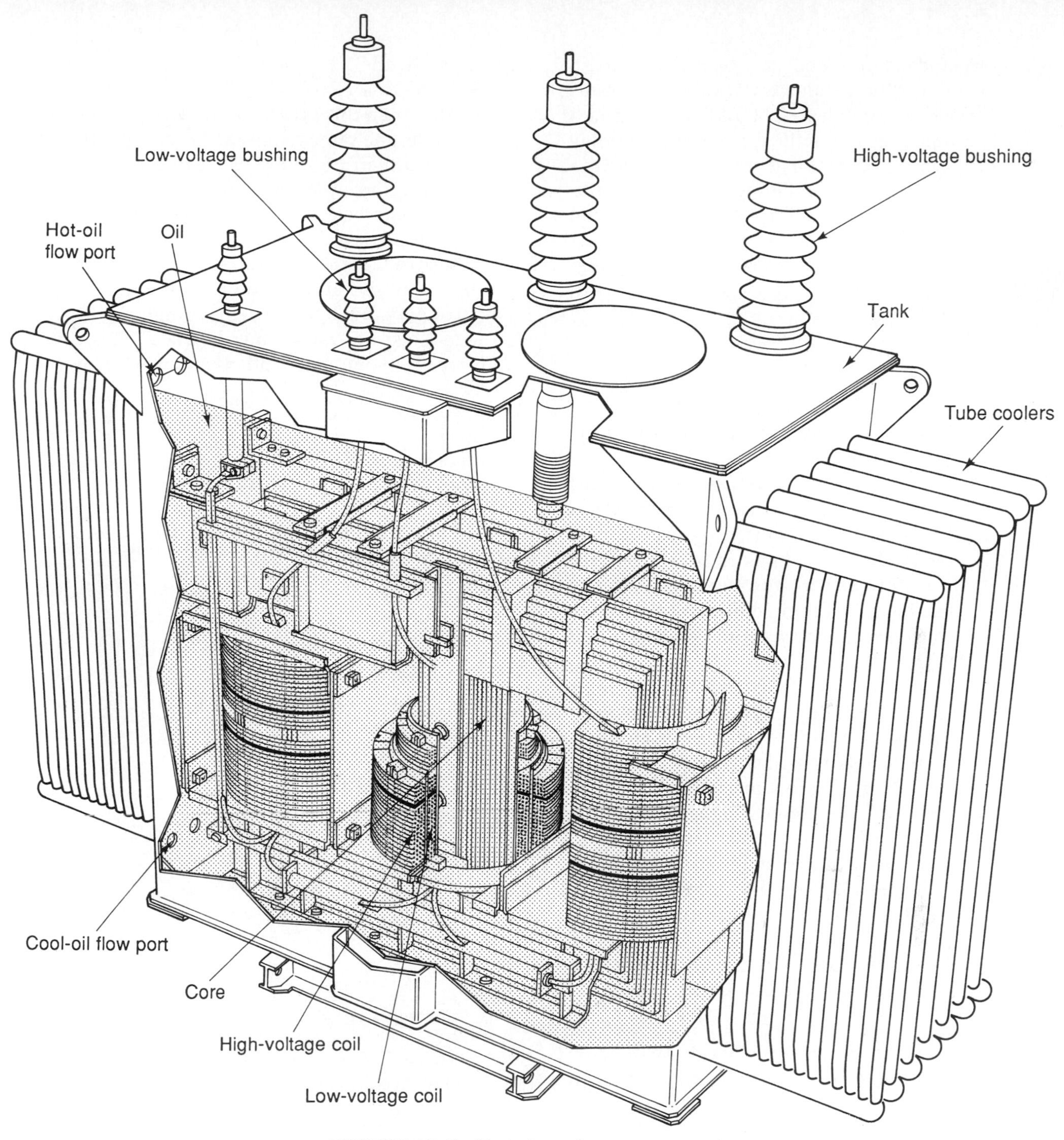

FIGURE 18–7 *Typical transformer, cutaway view.*

1977—U.S. Environmental Protection Agency (EPA) published the "Final Rule for Polychlorinated Biphenyls (PCBs) Manufacturing, Processing, Distribution in Commerce and Use Prohibitions."

1982—Electrical Use Rule—Required periodic inspection and records of PCB leaks. Also established new classification for PCB transformers which pose an exposure risk to food plants and animal feeds.

1985—(July) Fire Hazard Rule imposed new restrictions due to PCB transformers involved in fires. Requires removal of high-risk transformers or installation of fault protection by October 1, 1990.

1985—(August 16) Fire Hazard Rule requires the reporting of any PCB transformer involved in a fire that caused tank to rupture. Details cleanup procedures.

1985—(October 1) Deadline for removal of PCB

transformers in food and feed plants. Also prohibited the installation of used PCB transformers into commercial buildings. Commercial buildings are defined as public assembly properties, educational properties, institutions, stores, office buildings, and transportation centers, such as airports and bus or train stations.

1985—(December 1) Fire Hazard Rule further requires the following regarding all PCB transformers:

1. Removal of all combustibles in and around the transformer.
2. Notify local fire department of location of PCB transformer.
3. Access ways to PCB transformers must be marked and identified for firefighters.
4. Owners of PCB transformers in or near commercial buildings must register the transformer with the building owner in writing with a complete description of the transformer and location. (''Near'' means within 30 meters.)

1990—(October 1) All PCB transformers in or near commercial buildings must be removed, retrofitted, or equipped with new fault protection based on their classification.

1. Transformer secondary voltage less than 480Y/277 V requires high fault current protection.
2. A radial transformer with a secondary voltage of 480Y/277 V or higher must be equipped with both high and low fault current protection.
3. Use of network PCB transformers is prohibited as of this date.

After reviewing the requirements above it should be evident that the best permanent solution is retrofit or transformer replacement, providing a new non-PCB transformer.

AUTOTRANSFORMERS: THREE- AND SINGLE-PHASE

The three-phase autotransformer is chosen for its economy and energy considerations. They are used by those who have a 480Y/277- or 208Y/120-V, three-phase, four-wire distributing system in the building (Fig. 18–8).

An autotransformer cannot be used on a 480- or 240-V, three-phase, three-wire delta system. A grounded neutral phase conductor must be available in accordance with the *National Electrical Code*®Article 210-9, Exception 1.

Other than this article, autotransformers are installed under the same requirements as any other dry-type transformer. Figure 18–9 shows a typical wiring diagram for the autotransformer. Table 18–1 lists some of the characteristics of these transformers at 30 to 300 kVA.

Since the source and load share a common winding there is no isolation of the load from the source. Where this is a requirement it is recommended that the shielded isolating transformer be used. This is especially true for computers and other sensitive loads.

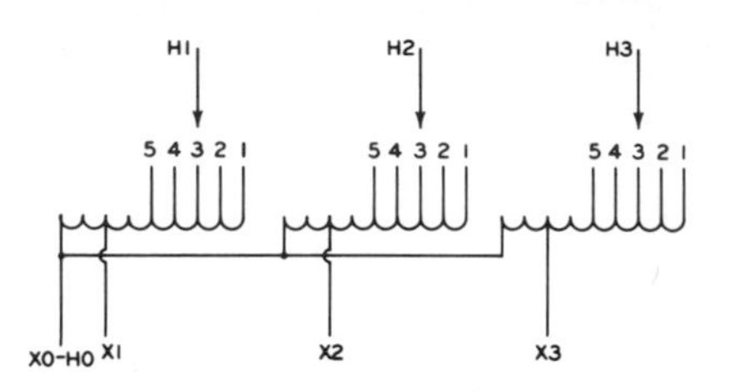

FIGURE 18–8 *Autotransformer. (Square D)*

FIGURE 18–9 *Typical wiring diagram for autotransformer. (Square D).*

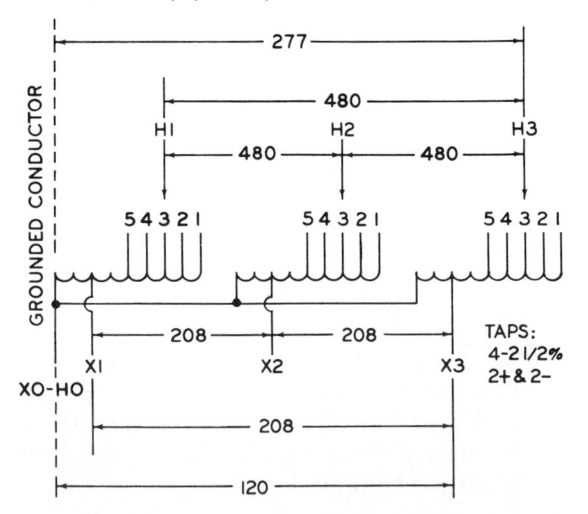

TABLE 18–1 AUTOTRANSFORMERS

kVA	Temp. Rise (°C)	Approx. Dimensions Height (in.)	Width (in.)	Depth (in.)	Weight (lb.)	Average X/R	% Z
30	150	23	22 1/4	15	250	1.4	2.1
45	150	23	22 1/4	15	275	1.0	3.3
75	150	30	30	20	425	1.2	3.7
112.5	150	30	30	20	605	1.0	2.4
150	150	42	36	24	750	1.5	3.5
225	150	42	36	24	1065	1.1	2.6
300	150	42	36	24	1375	2.6	3.5

Source: Courtesy of Square D.

Impedances

Impedances of autotransformers are quite a bit lower than comparable two-winding transformers. Although this provides better voltage regulation, it must be considered when making short-circuit studies for proper protective devices. Note the X/R and % Z values in Table 18–1. Table 18–2 shows impedances at 170°C.

Third Harmonics

Third harmonics are always present in wye–wye connections, but are kept at a minimum by using a three-legged core construction. For general use this presents no problem. Sound levels are indicated in Table 18–3.

Grounding

Article 250-26 of the NEC deals with grounding separately derived alternating-current systems. It is best to study the requirements of Article 250-26, 250-79(c) and Section 250-5 before connecting the transformers. Figure 18–10 shows the typical drawing for grounding a two-winding transformer, and Fig. 18–11 shows how the common grounded neutral is installed for autotransformers as per NEC 210-9.

In the case of the autotransformer, the grounded conductor of the supply, whether it be a 480Y/277- or

TABLE 18–2 IMPEDANCE AT 170°C

kVA	Impedance (%) Autotransformers	Insulating Transformers
30	2.1	5.5
45	3.3	5.7
75	3.7	5.2
112	2.4	6.9
150	3.5	6.7
225	2.6	6.6
300	3.5	3.7

Source: Courtesy of Square D.

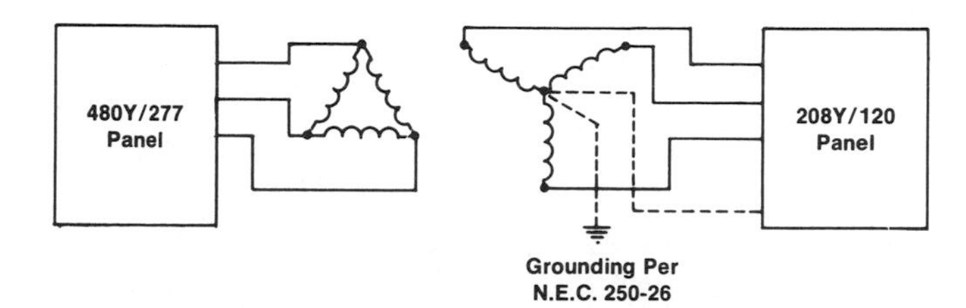

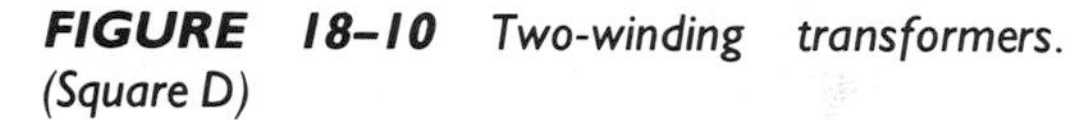
FIGURE 18–10 *Two-winding transformers. (Square D)*

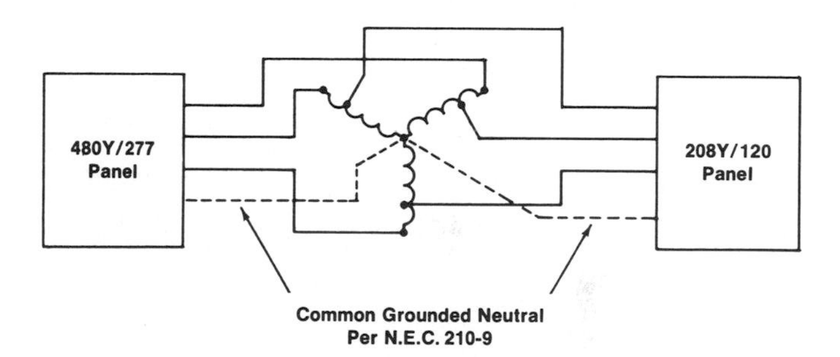

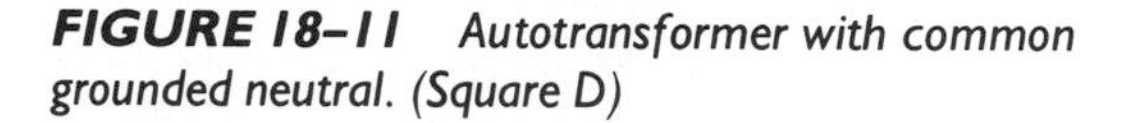
FIGURE 18–11 *Autotransformer with common grounded neutral. (Square D)*

TABLE 18–3 SOUND LEVELS FOR TRANSFORMERS

kVA	Sound Level (dB)[a] Design Levels Auto	Insulating	NEMA Standard
30	43	43	45
45	44	44	45
75	44	47	50
112	44	49	50
150	50	50	50
225	50	51	55
300	52	54	55

Source: Courtesy of Square D.
[a]dB, decibel. A 1-dB change in sound is probably the smallest change the trained ear can detect. Mathematically, it is equal to $10 \times \log_{10}$.

208Y/120-V system, is brought into this transformer to the common HO-XO terminal and the ground is established to satisfy the NEC requirement. Running the fourth wire is usually a negligible expense since in most cases the transformer is very close to its supply. The autotransformer connection is shown in Fig. 18–11.

BUCK AND BOOST TRANSFORMERS

Buck and boost transformers are used to increase or decrease voltage. They are insulating transformers that have 120 × 240 V primaries and either 12/24- or 16/32-V secondaries. When used as isolating transformers, they carry the rated kVA stated on the nameplate. Their prime use and value, however, lies in the fact that the primary and secondary of a buck and boost transformer can be interconnected and the unit used as an autotransformer. By varying the manner in which the two primaries and two secondary windings are connected, numerous ratios and current ratings can be obtained.

There are many applications where a slight adjustment in voltage (either upward or downward) is desirable or necessary. The use of a buck and boost transformer is one of the most economical and compact means of accomplishing adjustments of this type.

When used as an autotransformer, a buck and boost unit it will carry loads in excess of its nameplate rating. The increased ampacity is dependent on the ratio and voltage to which the transformer is subjected.

The application of buck and boost transformers as autotransformers has long been a source of confu-

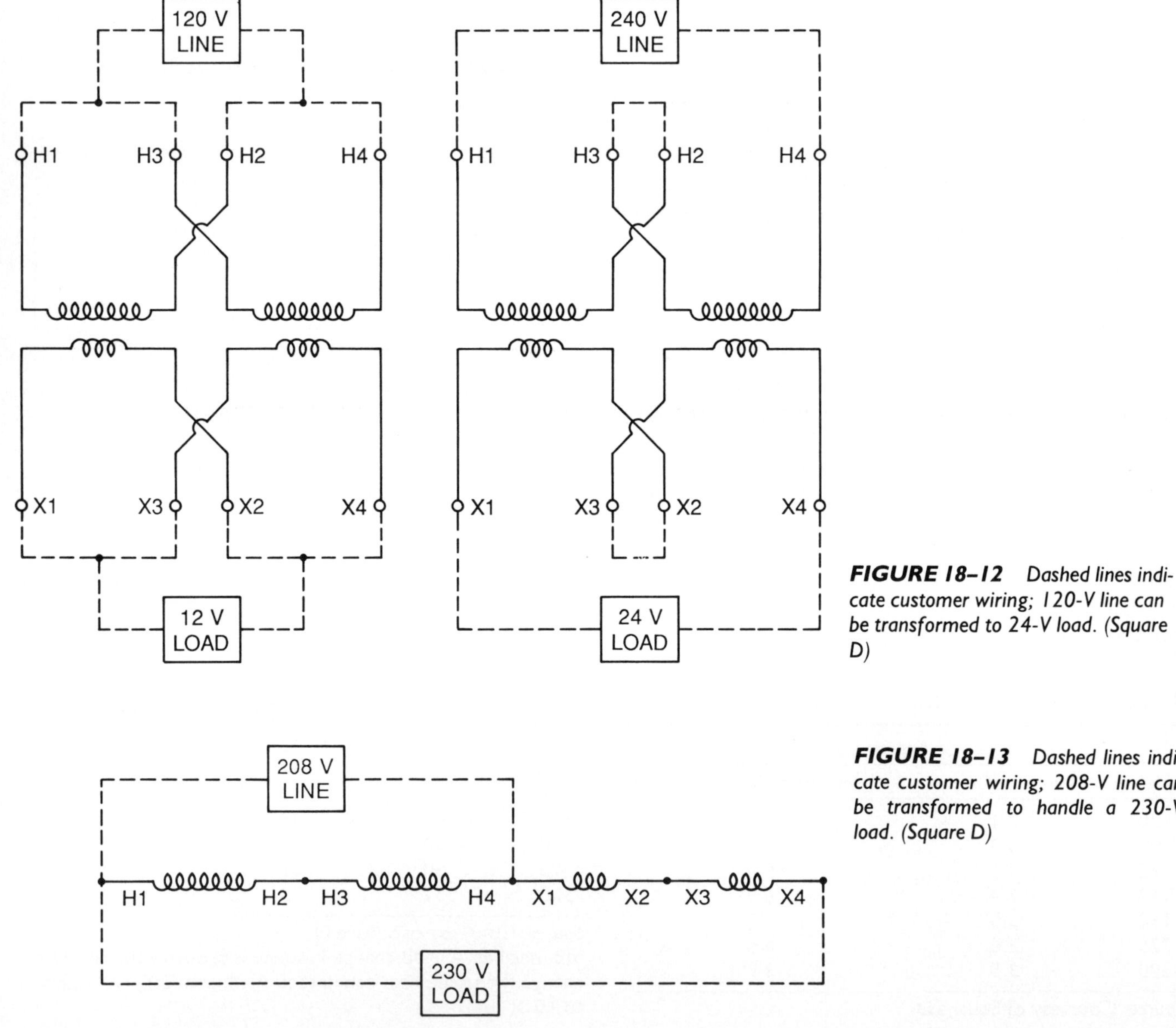

FIGURE 18–12 *Dashed lines indicate customer wiring; 120-V line can be transformed to 24-V load. (Square D)*

FIGURE 18–13 *Dashed lines indicate customer wiring; 208-V line can be transformed to handle a 230-V load. (Square D)*

sion to users. The key to this confusion undoubtedly lies in the fact that when used as an autotransformer the buck and boost unit carries loads in excess of its nameplate rating.

The nameplate rating shows the load that can be carried continuously by the unit when used as an isolation transformer, with a line voltage of 120 V or 240 V (Fig. 18–12). There is no direct electrical connection between the primary windings (letters H) and the secondary windings (letters X). When used in this way, each winding must carry full nameplate load.

Figure 18–13 shows a buck and boost transformer connected as an autotransformer. Note that in this case the primary windings (letters H) and the secondary windings (letters X) are in direct electrical contact at junction H1–X1. Because the primary and secondary windings are interconnected, they share the load, rather than each having to carry the full load. Consequently, the transformer can carry a load greater than on the nameplate. Winding section H1 to H4 carries only the difference of the primary and secondary currents, thus permitting the higher-load kVA ratings.

Calculating the Load

Increasing or decreasing line voltage by small percentages can be accomplished easily and economically by the proper use of buck and boost units as autotransformers. Depending on the percentage of change desired and the base voltage, the proper connections must be used; see Fig. 18–14, schematics 1, 2, 3, and 4.

FIGURE 18–14 *Wiring the buck and boost transformer. (Square D)*

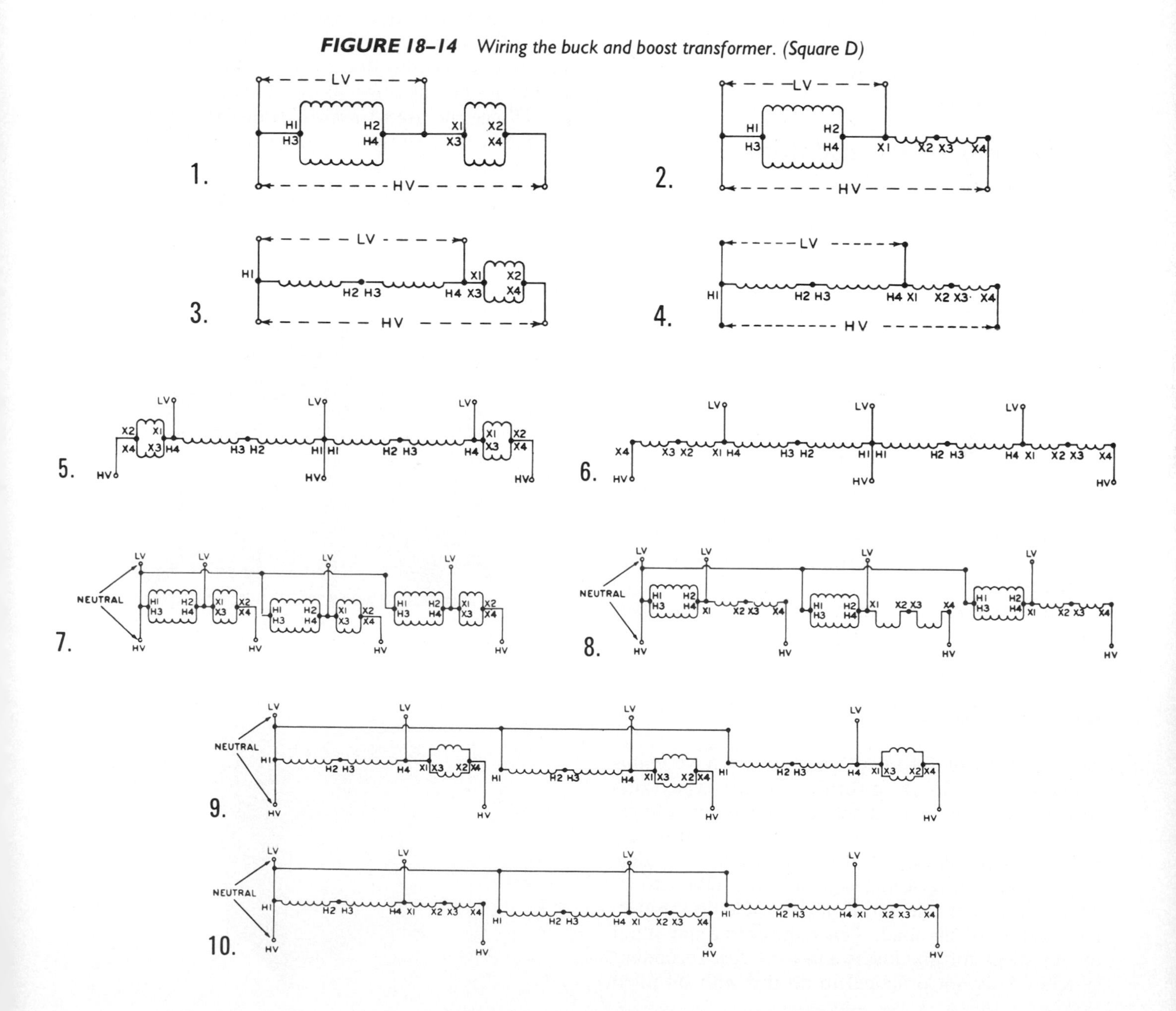

The percentage of voltage change is equal to:

Boosting:

$$\frac{\text{high voltage} - \text{low voltage}}{\text{low voltage}} \times 100$$

Bucking:

$$\frac{\text{high voltage} - \text{low voltage}}{\text{high voltage}} \times 100$$

Example:

Boosting 207 V to 230 V:

$$\% \text{ change} = \frac{230 - 207}{207} \times 100$$

$$\frac{23}{207} \times \frac{100}{1} = 11.1\%$$

Calculating single-phase kVA:

$$\frac{\text{volts} \times \text{load amperes}}{1000}$$

Three-Phase Buck and Boost Transformers

So far the discussion has centered on single-phase buck and boost transformers. Figure 18–14, schematics 5 to 10, show various means of connecting the transformers in three-phase arrangements.

Calculating three-phase kVA:

$$\frac{\text{volts} \times \text{load amperes} \times 1.73}{1000}$$

Single-phase uses one transformer unit. Three-phase uses two or three transformers to boost or buck. Three-phase loads can be served by using two single-phase units connected in open delta form. Three-phase loads can also be served by using three single-phase units connected in wye.

Before using autotransformers, check local codes for any restrictions pertaining to their use.

DRY-TYPE TRANSFORMERS

Newer makes of oil-filled transformers have eliminated the use of PCBs in their tanks. Dry-type transformers do have some advantages and are used to replace the older PCB types. Dry-type transformers increase the efficiency of electrical systems considerably by permitting voltages greater than 600 V to be conducted, as near as physically possible, to the electrical center of the load. This reduces to a minimum money spent for line losses and for larger secondary systems. It is not practical to do this with oil-filled transformers, which must be installed outdoors at a safe distance from a building, or in a fireproof vault which usually cannot be located for the most efficient power distribution. Although recent NEC changes permit vaults with lower fire rating if they are sprinkler protected, many engineers and users object to installing any water pipes in electrical equipment rooms. Some electrical equipment rooms also contain an emergency distribution system, including the generator. A water leak or sprinkler head malfunction could disable the entire electrical system.

A second type of application, which is quite common, is the use of dry-type transformers on a roof. Since they are usually lighter weight than a comparable oil-filled transformer, a lighter roof structure is required. Also, they do not require any special provisions for containing the oil in case of a leak or fire.

Another major area of use has been in high-rise buildings, permitting the economical distribution of power to the various floors using voltages up to 15 kV, and then stepping down to the utilization voltage.

Since the use of *askarel* transformers was prohibited because of the environmental concerns raised

FIGURE 18–15 *Dry-type transformers: (A) open cabinet; (B) closed cabinet. (Square D)*

(A)

(B)

by PCBs, new liquids with suitable cooling and insulating qualities have been developed and applied to transformer use. The NEC classifies them as "less flammable." This means that they can burn but are less combustible than mineral oil, or "nonflammable," which means that they do not have a flash point or fire point. Because this is a new application for most of these liquids, there are few historical field data to determine if there are some unknown, undesirable characteristics that might develop in some of these liquids over a period of years. Figure 18–15 shows dry-type transformers with and without a cabinet.

Cooling

The addition of cooling fans increases the capacity of a transformer by $33\frac{1}{3}\%$. This is an economical way to handle short-time load peaks or emergency overloads. Fans can be provided at the time the transformer is purchased, or provision can be made for the future addition of fans in the field. Fan cooling is generally economical to provide on transformers 300 kVA and larger. Automatic fan cooling systems have heat-sensing thermistors in each phase of a three-phase or single-phase transformer.

Insulation and Temperature Rise

Most dry-type transformer manufacturers use UL-component recognized 220°C insulation systems for 30 kVA and above. The 220°C represents the ultimate temperature of the winding and is the summation of the permissible rise (by resistance measurement) of the windings (150°C), the hot differential allowance (30°C), and the ambient allowance (40°C). This 220°C insulation system was formerly called the H Insulation System by NEMA and ANSI. The letter designation H was dropped due to the confusion caused by UL calling this same 220°C system a C Insulating System. H is still used in some catalog numbers.

Taps

Taps are made standard with two $2\frac{1}{2}\%$ above and below normal primary voltage. Four $2\frac{1}{2}\%$ taps below normal primary voltage can be obtained from most manufacturers.

SPECIAL TRANSFORMERS

Outdoor transformers of the ventilated dry-type are tamper-proof with special baffled louvers, and are constructed of heavy-gage sheet metal that has been cleaned and painted. Periodic inspections are recommended, but there is no oil or filter to check. The transformers are available in all voltages up to 15 kV and sizes through 1500 kVA (Fig. 18–16).

Pad-mounted transformers (dry-type; Fig. 18–17) are ventilated and have a 4160-V delta primary and 208Y/120-V secondary with two $2\frac{1}{2}\%$ taps (Fig. 18–17). This type of housing may also contain a primary hook-stick-operated switch, transformer, and a limited amount of secondary breakers and switches. All outdoor transformers over 600 V have lighting arresters as standard equipment for extra protection against lightning or switching surges.

Epoxy units are made for substations and are particularly suited for applications requiring a dry-type transformer with good performance characteristics. The windings are completely impregnated with epoxy resin, which together with fiberglass cloth and tape, form the solid dielectric system. The solid dielectric system protects the windings from its environment with respect to moisture and airborne contaminants. This solid dielectric system also provides the ability to withstand thermal shock and the mechanical forces of a short circuit. They are ideal replacements for PCB units (Fig. 18–18).

LIQUID-FILLED TRANSFORMERS

The transformer coils are immersed in an oil or silicone to increase the efficiency of the transformer. The liquid has the ability to transfer heat from the coil to the outside of the case efficiently. The liquid also has insulating qualities that are very desirable at high voltages. Now that PCBs are eliminated from the manufacture of transformers it is necessary to make sure that the replacements are environmentally safe. So far the oils used are as fire resistant as PCB oil.

Safety Notice: Keep in mind that to work with transformers, adequate training is needed. Extremely hazardous voltages are present in energized units. Adequate training and safe working procedures for this and related high-voltage equipment is needed to make sure that you can work with the equipment safely.

Working with and inspecting the transformer tank and its contents requires special instructions on its contents and operation. Some of these are mentioned here to inform those who may wish to obtain further training before work on the equipment.

SUBMERSIBLE TRANSFORMER

Some manufacturers use gray paint for the exterior of the transformer and others use green (Fig. 18–19). These three-phase transformers consist of a core and

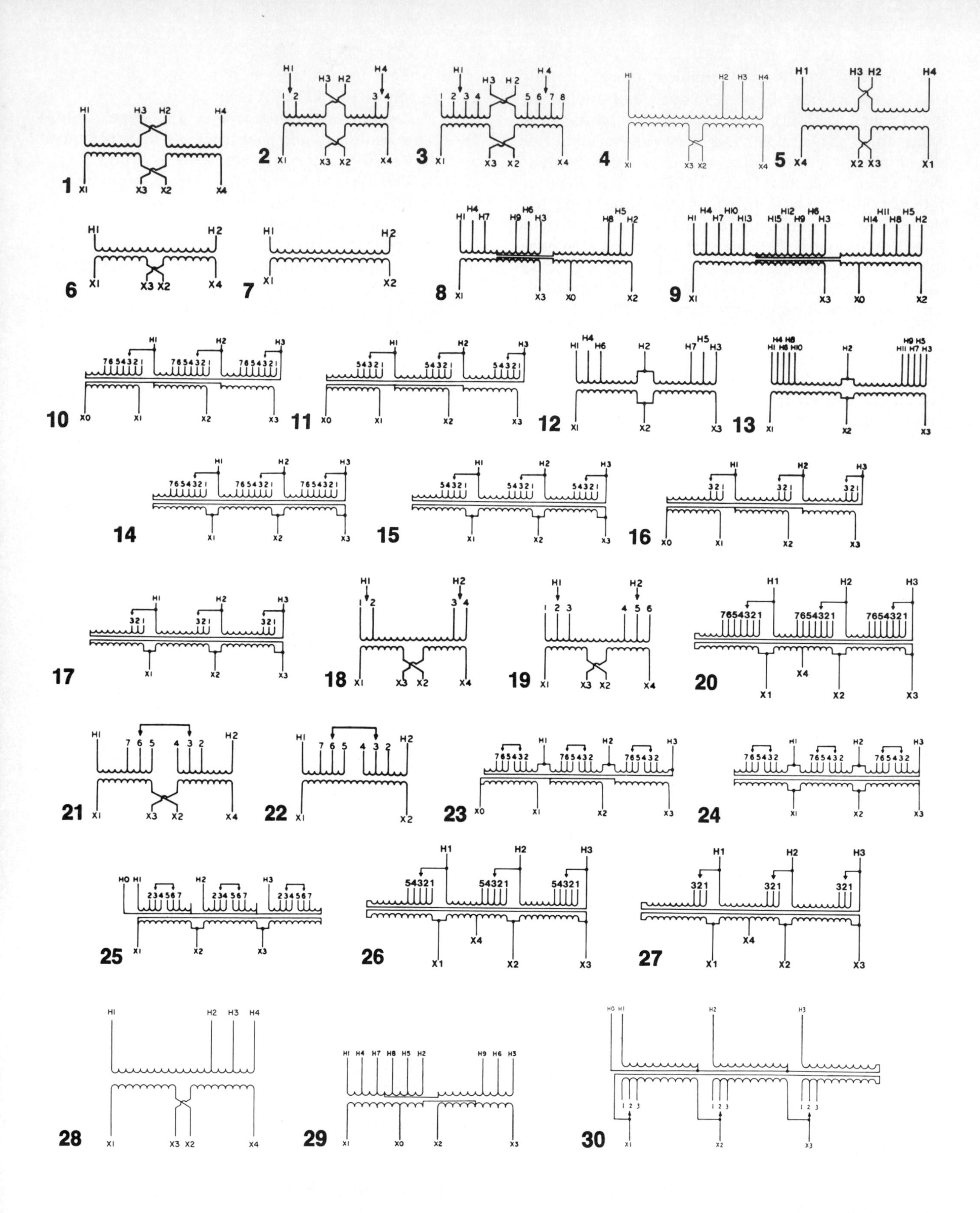

FIGURE 18–16 *Transformer wiring diagrams. (Square D)*

FIGURE 18–17 *Pad-mounted dry-type transformer. (Square D)*

FIGURE 18–18 *Three-phase cast-epoxy unit substation transformer. (Square D)*

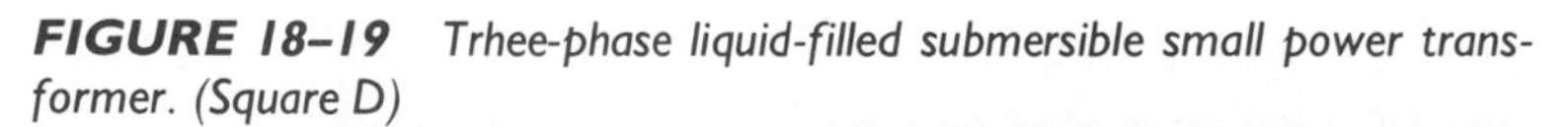

FIGURE 18–19 *Trhee-phase liquid-filled submersible small power transformer. (Square D)*

SQUARE D COMPANY

UNDERGROUND DISTRIBUTION TRANSFORMER

KVA ______ Continuous
Pri. Volts ______
Sec. Volts ______
Instr. Book ______
Impulse Level Full Wave
Test Pri Volts ______ KV Bil
Sec. Volts ______ KV Bil
Liquid Type ______
Max Operating Pressure: ______ psi
______ Gals ______ Lbs
Core & Coils ______ Lbs
Tank ______ Lbs
Total Weight ______ Lbs
Untanking Hgt ______ Ins
Mfg Date ______

Class ______ ______ °C Rise
Three Phase ______ Hertz
Serial Number ______
Impedance At ______ °C
______ %OA Rating
Ld Brk Sw Volts ______ Bil ______
Sw Continuous Amps ______
Sw Momentary Amps ______
Close & Hold Amps ______

Tap	Connects			Volts	Amps
A	1	TO	2		
B	2	TO	3		
C	3	TO	4		
D	4	TO	5		
E	5	TO	6		

CAUTION
DO NOT USE LIQUID OTHER THAN SPECIFIED ABOVE
DE-ENERGIZE TRANSFORMER BEFORE CHANGING TAPS

H2 H1 H3 X2 X0 X1 X3
H2 H1 H0 H3 X2 X1 X0 X3
H1B H2B H3B H0
H1A H2A H3A

MADE IN U.S.A. 43504-004-05

FIGURE 18–20 *Pad-mounted liquid-filled compartmental transformer. (Square D)*

FIGURE 18–21 *Three-phase, liquid-filled substation transformer. (Square D)*

coil assembly designed to reduce operating losses and to provide adequate mechanical strength should a system fault occur. The 75- to 5000-kVA core–coil assemblies are of the five-legged design with wound cores (Fig. 18–20). By using the five-legged design, the possibility of tank heating due to stray flux paths under short-circuit conditions is eliminated. Computer design techniques produce a core-coil assembly with high efficiency and low losses for the most economical operating characteristics. Strip aluminum secondary windings offer high axial short-circuit strength and fast coil heat dissipation.

Maintenance

Oil leaks are rare, but if detected, must be repaired at once to avoid the liquid level dropping below energized parts, creating a possibility of flashover or tank overheating. If required, the transformer must be refilled to its proper operating level. Small pinhole leaks in the exterior metal, weld seam, or other locations, resulting in slow dripping, can be repaired with a lifetime durable epoxy patch kit. The transformer must first be deenergized. Then a temporary oil stick is applied. Next the epoxy is applied. This usually eliminates the need for a vacuum pump to stop the oil leak while the epoxy is being cured. Larger leaks may require the use of a vacuum pump. The main reason for maintaining the level and preventing the oil to leak out is because of the entry of moisture into the tank. The moisture can cause problems with high voltages and can also cause the oil to become contaminated necessitating its removal and replacement.

Figure 18–21 shows all the parts of a liquid-filled substation transformer. Note the location of the pressure relief device, fill plug, liquid-level gage, dial-type thermometer, nameplate, and drain valve. The fins on the back are used to cool the liquid as it passes through the tubes from top to bottom. Hot oil rises and enters the tubes at the top of the fins. As it trickles down, it is cooled by the outside air. Once cooled it reenters the tank at the bottom to once again pick up excess heat and conduct it to the outside of the tank and away from the coils inside.

TRANSFORMER APPLICATIONS

Transformers are used in business, industry, and commerce as well as for schools and homes. They come in many sizes and shapes, as has been seen so far in this chapter. Six typical installations and applications for industrial work are discussed in the next part of this chapter.

Distributing Power at High Voltages

Transformers are used for distributing power at 480 or 600 V and stepping down voltage at the point of use (for 240-V motors, or 120-V equipment and lights). This results in better regulation of voltage and minimized line loss and reduces wiring costs. See Fig. 18–22 for a diagram illustrating these points.

Double-Wiring Elimination

Transformers are used to eliminate double wiring. For maximum safety, 120-V lighting and control circuits may be obtained from 240-, 480-, or 600-V power circuits by installing dry-type transformers at the most convenient location to the load. This eliminates separate circuits and independent metering for power and light and often results in large savings. See Fig. 18–23 for examples of this.

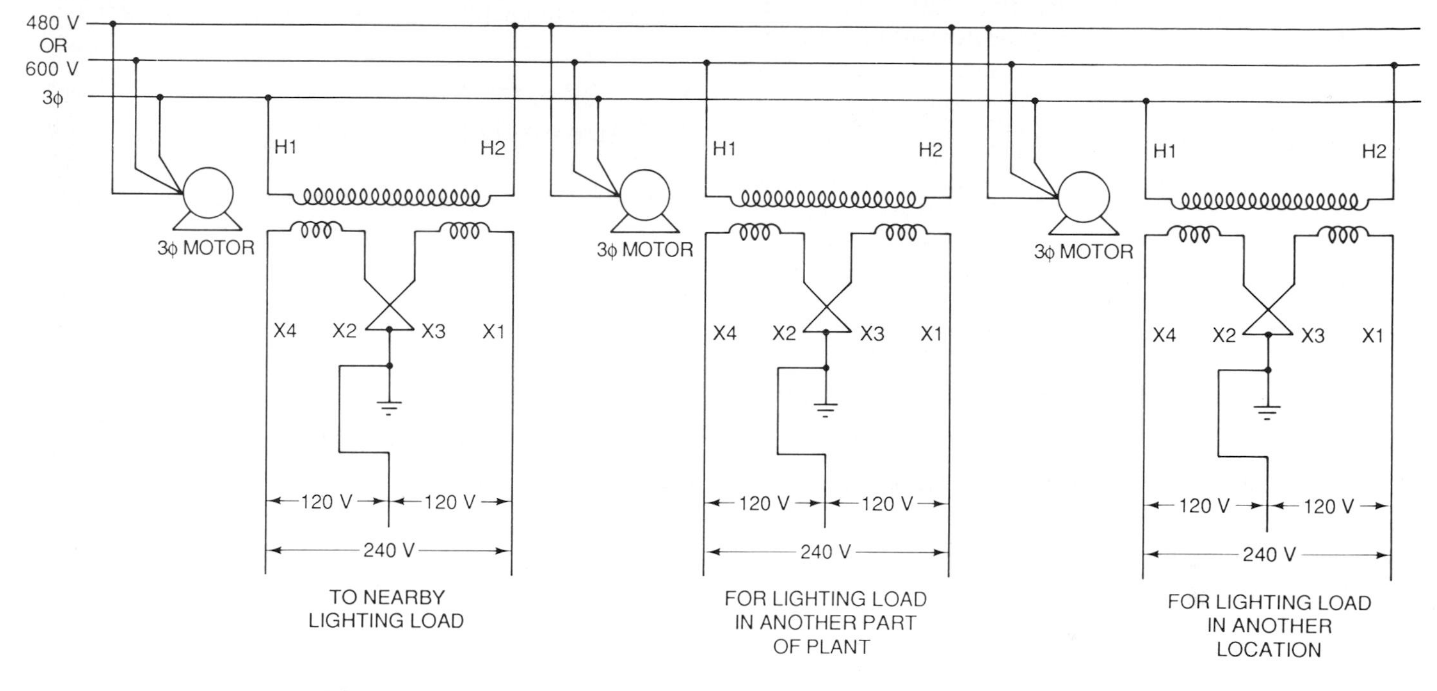

FIGURE 18–22 *120/240-V single-phase from three-phase high-voltage source.*

FIGURE 18–23 *120-V lighting and control circuits from three-phase source.*

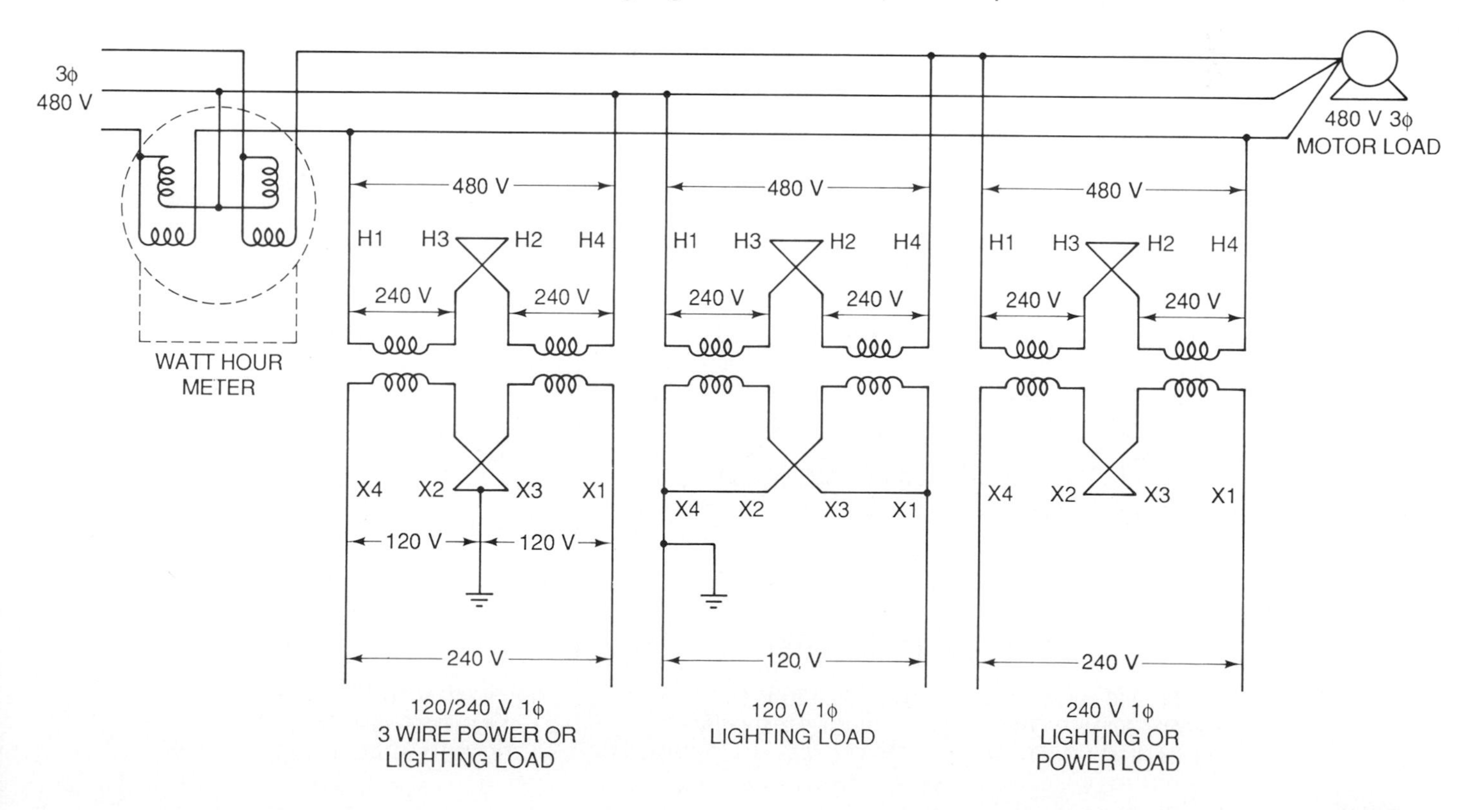

Operating 120/240-Volt Equipment from Power Circuits

Transformers are used to operate portable tools, electrical control devices, alarms, relays, soldering irons, heating pots, small heat-treating furnaces, bench welders, and other high-current devices more economically from 120- or 240-V circuits supplied through a dry-type transformer from a high-voltage power circuit (Fig. 18–24).

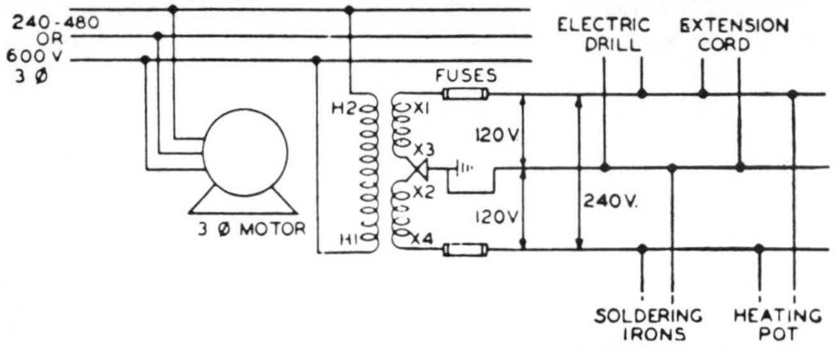

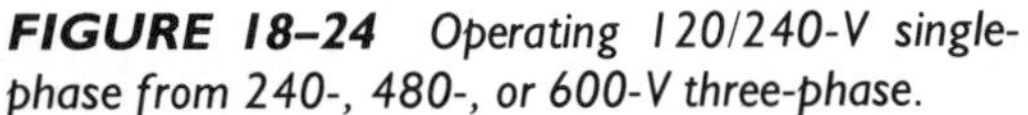

FIGURE 18–24 *Operating 120/240-V single-phase from 240-, 480-, or 600-V three-phase.*

Isolating Circuits

Transformers are used because of their ability to isolate one circuit from another. They can be air-cooled transformers and used as a means of subdividing circuits to accommodate independent demand. They can be connected to a three-phase, 480-V circuit, dry-type transformer to provide 120/240 V, three-wire, single-phase power for lighting loads. They can provide 120-V single-phase lighting loads or 240-V single-phase lighting or power loads. Transformers permit grounding of each low-voltage circuit (Fig. 18–25).

Changing from Four-wire to Three-wire Circuits

Transformers can be used to produced a three-wire, 120/240-V, single-phase circuit from the 120-V, two-wire circuit of a four-wire, 208Y/120-V, three-phase source. A three-phase primary of 240, 480, or 600 V may be used to provide a three-wire, 120/240-V, single-phase circuit (Figure 18–26).

Stepping-Up or Stepping-Down Voltage

Transformers can be used to step-up or step-down voltages simply by virtue of the turns ratio in the primary as compared to the secondary (Fig. 18–27). Whenever the voltage source is lower or higher than the nominal required by the equipment load, a buck and boost transformer may be used.

FIGURE 18–25 *Subdividing circuits.*

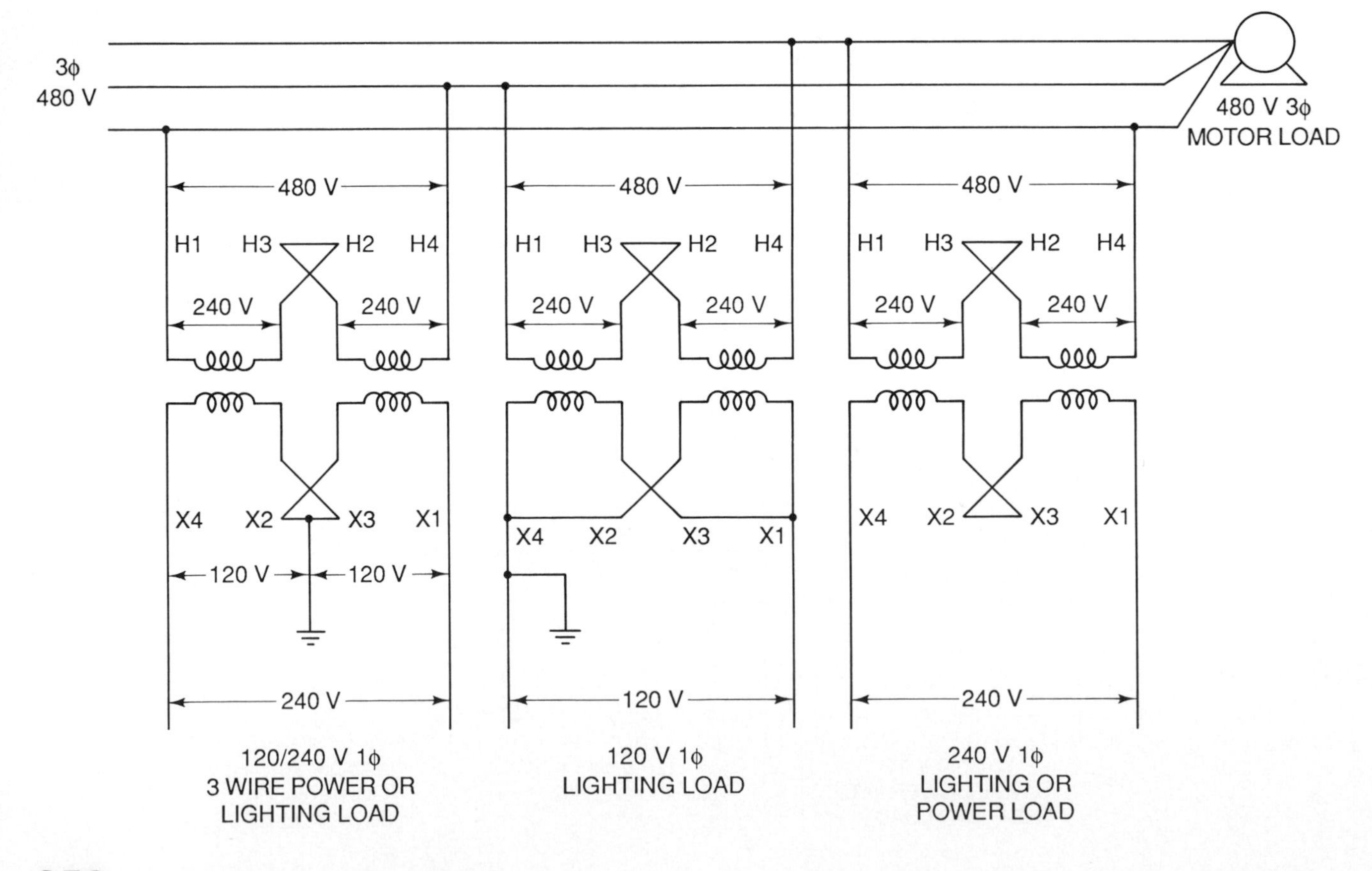

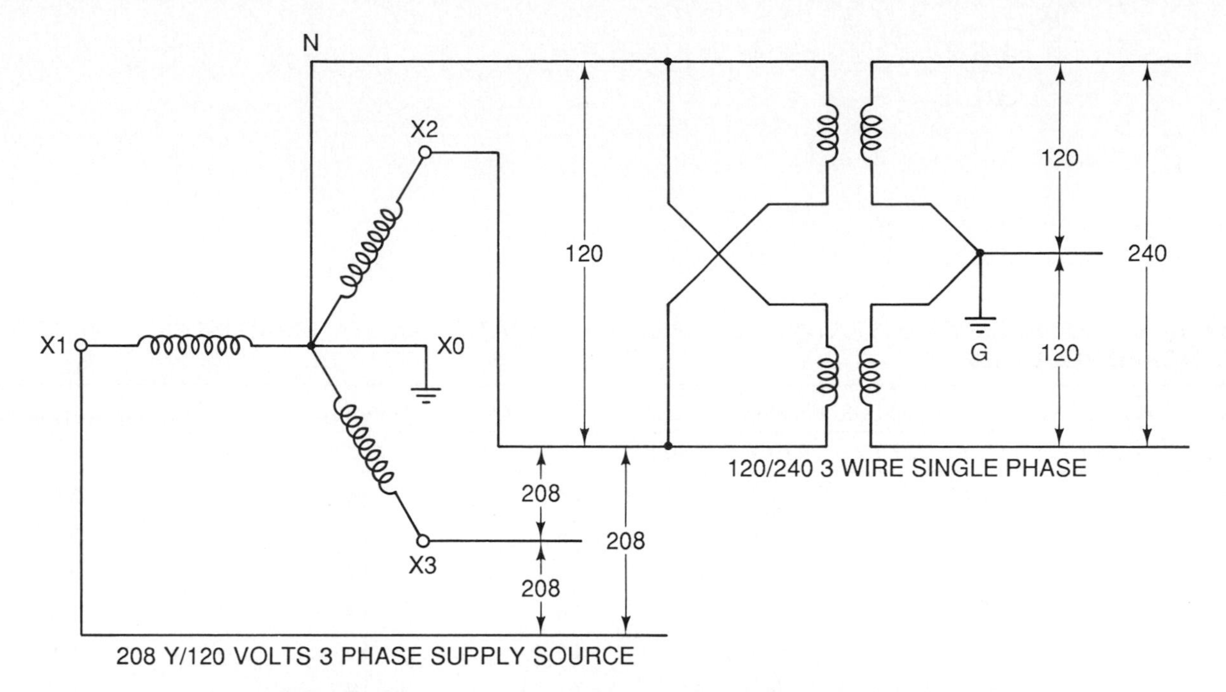

FIGURE 18–26 *Three-wire secondary circuit from four-wire three-phase.*

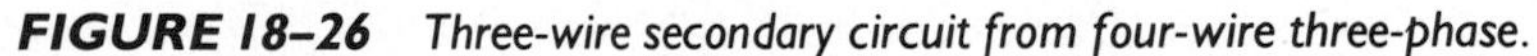

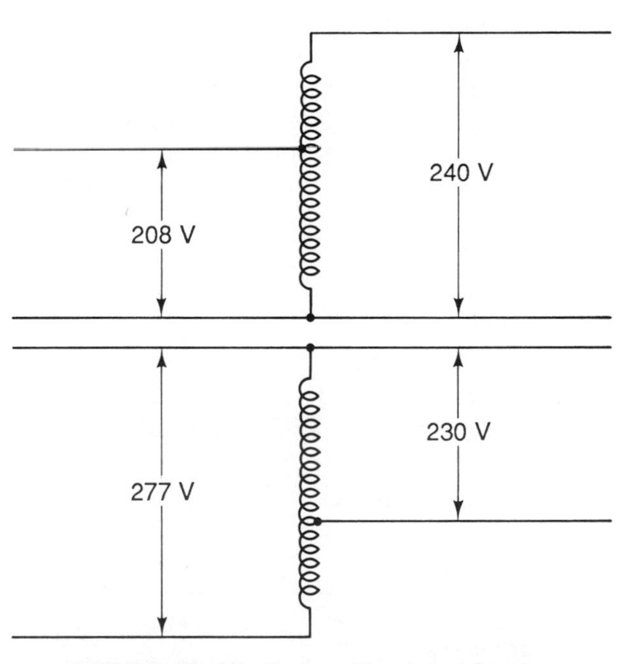

FIGURE 18–27 *Buck and boost transformers.*

MOTOR TRANSFORMERS

Transformers are very necessary to obtain the proper voltage for the operation of motors. Table 18–4 shows transformer ratings required for the operation of standard induction motors at standard voltages and at 60 Hz. Table 18–5 shows how to convert kilowatts to horsepower or horsepower to kilowatts, and Table 18–6 shows what can happen with improper voltage levels in a plant.

TABLE 18–4 TRANSFORMER kVA RATING REQUIRED FOR OPERATION OF STANDARD INDUCTION MOTORS AT STANDARD VOLTAGES

AC Motors	Horse-power	*Full-Load Amperes*				*Minimum Wire Size*[a]				*Minimum Transformer (kVA)*[b]
		115 V	*220 V*	*230 V*	*440 V*	*115 V*	*220 V*	*230 V*	*440 V*	
Single-phase	$\frac{1}{6}$	4.4		2.2		14		14		0.53
	$\frac{1}{4}$	5.8		2.9		14		14		0.70
	$\frac{1}{3}$	7.2		3.6		14		14		0.87
	$\frac{1}{2}$	9.8		4.9		14		14		1.18
	$\frac{3}{4}$	13.8		6.9		14		14		1.66
	1	16		8		12		14		1.92
	$1\frac{1}{2}$	20		10		12		14		2.4
	2	24		12		10		14		2.88
	3	34		17		8		12		4.1
	5	56		28		6		10		6.72
	$7\frac{1}{2}$	60		40		3		8		9.6
	10	100		50		1		6		12
Three-phase	$\frac{1}{2}$			2	1		14		14	0.9
	$\frac{3}{4}$			2.8	1.4		14		14	1.2
	1			3.5	1.8		14		14	1.5
	$1\frac{1}{2}$			5	2.5		14		14	2.1
	2			6.5	3.3		14		14	2.7
	3			9	4.5		14		14	3.8
	5			15	7.5		14		14	6.3
	$7\frac{1}{2}$			22	11		10		14	9.2
	10			27	14		10		14	11.2
	15			40	20		8		12	16.6
	20			52	26		6		10	21.6
	25			64	32		4		8	26.6
	30			78	39		3		8	32.4
	40			104	52		1		6	43.2
	50			125	63		0		4	52.0

Source: Courtesy of Square D.

[a]Not more than three conductors in cable or raceway.

[b]Allow 20% additional kVA if motors are started more than once per hour. Data above computed from standard motor data as listed in *National Electrical Code*®. For estimating only. For OEM application, check exact requirements with factory.

TABLE 18–5 CONVERSION TABLES FOR HORSEPOWER TO KILOWATTS AND KILOWATTS TO HORSEPOWER FOR ELECTRIC MOTORS

Kilowatts to Horsepower				*Horsepower to Kilowatts*			
kW	*hp*	kW	*hp*	*hp*	kW	*hp*	kW
1	1.341	55	73.733	1	.746	55	41.03
2	2.681	60	80.436	2	1.492	60	44.76
3	4.022	65	87.139	3	2.238	65	48.49
4	5.363	70	93.842	4	2.984	70	52.22
5	6.703	75	100.545	5	3.730	75	55.95
6	8.044	80	107.248	6	4.476	80	59.68
7	9.384	85	113.951	7	5.222	85	63.41
8	10.725	90	120.654	8	5.968	90	67.14
9	12.065	95	127.357	9	6.714	95	70.87
10	13.406	100	134.048	10	7.460	100	74.60
11	14.747	110	147.47	11	8.206	110	82.06
12	16.087	120	160.87	12	8.952	120	89.52
13	17.428	130	174.28	13	9.698	130	96.98
14	17.768	140	187.68	14	10.444	140	104.44
15	20.109	150	201.09	15	11.190	150	111.90
16	21.450	160	214.50	16	11.936	160	119.36
17	22.790	170	227.90	17	12.682	170	126.82
18	24.131	180	241.31	18	13.428	180	134.28
19	25.471	190	254.71	19	14.174	190	141.74
20	26.812	200	268.12	20	14.920	200	149.20
22	29.493	220	294.93	22	16.412	220	164.12
24	32.174	240	321.74	24	17.904	240	179.04
26	34.856	260	348.56	26	19.396	260	193.96
28	37.537	280	375.37	28	20.888	280	208.88
30	40.218	300	402.18	30	22.380	300	233.80
32	42.899	325	435.69	32	23.872	325	242.45
34	45.580	350	469.21	34	25.364	350	261.1
36	48.261	400	436.24	36	26.856	400	298.4
38	50.943	450	603.27	38	28.348	450	335.7
40	53.624	500	670.30	40	29.840	500	373.0
42	56.305	600	804.36	42	21.332	600	447.6
44	58.986	700	938.42	44	32.824	700	522.2
46	61.667	800	1072.48	46	34.316	800	596.8
48	64.349	900	1206.54	48	35.808	900	671.4
50	67.030	1000	1340.60	50	37.300	1000	746.0

Source: Courtesy of Square D.

Note: Conversions not given may be obtained by adding appropriate values given.

TABLE 18-6 IMPROPER VOLTAGE LEVELS AFFECT PERFORMANCE OF PLANT EQUIPMENT

Motors	Torque reduced 19% by 10% undervoltage, and motor temperature increases and shortens life of motor.	Starting current (inrush current causing voltage dips) is increased 12% and power factor decreased 5% when 10% overvoltage exists.
Rectifier loads: Electroplaters, battery chargers, static dc supplies for cranes, dc motor drives, magnetic chucks, precipitators	With a 10% undervoltage electroplating deposition rate drops 10 to 20%; battery charging rate falls 15 to 25%; precipitator cleaning power drops 20%; magnetic chuck holding power is reduced 19%.	Metallic rectifiers withstand 50% less transient surge when operated at 10% overvoltage.
Magnetic devices: Solenoids for clamping and ejecting, vibrating feeders, magnetic brakes, solenoid valves, motor starter contactors, ac relays	Solenoids take longer to open a valve, eject a part, close a relay, or close a starter. Holding power of relays varies as the square of the voltage, and at 10% undervoltage is so reduced that vibration or minor voltage dips will drop out contactors.	Wear and distortion of solenoid surfaces is substantially greater. Saturation of solenoid cores with associated drastic increase of operating current and heating takes place.

Source: Courtesy of Square D.

QUESTIONS

1. How efficient are transformers?
2. What is a buck and boost transformer?
3. Why are eddy currents undesirable in a transformer?
4. What is hysteresis?
5. How are hysteresis losses minimized?
6. How are transformer copper losses minimized?
7. Why were PCBs used in transformers?
8. What took place October 1, 1990 that affected transformers?
9. What is a third harmonic?
10. Why are dry-type transformers better in some cases than oil-filled?
11. How can a transformer's capacity be increased by 33%?
12. How are epoxy transformers used?
13. Which is the primary of a transformer?
14. What is induced voltage?
15. What is mutual inductance?
16. If you know the secondary voltage, primary voltage, and the number of turns in the primary, how do you find the number of turns in the secondary of a transformer?
17. What is an autotransformer?
18. What limits the use of autotransformers?
19. What generates the greatest losses in a transformer?
20. What circulates through tube coolers?

REVIEW PROBLEMS

Transformers are a very important part of any electrical system. They are utilized to make sure that the correct voltage and current are available where needed. A quick review of the turns ratio and some of the factors related to transformers will be useful.

1. When connected to a 60 Hz 120-V circuit, a 24-V transformer delivers up to 4 A to a 24-V solenoid. How much current does the transformer demand from the primary circuit when the full 4 A flows in its secondary?
2. How much current flows in the primary of a transformer whose primary is connected to a source of 120 V and whose secondary provides 12 A at 12 V? Assume 98% efficiency for the transformer.
3. If the turns ratio of a step-up transformer is 1:5, what is the voltage of the secondary if the primary is connected to a source of 120 V, 60 Hz?
4. A 16-V transformer delivers 1.0 A to a chime. If the primary voltage is 120, what is the current through the primary?
5. How much voltage do you get from a 64-VA transformer if the current rating of the secondary is 4 A?
6. A step-up transformer has 600 V output at 250 mA. The input voltage is 120. What is the primary cur-

rent needed to produce the 250 mA in the secondary?

7. If a transformer with 600 turns on the primary and operates at 120 V ac, what is the turns-per-volt value needed for determining the secondary windings to obtain a specific voltage from the secondary?
8. A transformer has 600 turns in the primary and uses 120 V for the input voltage. What would be the output voltage if the number of turns in the secondary is 300?
9. If a 660-V ac power source is connected to a transformer with 1800 turns in the primary and 300 turns in the secondary, what would be the resultant voltage available at the secondary?
10. If an autotransformer has 1800 turns on the entire winding but is tapped at 300 turns, what would be the output voltage when 240 V is applied to the primary or the entire 1800 turns?

19

Power Generation

Objectives

After studying this chapter, you will be able to:

1. Draw a sine wave.
2. Identify parts of an ac generator.
3. Draw output waveforms for single-phase, two-phase, and three-phase generators.
4. Explain why exciters are needed.
5. Draw wye-to-wye connections.
6. Describe how output frequency is determined for an alternator.
7. Draw a transfer switching circuit.
8. Describe how an automatic transfer switch operates.
9. Explain the difference between the two types of UPS systems.
10. Describe systems for paralleling multiple power sources.
11. Define cogeneration.
12. Explain peak-load shaving.

The basic principle for the generation of an EMF in an ac generator or alternator is the same as in a dc generator. The generation of an EMF in an armature conductor depends solely on a relative motion between the conductor and the magnetic field. Two constructions are possible. The magnetic field may be stationary and the armature may rotate. In this case, the magnetic field is called the stator and the armature is called the rotor. Or, the magnetic field may rotate and the armature may be stationary, in which case the magnetic field is called the rotor and the armature is called the stator.

In almost all dc generators, the field is stationary and the armature is rotated. But in almost all ac generators, the armature is stationary and the field is rotated. The latter type of construction provides some advantages. A rotating armature requires slip rings to carry current to the external load. Such rings are difficult to insulate. They are frequent sources of trouble, often causing open and short circuits. A stationary armature needs no slip rings. Thus armature leads can be continuously insulated conductors from the armature coils to the busbars. It is more difficult to insulate conductors in a rotating armature than in a stationary armature because of the centrifugal force that results from rotation. Also, the stationary armature allows alternators to operate with higher voltages than those in dc generators.

Inasmuch as an ac motor will not operate without a source of alternating current, it is important that a closer examination of the device that provides the power for the operation of all ac motors be made. Figure 19–1 shows a simplified drawing of an ac generator. Certain features of this generator are basic to the

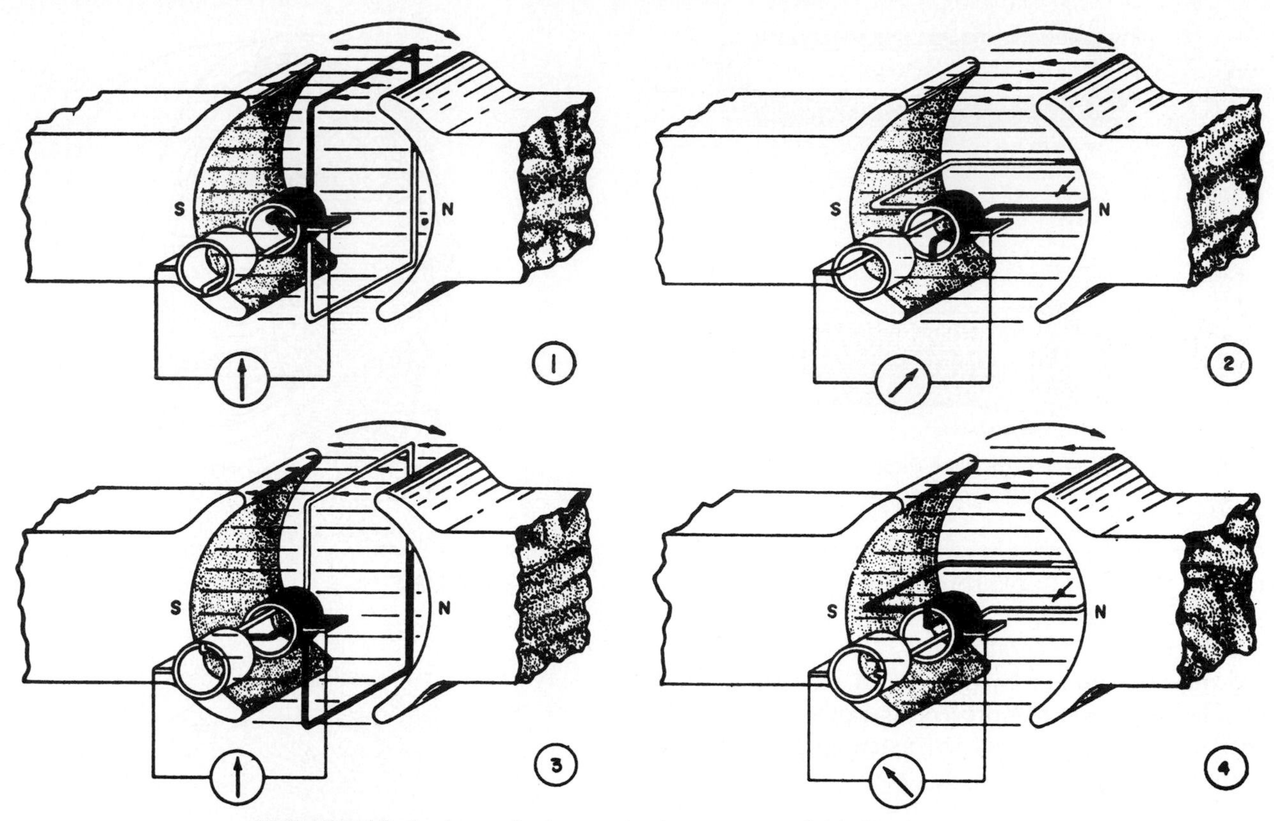

FIGURE 19–1 *Loop of wire rotating in a magnetic field. This basic alternator produces the waveform shown in Fig. 19–2.*

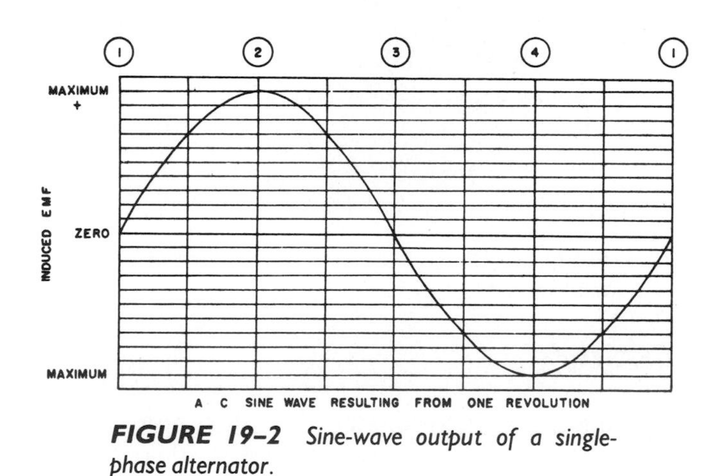

FIGURE 19–2 *Sine-wave output of a single-phase alternator.*

design of all ac generators. Note how the loop of wire rotating in a magnetic field creates the sine-wave output.

Sine Wave. The voltage produced by this generator is an alternating voltage. One complete revolution of the coil produces 1 Hz of voltage. That is, the voltage builds up from zero to a maximum, then drops to zero, then builds up again in the opposite direction to a maximum. Finally, the voltage drops to zero to complete the hertz or cycle. Such a hertz of alternating current or voltage is represented by a sine wave (Fig. 19–2).

COMPONENTS OF THE ALTERNATOR

Rotors

The rotating part of the generator is called the rotor. The field of the ac generator is placed on the rotor. It is either a salient-pole type or a turbo type (Fig. 19–3).

FIGURE 19–3 *Salient-pole rotor for an alternator.*

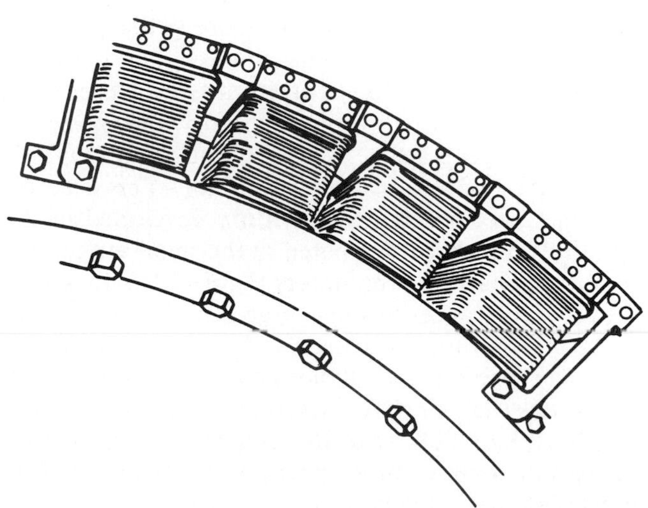

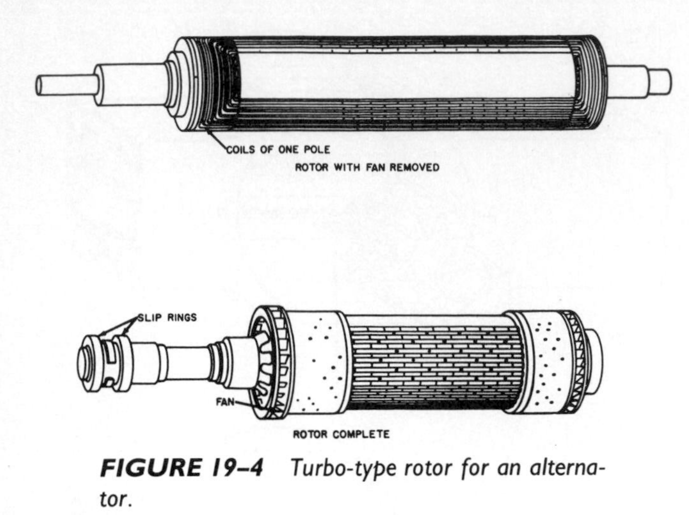

FIGURE 19–4 *Turbo-type rotor for an alternator.*

Compare the salient-pole rotor and Fig. 19–4, which has a turbo-type rotor.

When the ac generator is to be driven by a slow-speed diesel engine or by a water turbine (up to 720 rpm), the salient-pole or projecting-pole rotor is used. The field poles are formed by fastening a number of steel laminations to a spoked frame or spider. The heavy pole pieces produce a flywheel effect on the slow-speed rotor. This helps to keep the angular speed constant. It also reduces variation in the voltage and frequency of the generator output. In high-speed alternators (up to 3600 rpm), the smooth-surface turbo-type rotor is used for two major reasons: (1) it has less air-friction (heating) loss, and (2) the windings can be placed so that they can withstand the centrifugal forces developed at high speeds. Turbo-type rotors are a solid-steel forging, a number of steel disks fastened together with the field coils locked in slots. These field coils are usually placed so that they distribute the field flux evenly around the rotor, as shown in Fig. 19–4.

Stators

In a rotating-field ac generator, the armature windings are stationary and are therefore the stator. The armature iron, being in a moving magnetic field, is laminated to reduce eddy current losses. A typical ac generator stator is shown in Fig. 19–5. In high-speed turbo-type generators, the stator laminations are ribbed to provide sufficient ventilation. This is necessary because the high temperature developed in the windings cannot be dissipated in the small air gap between the rotor and the stator. Figure 19–6 illustrates the close tolerance. In some large installations, the alternators are totally enclosed and cooled by hydrogen gas under pressure, which has greater heat-dissipating properties than air. Stator coils in high-speed alternators must be well braced. Bracing prevents coils from being pulled out of place when the alternator is operating with heavy loads.

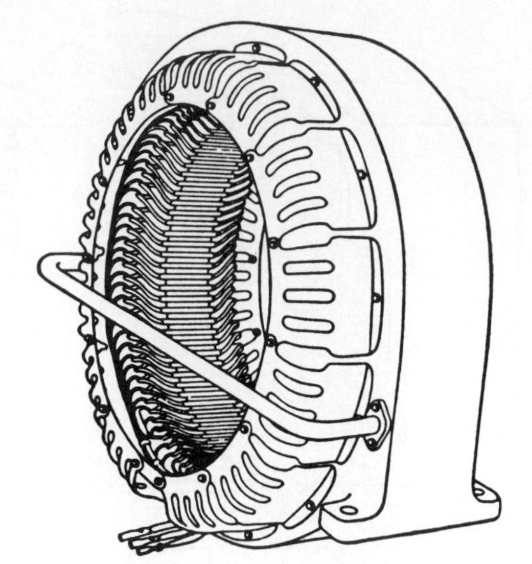

FIGURE 19–5 *Typical alternator stator.*

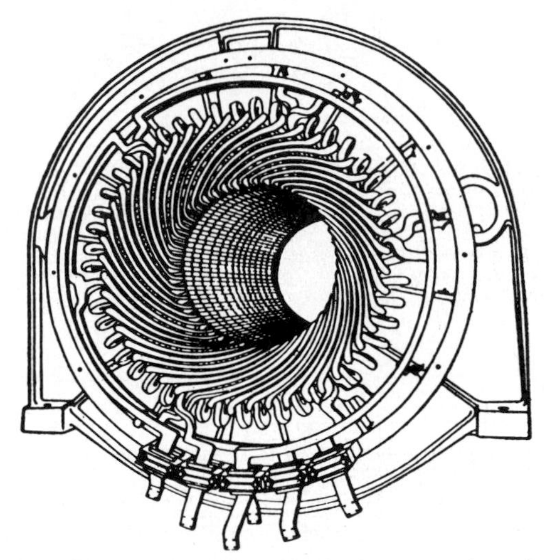

FIGURE 19–6 *Note how close together the windings are.*

Exciters

Like many dc generators, ac generators need a separate dc source for their fields. This dc field current must be obtained from an external source called an exciter. The exciter used to supply this current is usually a flat, compound-wound dc generator designed to furnish from 125 to 250 V. The exciter armature may be

FIGURE 19–7 *Exciter armature and generator field mounted on the same shaft.*

FIGURE 19–8 *Brushless rotor.*

mounted directly on the rotor shaft of the ac generator, or it may be belt driven. Figure 19–7 shows an exciter armature and generator field mounted on the same shaft.

Brushless exciters are also used to provide the dc fields. The brushless exciter is an ac generator that converts the ac power to dc. It does so by means of a diode rectifier assembly which is attached to, but insulated from, the generator shaft. The brushless exciter has no friction-producing parts, such as brushes, brush holders, commutator, or slip rings. It needs very little maintenance (Figure 19–8).

Static Exciters. Another method of field excitation commonly used is the static exciter. It is called a static exciter because it contains no moving parts. A portion of the ac current from each phase of the generator output is fed back through a system of transformers, rectifiers, and reactors to the field windings as dc excitation current. With this system an external source of dc current is necessary for initial excitation of the field windings. On engine-driven generators, the initial "field flash" may be obtained from storage batteries, which are also used to start the engine.

Frame and Shaft

The frame and shaft of an alternator serve the same purpose as in the dc generator. The frame completes the magnetic circuit of the field. It also supports the parts and windings. The shaft on which the rotor turns is supported by the end bells or end frames.

TYPES OF ALTERNATORS

Single-Phase

This type of alternator is seldom used except for special purposes. It is used as an emergency generator and for construction crews. As a rule, this type of alternator is low powered and self-excited. Figure 19–9 shows a typical alternator. Its construction is similar to a dc generator with an auxiliary ac winding on the dc armature. The dc winding on the rotating member is of the usual lap or wave-type construction. The winding is connected to the commutator bars in the usual way.

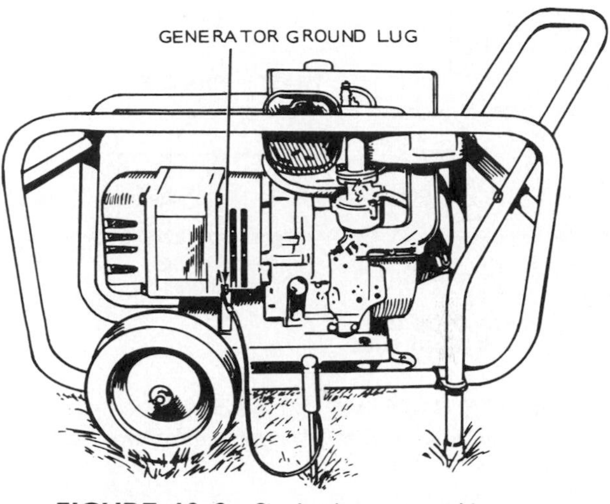

FIGURE 19–9 *Single-phase, portable generator.*

The dc winding output provides the current for the dc field excitation and other dc power applications. A second open wave winding is laid in the slots of the rotating member on top of the dc winding. This second winding, which is connected to slip rings, supplies the ac output.

Two-Phase

This type of alternator is called a multiphase or polyphase ac generator. It has two or more single-phase windings symmetrically spaced around the stator. In a two-phase alternator, there are two single-phase windings physically spaced so that the ac voltage induced in one is 90° out of phase with the voltage induced in the other. When one winding is being cut by maximum flux, the other is being cut by no flux.

Figure 19–10 shows a schematic diagram of a two-phase, four-pole alternator. This stator consists of two single-phase windings (phases) separated from each other. Each phase consists of four windings. These windings are connected in series so that their voltages add. The rotor is identical to the rotor used in the single-phase alternator. Note in Fig. 19–10B the waveforms produced by this type of generator.

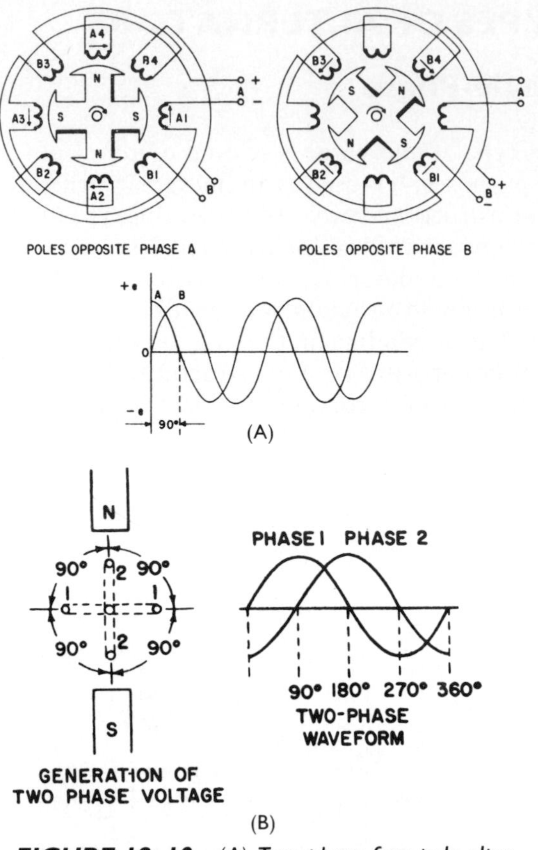

FIGURE 19–10 (A) Two-phase four-pole alternator and its waveform; (B) generation of two-phase voltage.

FIGURE 19–11 (A) The two phases of a two-phase alternator can be connected together to produce single-phase power; (B) two-phase power brought out with all three connections available for connection to consuming devices; (C) waveforms produced by a two-phase alternator with all three connections brought out separately.

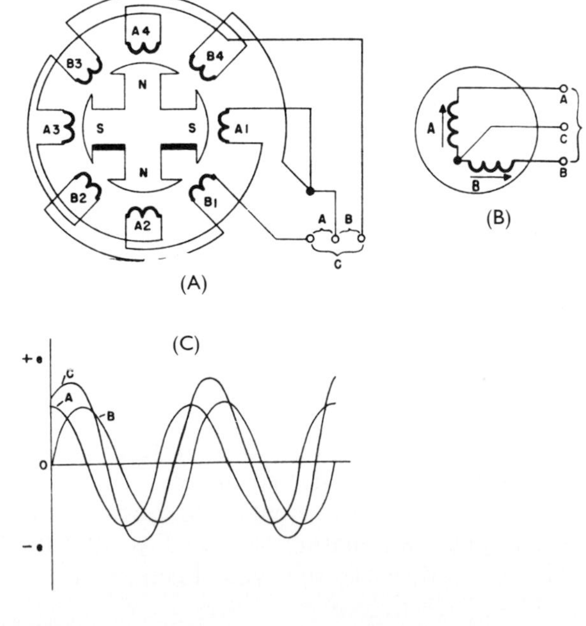

The two phases of a two-phase alternator can be connected together as shown in Fig. 19–11. Only three leads are brought out from the generator for connection to a load. This type of power is seldom used. However, in some parts of Europe this type of power is available for home and commercial use. It has an advantage when starting motors: ac motors that use two-phase power do not need a start winding or a switch to remove the windings from the circuit once the motor has reached operating speed, as is the case with single-phase motors.

Three-Phase

As the name indicates, this type of alternator has three single-phase windings. These windings are spaced so that the voltage induced in each winding is 120° out of phase with the voltage in the other two windings. A schematic diagram of a three-phase generator showing all three coils is complex. Figure 19–12A shows how some of the various load options can be connected on three-phase power.

Figure 19–12B shows the output waveform of the alternator. Note that there are 120° of separation between each phase of the output. Electrical power

FIGURE 19–12 (A) Various load options for three-phase alternator; (B) generation of three-phase voltage and the resultant waveforms.

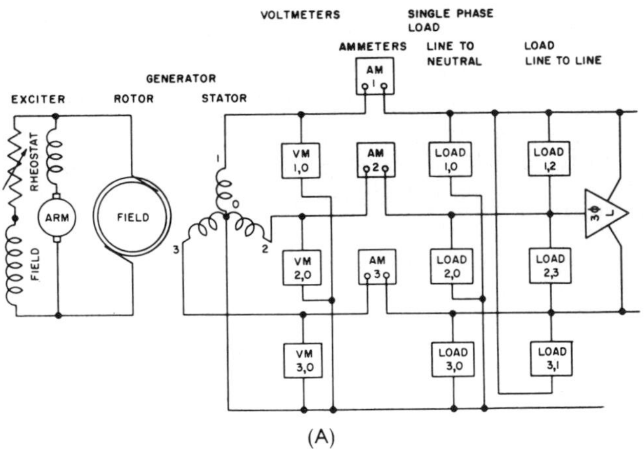

(B)

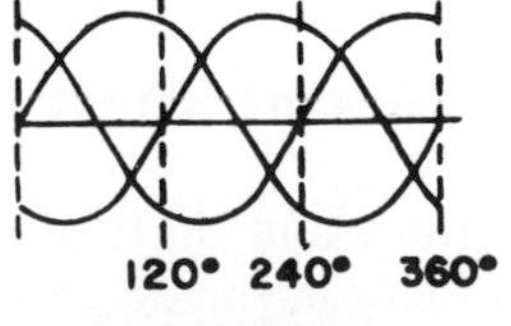

N
120°
120°
S
GENERATION OF
THREE-PHASE VOLTAGE

generated by power companies for use in homes and business is all produced as three-phase. The three phases are then divided by three separate transformers into single-phase power for three different subdivisions or three different customers. Some three-phase power is used by businesses to drive large motors. Three-phase motors do not require as much maintenance as single-phase motors.

WYE AND DELTA CONFIGURATIONS

Wye Connection

Instead of six leads coming out of the three-phase alternator, one of the leads from each phase may be connected to form a common junction. The stator is then wye- or star-connected. Figure 19–13 shows a wye connection. The common lead may or may not be brought out of the machine. If it is brought out, it is called the *neutral*. One advantage of the neutral is the balancing of the load between or among all coils. The neutral serves as a common return circuit from all three phases. It maintains a voltage balance across the loads. No current flows in the neutral when the loads are balanced. The three-phase, four-wire system is widely used in industry and for aircraft ac power systems.

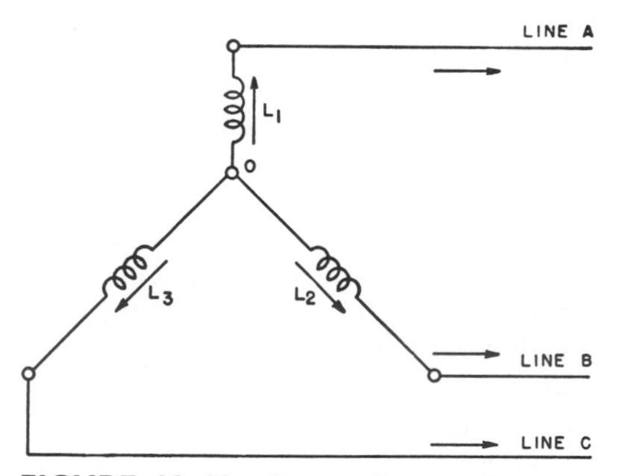

FIGURE 19–13 *Current flow in three-phase windings, wye-connected.*

Delta Connection

A three-phase stator may also be connected in a delta configuration. In a delta-connected alternator, one phase winding and the start end of another are connected to the finish end of the third. The start end of the third is connected to the finish end of the first. The three junction points are connected to the line wires leading to the load. Figure 19–14 shows a delta connection. When the generator phases are properly connected in delta, no appreciable current flows within the delta loop when there is no external load connected to the alternator. If any one of the phases is reversed with respect to its correct connection, a short-circuit current flows within the windings on no-load. This causes damage to the windings.

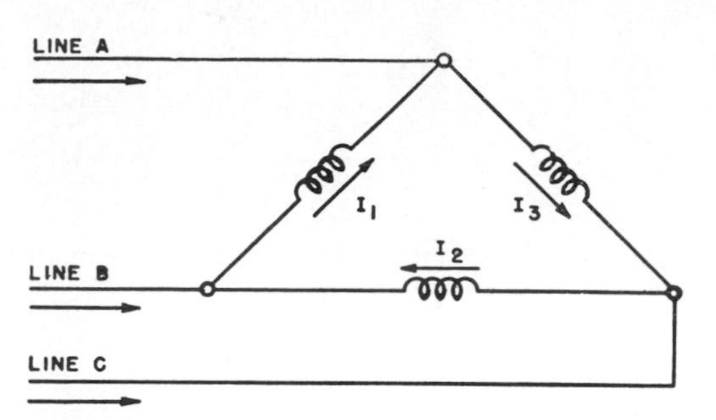

FIGURE 19–14 *Current flow in a delta-connected, three-phase alternator windings.*

To avoid connecting a phase in reverse, it is necessary to test the circuit before closing the delta. This is done by connecting a voltmeter or fuse wire between the two ends of the delta loop before closing the delta. The two ends of the delta loop are never connected if there is an indication of any appreciable current or voltage between them when no load is connected to the alternator.

Power in a Balanced Wye

The power delivered by a balanced three-phase wye-connected system is equal to three times the power delivered by each phase. The total true power is

$$P_t = \sqrt{3} \times E_{\text{phase}} \times I_{\text{phase}} \times \cos \angle\theta$$

Since

$$E_{\text{phase}} = \frac{E_{\text{line}}}{\sqrt{3}} \quad \text{and} \quad I_{\text{phase}} = I_{\text{line}}$$

the total true power is

$$P_t = \frac{E_{\text{line}}}{\sqrt{3}} \times I_{\text{line}} \times \cos \angle\theta$$

Power in a Balanced Delta

The power delivered by a balanced three-phase delta-connected system is also three times the power delivered by each phase (Fig. 19–15):

$$E_{\text{phase}} = E_{\text{line}} \quad \text{and } I_{\text{phase}} = \frac{I_{\text{line}}}{\sqrt{3}}$$

The total true power is

$$P_t = 3\,E_{\text{line}} \times \frac{I_{\text{line}}}{\sqrt{3}} \times \cos \angle\theta$$

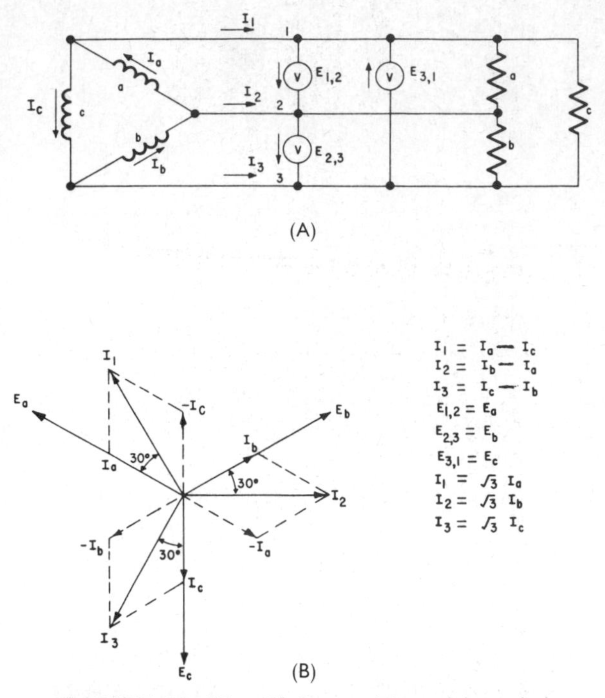

FIGURE 19–15 *(A) Power delivered by a balanced three-phase, delta-connected system; (B) current and voltage relationships in a three-phase alternator.*

Thus the expression for three-phase power delivered by a balanced delta-connected system is the same as the expression for three-phase power delivered by a balanced wye-connected system.

FREQUENCY

The frequency of the alternator voltage depends on the speed of rotation of the rotor and the number of poles. The higher the frequency needed, the faster the alternator must turn. The lower the speed, the lower the frequency. The more poles on the rotor, the higher the frequency for a given speed. When a rotor has rotated through an angle such that two adjacent poles (a north and a south pole) have passed one winding, the voltage induced in that winding has varied through one complete cycle or hertz.

$$f = \frac{P}{2} \times \frac{N}{60} = \frac{PN}{120}$$

P = number of poles,
N = speed in rpm, and
f = frequency.

LOAD CHANGES

When the load of an alternator is changed, the terminal voltage carries the load. The amount of variation depends on the design of the generator and the power factor of the load. With a load having a lagging power factor (one with inductance dominating), the drop in terminal voltage with increased load is greater than for unity (1.00) power factor (that is, a totally resistive load). With a load, a power factor that is leading the terminal voltage tends to rise. The causes of terminal voltage changes with load changes are the armature resistance, armature reactance, and armature reaction.

Armature Resistance

When current flows through a generator armature winding, there is an *IR* drop (voltage drop) due to the resistance of the winding. This drop increases with load. Thus the terminal voltage is reduced. The armature resistance drop is small because the resistance is low.

Armature Reactance

The armature current in an alternator varies approximately as a sine wave. The continuously varying current in the generator armature is accompanied by an IX_L voltage drop in addition to the *IR* drop. Armature reactance in an alternator may be from 30 to 50 times the value of armature resistance. This is because of the relatively large inductance of the coils in comparison with its resistance.

Armature Reaction

When an alternator supplies no load, the dc field flux is distributed uniformly across the air gap. When an alternator supplies a reactive load, however, the current flowing through the armature conductors produces an armature magnetomotive force (MMF). That force influences the terminal voltage by changing the magnitude of the field flux across the air gap. When the load is inductive, the armature MMF opposes the dc field and weakens it. Thus the terminal voltage decreases. When a leading current flows in the armature, the dc field is aided by the armature MMF. The flux across the air gap is increased. Thus the terminal voltage increases.

VOLTAGE REGULATION

Voltage regulation of an alternator is the change of voltage from full-load to no-load. This is expressed in percentage of full-load volts with a constant speed and dc field current. For example, the no-load voltage of an alternator is 250 V. Its full-load voltage is 220 V. What is the percent of regulation?

$$\frac{250 - 220}{220} = 13.6\%$$

STANDBY OR EMERGENCY POWER SOURCES

The loss of power can be very dangerous—catastrophic in some cases. When the power goes off it can interrupt ventilation fans, water pumps, milking machines, mechanical feeders, fallout shelters, refrigeration, furnace controls, and other vital modern production equipment that requires continuous electric service.

Storms, accidents, and equipment breakdown can all cause an interruption of electrical service. If a power outage lasts for any length of time, serious problems, such as animal suffocation in windowless animal shelters, food spoilage, frozen water pipes, or loss of production, can result. An ever-increasing dependence on a constant supply of electrical power causes increased interest in standby equipment for the generation of electricity.

Farms, hospitals, schools and businesses, and industry are all interested in a constant supply of electrical energy. With the widespread use of computers in industry, commerce, and education, the lack of a dependable source of electricity is even more important.

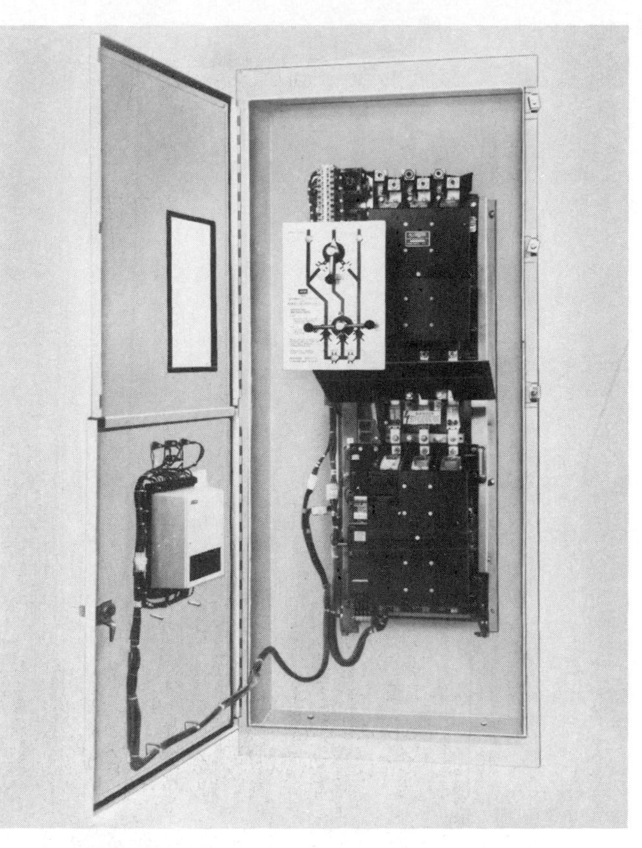

FIGURE 19–17 Automatic transfer and bypass isolation switch. (Automatic Switch)

Transfer Switch

One of the most important parts of an emergency power source is the transfer switch. The *National Electrical Code®* requires that a standby generator be connected so as to prevent the inadvertent interconnection of the two power sources. A double-pole, double-throw switch is usually installed between the power supplier's meter and the service entrance. If current transformers are used for metering, a pole-top transfer switch may be used. Figure 19–16 shows the typical transfer switch wiring.

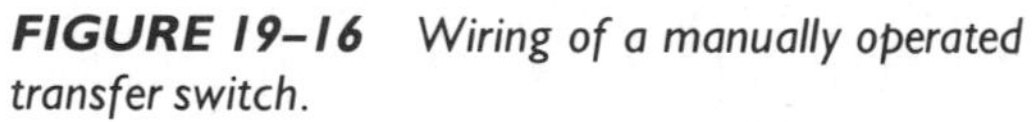

FIGURE 19–16 Wiring of a manually operated transfer switch.

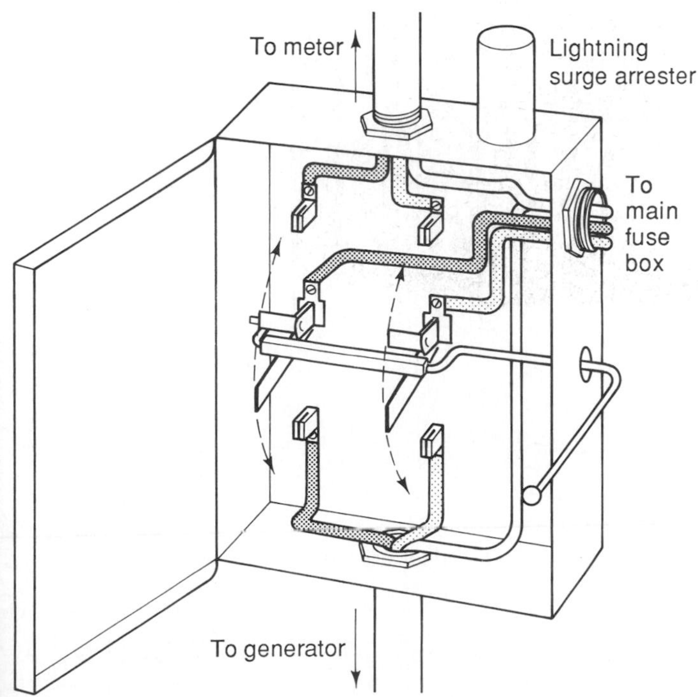

The use of a double-pole, double-throw switch prevents power from feeding back into the power supplier's line and endangering the lives of linemen who may be working to restore power. It also prevents accidental energizing of the farm, business, hospital, or other system and consequent burnout of the generator when regular power service is restored. Most standby equipment guarantees are voided if the transfer switch is not used.

The transfer switch shown in Fig. 19–16 is used at a home or farm. It shows the basic method used to transfer power from a generator to a home or farm system and then back to the power company's lines. Figure 19–17 shows the automatic transfer and bypass isolation switch.

Types of Standby Generators

Standby generators can be divided into two types: engine driven and tractor driven. (Fig. 19–18). Tractor-driven units can be stationary or portable, as a trailer-

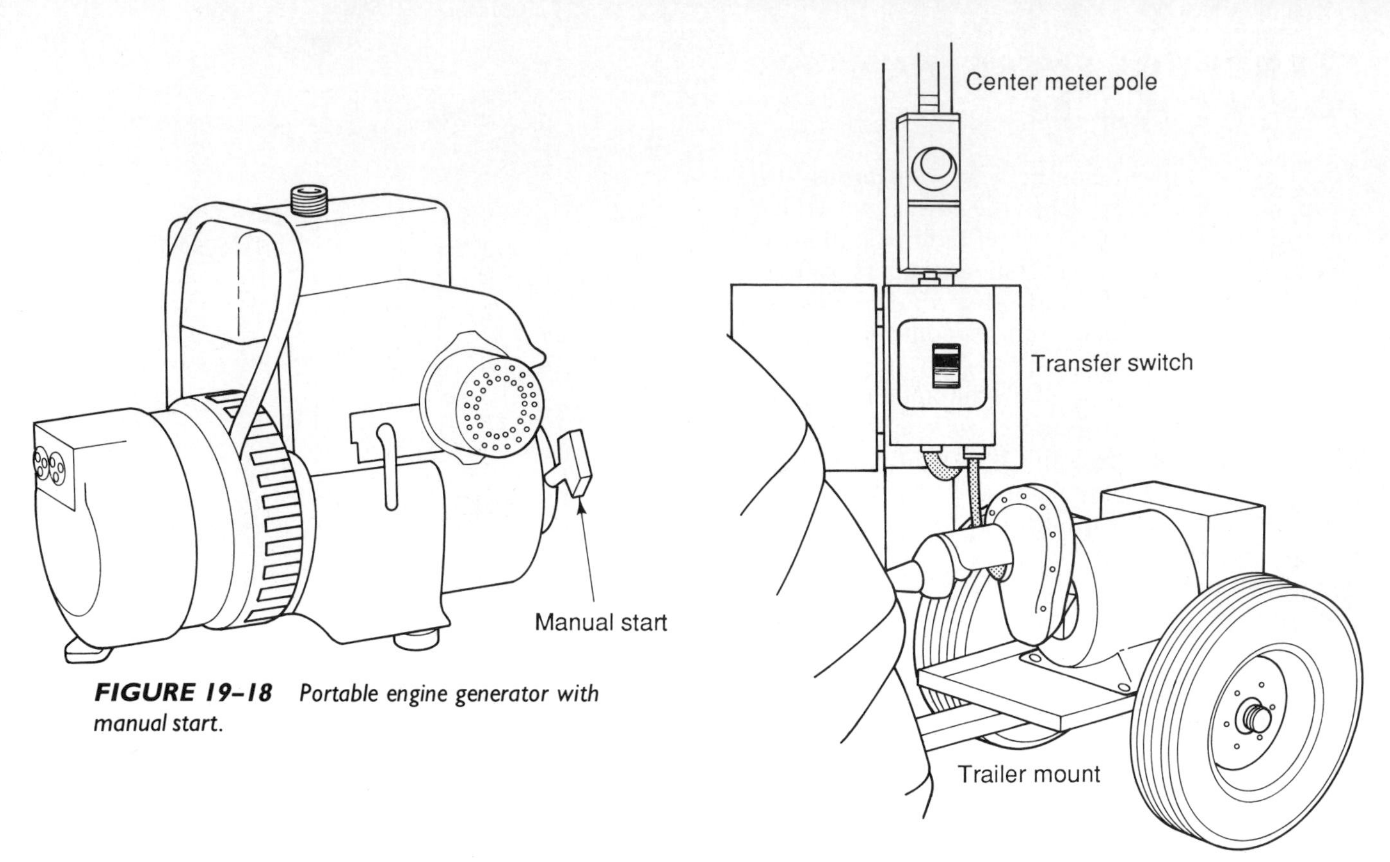

FIGURE 19–18 *Portable engine generator with manual start.*

FIGURE 19–19 *Tractor-driven generator.*

FIGURE 19–20 *Typical indoor installation for a large diesel-engine system.*

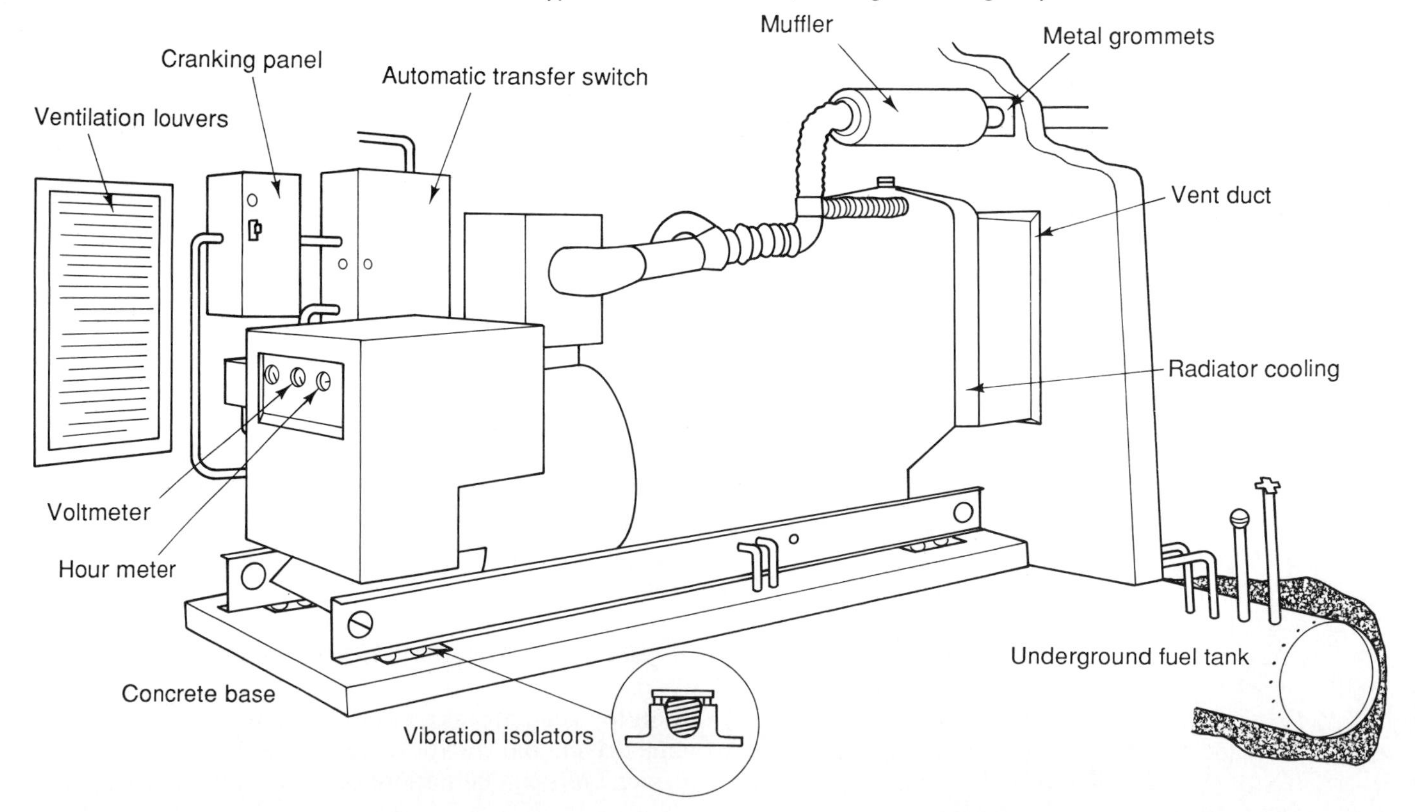

mounted unit (Fig. 19–19). Engine-driven units can also be stationary or portable and can be either manual start or automatic start. Standby generators are available to operate either as single- or three-phase. Some units are wired with four lead wires so that they can be operated either single- or three-phase. Generators must be matched to the power, voltage, and frequency used from the power company's lines. This is usually 120/240-V, single-phase, 60-Hz ac.

There are no standard ratings for standby generators. That means overload or maximum capacity, limited to short intermittent periods, must be considered in selecting the size of generator. Manufacturers' ratings can vary from zero overload capacity to 100%. A large overload capacity permits a smaller unit, particularly where large electric motors are involved. Figure 19–20 shows a generator design for indoor installation. Larger units usually rely on a diesel engine for power.

Automatic Transfer Switches

An automatic transfer switch can monitor the normal power source (Fig. 19–17). If there is a power outage, the switch signals the engine generator to start. When the generator reaches proper voltage and frequency, it transfers selected loads to it. When the normal source is then restored, the switch retransfers the load to it. It then shuts down the engine after a cool-down period.

At times a transfer switch handles more than its normal or continuous current rating. A reliable transfer switch must be able to handle all situations with no harmful effects on the switch. Motor starting current is one of these situations. When a motor starts up, it can draw as much as six times its running current. It can draw that much if it stalls while running. If the transfer switch operates at either of these times, it will have to interrupt that six times current. Thus the switch must be able to handle it; if not, the switch could be permanently damaged (Fig. 19–21).

All current goes through the transfer switch. Thus a short circuit on the load side will cause the maximum available current to go through the transfer switch until the circuit breaker opens the line. Therefore, the transfer switch must also withstand short-circuit loads.

Magnetic forces are so great during a short circuit that they can cause the contacts of an inadequately designed switch to open. Thus the switch must be able to lock its contacts closed until the circuit breaker operates.

Tungsten lamps can draw up to 16 times more current when they are cold. This means that a transfer switch feeding power to tungsten lamps must be able to handle this above-normal inrush current.

Motors can draw up to 15 times as much inrush current on starting as when they are running normally. The inrush of 15 times results when a motor is still running and is reconnected to a new normal or emergency source that is not synchronized with the motor. This can happen if the motor is as much as 180° out of phase with the new source (Fig. 19–22).

FIGURE 19–21 *Transfer switch, 150 A. (Automatic Switch)*

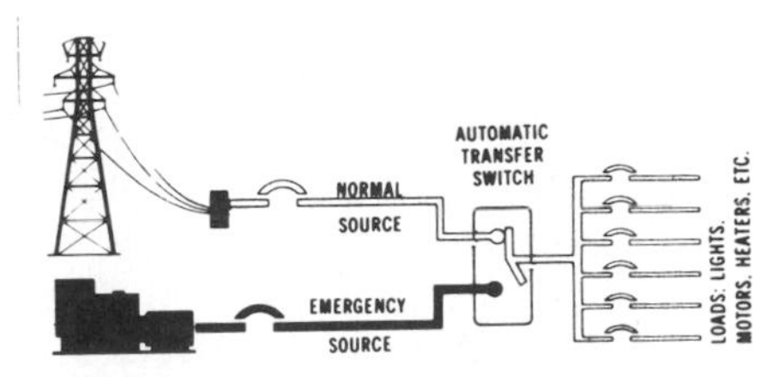

FIGURE 19–22 *The automatic transfer switch maintains power to selected loads from either the normal or emergency power source. (Automatic Switch)*

Electronic Loads

Almost every power system has some electronic load. Increased automation and the economies of solid-state control suggest that even larger percentages of power system loads will be electronic in the future. Typical examples of electronic loads include data processing equipment, intensive-care-unit monitoring equipment, numerical control machinery, and security equipment. There is virtually no commercial or industrial facility in which electronic loads are not becoming a critical part of the overall load profile.

Uninterruptible Power Systems

Electronic loads are very susceptible to voltage and frequency variations. As a typical example, real-time access computers require an ac power system that does not deviate from nominal voltage by more than +8% or less than −10%. Allowable frequency deviations are typically ± 0.5 Hz. In addition to the deviation stated, some data processing loads are affected by rate change of deviation as well. To protect computer systems, uninterruptible power systems (UPS) have been developed.

Essentially, there are two types of UPS systems: the motor-generator *flywheel set* and the *static inverter* with battery backup. Due to advances in solid-state equipment, the more commonly used UPS is now the static inverter with battery backup. This is due primarily to the lower cost per kVA and the elimination of a high-starting kVA requirement.

A solid-state UPS is made up of three major sections. The first section is the rectifier that converts the ac input into dc. The second section is a dc bus on which floats a battery system. The third section is the static inverter that converts the dc back to a clean ac sine wave. The principal function of the UPS is to isolate the protected electronic load from power deviations that would affect the operation of the connected electronic load. Any short-term outage or transient that occurs on the ac power bus is filtered out or overridden by the battery in the dc bus section providing constant power into the inverter.

Sometimes the battery supply of the UPS is required to carry the output of the inverter for extended periods of time (in excess of 3 to 5 minutes). Upon reconnection of the ac power source to the rectifier, the UPS draws power not only to carry the output load, but also to recharge the batteries.

Multiple-Engine Generator Sets

Where there are two critical loads with one more critical than the other, the three source priority load system can be used. This system supplies the priority load first, then the nonpriority load (Fig. 19–23).

The priority load system operates the same as the three-source system, with the following exceptions. The first engine generator set to reach acceptable output is connected to the emergency terminals of the automatic transfer switch 1 (ATS 1) by ATS 2, which is a six-pole, double-throw switch. The double-throw action of ATS 2 simultaneously connects the emergency terminals of ATS 3 to the other engine generator set. When that set reaches acceptable output, ATS3 transfers the nonpriority load to it. If the set carrying the priority load malfunctions, ATS 2 transfers the other set to the priority load. Other features built into the system provide optimum protection for the nonpriority load.

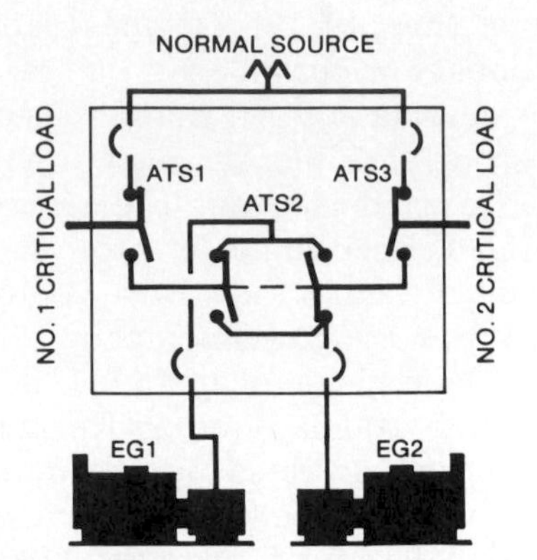

FIGURE 19–23 *Three-source priority load system. (Automatic Switch)*

Systems for Paralleling Multiple Power Sources

When it is necessary to start and parallel two or more engine generator sets to supply emergency power on a common bus, a system is needed that includes reverse power monitors, synchronizers, load sequencers, and other components to handle parallel-related operations.

A *two-engine system* is shown in Fig. 19-24. If the normal source fails or drops below acceptable levels, the controls start both engine generator sets. The first to reach adequate output is put onto the emergency bus. Assuming that ATS 1 is feeding the more critical load, it will transfer its load to the emergency bus as soon as the first engine is connected to the bus. When the other set reaches adequate output, the controls bring into synchronism with the first set, parallel it onto the same bus, and ATS 2 transfers the secondary load. Furthermore, the controls will operate to maintain power to the more critical load as long as power is available from any source.

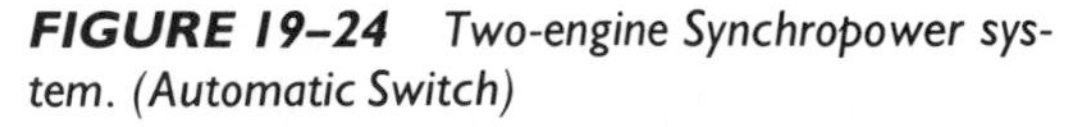

FIGURE 19–24 *Two-engine Synchropower system. (Automatic Switch)*

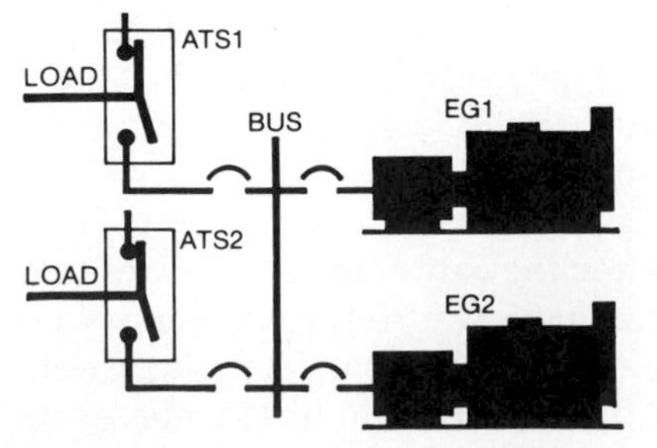

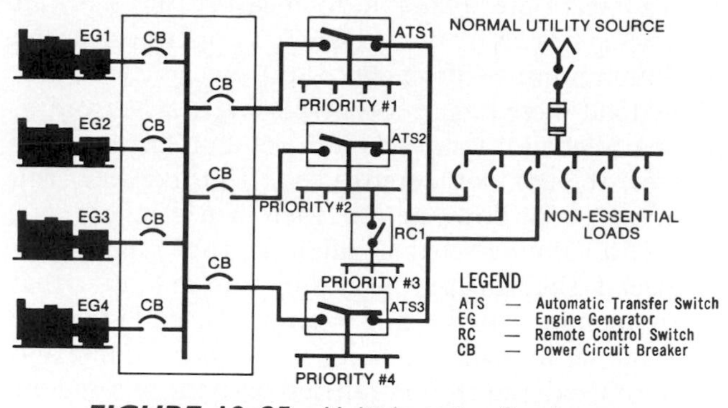

FIGURE 19–25 *Multiple-engine Synchropower system. (Automatic Switch)*

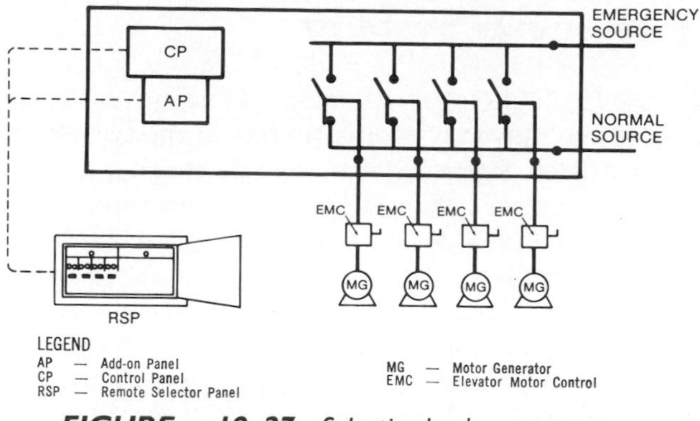

FIGURE 19–27 *Selective-load emergency power transfer system. (Automatic Switch)*

Figure 19–25 shows a *multiple-engine system.* Multiple-engine generator sets are used for various reasons, such as economy, reliability, or to minimize downtime. A multiple-engine system operates the same way as a two-engine system with additional controls to handle required number of engine generator sets. Figure 19–26 shows a power control system.

Selective Load Transfer Systems

The selective load transfer system is dependable and economical. It is also a limited supply of emergency power to be channeled to selected loads one at a time. Some of the applications for this system are: elevators, production processes, multiple-pump systems in sewage treatment plants, boiler feedwater pumps, HVAC chillers, chilled-water circulating pumps, equipment bays, and workstations.

The system is most frequently applied to elevator systems in both new construction and retrofit. In these cases, one elevator at a time can be operated when normal power fails. It uses the minimum amount of auxiliary power for operation because the standby generator can be sized for the necessary emergency load plus only one elevator. Once the first elevator reaches the main or selected floor, another elevator can be connected to the emergency source. This sequence continues until all elevators have safely been brought down—one at a time. Once the normal power is restored, all elevators automatically resume normal operation. Figure 19–27 shows a selective load emergency power transfer system.

PRIME MOVER SYSTEMS

To meet increased demands for clean and continuous reliable electric power, many installations are being supplied with on-site power generation. In some a cogeneration concept is employed where the heat from the power sources is used for heating and air conditioning.

Prime mover systems can consist of any number of engine generator sets. The number depends on the size and number of loads, and other factors, depending on individual applications. Basically, however, prime power applications fall into two categories: two-engine systems and multiple-engine systems.

FIGURE 19–26 *Power control system. (Automatic Switch)*

Two-Source Systems

Figure 19–28 shows a two-engine prime mover system. The prime mover system is similar to the two-engine generator emergency system except that typically in the prime power system either engine generator set can supply the total load. As a rule the sets are alternated on a weekly basis. That allows the idle set to be serviced easily. Changeover from one set to another can be initiated manually or automatically. The idle set is started, automatically synchronized, paralleled, and run together for a warm-up and stabilization period. Then the other set is disconnected, cooled down, and turned off. There is no interruption in power continuity during changeover.

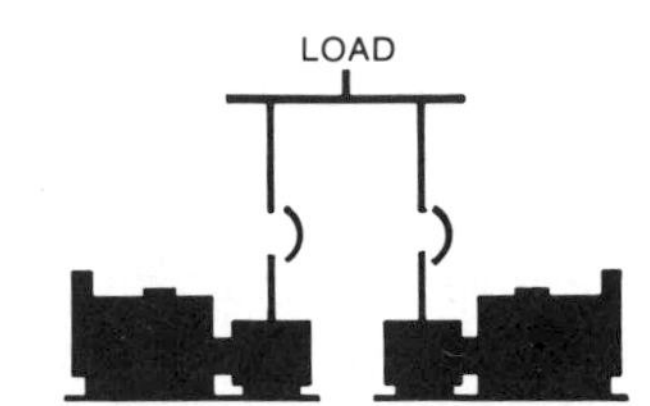

FIGURE 19–28 *Two-engine prime power system. (Automatic Switch)*

If a set malfunctions while in service, the other is started and put on the line automatically. There will be continuity of power for as long as it takes to start and connect the idle engine—usually less than 10 seconds.

Multiple-Source Systems

Figure 19–29 shows a multiple-engine generator prime power system. More than two engine generator sets may be required, for reasons mentioned previously. However, rarely do they require continuous operation of all sets on line at the same time. Load requirements vary throughout the day and with the season. Furthermore, fuel is saved and engine wear is reduced by operating only enough sets to carry the load and have some on-line reserve capacity.

FIGURE 19–29 *Multiple-engine generator set prime mover system. (Automatic Switch)*

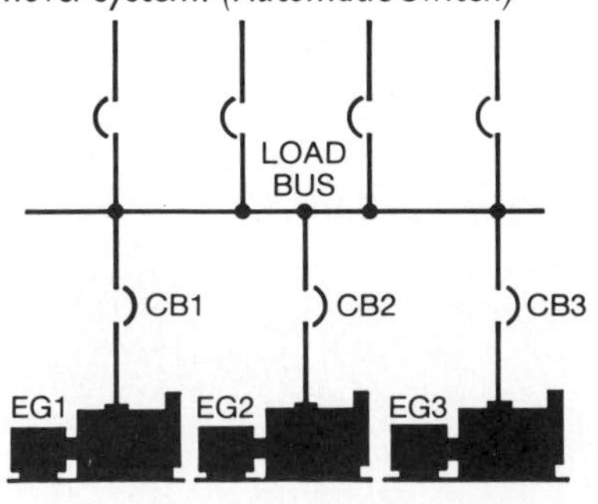

Systems are available to measure the kilowatt load being drawn from the bus. They operate only the minimum number of sets necessary to carry the load. If the load increases, the controls start, synchronize, and parallel additional sets to meet it. If the load decreases, the controls remove sets from the line, run them for a cool-down period, then turn them off.

If a running set malfunctions, the controls disconnect it, shed an appropriate amount of load so that the remaining sets are not overloaded, and start the next idle set in turn. When its output is adequate and synchronized, the set is paralleled onto the bus and the shed load is reconnected. As a rule the total time from malfunction to reconnecting the shed load will not exceed 10 seconds.

POWER MANAGEMENT SYSTEMS

A *cogeneration system* is defined as any system where a single source of thermal energy (fuel) drives two processes, the second process being driven by waste heat from the first process. Most power management systems for cogeneration usually involve the recovery and use of rejected heat (second process) produced in the generation of electricity (first process).

The electrical generation portion of the cogeneration system can operate in either of two ways: isolated from the utility or in parallel with the utility. When operating in parallel with the utility, the design must incorporate protective relaying, as required in guidelines issued by the local utility, in addition to the standard controls. Optimum cost-effectiveness, energy conversion efficiency, and system performance are achieved when the power management system operates in parallel with the utility system. These systems incorporate control functions that cause the on-line engines to produce only that amount of electricity which would provide recoverable heat to satisfy the heat load demand. Thus fuel consumption is minimized and efficiency is maximized.

Figure 19–30 shows a typical cogeneration system. Note that this diagram illustrates a cogeneration system operating in parallel with the utility. The recoverable heat may be applied to absorption chilling, as is usually the case in commercial buildings. In industrial buildings the recovered heat is more likely used for process heat.

As indicated, cogenerating systems require control functions in addition to those controls supplied as standard in a prime power system. To ensure that these system controls meet all the specialized requirements of cogeneration, including operating in parallel with the utility, contact the manufacturer of such systems during the preliminary design stages for sugges-

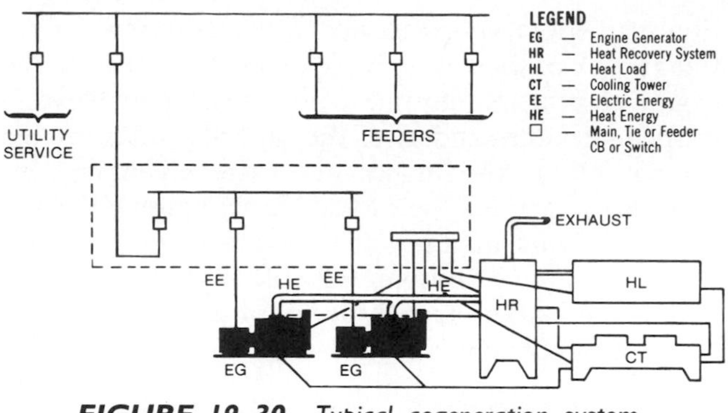

FIGURE 19-30 *Typical cogeneration system. (Automatic Switch)*

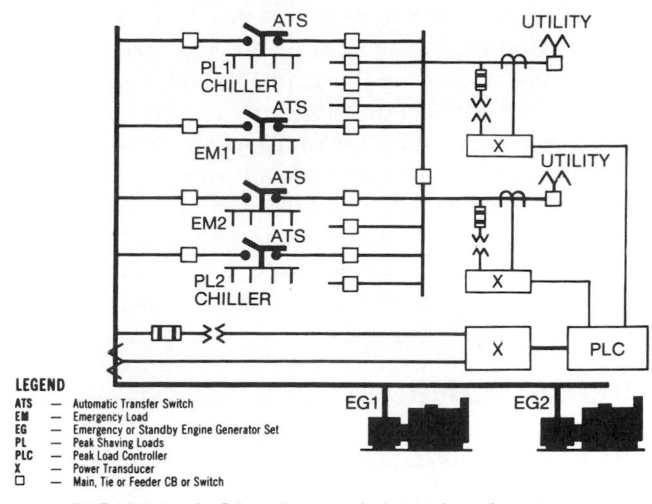

FIGURE 19-31 *Typical dedicated load-peaking transfer system. (Automatic Switch)*

tions in selecting appropriate controls and to assure proper coordination.

Peak-load shaving systems (peaking systems) allow designated building loads to be powered by on-site power generation whenever the building's total load demand exceeds a predetermined level. While an on-site generation system usually supplies emergency or standby power as its major function, such a system can be modified with additional controls to provide peaking service. These systems can be grouped into two categories: (1) isolated from the utility service, or (2) paralleled with the utility service.

Utility Isolated Peaking Transfer System

When on-site generation is isolated from the utility service, it is necessary to include transfer switches for the peaking loads (shown in PL in Fig. 19-31) when they are not the emergency loads. The transfer switches allow the designated peaking loads to draw power from either the utility or the engine generators as a function of total building demand. Only after the present level of demand is reached are the designated loads transferred to the generator sets. Thus the higher demand charges are avoided. When building demand falls below the retransfer level, the peak load is retransferred back to the utility service and the engine generators are signaled to shut down.

When two or more generator sets provide power in a power management system, generators are brought on-line only as required. When the demand limit from the utility is exceeded, the first generator is started and one of the peaking loads is transferred to it. If the limit is again exceeded, the second generator is started, synchronized, and paralleled with the first, and the second peak load is transferred. This procedure can be repeated until all on-site generation is on-line, provided that sufficient peak loads are fed by transfer switches.

Priority interrupt logic is necessary in peak-shaving transfer systems because peak loads are sometimes not the same loads as emergency loads, as indicated in Fig. 19-31. This interrupt logic automatically suspends peak shaving upon sensing loss of adequate power to emergency loads. The logic initiates the retransfer or disconnect of the peak-shaving loads from the standby source to enable the immediate transfer of the emergency loads to the standby source. The emergency loads are supplied with power without the usual startup delay of the generator sets since these sets are already up and on-line.

Utility Paralleled Peaking Systems

An optional form of peak shaving is shown in Fig. 19-32. The on-site generation is paralleled with the

FIGURE 19-32 *Typical utility paralleled peaking system. (Automaltic Switch)*

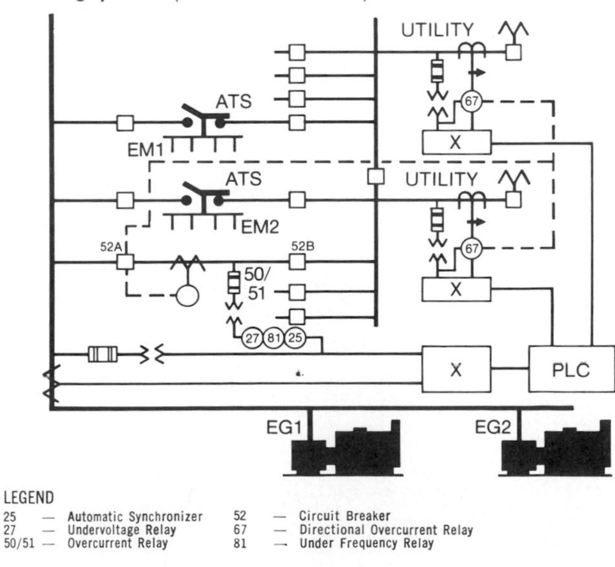

utility. In this case it is not necessary to designate particular loads as peak loads and add peak-shaving transfer switches since the utility and the standby generators share the building load.

As in the peaking transfer system, interrupt logic automatically suspends peak shaving upon loss of power to the emergency loads. The operation of the generator controls, however, is very different. The governors and voltage regulators are controlled to cause the engine generators to assume only that portion of the load in excess of the demand limit, not a fixed block of load as with peak transfer systems. The engine generators continue to assume all load in excess of the limit point and up to the capacity of the engine generator sets. Any further excess after all engine generators are carrying their rated load is assumed by the utility. In this manner, the on-site-generated kilowatthour value is minimized while kilowatt peak demand limiting is effectively accomplished. This results in the most economical mode of peaking operation.

QUESTIONS

1. How many revolutions of the alternator does it take to produce 1 Hz?
2. What type of rotor is used for a slow-speed alternator?
3. What is the purpose of brushless exciters?
4. In three-phase, what is the angular separation between phases?
5. What is the advantage of the neutral common in three-phase?
6. What determines the frequency of the alternator voltage?
7. What effect on output voltage does armature resistance have?
8. How is percent of voltage regulation of an alternator found?
9. What does a standby generator do? When?
10. What is a transfer switch?
11. What is a static inverter?
12. What does *UPS* stand for?
13. What is a cogeneration system?
14. Why is a peak-load shaving system used?
15. What is interrupt logic?
16. Why are selective load transfer systems used?
17. What is a prime mover?
18. What determines how many generator sets are needed?
19. What are the two categories of prime power?
20. How can changeover from one set to another be initiated?

REVIEW PROBLEMS

Power factor plays an important role wherever electric motors are connected to an ac line. The windings have inductance and resistance. A quick review of the way in which power factor is found mathematically will aid in the understanding of why this unit keeps showing up in the power calculations for motors. When reactance and resistance both exist in an ac circuit, the phase difference between the current and the voltage is no longer 90° as it is with reactance alone. It is some quantity less than 90°. The phase angle (θ), the angular difference between the voltage and the current, can quickly be converted to a power factor.

1. What is the phase angle of a circuit when the resistance is 100 Ω and the impedance is 100 Ω? What does this condition represent?
2. What is the phase angle of a circuit if the resistance is 100 Ω and the impedance is 150 Ω?
3. If a resistor with 500 Ω of resistance is placed in series with an inductor of 1 H on a 120-V line with 60 Hz, what is the phase angle?
4. What is the phase angle if the circuit is a series combination of resistance (1000 Ω) and inductance with an inductive reactance of 628 Ω in a circuit where 1.660 H and a frequency of 60 Hz are present?

Power Distribution Systems

Objectives

After studying this chapter, you will be able to:

1. Identify components of a power generation/distribution system.
2. List present-day commercial generating systems.
3. Describe the following radial power distribution systems: looped-primary, banked-secondary, primary-selective, secondary-selective.
4. Explain how a simple network distribution system operates.
5. Explain how fault currents are calculated.
6. Draw wattmeter schematic hookups for three-phase power circuits.
7. Explain the role of the transformer in power distribution systems.
8. Draw schematic drawings of transformer combinations used for power distribution.
9. Explain how switching transfers loads.
10. Describe a cable tray system of power distribution.

SOURCES OF COMMERCIAL AC

The use of magnetism is the most common method of generating electricity in large quantities for businesses, homes, industry, hospitals, and other institutions. A simple ac generator consists of a single loop of wire wrapped around an iron core. As the loop rotates and cuts lines of force, it has an EMF induced in it according to the strength of the magnet and the polarity of the lines being cut. This has been discussed in Chapter 19. What we are interested in here is the way the ac is generated by large devices called alternators.

Falling-Water Generators

The principle of electricity generation is the same whether it uses mechanical energy produced by gasoline or diesel engines, or falling water or nuclear energy. The Niagara Power Project is an example of falling water used to generate huge amounts of electrical energy. It is one of the world's largest facilities of this type. Total installed power at Niagara's two generating plants is 2190 MW. There are 340 miles of transmission lines to interconnect this project with another at the dam on the St. Lawrence.

The generating section of the Robert Moses plant in Niagara Falls consists of 13 unit blocks. Each of these unit blocks consists of a Francis-type hydraulic (water-driven) turbine rated at 200,000 hp at 300 feet net head. The 150,000-kW generators operate at 120 rpm, three-phase, 60 Hertz. They are mounted below the deck under removable hatch covers, as can be seen in Fig. 20–1. The crane travels along the top of them and can be positioned over the generator pits for handling the rotating parts of the turbine and generator assemblies. Transformers are located on the deck behind the generators and under what appears to be black rectangles with puffed-out pointed sides.

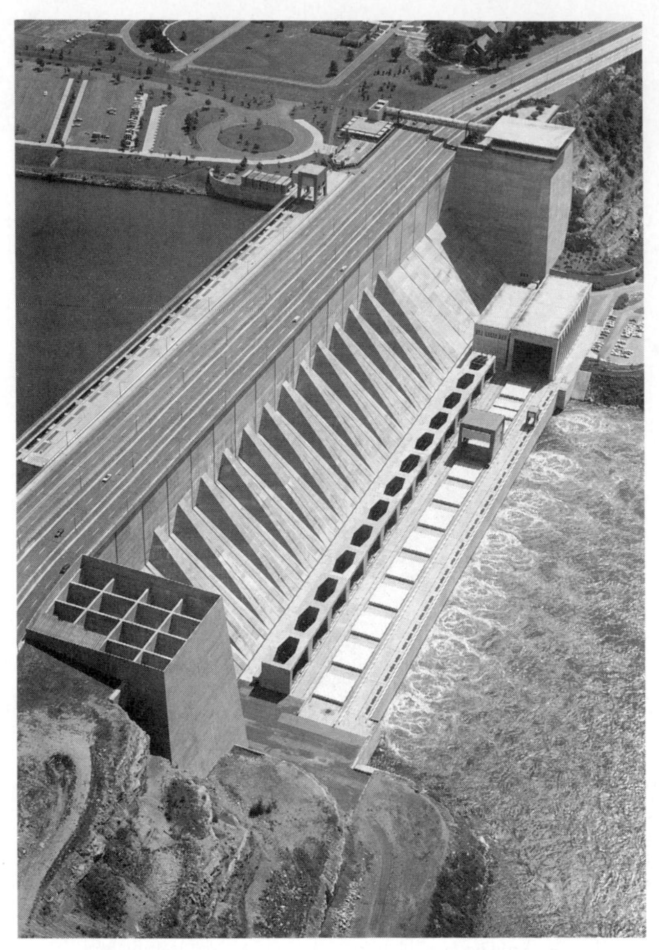

FIGURE 20–1 Robert Moses power plant, Niagara Falls, New York.

Switchyard

The Niagara switchyard is situated on a 35-acre site south of the power canal, halfway between the Robert Moses and Lewiston power plants. Purpose of the switchyard is to collect and meter power from the generators and send it out over transmission lines. The switchyard has three voltage sections: 115,000, 230,000, and 345,000 V. Power enters the switchyard from the Robert Moses plant through seven 115,000-V cable circuits and six 230,000-V cable circuits. The cables are installed in underground power tunnels. These cable circuits, 61,620 feet in length, are filled with a total of 138,000 gallons of a special insulating oil (Fig. 20–2). Electrical power leaves the switchyard at 115,000, 230,000, and 345,000 V for use in many locations in New York State and as far north as Montpelier, Vermont. The power is generated at 13,800 V and stepped up by the generators located nearby to 138,000 and 230,000 V before it is sent to the switchyard.

Nuclear Power Generators

Nuclear plants are similar to the fossil-fuel burning plants. The chief difference is in the way the heat is generated, controlled, and used to produce steam to

FIGURE 20–2 Switchyard.

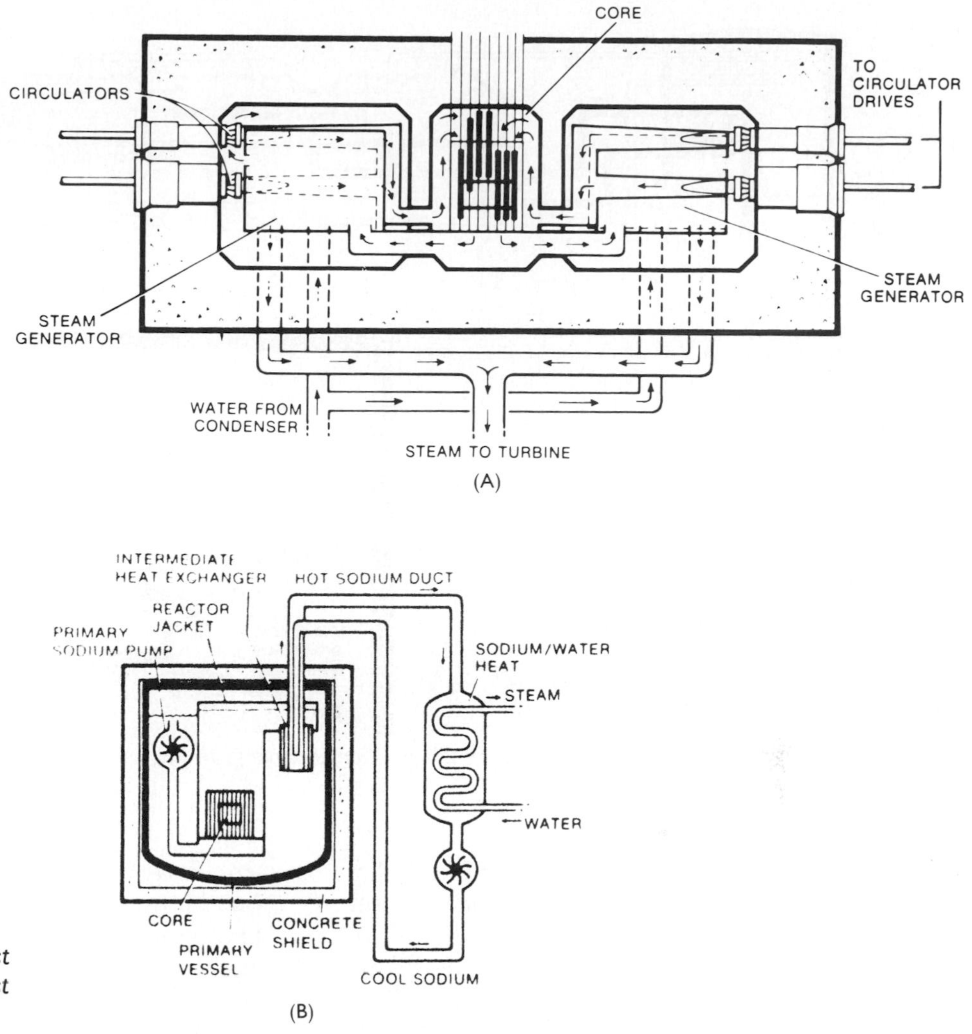

FIGURE 20–3 *(A) Gas-cooled fast breeder reactor; (B) liquid-metal fast breeder reactor.*

run turbine generators. In a nuclear power plant, the furnace for burning coal, oil, or gas is replaced by a reactor that contains a core of nuclear fuel. Energy is produced in the reactor by a process called nuclear fission. This fission process splits the center, or nucleus, of certain atoms when they are struck by a subatomic particle called a neutron. The resulting fragments, or fission products, then fly apart at great speed, generating heat as they collide with surrounding matter.

Figure 20–3A shows the gas-cooled, fast-breeder reactor used in some of today's older nuclear plants. Figure 20–3B shows the liquid-metal fast breeder reactor used in some other types of generating plants.

Fossil-Fuel Power Generators

Steam needed for driving turbines, which in turn drive electric generators, must be produced by heat. The method of heat production often becomes a rather difficult engineering problem. With the development of some dependable sources of heat from fossil fuel, former design problems have been simplified. Almost any substance may be used as a fuel. If it can be pulverized and fed into a furnace with extremely high temperatures it will burn (Fig. 20–4).

Figure 20–5 shows a steam generator (side elevation) with tangential firing system and natural circulation. Of great concern today are power plants that use inexpensive fuels that pollute the atmosphere. Sulfur contained in coal combine with the moisture in clouds to produce acid rain, which does a great deal of damage to both trees and streams.

Electricity may be generated by fossil-fuel-powered generators, by atomic-powered generators, or by the use of falling water to drive generators. It is also possible to generate power by engine-driven generators. In Alaska, for example, where long distribution lines are impractical due to ice and wind conditions that result in line damage, engine-driven generators are common.

DISTRIBUTION

Once electricity is generated in sufficient amounts for consumption in large quantities, the second necessary

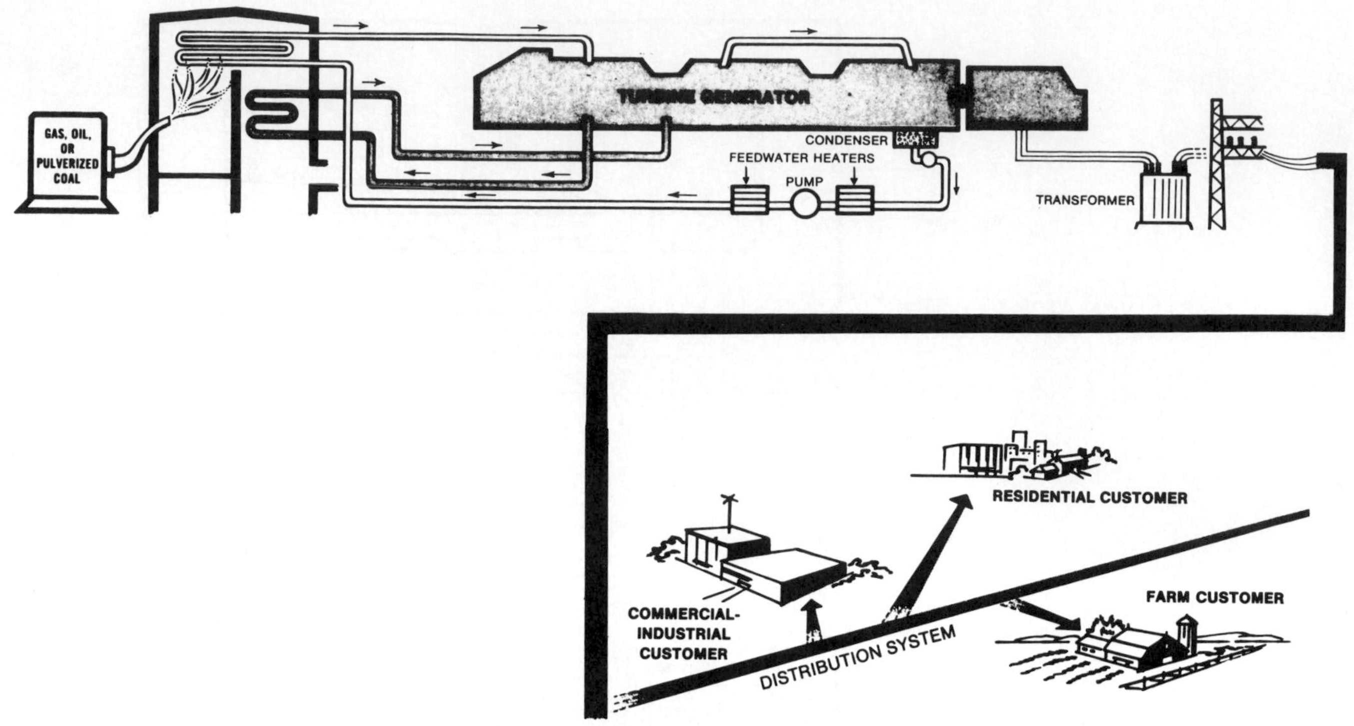

FIGURE 20–4 *Conventional fossil-fuel power plant.*

step is to get the energy to the consumer. Herein lies a distribution problem: that of stringing and maintaining long lines (Fig. 20–6).

The best distribution system is one that will most economically and safely supply adequate electric service to both present and future probable loads. The best system for a given building or industrial plant is one that takes into consideration the needs and requirements of the electrical equipment. Various types of systems have been designed to furnish the proper power when needed to different types of loads.

Types of Distribution Systems

The simplest type of system is the radial. However, there are other modifications that have specific advantages when special problems or situations are present. In the great majority of cases, power is supplied to a building at the utilization voltage. In practically all of these cases, the distribution of power within the building is achieved through the use of a simple radial system. In some cases where service is available at the building at a voltage higher than the utilization voltage, there are a number of systems that may be used.

Simple Radial System. The conventional simple radial system receives power at the utility supply voltage at a single substation and steps the voltage down to the utilization level. In some cases the utility supply is at the utilization voltage, and the substation then becomes a main distribution switchgear or switchboard.

Low-voltage feeder circuits run from the substation bus to switchgear or switchboard assemblies and panelboards that are located with respect to the loads as shown in Fig. 20–7. Each feeder is connected to the substation bus through a circuit breaker or other overcurrent protective device. Relatively small circuits are used to distribute power to the loads from the switchgear or switchboard assemblies and panelboards.

The *modern* simple-radial system is an improvement over the simple radial system. It distributes power at a primary voltage. The voltage is stepped down to utilization level in the several load areas within the building through power center transformers. The transformers are usually connected to their associated load bus through a circuit breaker (Fig. 20–8). Each power center is a factory-assembled unit substation consisting of a three-phase liquid-filled or air-cooled transformer, an integrally mounted primary switch, and low-voltage circuit breakers. Circuits are run to the load from these circuit breakers.

An improved version of the simple-radial system is the modified modern simple-radial system (Fig. 20–9). In this arrangement one primary feeder is used per transformer. The system is comparable in service continuity to the single substation form of the simple radial system. This system arrangement is usually very expensive. The cost can be materially reduced, however, by replacing automatic feeder circuit breakers with load-break switches, fused or unfused, and backing up a number of these switches with one automatic

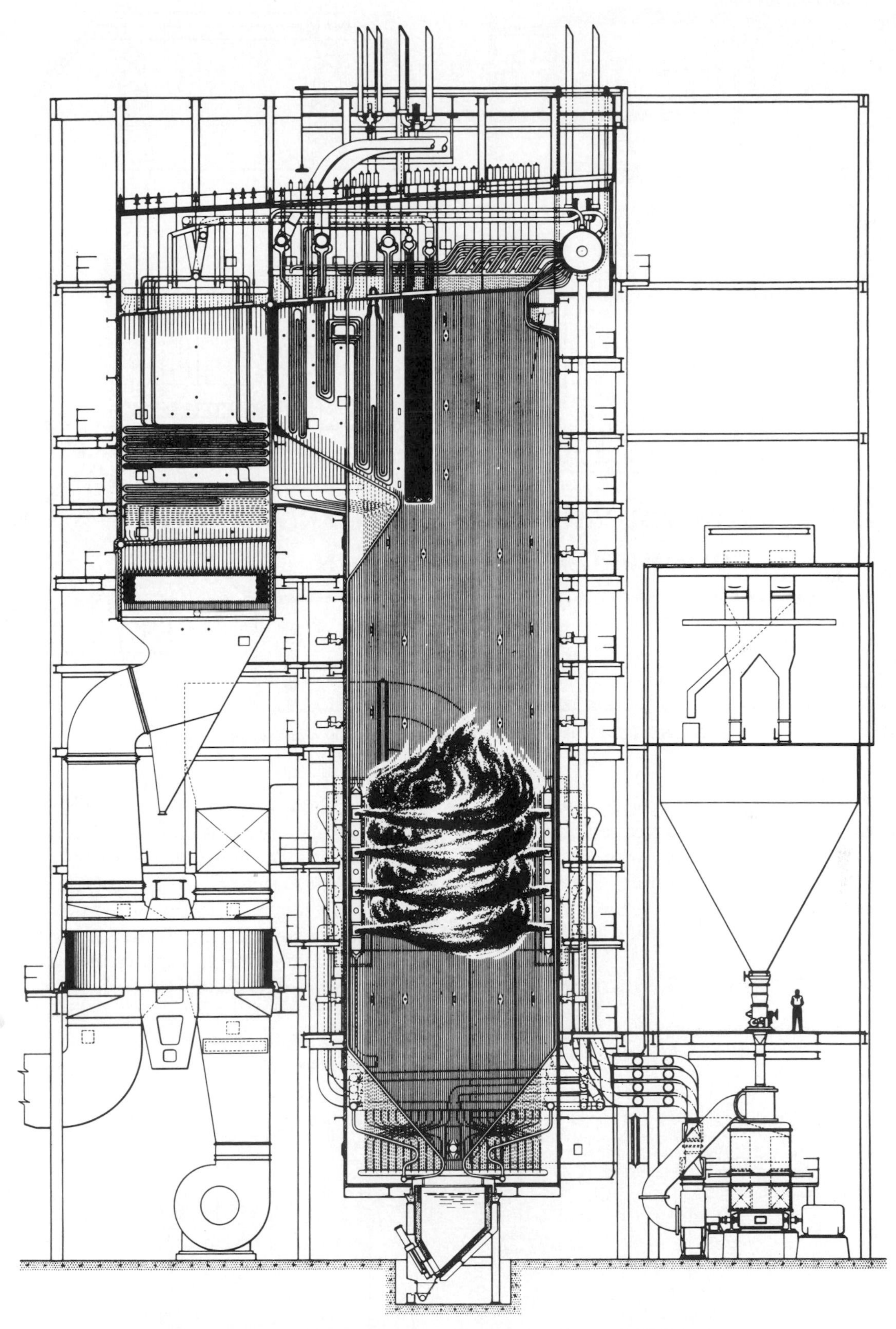

FIGURE 20–5 *Steam generator, side elevation. Tangential firing system.*

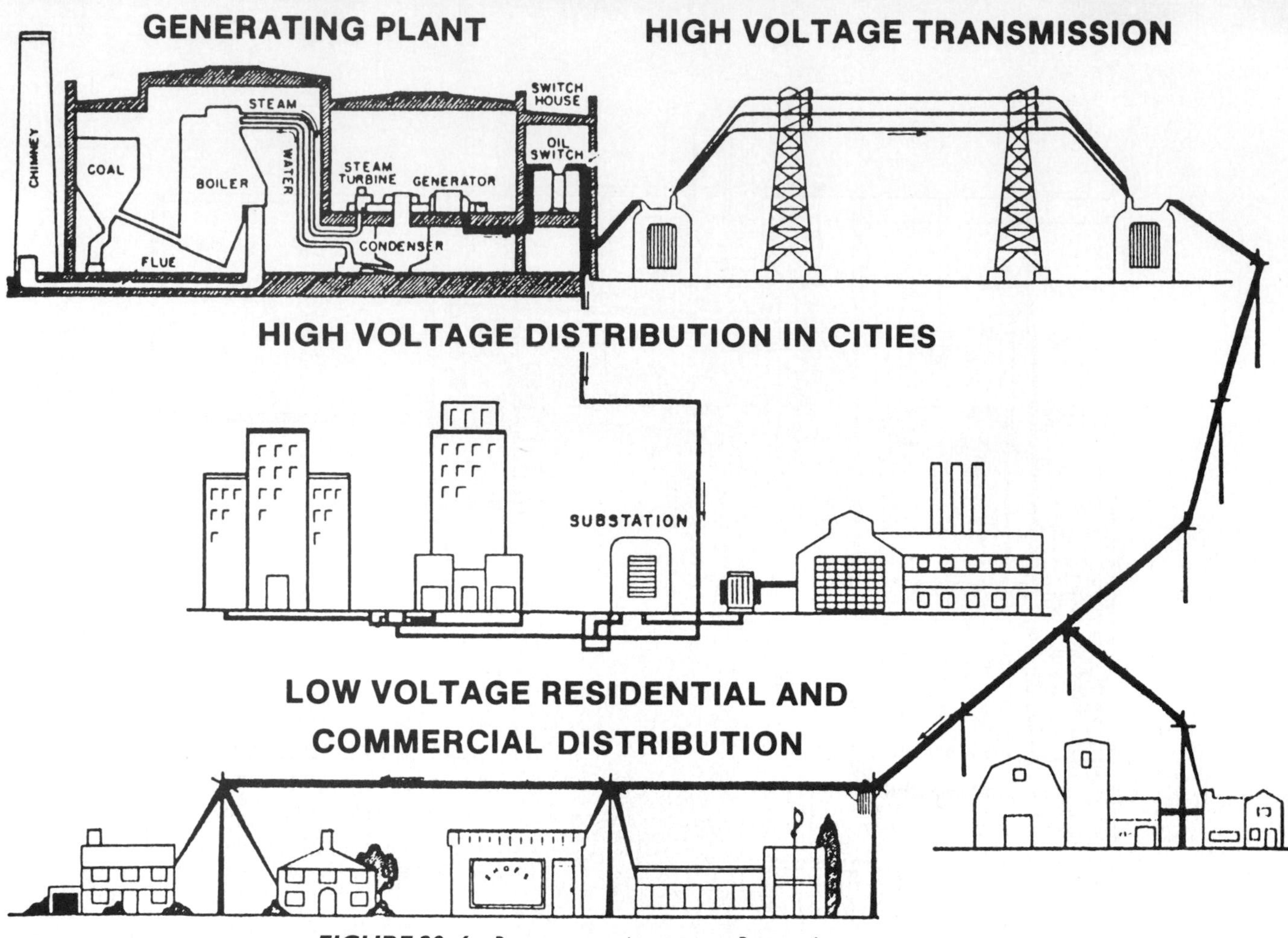

FIGURE 20–6 *Power generation system. Power plant to consumer.*

circuit breaker. In this system a primary feeder fault interrupts service to all loads. Service can be restored to all loads except those associated with the fault system element by opening the load-break switch on the faulted circuit and closing the circuit breaker. Service can be restored to the remaining load when the necessary repairs are made.

Loop Primary Radial System. This system is similar to the modern form of a simple radial system. It provides for quick restoration of service when a primary feeder or transformer fault occurs, as does the modified modern simple radial system shown in Fig. 20–9, but at a lower cost. A sectionalized primary loop controlled by a single primary feeder breaker, rather than a radial primary feeder, is shown in Fig. 20–10.

Banked-Secondary Radial System. This system per-

FIGURE 20–7 *Conventional simple-radial system. (Westinghouse)*

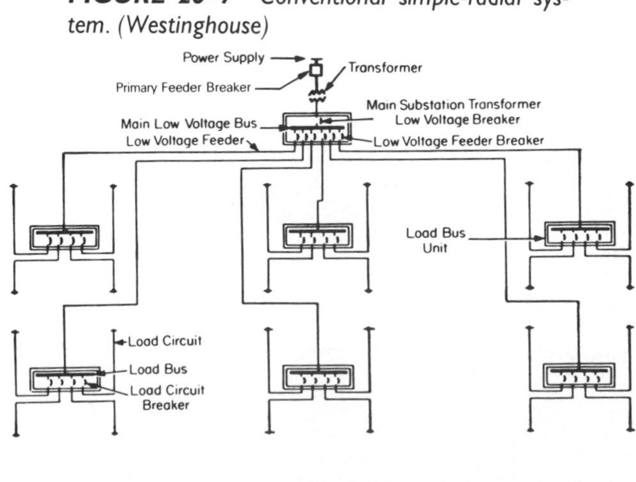

FIGURE 20–8 *Modern simple-radial system. (Westinghouse)*

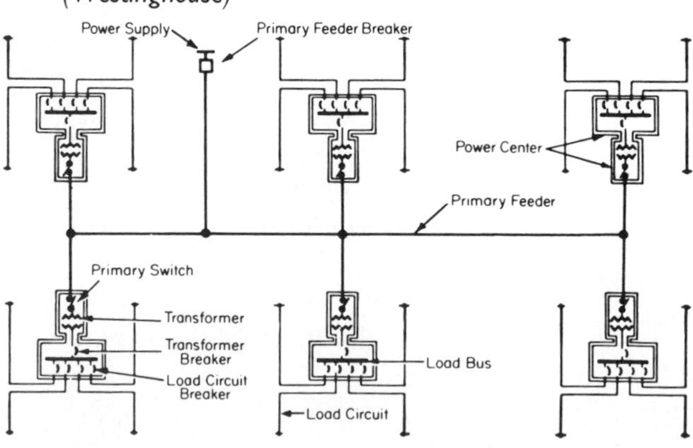

mits quick restoration of service to all loads following a primary feeder or transformer fault. It uses a secondary loop to provide an emergency supply when a fault occurs in a transformer or a section of the primary loop (Fig. 20–11).

The secondary loop gives a number of important advantages other than providing an emergency supply to restore service. It helps equalize the loads on all transformers and thus makes it unnecessary to match the transformer capacity at each load center to the load connected to each load bus. This system is better adapted for across-the-line starting of relatively large motors than is any other type of radial system because the starting current is supplied through several transformers in parallel rather than through a single transformer. Thus if the starting of large motors is involved, the use of this system may result in a savings in the cost of motor starting equipment. It is also the most satisfactory type of radial system for combined light and power secondary circuits.

Primary-Selective Radial System. This type of system uses at least two primary feeder circuits in each load area (Fig. 20–12). It is designed so that when one

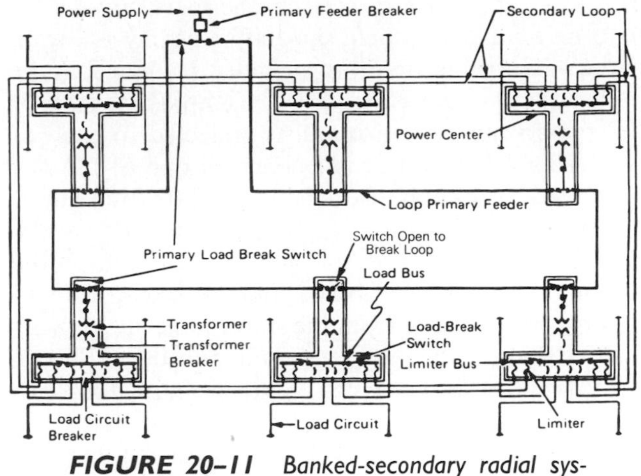

FIGURE 20–11 *Banked-secondary radial system. (Westinghouse)*

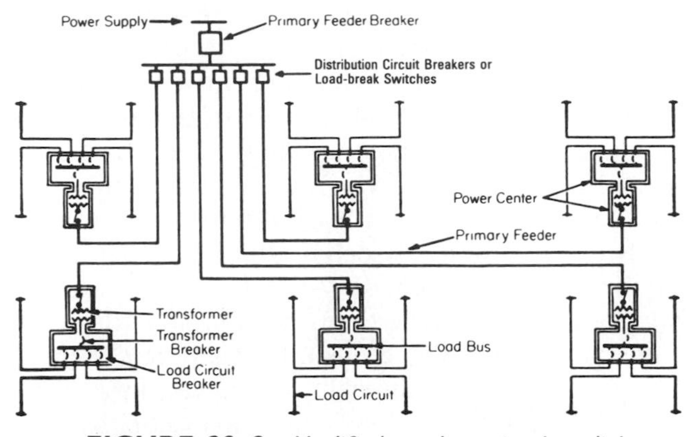

FIGURE 20–9 *Modified modern simple-radial system. (Westinghouse)*

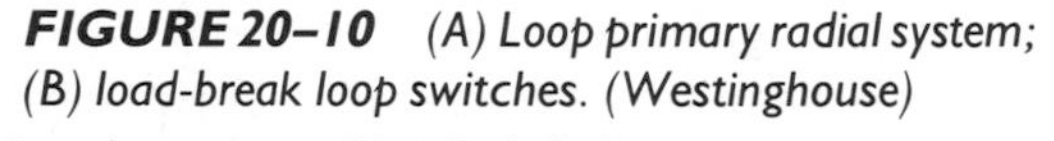

FIGURE 20–10 *(A) Loop primary radial system; (B) load-break loop switches. (Westinghouse)*

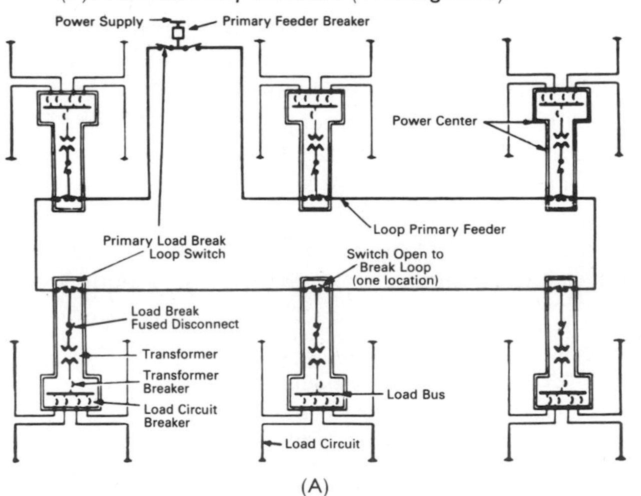

(A)

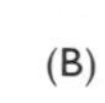

(B)

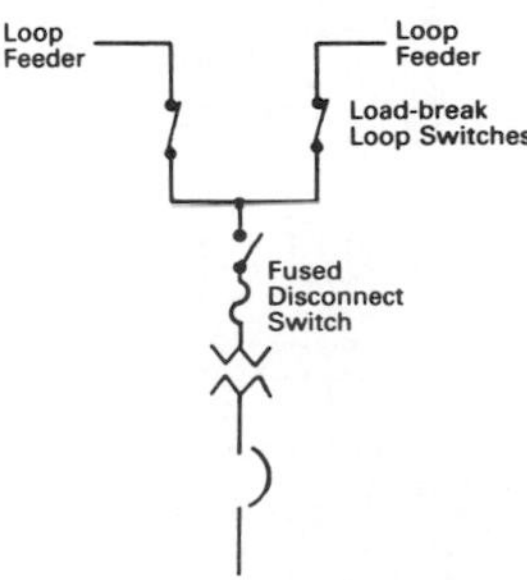

FIGURE 20–12 *Primary-selective radial system. (Westinghouse)*

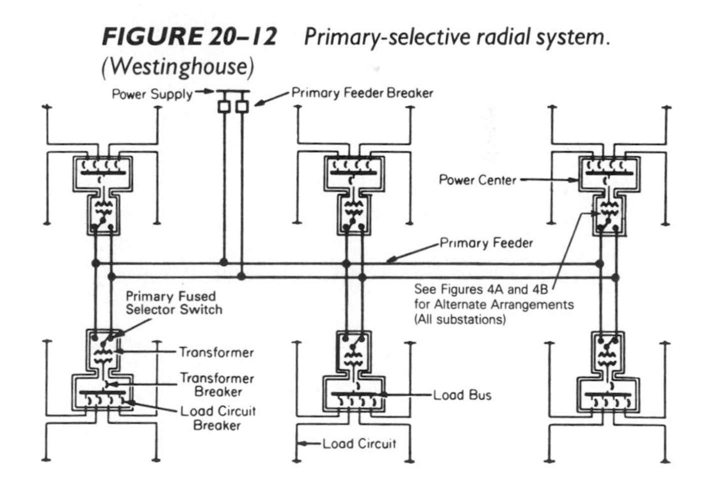

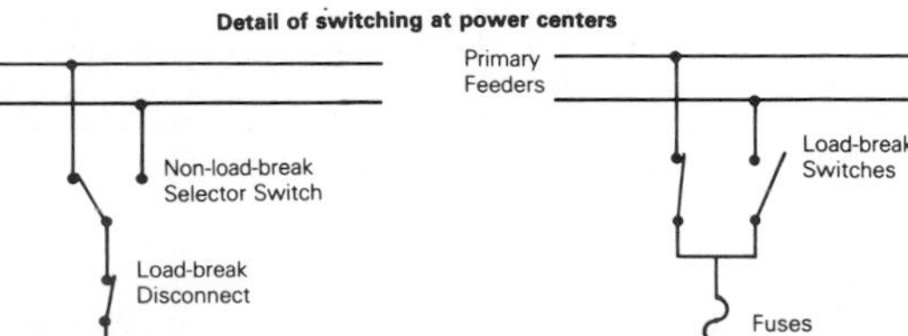

primary circuit is out of service, the remaining feeder or feeders have sufficient capacity to carry the total load. While three or more primary feeders may be used, usually only two feeders are employed. Half of the transformers are normally connected to each of two feeders. When a fault occurs on one of the primary feeders, only half of the load in the building is dropped. In the systems discussed previously a primary feeder fault causes an outage to all loads in the building. This system is not as good as the banked-secondary radial system from the standpoint of restoring service to the loads after a primary feeder or transformer fault. It is only a little better than the loop-primary radial system in this respect. In some cases, particularly in fairly large buildings, with medium- or light-load density, about the same quality of service can be rendered at less cost by using the loop primary radial system employing two separate loops instead of one. In this case, half of the transformers are connected to each loop of the system.

Secondary-Selective Radial System. This system uses the same principle of duplication feed from the power supply point as the primary-selective radial system. In this system, however, the duplication is carried all the way to each load bus on the secondary side of the transformers instead of just to the primary terminals of the transformers. This arrangement permits quick restoration of service to all loads when a primary feeder or transformer fault occurs, as does the banked-secondary radial system.

The usual form of secondary-selective radial system is shown in Fig. 20–13. Each load area in the building is supplied over two primary feeders and through two transformers.

A bus-tie breaker is provided for connecting the two secondary or load bus sections at each load center. A primary feeder fault causes half of the load in the building to be dropped.

A fault in a transformer causes the associated primary feeder breaker to trip and interrupts service to half the building load. Service can be restored by opening the disconnecting switch on the primary and the circuit breaker on the secondary of the fault transformer, closing the associated bus-tie breaker, and reclosing the primary feeder breaker. This manual switching restores the system to normal operating conditions except that the faulted transformer is de-energized and its load is being supplied through the adjacent transformer.

From the standpoint of voltage fluctuation when a load such as a relatively large motor is thrown on the system, this system is not as good as the banked-secondary radial system. The advantage that the secondary selective radial system offers over the banked-secondary radial system is that a primary feeder or transformer fault causes only one-half the load to be dropped instead of all the loads in the area. Both systems permit quick restoration of service to all loads when a primary feeder or transformer fault occurs.

Modified Secondary-Selective Radial System. A modified form of secondary-selective radial system will often be less costly than the common form described previously (Fig. 20–14). In this system there is only one transformer at each load center, instead of two. Pairs of adjacent load buses connected by secondary cables or bus permit picking up the load of any bus when a primary feeder or transformer fault occurs.

Each tie circuit is connected to each of its two load buses through a secondary-tie circuit breaker. Each secondary-tie breaker is interlocked so that it cannot be closed unless one of the two transformer breakers is open.

Basically, the single-transformer-substation form of secondary-selective radial system, regardless of the number and size of the transformers used, functions as described in connection with the more usual form of this system.

Simple Network System. The ac secondary network system is the system that has been used for many years

FIGURE 20–13 Secondary-selective radial system. (Westinghouse)

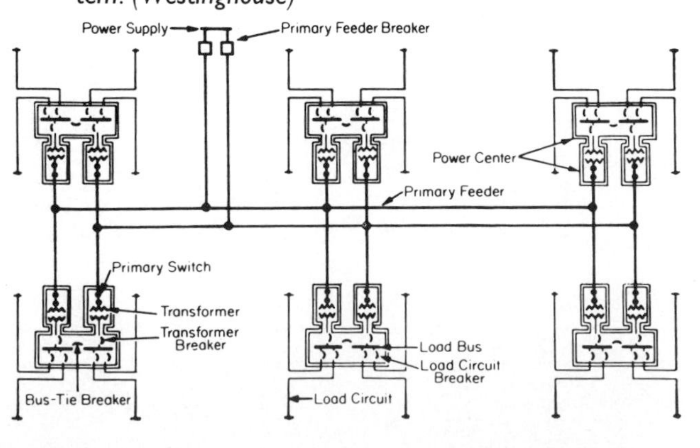

FIGURE 20–14 Modified secondary-selective radial system. (Westinghouse)

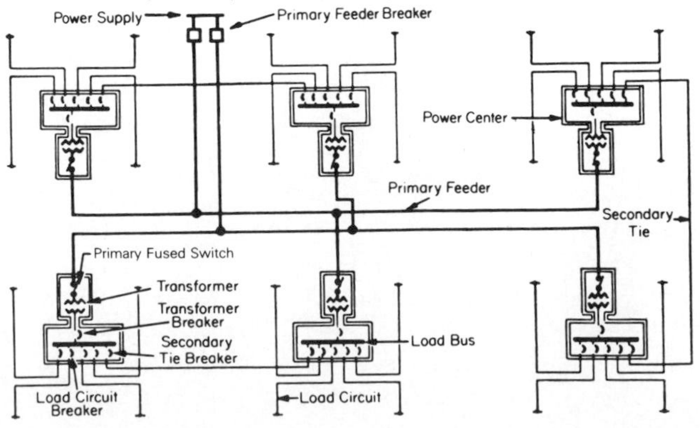

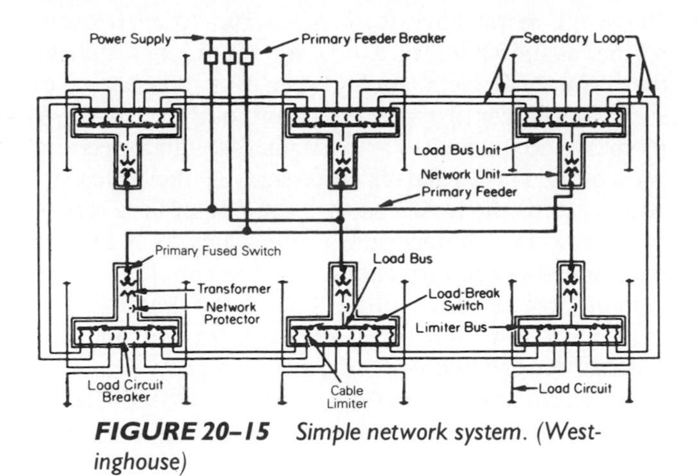

FIGURE 20–15 *Simple network system. (Westinghouse)*

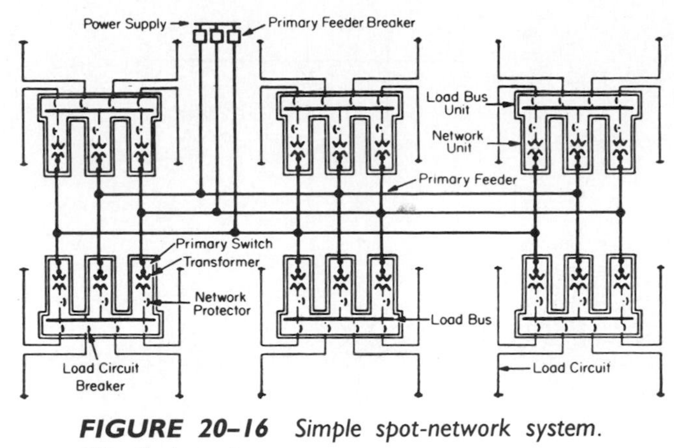

FIGURE 20–16 *Simple spot-network system. (Westinghouse)*

to distribute electrical power in the high-density downtown areas of cities. Modifications of this type of system make it applicable to serve loads within buildings (Fig. 20-15).

The best known advantage of the secondary network system is continuity of service. No single fault anywhere on the system will interrupt service to more than a small part of the system load. Most faults will be cleared automatically without interrupting service to any load. Another outstanding advantage that the network system offers is its flexibility to meet changing and growing load conditions at minimum cost and with minimum interruption of service to the other loads. In addition to flexibility and service reliability, the secondary network system provides exceptionally uniform and good voltage regulation, and its high efficiency materially reduces the cost of losses.

The chief purpose of the secondary loop is to provide an alternative supply to any load bus when the primary feeder that normally supplies it is deenergized. This prevents any service interruption when a transformer or feeder fault occurs. The secondary loop makes it unnecessary to match the load supplied from any load bus with the transformer capacity at that point. Loads can be served from the loop at load buses located between network transformers and the load circuits can be run directly to the loads from the nearest load bus. This permits the use of very short radial circuits and result in considerable saving in secondary cable and conduit compared with the load circuits of a simple radial system. The voltage regulation on a network system is such that both lights and power can be fed from the same load bus. Much larger motors can be started across the line than on a simple radial system. This often results in greatly simplified motor control and permits the use of relatively large low-voltage motors with their less expensive control.

To obtain satisfactory selectivity between limiters under all fault conditions, a minimum of two similar single-conductor cables per phase should be used in the secondary loop. When a fault occurs on a primary feeder or in a transformer, the fault is isolated from the system through automatic tripping of the primary feeder circuit breaker and all of the network protectors associated with that feeder circuit. This operation does not interrupt service to any load. When the necessary repairs have been made, the system is restored to normal operating condition by closing the feeder circuit breaker. All network protectors associated with that feeder will close automatically.

The *simple spot-network system* resembles the secondary-selective radial system in that each load area is supplied over two or more primary feeders through two or more transformers. In this system, however, the transformers are connected through network protectors to a single load bus, as shown in Fig. 20-16.

Simple spot-network systems are more economical than other forms of network systems for buildings where there are heavy concentrations of load covering small areas, with considerable distances between these concentrations and very little load in areas between them. They are commonly used in high-rise office buildings. The chief disadvantage of these systems is that they are not nearly as flexible for growing and shifting loads as are the other forms of network systems. Simple spot-network systems are used where a high degree of service continuity is required, flexibility is of minor importance, and their cost is less than the network systems using a secondary loop. The simple spot-network system is most economical where three or more primary feeders are required. This is because supplying each load bus through three or more transformers greatly reduces the spare cable and transformer capacity that is required.

Primary-Selective Network. This is the most generally applicable and widely used form of industrial secondary network system. Each transformer in the primary-selective network system is equipped with a primary selector switch arrangement (Figs. 20-17 and

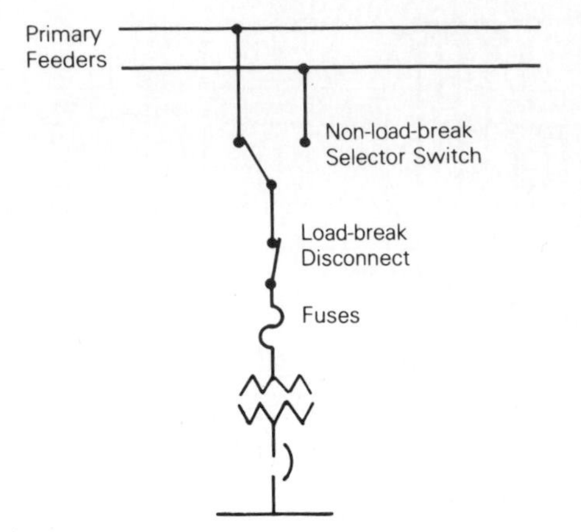

FIGURE 20–17 *Nonload break selector switch location. (Westinghouse)*

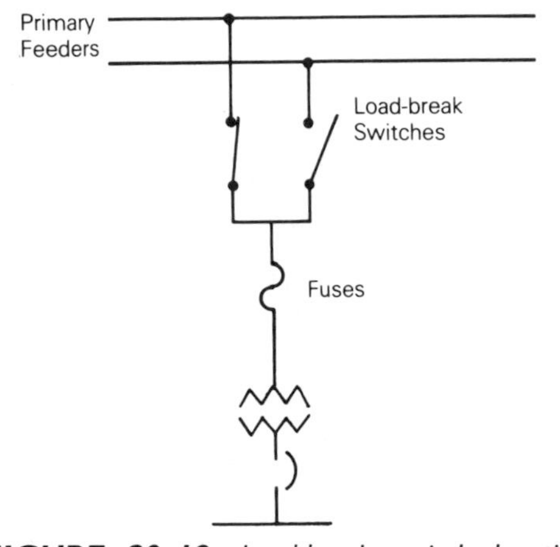

FIGURE 20–18 *Load-break switch location. (Westinghouse)*

20–18). Two primary feeders are run to each transformer as shown in Fig. 20–19. In a building requiring two primary feeders, each feeder must be capable of supplying the entire load in the building. Each transformer is connected to a load bus through a network protector. The radial circuits serving the loads are connected to the load bus through circuit breakers or fused switches. A secondary loop, such as is used in the banked-secondary radial and the simple network systems, connects each load bus to the two adjacent load buses. One half of the network transformers are normally connected to each primary feeder.

A primary feeder fault causes the faulty feeder and its transformers to be disconnected from the system by automatic operation of the primary feeder breaker and associated network protectors. The entire building load is then carried over the remaining primary feeder and one-half of the network transformers. The transformers associated with the faulty feeder can be connected to the good feeder by manually operating their selective selector switches. This relieves the overload on the transformers normally associated with the good feeder.

A transformer fault is isolated from the system by operation of the primary feeder breaker and the network protectors associated with the primary feeder. The defective transformer can be disconnected from the primary feeder by opening its selector switch. The feeder and its good transformer can be returned to service immediately. A primary feeder or transformer fault will not cause any interruption of service to the load.

As with the simple network system, the primary-selective network may also take the form of a spot-network system. Each transformer in the primary-selective network system is equipped with a primary selector switch and arranged for connection to either of two primary feeders (Fig. 20–20). This largely eliminates the necessity for spare transformer capacity and

FIGURE 20–19 *Primary-selective network system. (Westinghouse)*

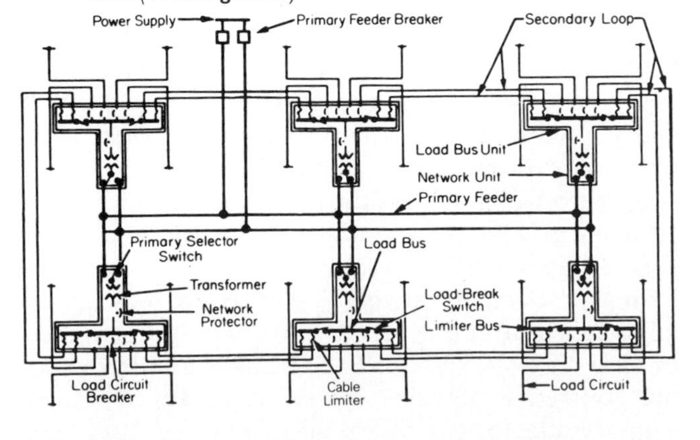

FIGURE 20–20 *Primary-selective spot-network system. (Westinghouse)*

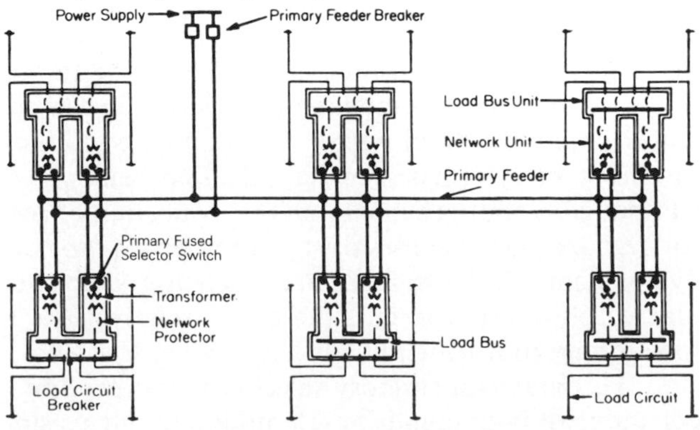

the system is ordinarily designed without providing for such capacity.

SYSTEMS ANALYSIS

A major consideration in the design of a distribution system is to ensure that it provides the required quality of service to the various loads. This includes serving each load under normal conditions and under abnormal conditions providing the desired protection to service the system apparatus so that the interruptions of service are minimized consistent with good economic and mechanical design. Short-circuit currents and fault currents are to be considered in any system.

Short-Circuit Currents

The amount of current available in a short-circuit fault is determined by the capacity of the system voltage sources and impedances of the system, including the fault. Constituting voltage sources are the power supply (utility or on-site generation) plus all rotating machines connected to the system at the time of the fault. A fault may be either an arcing or bolted fault. In all arcing faults, part of the circuit voltage is consumed across the fault and the total fault current is somewhat smaller than for a bolted fault, so the latter condition is the value sought in the fault calculations.

Basically, the short-circuit current is determined by Ohm's law except that the impedance is not constant since some reactance is included in the system. The effect of reactance in an ac system is to cause the initial current to be high and then decay toward a steady-state or Ohm's law value. The fault current consists of an exponentially decreasing direct-current component superimposed on a decaying alternating current. The rate of decay of both the dc and ac components depends on the ratio of reactance to resistance, X/R, of the circuit. The greater this ratio, the longer the current remains higher than the steady state that it would eventually reach.

Fault Currents

The calculation of asymmetrical currents is a laborious procedure since the degree of asymmetry is not the same on all three phases. It is common practice to calculate the rms symmetrical fault current, with the assumption being made that the dc component has decayed to zero, and then apply a multiplying factor to obtain the first half-cycle rms asymmetrical current, which is called the *momentary current*.

To determine motor contribution to the first half-cycle fault current when the system motor load is known, the following assumptions generally are made:

1. 208Y/120-V systems:
   a. Assume 50% lighting and 50% motor load.
   b. Assume motor feedback contribution of twice full-load current of transformer.
2. 240/480/600-V three-phase, three-wire systems:
   a. Assume 100% motor load.
   b. Assume motor feedback contribution of four times full load current of transformer.
3. 480Y/277-V systems in commercial buildings:
   a. Assume 50% motor load.
   b. Assume 100% induction motors.
   c. Assume motor feedback contribution of two times full load current of transformer source.
4. For industrial plants, make same assumptions as for three-phase, three-wire systems.

Three-Phase Power

Three-phase power can be generated by a wye-connected generator or a delta-connected generator. Instead of having six leads come out of the three-phase ac generator, one of the leads from each phase can be connected to form a common junction. The stator is then called a *wye*. Sometimes the wye connection is also called a *star* connection. The common lead may not be brought out of the generator. If it is brought out, it is called the *neutral* (Fig. 20–21). If the wye connection has the neutral brought out of the generator, it is called a four-wire, three-phase system.

The wye connection provides 1.73 times the phase voltage for any two of the three wires connected. The line currents are equal to the current in any phase winding. The advantage of a wye connection is its ability to produce more voltage. Note in Fig.

FIGURE 20–21 *Single-phase voltage from a wye connection.*

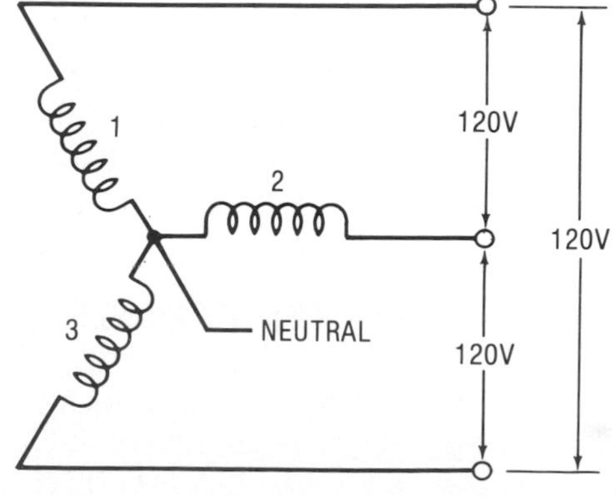

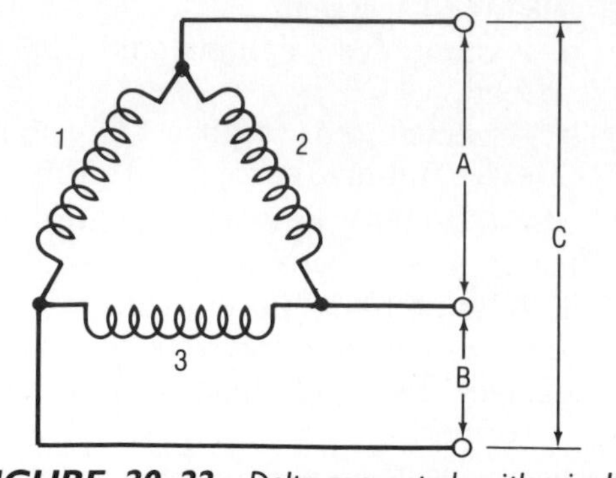

FIGURE 20–22 Delta-connected with single-phase voltages identified.

20–21 that windings 1 and 2 are in series with each other. Windings 2 and 3 are in series with each other also. If windings 1 and 3 are used for connections, they, too, are in series. Thus no matter which connections are used, there are two coils in series to produce the single-phase power needed. The current provided, however, is as in any series connection. It is that which can be provided by only one winding.

The true power delivered by a wye-connected generator is the same as that of a delta-connected generator. The total true power is

$$P_t = 1.73E_{\text{line}} \times I_{\text{line}} \times \cos \angle\theta$$

The delta connection provides 1.73 times the phase current for any two connections (Fig. 20–22). The line voltages are equal to the voltage in any phase winding. The advantage to the delta connection is its ability to produce more current. Note that winding 1 is in parallel with windings 2 and 3, and windings 2 and 3 are in series with each other. This produces the output labeled C. Winding 2 is in parallel with windings 1 and 3, and windings 1 and 3 are in series with each other. They produce output A. Winding 3 is in parallel with windings 1 and 2, and windings 1 and 2 are in series with each other. This combination produces output B. With delta connection, the current is the advantage. With the wye connection, voltage is the advantage. In summary, the delta connection produces line voltage equal to the phase voltage and the line current equal to the square root of 3 or 1.73 times the phase current. True power is the same in the delta connection as in the wye connection. In both cases it is merely a matter of multiplying the line current times the line voltage times the 1.73 factor and the cosine of the angle θ.

MEASUREMENT OF POWER

The wattmeter connections for measuring the true power in a three-phase system are shown in Fig. 20–23. The method shown in Fig. 20–23A uses three wattmeters with their current coils inserted in series. The line wires and their potential coils are connected between line and neutral wires. The total true power is equal to the arithmetic sum of the three wattmeter readings.

In Fig. 20–23B two wattmeters are used with their potential coils connected between these line wires and the common or third wire that does not contain the current coils. The total true power is equal to the algebraic sum of the two wattmeter readings. If one meter reads backward, its potential coil connections are reversed. Note the polarity indications placed there to prevent incorrect connections. The incorrect connection causes the meter to read up-scale. The total true power is then equal to the difference in the two wattmeter readings. If the load power factor is less than 0.5 and the loads are balanced, the true power will equal the difference between the two wattmeter readings. If the load power factor is 0.5, one meter will indicate zero. If the load factor is above 0.5, the total true power is the sum of the two wattmeter readings.

TRANSFORMERS

Transformers are important in any power distribution system. A quick review of their basic principles will aid in understanding their use in various parts of a distribution system.

A transformer is a device that has no moving parts. It transfers energy from one circuit to another by electromagnetic induction. The energy is always transferred without a change in frequency. Usually, changes in voltage and current are evident. A step-up transformer receives electrical energy at one voltage and delivers it at a higher voltage. Conversely, a step-down transformer receives energy at one voltage and delivers it at a lower voltage. Transformers require little care and maintenance because they are simple, rugged, and durable in construction. Since there are no moving parts, there is little to wear out. The efficiency can be more than 99%. Transformers are responsible for the use of ac today in every phase of life.

Transformer Construction

A transformer in its simplest form has two windings insulated electrically from each other. These windings are wound on a common magnetic circuit built of laminated steel sheet. The principal parts are the following:

1. *Core:* provides a circuit of low reluctance for the magnetic flux
2. *Primary winding:* receives the energy from the ac source

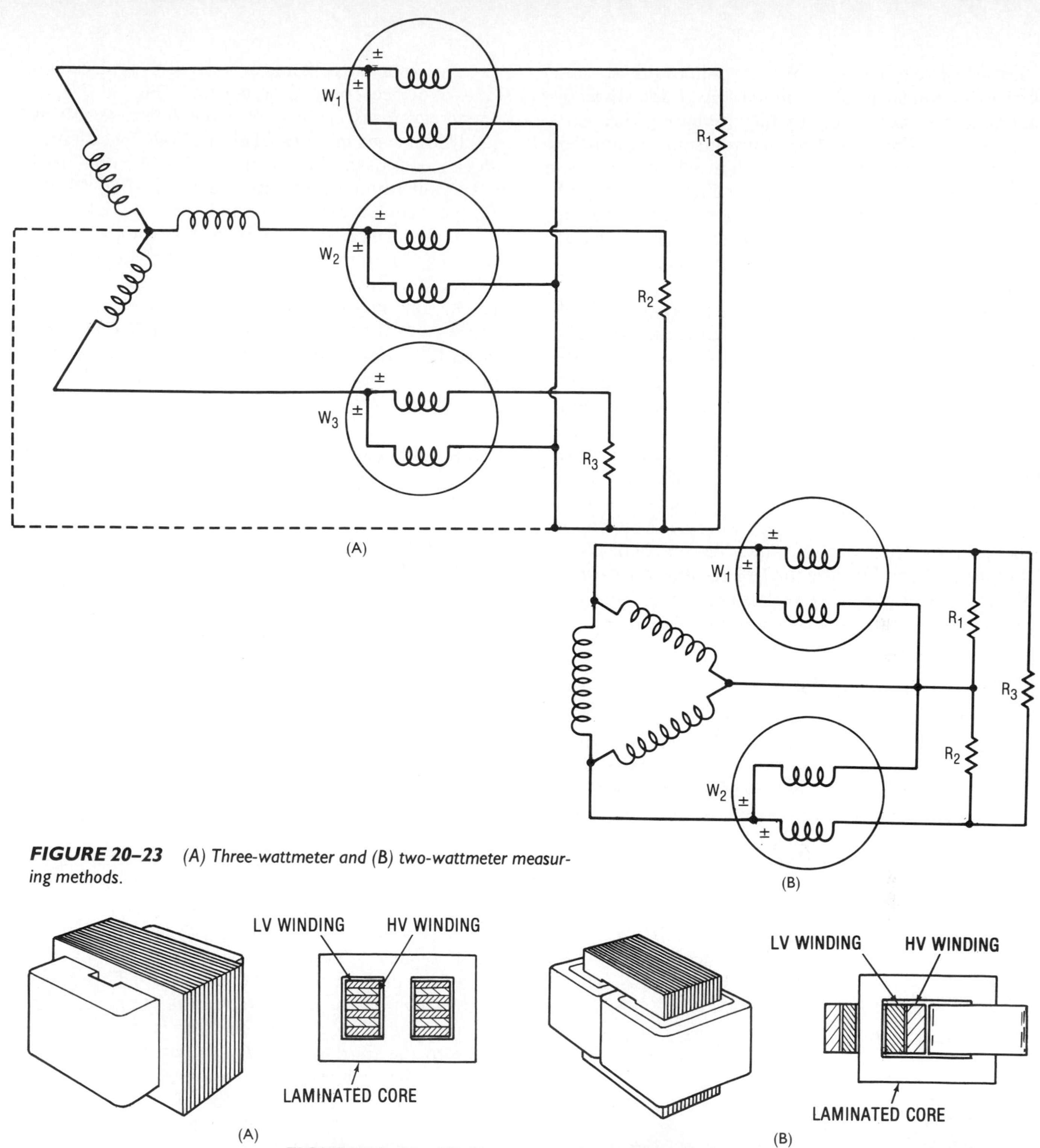

FIGURE 20–23 *(A) Three-wattmeter and (B) two-wattmeter measuring methods.*

FIGURE 20–24 *(A) Core-type and (B) shell-type transformers.*

3. *Secondary winding:* receives the energy by mutual induction from the primary and delivers it to the load
4. *Enclosure:* prevents damage when the transformer is overloaded

When a transformer is used to step up voltage, the low-voltage winding is the primary. When a transformer is used to step down the voltage, the high-voltage winding is the primary. The primary is always connected to the source of the power. The secondary is always connected to the load. It is common practice to refer to the windings as the primary and secondary rather than the high-voltage and low-voltage windings.

The principal types of transformer construction are the *core* type and the *shell* type. Figure 20–24 shows the two types. The cores are built of thin stampings of silicon steel called *laminations.* Eddy currents are generated in the core by the alternating flux as it cuts through the iron. These currents are minimized by using the thin laminations with insulating varnish. Hysteresis losses are caused by the friction developed between magnetic particles as they rotate through each cycle of magnetization. These losses are minimized by using a special grade of heat-treated grain-oriented silicon-steel laminations.

In the core-type transformer, the copper windings surround the laminated iron core. In the shell-type transformer the iron core surrounds the copper windings. Distribution transformers are generally of the core type. Some of the largest power transformers are of the shell type.

If the windings of a core-type transformer were placed on separate legs of the core, a relatively large amount of flux produced by the primary windings would fail to link the secondary winding. This would cause a large leakage flux. The effect of the leakage flux would be to increase the leakage reactance drop, X_L, in both windings. To reduce the leakage flux and reactance drop, the windings are subdivided. Half of each winding is placed on each leg of the core. The windings may be cylindrical in form and placed one inside the other with the necessary insulation. The low-voltage winding is placed with a large part of its surface area next to the core. The high-voltage winding is placed outside the low-voltage winding to reduce the insulation requirements of the two windings. If the high-voltage winding were placed next to the core, two layers of high-voltage insulation would be required. One would be needed next to the core; the other would be needed between the two windings. In another method, the windings are built up in thin flat sections called *pancake coils.* The pancake coils are sandwiched together with the required insulation between them.

FIGURE 20–25 *75 VA for 12 V output from 120-V line.*

The complete core and coil assembly is placed inside a set of steel covers or a steel tank (Fig. 20–25). In commercial transformers, the complete assembly is usually immersed in a special mineral oil. This oil provides a means of insulation and cooling. No oil is used in the transformer enclosures shown in Fig. 20–25. This type requires air circulation to keep it cool.

Transformers are built in both single-phase and three-phase units. A three-phase transformer consists of separate insulated windings for the different phases. The windings are wound on a three-legged core capable of establishing three magnetic fluxes displaced 120° in time phase. Three-phase transformers are not used in locations where the operation of a single unit may affect the others.

Three-Phase Transformer Connections

Power may be supplied through three-phase circuits containing transformers in which the primaries and secondaries are connected in various wye and delta combinations. For example, three single-phase transformers may supply three-phase power with four possible combinations of their primaries and secondaries.

Possible combinations are:

1. Primaries in delta and secondaries in delta
2. Primaries in wye and secondaries in wye
3. Primaries in wye and secondaries in delta
4. Primaries in delta and secondaries in wye

If the primaries of three single-phase transformers are properly connected (either in wye or delta) to a three-phase source, the secondaries may be connected in delta (Fig. 20–26). A topographic vector diagram of the three-phase voltages is shown in Fig. 20–26A. The vector sum of these three voltages is zero. This can be seen by combining any two vectors. For example, check E_A and E_B. Note that their sum is equal and opposite to the third vector, E_C. A voltmeter inserted within the delta will indicate zero voltage (Fig. 20–26B).

Assuming that all three transformers have the same polarity, the delta connection consists of the X_2 lead of winding A to the X_1 lead of B, the X_2 lead of B to X_1 of C, and the X_2 lead of C to X_1 of A. If any one of the three windings is reversed with respect to the other two windings, the total voltage within the delta will equal twice the value of one phase. If the delta is closed on itself, the resulting current will be of short-circuit magnitude. The result is damage to the transformer windings and cores. The delta should never be closed until a test is made to determine that the voltage within the delta is zero or nearly zero. This may be done with a voltmeter, fuse wire, or a test lamp. In Fig. 20–26B, when the voltmeter is inserted between

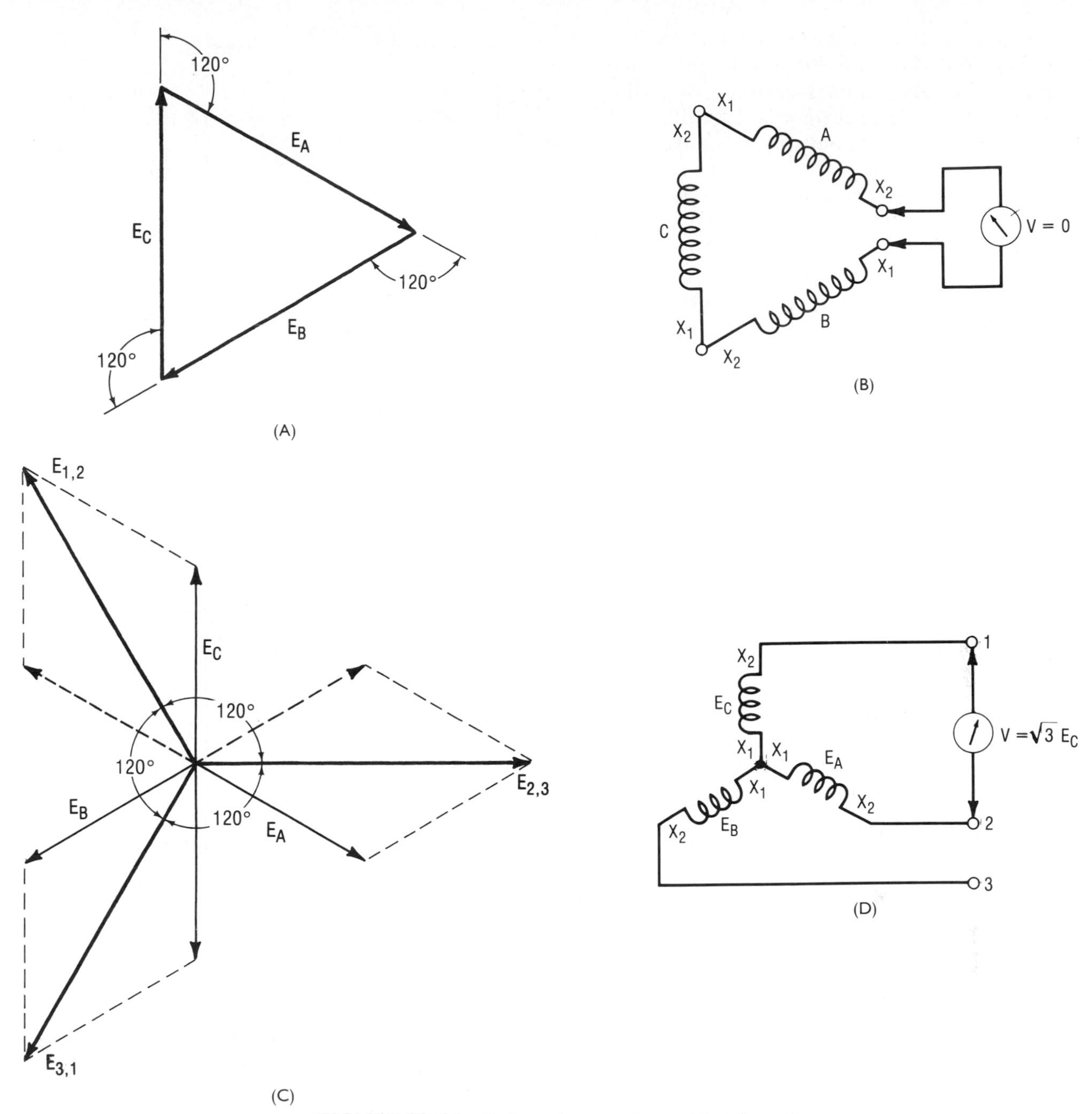

FIGURE 20–26 *Delta and wye vectors and test for voltages.*

the X_2 lead of A and the X_1 lead of B, the delta circuit is completed through the voltmeter. The indication should be approximately zero. Then the delta is completed by connecting the X_2 lead of A to the X_1 lead of B.

If the three secondaries of an energized transformer bank are properly connected in delta and are supplying a balanced three-phase load, the line current will be equal to 1.73 times the phase current. If the rate current of a phase (winding) is 100 A, the rated line current will be 173 A. If the rated voltage of a phase is 120 V, the voltage between any two line wires will be 120 V.

The three secondaries of the transformer bank may be reconnected in wye to increase the output voltage. The voltage vectors are shown in Fig. 20–26C. If the phase voltage is 120 V, the line voltage will be 1.73 times 120 = 208 V. The line voltages are represented by vectors, E1,2, E2,3, and E3,1. A voltmeter test for the line voltage is represented in Fig. 20–26D. If the three transformers have the same polarity, the proper connections for a wye-connected secondary bank are indicated in the figure. The X_1 leads are connected to form a common or neutral connection. The X_2 leads of the three secondaries are brought out to the line leads. If the connections of any one winding are reversed, the voltages between the three line wires will become unbalanced. Thus the loads will not receive their proper magnitude of load current. Also, the phase angle between the line currents will be changed. They will no longer be 120° out of phase with each other.

Therefore, it is important to connect the transformer secondaries correctly to preserve the symmetry of the line voltages and currents.

Transformer installations with both primary and secondary windings delta connected are shown in Fig. 20–27. The H_1 lead of one phase is always connected to the H_2 lead of an adjacent phase. The X_1 lead is connected to the X_2 terminal of the corresponding adjacent phase, and so on. The line connections are made at these junctions. This arrangement assumes that the three transformers have the same polarity.

An open-delta connection results when any one of the three transformers is removed from the delta-connected transformer bank without distributing the three-wire, three-phase connections to the remaining two transformers. These transformers will maintain the correct voltage and phase relations on the secondary to supply a balanced three-phase load. An open-delta connection is shown in Fig. 20–28. The three-phase source supplies the primaries of the two transformers. The secondaries supply a three-phase voltage to the load. The line current is equal to the transformer phase current in the open-delta connection. In the closed-delta connection, the transformer phase current, $I_{\text{phase}} = I_{\text{line}}/\sqrt{3}$. Thus when one transformer is removed from a delta-connected bank of three transformers, the remaining two transformers will carry a current to $\sqrt{3}\, I_{\text{phase}}$. This value amounts to an overload current on each transformer of 1.73 times the rated current, or an overload of 73.2%.

Thus to prevent overloading transformers in an open-delta connection, the line current must be re-

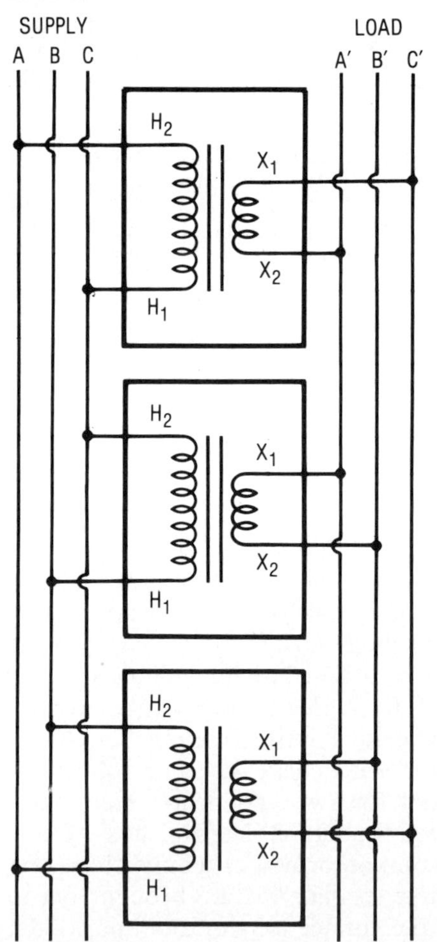

FIGURE 20–27 *Connections for a three-transformer bank.*

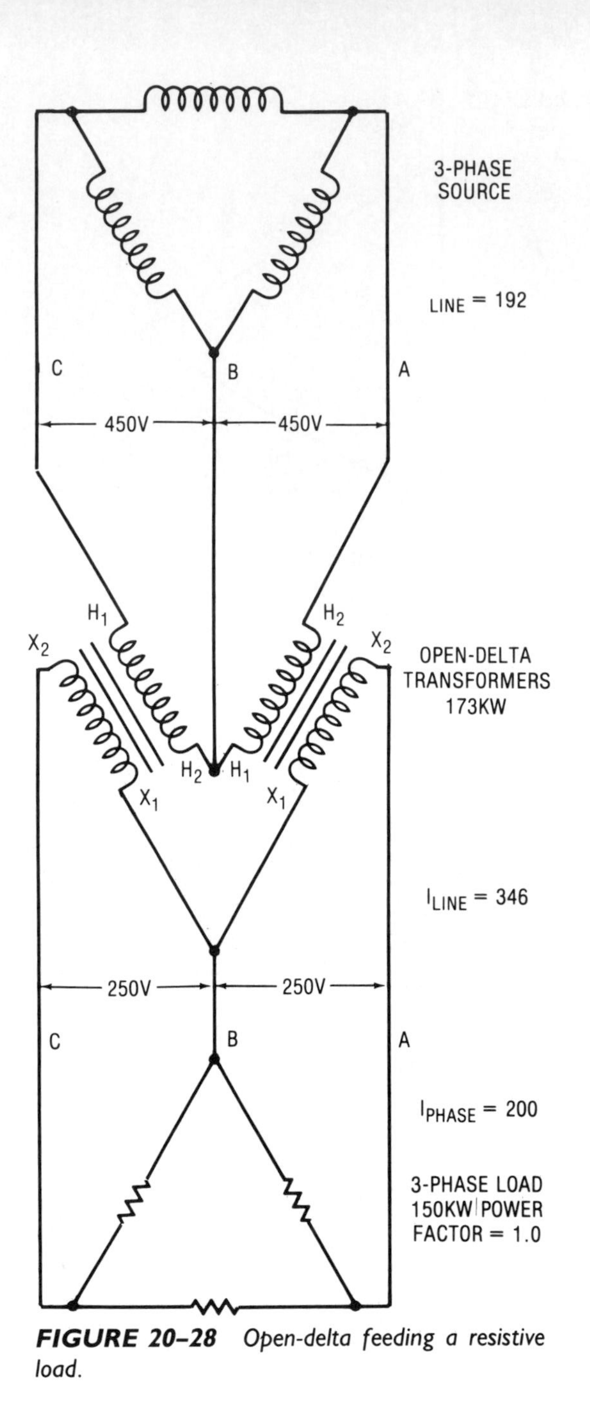

FIGURE 20–28 *Open-delta feeding a resistive load.*

duced so that the rated current of the individual transformers is not exceeded. The open-delta connection results in a reduction in system capacity. The full-load capacity in a delta connection at unit power factor is

$$P_{\Delta} = \sqrt{3}\, E_{\text{phase}} \quad I_{\text{phase}} = \sqrt{3}\, E_{\text{line}}\, I_{\text{line}}$$

In an open-delta connection, the line current is limited to the rate phase current of $I_{\text{line}}/\sqrt{3}$. The full-load capacity of the open-delta or V-connected system is

$$P_V = \sqrt{3}\, E_{\text{line}} \frac{I_{\text{line}}}{\sqrt{3}} = E_{\text{line}}\, I_{\text{line}}$$

The ratio of the load that can be carried by two transformers connected in open-delta to the load that can be carried by three transformers in closed-delta is

$$\frac{P_V}{P} = \frac{E_{\text{line}}\, I_{\text{line}}}{\sqrt{3}\, E_{\text{line}}\, I_{\text{line}}} = \frac{1}{\sqrt{3}} = 0.577$$

or 57.7% of the closed-delta rating.

For example, a 150-kW, three-phase balanced load operating at unity power factor is supplied at 250 V. The rating of each of three transformers in closed-delta is 150/3 = 50 kW. The phase current is 50,000/250 = 200 A. The line current is $200 \times \sqrt{3} =$ 346 A. The removal of one transformer from the bank leaves two transformers that would be overloaded 346 − 200 = 146 A, or (146/200) × 100 = 73%. To prevent overload on the remaining two transformers, the line current must be reduced from 346 A to 200 A. Also, the total load must be reduced to

$$\frac{\sqrt{3} \times 250 \times 200}{1000} = 86.5 \text{ kW}$$

$$\frac{86.5}{150} \times 100 = 57.7\% \text{ of the original load.}$$

The rating of each transformer in an open-delta necessary to supply the original 150-kW load is

$$\frac{E_{\text{phase}}\, I_{\text{phase}}}{1000} \quad \text{or} \quad \frac{250 \times 346}{1000} = 86.5 \text{ kW}$$

Two require a total rating of 2 × 86.5 = 173.0 kW. This compares with 150 kW for three transformers in closed-delta. Assume that two transformers are used in open-delta to supply the same load as three 50-kW transformers in closed-delta. Then the required increase in transformer capacity is

$$173.0 - 150 = \frac{23.0}{150} \text{ kW}$$

$$\text{or} \quad \frac{23.0}{150} \times 100 = 15.3\%$$

Three single-phase transformers with both primary and secondary windings wye-connected are shown in Fig. 20–29. Only 57.7% of the line voltage ($E_{\text{line}}/\sqrt{3}$) is impressed across each winding. But full-line current flows in each transformer winding.

Three-phase transformers delta-connected to the primary circuit and wye-connected to the secondary circuit are shown in Fig. 20–30. This connection provides four-wire, three-phase service with 208 V between lines wires $A'B'C'$. There is $208/\sqrt{3}$ or 120 V between each line wire and neutral, N.

The wye-connected secondary is desirable when a large number of single-phase loads are to be supplied from a three-phase transformer bank. The neutral, or grounded, wire is brought out from the midpoint of the wye connection. This permits the single-phase loads to be distributed evenly across the three phases, and three-phase loads can be connected directly across the line wires. The single-phase loads have a voltage rating of 120 V. The three-phase loads are rated at 208 V. This connection is often used in high-voltage power supply transformers for radar installations. The phase voltage is 1/1.73, or 0.577 (57.7%) of the line voltage.

Three single-phase transformers with wye-connected primaries and delta-connected secondaries are shown in Fig. 20–31. This arrangement is used for

FIGURE 20–29 *Connections for a three-transformer bank.*

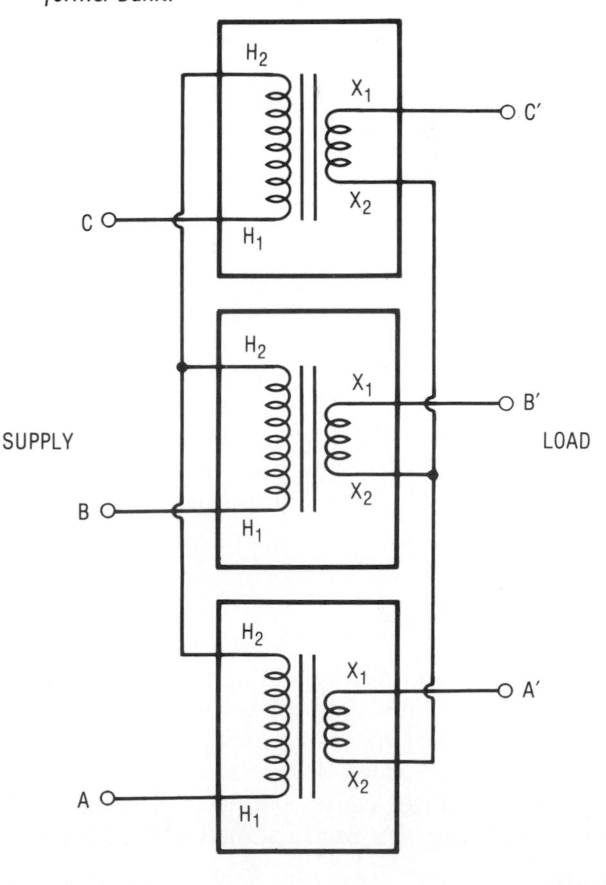

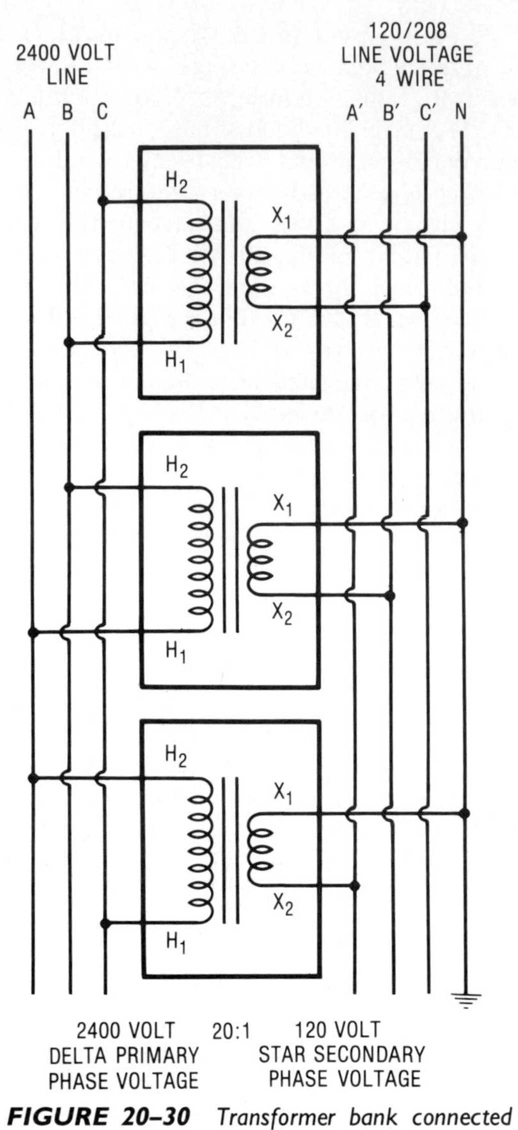

FIGURE 20–30 *Transformer bank connected delta to wye.*

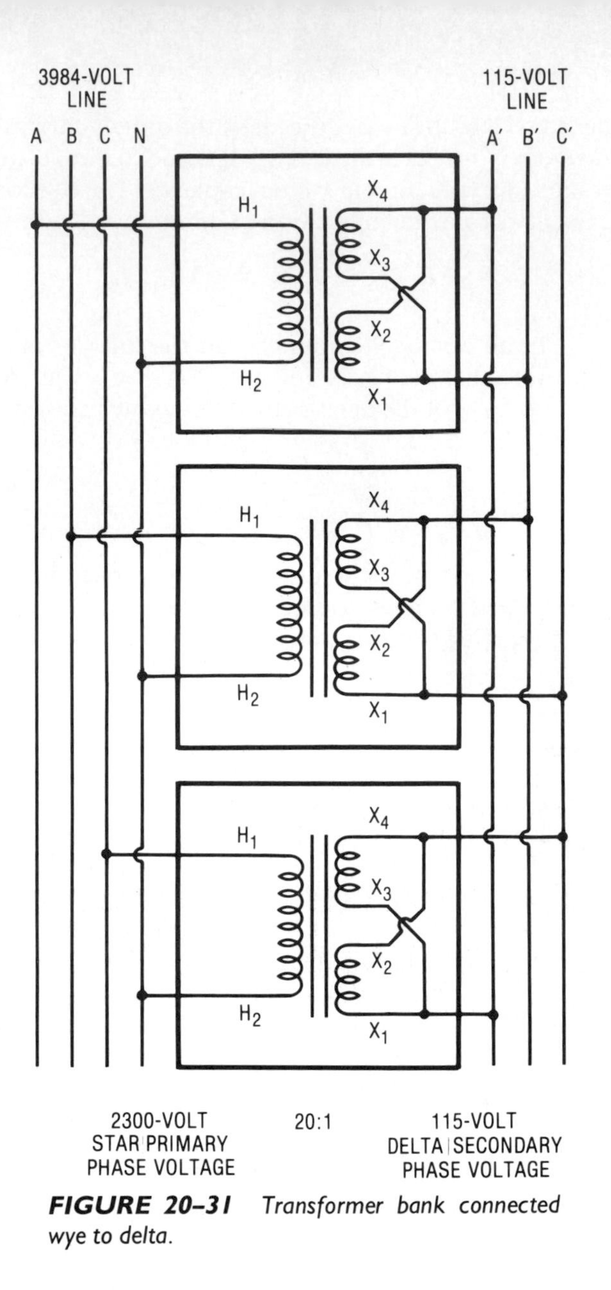

FIGURE 20–31 *Transformer bank connected wye to delta.*

stepping down the voltage from approximately 4000 V between line wires on the primary side to either 115 or 230 V, depending on whether the secondary windings of each transformer are connected in parallel, and the secondary output voltage is 115 V. There is an economy in transmission with the primary in wye. The line voltage is 73% higher than the phase voltage, and the line current is accordingly less. Thus line losses are reduced and the efficiency of transmission is improved.

SWITCHING

Electrical distribution systems are often quite complicated. They cannot be made absolutely fail safe. Circuits are subject to destructive overcurrents. Harsh environments, general deterioration, accidental damage from natural causes, excessive expansion, and overloading of the electrical distribution system are factors that contribute to the occurrence of such overcurrents. Reliable protective devices prevent or minimize costly damage to transformers, conductors, motors, and many other components and loads that make up the complete distribution system. Reliable circuit protection is essential to avoid the severe momentary losses that can result from power blackouts and prolonged downtime of facilities. Switching is one of the means utilized to be able to disable a circuit while the fuses or circuit breakers are replaced or transformers or other devices are repaired or maintained.

Automatic Emergency Power Switching

When the utility power or the normal source fails, controls for automatic handling of the emergency

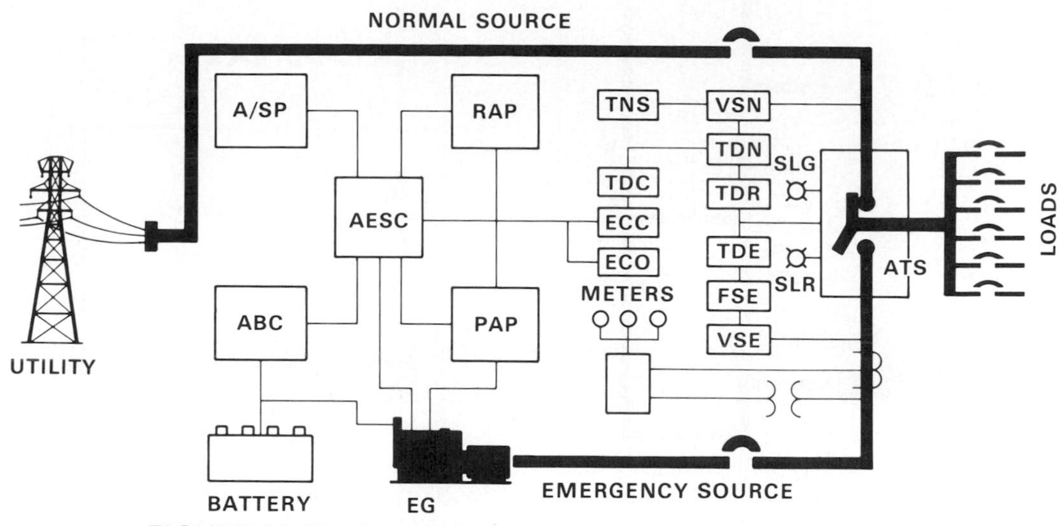

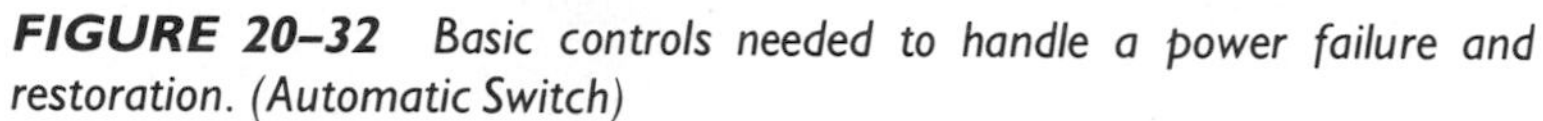

FIGURE 20–32 *Basic controls needed to handle a power failure and restoration. (Automatic Switch)*

power and the restoration of utility power are needed (Fig. 20–32). Controls are needed to start the engine and transfer the loads to the generator when it reaches proper voltage and frequency. Then, when the utility power is restored, the controls are needed to sense that power is restored and a need to retransfer the loads to the utility is evident. Allowing the engine to cool off before shutting down also needs to be handled automatically. The basic controls needed to accomplish this job are shown in Fig. 20–32.

All the abbreviations are defined below for ease in understanding some of the complexity of a system with an automatic emergency power source.

ATS (Automatic Transfer Switch): switches the loads from the utility power (normal source) when it fails, to the engine generator (emergency source). Then retransfers the loads back to the utility when its power is restored. Switching is done automatically at the proper times under the supervision of the appropriate sensors, time delays and relays.

ECC (Engine Control Contact): closes when the normal source fails and thereby initiates engine operation through the Automatic Engine-Starting Controls (AESC).

ECO (Engine Control Contact): opens when the normal source fails. This may be used to trigger the Automatic Engine-Starting Controls, depending on the type of controls being used.

FSE (Frequency Sensor, Emergency): senses the frequency of the power from the engine generator.

PAP (Prealarm Panel): alerts personnel that engine oil pressure is dropping or that water temperature is rising *before* they reach a critical point. This panel also indicates low fuel level, low water temperature. In addition, it shows the system status and sounds an alarm when the engine is shut down from overcrank, overspeed, low oil pressure, and high water temperature.

RAP (Remote Alarm Panel): located in an occupied area such as the maintenance engineer's or building security office, this panel alerts personnel who are not in the vicinity of the power system and engine of the status of the complete system.

SLG (Signal Light, Green): when lighted, indicates that the automatic transfer switch is connected to the utility (normal source).

SLR (Signal Light, Red): when lighted, indicates that the automatic transfer switch is connected to the engine–generator set (emergency source).

TDC (Time Delay, Cool-down): delays the engine from shutting down immediately when the load is retransferred to allow the engine to cool down from operating temperature.

TDE (Time Delay, Emergency): delays the transfer switch from switching loads from the failed utility to the engine generator. This delay is usually set at zero unless there is more than one transfer switch handling loads for one engine generator; in which case, you may want to set the time delays so that all switches do not transfer their loads at the same time.

TDN (Time Delay, Normal): delays starting the engine when the utility power dips momentarily, then comes back up. This time delay avoids nuisance starts on the engine.

TDR (Time Delay, Retransfer): delays the transfer switch from retransferring the loads from

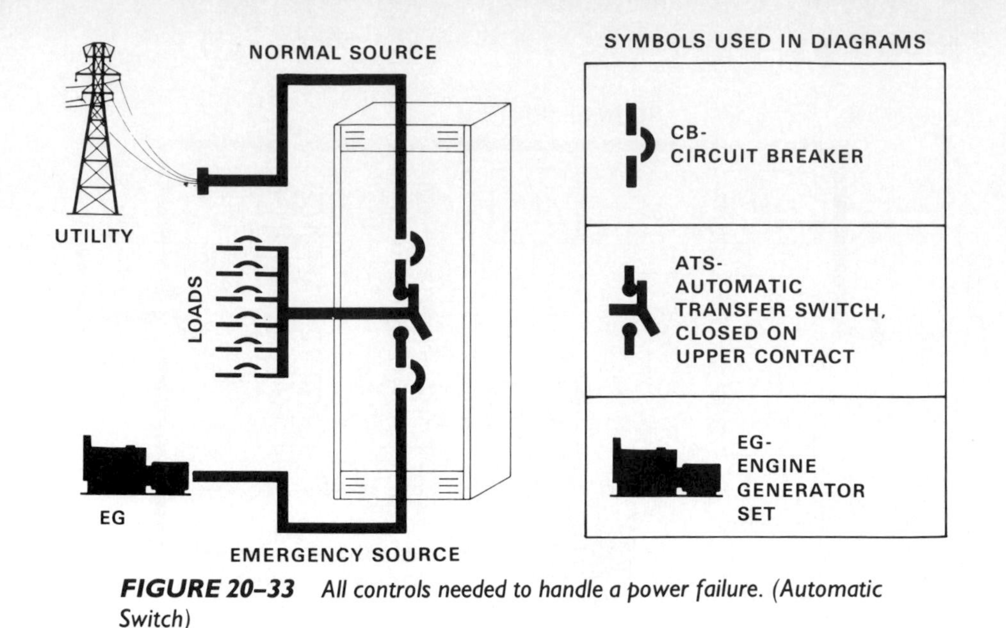

FIGURE 20–33 *All controls needed to handle a power failure. (Automatic Switch)*

the generator (emergency source) back to the utility (normal source) as soon as the utility power has been restored. This delay gives the utility time to establish itself.

TNS (Test Normal Switch): simulates a failure of the utility power. When this switch is operated, the engine generator starts and runs. Then the transfer switch transfers the loads to the engine generator at the proper time. After a prescribed amount of engine-running time determined by the Time Delay Retransfer (TDR) setting, the transfer switch retransfers the loads back to the utility power and shuts down the engine. If it is important that *loads not be interrupted when testing,* a combination automatic transfer and bypass-isolation switch should be used.

VSE (Voltage Sensor, Emergency): senses when the engine generator is producing acceptable voltage, and if the frequency (see FSE) is also acceptable, signals the transfer switch to transfer the loads to the generator.

VSN (Voltage Sensor, Normal): senses when the utility power (normal source) drops below an acceptable value on any phase and, through the Engine Control Contact, triggers the engine-starting controls. Also senses when the utility power is again acceptable and signals the transfer switch to return the loads back to the restored utility. Then signals the engine controls to begin cool-down and shutdown.

In addition to the items mentioned above, which are basic for a reliable automatic emergency power control system, other lights, meters, and specialized accessories may be added, depending on the installation. Accessories may include additional time delays, manual controls, engine–generator control accessories, and indicators. However, the items listed, if properly designed and coordinated, will provide reliable control for automatic transfer of two sources of power: one normal, the other emergency. Figure 20–33 has all the controls necessary to handle a power failure and restoration in a facility having one normal and one emergency source of power. One enclosure houses all the controls for switching purposes.

It is important to test and inspect the automatic transfer switch because it is the heart and brains of the emergency power system. It senses the power failure, signals the engine to start, transfers the loads and then

FIGURE 20–34 *Automatic transfer and bypass-isolation switch. (Automatic Switch)*

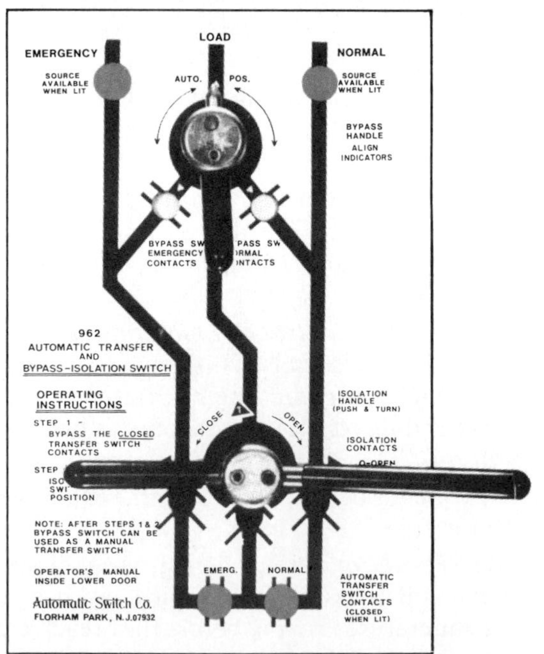

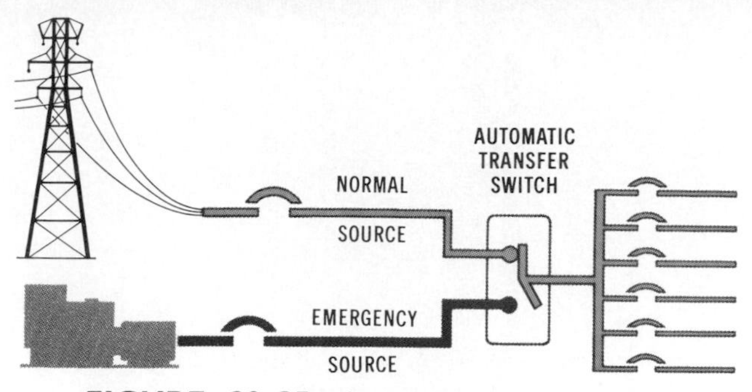

FIGURE 20–35 Automatic transfer switch, basic system. (Automatic Switch)

retransfers the load back to normal, and shuts down the engine. If the automatic transfer switch fails, nothing else in the system will respond to the need.

However, in hospitals and other facilities where an uninterrupted supply of power is vital to human life—places where the automatic transfer switches must be periodically tested—there is a reluctance to do so because maintenance personnel do not want to interrupt the power even momentarily. For facilities such as these where power interruption and downtime cannot be tolerated, an automatic *transfer and bypass isolation switch* (Fig. 20–34) solves the problem. The automatic transfer and isolation switch unit goes into the same location in the electrical circuit as the automatic transfer switch (Fig. 20–35).

Switchboards

There are a number of manufacturers of switchboards. Each does the same thing but in a slightly different way. Square D switchboards with 200,000-A short-circuit rating are referred to as QMB. This type of switchboard uses fusible switches. The switches are of modular design (Fig. 20–36). Red and black operating handles and large on–off nameplates clearly iden-

FIGURE 20–36 Fusible switches QMB. (Square D)

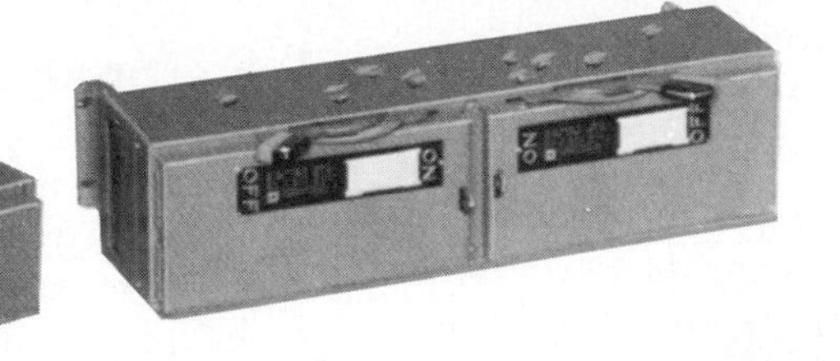

60 Ampere
2 or 3-Pole
240V. or 600V. ac, 250V. dc
Twin branch switch units

30 Ampere
2 or 3-Pole
240V. or 600V. ac, 250V. dc
Twin branch switch units

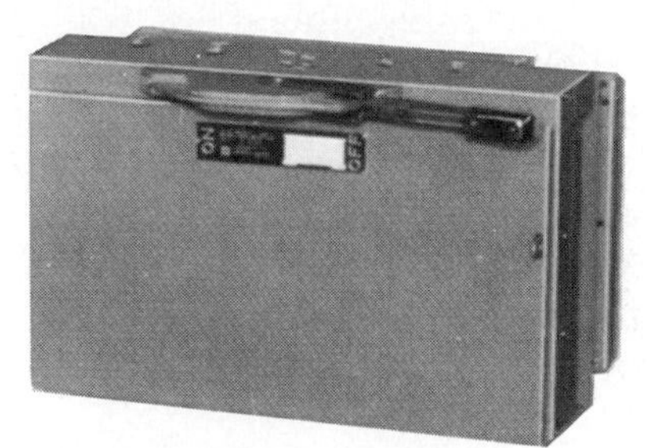

400, 600 or 800 Ampere
2 or 3-Pole
480Y/277V. ac
Single branch switch units

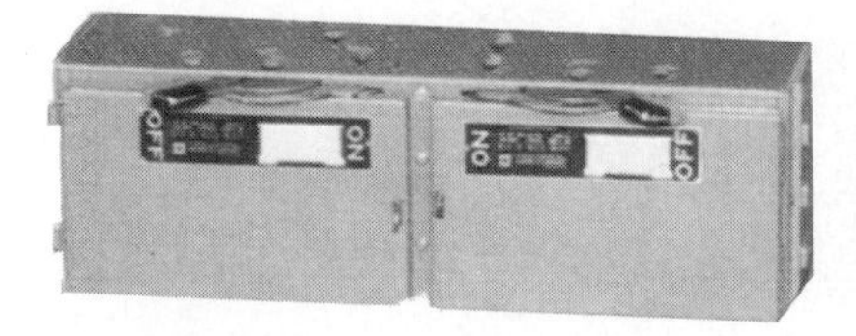

100 Ampere
2 or 3-Pole
240V. or 600V. ac, 250V. dc
Twin branch switch units

200 Ampere
2 or 3-Pole
240V. or 600V. ac, 250V. dc
Single branch switch units

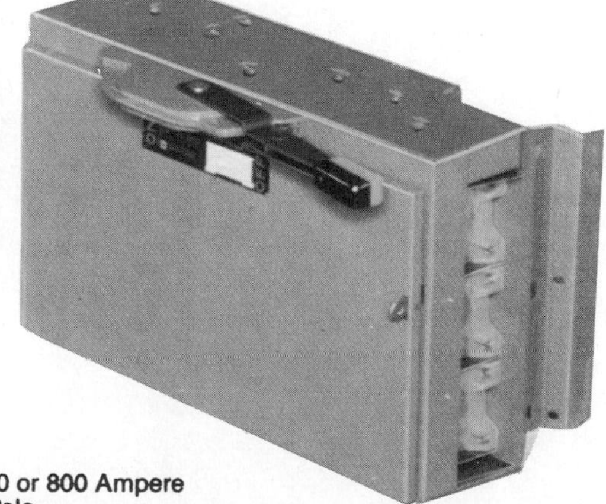

400, 600 or 800 Ampere
2 or 3-Pole
240V. or 600V. ac, 250V. dc
Single branch switch units

FIGURE 20–37 *(A) Powerstyle switchboard; (B) Speed-D switchboard. (Square D)*

tify the position of the switchblades. When the switch door is open, the switch blades are completely visible and the entire length of the fuse is exposed for circuit testing or installation and replacement of fuses. They also have built-in fuse pullers (through 100 A) with dual cover interlocks and positive lock-off means. These switches are rated from 30 to 800 A. Switchboards are shown in Fig. 20–37. Circuit breaker units for motor starter applications are available with either thermal-magnetic circuit breakers or magnetic trip circuit breakers (Fig. 20–38).

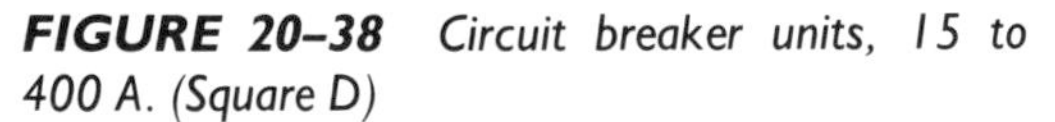

FIGURE 20–38 *Circuit breaker units, 15 to 400 A. (Square D)*

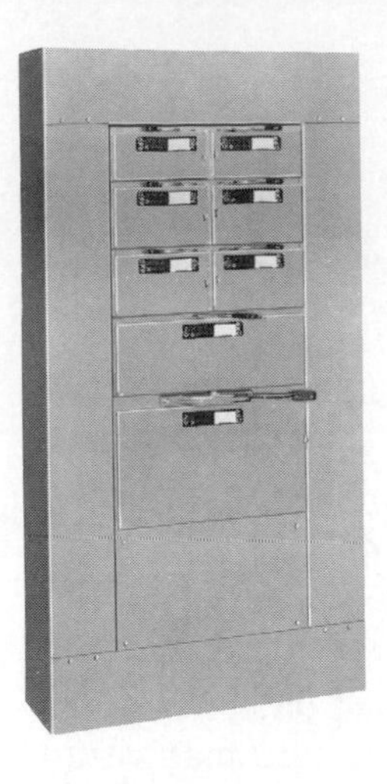

FIGURE 20–39 *Motor starter center. (Square D)*

Motor starter centers are available for 120 to 600 V with reversing or nonreversing types. They have pushbuttons, pilot lights, control voltage transformers, and fuse blocks (Fig. 20–39).

Panelboards

Panelboards are available from many different manufacturers. A wide range of panel types is available for applications to 600 V ac with 10,000 through 200,000 A maximum short-circuit current rating. Plug-in or bolt-on branch circuit breakers are available. Plug-in circuits are locked in position with dead front-panel cover, assuring positive contact.

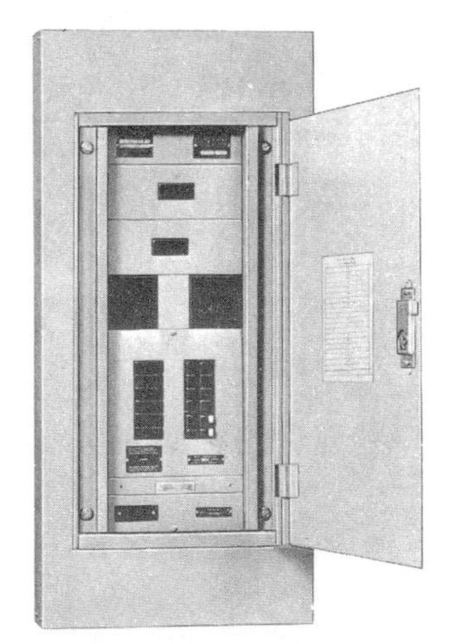

FIGURE 20–40 *Power panelboard. (Westinghouse)*

Power panelboards have a wide range of mains capacities and circuit flexibility (Fig. 20–40). Main lugs are available from 225 to 1600 Amperes, main breakers or main fusible switches to 1200 Amperes. Branch circuit breakers or fusible switches are rated 15 to 1200 A, 600 V ac–250 V dc maximum. Figure 20–41

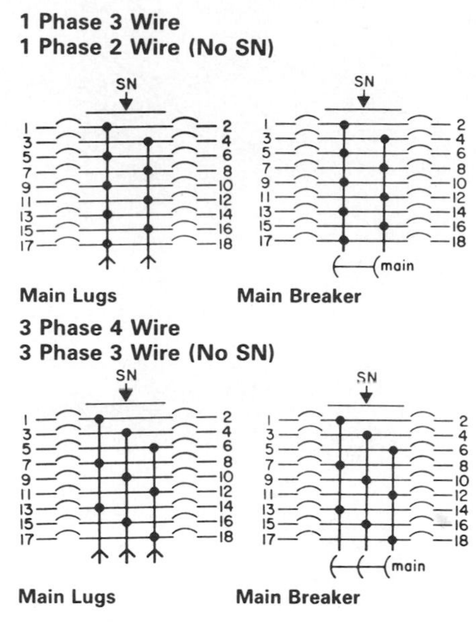

FIGURE 20–41 *Typical panel wiring diagrams. (Westinghouse)*

shows the typical panelboard wiring diagrams for single-phase and three-phase.

Raceways

In the distribution of electrical power to various locations within a building, a number of methods have been devised by various manufacturers; it would require an entire book just to give examples. However, the main purpose is to distribute the power to the motor or device that needs it and do the job without exposing people or equipment to the dangers of high-voltage and high-current-carrying cables.

Aluminum Lay-In Wall Duct and Floor Trench Duct

Evolving technology in medical care facilities has identified the need for a nonferrous raceway system. Magnetic resonance imaging (MRI) is a new diagnostic procedure that utilizes magnetic field rather than x-rays. Because the equipment uses magnets, it is important that all or as much ferrous material as possible be eliminated from the room (Fig. 20–42). These rooms will normally utilize a raised floor in areas with the magnetic and electronic equipment. Along with the raised floor, cables must be routed back to control rooms. This can be accomplished with aluminum conduit, but aluminum lay-in wall duct and in some cases aluminum floor trench are preferred.

Cable Tray

An economical raceway system designed to support and protect electrical wire and cable is available in aluminum and two types of galvanized steel for outdoor or indoor applications (Fig. 20–43).

Cable trays are not raceways. They are covered

FIGURE 20–42 *Aluminum lay-in wall duct and floor trench duct. (Square D)*

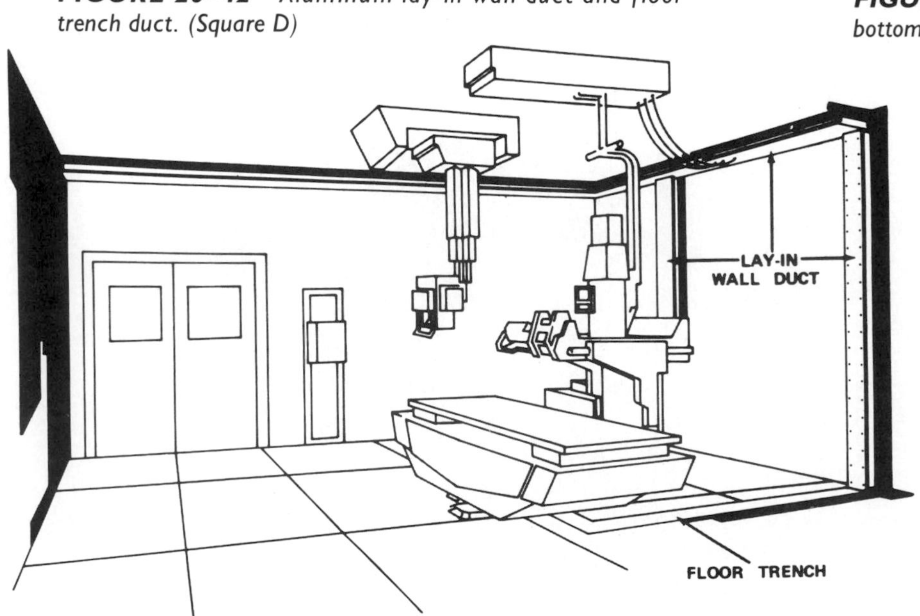

FIGURE 20–43 *Cable tray, ladder, trough, solid bottom channel. (Square D)*

Ladder

Used for support of larger power cables, ladder type provides maximum conductor ventilation.

Trough

Provides additional support for smaller cables. Trough has smooth surfaces and adequate openings for cable dropouts.

Solid Bottom

This tray provides shielding for sensitive communication and signal circuitry.

Channel

Used to carry single feeder or small branch cables to termination points. Channel offers a low cost alternative to conduit.

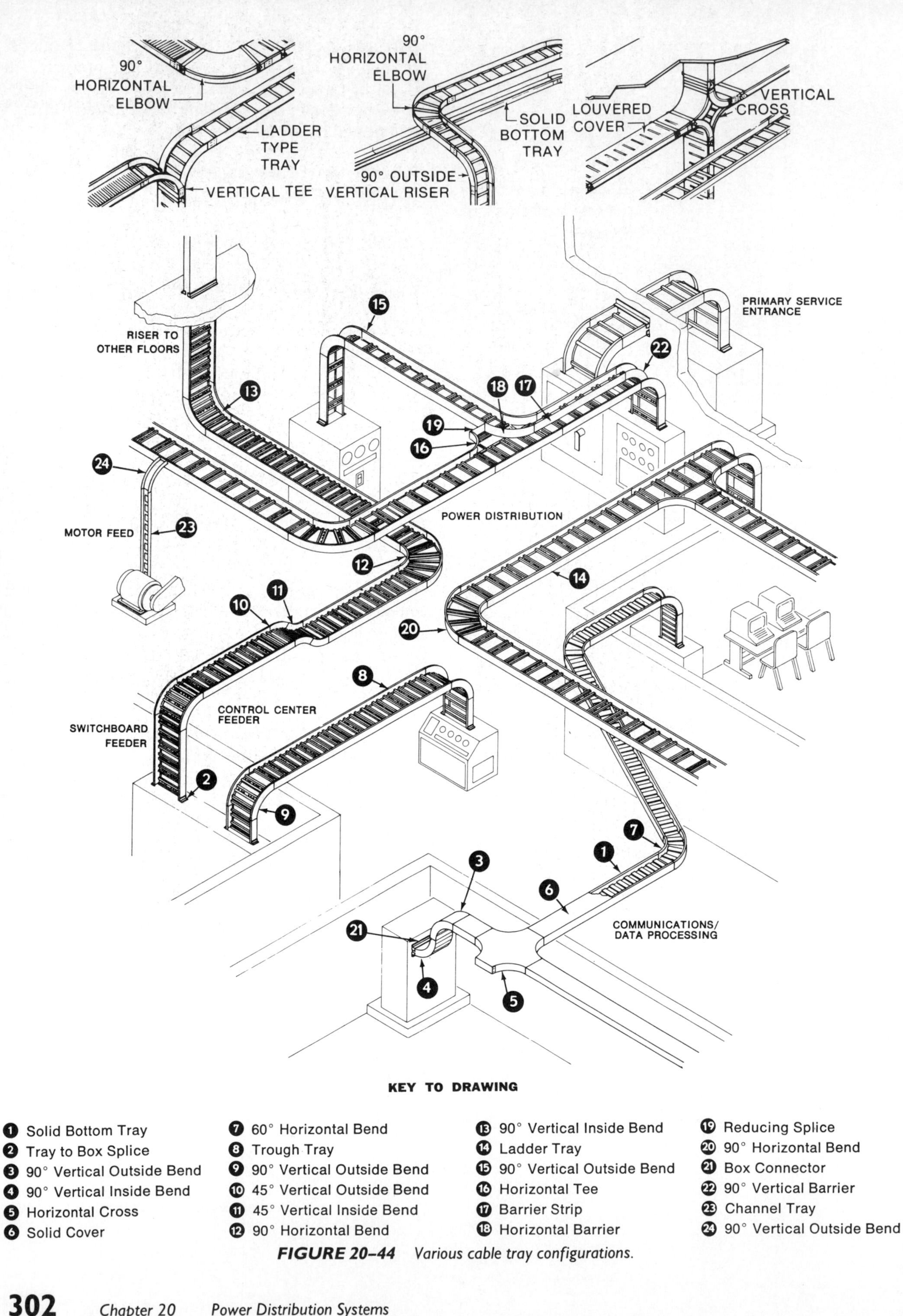

FIGURE 20–44 *Various cable tray configurations.*

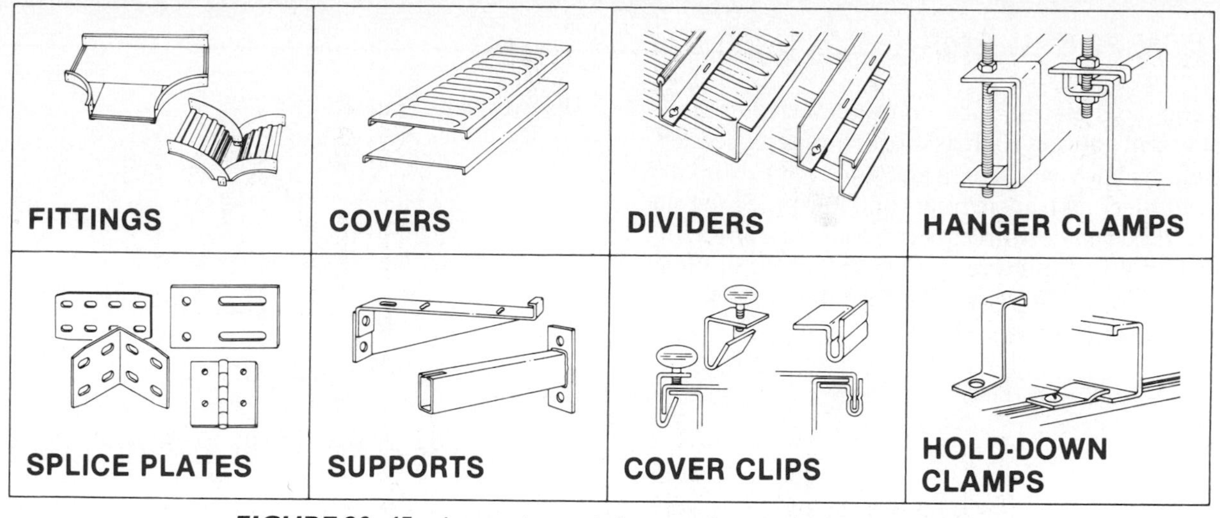

FIGURE 20–45 *Accessories needed to complete a cable tray installation.*

by Article 318 of the *National Electrical Code®*. Cable trays are open raceway-like assemblies made of steel, aluminum, or a suitable nonmetallic material. They are used in buildings to route cables and support them out of the way of normal building activities. A strong, sturdy support for cables is provided through troughs and ladders to route the cables to their destination or termination.

Figure 20–44 provides examples of how the trays can be routed and used to support heavy cables. The trays are made in straight sections, with matching fittings to accommodate all changes of direction or quantity of cables. They are usually made of aluminum or zinc-coated steel.

Trough-type trays protect cables from damage and give good support and ample ventilation. Solid-bottom fittings generally create no ventilation problems since they are a small part of the system. Cables are adequately ventilated through straight sections. Ladder trays provide maximum ventilation to power cables and other heat-producing cables.

However, cables are vulnerable to damage and covers are available. Various parts are needed to support the trays and covers (Fig. 20–45). Cables suitable for use in cable trays are marked CT (cable tray) on the outside of the jacket.

The cable system must be complete. It must be used as a complete system of straight sections, angles, offsets, saddles, and other associated parts to form a cable support system that is continuous and grounded as required by the NEC in Section 318-6(a). The system must be grounded as any raceway system must also be grounded. The Code treats the cable tray as a raceway and a wiring method. Limitations are placed on the number, size, and placement of conductors inside the tray. These limitations can be obtained by checking the Code.

QUESTIONS

1. What is the simple radial system?
2. What is the loop primary radial system?
3. What is the banked-secondary radial system?
4. What is the primary-selective radial system?
5. What is the secondary-selective radial system?
6. What is the modified secondary-selective radial system?
7. What is a simple network system?
8. What is a simple spot-network system?
9. What is a primary selective network?
10. What are fault currents?
11. What is the most generally applicable and widely used form of industrial secondary network system?
12. Differentiate between a core and a shell transformer.
13. Describe an automatic emergency power switching system.
14. Why is periodic testing of an emergency power system necessary?
15. What is a raceway?
16. What is a panelboard?
17. Are cable trays classified as raceways?
18. How are cables for cable trays marked?
19. Why is reliable circuit protection needed?
20. What does *VSE* stand for in emergency power service?

REVIEW PROBLEMS

Delta and wye circuits are commonplace in three-phase current sources such as most commercially generated electrical power. Delta and wye connections are used in connecting transformers and in connecting the loads to these power sources. Formulas for converting from delta to wye and wye to delta are useful in figuring three-phase resistances. The formulas shown here can be used for purposes of review.

Delta-to-Wye (Fig. P–1):

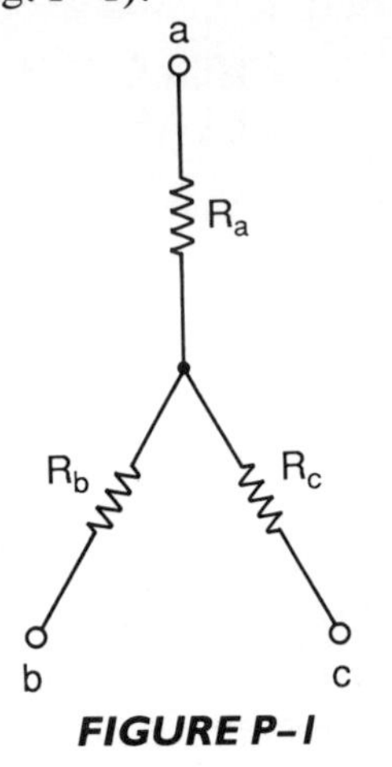

FIGURE P–1

$$R_a = \frac{R_y \times R_z}{R_x + R_y + R_z}$$

$$R_b = \frac{R_x \times R_z}{R_x + R_y + R_z}$$

$$R_c = \frac{R_x \times R_y}{R_x + R_y + R_z}$$

Wye-to-Delta (Fig. P–2):

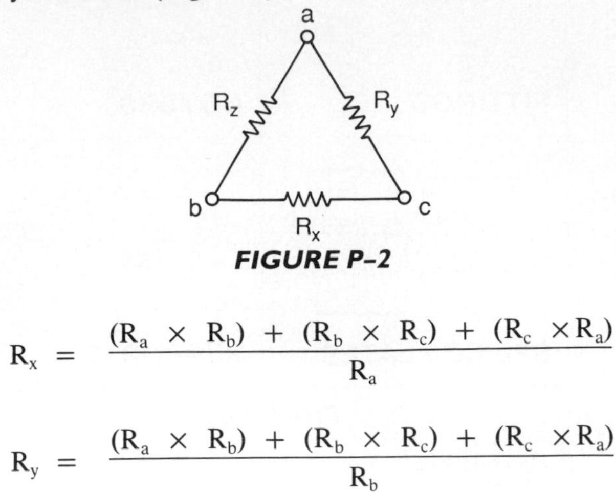

FIGURE P–2

$$R_x = \frac{(R_a \times R_b) + (R_b \times R_c) + (R_c \times R_a)}{R_a}$$

$$R_y = \frac{(R_a \times R_b) + (R_b \times R_c) + (R_c \times R_a)}{R_b}$$

$$R_z = \frac{(R_a \times R_b) + (R_b \times R_c) + (R_c \times R_a)}{R_c}$$

1. In the circuit shown in Fig. P–1, R_a is 6000 Ω, R_b is 2000 Ω, and R_c is 3000 Ω. What is the value of R_y in the equivalent delta circuit?
2. Find the value of R_x in problem 1.
3. Find the value of R_z in problem 1.
4. If the delta circuit shown in Fig. P–2, has all resistors of the same size (12,000 Ω). Find the value of the resistors in the equivalent wye circuit.
5. In the circuit shown in Fig. P–2, R_x is 20,000 Ω, R_y is 10,000 Ω, and R_z is 30,000 Ω. What is the value of R_c in the equivalent wye circuit?

21

Programmable Controllers

Objectives

After studying this chapter, you will be able to:

1. Describe solid-state controllers that use PWM for operation.
2. List standard electronic features in controls.
3. Identify parts of a PLC system.
4. Define interfacing and input–output.
5. Differentiate between parallel and serial ports on a PLC.
6. Explain the RS-232C standard.
7. Identify PLC problems with electrical noise.
8. Explain solid-state reliability in PLCs.
9. List at least five processors available with Square D PLC systems.
10. List three types of display systems used with PLCs.
11. Explain the operation of a cell controller.
12. Explain the difference between a microcell and a minicell.
13. Discuss the future of PLCs.

The National Electrical Manufacturers Association gives the following definition for a programmable controller (PLC): *a digitally operating electronic apparatus which uses a programmable memory for the internal storage of instructions for implementing specific functions such as logic, sequencing, timing, counting and arithmetic to control, through digital or analog input/output modules, various types of machines or processes.*

SOLID-STATE ELECTRONICS

Solid-state electronics have gradually been making the control of motors easier and predictable in terms of load requirements and changes in loads. Electronics devices have been used to detect phase failure, phase reversals, open circuits, and short circuits. Electronic devices have also been used to provide reduced starting current and high starting torque. They also reduce the voltage to lightly loaded motors. Reducing the voltage in these applications can save energy.

Typical applications for controllers are in motors used in food-processing facilities, beverage bottling facilities, textile machinery, cranes, belt-driven equipment, conveyors, general machinery, materials-handling facilities, compressors, machine tools, woodworking equipment, and water treatment facilities. Figure 21–1 shows a solid-state ac motor control circuitry. Note the control logic and overload protection module. This type of electronics is utilized in the soft start and voltage reduction during light-loads situations.

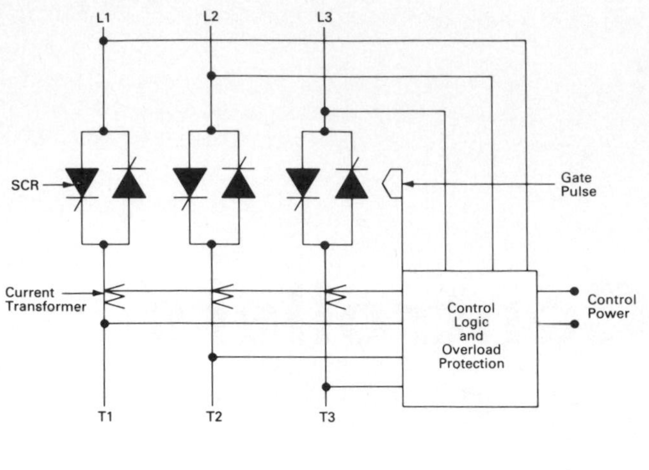

FIGURE 21-1 *Solid-state ac motor control circuitry. (Westinghouse)*

The conversion of the fixed-frequency input to an adjustable-frequency output is also possible using electronics for adjustable-frequency drives. The rectifier converts incoming ac supply voltage to a fixed-potential dc bus level. The dc voltage is in turn inverted by a three-phase, pulse-width modulated (PWM) inverter section to an adjustable frequency output whose voltage is also adjusted proportionately to the frequency to provide constant voltage per hertz excitation to the motor terminals up to 60 Hz. Above 60 Hz, the voltage remains constant at nominal motor full-voltage rating. In this way energy-efficient speed control is obtained in the range from 6 to 120 Hz.

Standard Electronic Features in Controls

Some of the standard protective features provided by electronics circuitry are instantaneous power failure protection. The controller trips if power outage exceeds 15 milliseconds. Electronic instantaneous overcurrent protection, inverse time overload protection, and undervoltage protection are provided by electronics. Dc bus overload protection and controller overtemperature protection are provided by electronics. Torque (current) limit protect, if necessary, is automatically extended to accelerate or decelerate. When running under stated state conditions, current limit will reduce output frequency when the adjustable value is exceeded. Electronic ground-fault protection is also available in the package. Surge protection from input ac line transients is provided by a snubber network and electrical isolation is provided between the power and the logic circuits. All this and more can be accomplished by a programmable controller.

THE PROGRAMMABLE CONTROLLER

The programmable controller (PLC)* is made up of a power supply, a processor unit, and I/O modules. Figure 21-2 shows a block diagram of a programmable controller system. The inputs can be a keyboard or limit switches, pressure switches, thermostats, or any number of devices that can provide an *on-off* status.

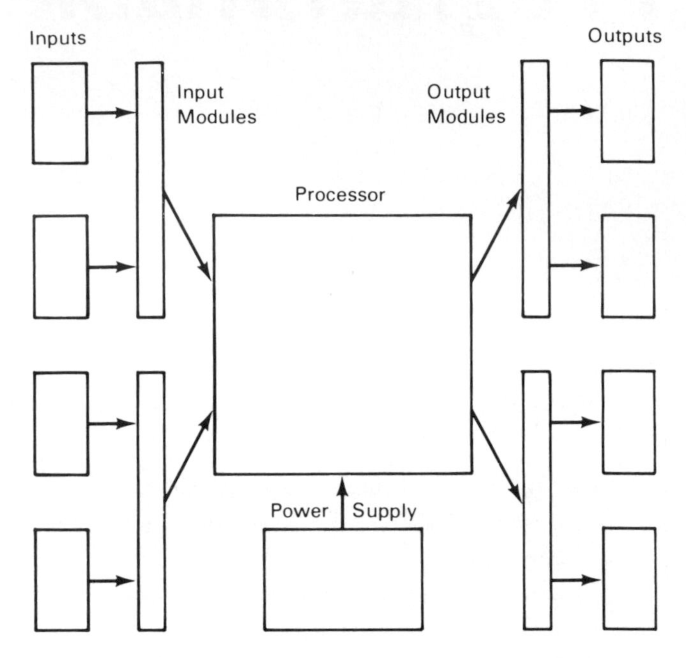

FIGURE 21-2 *Block diagram for PLC system, including I/Os.*

The controller has a memory so that it can be programmed and can retain the information presented to it and compare it to what has been programmed and then make a decision and send the required signal by way of the output modules to the motor it is controlling.

Input-Output

The connections that interface with the outside world of the programmable controller are called input-output (I/O) ports. The input port allows data from a keyboard or other input device to be taken into the controller. The output port is used to send data to an output device such as a motor. Bus lines carry the signals to and from the major parts of the system.

*Inasmuch as you see the abbreviation PC used to mean *personal computer,* it is now suggested that PLC (for *programmable logic controller*) be used for programmable controller to differentiate it from the PC.

Interfacing

Interfacing is used to describe what happens when the controller communicates with the motor it is controlling and the switches or sensing devices that sense require information needed for the process to be completed. The programmable controller is able to communicate with the rest of the equipment around it by being connected through *ports.* It is necessary to input information to the controller and is necessary to receive information from the controller after it has processed it according to its program.

In some instances the programmable controller can be programmed using a computer keyboard and the information can be displayed on a cathode ray tube (CRT) and checked for accuracy before being stored in the controller's memory.

ASCII Code

Digital information is processed and communicated by using the ASCII Code (Table 21–1). To communicate with the controller, it is necessary to be able to input instructions to the microprocessor or chip that makes up the processing unit. A special code has been designed so that the regular typewriter type of keyboard can be used to type in instructions. However, the keys of the keyboard are switches and they send a pulse to a decoder which generates a special binary code. That code is most often the American Standard Code for Information Interchange (ASCII).

TABLE 21–1 SEVEN-BIT ASCII CODE FOR DIGITAL USE

A	1 000 001	a	1 100 001	0	0 110 000
B	1 000 010	b	1 100 010	1	0 110 001
C	1 000 011	c	1 100 011	2	0 110 010
D	1 000 100	d	1 100 100	3	0 110 011
E	1 000 101	e	1 100 101	4	0 110 100
F	1 000 110	f	1 100 110	5	0 110 101
G	1 000 111	g	1 100 111	6	0 110 110
H	1 001 000	h	1 101 000	7	0 110 111
I	1 001 001	i	1 101 001	8	0 111 000
J	1 001 010	j	1 101 010	9	0 111 001
K	1 001 011	k	1 101 011	SP	0 100 000
L	1 001 100	l	1 101 100	I	0 100 001
M	1 001 101	m	1 101 101	‘‘	0 100 010
N	1 001 110	n	1 101 110	#	0 100 011
O	1 001 111	o	1 101 111	$	0 100 100
P	1 010 000	p	1 110 000	%	0 100 101
Q	1 010 001	q	1 110 001	&	0 100 110
R	1 010 010	r	1 110 010	•	0 100 111
S	1 010 011	s	1 110 011	(	0 101 000
T	1 010 100	t	1 110 100	)	0 101 001
U	1 010 101	u	1 110 101	*	0 101 010
V	1 010 110	v	1 110 110	+	0 101 011
W	1 010 111	w	1 110 111	'	0 101 100
X	1 011 000	x	1 111 000	-	0 101 101
Y	1 011 001	y	1 111 001	·	0 101 110
Z	1 011 010	z	1 111 010	/	0 101 111

The ASCII code is made up of seven binary bits, so there are 128 possible combinations. This is obtained when you take 2 to the seventh power (2^7). The 128 (2^7) possible combinations of 1's and 0's can represent all the letters of the alphabet, both upper and lower case, as well as the numbers 0 through 9 and several special codes that include punctuation and machine control information.

Parallel Ports. Parallel ports are the outputs of the microprocessor or computer. The parallel port has a flat cable connected to it with eight conductors. Seven of these wires or conductors carry the information just mentioned. The eighth conductor carries the *strobe line,* the line that prevents the problem of switch bounce. When a switch is closed it also bounces or allows more than the on and off information to be given. It is necessary to eliminate this noise or incorrect signal information from being transmitted from the keyboard to the controller and then to the motor. It may cause the motor to make some incorrect on–off moves.

Serial Ports. The serial format may also be used to transmit data in the ASCII code. The serial format allows the information to be transmitted along two wires. This is very convenient when transmitting over long distances. The parallel format is very good for short distances or connection between machines with the same work cell, but if the information has to be sent for a greater distance, it is better to send it by serial formatting.

The information may be transmitted as changes in voltage or changes in current. There are standards for both. In fact, there are two standards for each. The two voltage standards are known as RS-232C and TTL. The two current standards are the 60-mA current loop and the 20-mA current loop. As you can see, the raw data or the 20 mA is not enough to operate a motor, so it must be used to control a circuit that will turn the motor on and off.

The *RS-232C standard* says that the voltage of the signal will be between −3 and −25 V to represent the logic 1 or "on" condition. A voltage between +3 and +25 V is used to represent the logic 0 or "off" position. This standard was developed by the Electronic Industry Association (EIA). There is an advantage to this standard inasmuch as the line noise will have to be very high to make any false signals and the voltage losses along the line will not affect the signal level as much as lower voltages do. It does have a disadvantage, inasmuch as it has to be converted to transistor-transistor logic (TTL) at the port of the computer or microprocessor.

The *TTL standard* specifies that 5 V presents a logic 1 and 0 volts represents a logic 0. TTL standard is compatible with TTL logic and interfaces directly. There are problems with any transmission of data over a distance. If there is a line voltage of at least $\frac{1}{2}$ V, there is a possibility of receiving incorrect data. Since the peak is only 5 V there is always the possibility of picking up a noise signal when a wire is spread over a distance and in an electrical noise generating environment.

The *60 mA standard* says that a current of 60 mA is logic 1 while zero represents logic 0. The main advantage is that the noise usually encountered over long-distance transmission lines does not affect the quality of the data being transmitted. However, the main disadvantage of this standard is that it has to be converted to voltage variations if used as inputs to a computer port.

The *20 mA standard* is basically the same as the 60 mA standard except that it is 20 mA. The 20 mA level represents logic 1 and zero current represents logic 0. The same advantage is experienced with this standard as with the 60 mA standard. It is also necessary to convert the current variations to voltage variations if used as inputs to a computer port.

THE ENVIRONMENT AFFECTS PERFORMANCE

Many solid-state controls are sensitive to various environment factors but not the same ones as those that generally affect electromechanical devices. Solid-state controls are generally less sensitive to shock and vibration since they contain no moving parts. At relatively high levels of shock and vibration, circuit boards may loosen, and crack or component leads may fail. Some electromechanical components are more susceptible to activation under shock and vibration. In this area solid-state controls may prove superior. Mounting position is also usually of little significance with the solid-state control, except in instances where airflow is required for cooling.

ELECTRICAL NOISE

This noise is defined as any unwanted electrical signal that enters the equipment through various means. Electrical noise is capable of causing various types of malfunctions of solid-state equipment. Electrical noise may cover the entire spectrum of frequencies and exhibit any waveshape. Solid-state devices are especially sensitive to noise since they operate at low signal levels. Also, the speed of solid-state components allows them to respond to relatively high frequencies.

Electrical noise entering a solid-state control is often incapable of damaging components directly unless extremely high energy and/or voltage levels are encountered. Most of the malfunctions due to noise are temporary *nuisance-type* occurrences or *operating errors,* but could result in hazardous machine operations under certain conditions (Fig. 21–3).

Input lines, output lines, and power supply lines are the most common sources of noise entering solid-state controls. Noise may be coupled into the lines electrostatically through capacitance between these lines and lines carrying other signals. A high potential is usually required or long closely spaced conductors are necessary. Magnetic coupling is also quite common when control lines are closely spaced to lines carrying large currents. The signals in this case are coupled through the mutual inductances as in a transformer. Electrostatic and magnetic noise may also be coupled directly into the control logic circuitry. This generally appears in unenclosed, unshielded control electronics and requires a strong noise field. Noise can occur in the form of electromagnetic radiation from remote sources.

Close coupling is not required and the various lines entering the control act as receiving antennas. Occasionally, the control circuitry itself is sufficiently sensitive to detect a radiated signal and respond to it. This type of noise is troublesome when it is encountered since it is usually of high frequency and occasionally difficult to filter and shield against on a generalized basis. A particular installation may require special treatment, since the various coupling elements, such as control lines, exhibit unpredictable characteristics, some of which could render built-in filters inef-

FIGURE 21–3 *Electric noise.*

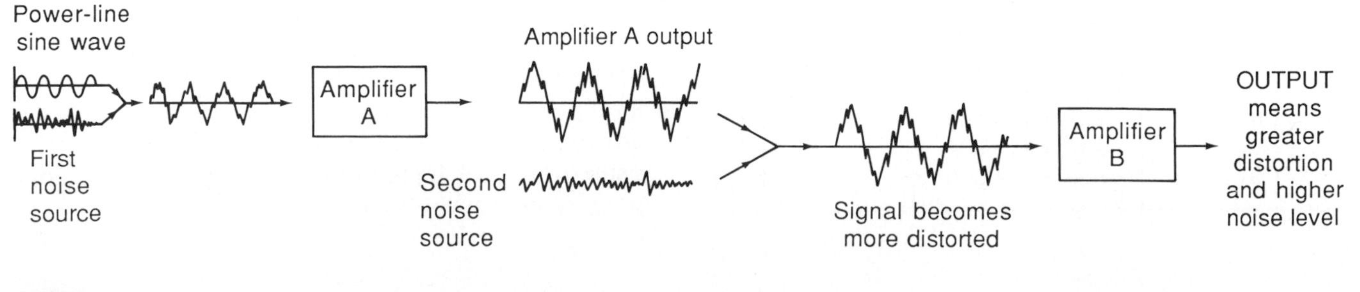

fective. Metal enclosures are effective shields where adequate electrical bonding around doors and bolted surfaces is provided.

Many designs have been tested and retested. Much effort has gone into filters, shielding, and the design of generally insensitive circuitry. It is, however, impossible to design a control that can handle every form of noise coupled into it, especially through the external wiring.

Installation Practices

Noise must be minimized as much as possible from entering the control by using the appropriate installation practices, especially when the anticipated noise signal has characteristics very similar to the desired control input signal.

Grounding

Grounding practices that are used have a significant effect on noise immunity. Each ground should be connected to its respective reference point by no more than one wire, called a single-point ground. Under no circumstance should two or more systems share a common single ground wire, either equipment ground or control common.

SOLID-STATE RELIABILITY

Solid-state components exhibit a high degree of reliability when operated within their ratings. For example, a *triac* might have an average life of 450,000 hours or 50 years under typical operating conditions, but it fails at random even when operated within its ratings. The lifetime above is an average. The time of failure of an individual device cannot be predicted by observation as in the case of a relay, where patterns of wear might be watched. It is thus advisable to provide some sort of independent check on the operation of individual devices when they are controlling a critical or potentially hazardous operation. In addition, the predominant mode of failure of a solid-state output is in the "on" or short-circuit mode. This must be considered in certain critical applications.

Backup Operations

For any size of solid-state control capable of performing potentially hazardous machine operations, *emergency* circuits for stopping operation must be routed outside the controller. For example, devices such as end-of-travel limit switches or emergency-stop pull cord switches should operate motor starters directly without being processed through controller logic. This forms a reliable means of control and should be implemented using a minimum number of simple, highly dependable components of an electromechanical nature, if possible. Thus, in the event of a complete controller failure, an independent rapid shutdown means is available. A convenient means of disconnecting the critical or potentially hazardous portions of a machine from the controller should be provided for use during troubleshooting or setup following maintenance.

CONTROLLERS

A number of manufacturers make programmable controllers. Each has slightly different electronics packages. However, to make sense of the process and

FIGURE 21–4 *Processor family. (Square D)*

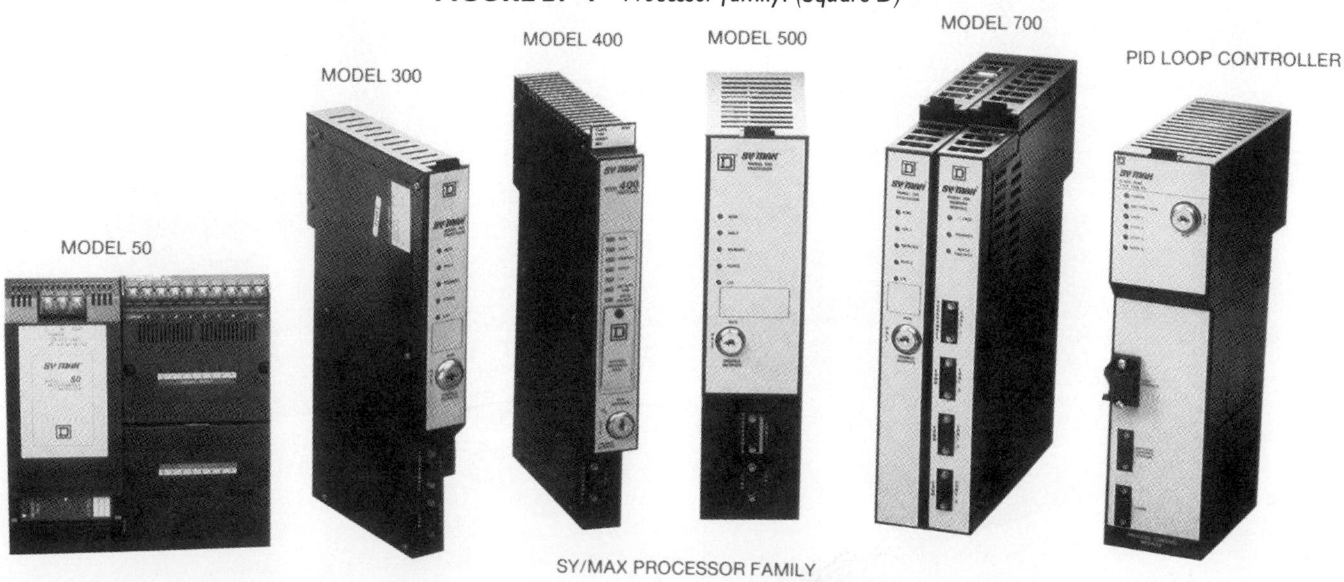

the equipment, one has been chosen for this discussion. It is the Sy/Max, made by Square D (Fig. 21–4). There are a number of processors for this particular manufacturer to suit as many of the requirements of customers as possible.

Programmable controllers are used in a variety of applications to replace conventional control devices, such as relays and solid-state logic. When compared with conventional control means, programmable controllers (PLCs) allow ease of installation, quick and efficient system modifications, more functional capability, troubleshooting diagnostics, and a high degree of reliability. Typical applications include automated material handling, machine tool, and assembly machine control, wood and paper processing control, injection molding machine control and process control applications such as film, chemical, food, and petroleum.

System Hardware and Programming Equipment

The controller family consists of two groups of equipment: system hardware and programming equipment. System hardware is used to control the actual operation, while the programming equipment is used to enter the user control program into the system hardware. Once the program is entered, the programming equipment can be used for monitoring, program alteration, or message displays but is not required for system operation (Fig. 21–5).

System hardware consists of a processor, one or more rack assemblies, power supplies, I/O modules, and various other modules that provide additional capabilities. The rack assemblies and associated I/O modules communicate with external I/O control devices such as limit switches, motor starters, and other devices.

Processors

There are five processors available with the Sy/Max system. The smallest or least expensive, Model 50, can be programmed with an IBM or compatible personal computer using Sy/Mate software in addition to its own hand-held programmer, or the process parameters can be fine tuned with the controller by using the control station shown in Fig. 21–6. Memory is available up to 4K of EPROM with a battery-backed RAM or E^2PROM. The I/O capacity is 256. To make a comparison of what is available, take a look at Model 300 processor. The Model 300 has an I/O capacity of 256. The processor has 128 internal relay equivalents and 96 four-digit storage registers for timers, counters,

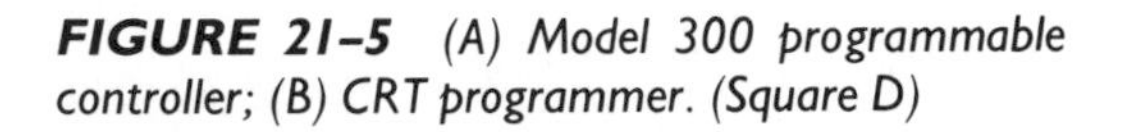

FIGURE 21–5 *(A) Model 300 programmable controller; (B) CRT programmer. (Square D)*

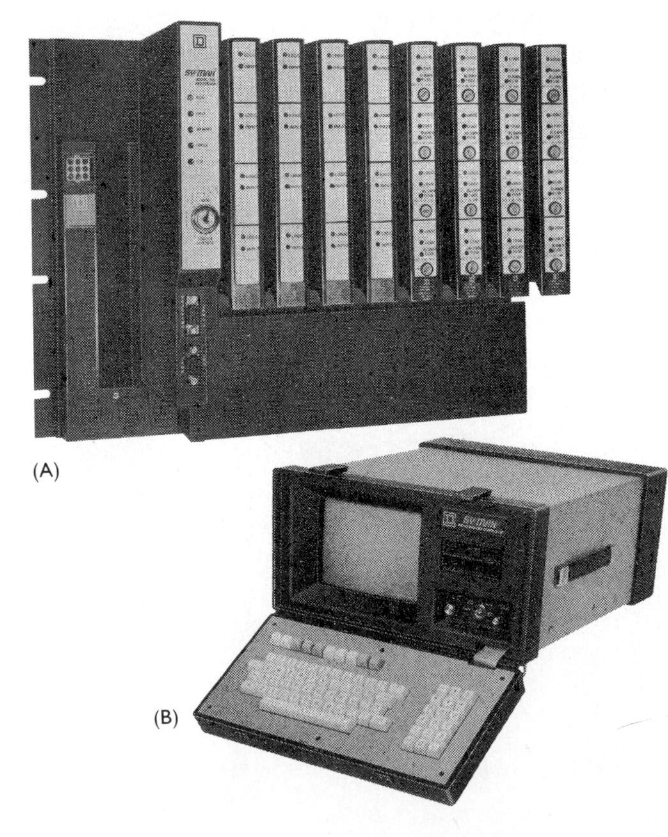

FIGURE 21–6 *PID loop control station. (Square D)*

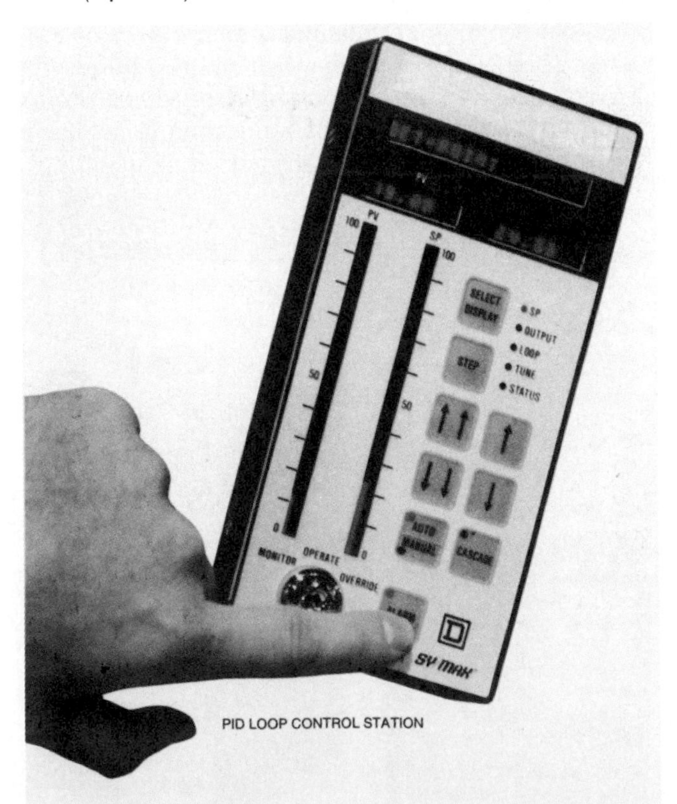

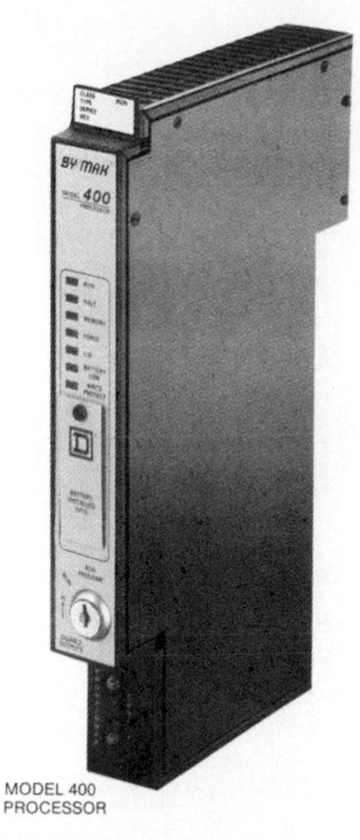

FIGURE 21-7 *Model 400 processor. (Square D)*

synchronous shift registers, and data storage. In addition to relay logic, counting, timing, and data manipulation, this processor offers modular construction for ease of installation and troubleshooting, four-function math, ASCII output to generate alarm messages and reports, plus peer-to-peer communication with other processors through either of two communication ports.

By upgrading to the Model 500 level, other features may be acquired. The 500 can perform square-root math functions, scan control, subroutines, timed interrupts, and matrix operations. It also has several levels of security that can prevent unauthorized access to data and program information through either of its two communications ports. It has up to 8K of battery-backed RAM or combination RAM/EPROM memory and has an I/O capacity of over 2000.

Another upgrade to Model 400 (See Fig. 21-7) produces a processor that can handle almost any application. In addition to Model 500 processor capabilities, the instruction set of the Model 400 includes trigometric, transcendental, and statistical functions and the ability to perform these functions in integer or floating-point (result to $10^{\pm 38}$) format. The PID control algorithm is also part of the 400 instruction set to simplify and speed implementation of process control applications. Another unique feature is the capability of an on-board battery to support the RAM memory and real-time clock upon removal of the processor from the rack. This model can also read ASCII data through either of its two independent communication ports, allowing it to interface directly with ASCII weight scales, bar code readers, and other such inputs for applications such as materials handling. It has up to 16K words of RAM or RAM/EPROM combination and has an I/O capacity of 4000.

Input–Output Modules

The I/O modules provides the interface between the processor and the field device that is being switched or controlled. The I/O modules shown in Fig. 21-8 are

FIGURE 21-8 *I/O systems. (Square D)*

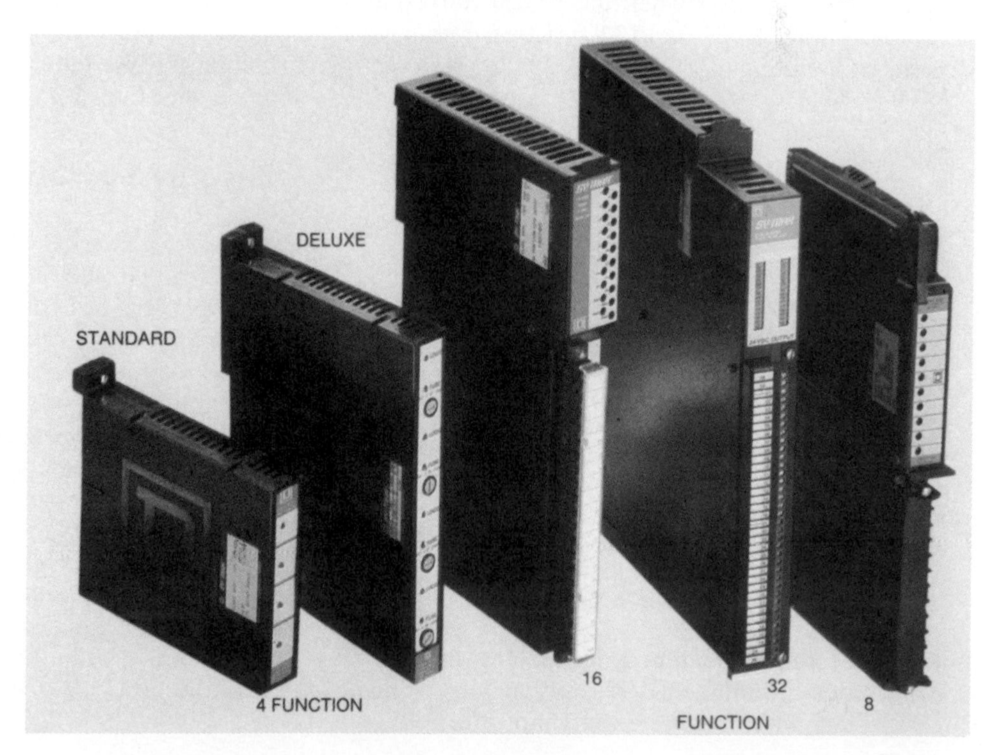

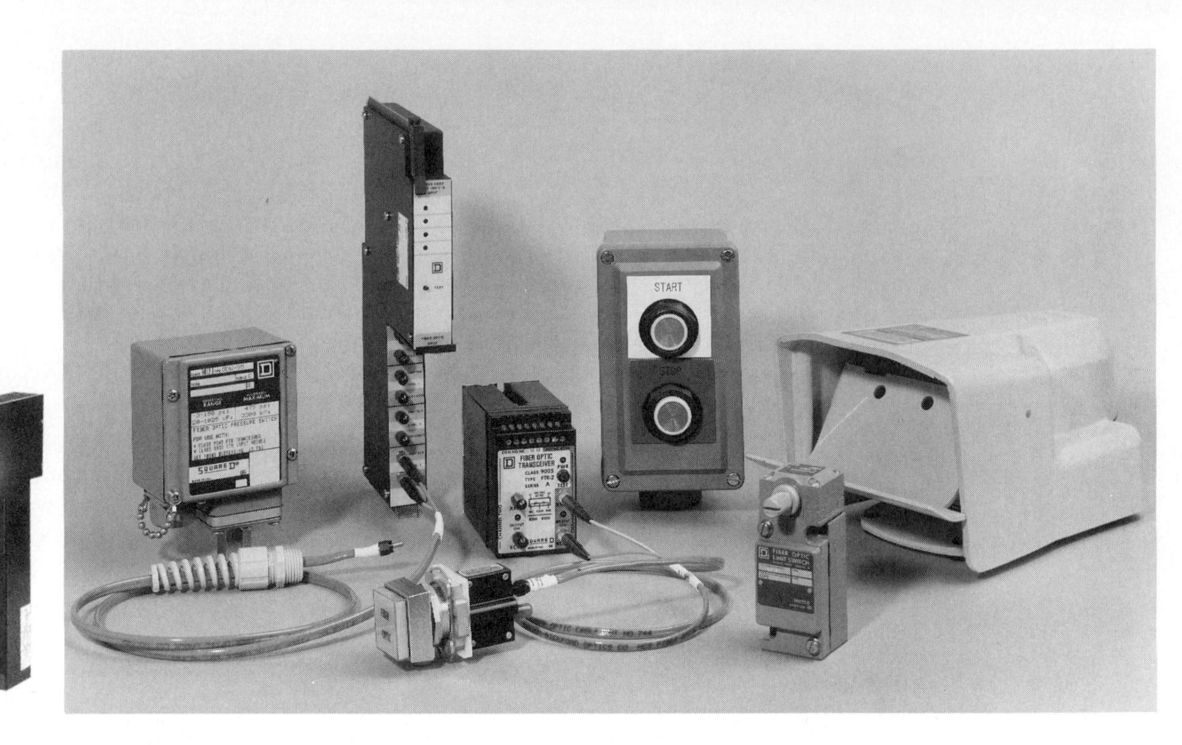

FIGURE 21–9 *Fiber-optic interface and input module with field devices. (Square D)*

available in five versions. The standard four-function covers the range of operating voltages from TTL to 240-V ac/250-V dc. Standard I/Os have a single diagnostic LED for each point. Output fuses are accessible from the side of the module. The deluxe four-function module is interchangeable and compatible with the standard four-function modules but has a high-power (5 A per output) dual-point module and individual power circuit isolation when used with the isolated I/O rack. The 4-, 8-, 16-, and 32-function modules are capable of handling most processes.

The fiber-optic input module has field devices (switches) designed for use with its circuitry (Fig. 21–9). The fiber-optic interface module converts all programmable controller differential communication (remote I/O, programming, etc.) to optical communication. Fiber-optic communications provides immunity from electromagnetic interference (EMI) and radio-frequency interference (RFI) and complete electrical isolation. It is intrinsically safe and uses lightweight, easily installed cable. Maximum length for the cable is 5 miles or 8 km. This type of communication system is useful in process control systems, petrochemical plants, utility power substations, and outdoor/underground installations.

Intelligent I/O Modules

Intelligent (register) I/O modules such as multiplexed binary-coded decimal (BCD) and analog I/O and high-speed counter input, stepping motor output, and speech output modules provide special functions to the PLC. On-board microprocessors in each of these devices allow information to be directly transferred and stored in processor data registers. A plant floor microcomputer allows for a production report. Graphic generation is also available.

The speech module is a synthesized speech/message annunciator. It can provide audible alarm annunciation, operator instructions, or directions to supplement or replace a visual display. The speech module is compatible with the processors and data controllers or any device that can generate ASCII output.

Display Systems

There are a number of display systems that can utilize the features of a programmable controller. Several different types of color graphic displays are available. Color graphics are available with the hardware and software systems that interface with a computer such as that in Fig. 21–10.

The CRT programmer is a portable device that can monitor, program, and document the control logic of any processor mentioned previously (Fig. 21–11). This programmer makes programming the processor easy. This is due to its set of multifunction soft keys. In addition, high-level functions such as math, shift registers, timers, and counters are programmed using simple fill-in-the-blank function boxes in many cases. Programming is fairly easy once you have learned the procedures.

FIGURE 21–10 *Programmers for Sy/Max. (Square D)*

FIGURE 21–11 *CRT display with function names. (Square D)*

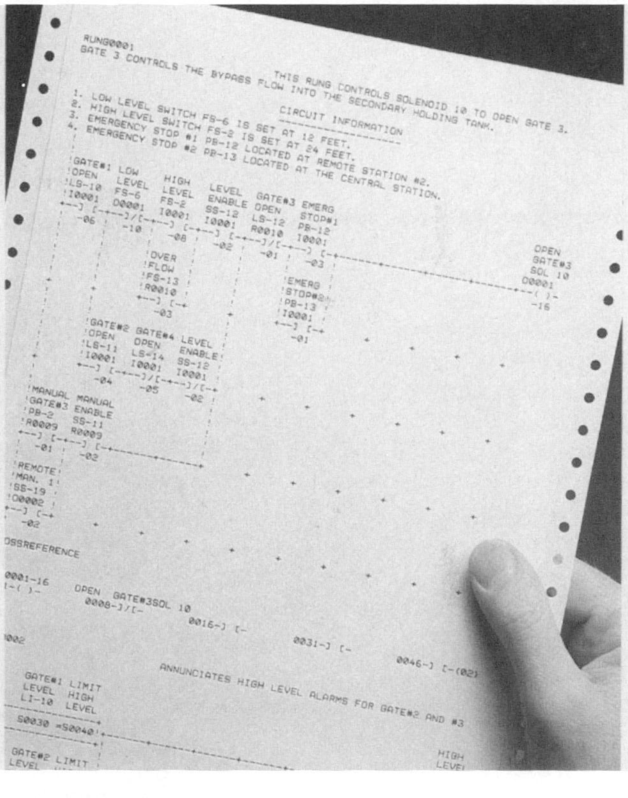

FIGURE 21–12 *Program documentation. (Square D)*

Off-line programming allows the processor control program to be developed, edited, and documented in a nonfactory environment without a processor. The control program can then be stored on tape and downloaded into the processor on the plant floor, reducing overall system development time.

I/O function names improve the ability for maintenance and other plant personnel to understand a detailed control program by displaying the name (up to 12 alphanumeric characters) of the I/O device. This enhances using the CRT to diagnose control system faults. Each I/O element can have up to an 18-character alphanumeric name along with its address. In addition, each logic run can have up to a full-page description of its operation. See the program in Fig. 21–12.

Cell Controllers

A cell controller is typically used to coordinate and manage the operation of a manufacturing cell, consisting of a group of automated programmable machine controls (programmable controllers, robots, etc.) designed to work together and perform a complete manufacturing or process-related task.

Microcell Controller

The *microcell* controller is a programmable multifunction, data and program storage device, designed for small cell control applications (Fig. 21–13).

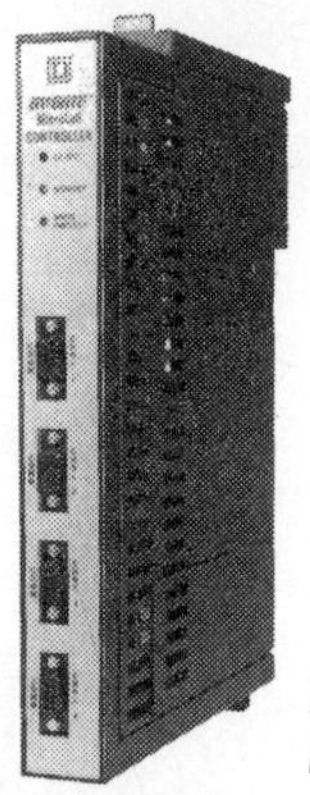

FIGURE 21–13 *Microcell controller. (Square D)*

FIGURE 21–14 *Minicell controller. (Square D)*

The *minicell* controller is the midrange member of the cell controller family. It is designed to perform basic control functions in addition to high-level functions such as data analysis, trending, statistical process control, statistical quality control, color graphic generation and serve as a communications gateway (Fig. 21–14).

Local Area Network

One of the advantages of keeping the same type of programmable controller within a plant is its ability to become part of a local area communication network (Fig. 21–15). The network can have up to 200 controllers and other devices communicate with each other. The network consists of twin-axial cable up to 10,000 feet long and up to 100 network interface modules. Two devices (PLCs, computers, CRTs, printers, etc.) can be connected to each network interface module. The network allows any programmer or programmable package of the acceptable type to be used on the network. Several versions of network interface modules are available.

Future of PLCs

The programmable controller (PLC) has a bright future inasmuch as it will probably be used in all new production facilities. In some instances it has found applications that were once thought to be robot jobs. The cost of robots and their limited re-programmable nature makes the inexpensive PLC the device for the future.

Each manufacturer has a different training program for its particular devices. As you have witnessed

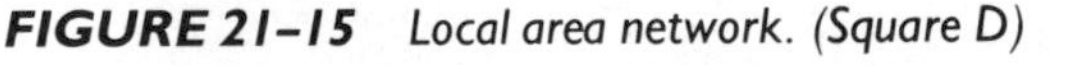
FIGURE 21–15 *Local area network. (Square D)*

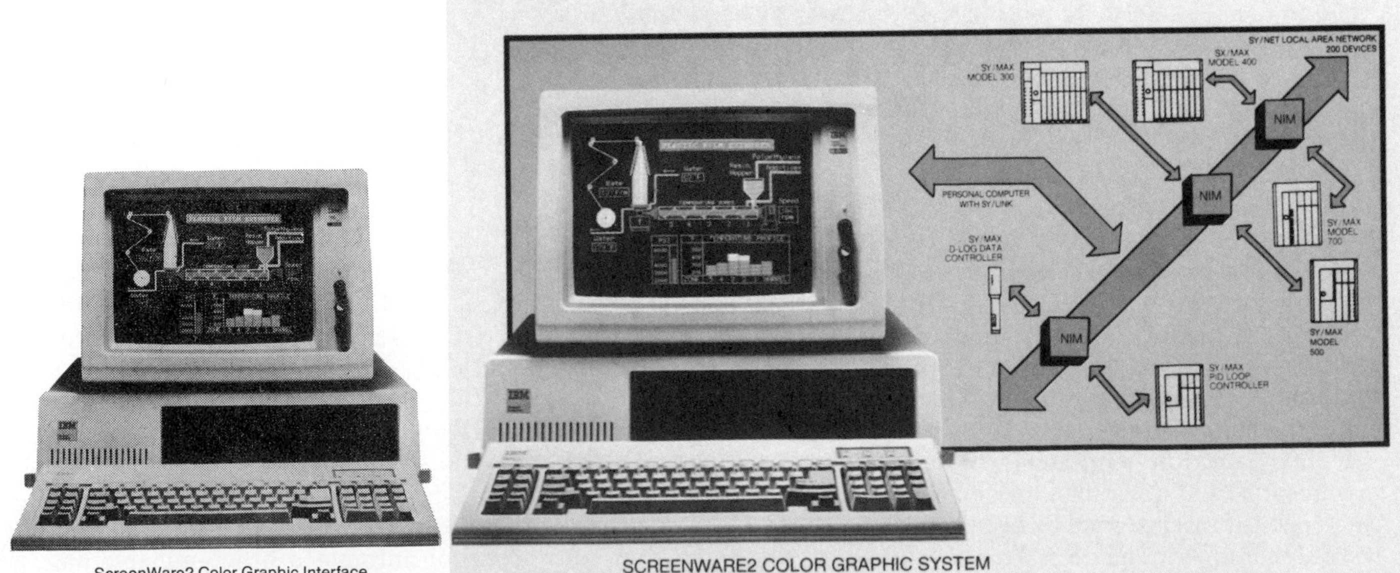

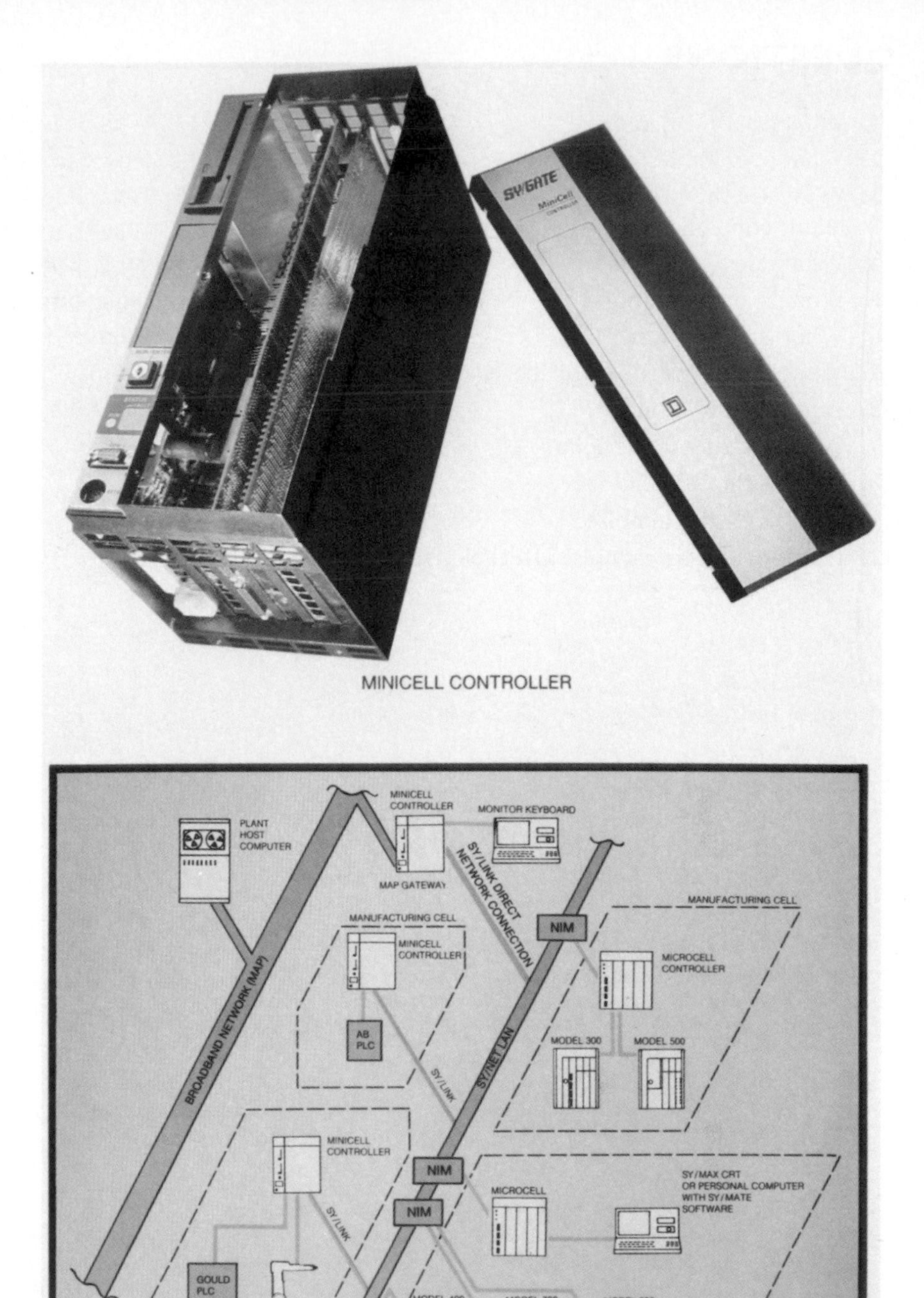

FIGURE 21–16 *Minicell controller. (Square D)*

here it is impossible to cover PLCs in one chapter of a book. It is more a subject for an entire book and training program.

What has been done here is an introduction to some of the concepts, ideas and equipment involved in a PLC system that is used to form part of a larger unit such as a manufacturing cell.

Much more time and effort will be needed by all those involved in electrical motor control to keep abreast of how electronics is doing the job and becoming less expensive and more reliable while doing so.

A good background in digital electronics is necessary to be able to understand how the programmable controller operates and functions in computer integrated manufacturing (CIM) as part of a computer-controlled manufacturing facility (Fig. 21–16).

QUESTIONS

1. What does *PC* denote?
2. What does *PLC* stand for?
3. What are the parts needed to make up a programmable controller?
4. What does *I/O* mean?
5. What is a parallel port?
6. What is a serial port?
7. What does the strobe line do?
8. What does *TTL* mean?
9. Who uses the ASCII code?
10. What is the RS232C code?
11. What is electrical noise?
12. How does electrical noise affect electronic devices?
13. What is EMI?
14. How can a PC be programmed?
15. What does *CRT* mean?
16. What is a cell controller?
17. What is the difference between microcell and minicell controllers?
18. What are electronic devices used for in terms of detecting problems?
19. What are some typical applications for motor controllers?
20. How are fixed-frequency power source inputs changed to adjustable-frequency outputs?

22

Troubleshooting and Maintenance

Objectives

After studying this chapter, you will be able to:

1. Explain how preventive maintenance can prolong trouble-free operation of a system.
2. Explain how to prevent accidental shock.
3. Troubleshoot ac and dc motors.
4. Troubleshoot power supply disturbances.
5. Troubleshoot circuits using a VOM.
6. Troubleshoot circuits using an oscilloscope.
7. Troubleshoot relays.
8. Troubleshoot solid-state motor control equipment.

Troubleshooting is another of the tasks performed by the electrician. It tests your ability to observe everything around you and your ability to understand how things work. One of the best ways to prevent trouble is to check off certain items as a routine procedure to catch trouble before it becomes a major item and causes fire, damage, and/or death. Electrical problems are many, and every connection and every device is a potential problem. Each device and service as well as circuits should have been properly wired, but that is not always the case.

One of the biggest problems is troubleshooting electric motors. A troubleshooting chart will aid in this task as well as some of the more obvious observations made by the person on the scene. All textbook troubleshooting can do is identify the logical problems. On-scene facts are not always detailed in textbooks, so a good observer must also be able to uncover the facts needed to make a diagnosis. Once the problem is found, it is usually easily corrected.

PREVENTIVE MAINTENANCE

Damp and Wet Areas

One of the areas that can cause problems in any home or shop wiring system is dampness and wetness. Watertight equipment should be installed wherever there is a danger of water coming in contact with live wires. One of the largest problems is the condensation of moisture inside the panelboard (Fig. 22–1). Moisture condenses where the warm, moist air in the basement moves up to come in contact with the cold air outside making the riser cold. In areas where this is a problem, either an underground entrance can be made or, an outside riser can be mounted alongside the building, which will make its entrance only when it reaches the panelboard. An entrance as low as possible is preferred so that any moisture that does condense will

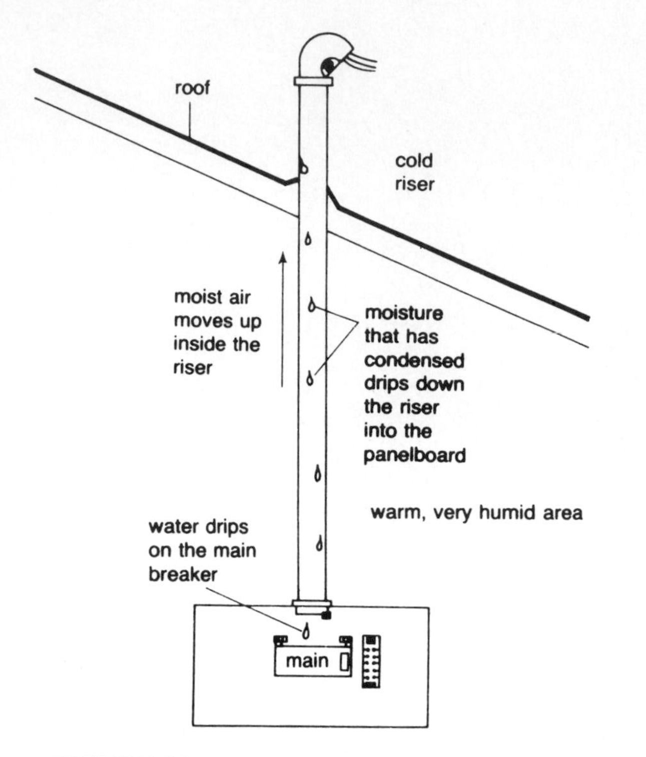

FIGURE 22–1 *Moisture in the panelboard can couse problems.*

easily drain out the bottom of the panelboard without contacting the hot side of the distribution panel.

Another problem is rust and corrosion. Anywhere there is moisture there is the possibility of rust and corrosion. Both rust and corrosion can cause contact problems with metals and remove or place high resistance in the path of a ground system. Removing a ground produces a situation that can be very hazardous (Fig. 22–2). In a single-phase system the current on the neutral of a properly installed 120/240-V system carries the difference between the current flowing on the hot lines. If the ground is removed by corrosion or rust preventing contact with the proper grounding lugs, it is the same as having an open ground.

One of the indications of this condition is that some lights in the building will appear very bright and others very dim. Turn off the main switch and locate the open or corroded ground connection before allowing continued operation. A situation of this sort will make it very dangerous for anyone who touches any of the conductors. That person or animal (in the case of barnyards) will complete the ground circuit and fatal shock may occur.

FIGURE 22–2 *Removing a ground.*

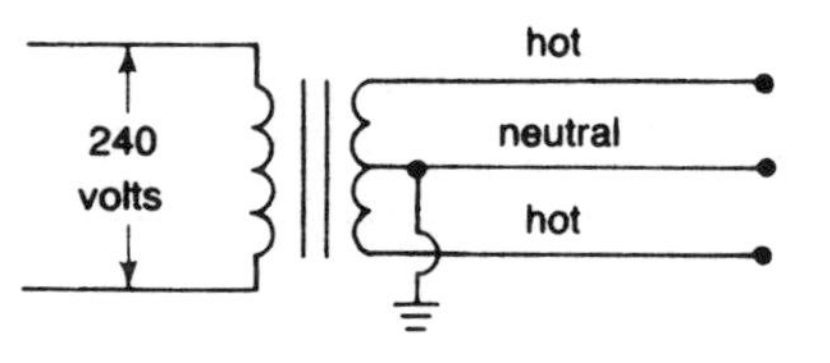

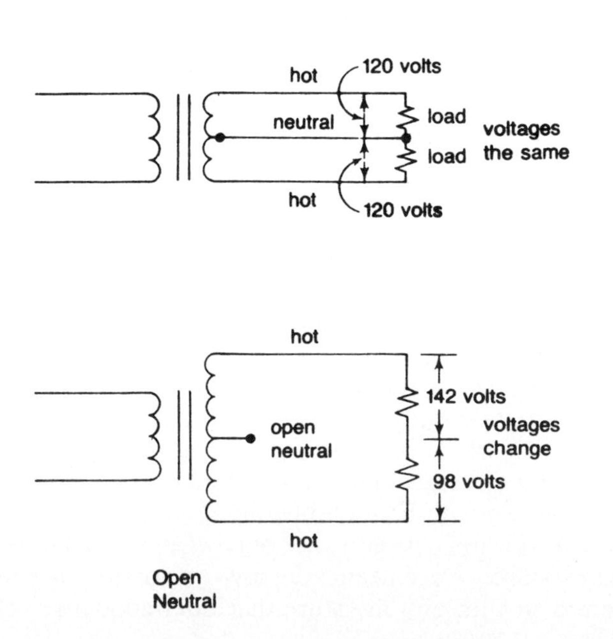

Prevention of Accidental Shock

The ground-fault circuit interrupter (GFCI) is one device used to prevent accidental shock. However, GFCI protection should never be a substitute for good grounding practice, but should support a well-maintained grounding system. Figure 22–3 shows a device that can be used in various locations for the prevention of shock by checking the grounding system. This device is used in homes, plants, and businesses where people are employed and use electrical equipment. It is often encountered by the electrician whether in the home or on the job.

The arrangement shown in Fig. 22–4A checks polarity and grounding. It also diagnoses five other incorrect wiring conditions with a plug-in tester. Figure 22–4B shows the same tester being used to check continuity of the ground path of a tool. This is very important since the hand drill has a metal handle.

In Fig. 22–4C the meter is a ground loop tester; it measures the ground loop impedance of live circuits. It can also be used to check for grounding of tools, piping systems, and other equipment. The meter in Fig. 22–4D is used to check the 500- and 1000-V dc insulation resistance of deenergized circuits and electrical equipment. It also checks for continuity in low-resistance circuits.

The ground-fault circuit interrupter shown in Fig. 22–4E mainly provides insurance that a tool will not develop a fault on the job, causing a serious personal injury. It tests tools to assure that any current leakage is below a hazard level. Keep in mind that nuisance tripping of a GFCI can be caused by a few drops

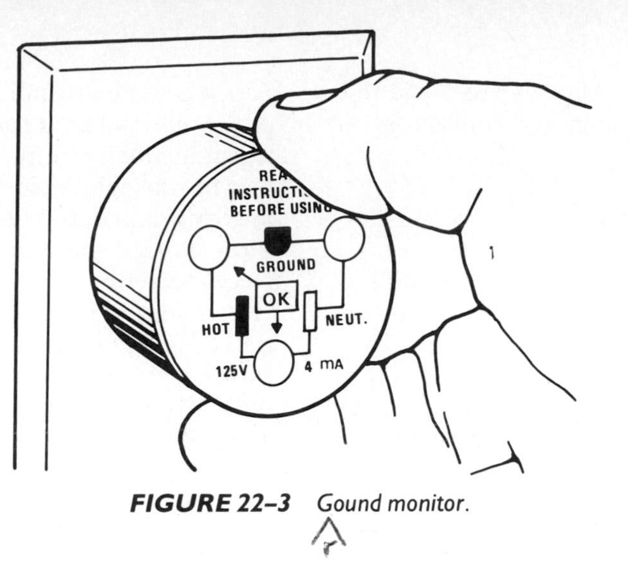

FIGURE 22–3 *Gound monitor.*

FIGURE 22–4 *(A) Ground monitor used to check polarity and grounding; (B) checking grounding path for a tool; (C) ground loop tester; (D) checking impedance of live circuit; (E) testing tools to ensure that leakage is below hazardous level.*

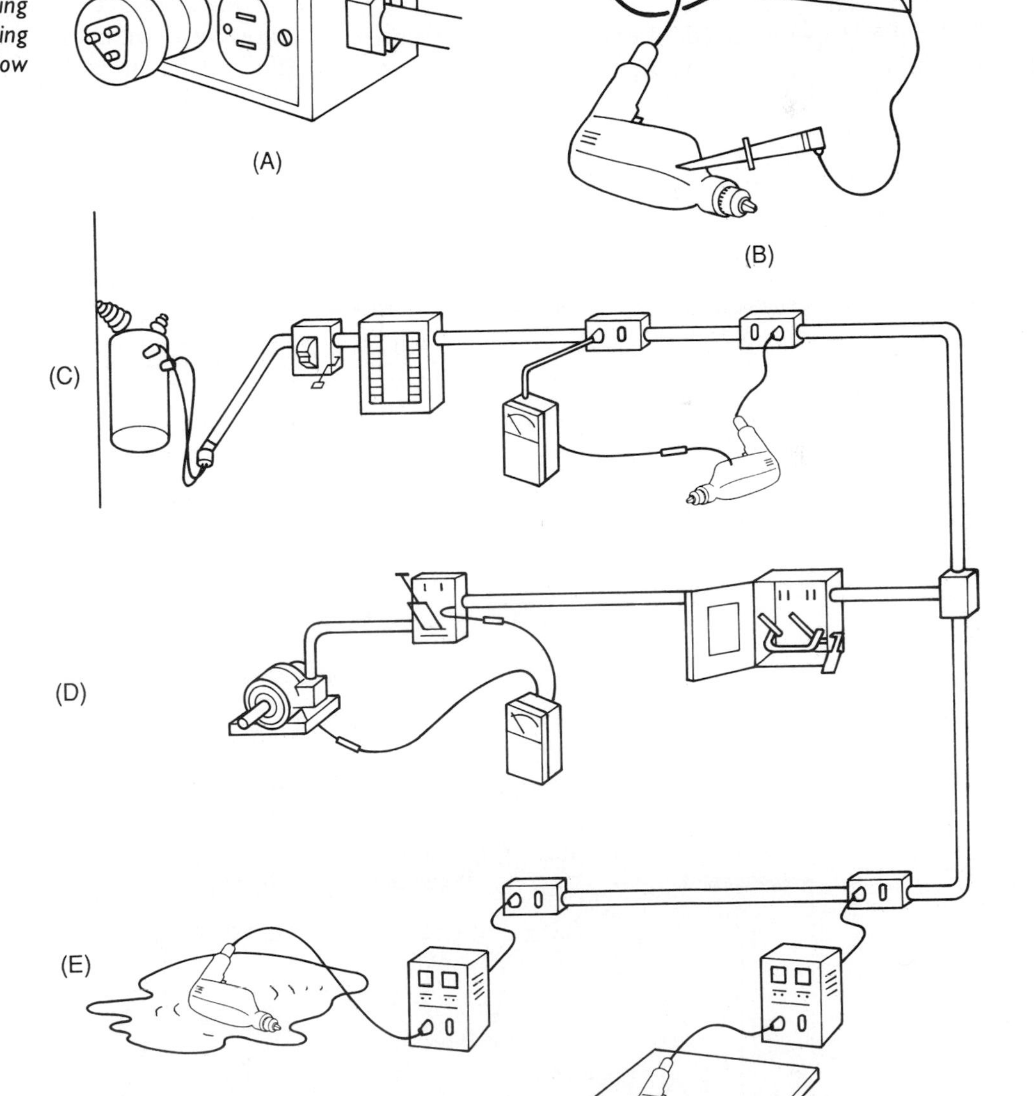

of moisture or flecks of dust. One way to avoid this problem is to use watertight plugs and connectors on extension cords.

Ground-Fault Receptacles

There are two different ways to wire up ground-fault receptacles (GFRs) (Figs. 22–5 and 22–6). The devices shown are not only GFCIs but also receptacles. They can be used, as shown in Fig. 22–6, to protect other downstream receptacles. This brings about problems in some places, inasmuch as the protected outlets are not always known by the persons using them, and when the GFR trips it takes them off-line also.

One way to check for a terminal installation is to check the red and gray wires. If they are capped with a wire nut, you know that GFR does not service any other outlets.

Wiring Devices

Using the proper wiring devices is a form of preventive maintenance inasmuch as it prevents problems later. Shock hazards are minimized by the dielectric strength of the material used for the molded interior walls and the individual wire pocket areas. Nylon seems to be best for this job. Nylon devices withstand high impact in heavy-duty industrial and commercial applications. Each molded piece has to support adjacent molded pieces to result in good resiliency and strength. Devices made of vinyl, neoprene, urea, or phenolic materials can crack or be damaged under pressure. Damage can be invisible and cause direct shorts and other hazards. Nylon also has the ability to withstand high voltages without breaking down.

MAINTENANCE OF SMALL ELECTRIC MOTORS

Small motors usually operate with so little trouble that they are apt to be neglected. They should be thoroughly inspected twice yearly to detect wear and to remove any conditions that might lead to further wear. Special care must be taken to inspect motor bearings,

FIGURE 22–5 *Ground-fault receptacle installed in a box.*

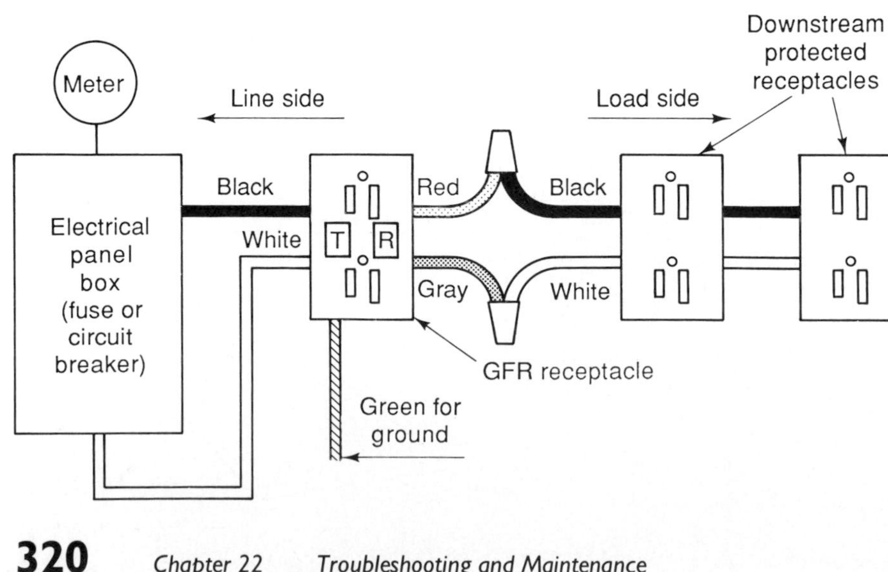

FIGURE 22–6 *Ground-fault receptacle wired to protect downstream devices.*

cutouts, and other wearing parts. Make sure that dirt and dust are not interfering with ventilation or clogging moving parts.

Adequate Wiring

When installing a new motor or transferring a motor from one installation to another, it is well to check the wiring. Be sure that adequate wire sizes are used to feed electrical power to the motor; in many cases, replacement of wires will prevent future breakdown. Adequate wiring assists in preventing overheating of motors and reduces electric power costs.

Check Internal Switches

Start winding switches usually give little trouble; however, regular attention makes them last even longer. Use fine sandpaper to clean contacts. Make sure that the sliding member on the shaft that operates the start winding switch moves freely. Check for loose screws.

Check Load Condition

Check the driven load regularly. Sometimes, additional friction develops gradually within the machine and imposes an overload on the motor, so watch the motor temperature. Protect motors with properly rated fuses or overload cutouts.

Extra Care in Lubrication

A motor running three times as much as usual will need three times as much attention to lubrication. Motors should be lubricated according to the manufacturer's recommendation. Provide enough oil, but do not *overdo* it.

Keep Commutators Clean

Do not allow a commutator on a dc motor to become covered with dust or oil. It should be wiped occasionally with a clean, dry cloth or one moistened with a solvent that does not leave a film. If necessary to use sandpaper, use No. 0000 paper or finer. Sandpaper or abrasive papers are available with ratings as high as 1500 grit.

Motors Must Have Proper Service Rating

Sometimes it is necessary to move a motor from one job to another or to operate a machine continuously when it has previously been running for short periods of time. Whenever a motor is operated under different conditions or on a new application, make sure that it is rated properly. A motor is rated for intermittent duty because the temperature rise within the motor will not be excessive when it is operated for short periods. Putting such a motor on a continuous-duty application will result in excessive temperature rise, which will cause the insulation to deteriorate or may even cause burnout.

Replace Worn Brushes

Brushes should be inspected at regular intervals so that replacements can be made if necessary. Whenever a brush is removed for inspection, be sure that it is replaced in the same axial position; that is, it must not be turned around in the brush holder when putting it back in the motor. If the contact surface, which has been "worn in" to fit the commutator, is not replaced in the same position, excessive sparking and loss of power will result. Brushes naturally wear down and should be replaced before they are less than $\frac{1}{4}$ in. in length. The commutator should also be inspected when brushes are removed. See the section "DC Motor Problems" later in the chapter.

MOTOR PROBLEMS

Certain danger signals are presented before a motor overheats or burns out.

Ball Bearing Motors

Danger Signals

1. A sudden increase in the temperature differential between the motor and bearing temperatures is an indication of malfunction of the bearing lubricant.
2. A temperature higher than that recommended for the lubricant warns of a reduction in bearing life. The rule of thumb is that grease life is halved for each 25°F increase in operating temperature.
3. An increase in bearing noise, accompanied by a bearing temperature rise, is an indication of a serious malfunction of the bearing.

Major Duties of Ball Bearing Lubricant

1. To dissipate heat caused by friction of bearing members under load.
2. To protect bearing members from rust or corrosion.
3. To offer maximum protection against entrance of foreign matter into the bearings.

Causes of Bearing Failures

1. Foreign matter in bearing from dirty grease or ineffective seals.
2. Deterioration of grease because of excessive temperature or contamination.
3. Overheated bearings as a result of too much grease.

Sleeve Bearing Motors

The lubricant used with sleeve bearings must actually provide an oil film that completely separates the bearing surface from the rotating shaft member, and ideally, eliminate metal-to-metal contact.

Lubricant. Oil, because of its adhesive properties and because of its viscosity or resistance to flow, is dragged along by the rotating shaft of the motor and forms a wedge-shaped film between the shaft and the bearing. The oil film forms automatically when the shaft begins to turn and is maintained by the motion. The forward motion sets up a pressure in the oil film, which in turn supports the load. This wedge-shaped film of oil is an absolutely essential feature of effective hydrodynamic, sleeve bearing lubrication. Without it no great load can be carried, except with high friction loss and resultant destruction of the bearing. When lubrication is effective and an adequate oil film is maintained, the sleeve bearing serves chiefly as a guide to ensure alignment. In the event of failure of the oil film, the bearing functions as a safeguard to prevent actual damage to the motor shaft.

Selection of Oil. The selection of the oil that will provide the most effective bearing lubrication and not require frequent renewal merits careful consideration. Good lubricants are essential to low maintenance cost. Top-grade oils are recommended, as they are refined from pure petroleum, are substantially noncorrosive as far as metal surfaces to be lubricated are concerned, are free from sediment, dirt, or other foreign materials, and are stable with respect to heat and moisture encountered in the motor. In performance terms, the higher-priced oils prove to be cheaper in the long run.

An oil film is built up of many layers or laminations that slide upon one another as the shaft rotates. The internal friction of the oil, which is due to the sliding action of the many oil layers, is measured as the *viscosity*. The viscosity of the oil chosen for a particular application should provide ample *oiliness* to prevent wear and seizure at ambient temperature, low speeds, and heavy loads, before the oil film is established and operating temperature is reached. Low-viscosity oils are recommended for use with fractional-horsepower motors, as they offer low internal friction, permit fuller realization of the motors efficiency, and minimize the operating temperature of the bearing.

Standard Oils. High ambient temperatures and high motor operating temperature will have a destructive effect on sleeve bearings lubricated with standard temperature range oil by increasing the bearing operating temperature beyond the oil's capabilities. Such destructive effects include reduction in oil viscosity, an increase in corrosive oxidation products in the lubricant, and usually a reduction in the quantity of the lubricant in contact with the bearing. Special oils are available, however, for motor applications at high temperatures and also for motor applications at low temperatures. The care exercised in selecting proper lubricant for expected extremes in bearing operating temperatures will have a decided influence on motor performance and bearing life.

Wear. Although sleeve bearings are less sensitive to a limited amount of abrasive or foreign materials than are ball bearings, owing to the ability of the relatively soft surface of the sleeve bearing to absorb hard particles of foreign materials, good maintenance practice recommends that the oil and bearing be kept clean. Frequency of oil changing will depend on local conditions such as severity and continuity of service and operating temperature. A conservative lubrication maintenance program should call for periodic inspections of the oil level and cleaning and refilling with new oil every 6 months.

Warning: Overlubrication should be avoided. Insulation damage by excess motor lubricant represents one of the most common causes of motor winding insulation failure in both sleeve and ball bearing motors.

COMMON MOTOR PROBLEMS AND THEIR CAUSES

Easy-to-detect symptoms, in many cases, indicate exactly what is wrong with a fractional-horsepower motor. However, where general types of trouble have similar symptoms, it becomes necessary to check each possible cause separately. Table 22–1 lists some of the more common ailments of small motors, together with suggestions as to possible causes. Most common motor problems can be checked by some test or inspection. The order of making these tests rests with the troubleshooter, but it is natural to make the simplest ones first. For instance, when a motor fails to start, you first inspect the motor connections, since this is an easy and simple thing to do.

Problem Diagnosis

In diagnosing problems, a combination of symptoms will often give a definite clue to the source of the trouble and hence eliminate other possibilities. For in-

TABLE 22–1 SQUIRREL-CAGE MOTOR PROBLEMS

Symptom and Possible Cause	*Possible Remedy*
Motor will not start	
(a) Overload control tripped	(a) Wait for overload to cool. Try starting again. If motor still does not start, check all the causes as outlined in the following.
(b) Power not connected	(b) Connect power to control and control to motor. Check clip contacts.
(c) Faulty (open) fuses	(c) Test fuses.
(d) Low voltage	(d) Check motor nameplate values with power supply. Also check voltage at motor terminals with motor under load to be sure wire size is adequate.
(e) Wrong control connections	(e) Check connections with control wiring diagram.
(f) Loose terminal lead connection	(f) Tighten connections.
(g) Driven machine locked	(g) Disconnect motor from load. If motor starts satisfactorily, check driven machine.
(h) Open circuit in stator or rotor winding	(h) Check for open circuits.
(i) Short circuit in stator winding	(i) Check for shorted coil.
(j) Winding grounded	(j) Test for grounded winding.
(k) Bearings stiff	(k) Free bearings or replace.
(l) Grease too stiff	(l) Use special lubricant for special conditions.
(m) Faulty control	(m) Check control wiring.
(n) Overload	(n) Reduce load.
Motor noisy	
(a) Motor running single phase	(a) Stop motor, then try to start. (It will not start on single phase.) Check for "open" in one of the lines or circuits.

stance, in the case just cited of a motor that will not start, if heating occurs, it offers the suggestion that a short or ground exists in one of the windings and eliminates the likelihood of an open circuit, poor line connection, or defective starter switch.

Centrifugal Switches

Centrifugal starting switches, found in many types of single-phase fractional horsepower motors, occasionally are a source of trouble. If the mechanism sticks in the running position, the motor will not start. On the other hand, if stuck in the closed position, the motor will not attain speed and the starting winding heats up quickly. The motor may also fail to start if the contact points of the switch are out of adjustment or coated with oxide. It is important to remember, however, that any adjustment of the switch or contacts should be made only at the factory or authorized service station.

Commutator-Type Motors

More maintenance is required by motors with commutators. High-speed series-wound motors should not be used on long, continuous-duty cycle applications because the commutator and brushes are a potential source of trouble. Gummy commutators and oil-soaked brushes can cause sluggish action and severe sparking. The commutator can be cleaned with fine sandpaper. However, if pitted spots still appear, the commutator should be reground.

TROUBLESHOOTING AIDS

Connection Diagrams

Figure 22–7 shows motor connection diagrams as an aid in troubleshooting. Knowing the arrangement of coils aids in checking out the shorts and grounds as well as opens.

Small Three-Phase Motor Rating Data

Knowing the current expected to be drawn in normal operation aids in troubleshooting. It is possible to use a clamp-on type meter and check the current drawn by the motor to see if it is excessive or incorrect. Table 22–2 shows the ampere rating of ac motors that operate on three-phase power.

Ampere ratings of motors vary somewhat depending on the type of motor. The values given below are for drip-proof class B-insulated (T frame) 1.15 service factor NEMA design B motors. These values represent an average full-load motor current that was calculated from the motor performance data published by several motor manufacturers. In the case of high-torque squirrel-cage motors, the ampere ratings will be at least 10% greater than the values given below.

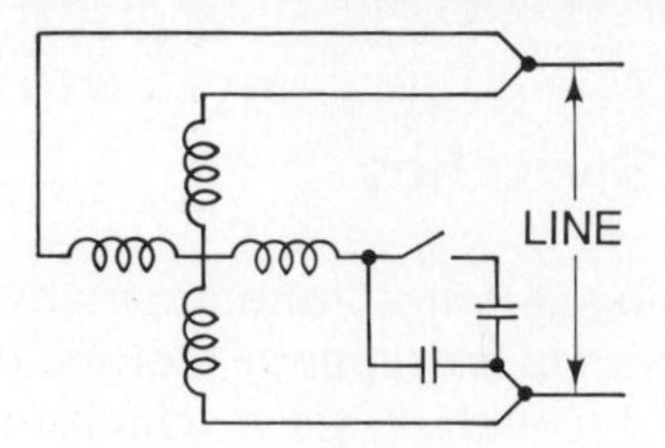

Reluctance synchronous two-value capacitor – Reversible only from rest (by transposing leads).

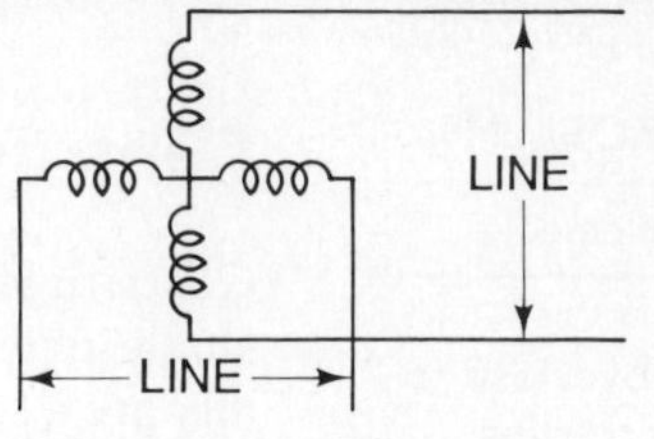

Reluctance synchronous two-phase 4-lead – Reversible by transposing either phase leads with line.

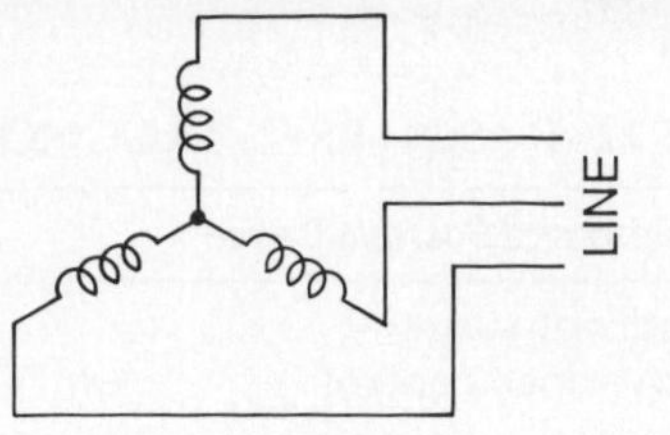

Reluctance synchronous three-phase – reversible by transposing any two leads.

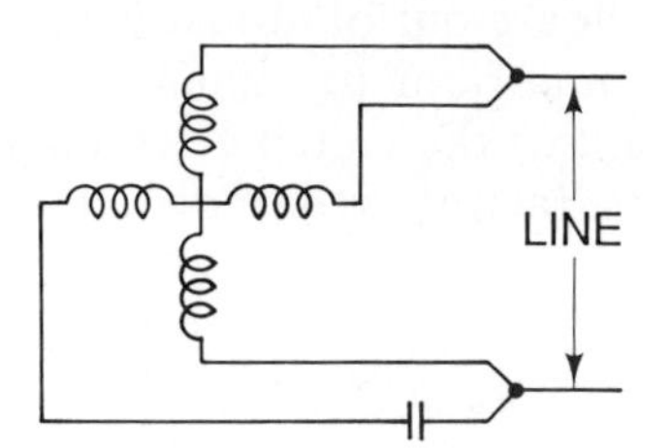

Hysteresis synchronous permanent-split capacitor – Reversible by transposing leads.

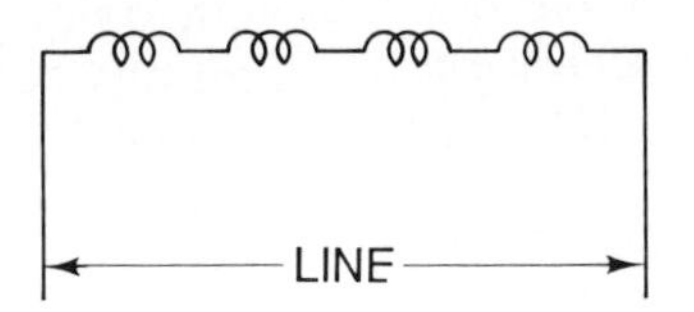

Shaded pole – Nonreversible.

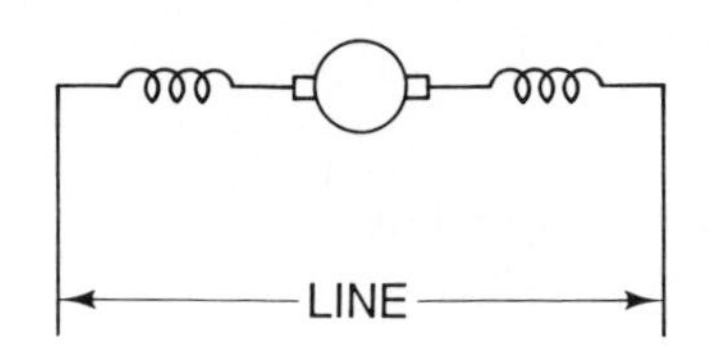

Series-wound, 2-lead – Nonreversible

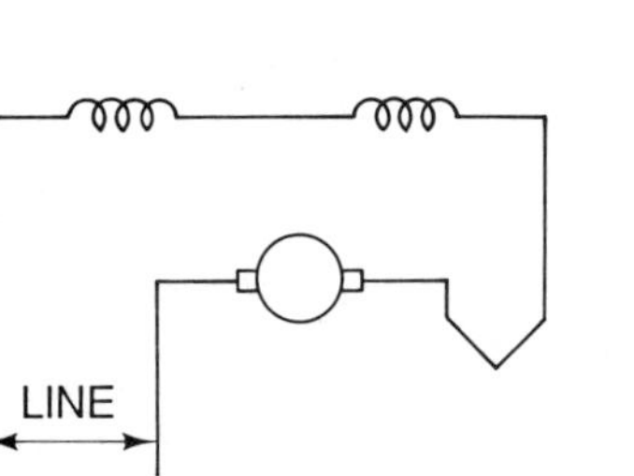

Series-wound, 4-lead – Reversible by transposing armature leads.

C. ROT.
C.C. ROT.
LINE

Series-wound split field – Reversible by connecting either field lead to line.

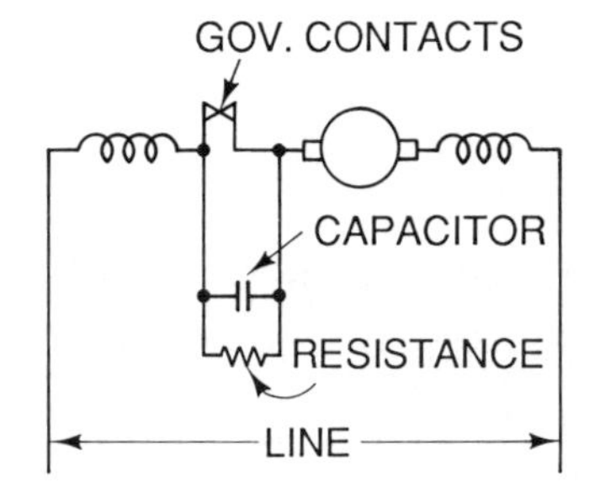

Electric governor controlled series-wound – Nonreversible.

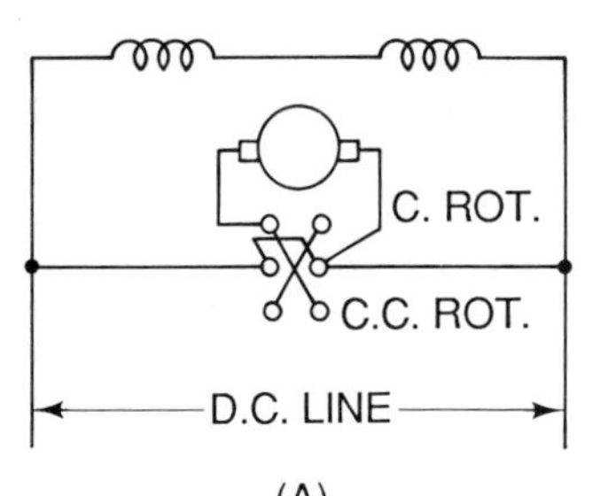

(A)
Shunt motor
4-wire reversible

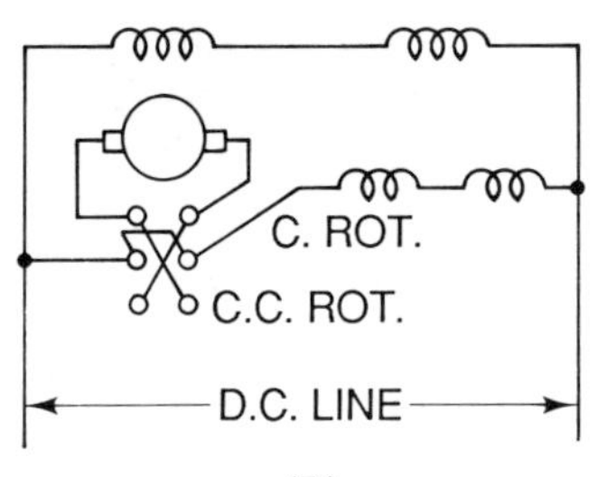

(B)
Compound motor
5-wire reversible

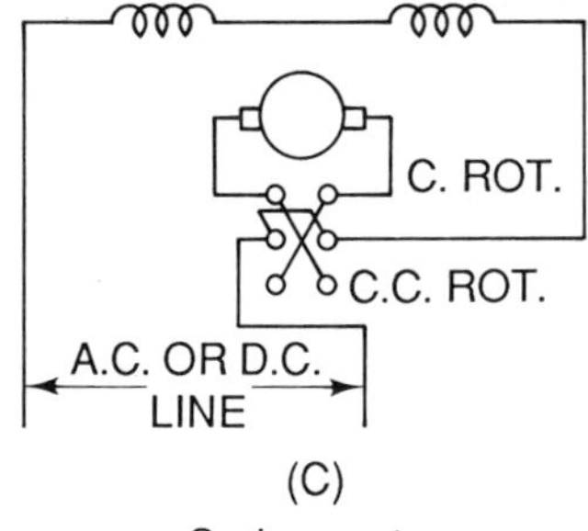

(C)
Series motor
4-wire reversible

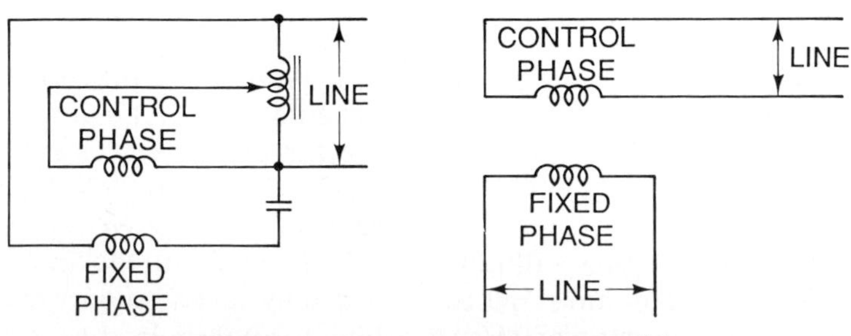

Two-phase servo-type control – Reversible.

FIGURE 22–7 *Connection diagrams.*

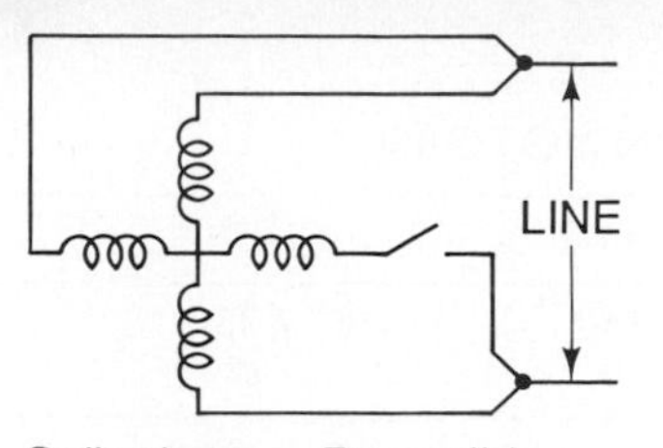

Split-phase – Reversible only from rest (by transposing leads).

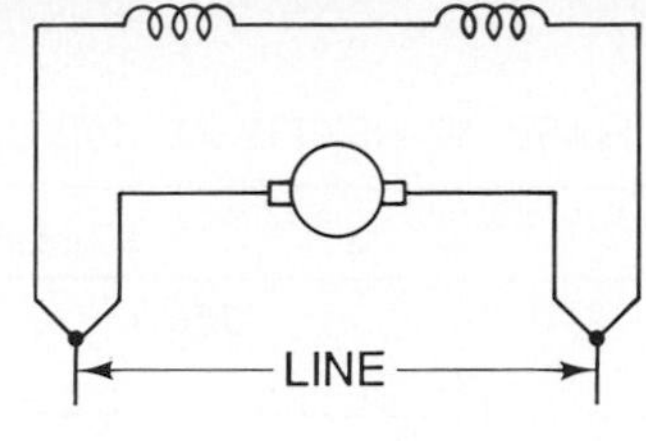

Shunt-wound – Reversible by transposing leads.

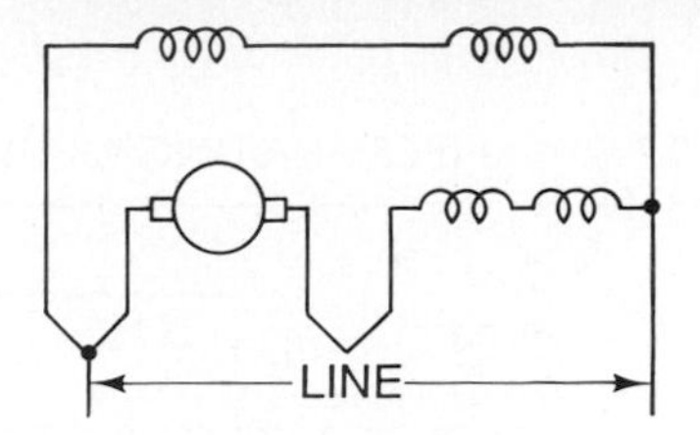

Compound-wound – Reversible by transposing armature leads.

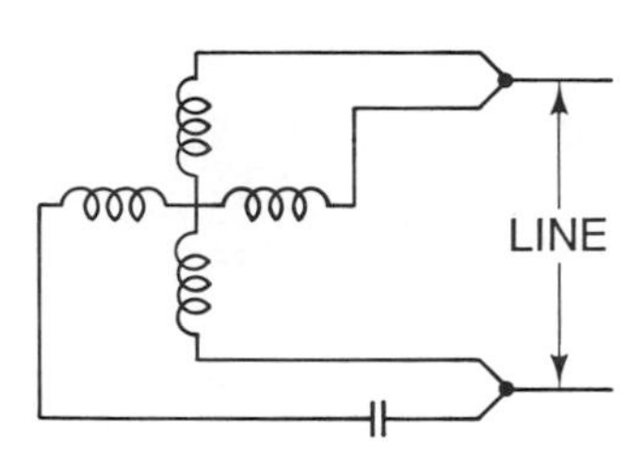

Permanent-split capacitor, 4-lead – Reversible by transposing leads.

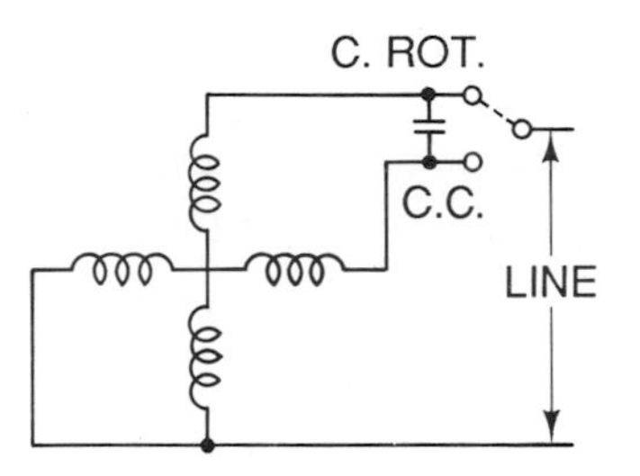

Permanent-split capacitor 3-lead – Reversible by connecting either side of capacitor to line.

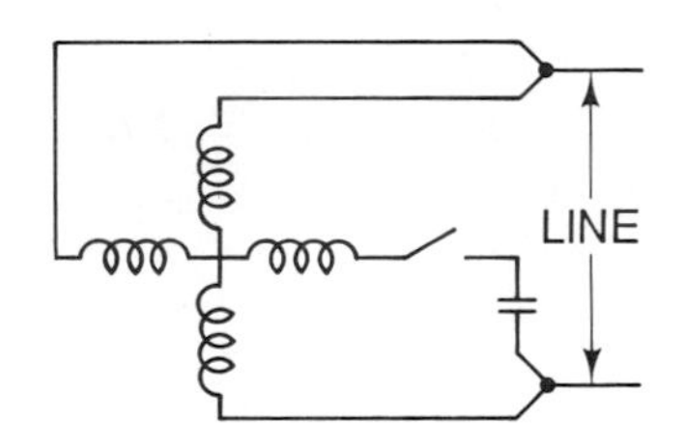

Capacitor-start – Reversible only from rest (by transposing leads).

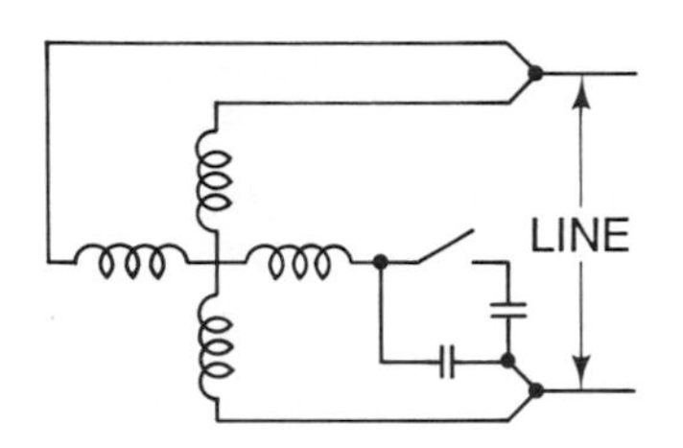

Two-value capacitor – Reversible only from rest (by transposing leads).

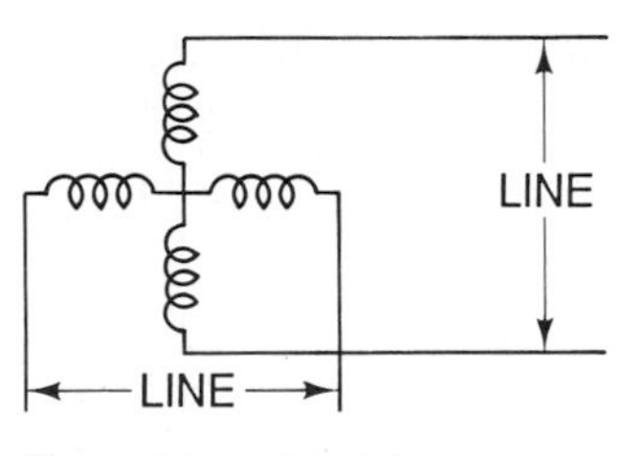

Two-phase 4-lead – Reversible by transposing either phase leads with line.

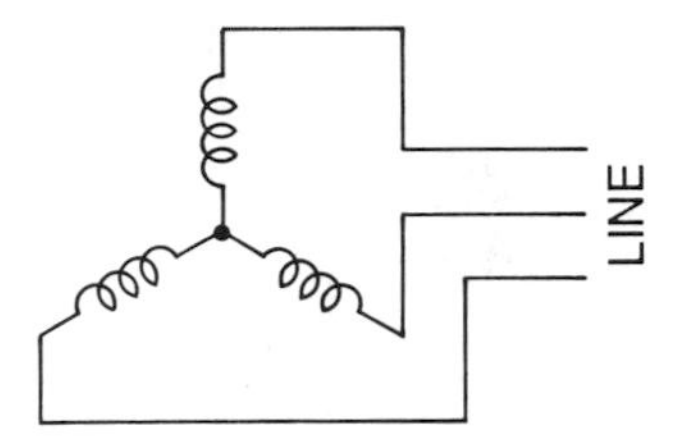

Three-phase, single voltage – Reversible by transposing any two leads.

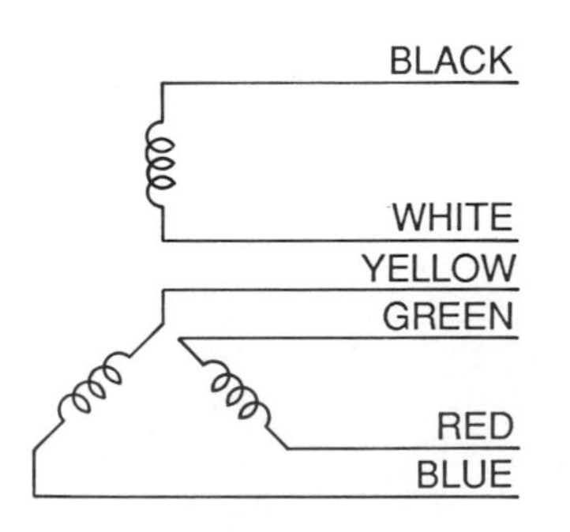

3-phase, star-delta 6-lead reversible – For 440 volts connect together white, yellow, and green: Connect to line black, red, and blue. To reverse rotation, transpose any two line leads. For 220 volts connect white to blue, black to green, and yellow to red: Then connect each junction point to line.

To reverse rotation, transpose any two junction points with line.

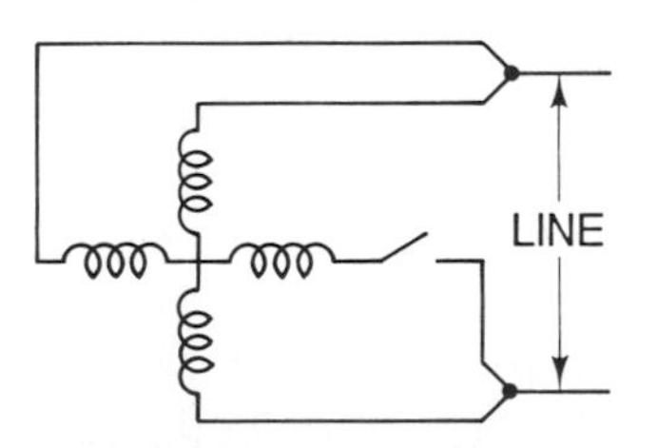

Reluctance synchronous split-phase – Reversible only from rest (by transposing leads).

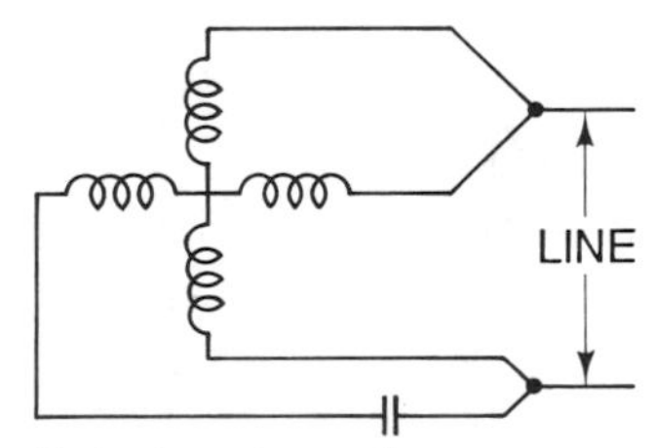

Reluctance synchronous permanent split-capacitor, 4-lead – Reversible by transposing lead.

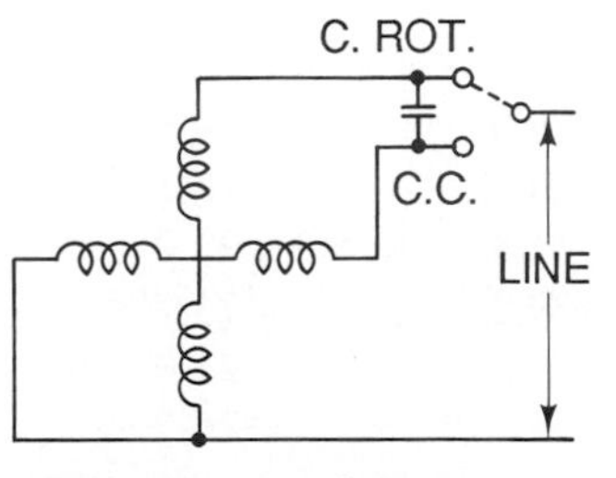

Reluctance synchronous permanent-split capacitor, 3-lead – Reversible by connecting either side of capacitor to line.

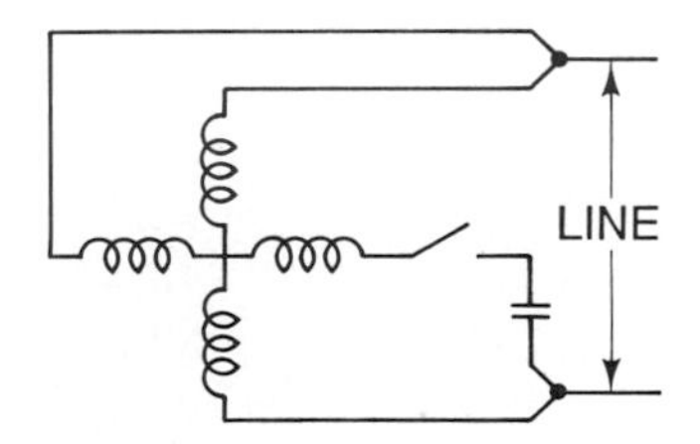

Reluctance synchronous capacitor-start – Reversible only from rest (by transposing leads).

FIGURE 22–7 *Continued*

TABLE 22–2 AMPERE RATING OF THREE-PHASE, 60-HERTZ, AC INDUCTION MOTORS[a]

Hp	Syn. Speed (rpm)	Current(A) 115 V	230 V	380 V	460 V	575 V	2200 V
1/4	1800	1.90	0.95	0.55	0.48	0.38	
	1200	2.80	1.40	0.81	0.70	0.56	
	900	3.20	1.60	0.93	0.80	0.64	
1/3	1800	2.38	1.19	0.69	0.60	0.48	
	1200	3.18	1.59	0.92	0.80	0.64	
	900	3.60	1.80	1.04	0.90	0.72	
1/2	1800	3.44	1.72	0.99	0.86	0.69	
	1200	4.30	2.15	1.24	1.08	0.86	
	900	4.76	2.38	1.38	1.19	0.95	
3/4	1800	4.92	2.46	1.42	1.23	0.98	
	1200	5.84	2.92	1.69	1.46	1.17	
	900	6.52	3.26	1.88	1.63	1.30	
1	3600	5.60	2.80	1.70	1.40	1.12	
	1800	7.12	3.56	2.06	1.78	1.42	
	1200	7.52	3.76	2.28	1.88	1.50	
	900	8.60	4.30	2.60	2.15	1.72	
1 1/2	3600	8.72	4.36	2.64	2.18	1.74	
	1800	9.71	4.86	2.94	2.43	1.94	
	1200	10.5	5.28	3.20	2.64	2.11	
	900	11.2	5.60	3.39	2.80	2.24	
2	3600	11.2	5.60	3.39	2.80	2.24	
	1800	12.8	6.40	3.87	3.20	2.56	
	1200	13.7	6.84	4.14	3.42	2.74	
	900	15.8	7.90	4.77	3.95	3.16	
3	3600	16.7	8.34	5.02	4.17	3.34	
	1800	18.8	9.40	5.70	4.70	3.76	
	1200	20.5	10.2	6.20	5.12	4.10	
	900	22.8	11.4	6.90	5.70	4.55	
5	3600	27.1	13.5	8.20	6.76	5.41	
	1800	28.9	14.4	8.74	7.21	5.78	
	1200	31.7	15.8	9.59	7.91	6.32	
	900	31.0	15.5	9.38	7.75	6.20	
7 1/2	3600	39.1	19.5	11.8	9.79	7.81	
	1800	43.0	21.5	13.0	10.7	8.55	
	1200	43.7	21.8	13.2	10.9	8.70	
	900	46.0	23.0	13.9	11.5	9.19	
10	3600	50.8	25.4	15.4	12.7	10.1	
	1800	53.8	26.8	16.3	13.4	10.7	
	1200	56.0	28.0	16.9	14.0	11.2	
	900	61.0	30.5	18.5	15.2	12.2	
15	3600	72.7	36.4	22.0	18.2	14.5	
	1800	78.4	39.2	23.7	19.6	15.7	
	1200	82.7	41.4	25.0	20.7	16.5	
	900	89.0	44.5	26.9	22.2	17.8	

TABLE 22–2 CONTINUED

Hp	*Syn. Speed (rpm)*	*Current(A)* 115 V	230 V	380 V	460 V	575 V	2200 V
20	3600	101.1	50.4	30.5	25.2	20.1	
	1800	102.2	51.2	31.0	25.6	20.5	
	1200	105.7	52.8	31.9	26.4	21.1	
	900	109.5	54.9	33.2	27.4	21.9	
25	3600	121.5	60.8	36.8	30.4	24.3	
	1800	129.8	64.8	39.2	32.4	25.9	
	1200	131.2	65.6	39.6	32.8	26.2	
	900	134.5	67.3	40.7	33.7	27.0	
30	3600	147.	73.7	44.4	36.8	29.4	
	1800	151.	75.6	45.7	37.8	30.2	
	1200	158.	78.8	47.6	39.4	31.5	
	900	164.	81.8	49.5	40.9	32.7	
40	3600	193.	96.4	58.2	48.2	38.5	
	1800	202.	101.	61.0	50.4	40.3	
	1200	203.	102.	61.2	50.6	40.4	
	900	209.	105.	63.2	52.2	41.7	
50	3600	241.	120.	72.9	60.1	48.2	
	1800	249.	124.	75.2	62.2	49.7	
	1200	252.	126.	76.2	63.0	50.4	
	900	260.	130.	78.5	65.0	52.0	
60	3600	287.	143.	86.8	71.7	57.3	
	1800	298.	149.	90.0	74.5	59.4	
	1200	300.	150.	91.0	75.0	60.0	
	900	308.	154.	93.1	77.0	61.5	
75	3600	359.	179.	108.	89.6	71.7	
	1800	365.	183.	111.	91.6	73.2	
	1200	368.	184.	112.	92.0	73.5	
	900	386.	193.	117.	96.5	77.5	
100	3600	461.	231.	140.	115.	92.2	
	1800	474.	236.	144.	118.	94.8	23.6
	1200	478.	239.	145.	120.	95.6	24.2
	900	504.	252.	153.	126.	101.	24.8
125	3600	583.	292.	176.	146.	116.	
	1800	584.	293.	177.	147.	117.	29.2
	1200	596.	298.	180.	149.	119.	29.9
	900	610.	305.	186.	153.	122.	30.9
150	3600	687.	343.	208.	171.	137.	
	1800	693.	348.	210.	174.	139.	34.8
	1200	700.	350.	210.	174.	139.	35.5
	900	730.	365.	211.	183.	146.	37.0
200	3600	904.	452.	274.	226.	181.	
	1800	915.	458.	277.	229.	184.	46.7
	1200	920.	460.	266.	230.	184.	47.0
	900	964.	482.	279.	241.	193.	49.4

TABLE 22–2 CONTINUED

Hp	Syn. Speed (rpm)	Current(A) 115 V	230 V	380 V	460 V	575 V	2200 V
250	3600	1118.	559.	338.	279.	223.	
	1800	1136.	568.	343.	284.	227.	57.5
	1200	1146.	573.	345.	287.	229.	58.5
	900	1200.	600.	347.	300.	240.	60.5
300	1800	1356.	678.	392.	339.	274.	69.0
	1200	1368.	684.	395.	342.	274.	70.0
400	1800	1792.	896.	518.	448.	358.	91.8
500	1800	2220.	1110.	642.	555.	444.	116.

Source: Courtesy of Bodine Electric Company, Chicago.

[a]Ampere ratings of motors vary somewhat. The values given here are for drip-proof, class B insulated (T frame) where available, 1.15 service factor, NEMA design B motors. The values represent an average full-load motor current that was calculated from the motor performance data published by several motor manufacturers. In the case of high-torque squirrel-cage motors, the ampere ratings will be at least 10% greater than the values shown.

POWER SUPPLY DISTURBANCES

Maintenance of equipment is affected by the quality of the power supplied to it. There are a number of problems associated with various line power disturbances. Three types of irregularity that affect power supply are voltage fluctuations, transients, and power outages.

Voltage Fluctuations

In many states the appropriate public service commissions establish allowable voltage tolerances for utilities. These tolerances are continually monitored, and in most instances every reasonable precaution is taken to stay within these limits. However, some equipment is so sensitive that fluctuations within the tolerance limits can still cause problems (Fig. 22–8).

Voltage fluctuations can usually be detected by visible flickering of lights. High- or low-voltage conditions can result in damage to equipment, loss of data, and erroneous readings in monitoring systems (Fig. 22–9).

Undervoltage can result from overloaded power circuits. Intermittent low voltage is typically caused by

FIGURE 22–8 *Normal power sine wave.*

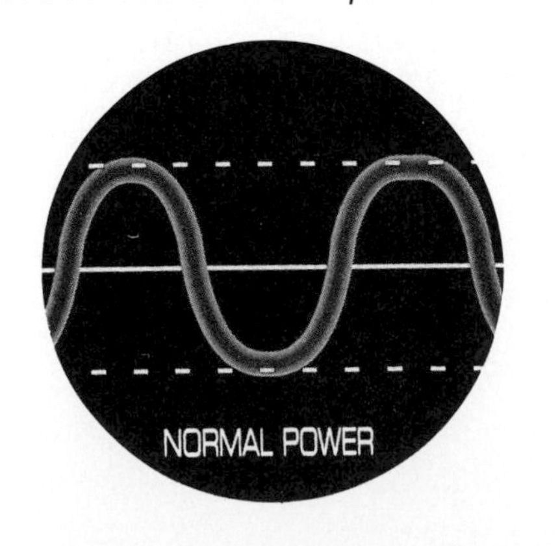

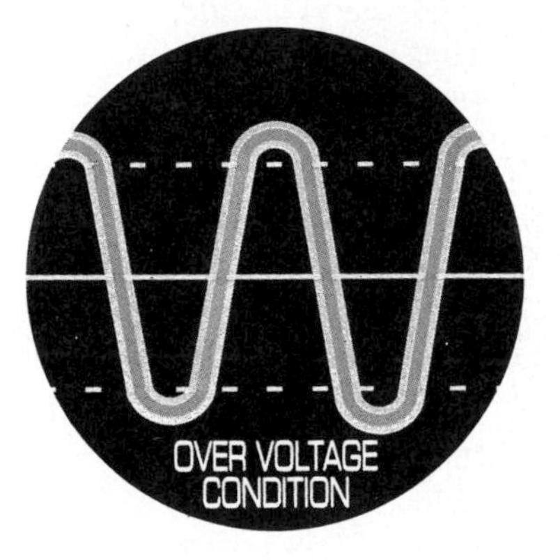

FIGURE 22–9 *Overvoltage condition.*

FIGURE 22–10 *Undervoltage condition.*

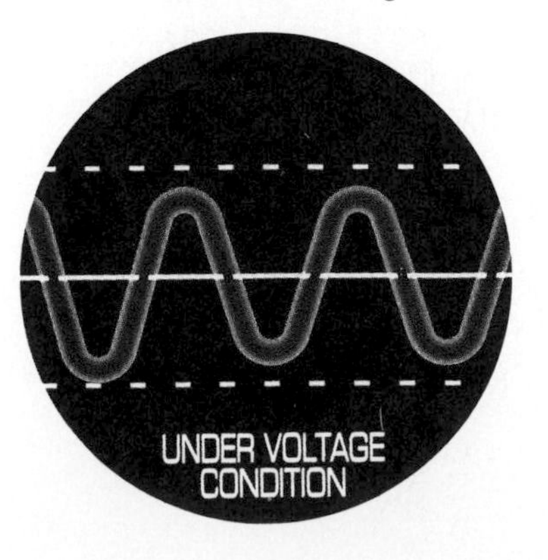

starting a large heavily loaded motor such as an air conditioner. Overvoltage conditions are less common but are more damaging and are seen frequently in facilities with rapidly varying loads (Fig. 22–10).

Transients

Voltage Spikes. Short-duration impulses in excess of the normal voltage are called spikes or surges. Although their duration is incredibly brief, a spike may exceed the normal voltage level five- or 10-fold. Spikes can wipe out data stored in memory, produce output errors, or cause extensive equipment damage. Besides the immediate damage, there are also harder-to-detect effects, particularly reduced service life. Subsequent random failures can be particularly annoying and expensive (Fig. 22–11).

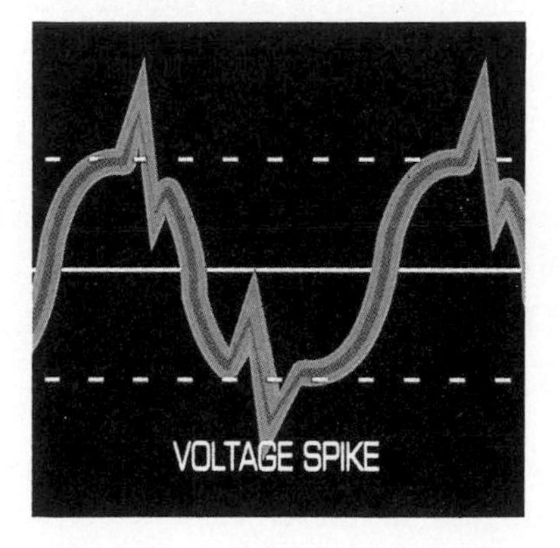

FIGURE 22–11 *Voltage spike on a sine wave.*

The leading day-to-day cause of small low-energy spikes is the switching on and off of an electrical motor (inductive load switching). Air conditioners, electrical power tools, furnace ignitions, electrostatic copy machines, arc welders, and elevators are particularly guilty of creating voltage spikes. The problems created by the inductive load switching are very common in industrial plants. Larger spikes are typically caused by lightning. A direct lightning hit, of course, is catastrophic but of very low probability. However, a distant lightning strike several miles away may be transmitted through utility power lines and show up as a voltage spike all along the line.

Electrical Noise. As contrasted to outright equipment damage, computer "glitches" are caused by electrical noise (Fig. 22–12). The same causes of voltage spikes can (at a lower voltage magnitude) cause noise interference. Other electrical noise generators include radio transmitters, fluorescent lights, computers, business machines, and electrical devices such as light sockets, wall receptacles, plugs, and loose electrical connections. Interaction between system components may generate sufficient noise to cause errors. Although most electronic equipment has some internal noise filtering, equipment located in severe noise environments may encounter some interference.

FIGURE 22–12 *Noise interference on a power-line sine wave.*

Transients are by far the most common sort of power disturbance and fortunately, often the easiest to correct. However, they may be difficult to detect since they last such a short time.

Power Outages

Power outages are a total interruption of power supply. An interruption of a mere 15 milliseconds is considered a blackout to sensitive equipment. Power outages can cause problems for equipment users, most critical of which are loss of valuable data and expensive time-consuming reprogramming.

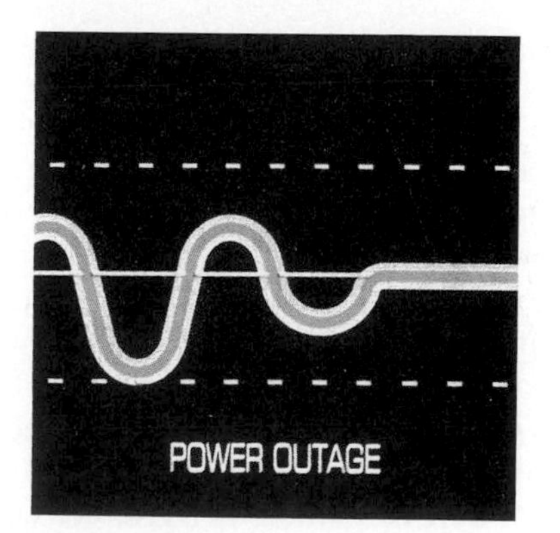

FIGURE 22–13 *Power outage with sine wave diminishing to zero.*

Outages tend to be caused by a larger-scale problem than a transient. Interruptions may be caused, for example, by utility or on-site load changes, on-site equipment malfunction, or by faults on the power system (Fig. 22–13).

Looking for Shorts

Shorted turns in the winding of a motor behave like a shorted secondary of a transformer. A motor with a shorted winding will draw excessive current while running at no load. Measurement of the current can be made without disconnecting lines. This means that you engage one of the lines with the split-core transformer of the tester. If the ammeter reading is much higher than the full-load ampere rating on the nameplate, the motor is probably shorted.

In a two- or three-phase motor, a partially shorted winding produces a higher current reading in the shorted phase. This becomes evident when the current in each phase is measured.

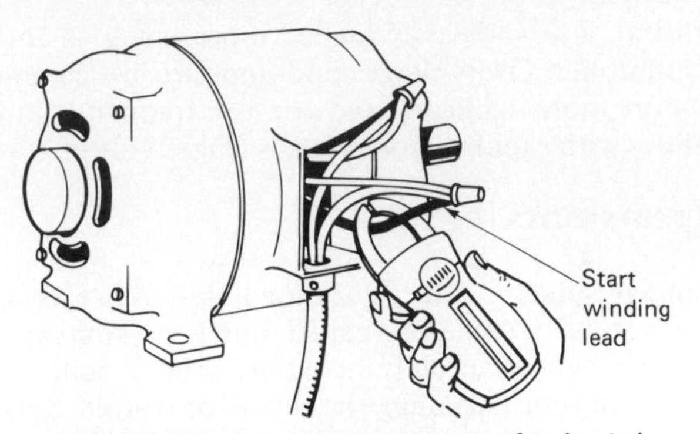

FIGURE 22–15 *Checking the centrifugal switch with a clamp-on meter. (Amprobe Instrument)*

MOTORS WITH SQUIRREL-CAGE ROTORS

Loss in output torque at rated speed in an induction motor may be due to opens in the squirrel-cage rotor.

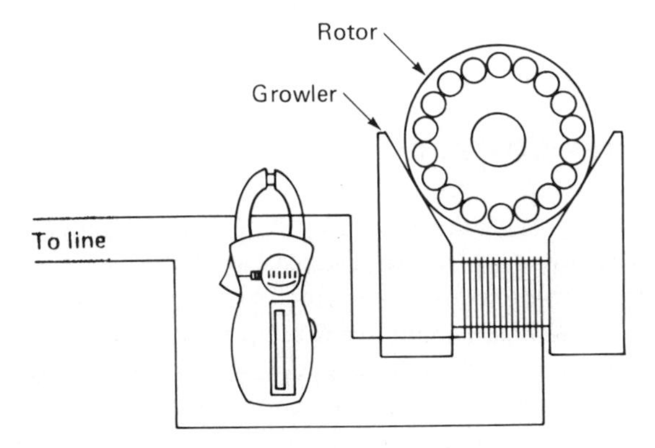

FIGURE 22–14 *Using a growler to test a rotor.*

To test the rotor and determine which rotor bars are loose or open, place the rotor in a growler. Engage the split-core ammeter around the lines going to the growler, as shown in Fig. 22–14. Set the switch to the highest current range. Switch on the growler and then set the test unit to the approximate current range. Rotate the rotor in the growler and take note of the current indication whenever the growler is energized. The bars and end rings in the rotor behave similarly to a shorted secondary of a transformer. The growler windings act as the primary. A good rotor will produce approximately the same current indications for all positions of the rotor. A defective rotor will exhibit a drop in the current reading when the open bars move into the growler field.

TESTING THE CENTRIFUGAL SWITCH IN A SINGLE-PHASE MOTOR

A defective centrifugal switch may not disconnect the start winding at the proper time. To determine conclusively that the start winding remains in the circuit, place the split-core ammeter around one of the start-winding leads. Set the instrument to the highest current range. Turn on the motor switch. Select the appropriate current range. Observe if there is any current in the start-winding circuit. A current indication signifies that the centrifugal switch did not open when the motor came up to speed (Fig. 22–15).

TESTING FOR SHORT CIRCUITS BETWEEN RUN AND START WINDINGS

A short between run and start windings may be determined by using the ammeter and line voltage to check for continuity between the two separate circuits. Disconnect the run- and start-winding leads and connect the instrument as shown in Fig. 22–16. Set the meter on voltage. A full-line voltage reading will be obtained if the windings are shorted to one another.

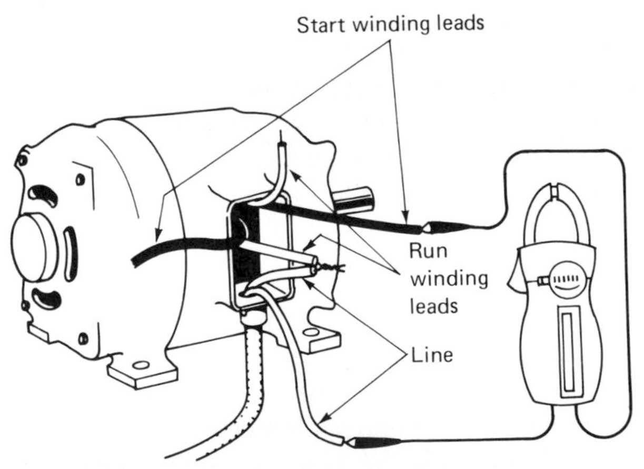

FIGURE 22–16 *Finding a shorted winding using a clamp-on meter. (Amprobe Instrument)*

CAPACITOR TESTING

Defective capacitors are very often the cause of trouble in capacitor-type motors. Shorts, opens, grounds, and insufficient capacity in microfarads are condi-

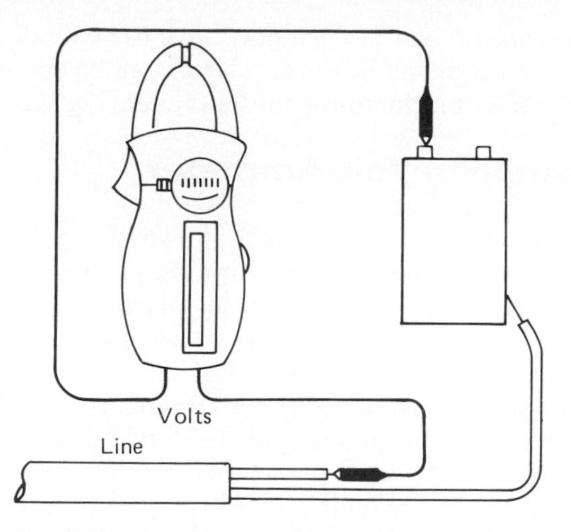

FIGURE 22–17 *Finding a grounded capacitor with a clamp-on meter. (Amprobe Instrument)*

tions for which capacitors should be tested to determine whether they are good. You can determine a grounded capacitor by setting the instrument on the proper voltage range and connecting it and the capacitor to the line as shown in Fig. 22–17. A full-line voltage indication on the meter signifies that the capacitor is grounded to the can. A high-resistance ground is evident by a voltage reading that is somewhat below the line voltage. A negligible reading or a reading of no voltage will indicate that the capacitor is not grounded.

Measuring Capacity of a Capacitor

To measure the capacity of the capacitor, set the test unit's switch to the proper voltage range and read the line-voltage indication. Then set to the appropriate current range and read the capacitor current indication. During the test, keep the capacitor on the line for a very short period of time. Keep the capacitor on the line for a very short period of time, because motor starting electrolytic capacitors are rated for intermittent duty (Fig. 22–18). The capacity in microfarads is then computed by substituting the voltage and current readings in the following formula, assuming that a full 60-Hz line is used:

$$\text{microfarads} = \frac{2650 \times \text{amperes}}{\text{volts}}$$

An open capacitor will be evident if there is no current indication in the test. A shorted capacitor is easily detected. It will blow the fuse when the line switch is turned on to measure the line voltage.

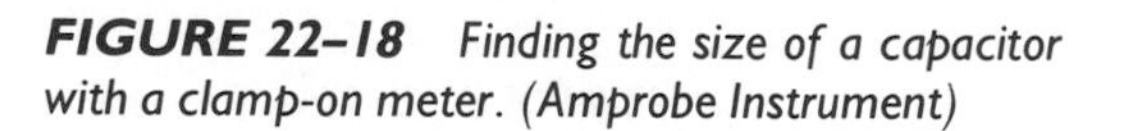

FIGURE 22–18 *Finding the size of a capacitor with a clamp-on meter. (Amprobe Instrument)*

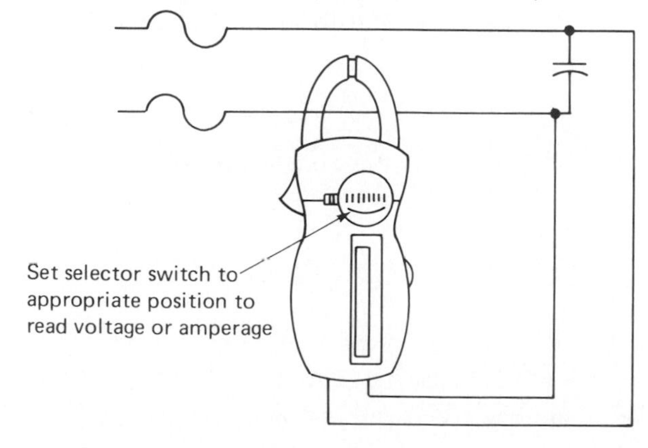

USING METERS TO CHECK FOR PROBLEMS

The voltmeter and the ohmmeter can be used to isolate various problems. You should be able to read the schematic and make the proper voltage or resistance measurements. An incorrect reading will indicate the possibility of a problem. Troubleshooting charts will aid in isolating the problem to a given system. Once you have arrived at the proper system that may be causing the symptoms noticed, you will then need to use the ohmmeter with the power off to isolate a section of the system. Once you have zeroed in on the problem, you can locate it by knowing what the proper reading should be. Deviation from a stated reading of over 10% is usually indicative of a malfunction, and in most cases the component part must be replaced to assure proper operation and no callbacks.

Using a Volt-Ammeter for Troubleshooting Electric Motors

Most electrical equipment will work satisfactorily if the line voltage differs ± 10% from the actual nameplate rating. In a few cases, however, a 10% voltage drop may result in a breakdown. Such may be the case with an induction motor that is being loaded to its fullest capacity on both start and run. A 10% loss in line voltage will result in a 20% loss in torque.

The full-load current rating on the nameplate is an approximate value based on the average unit coming off the manufacturer's production line. The actual current for any unit may vary as much as ± 10% at rated output. However, a motor whose load current exceeds the rated value by 20% or more will reduce the life of the motor due to higher operating temperatures, and the reason for excessive current should be determined. In many cases it may simply be an overloaded motor. The percentage increase in load will not correspond with the percentage increase in load current. For example, in the case of a single-phase induc-

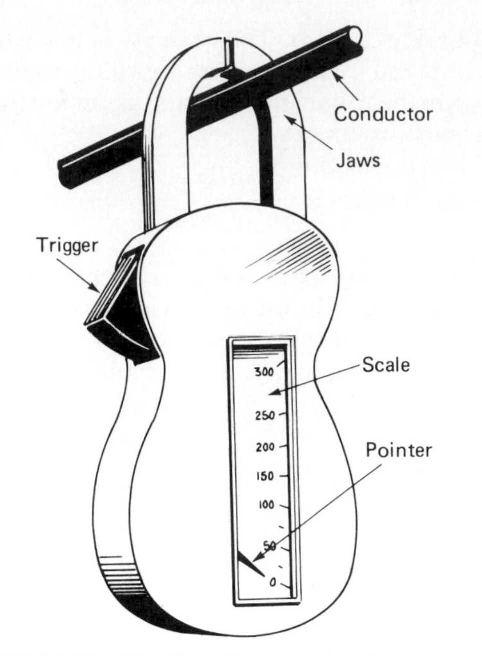

FIGURE 22–19 *Clamp-on volt-ammeter. (Amprobe Instrument)*

tion motor, a 35% increase in current may correspond to an 80% increase in torque output.

Operating conditions and behavior of electrical equipment can be analyzed only by actual measurement. A comparison of the measured terminal voltage and current will check whether the equipment is operating within electrical specifications.

A voltmeter and an ammeter are needed for the two basic measurements. To measure voltage, the test leads of the voltmeter are in contact with the terminals of the line under test. To measure current, the conventional ammeter must be connected in series with the line so that the current will flow through the ammeter.

To insert the ammeter means that you have to shut down the equipment, break open the line, and connect the ammeter and then start up the equipment to read the meter. You have to do the same to remove the meter once it has been used. Then there are other time-consuming tests that may have to be made to locate the problem. However, all this can be eliminated by the use of a clamp-on volt-ammeter (Fig. 22–19).

Clamp-On Volt-Ammeter

The pocket-size volt-ammeter shown in Fig. 22–19 is the answer to most troubleshooting problems on the job. The line does not have to be disconnected to obtain a current reading. The meter works on the transformer principle that picks up the magnetic lines surrounding a current-carrying conductor and then presents this as a function of the entire amount flowing through the line. Remember that earlier we discussed how the magnetic field strength in the core of the transformer determines the amount of current in the secondary. Well, the same principle is used here to detect the flow of current and how much.

To get transformer action, the line to be tested is encircled with the split-type core simply by pressing the trigger button. Aside from measuring terminal voltages and load currents, the split-core ammeter-voltmeter can be used to track down electrical difficulties in electric motor repair.

Looking for Grounds

To determine whether a winding is grounded or has a very low value of insulation resistance, connect the unit and test leads as shown in Fig. 22–20. Assuming that the available line voltage is approximately 120 V, use the unit's lowest voltage range. If the winding is grounded to the frame, the test will indicate full-line voltage.

A high-resistance ground is simply a case of low insulation resistance. The indicated reading for a high-resistance ground will be a little less than line voltage. A winding that is not grounded will be evidenced by a small or negligible reading. This is due mainly to the capacitive effect between the windings and the steel lamination.

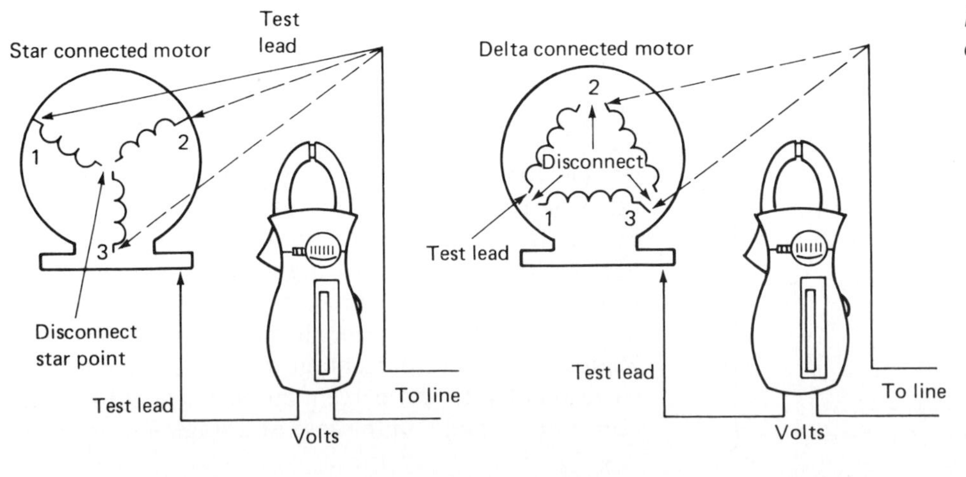

FIGURE 22–20 *Grounded phase of a motor.*

To locate the grounded portion of the windings, disconnect the necessary connection jumpers and test. Grounded sections will be detected by a full-line voltage indication.

Looking for Opens

To determine whether a winding is open, connect test leads as shown in Figs. 22-21 and 22-22. If the winding is open, there will be no voltage indication. If the circuit is not open, the voltmeter indication will read full-line voltage.

TROUBLESHOOTING GUIDE

One of the quickest ways to troubleshoot is to check out symptoms and possible causes in a chart. This allows for quick isolation of the cause and suggests possible corrections. Both three-phase motors and their starters can be checked quickly this way. Tables 22-3 and 22-4 will aid in troubleshooting motors and starters.

MOTOR LIFE

The stator windings of integral horsepower ac motors are capable of full-power operation for many years. However, winding life can be shortened by any combination of the following:

1. *Mechanical damage* produces weak spots in the insulation. It can occur during maintenance of the motor or result from such operating problems as severe vibration.
2. *Excessive moisture* encountered in service causes deterioration of the insulation.
3. *High dielectric stress,* such as voltage surges or excess input current, can cause overheating and insulation deterioration.
4. *High temperature* reduces the ability of the insulation system to withstand mechanical or electrical abuse. Overtemperature is usually a result of poor installation or misapplication of the motor.

Regardless of the reason for failures, the obvious result is thermal degradation of the insulation—or

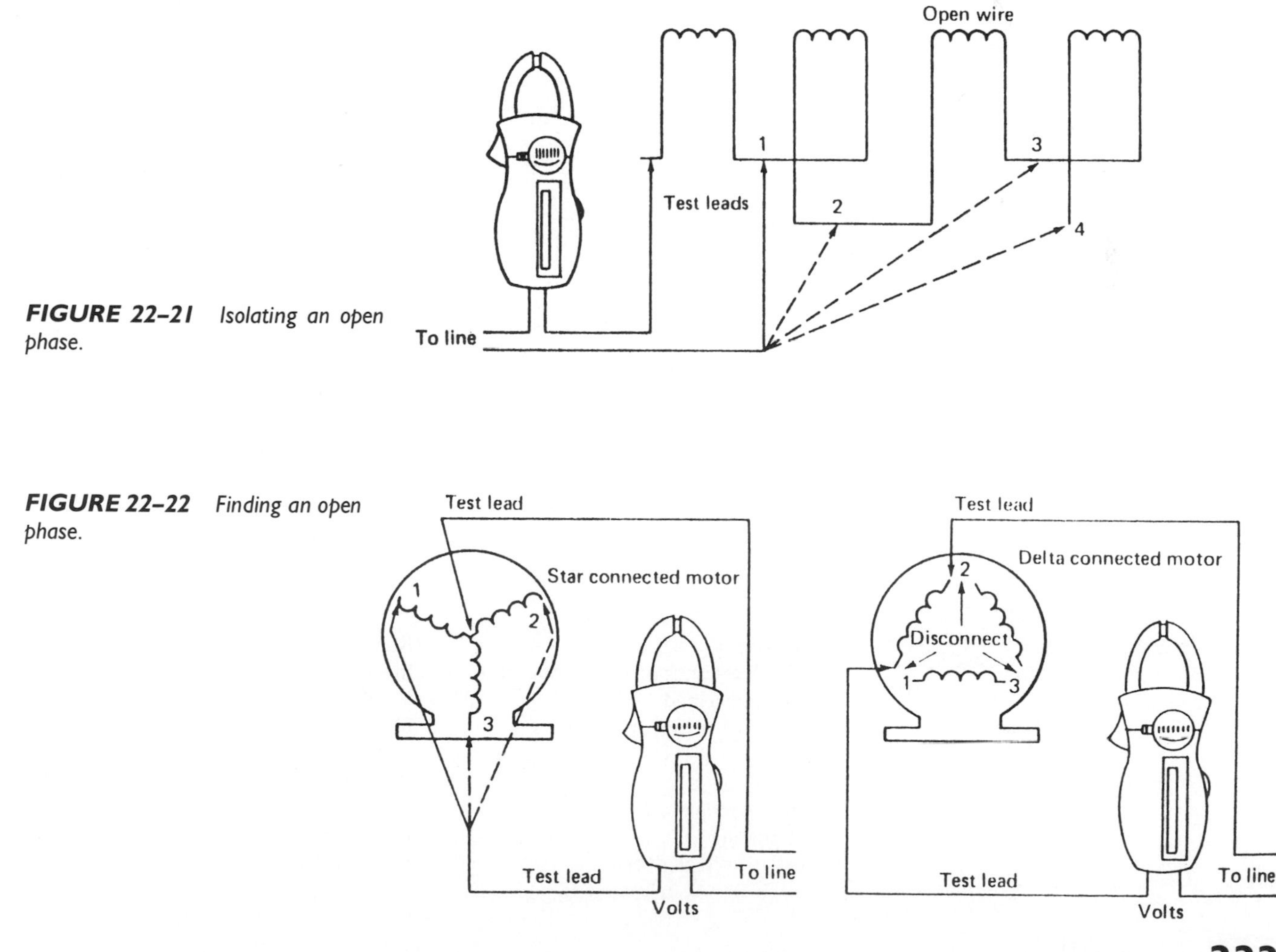

FIGURE 22-21 *Isolating an open phase.*

FIGURE 22-22 *Finding an open phase.*

Table 22–3 Three-Phase Motor Troubleshooting Guide

Symptom	*Possible Causes*	*Correction*
High input current (all three phases)	Accuracy of ammeter readings	First check accuracy of ammeter readings on all three phases.
Running idle (discounted from load)	High line voltage: 5 to 10% over nameplate	Consult power company—possibly decrease by using lower transformer tap.
Running loaded	Motor overloaded	Reduce load or use larger motor.
	Motor voltage rating does not match power system voltage	Replace motor with one of correct voltage rating. Consult power company—possibly correct by using a different transformer tap.
Unbalanced input current (5% or more deviation from the average input current)	Unbalanced line voltage due to: a. Power supply b. Unbalanced system loading c. High-resistance connection d. Undersized supply lines	Carefully check voltage across each phase *at the motor terminals* with good, properly calibrated voltmeter.
Note: A small voltage imbalance will produce a large current imbalance. Depending on the magnitude of imbalance and the size of the load, the input current in one or more of the motor input lines may greatly exceed the current rating of the motor.	Defective motor	If there is doubt as to whether the trouble lies with the power supply or the motor, check as follows: Rotate *all three input power lines* to the motor by one position (i.e., move line 1 to motor lead 2, line 2 to motor lead 3 and 3 to motor lead 1. a. If the unbalanced current pattern follows the *input power lines*, the problem is in the power supply b. If the unbalanced current pattern follows the *motor leads*, the problem is in the motor. Correct the voltage balance of the power supply or replace the motor, depending on the answer to a and b above.
Excessive voltage drop (more than 2 or 3% of nominal supply voltage)	Excessive starting or running load	Reduce load.
	Inadequate power supply	Consult power company.
	Undersized supply lines	Increase line sizes.
	High-resistance connections	Check motor leads and eliminate poor connections.
	Each phase lead runs in separate conduits	All three-phase leads must be in a single conduit, according to the *National Electrical Code*®. (This applies only to metal conduit with magnetic properties.)
Overload relays tripping upon starting (see also "Slow Starting"	Slow starting (10–15 seconds or more) due to high-inertia load	Reduce starting load. Increase motor size if necessary.
	Low voltage at motor terminals	Improve power supply and/or increase line size.
Running loaded	Overload	Reduce load or increase motor size.
	Unbalanced input current	Balance supply voltage.
	Single phasing	Eliminate.
	Excessive voltage drop	Eliminate (see above).

Table 22-3 CONTINUED

Symptom	*Possible Causes*	*Correction*
	Too frequent starting or intermittent overloading	Reduce frequency of starts and overloading or increase motor size.
	High ambient starter temperatures	Reduce ambient temperature or provide outside source of cooler air.
	Wrong-size relays	Correct size per nameplate current of motor. Relays have built in allowances for service factor current. Refer to *National Electrical Code*®.
Motor runs excessively hot	Overloaded	Reduce load or load peaks and number of starts in cycle or increase motor size.
	Blocked ventilation a. TEFCs b. ODPs	Clean external ventilation system; check fan. Blow out internal ventilation passages. Eliminate external interference to motor ventilation.
	High ambient temperature over 40°C or 105°F	Reduce ambient temperature or provide outside source of cooler air.
	Unbalanced input current	Balance supply voltage. Check motor leads for tightness.
	Single phased	Eliminate.
Won't start (just hums and heats up)	Single phased	Shut power off. Eliminate single phasing. Check motor leads for tightness.
	Rotor or bearings locked	Shut power off. Check shaft for freeness of rotation. Be sure proper-sized overload relays are in *each of the three phases* of starter. Refer to *National Electrical Code*®.
Runs noisy under load (excessive electrical noise or chatter under load)	Single phased	Shut power off. If motor cannot be restarted, it is single phased. Eliminate single phasing. Be sure that proper-sized overload relays are in *each of the three phases* of the starter. Refer to *National Electrical Code*®.
Slow starting (10 or more seconds on small motors; 15 or more seconds on large motors)		
Across the line start	Excessive voltage drop (5–10% voltage drop causes 10–20% or more drop in starting torque)	Consult power company; check system. Eliminate voltage drop.
	High-inertia load	Reduce starting load or increase motor size.
Reduced voltage start	Excessive voltage drop	Check and eliminate.
	Loss of starting torque	
Wye–delta	Starting torque reduced to 33%	Reduce starting load or increase motor size.
PWS	Starting torque reduced to 50%	Choose starting method with higher starting torque.
Autotransformer	Starting torque reduced 25 to 64%	Reduce time delay between first and second steps on starter; get motor across the line sooner.

Table 22–3 CONTINUED

Symptom	*Possible Causes*	*Correction*
Load speed appreciably below nameplate speed	Overload Excessively low voltage	Reduce load or increase voltage. *Note:* A reasonable overload or voltage drop of 10–15% will reduce speed only 1–2%. A report of any greater drop would be questionable.
	Wrong nameplate	If speed is off appreciably (i.e., from 1800 to 1200 rpm, check Lincoln code stamp (on top of stator) with nameplate. If codes do not agree, replace with motor of proper speed.
	Inaccurate method of measuring rpm	Check meter using another device or method.
Excessive vibration (mechanical)	Out of balance a. Motor mounting	Be sure motor mounting is tight and solid.
	b. Load	Disconnect belt or coupling; restart motor. If vibration stops, the unbalance was in load.
	c. Sheaves or coupling	Remove sheave or coupling; securely tape $\frac{1}{2}$ key in shaft keyway and restart motor. If vibration stops, the imbalance was in the sheave or coupling.
	d. Motor	If the vibration does not stop after checking a, b, and c above, the imbalance is in the motor; replace the motor.
	e. Misalignment on close-coupled application	Check and realign motor to the driven machine.
Noisy bearings (listen to bearings)		
Smooth midrange hum	Normal fit	Bearing OK.
High whine	Internal fit of bearing too tight	Replace bearing; check fit.
Low rumble	Internal fit of bearing too loose	Replace bearing; check fit.
Rough clatter	Bearing destroyed	Replace bearing; avoid: a. Mechanical damage b. Excess greasing c. Wrong grease d. Solid contaminants e. Water running into motor f. Misalignment on close-coupled application g. Excessive belt tension
Mechanical noise	Driven machine or motor noise?	Isolate motor from driven machine; check difference in noise level.
	Motor noise amplified by resonant mounting	Cushion motor mounting or dampen source of resonance.
	Driven machine noise transmitted to motor through drive	Reduce noise of driven machine or dampen transmission to motor.
	Misalignment on close-coupled application	Improve alignment.

Source: Courtesy of Lincoln Electric Co.

Table 22–4 TROUBLESHOOTING MOTOR STARTERS

Symptoms	*Possible Causes*	*Correction*
	Magnetic and Mechanical Parts	
Noisy magnet Humming	Misalignment or mismating of magnet pole faces	Realign or replace magnet assembly.
	Foreign matter on pole face (dirt, lint, rust, etc.)	Clean (but do not file) pole faces and realign if necessary.
	Low voltage applied to coil	Check system and coil voltage. Observe voltage variations during startup time.
Loud buzz	Broken shading coil	Replace shading coil and/or magnet assembly.
Failure to pick up and seal in	Low voltage	Check system voltage, coil voltage, and watch for voltage variations during start.
	Wrong magnet coil or wrong connection	Check wiring, coil nomenclature, etc.
	Coil open or shorted	Check with an ohmmeter and when in doubt, replace.
	Mechanical obstruction	Disconnect power and check for free movement of magnet and contact assembly.
Failure to drop out	"Gummy" substance on pole faces or magnet slides	Clean with nonvolatile solvent, degreasing fluid, possibly gasoline (with caution).
	Voltage to coil not removed	Shorted seal-in contact (exact cause found by checking coil circuit).
	Worn or rusted parts causing binding	Clean or replace worn parts.
	Residual magnetism due to lack of air gap in magnet path	Replace any worn magnet parts or accessories.
	Contacts	
Contact chatter (source is probably from magnetic assembly)	Broken shading coil	Replace assembly.
	Poor contact continuity in control circuit	Improve contact continuity or use holding-circuit interlock (three-wire control).
	Low voltage	Correct voltage condition. Check momentary voltage dip during start.
Welding	Abnormal inrush of current	Use larger contactor or check for grounds, shorts, or excessive motor load current.
	Rapid jogging	Install larger device rated for jogging service or caution operator.
	Insufficient tip pressure	Replace contact springs; check contact carrier for deformation or damage.
	Low voltage preventing magnet from sealing	Correct voltage condition. Check momentary voltage dip during starting.
	Foreign matter preventing contacts from closing	Clean contacts with nonvolatile solvent. Contactors, starters, and control accessories used with very small current or low voltage should be cleaned with solvent and then with acetone to remove the solvent residue.
	Short circuit	Remove short fault and check to be sure that fuse or breaker size is correct.
Short contact life or overheating	Filing or dressing	Do not file silver contacts. Rough spots or discoloration will not harm them or impair their efficiency.
	Interrupting excessively high currents.	Install larger device or check for grounds, shorts, or excessive motor currents.

Note: Any contact replacement should include a complete set replacement, including support springs, screws, etc.

Table 22-4 CONTINUED

Symptom	*Possible Causes*	*Correction*
	Coils	
Open circuit	Mechanical damage	Handle and store coils carefully. Replace coil.
Cooked coil (overheated)	Overvoltage or high ambient temperature	Check application and circuit. Coils will operate over a range of 85–110% rated voltage.
	Incorrect coil	Check rating and, if incorrect, replace with proper coil.
	Shorted turns caused by mechanical damage or corrosion	Replace coil.
	Undervoltage, failure of magnet to seal in	Correct system voltage.
	Dirt or rust on pole faces, increasing air gap	Clean pole faces.
	Sustained low voltage	Remedy according to local code requirements, low voltage system protection, etc.
	Overload Relays	
Nuisance tripping	Sustained overload	Check for motor or electrical equipment grounds and shorts, as well as excessive motor currents due to overload. Check motor winding resistance to ground.
	Loose connections	Clean connections and tighten. This includes load wires and heater element mounting screws.
	Incorrect heater	Check heater sizing and ambient temperature.
Failure to trip out (causing motor burnout)	Mechanical binding, dirt, corrosion, etc.	Clean or replace.
	Incorrect heater or heaters omitted and jumper wires used instead	Recheck ratings and heater size. Correct if necessary.
	Wrong calibration adjustment	Consult factory. Calibration adjustment is not normally recommended unless factory supervised. It is customary to return units to factory for check and calibration.
	Wrong calibration adjustment	Consult factory. Calibration adjustment is not normally recommended unless factory supervised. It is customary to return units to factory for check and calibration.
	Manual Starters	
Failure to operate (mechanically)	Mechanical parts, including springs, worn or broken	Replace parts as needed.
	Welded contacts due to misapplication or other abnormal cause	Replace contacts and recheck operation.
Trips out prematurely	Motor overload, incorrect heaters, or misapplication	Check conditions and replace or adjust as needed.
	Pushbuttons	
Button inoperable		
Mechanical	Shaft binding due to dirt or residue	Check, clean, and clear.
	Contact board spring broken	Replace contact board.
Electrical	Contaminated contacts and corrosion	Clean.
	Excessive jogging	Install larger device or check rated for jogging or caution operator.
	Weak contact pressure	Replace contact springs; check contact carrier for deformation or damage.

Table 22–4 CONTINUED

	Dirt or foreign matter on contact surface	Clean contacts with nonvolatile solvent.
	Short circuits	Remove short fault and check to be sure fuse or breaker size is correct.
	Loose connection	Clean and tighten.
	Sustained overload	Install larger device or check for excessive motor load current.
	Excessive wear	Higher-than-normal voltage will cause unnecessary forces that may result in mechanical wear.
Contacts, supports, discoloring	Loose connections	Tighten hardware or replace.

Source: Courtesy of Square D.

burnouts. The rate of insulation degradation is increased by higher temperature. In fact, insulation life is reduced by about half for each 10°C increase of winding temperature. Therefore, long winding life requires normal operating temperatures.

Ventilation

Forced ventilation is generally an inherent design feature of induction motors. Decreased cooling air volume caused by blocked air passages, blower failure, or low air density at higher altitudes leads to overheating and shortened winding life.

Ambient Temperature

The insulation system of motors is usually designed to operate at a maximum 40°C (104°F) ambient temperature. Any increase of ambient over 40°C requires derating the motor or its expected life will be shortened. Factors that raise input air temperature include placement of motors in discharge airstreams from other equipment and high-temperature locations.

To calculate the derating required for high ambient temperatures, multiply the ambient factor obtained from Fig. 22–23 by the rated horsepower of the motor. The ambient factor from Fig. 22–23 can also be used to *uprate* motors used in ambient temperatures under 40°C.

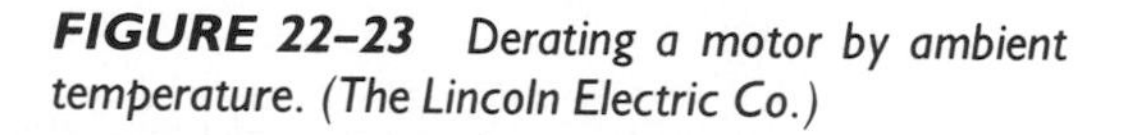

FIGURE 22–23 *Derating a motor by ambient temperature. (The Lincoln Electric Co.)*

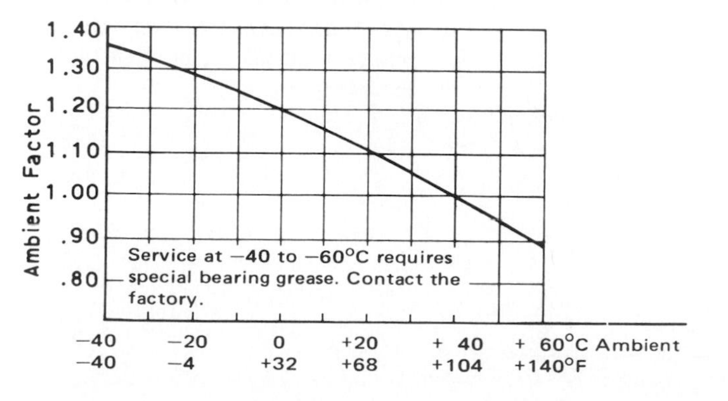

Whenever a motor is derated or uprated, the starting, pull-up, and breakdown torques remain the same as the nameplate rating, but the bearings, shaft, and other components may be subjected to life that is a function of the new rating.

PERFORMANCE CHARACTERISTICS

Performance characteristics of motors change when they are operated on high or low voltages at rated frequency. Table 22–5 shows the characteristic and the change when operated 10% above rated voltage and 10% below rated voltage. Winding failures resulting from extreme voltage variation are identical to those of overloads because the input current is uniformly excessive.

Voltage Unbalance

Voltage unbalance occurs when the phase voltages differ from one another. At one extreme this phenomenon occurs as single phasing. It can also occur in a more subtle form as voltage differences among the three phases.

Single Phasing. A single-phase motor at standstill will not start and the large inrush current overheats the windings rapidly. If the single phasing occurs when running, the motor will continue to supply torque, but the input current to the remaining two phases increases, causing overheating.

Voltage Unbalance. By definition

$$\text{voltage unbalance (\%)} = 100 \times \frac{\text{maximum voltage deviation from average voltage}}{\text{average voltage}}$$

TABLE 22–5 PERFORMANCE CHARACTERISTIC CHANGES WHEN MOTORS ARE OPERATED ON HIGH OR LOW VOLTAGES (AT RATED FREQUENCY)

	Change, Relative to Performance at Rated Voltage	
Performance Characteristic	*When actual voltage is 10% above rated voltage*	*When actual voltage is 10% below rated voltage*
Starting–Pull-Up–Breakdown		
Current	Increases 10–12%	Decreases 10–12%
Torque	Increases 21–25%	Decreases 19–23%
	Idle	
Current		
1800/1200/900-rpm motors	Increases 12–39%	Decreases 10–21%
3600-rpm motors	Increases 28–60%	Decreases 21–34%
	Rated Load	
Current (1800- and 3600-rpm motors only)		
143T–182T	Varies −4 to +11%	Varies −11 to +4%
184T–256T	Decreases 1–10%	Increases 1–10%
284T–445T	Decreases 0–7%	Increases 0–7%
Current (1200- and 900-rpm motors only): These motors do not follow the pattern above. Most will respond to a 10% overvoltage with an increase (as much as 15%) in input current. A similar reduction in input current is characteristic of a 10% reduction in input voltage.		
Power factor[a]	Decreases 5–8%	Increases 2%
Efficiency[a]	Little change	Decreases 2%
Speed	Increases 1%	Decreases 1.5%
Percent slip	Decreases 1.0	Increases 1.5

[a]Note at $\frac{3}{4}$ and $\frac{1}{2}$ load, these changes are approximately the same.

When the voltage varies among the three phases, the unbalanced voltage raises the current in one or two phases, causing overheating.

Voltage unbalances are caused by the following conditions:

1. Unequal loading of the three-phase system
2. Unequal tap settings on transformers
3. Poor connections in the power supply
4. Open delta transformer systems
5. Improper function of capacitor banks

The winding failure pattern from either single-phasing or voltage unbalance will be similar. Figure 22–24 shows two phases that are overheated in a Y-connected motor. A delta-connected motor will have one overheated phase.

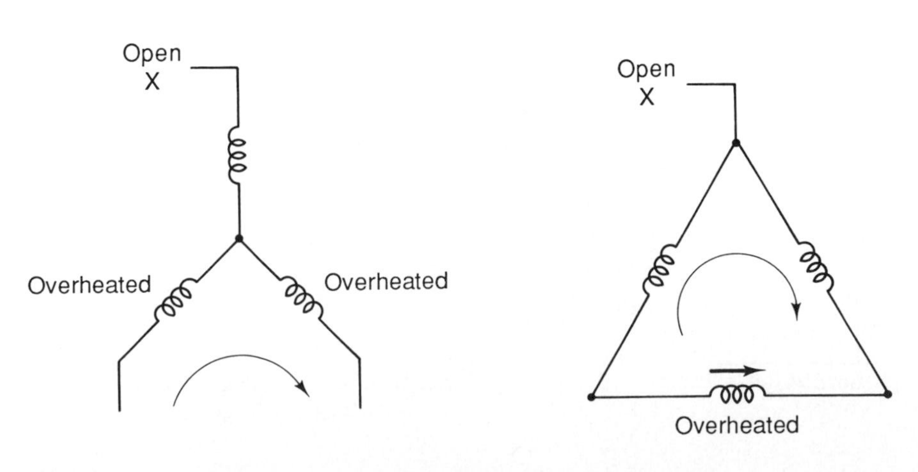

FIGURE 22–24 *Heating effects on wye- or delta-connected windings when voltage is unbalanced or single phasing occurs.*

MOTOR PROTECTION

In each of the abnormal operating conditions listed—overloading, voltage regulation, and voltage unbalance—the risk of excessive winding temperature exists because input current is higher than nameplate current. Motor controls and protective devices must prevent input current from exceeding nameplate current for extended periods.

The *National Electrical Code®* has established standards that apply to the control and protection of motors and associated circuits. For example, Fig. 22–25 is a general circuit for the installation of squirrel-cage inductor motors. Consult the *National Electrical Code®* for complete requirements for all installations.

Motor controllers and their associated circuits will vary among different models as well as different controller manufacturers. Figure 22–26 is a functional schematic diagram of a motor controller that incorporates overload protection. Good engineering practice emphasizes using three overload relays, one in each phase line of the motor, to give protection from voltage unbalance conditions. The size of the controller and overload protection relays must be in accord with the controller manufacturer's specifications as well as the *National Electrical Code®*.

As a further safeguard against motor failure due to prolonged locked rotor conditions, the overload protection should trip out in 15 seconds or less at locked-rotor current.

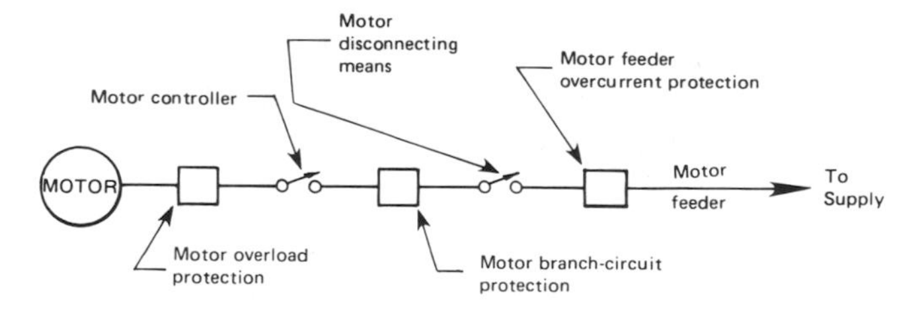

FIGURE 22–25 *Diagram showing motor protective devices.*

FIGURE 22–26 *A motor control consists of a contactor and overload relays.*

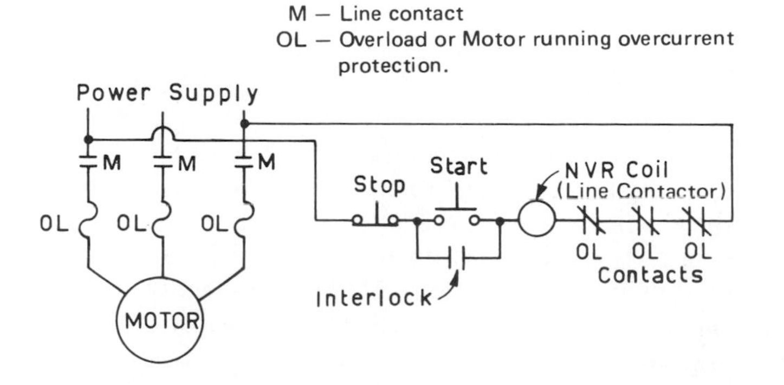

DC MOTOR PROBLEMS

One of the problems associated with dc motors is high maintenance costs in both time and materials. They do have a tendency to need attention at the commutator and brush level. Figure 22–27 shows some of the problems associated with worn or damaged commutators.

Figure 22–27A. The pitch bar-marking produces low or burned spots on the commutator surface that equals half or all the number of poles of the motor.

Figure 22–27B. Streaking on the commutator surface denotes the beginning of serious metal transfer to the carbon brush.

Figure 22–27C. Heavy slot bar-marking involves etching of the trailing edge of the commutator bar in relation to the numbered conductors per slot.

Figure 22–27D. Threading of the commutator with fine lines is a result of excessive metal transfer leading to resurfacing and excessive brush wear.

Figure 22–27E. Copper drag is an abnormal amount of excessive commutator material at the trailing edge of the bar. Although rare, flashover may occur if not corrected.

Figure 22–27F. Grooving is caused by an abrasive material in the brush or the atmosphere.

Satisfactory commutator surfaces are shown in Fig. 22–28. Determine desirable and undesirable commutator surfaces by checking these with those shown in the previous figure. Figure 22–29 shows what can be done to correct the situation.

Figure 28–30A. Light tan film over the entire commutator surface is a normal condition.

Figure 28–30B. Mottled surface with random film patterns is satisfactory.

Figure 28–30C. Slot bar markings appearing on bars in a definite pattern depicts normal wear.

Figure 28–30D. Heavy film with uniform appearance over entire commutator surface is acceptable.

Figure 22–30 shows 30 illustrations of brushes used in dc motors and generators.

SOLID-STATE EQUIPMENT TROUBLESHOOTING

Troubleshooting is easier with solid-state equipment since it is usually made with printed circuit boards or modules that can be identified as the cause of the problem and removed to be exchanged by another board or circuit module of the same type. Many are removed by pulling the card from clips and others have plugs that have to be disconnected.

Most of the devices are self-diagnosing. LEDs are used to indicate possible problems not only in the

(A)

PITCH BAR-MARKING produces low or burned spots on the commutator surface that equals half or all the number of poles on the motor.

(B)

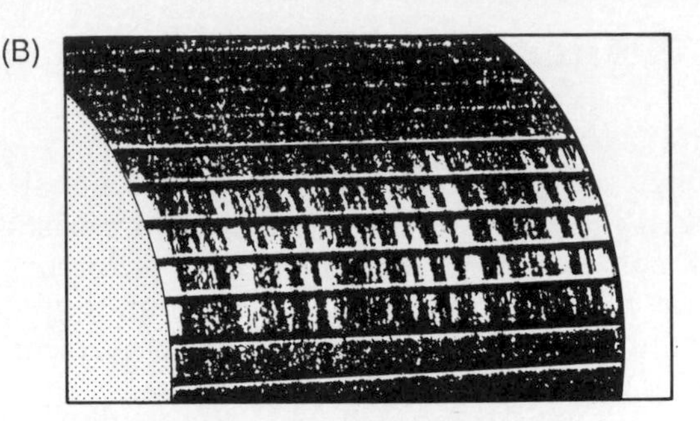

STREAKING on the commutator surface denotes the beginning of serious metal transfer to the carbon brush.

(C)

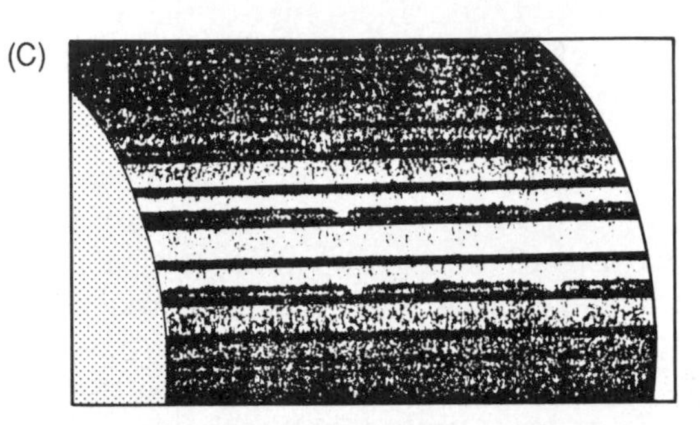

HEAVY SLOT BAR-MARKING involves etching of the trailing edge of commutator bar in relation to the numbered conductors per slot.

(D)

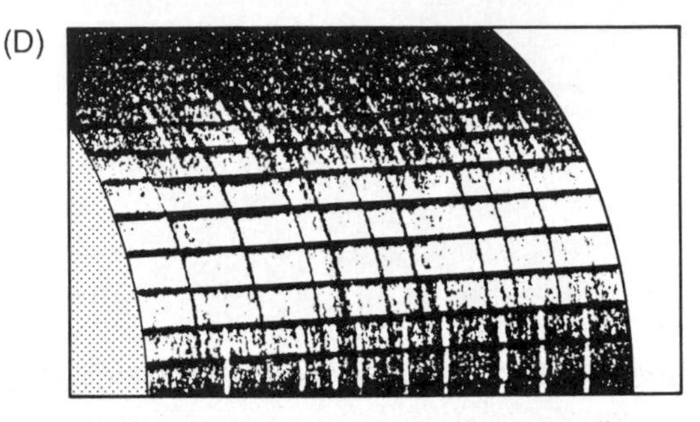

THREADING of commutator with fine lines is a result of excessive metal transfer leading to resurfacing and excessive brush wear.

(E)

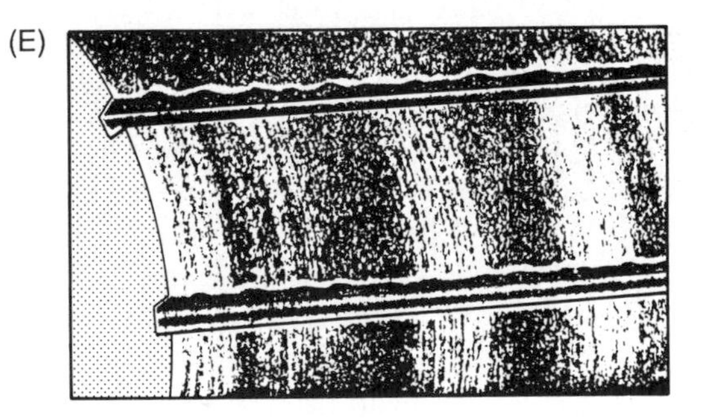

COPPER DRAG is an abnormal amount of excessive commutator material at the trailing edge of bar. Even though rare, flashover may occur if not corrected.

(F)

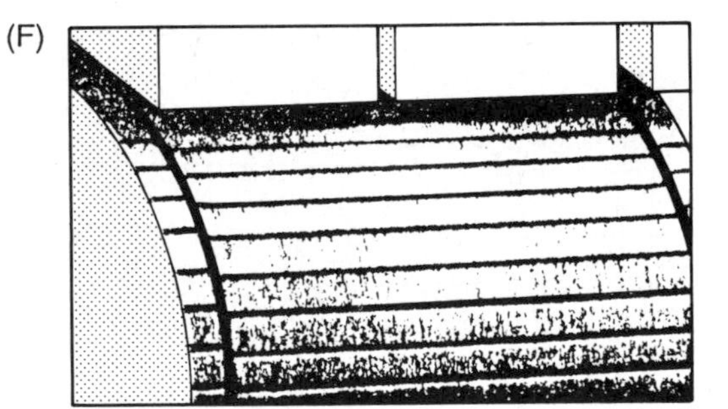

GROOVING is caused by an abrasive material in the brush or atmosphere.

FIGURE 22–27 *Worn or damaged commutators. (Reliance)*

circuitry or equipment as a whole but also in individual devices. In fact, many pieces of equipment are self-diagnosing to the point of telling you what to replace to repair the device.

In many of these devices a diode is either shorted or open. You can check the diode using a VOM and changing the probes across the diode. The diode, if good, will read high resistance in one direction and low resistance in the other. If the diode is shorted, there will be a low-resistance reading in both directions. If the diode is open, there will be an infinite reading in both directions. If a diode is found open or shorted, it must be replaced with one of the same size and rating.

(A)

LIGHT TAN FILM over entire commutator surface is a normal condition.

(B)

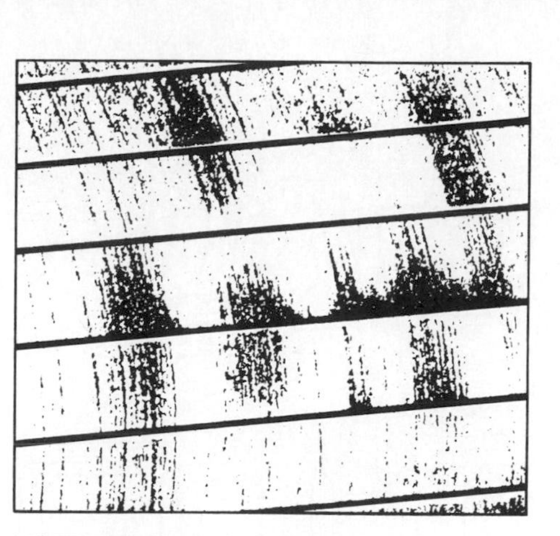

MOTTLED SURFACE with random film patterns is satisfactory.

(C)

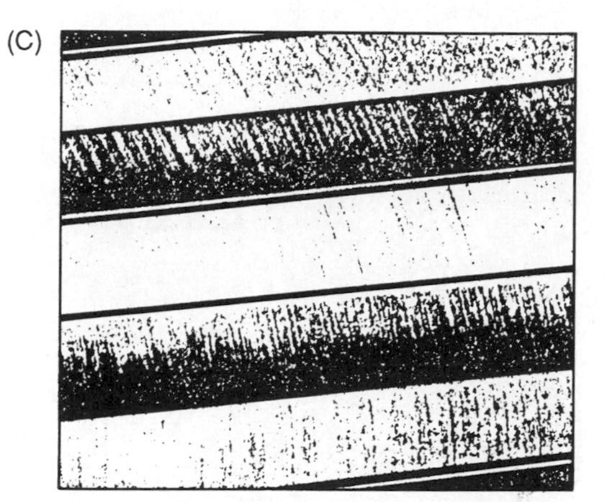

SLOT BAR MARKINGS appearing on bars in a definite pattern depicts normal wear.

(D)

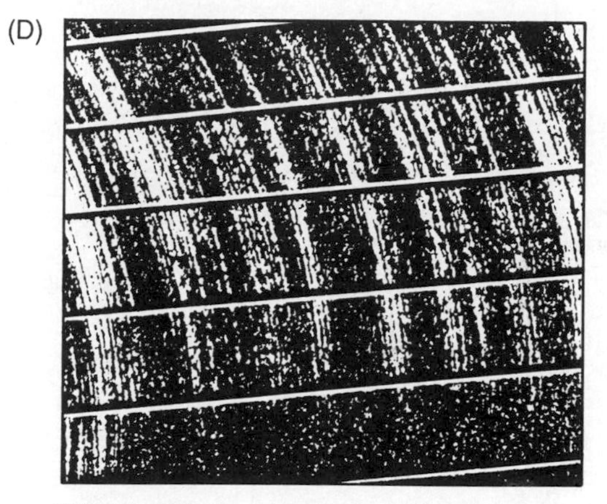

HEAVY FILM with uniform appearance over entire commutator surface is acceptable.

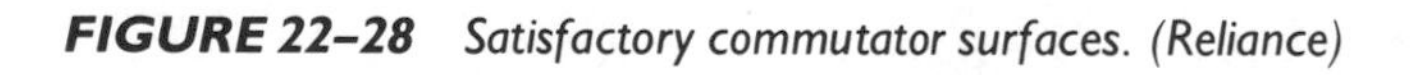

FIGURE 22–28 *Satisfactory commutator surfaces. (Reliance)*

FIGURE 22–29 *Troubleshooting causes of worn commutators. (Reliance)*

								Contamination		Type of Brush In Use	
	Vibration	Brush Pressure (light)	Unbalanced Shunt Field	Armature Connection	Light Electrical Load	Electrical Overload	Electrical Adjustment	Gas	Abrasive Dust	Abrasive Brush	Porous Brush
Pitch bar-marking	■	■	■	■						■	
Slot bar-marking						■	■	■			
Copper Drag	■	■						■		■	
Streaking		■			■			■	■	■	■
Threading		■			■			■			■
Grooving									■	■	

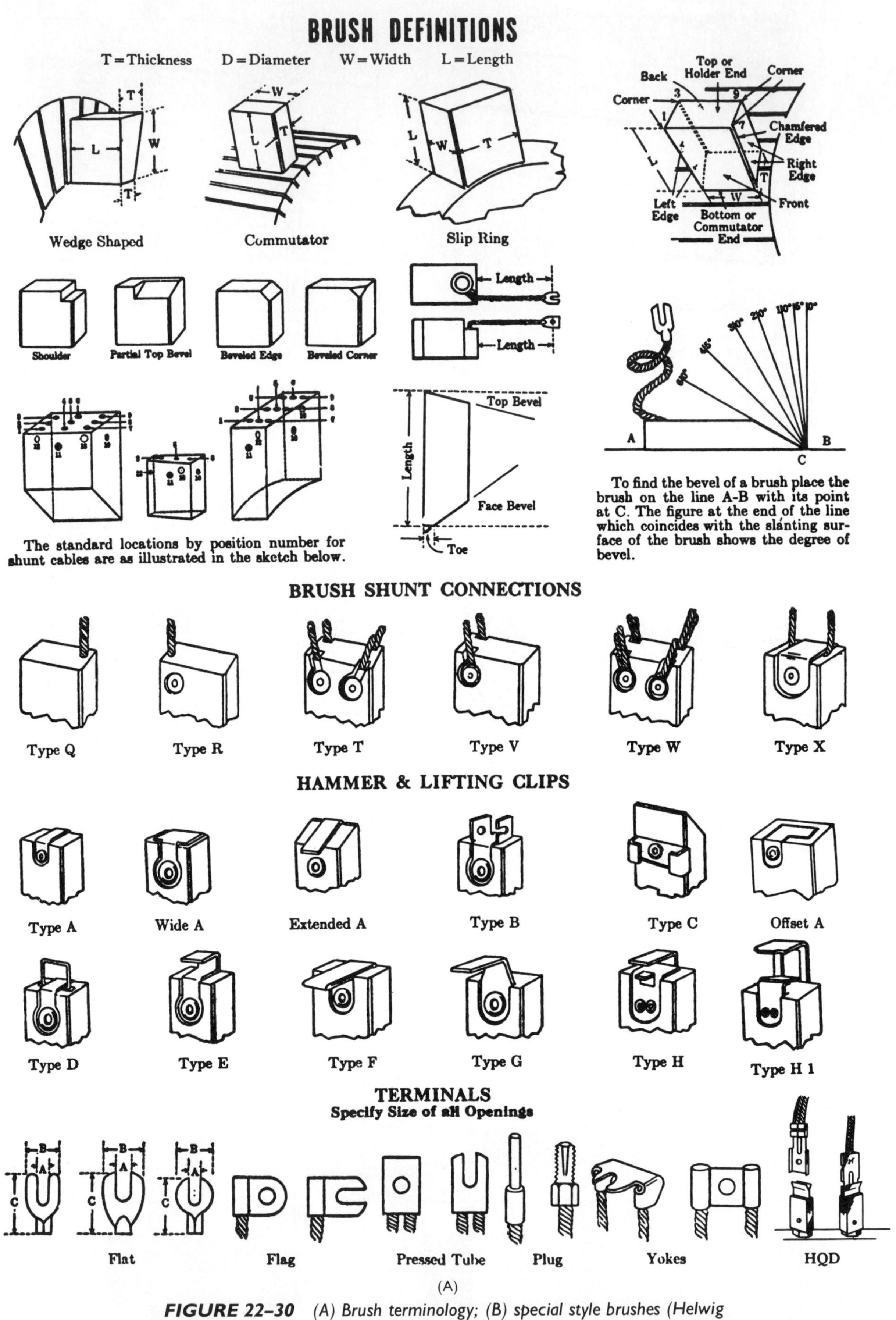

(A)

FIGURE 22–30 *(A) Brush terminology; (B) special style brushes (Helwig Carbon)*

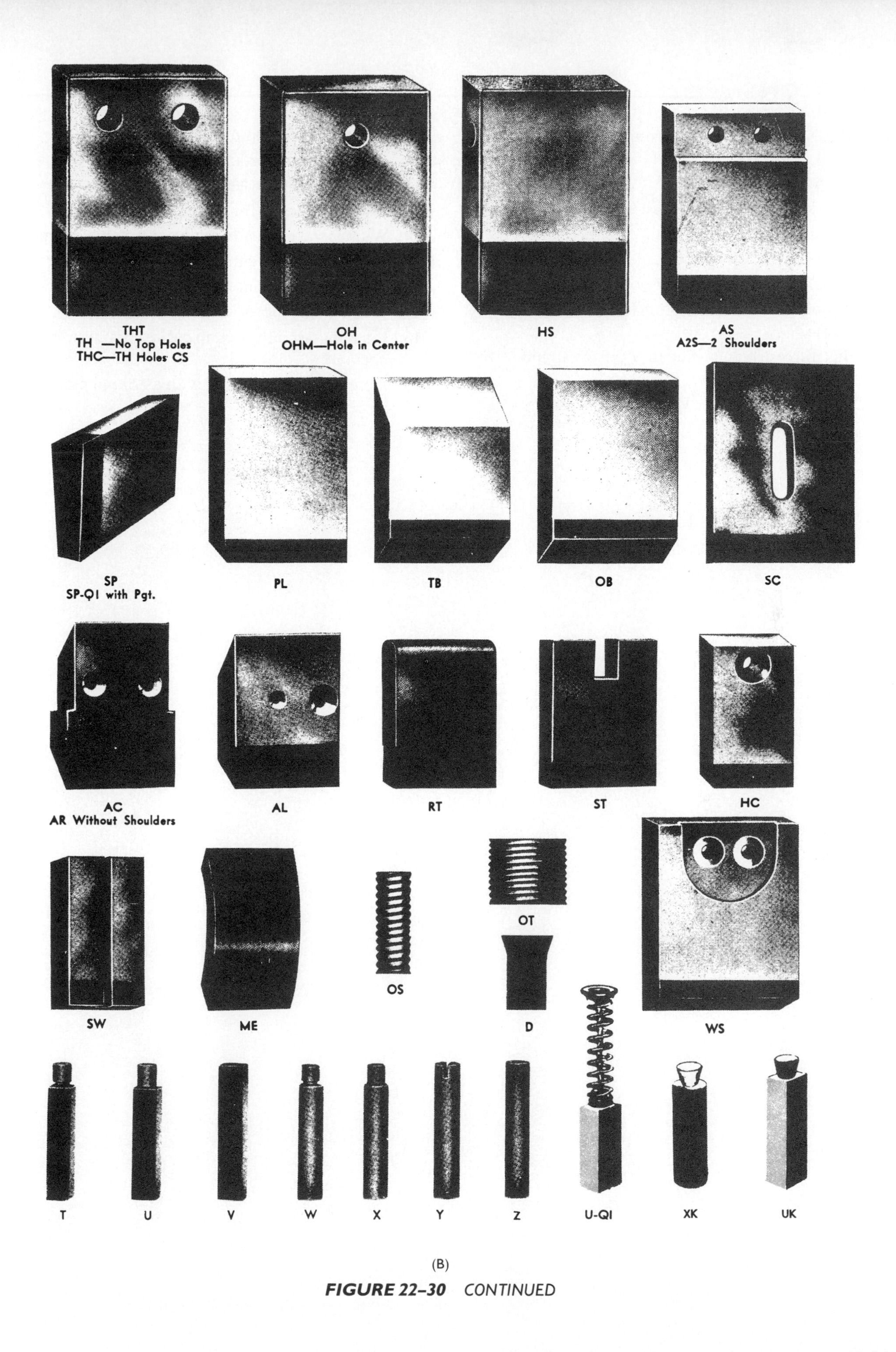

(B)

FIGURE 22–30 *CONTINUED*

QUESTIONS

1. What is an indication of losing a ground in an electrical system?
2. How does the GFCI work?
3. What is the main duty of ball bearing lubrication?
4. What causes bearing failure?
5. Why is oil viscosity important to a motor?
6. How often should motors be oiled?
7. Why are commutator motors in need of more maintenance?
8. What three irregularities in a power supply affect motor power supplies?
9. Define voltage spikes. What causes them?
10. What is electrical noise? What is its source?
11. What are transients? What is their source?
12. What does a growler do?
13. How do you test for a grounded capacitor?
14. What is the main advantage of a clamp-on meter?
15. What does high temperature do to a motor?
16. What does ambient temperature mean?
17. What causes motor winding failures?
18. Is a light tan film over an entire commutator surface normal?
19. How is a diode checked with an ohmmeter?

Glossary

Ac contactor An alternating-current (ac) contactor is designed for the specific purpose of establishing or interrupting an ac power circuit.

Across-the-line starter Device consisting of contactor and overload relays to start–stop a motor from rest to normal speed and to protect motor from overload.

Actuator Cam, arm, or similar mechanical piece used to actuate a device.

Adapter module Printed circuit card that provides communications between an I/O rack and the processor. It transmits I/O rack input terminal status to, and receives output data from, the processor.

Adjustable speed Concept of varying the speed of a motor, either manually or automatically. The desired operating speed (set speed) is relatively constant regardless of load.

Adjustable-speed drive (electrical) An adjustable-speed drive is comprised of a motor, drive controller, and operator's controls (either manual or automatic).

Air (as a prefix) The prefix "air" applied to a device that interrupts an electric circuit indicates that the interruption occurs in air.

Ambient temperature Temperature of air, water, or a surrounding medium where equipment is operated or stored.

Ambient temperature compensated Device, such as an overload relay, which is not affected by the temperature surrounding it.

Ampacity Current-carrying capacity expressed in amperes.

Ampere Measurement of intensity or rate of flow of electrons in an electric circuit. An ampere is the amount of current that will flow through a resistance of 1 ohm under a pressure of 1 volt.

Ampere rating Current-carrying capacity of a fuse. When a fuse is subjected to a current above its ampere rating, it will open the circuit after a predetermined period of time.

Ampere squared seconds, I^2t Measure of heat energy developed within a circuit during the fuse's clearing. It can be expressed as "melting I^2t", "arcing I^2t" or the sum of them as "clearing I^2t". I stands for effective let-through current (rms), which is squared, and t stands for time of opening, in seconds.

Analog Expression of values that can vary continuously (e.g., translation, rotation, voltage, or resistance). (Contrasted with *digital.*)

Antiplugging protection Effect of a control function or a device that operates to prevent application of countertorque by the motor until the motor speed has been reduced to an acceptable value.

Arcing time Amount of time from the instant the fuse link has melted until the overcurrent is interrupted or cleared.

Armature Rotating part of a dc motor.

Armature control Abbreviated term for *armature voltage control* of a dc motor, which describes the usual method of changing the speed of a dc motor by controlling the magnitude of applied armature voltage.

Armature current Dc current required by a dc motor to produce torque and drive a load. The maximum safe, continuous current is stamped on the motor nameplate. This can be exceeded only for initial acceleration and for short periods of time. Armature current is proportional to the amount of torque being produced; therefore, it rises and falls as the torque demand rises and falls.

Armature voltage feedback Armature voltage can be used as the speed feedback signal to an electronic speed regulator. This voltage is about directly proportional to motor speed, assuming a constant motor field and ignoring IR drop. Armature voltage feedback is used where the expense of a tachometer-generator for speed feedback is not justified and a regulation accuracy of 2–5% is adequate.

ASCII Acronym for *American Standard Code for Information Interchange,* an eight-bit (7 bits plus a parity bit) code for representing alphanumerics, punctuation marks, and certain special characters for control purposes.

Attenuation (also called ***suppression*** or ***rejection***) Amount of signal loss in a system. In a device such as an isolation transformer, the degree of reduction of unwanted spikes and signals; usually expressed either as a voltage ratio or in decibels.

Auxiliary contacts (switching device) Contacts in addition to the main-circuit contacts and function with the movement of the latter.

Auxiliary device Any electrical device other than motors and motor starters necessary to fully operate the machine or equipment.

Axial approach When the target to be detected approaches the sensing face head-on.

Axis Principal direction along which movement of a tool or workpiece occurs; also refers to one of the reference lines of a coordinate system.

Back of a motor End that carries the coupling or driving pulley. This is sometimes called the drive end (DE) or pulley end (PE).

Bandwidth Generally, frequency range of system input over which the system will respond satisfactorily to a command.

Base speed Manufacturer's nameplate rating where the motor will develop rated horsepower at rated load and voltage. With dc drives, it is commonly the point where full armature voltage is applied with full-rated field excitation. With ac systems, it is commonly the point where 60 Hz is applied to the induction motor.

Baud (1) Unit of data transmission speed equal to the number of code elements (bits) per second; (2) unit of signaling speed equal to the number of discrete conditions or signal events per second.

Bearing (ball) Ball-shaped component that is used to reduce friction and wear while supporting rotating elements. For a motor, this type of bearing provides a relatively rigid support for the output shaft.

Bearing (roller) Special bearing system with cylindrical rollers capable of handling belted load applications, too large for standard ball bearings.

Binary-coded decimal (BCD) Method used to express individual decimal digits (0–9) in 4-bit binary notation (e.g., the number 23 is represented as 0010 0011 in BCD notation).

Bit (1) Acronym for *binary digit,* the smallest unit of information in the binary numbering system, represented by the digits 0 and 1; (2) smallest division of a PC word.

Block diagram Diagram showing the relationship of separate subunits (blocks) in a control system.

Brake—magnetic Friction brake controlled electromagnetically.

Braking Means of stopping an ac or dc motor.

Branch circuit That portion of a wiring system extending beyond the final overcurrent device protecting the circuit. (A device not approved for branch-circuit protection, such as a thermal cutout or motor overload protective device, is not considered as the overcurrent device protecting the circuit.)

Break (of a circuit-opening device) Minimum distance between the stationary and movable contacts when these contacts are in the open position.

Breakaway torque Torque required to start a machine from standstill. It is always greater than the torque needed to maintain motion.

Breakdown torque The breakdown torque of an ac motor is the maximum torque that it will develop with rated voltage applied at rated frequency.

Bridge rectifier Full-wave rectifier that conducts current in only one direction of the input current. Ac applied to the input results in approximate dc at the output.

Bridge rectifier (diode, SCR) A diode bridge rectifier is a noncontrolled full-wave rectifier that produces a constant rectified dc voltage. An SCR bridge rectifier is a full-wave rectifier with an output that can be controlled by switching on the gate control element.

Brush Conductor, usually composed of some element of carbon, serving to maintain an electrical connection between stationary and moving parts of a machine (commutator of a dc motor). The brush is mounted in a spring-loaded holder and positioned tangent to the commutator segments against which it "brushes." Pairs of brushes are equally spaced around the circumference of the commutator.

Bus Set of power supply leads or a conductor providing for multiple connections.

Cam Device with one or more lobes (projections) which, as it moves, operates levers or switches that cause mechanical or electrical functions.

CAM Acronym for *computer-aided manufacturing.*

Captive screw Screw-type fastener that is retained in some manner when unscrewed and cannot easily be separated from the part it secures.

Cartridge fuse Fuse consisting of a current-responsive element inside a fuse tube with terminals on both ends.

CEMF Abbreviation for *counter electromotive force,* the product of a motor armature rotating in a magnetic field. This generating action takes place whenever a motor is rotating. Under stable motoring conditions the generated voltage (CEMF) is equal to the voltage supplied to the motor minus small losses. However, the polarity of the CEMF is opposite to that of the power being supplied to the armature.

Central processing unit (CPU) Another term for *processor.* It includes the circuits controlling the interpretation and execution of the user-inserted program instructions stored in the PC memory.

C face (motor mounting) Mounting used to close couple pumps and similar applications where the mounting holes in the face are threaded to receive bolts from the pump. Normally, a C face is used where a pump or similar item is to be overhung on the motor. This type of mounting is a NEMA standard design and available with or without feet.

Circuit breaker Device designed to open and close a circuit by nonautomatic means, and to open the circuit automatically on a predetermined overload of current, without injury to itself when properly applied within its rating.

Circuit interrupter Nonautomatic manually operated device designed to open, under abnormal conditions, a current-carrying circuit without injury to itself.

Class CC fuses 600-V, 200,000-A interrupting rating, branch-circuit fuses with overall dimensions of $\frac{13}{32} \times 1\frac{1}{2}$ in. Their design incorporates a rejection feature that allow them to be inserted into rejection fuse holders and fuse blocks that reject all lower voltage, lower interrupting rating $\frac{13}{32} \times 1\frac{1}{2}$ in. fuses. They are available from $\frac{1}{10}$ to 30 A.

Class G fuses 300-V, 100,000-A interrupting rating branch-circuit fuses that are size rejecting to eliminate overfusing. The fuse diameter is $\frac{13}{32}$ in. and the length varies from $1\frac{5}{16}$ to $2\frac{1}{4}$ in. These are available in ratings from 1 to 60 A.

Class H fuses 250-V and 600-V, 10,000-A interrupting rating branch-circuit fuses that may be renewable or nonrenewable. These are available in ampere ratings of 1 to 600 A. (*See also* One-time fuses *and* Renewable fuses.)

Class J fuses Fuses rated to interrupt 200,000 A ac. They are UL labeled as "current limiting," are rated for 600 V ac, and are not interchangeable with other classes.

Class K fuses Fuses listed by UL as K-1, K-5, or K-9. Each subclass has designated I^2t and I_p maximums. These are dimensionally the same as class H fuses (NEC dimensions) and they can have interrupting ratings of 50,000, 100,000, or 200,000 A. These fuses are current limiting; however, they are not marked "current limiting" on their label since they do not have a rejection feature.

Class L fuses Fuses rated for 601 to 6000 A and rated to interrupt 200,000 A ac. They are labeled "current limiting" and are rated for 600 V ac. They are intended to be bolted into their mountings and are not normally used in clips. Some class L fuses have designed in time-delay features for all-purpose use.

Class R fuses High-performance fuses rated $\frac{1}{10}$ to 600 A in 250- and 600-V ratings. All are marked "current limiting" on their label and all have a 200,000-A interrupting rating. They have identical outline dimensions with the NEC fuses (class H) but have a rejection feature that prevents the user from mounting a fuse of lesser capabilities (lower interrupting capacity) when used with special class R clips. Class R fuses will fit into either rejection or nonrejection clips.

Class T fuses A UL classification of fuses in 300- and 600-V ratings from 1 to 1200 A. They are physically very small and can be applied where space is at a premium. They are fast-acting fuses, with an interrupting rating of 200,000 A rms.

Clearing time Total time between the beginning of the overcurrent and the final opening of a circuit at rated voltage by an overcurrent protective device. Clearing time is the total of the melting time and the arcing time.

Closed-circuit transition As applied to reduced-voltage controllers, a method starting by which the power to the motor is not interrupted during a normal starting sequence.

Closed loop Regulator circuit in which the actual value of the controlled variable (e.g., speed) is sensed and a signal proportional to this value (feedback signal) is compared with a signal proportional to the desired value (reference signal). The difference between these signals (error signal) causes the actual value to change in the direction that will reduce the difference in signals to zero.

CMOS Acronym for *complementary metal-oxide semiconductor circuitry,* an integrated-circuit family that has high threshold logic and low power consumption, thus making it especially useful in remote applications where supplying power becomes expensive.

Cogging Condition in which a motor does not rotate smoothly but "steps" or "jerks" from one position to another during shaft revolution. Cogging is most pronounced at low motor speeds and can cause objectionable vibrations in the driven machine.

Combination starter Magnetic starter having a manually operated disconnecting means built into the same enclosure with the magnetic contractor or starter.

Common-mode noise Signals or spikes impressed from line-to-ground in a power distribution system.

Commutator Cylindrically shaped assembly that is fastened to a motor shaft and is considered part of the armature assembly. It consists of segments or "bars" that are electrically connected to two ends of one (or more) armature coils. Current flows from the power supply through the brushes, to the commutator, and hence through the armature coils. The arrangement of commutator segments is such that the magnetic polarity of each coil changes a number of times per revolution (the number of times depends on the number of poles in the motor).

Commutation (inverter) Process by which forward current is interrupted or transferred from one switching device to the other. In most circuits where power is supplied from an ac source, turn-on control is adequate and turn-off occurs naturally when the ac cycle causes the polarity across a given device to reverse.

Commutation (dc motor) Reversing the current in an armature coil when the coil (ends) moves from one side of the brush to the other side of the same brush. This completes the connection between the armature winding and the external circuit.

Comparator Device that compares one signal to another, usually the process signal compared to the set point or command signal.

Complementary outputs Sensors with both NO and NC outputs which change state simultaneously.

Computerized numerical control (CNC) Numerical control system where a computer is used to perform some or all of the basic numerical control functions.

Conditional ignore zones Distinct program areas that control the same outputs, through separate rungs, at different times. Each conditional ignore zone is delimited and controlled by fence codes. For any grouping of outputs, the user's program must enable only one conditional ignore zone at a time. (If all zones are disabled, these outputs would remain in their last states.)

Conduit, flexible metal Flexible raceway of circular cross section especially constructed for the purpose of the pulling in or withdrawing of wires or cables after the conduit and its fittings are in place.

Conduit, flexible nonmetallic Flexible raceway of circular cross section especially constructed for the purpose of the pulling in or withdrawing of wires or cables after the conduit and its fittings are in place.

Conduit, rigid metal Raceway especially constructed for the purpose of the pulling in or winding of wires or cables after the conduit is in place, made of metal pipes of standard weight and thickness, permitting the cutting of standard threads.

Connector Plug or receptacle for electrically interconnecting one or more cables or electronic circuits.

Constant-horsepower range Range of motor operation where motor speed is controlled by field weakening. In this range, motor torque decreases as speed increases. Since horsepower is speed times torque (divided by a constant), the value of horsepower developed by the motor in this range is constant.

Constant-torque range Speed range in which the motor is capable of delivering a constant torque, subject to cooling limitations of the motor.

Constant-voltage range (ac drives) Range of motor operation where the drive's output voltage is held constant as output frequency is varied. This speed range produces motor performance similar to a dc drive's constant-horsepower range.

Constant volts per hertz (V/Hz) Relationship that exists in ac drives where the output voltage is varied directly proportional to frequency. This type of operation is required to allow the motor to produce constant-rated torque as speed is varied.

Contactor Device for repeatedly establishing and interrupting an electric power circuit.

Contactor reversing Method of reversing motor rotation by the use of two separate contactors, one of which produces rotation in one direction and the other which produces rotation in the opposite direction. The contactors are electrically (and mechanically) interlocked so that both cannot be energized at the same time.

Contacts Connecting parts that co-act to complete or to interrupt a circuit.

Contact wear Total thickness of material that may be worn away before the co-acting contacts cease to perform adequately.

Continuous rating Rating that defines the substantially constant load which can be carried for an indefinitely long time.

Contouring Operation in which simultaneous control of more than one axis is accomplished.

Control circuit The control circuit of the control apparatus or system is the circuit that carries the signals directing the performance of the controller.

Control circuit transformer Voltage transformer utilized to supply a voltage suitable for the operation of control devices.

Control circuit voltage Voltage provided for the operation of the coil of magnetic devices.

Control compartment Space within the base, frame, or column of the machine used for mounting the control panel.

Control, three-wire Control function that utilizes a momentary-contact pilot device and a holding circuit contact to provide undervoltage protection. Upon drop in control voltage to the dropout voltage of the device coil, the line contacts open and upon restoration of pickup voltage to the coil, the contacts will not close until the momentary-contact pilot service is actuated.

Control, two-wire Control function that utilizes a maintained-contact type of pilot device to provide undervoltage release. Upon a drop in control voltage to the dropout voltage of the device coil the line contacts will open, and upon the restoration of pickup voltage the contacts will close without manual operation of the pilot device.

Controller Device, or group of devices, that serves to govern, in a predetermined manner, the electric power delivered to the apparatus to which it is connected.

Controller, definite purpose Any controller having ratings, operating characteristics, or mechanical construction for use under service conditions other than usual or for use on a definite type of application, such as air-conditioner compressors.

Controller, drum Electric controller made of stationary contacts connected in the circuit by the rotation of a rotary group of movable contacts.

Controller, general purpose Any controller having ratings, characteristics, and mechanical construction for use under usual service conditions in accordance with NEMA standards for industrial control and systems.

Controller, manual Electric controller having all of its basic functions performed by devices operated by hand.

Converter Process of changing ac to dc. This is accomplished through the use of a diode rectifier or thyristor rectifier circuit. The term may also refer to the process of changing ac to dc to ac (e.g., adjustable frequency drive). A frequency converter, such as that found in an adjustable-frequency drive, consists of a rectifier, a dc intermediate circuit, an inverter, and a control unit.

Core memory Type of memory used to store information in ferrite cores. Each may be magnetized in either polarity, to represent a logical "1" or "0." This type of memory is nonvolatile.

Current limitation Fuse operation relating to short circuits only. When a fuse operates in its current-limiting range, it will clear a short circuit in less than a half-cycle. It will also limit the instantaneous peak let-through current to a value substantially less than that obtainable in the same circuit if that fuse were replaced with a solid conductor of equal impedance.

Current limiting Electronic method of limiting the maximum current available to the motor. This is adjustable so that the motor's maximum current can be controlled. It can also be preset as a protective device to protect both the motor and control from extended overloads.

Current-sinking sensor (NPN transistor) A current-sinking sensor "sinks" current *from* the load to the negative terminal (−) of the dc voltage supply.

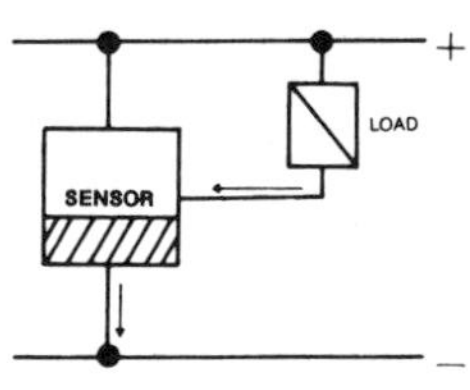

Current-sourcing sensor (PNP transistor) A current-sourcing sensor "sources" current from the positive terminal (+) of the dc voltage supply *to* the load.

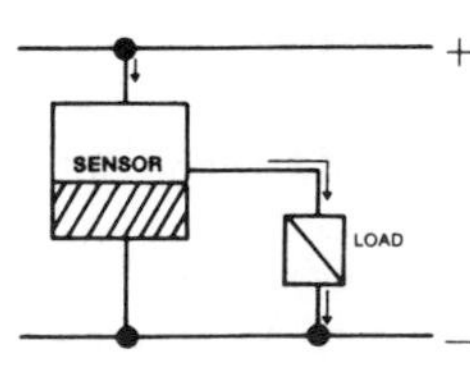

Damping Reduction in amplitude of an oscillation in the system.

Data General term for any type of information.

Data link Equipment, especially transmission cables and interface modules, which permits the transmission of information.

DC contactor Contactor specifically designed to establish or interrupt a direct-current power circuit.

Dead band Range of values through which a system input can be changed without causing a corresponding change in system output.

Decibel Logarithmic unit. The fundamental logarithmic unit is the bel, which is the common logarithm of the ratio of two values of power. The practical unit is the decibel (dB), which is 1/10 bel.

$$\text{Number of decibels} = 10 \log_{10} \frac{W_2}{W_1}$$

W_2 and W_1 = values of power expressed in the same units. A power ratio of 2 corresponds to 3 dB. The total number of decibels is three times the number of times that 2 is contained as a factor in the ratio. The decibel may also be used to express voltage or current ratios.

$$\text{Number of decibels} = 20 \log_{10} \frac{E_2}{E_1} \text{ or } 20 \log_{10} \frac{I_2}{I_1}$$

The factor 10 in the power equation is multiplied by 2 because the power varies as the square of the current or voltage ratio. Thus a voltage ratio of 2 corresponds to 6 dB.

Definite-purpose motor Any motor design listed and offered in standard ratings with standard operating characteristics from a mechanical construction for use under service conditions other than usual or for use on a particular type of application (NEMA).

Demagnetization (current) When a permanent-magnet dc motor is subjected to high current pulses, the magnets may become slightly demagnetized, resulting in a lower torque constant. Pulse currents of 7 to 10 times rated values can usually be applied without demagnetization.

Deviation Difference between an instantaneous value of a controlled variable and the desired value of the controlled variable corresponding to the set point. Also called *error.*

D flange (motor mounting) This type of motor mounting is used when a motor is to be built as part of a machine. The mounting holes of the flange are not threaded. The bolts protrude through the flange from the motor side. Normally, D-flange motors are supplied without feet since the motor is mounted directly to the driven machine.

di/dt Rate of change in current versus a rate of change in time. Line reactors and isolation transformers can be used to provide the impedance necessary to reduce the harmful effects that unlimited current sources can have on phase-controlled rectifiers (SCRs).

Differential (hysteresis) Distance between the operating point where the target enters the sensing field (sensor energizes) to the release point where the target leaves the sensing field (sensor deenergizes).

Digital Representation of numerical quantities by means of discrete numbers. It is possible to express in binary digital form all information stored, transferred, or processed by dual-state conditions (e.g., on–off, open–closed, octal, and BCD values). (Contrasted with *analog.*)

Diode Device that passes current in one direction but blocks current in the reverse direction.

Drift Deviation from the initial set speed with no load change over a specific time period. Normally, the drive must be operated for a specified warm-up time at a specified ambient temperature before drift specifications apply. Drift is normally caused by random changes in operating characteristics of various control components.

Drive controller (also called ***variable-speed drive***) Electronic device that can control the speed, torque horsepower, and direction of an ac or dc motor.

Dual-element fuse Fuse with a special design that utilizes two individual elements in series inside the fuse tube. One element, the spring-actuated trigger assembly, operates on overloads up to five to six times the fuse current rating. The other element, the short-circuit section, operates on short circuits up to their interrupting rating.

Duplex Means of two-way data communication. *See also* Full duplex.

dv/dt Rate of changes in voltage versus a rate of change in time. Specially designed resistor–capacitor networks can help protect the SCRs from excessive *dv/dt,* which can result from line-voltage spikes, line disturbances, and circuit configurations with extreme forward conducting or reverse blocking requirements.

Dwell Time spent in one state before moving to the next. In motion control applications, for example, a dwell time may be programmed to allow time for a tool change or part clamping operation.

Eddy current Currents induced in motor components from the movement of magnetic fields. Eddy currents produce waste heat and are minimized by lamination of the motor poles and armature.

Efficiency Ratio of mechanical output to electrical input indicated by a percent. In motors, it is the effectiveness with which a motor converts electrical energy into mechanical energy.

Electrical load That part of the electrical system which actually uses the energy or does the work required.

Electrical-optical isolator Device that couples input to output using a light source and detector in the same package. It is used to provide electrical isolation between input circuitry and output circuitry.

Electrostatic shield Typically, one turn of a thin sheet of aluminum or copper, extending over the full width of the windings of a transformer, usually located between the primary and secondary windings.

EMF Abbreviation for *electromotive force,* another term for voltage or potential difference. In DC adjustable-speed drives, voltage applied to the motor armature from a power supply is the EMF and the voltage generated by the motor is the counter EMF or CEMF.

Enable To allow an action or acceptance of data by applying an appropriate signal to the appropriate input.

Enclosure Housing in which the control is mounted. Enclosures are available in designs for various environmental conditions.

Encoder Electromechanical transducer that produces a serial or parallel digital indication of mechanical angle or displacement. Essentially, an encoder provides high-resolution feedback data related to shaft position and is used with other circuitry to indicate velocity and direction. The encoder produces discrete electrical pulses during each increment of shaft rotation.

Error Difference between the set-point signal and the feedback signal. An error is necessary before a correction can be made in a controlled system.

External control devices All control devices mounted external to the control panel.

Eyelet Used on printed circuit boards to make electrical connections from one side of the board to the other side.

Fail-safe operation Electrical system so designed that the failure of any component in the system will prevent unsafe operation of the controlled equipment.

Fast-acting fuse Fuse that opens very quickly on overloads and short circuits. This type of fuse is not designed to withstand temporary overload currents associated with some electrical loads (inductive loads).

FCC (CFR47) Defines limits of radiated energy (RFI) from computing devices to protect a TV receiver located at least 10 meters from the RFI source.

Feedback Element of a control system that provides an actual operation signal for comparison with the set point to establish an error signal used by the regulator circuit.

Feeder Circuit conductors between the service equipment, or the generator switchboard of an isolated plant, and the branch-circuit overcurrent device.

Field Stationary electrical part of a dc motor.

Field control Method of controlling dc motor speed by varying the field current in the shunt field windings.

Field economy Circuit design feature of a dc motor shunt field supply that reduces the supply voltage output after a predetermined period of time. On many field supplies, this means a 50% reduction in output voltage 2 to 3 minutes after machine shutdown (idle). A field economy circuit serves to reduce standby power consumption and prolong the insulation life of the motor field windings.

Field forcing Temporarily overexciting a motor shunt field to overcome the L/R time constant, increase the rate of flux change, and rapidly reverse the direction of shunt motor field current.

Field range Range of motor speed from base speed to the maximum rated speed.

Field reversing One method for producing regeneration. It is accomplished by changing the direction of current through the motor field, which reverses the polarity of the motor CEMF to account for generator action.

Field weakening Action of reducing the current applied to a dc motor shunt field. This action weakens the strength of the magnetic field and thereby increases the motor speed.

Filter Device that passes a signal or a range of signals and eliminates all others.

Floating ground Circuit whose electrical common point is not at earth ground potential or the same ground potential as circuitry with which it is associated. A voltage difference can exist between the floating ground and earth ground.

Float switch Switch that is operated by the buoyant constituent part and is responsive to the level of liquid.

Foot switch Switch that is suitable for operation by an operator's foot.

Force Tendency to change the motion or position of an object with a push or pull. Force is measured in ounces or pounds.

Form factor Figure of merit that indicates how much rectified current deviates from pure (nonpulsating) dc. A large departure from unity form factor (pure dc) increases the heating effect of the motor. Mathematically, it is expressed as I_{rms}/I_{av} (motor heating current/torque producing current).

FORTRAN Acronym for *formula translation,* a scientific programming language.

Four-quadrant operation The four combinations of forward and reverse rotation and forward and reverse torque of which a regenerative drive is capable. The four combinations are: (1) forward rotation/forward torque (motoring), (2) forward rotation/reverse torque (regeneration), (3) reverse rotation/reverse torque (motoring), and (4) reverse rotation/forward torque (regeneration).

Frame size Physical size of a motor, usually consisting of NEMA defined D and F dimensions at a minimum. The D dimension is the distance in quarter-inches from the center of the motor shaft to the bottom of the mounting feet. The F dimension relates to the distance between the centers of the mounting-feet holes.

Frequency Hertz measurement of a circuit.

Front of a motor End opposite the coupling or driving pulley. This is sometimes called the opposite pulley end (OPE) or commutator end (CE).

Full duplex (FDX) Mode of communications in which data may be transmitted and received simultaneously by both ends of the circuitry.

Full-load torque Torque necessary to produce rated horsepower at full-load speed.

Fuse Overcurrent protective device with a circuit opening fuseable member that is heated and severed by passage of overcurrent through it.

Gate Control element of an SCR (silicon-controlled rectifier), commonly referred to as a *thyristor.* When a small positive voltage is applied to the gate momentarily, the SCR will conduct current (when the anode is positive with respect to the cathode of the SCR). Current conduction will continue even after the gate signal is removed.

General-purpose motor This motor has a continuous class B rating and design, listed and offered in standard ratings with standard operating characteristics and mechanical construction for use under usual service conditions without restriction to a particular application or type of application.

Grounded Connected to earth or to some conducting body that serves in place of the earth.

Grounded circuit Circuit in which one conductor or point (usually, the neutral or neutral point of transformer or generator windings) is intentionally grounded (earthed), either solidly or through a grounding device.

Grounding conductor Conductor which, under normal conditions, carries no current but serves to connect exposed metal surfaces to an earth ground, to prevent hazards in case of breakdown between current-carrying parts and exposed surfaces. If insulated, the conductor is colored green, with or without a yellow stripe.

GTO Gate turn-off or gate turn-on power semiconductor device.

Guarded Covered, shielded, fenced, enclosed,or otherwise protected by means of suitable covers or casings, barriers, rails or screens, mats or platforms to remove the likelihood of dangerous contact or approach by persons or objects to a point of danger.

Hardware Mechanical, electrical, and electronic devices that comprise a programmable controller and its application.

Head Measurement of pressure, usually in feet of water. A 30-foot head is the pressure equivalent to the pressure found at the base of a column of water 30 feet high.

Horsepower Measure of the amount of work that a motor can perform in a given period of time.

Hunting Undesirable fluctuations in motor speed that can occur after a step change in speed reference (either acceleration or deceleration) or load.

Hysteresis loss Resistance offered by materials to becoming magnetized results in energy being expended and corresponding loss. Hysteresis loss in a magnetic circuit is the energy expended to magnetize and demagnetize the core.

IEEE STD 519 *Guide for Harmonic Control and Reactive Compensation of Static Power Converters.* Recommends limits of disturbances to the ac power distribution system which affect other equipment and communications.

Induction motor Alternating-current motor in which the primary winding on one member (usually the stator) is connected to the power source. A secondary winding on the other member (usually the rotor) carries the induced current. There is no physical electrical connection to the secondary winding; its current is induced.

Inductive load Electrical load that pulls a large amount of current—an inrush current—when first energized. After a few cycles or seconds the current "settles down" to the full-load running current.

Inertia Measure of a body's resistance to changes in velocity, whether the body is at rest or moving at a constant velocity. The velocity can be either linear or rotational. The moment of inertia (WK^2) is the product of the weight (W) of an object and the square of the radius of gyration (K^2). The radius of gyration is a measure of how the mass of the object is distributed about the axis of rotation. WK^2 is usually expressed in units of lb-ft^2.

Instability State or property of a system where there is an output but no corresponding input.

Integral horsepower motor Motor built in a frame having a continuous rating of 1 hp or more.

Intermittent duty (INT) Motor that never reaches equilibrium temperature (equilibrium), but is permitted to cool down between operations. For example, a crane, hoist, or machine tool motor is often rated for 15 or 30 duty.

Interrupting rating Defines a fuse's ability to interrupt and clear short circuits *safely.* This rating is much greater than the ampere rating of a fuse. The NEC defines *interrupting rating* as "the highest current at rated voltage that an overcurrent protective device is intended to interrupt under standard test conditions."

Inverter Term commonly used for an ac adjustable-frequency drive. Also used to describe a particular section of an ac drive; this section uses the dc voltage from a previous circuit stage (intermediate dc circuit) to produce an ac current or voltage having the desired frequency.

I/O scan time Time required for the PC processor to monitor all inputs and control all outputs. The I/O scan repeats continuously.

IR compensation A way to compensate for the voltage drop across resistance of the ac or dc motor circuit and the resultant reduction in speed. This compensation also provides a way to improve the speed regulation characteristics of the motor, especially at low speeds. Drives that use a tachometer-generator for speed feedback generally do not require an *IR* compensation circuit because the tachometer will inherently compensate for the loss in speed.

Isolated I/O module Module that has each input or output electrically isolated from every other input or output on that module. That is, each input or output has a separate return wire.

Isolation transformer Transformer that electrically separates the drive from the ac power line. An isolation transformer provides the following advantages: (1) in dc motor applications, it guards against inadvertent grounding of plant power lines through grounds in the dc motor armature circuit; (2) enhances protection of semiconductors from line-voltage transients; (3) reduces disturbances from other solid-state control equipment, such as drives without isolation transformers, time clock systems, electronic counters, and so on.

Jogging Means of accomplishing momentary motor movement by repetitive closure of a circuit using a single pushbutton or contact element.

Kinetic energy Energy of motion possessed by a body.

Knockout Portion of the wall of a box or cabinet so fashioned that it may be removed readily by hammer, screwdriver, and pliers.

Ladder diagram programming Method of writing a user's PC program in a format similar to a relay ladder diagram.

Language Set of symbols and rules for representing and communicating information (data) among people, or between people and machines.

Large-scale integration (LSI) Any integrated circuit that has more than 100 equivalent gates manufactured simultaneously on a single slice of semiconductor material. (*See also* Medium-scale integration *and* Small-scale integration.)

Latching relay Relay that can be mechanically latched in a given position manually or when operated by one element, and released manually or by the operation of a second element.

Lateral approach When the target to be detected approaches the sensing face from the side (slide-by).

LED Acronym for *light-emitting diode.*

Legend plates Identify the function of operator controls, indicating lights, and the like.

Limit switch Switch operated by some part or motion of a power-driven machine or equipment to alter the electric or electronic circuits associated with the machine or equipment.

Linear acceleration/deceleration (LAD) Circuit that controls the rate at which the motor is allowed to accelerate to a set speed or decelerate to zero speed. On most drives, this circuit is adjustable and can be set to accommodate a particular application.

Linearity Measure of how closely a characteristic follows a straight-line function.

Line-powered sensor (three- or four-wire) Sensor that draws its operating current (burden current) directly from the line. Its operating current does not flow through the load, and a minimum of three connections (three-wire) are required. A four-wire sensor has complementary outputs and requires four connections.

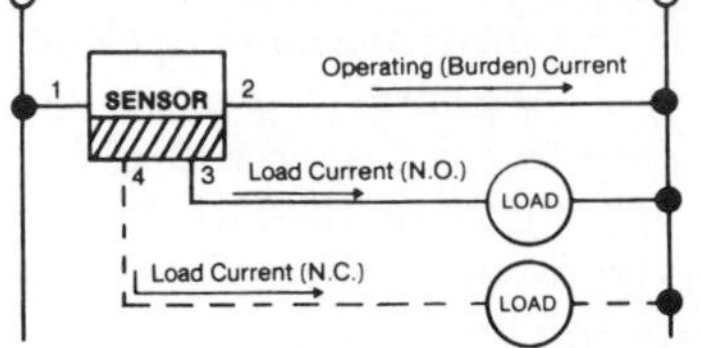

Load-on LED Indicates when output is conducting.

Load-powered sensor (two-wire) Sensor that draws its operating current (leakage current) through the load. The sensor is always in series with the load and only two connections are required.

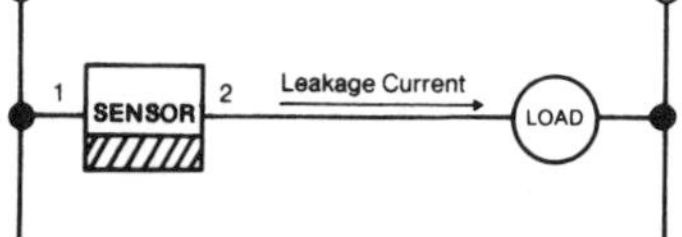

Load resistor Resistor connected in parallel with a high-impedance load so that the output circuit driving the load can provide at least the minimum current required for proper operation.

Locked-rotor current Steady-state current taken from the line with the rotor at standstill (at rated voltage and frequency). This is the current when starting the motor and load.

Locked-rotor torque Minimum torque that a motor will develop at rest for all angular positions of the rotor (with rated voltage applied at rated frequency).

Magnetic device Device actuated by electromagnetic means.

Magnetic starter Starter actuated by electromagnetic means.

Manual controller Device manually closed or opened.

Manual reset Device that requires manual action to reengage the breaking surfaces.

Master terminal box Main enclosure on the equipment containing terminal blocks for the purpose of terminating conductors for the control enclosure. (Normally associated with equipment requiring a separately mounted control enclosure.)

Medium-scale integration (MSI) Any integrated circuit which has between 12 and 100 equivalent gates. (*See also* Large-scale integration *and* Small-scale integration.)

Megger test Test used to measure an insulation system's resistance. This is usually measured in megohms and tested by passing a high voltage at low current through the motor windings and measuring the resistance of the various insulation systems.

Melting time Amount of time required to melt the fuse link during a specified overcurrent. *See also* Arcing time *and* Clearing time.

Memory Grouping of circuit elements that has data storage and retrieval capability.

Metal-oxide semiconductor (MOS) Electronic circuit in which the active region is a metal-oxide semiconductor sandwich. The oxide acts as the dielectric insulator between the metal gate and the conducting channel.

Microprocessor Electronic computer processor section implemented in relatively few IC chips (typically, LSI) which contain arithmetic, logic, register, control, and memory functions. The microprocessor is characterized by having instructions that reference micro operations. Functional equivalence of minicomputer instructions accomplished by programming series of micro instructions.

Modem Acronym for *modulator/demodulator,* a device used to transmit and receive data by frequency-shift keying (FSK). It converts FSK tones into their digital equivalent, and vice versa.

Most significant digit (MSD) Digit representing the greatest value.

Motor junction (conduit) box Enclosure on a motor for the purpose of terminating a conduit run and joining motor to power conductors.

Multiplexing Time-shared scanning of a number of data lines into a single channel. Only one data line is enabled at any instant.

Multispeed motor Induction motor that can obtain two, three, or four discrete (fixed) speeds by the selection of various stator winding configurations.

NEC The *National Electrical Code*®consists of recommendations by the National Fire Protection Association and is revised every three years. City or state regulations may differ from code regulations and take precedence over NEC rules.

NEC dimensions Dimensions once referenced in the *National Electrical Code*®. They are common to class H and K fuses and provide interchangeability between manufacturers for fuses and fusible equipment of given ampere and voltage ratings.

Negative feedback Condition where feedback is subtractive to the input reference signal. Negative feedback forms the basis for automatic control systems.

NEMA The National Electrical Manufacturers Association is a nonprofit organization organized and supported by manufacturers of electrical equipment and supplies. Some of the standards NEMA specifies are: hp ratings, speeds, frame sizes and dimensions, torques, and enclosures.

Noise Extraneous signals; any disturbance that causes interference with the desired signal or operation.

Nonautomatic Implied action required personal intervention for its control.

Nonreversing Control function that provides for operation in one direction only.

Nonvolatile memory Memory designed to retain its information while its power supply is turned off.

Normally closed (NC) output The sensor opens a circuit to the load when a target is detected.

Normally open (NO) output The sensor closes a circuit to the load when a target is detected.

Octal numbering system System that uses base 8 (e.g., the decimal number 324 would be written in octal notation as 504_8). Only the digits 0–7 are used.

Odd parity Condition existing when the sum of the number of 1's in a binary word is always odd.

Off-delay timer That function of an element whereby an output is discontinued following a definite intentional time delay after the input is deenergized.

Offset Steady-state deviation of a controlled variable from a fixed set point.

Ohm Unit of measure for electric resistance. An ohm is the amount of resistance that will allow 1 ampere to flow under a pressure of 1 volt.

Ohm's law Relationship between electrical pressure, resistance, and current flow in a circuit $E = IR$.

One-shot Programming technique that energizes (sets to "1" or ON) a storage bit or output for only one program scan. This pulsing effect occurs each time the PRECONDITIONS for the storage bit change from FALSE to TRUE. The storage bit is de-energized ("0" or OFF) at all other times. (The one-shot technique may be used in routines for shift registers, data manipulation, data file transfer, and report generation.)

One-time fuses Generic term used to describe a class H nonrenewable cartridge fuse with a single element.

On-line Describes equipment or devices connected to the communications line.

Op amp (Operational Amplifier) Usually a high-gain dc amplifier designed to be used with external circuit elements.

Open loop Control system that lacks feedback.

Open machine (motors) Machine having ventilating openings that permit passage of external cooling air over and around the windings of the machine.

- **A.** *Dripproof machine* is an open-type machine in which the ventilating openings are so constructed that successful operation is not interfered with when drops of liquid or solid particles strike or enter the enclosure at any angle from 0 to 15° downward from vertical.
- **B.** *Splashproof* is an open machine in which the ventilating openings are so constructed that successful operation is not interfered with when drops of liquid or solid particles strike or enter the enclosure at any angle not greater than 100° downward from the vertical.
- **C.** *Semiguarded* is an open machine in which part of the ventilating openings in the machine, normally in the top half, are guarded, as in the case of a "guarded machine," but the others are left open.
- **D.** *Guarded machine (NEMA Standard)* is an open machine in which all openings giving direct access to live metal or rotating parts (except smooth rotating surfaces) are limited in size by the structural parts or by the screens, baffles, grilles, expanded metal, or other means to prevent accidental contact with hazardous parts. Openings giving direct access to such live or rotating parts should not permit the passage of a cylindrical rod 0.75 in. in diameter.
- **E.** *Drip-proof guarded machine* is a drip-proof machine whose ventilating openings are guarded in accordance with the definition of a guarded machine.
- **F.** *Open externally ventilated machine* is one ventilated by means of a separate motor-driven blower mounted on the machine enclosure. This machine is sometimes known as a blower-ventilated or a force-ventilated machine.
- **G.** *Open pipe ventilated machine* is basically an open machine except that openings for admission of ventilating air are so arranged that inlet ducts or pipes can be connected to them. Air may be circulated by means integral with the machine or by means external to the machine (separately or forced ventilated).
- **H.** *Weather-protected machine* is an open enclosure divided into two types: (1) *Type 1* enclosures have ventilating passages constructed to minimize the entrance of rain, snow, and airborne particles and prevent passage of a 0.75-in. diameter cylindrical rod. (2) *Type 2* enclosures provide additional protection through the design of their intake and exhaust ventilating passages. The passages are so arranged that wind and airborne particles blown into the machine can be discharged without entering directly into the electrical parts of the machine. Additional baffling is provided to minimize the possibility of moisture or dirt being carried inside the machine.

Operating distance Distance from the sensing face to the plane of the target's path once it reaches the operating point.

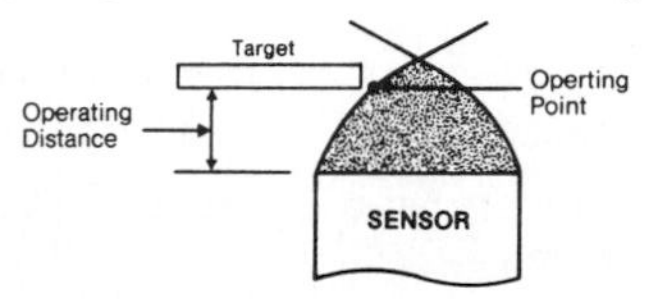

Operating overload Overcurrent to which electric apparatus is subjected in the course of normal operating conditions that it may encounter. (The maximum operating overload is considered to be six times normal full-load current for alternating-current industrial motors and control apparatus; four times normal full-load current for direct-current industrial motors and control apparatus used for reduced-voltage starting; and 10 times normal full-load current for direct-current industrial motors and control used for full-voltage starting. It should be understood that the overloads are currents that may persist for a very short time only, usually a matter of seconds.)

Operating/service deviation Means of specifying the speed regulating performance of a drive controller, generally in percent of base speed.

- **A.** *Operating deviation* defines speed change due to load change and typically assumes: (1) a change from one steady-state load value to another (not transient), and (2) a 95% maximum load change.
- **B.** *Service deviation* defines speed change due to changes in ambient conditions greater than these typical variations:

Condition	Change
Ac line voltage	+10 to −5%
Ac line frequency	+3 to −3%
Ambient temperature	15°C

Operator's control station (pushbutton station) Unit assembly of one or more externally operable pushbutton switches, sometimes including other pilot devices, such as indicating lights or selector switches, in a suitable enclosure.

Outline drawing Drawing showing approximate overall shape with no detail.

Overcurrent In an electric or electronic circuit, that current which will cause an excessive or dangerous temperature in the conductor or conductor insulation.

Overcurrent protective device Device operative on excessive current which causes and maintains the interruption of power in the circuit.

Overlapping contacts Combinations of two sets of contacts, actuated by a common means, each set closing in one of two positions, and so arranged that the contacts of one set opens after the contacts of the other set have been closed.

Overload Can be classified as an overcurrent that exceeds the normal full-load current of a circuit. Also characteristic of this type of overcurrent is that it does not leave the normal current-carrying path of the circuit—that is, it flows from the source, through the conductors, through the load, back through the conductors, to the source again.

Overload capacity Ability of the drive to withstand currents beyond the systems continuous rating. It is normally specified as a percentage of full-load current for a specified time period. Overload capacity is defined by NEMA as 150% of rated full-load current for 1 minute for standard industrial dc motors.

Overload relay Device that provides overload protection for electrical equipment.

Overshoot Amount that a controlled variable exceeds desired value after a change of input.

Parallel operation Type of information transfer whereby all digits of a word are handled simultaneously.

Parity Method of testing the accuracy of binary numbers used in recorded, transmitted, or received data.

Parity bit Additional bit added to a binary word to make the sum of the number of 1's in a word always even or odd.

Parity check Check that tests whether the number of 1's in an array of binary digits is odd or even.

PC (programmable controller) Solid-state control logic for machines and processes where a sequence of operations can be changed easily with programming (software).

Peak let-through current, I_p Instantaneous value of peak current let through by a current-limiting fuse when it operates in its current-limiting range.

Peripheral equipment Units that may communicate with the programmable controller but are not part of the programmable controller (e.g., teletype, cassette recorder, tape reader, etc.).

Phase-sequence reversal Reversal of the normal phase sequence of the power supply. For example, the interchange of two lines on a three-phase system will give a phase-sequence reversal.

Phase-sequence-reversal protection (phase-reversal protection) Effect of a device to prevent energization of the protected equipment on reversal of the phase sequence in a polyphase circuit.

Photo-isolator Solid-state device that allows complete electrical isolation between the field wiring and the controller.

Pickup and seal voltage Minimum voltage, suddenly applied, at which the device moves from its deenergized into its fully energized position.

Pickup voltage or current Voltage or current, suddenly applied, at which the device starts to operate.

Pilot duty rating *See* Rating, pilot duty.

PLC (see **PC**).

Plugging Control function that provides braking by reversing the motor line-voltage polarity or phase sequence so that the motor develops a countertorque that exerts a retarding force.

Position stopping Control function that provides for stopping the driven equipment at a preselected position.

Position transducer Electronic device (e.g., encoder or resolver) that measures actual position and converts this measurement into a feedback signal convenient for transmission. This signal may then be used as an input to a programmable controller that controls the parameters of the positioning system.

Positive feedback Condition where the feedback is additive to the input signal.

Power Work done per unit of time. Measured in horsepower or watts: 1 hp = 33,000 ft-lb/min = 746 watts.

Power factor Measurement of the time phase difference between the voltage and current in an ac circuit. It is represented by the cosine of the angle of this phase difference. Power factor is the ratio of real power (kW) to total kVA or the ratio of actual power (W) to apparent power (volt-amperes).

Preset speed One or more fixed speeds at which the drive will operate.

Pressure transducers

BASIC TERMINOLOGY

Transducer terminology can be divided into four categories: performance, electrical, temperature, and physical.

Performance

Pressure reference—All pressure measurements are made with respect to a pressure reference. Most manufacturers use the following types of references:

Absolute pressure (psia)—Pressure measured relative to a vacuum (absolute zero).

Gauge pressure (psig)—A measure of pressure reference to atmospheric pressure.

Sealed gauge pressure (psis)—A measure of pressure with respect to an internal reference chamber sealed at atmospheric pressure.

Pressure range—The minimum and maximum pressure limits over which a specified transducer is designed to operate.

Proof pressure (overload pressure)—The maximum pressure that can be applied to the sensing element of a transducer without causing a permanent change in the output specifications.

Burst pressure—The maximum pressure that can be applied to a transducer without rupture of either the sensing element or the transducer case.

Accuracy—The total of all deviations from a specified straight line, including the sum of nonlinearity, repeatability, and hysteresis expressed as a percent of full-scale output.

Best-fit straight line (BFSL)—The best straight line chosen such that the transducer response curve contains points of equal maximum deviations.

Error—The difference between the value indicated by the transducer and the true value of the pressure being sensed.

(Non)linearity—The error defined by the maximum deviation of measured output from a best-fit straight line during any one calibration cycle.

Hysteresis—The error defined by the difference in output when the pressure value is first approached with increasing pressure and then with decreasing pressure.

(Non)repeatability—The error defined by the ability of a transducer to reproduce an output signal when the same pressure value is applied to it consecutively, under the same conditions, and in the same direction.

Stability—The ability of a transducer to maintain all of its performance specifications throughout its life.

Response time—The length of time required for an output of a transducer to rise to a specific percentage of its final value as a result of a step change in input.

Electrical

Output—The electrical signal that is produced by an applied input to the transducer.

Span (full-scale output or FSO)—The algebraic difference in the electrical output when the maximum and minimum pressure is applied to the input.

Zero (balance)—The output of the transducer when zero input pressure is applied.

Excitation voltage—The external application of an electrical voltage source applied to a transducer for normal operation.

Signal conditioning—To process the form or mode of a signal so as to make it electrically compatible with a given device.

Insulation resistance—The resistance measured between two insulated points on a transducer when a specified dc voltage is applied.

Temperature

Operating temperature range—The range of ambient temperatures within which the transducer may be operated and not suffer damage or a permanent change in its specifications.

Compensated temperature range—The range of ambient temperatures within which the transducer will perform according to its output specifications.

Thermal effect on zero—The error defined by the maximum deviations from the zero due to changes in ambient temperature, within the compensated temperature range and no pressure applied.

Thermal effect on span—The error defined by the change in sensitivity due to a change in ambient temperature within the compensated temperature range.

SELECTION FACTORS

The following criteria should be considered carefully to determine the type of pressure transducers needed for your application:

Performance

1. Determine the pressure range that will be applied to the transducer. Consider the maximum operating pressure you will experience.
2. Consider if your application will produce any pressure transients or pulsations. Make certain that these pressures are normally within the specified proof (overload) pressure rating. The application of transient pressures requires special consideration.
3. Select a transducer that can measure your pressure range accurately for your particular application.

Electrical

1. Determine the output signal most appropriate for your application. Be sure to consider both zero and full-scale output (FSO) when selecting an output type.
2. Evaluate the excitation voltage that will be necessary for a particular transducer. Your power source will need to provide a voltage within the specified range.

Temperature

1. Variations in ambient temperature can have significant effects on transducer performance. Consider the compensated temperature range as well as thermal effects on zero and span in regard to your transducer selection. Pressure transducers are compensated over a specified temperature range to provide stable operation even when the temperature varies. However, operation outside the compensated temperature range will reduce transducer accuracy.

Physical

1. Select a pressure transducer whose pressure fitting (port) is an appropriate interface for your pressure system.
2. Determine how the connection of the transducer to the electrical system will be made. Choose an electrical termination suited for your installation.
3. Consider the total size (dimensions, weight) of the transducer package.
4. Evaluate the environment that the transducer will be used in and consider both the enclosure and the wetted materials of the transducer.

Process (1) Continuous and regular production executed in a definite uninterrupted manner. (2) A PLC application that primarily requires data comparison and manipulation. The PLC monitors the input parameters in order to vary the output values. (As generally contrasted with a machine, a process does not cause mechanical motion.)

Processor Unit in the programmable controller that scans all the inputs and outputs in a predetermined order. The processor monitors the status of the inputs and outputs in response to the user-programmed instructions in memory, and it energizes or deenergizes outputs as a result of the logical comparisons made through these instructions.

Program (as a verb) To prepare a program for the series of operations that a machine is to perform.

Programmable controller Solid-state control system that has a user-programmable memory for storage of INSTRUCTIONS to implement specific functions such as: I/O control logic, timing, counting, arithmetic, and data manipulation. A PLC consists of a central processor, input–output interface, memory, and programming device which typically uses relay-equivalent symbols. A PLC is purposely designed as an industrial control system that can perform functions equivalent to those performed by a relay panel or a wired solid-state logic control system.

Programmed control Control system in which the operations are determined by a predetermined input program from cards, tape, plug boards, cams, and so on.

Programming, automatic Any programming technique

in which the computer is used to help the human programmer convert a problem into machine language.

PROM Acronym for *programmable read-only memory,* a type of ROM that requires an electrical operation to generate the desired bit or word pattern. In use, bits or words are accessed on demand, but not changed.

Proof (as a suffix) So constructed, protected, or treated that successful operation of the apparatus is not interfered with when subjected to the specified material or condition.

Proportional, integral, derivative module Optional PLC processor module that provides automatic closed-loop operation of multiple continuous process control loops. For each loop, this module can perform any or all of the following control actions: (1) proportional control causes the output signal to change as a direct ratio of input signal variation; (2) integral control causes the output signal to change according to the summation of input signal values sampled up to the present time; (3) derivative control causes the output signal to change according to the rate at which input signal variations occur during a certain time interval.

Protection, antiplugging Effect of a control function or a device that operates to prevent the application of countertorque by the motor until the motor speed has been reduced to a specific value.

Protection, overload Effect of a device operative on excessive current, but not necessarily on short circuit, to cause and maintain the interruption of current flow to the device governed.

Protection, overspeed Effect of a device operative whenever the speed rises above a preset value to cause and maintain an interruption of power to the protected equipment or a reduction of its speed.

Protection, overvoltage Effect of a device operative on excessive voltage to cause and maintain the interruption of power in the circuit or reduction of voltage to the equipment governed.

Protocol Defined means of establishing criteria for receiving and transmitting data through communication channels.

Proximity switch *See* Switch, proximity.

Pull-in torque (synchronous motors) Maximum constant torque that a synchronous motor will accelerate into synchronism at rated voltage and frequency.

Pull-out torque (synchronous motors) Maximum running torque of a synchronous motor.

Pull-up torque Torque required to accelerate a load from standstill to full speed (where breakdown torque occurs), expressed in percent of running torque. It is the torque required not only to overcome friction, windage, and product loading but also to overcome the inertia of the machine. The torque required by a machine may not be constant after the machine has started to turn. This load type is characteristic of fans, centrifugal pumps, and certain machine tools.

Pushbutton Master switch having a manually operable plunger or button for actuating the switch.

Pushbutton station Unitassembly of one or more externally operable pushbutton switches, sometimes including other pilot devices, such as indicating lights or selector switches, in a suitable enclosure.

PWM Type of ac adjustable-frequency drive that accomplishes frequency and voltage control at the output section (inverter) of the drive. The drive's output voltage is always a constant amplitude and by ''chopping'' (pulse-width modulating) the average voltage is controlled.

Rainproof So constructed, protected, or treated as to prevent rain under specified test conditions from interfering with successful operation of the apparatus.

Raintight So constructed or protected as to exclude rain under specified test conditions.

RAM Acronym for *random-access memory,* memory that can be accessed (read from) or loaded (written into) depending on the particular addressing and operation codes generated internally in the PLC.

Rating Designated limit of operating characteristics based on definite conditions. Such operating characteristics as load, voltage, and frequency may be given in the rating.

Rating (of a Controller) Arbitrary designation of an operating limit. It is based on the power governed and on the duty and service required. A rating is arbitrary in the sense that it must necessarily be established by definite fixed standards and cannot, therefore, indicate the safe operating limit under all conditions which any occur.

Rating, break Value of current for which a contact assembly is rated for opening a circuit repeatedly at a specified voltage and under specified operating conditions.

Rating, continuous Rating that defines the substantially constant load which can be carried for an indefinitely long time.

Rating, eight-hour of a magnetic contactor Rating based on its ampere carrying capacity for 8 hours, starting with new clean contact surfaces, under conditions of free ventilation, with fully rated voltage on the operating coil and without causing any of the established limitations to be exceeded.

Rating, make Value of current for which a contact assembly is rated for closing a circuit repeatedly under specified operating conditions.

Rating, pilot duty Generic term formerly used to indicate the ability of a control circuit device to control other devices.

Reactance Any force that opposes changes in current voltage. The inertia of electrons causes them to oppose sudden changesin current flow or voltage.

Reactor, saturable Inductor having means to change the degree of magnetic saturation of its core(s), thereby controlling the magnitude of the alternating current.

Read/write memory Memory in which data can be placed (write mode) or accessed (read mode). The write mode destroys previous data; the read mode does not alter stored data.

Rectifier Device that transforms alternating current to direct current.

Redundancy Technique that provides a secondary method of control for certain critical functions. This is also called backup control.

Regeneration Characteristic of a motor to act as a generator when the CEMF is larger than the drive's applied voltage (dc drives) or when the rotor synchronous frequency is greater than the applied frequency (ac drives).

Regenerative braking Technique of slowing or stopping a drive by regeneration. *See also* Braking.

Regenerative control A regenerative drive contains the inherent capability and/or power semiconductors to control the flow of power to and from the motor.

Register Memory word or area used for temporary storage of data used within mathematical, logical, or transferral functions.

Regulation Ability of a control system to hold a speed once it has been set. Regulation is given in percentages of either base speed or set speed. Regulation is rated upon two separate sets of conditions:

- **A.** *Load regulation (speed regulation)* is the percentage of speed change with a defined change in load, assuming all other parameters to be constant. Speed regulation values of 2% are possible in drives utilizing armature voltage feedback, while regulation of 0.01% is possible using digital regulator schemes.
- **B.** *Line regulation* is the percentage of speed change with a given line voltage change, assuming all other parameters to be constant.

Relay Device that is operative by a variation in the conditions of one electric circuit to effect the operation of other devices in the same or another electric circuit. Where relays operate in response to changes in more than one condition, all functions should be mentioned.

Relay, current Relay that functions at a predetermined value of current. It may be an overcurrent relay, an undercurrent relay, or a combination of both.

Relay, latching Relay in which the contact assembly moves to and latches in a given position when operated manually or by a first element. The contact assembly can be returned to its original position either manually or by operation of the same or a second element.

Relay, magnetic control Relay actuated by electromagnetic means. When not otherwise qualified, the term refers to a relay intended to be operated by the opening and closing of its coil circuit, and having contacts designed for energizing and/or deenergizing the coils of magnetic contactors or other magnetically operated device.

Relay, magnetic overload Overcurrent relay the electrical contacts of which are actuated by the electromagnetic force produced by the load current or a measure of it.

Relay, open-phase Relay that functions by reason of the opening of one or more phases of a polyphase circuit.

Relay, overload Overcurrent relay that functions at a predetermined value of overcurrent to cause disconnection of the load from the power supply. An overload relay is intended to protect the load (e.g., motor armature) or its controller, and does not necessarily protect itself.

Relay, undervoltage or low-voltage protection Relay that functions to provide undervoltage or low-voltage protection.

Relay, voltage Relay that functions at a predetermined value of voltage. It may be an overvoltage relay, an undervoltage relay, or a combination of both.

Renewable fuses (600V and below) Fuses in which the element, typically a zinc link, may be replaced after the fuse has opened, and then reused. Renewable fuses are made to class H standards.

Repeat accuracy (repeatability) Measure of variation in operating distance between successive operations under constant operating conditions. This measurement is often expressed as a maximum percentage of the "operating distance" (e.g., 5%).

Reset To restore a mechanism, storage, or device to a prescribed state.

Reset, automatic Function that operates to reestablish specific conditions automatically.

Reset, manual Function that requires a manual operation to reestablish specific conditions.

Resistance Nonreactive opposition that a device or material offers to the flow of direct or alternating current. The opposition results in production of heat in the material carrying the current. Resistance is measured in ohms and is usually designated by the letter R.

Resistance starting Form of reduced-voltage starting employing resistances that are short-circuited in one or more steps to complete the starting cycle.

Resistant (as a suffix) So constructed, protected, or treated that the apparatus will not be damaged when subjected to the specified material or conditions for a specified time.

Resistive load Electrical load that is characteristic of not having any significant inrush current. When a resistive load is energized, the current rises instantly to its steady-state value without first rising to a higher value.

Resistor Device, the primary purpose of which is to introduce resistance into an electric circuit.

Resolution Smallest distinguishable increment into which a quantity can be divided (e.g., position or shaft speed). It is also the degree to which nearly equal values of a quantity can be discriminated. For encoders, it is the number of unique electrically identified positions occurring in 360° of input shaft rotation.

Response time Time delay from when a target reaches the operating point to when an output actually occurs. A target must stay within the sensing field for at least the response time or the sensor will not change output status. (*Note:* The target must also remain within the sensing field long enough to allow the load sufficient time to respond to the output signal of the sensor.)

Reversing Changing the direction of rotation of the motor armature or rotor. A dc motor is reversed by changing the polarity of the field or the armature, but not both. An ac motor is reversed by reversing the connections of one leg on the three-phase power line. The reversing function can be performed in one of the following ways:

- **A.** *(Dc) Contactor reversing* is done by changing the phase rotation of an ac motor or the polarity to a dc motor armature with switching contactors. The contactors are operated by momentary pushbuttons and/or limit switches to stop the motor and change directions. A zero-speed (antiplugging) circuit is associated with this system to protect the motor and control.
- **B.** *(Dc) Field reversing* is accomplished by changing the dc polarity to the motor shunt field. This type of reversing can be accomplished with dc-rated contactors or by means of an electronically controlled solid-state field supply.

C. *(Dc) Manual reversing* is the act of reversing the dc polarity to the motor armature by changing the position of a single switch. The switch is usually detented to give a degree of mechanical antiplugging protection. Limit switches and remote stations cannot be used with this system. Dynamic braking is recommended.

D. *(Ac or dc) Static reversing* is the act of reversing the dc polarity of the dc motor armature or phase rotation of an ac motor with no mechanical switching. This is accomplished electronically with solid-state devices. Solid-state antiplugging circuitry is generally a part of the design.

Rheostat Adjustable resistor so constructed that its resistance may be changed without opening the circuit in which it may be connected.

RMS current Effective value of an alternating-current sine wave calculated as the square root of the average of the squares of all the instantaneous values of the current throughout one cycle. Rms alternating current is that value of an alternating current which produces the same heating effect as a given direct-current value.

ROM Acronym for *read-only memory*. Solid-state digital storage memory whose contents cannot be altered by the PLC.

Scan time Time necessary to execute the entire PLC program completely one time.

Sealing, voltage or current Voltage or current necessary to seat the armature of a magnetic circuit closing device from the position at which the contacts first touch each other.

Selectable (NO/NC) output Convertible from NO to NC.

Sensing range Maximum operating range at which the sensor will reliably detect a standard target under conditions of nominal voltage and temperature.

Serial operation Type of information transfer within a programmable controller whereby the bits are handled sequentially rather than simultaneously, as they are in parallel operation. Serial operation is slower than parallel operation for equivalent clock rate. However, only one channel is required for serial operation.

Service (of a controller) Is the specific application in which the controller is to be used; for example, general purpose or definite purpose (crane and hoist, elevator, machine tool, etc.).

Service factor When used on a motor nameplate, a number that indicates how much above the nameplate rating a motor can be loaded without causing serious degradation (i.e., a motor with 1.15 S-F can produce 15% greater torque than one with 1.0 S-F). When used in applying motors or gearmotors, it is a figure of merit used to adjust measured loads in an attempt to compensate for conditions that are difficult to measure or define.

Set speed Desired operating speed.

Shielded Proximity switch that can be flush mounted (embedded) in metal without affecting the sensing range.

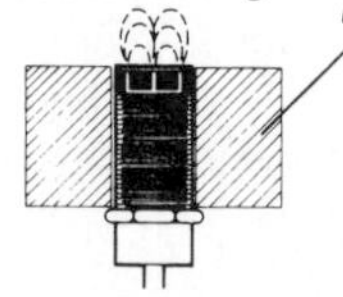

Shift register Program, entered by the user into the memory of a programmable controller, in which the information data (usually single bits) are shifted one or more positions on a continual basis. There are two types of shift registers: asynchronous and synchronous.

Shock load Load seen by a clutch, brake, or motor in a system that transmits high peak loads. This type of load is present in crushers, separators, grinders, conveyors, winches, and cranes.

Short circuit Can be classified as an overcurrent that exceeds the normal full-load current of a circuit by a factor many times (tens, hundreds, or thousands) greater. Also characteristic of this type of overcurrent is that it leaves the normal current-carrying path of the circuit—it takes a shortcut around the load and back to the source.

Signal Information about a variable that can be transmitted in a system.

Significant digit Digit that contributes to the precision of a number. The number of significant digits is counted beginning with the digit contributing the most value, called the most significant digit, and ending with the one contributing the least value, called the least significant digit.

Silicon-controlled rectifier (SCR) Solid-state switch, sometimes referred to as a *thyristor*. The SCR has an anode, cathode, and control element called a *gate*. The device provides controlled rectification since it can be turned on at will. The SCR can rapidly switch large currents at high voltages. They are small in size and low in weight.

Single phasing Condition that occurs when one phase of a three-phase system opens, in either a low-voltage (secondary) or high-voltage (primary) distribution system. Primary or secondary single phasing can be caused by any number of events. This condition results in unbalanced loads in polyphase motors, and unless preventative measures are taken, causes overheating and failure.

Skew Arrangement of laminations on a rotor or armature to provide a slight angular pattern of their slots with respect to the shaft axis. This pattern helps to eliminate low-speed cogging in an armature and minimize induced vibration in a rotor as well as to reduce associated noise.

Skewing Time delay or offset between any two signals in relation to each other.

Slewing Incremental motion of the motor shaft or machine table from one position to another at maximum speed without losing position control.

Slip Difference between rotating magnetic field speed (synchronous speed) and rotor speed of ac induction motors. Usually expressed as a percentage of synchronous speed.

Small-scale integration (SSI) Any integrated circuit which has fewer than 12 equivalent gates. (*See also* Large-scale integration *and* Medium-scale integration.)

Snap action Rapid motion of the contacts from one position to another position, or their return. This action is relatively independent of the rate of travel of the actuator.

Solid-state devices (semi-conductors) Electronic components that control electron flow through solid materials (e.g., transistors, diodes, integrated circuits).

Special-purpose motor Motor with special operating characteristics or special mechanical construction, or both, designed for a particular application and not falling within the definition of a general-purpose or definite-purpose motor.

Speed range Speed minimum and maximum at which a motor must operate under constant- or variable-torque load conditions. A 50:1 speed range for a motor with a top speed of 1800 rpm means that the motor must operate as low as 36 rpm and still remain within regulation specifications. Controllers are capable of wider controllable speed ranges than motors because there is no thermal limitation, only electrical. Controllable speed range of a motor is limited by the ability to deliver 100% torque below base speed without additional cooling.

Speed regulation Numerical measure, in percent, of how accurately motor speed can be maintained. It is the percentage of change in speed between full load and no load.

Stability Ability of a drive to operate a motor at constant speed (under varying load), without "hunting" (alternately speeding up and slowing down). It is related to both the characteristics of the load being driven and electrical time constants in the drive regulator circuits.

Standard target Object used to determine sensing range. This is normally a square mild steel plate 1 mm thick, with the length of each side equal to the diameter of the sensing face or three times the nominal sensing distance of the sensor.

Starter Electric controller for accelerating a motor from rest to normal speed. A device designed for starting a motor in either direction of rotation; includes the additional function of reversing and should be designated a controller.

Starter, automatic Starter that automatically controls the acceleration of a motor.

Starter, autotransformer Starter provided with an autotransformer that furnishes a reduced voltage for starting. It includes the necessary switching mechanism and is frequently called a *compensator* or *autostarter.*

Starter, increment (network) Starter that applies starting current to a motor in a series of increments of predetermined value and apredetermined time intervals in a closed-circuit transition for the purpose of minimizing line disturbance. One or more increments may be applied before the motor starts.

Starter, part-winding Starter that applies voltage successively to the partial sections of the primary winding of an alternating-current motor.

Starter, primary reactor Starter that includes a reactor connected in series with the primary winding of an induction motor to furnish reduced voltage for starting. It includes the necessary switching mechanism for cutting out the reactor and connecting the motor to the line.

Starter, primary resistor Starter that includes a resistor connected in series with the primary winding of an induction motor to furnish reduced voltage for starting. It includes the necessary switching mechanism for cutting out the resistor and connecting the motor to the line.

Starting, slow-speed Control function that provides for starting an electric drive only at the minimum speed setting.

Stepping motor Bidirectional, permanent-magnet motor that turns through one angular increment for each pulse applied to it.

Stiffness Ability of a device to resist deviation due to load change.

Submersible So constructed as to exclude water when submerged in water under specified test conditions of pressure and time.

Surface-mounted (type) Designed to be secured to and to project from a flat surface.

Surge protection Process of absorbing and clipping voltage transients on an incoming ac line or control circuit. MOVs (metal-oxide varistors) and specially designed *RC* networks are usually used to accomplish this.

Switch Device for making, breaking, or changing the connections in an electric circuit.

Switch, cam-operated Switch in which the electrical contacts are opened and/or closed by a mechanical action of a cam or cams.

Switch, control-circuit limit Limit switch whose contacts are connected only into the control circuit.

Switch, control cutout Switch that interrupts and isolates the control circuit of an electric controller.

Switch, drum Switch in which the electric contacts are made on segments or surfaces on the periphery of a rotating cylinder or sector or by the operation of a rotating cam.

Switch, float Switch operated by a buoyant constituent part and responsive to the level of a liquid.

Switch, foot Switch suitable for operation by an operator's foot.

Switch, general-use Switch extended for use in general distribution and branch circuits. It is rated in amperes and is capable of interrupting the rated current at the rated voltage.

Switch, isolating Switch intended for isolating an electric circuit from the source of power. It has no interrupting rating and is intended to be operated only after the circuit has been opened by some other means.

Switch, limit Switch operated by some part or motion of a power-driven machine or equipment to alter the electrical circuit associated with the machine or equipment.

Switch, master Switch that dominates the operation of contactors, relays, or other remotely operated devices. *See also* Pushbutton.

Switch, motor-circuit Switch intended for use in a motor branch circuit. It is rated in horsepower and is capable of interrupting the maximum operating overload (*see* Operating Overload) current of a motor of the same rating at the rated voltage.

Switch, power-circuit limit Limit switch whose contacts are connected into the power circuit.

Switch, pressure Switch operated by a constituent part and responsive to fluid (gas or liquid) pressure.

Switch, proximity Device that reacts to the proximity of an actuating means without physical contact or connection therewith.

Switch, selector Manually operated multiposition switch for selecting an alternative control circuit.

Switch, speed Speed-responsive switch actuated when the equipment by which it is driven attains a predetermined speed.

Switch, spring-return Multiposition switch in which the self-returning function is affected by the action of a spring.

Synchronous shift register Shift register that uses a clock for timing of a system operation and where only one state change per clock pulse occurs.

Synchronous speed Speed of an ac induction motor's rotating magnetic field. It is determined by the frequency applied to the stator and the number of magnetic poles present in each phase of the stator windings. Mathematically, it is expressed as: sync speed (rpm) $= 120 \times$ applied frequency (Hz)/number of poles per phase.

System Collection of units combined to work as a larger integrated unit having the capabilities of all the separate units.

Tachometer-generator (tach) Small generator normally used as a rotational speed-sensing device. Tachometers are typically coupled to the shaft of dc or ac motors requiring close speed regulation. The tach feeds a signal to a controller, which then adjusts the output voltage or frequency to the motor.

Target Object to be detected.

Target detection LED Indicates when target is present.

Temperature, ambient Temperature of the medium, such as air, water, or earth, into which the heat of the equipment is dissipated. For self-ventilated equipment, the ambient temperature is the average temperature of the air in the immediate neighborhood of the equipment. For air- or gas-cooled equipment with forced ventilation or secondary water cooling, the ambient temperature is taken as that of the ingoing air or cooling gas. For self-ventilated enclosed (included oil-immersed) equipment considered as a complete unit, the ambient temperature is the average temperature of the air outside the enclosure in the immediate neighborhood of the equipment.

Terminal board (block or strip) Insulating base or slab equipped with one or more terminal connectors for the purpose of making electrical connections thereto.

Terminal connector Connector for attaching a conductor to a lead, terminal block, or stud of electric apparatus.

Tests, application Tests performed by a manufacturer to determine those operating characteristics that are not necessarily established by standards but which are of interest in the application of devices.

Tests, dielectric Tests consisting of the application of a voltage higher than the rated voltage for a specified time.

Tests, production Tests made at the discretion of the manufacturer on some or all production units for the purpose of maintaining quality and performance.

Thermal cutout Overcurrent protective device that contains a heater element in addition to and affecting a fusible member that opens the circuit.

Thermal protector (as applied to motors and generators) Protective device intended for assembly as an integral part of the machine, which, when properly applied, protects the machine against dangerous overheating due to overload and, in a motor, failure to start.

Thread speed Fixed low speed, usually adjustable, supplied to provide a convenient method for loading and threading machines. May also be called a *preset speed.*

Threshold current Symmetrical rms available current at the threshold of the current-limiting range, where the fuse becomes current limiting when tested to the UL standard. This value can be read off a peak let-through chart where the fuse curve intersects the *A-B* line. A threshold ratio is the relationship of the threshold current to the fuse's continuous current rating.

Tight (as a suffix) So constructed that the enclosure will exclude the specified material under specified conditions.

Time, accelerating Time in seconds for a change of speed from one specified speed to a higher specified speed while accelerating under specified conditions.

Time, decelerating Time in seconds for a change of speed from one specified speed to a lower specified speed while decelerating under specified conditions.

Time delay A time interval is purposely introduced in the performance of a function.

Time-delay fuse Fuse with a built-in delay that allows temporary and harmless inrush currents to pass without opening, but is designed to open on sustained overloads and short circuits.

Time, inverse Qualifying term applied to a relay, indicating that its time of operation decreases as the magnitude of the operating quantity increases.

Time response Output, expressed as a function of time (curve), resulting from the application of a specified input under specified operating conditions.

Time, setting Time required, following the initiation of a specified stimulus to a system, for a specified variable to enter and remain within a specified narrow band centered on its final value. Setting time is expressed in seconds.

Torque Turning force applied to a shaft, tending to cause rotation. Torque is normally measured in ounce-inches or pound-feet and is equal to the force applied times the radius through which it acts.

Torque constant Motor parameter that provides a relationship between input current and output torque. For each ampere of current applied to the rotor, a fixed amount of torque will result.

Torque control Method of using current-limit circuitry to regulate torque instead of speed.

Totally enclosed machine (motor) Machine so enclosed as to prevent the free exchange of air between the inside and the outside of the case but not sufficiently enclosed to be termed airtight.

A. *Totally enclosed fan-cooled* is a totally enclosed machine equipped for exterior cooling by means of a fan or fans integral with the machine but external to the enclosing parts.

B. *Explosion-proof machine* is a totally enclosed machine whose enclosure is designed and constructed to withstand an explosion of a specified gas or vapor that may occur within it and to prevent the ignition of the specified gas or vapor surrounding the machine by sparks, flashes, or explosions of the specified gas or vapor that may occur within the machine casing.

C. *Dust-ignition-proof machine* is a totally enclosed machine whose enclosure is designed and constructed in a manner that will exclude ignitable amounts of dust or amounts that might affect performance or rating, and that will not permit arcs, sparks, or heat otherwise generated or liberated inside the enclosure to cause ignition of exterior accumulations or atmospheric suspensions of a specific dust on or in the vicinity of the enclosure.

D. *Waterproof machine* is a totally enclosed machine so constructed that it will exclude water applied in the form of a stream from a hose, except that leakage may occur around the shaft provided that it is prevented from entering the oil reservoir and provision is made for draining the machine automatically. The means for automatic draining may be a check valve or a tapped hole at the lowest part of the frame that will serve for application of a drain pipe.

E. *Totally enclosed water-cooled machine* is a totally enclosed machine that is cooled by circulating water, the water or water conductors coming in direct contact with the machine parts.

F. *Totally enclosed water-air-cooled machine* is a totally enclosed machine that is cooled by circulating air, which, in turn, is cooled by circulating water. It is provided with a water-cooled heat exchanger for cooling the internal air and a fan or fans, integral with the rotor shaft or separate, for circulating the internal air.

G. *Totally enclosed air-to-air-cooled machine* is a totally enclosed machine that is cooled by circulating the internal air through a heat exchanger which, in turn, is cooled by circulating external air. It is provided with an air-to-air heat exchanger for cooling the internal air and a fan or fans, integral with the rotor shaft or separate, for circulating the internal air and a separate fan for circulating the external air.

H. *Totally enclosed fan-cooled guarded machine* is a totally enclosed fan-cooled machine in which all openings giving direct access to the fan are limited in size by the design of the structural parts or by screens, grilles, expanded metal, and so on, to prevent accidental contact with the fan. Such openings must not permit the passage of a cylindrical rod 0.75 in. in diameter, and a probe should not contact the blades, spokes, or other irregular surfaces of the fan.

I. *Totally enclosed air-over machine* is a totally enclosed machine intended for exterior cooling by a ventilating means external to the machine.

Transducer Device that converts one energy form to another (e.g., mechanical to electrical). Also, a device that when actuated by signals from one or more systems or media, can supply related signals to one or more other systems or media.

Transient Momentary deviation in an electrical or mechanical system.

Transistor Solid-state three-terminal device that allows amplification of signals and can be used for switching and control. The three terminals are called the emitter, base, and collector.

Triac Solid-state component capable of switching alternating current.

TTL Acronym for *transistor-transistor logic*. Family of integrated-circuit logic. (Usually, 5 V is high or "1" and 0 volt is low or "0.")

UL classes Underwriters' Laboratories has developed basic physical specifications and electrical performance requirements for fuses with voltage ratings of 600 V or less. These are known as UL standards. If a type of fuse meets the requirements of a standard, it can fall into that UL class. Typical UL classes are K, RK1, RK5, G, L, H, T, CC, and J.

Unshielded Proximity switch that can sense a metal target at a greater distance, but due to its side sensitivity may detect surrounding metal.

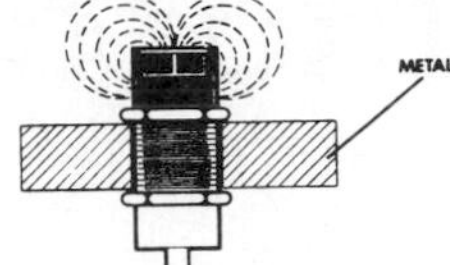

UV erasable PROM Erasable PROM that can be cleared (set to "0") by exposure to intense ultraviolet light. After being cleared, it may be reprogrammed. (*See also* PROM.)

Vector Quantity that has magnitude, direction, and sense. This quantity is commonly represented by a directed line segment whose length represents the magnitude and whose orientation in space represents the direction.

Volatile memory Memory that loses its information if the power is removed from it.

Voltage rating Maximum value of system voltage in which a fuse can be used, yet safely interrupt an overcurrent. Exceeding the voltage rating of a fuse impairs its ability to clear an overload or short circuit safely.

VVI Type of ac adjustable-frequency drive that controls the voltage and frequency to the motor to produce variable-speed operation. VVI-type drive controls the voltage in a section other than the output section where frequency generation takes place. The frequency control is accomplished by an output bridge circuit that switches the variable voltage to the motor at the desired frequency.

Watertight So constructed as to exclude water applied in the form of a hose stream under specified test conditions.

Withstandability, fault Ability of electrical apparatus to withstand the effects of specified electrical fault current conditions without exceeding specified damage criteria.

Word Grouping or a number of bits in a sequence that is treated as a unit.

Work Force moving an object over a distance. Measured in inch-ounces (in-oz) or foot-pounds (ft-lb). Work = force times distance.

X axis Axis of motion that is always horizontal and parallel to the work holding surface.

Y axis Axis of motion that is perpendicular to both the *X* and *Z* axes.

Z axis Axis of motion that is always parallel to the principal spindle of the machine.

I

DC Motor Trouble Chart

Symptom and Possible Cause	Possible Remedy
Motor will not start	
(a) Open circuit in controller	(a) Check controller for open starting resistor, open switch, or open fuse.
(b) Low terminal voltage	(b) Check voltage with nameplate rating.
(c) Bearing frozen	(c) Recondition shaft and replace bearing.
(d) Overload	(d) Reduce load or use larger motor.
(e) Excessive friction	(e) Check bearing lubrication to make sure that the oil has been replaced after installing motor. Disconnect motor from driven machine and turn motor by hand to see if trouble is in motor. Strip and reassemble motor; then check part by part for proper location and fit. Straighten or replace bent or sprung shaft (machines under 5 hp [3.73 kW]).
Motor stops after running short time	
(a) Motor is not getting power	(a) Check voltage at the motor terminals; also fuses, coils, and overload relay.
(b) Motor is started with weak or no field	(b) If adjustable-speed motor, check rheostat for correct setting. If correct, check condition of rheostat. Check field coils for open winding. Check wiring for loose or broken connection.
(c) Motor torque insufficient to drive load	(c) Check line voltage with nameplate rating. Use larger motor or one with suitable characteristic to match load.
Motor runs too slow under load	
(a) Line voltage too low	(a) Check and remove any excess resistance in supply line, connections, or controller.
(b) Brushes ahead of neutral	(b) Set brushes on neutral.
(c) Overload	(c) Check to see that load does not exceed allowable load on motor.
Motor runs too fast under load	
(a) Weak field	(a) Check for resistance in shunt-field circuits.
(b) Line voltage too high	(b) Correct high-voltage condition.
(c) Brushes back of neutal	(c) Set brushes on neutral.
Sparking at brushes	
(a) Commutator in bad condition	(a) Clean and reset brushes.
(b) Commutator eccentric or rough	(b) Grind and true commutator. Undercut mica.
(c) Excessive vibration	(c) Balance armature. Check brushes to make sure they ride freely in the holders.
(d) Broken or sluggish brush-holder spring	(d) Replace spring and adjust pressure to manufacturer's recommendations.
(e) Brushes too short	(e) Replace brushes.

11

Wound-Rotor Motor Trouble Chart

Symptom and Possible Cause	*Possible Remedy*
Motor runs at low speed with external resistance cut out	
(a) Wires to the control unit too small	(a) Use larger cable to the control unit.
(b) Control unit too far from motor	(b) Bring control unit nearer motor.
(c) Open circuit in rotor circuit (including cable to the control unit)	(c) Test by ringing out circuit and repair.
(d) Dirt between brush and ring	(d) Clean rings and insulation assembly.
(e) Brushes stuck in holders	(e) Use right-size brush.
(f) Incorrect brush tension	(f) Check brush tension and correct.
(g) Rough collector rings	(g) File, sand, and polish.
(h) Eccentric rings	(h) Turn down on lathe, or use portable tool to true-up rings without disassembling motor.
(i) Excessive vibration	(i) Balance motor.
(j) Current density of brushes too high (overload)	(j) Reduce load. If brushes have been replaced, make sure they are of the same grade as originally furnished.

III

Fractional-Horsepower Motor Trouble Chart

Symptom and Possible Cause	*Possible Remedy*
	Split-Phase Induction Motor
Failure to start	
(a) No voltage	(a) Check for voltage at motor terminals with test lamp or voltmeter. Check for blown fuses or meter. Check for blown fuses or open overload device in starter. If motor is equipped with a slow-blow fuse, see that the fuse plug is not open and that it is screwed down tight.
(b) Low voltage	(b) Measure the voltage at the motor terminals with the switch closed. Voltage should read within 10% of the voltage stamped on the motor nameplate. Overload transformers or circuits may cause low voltage. If the former, check with the power company. Overloaded circuits in the building can be found by comparing the voltage at the meter with the voltage at the motor terminals with the switch closed.
(c) Faulty cutout switch operation	(c) Cutout switch operation may be observed by removing the inspection plate in the front end bracket. The mechanism consists of a cutout switch mounted on the front end bracket and a rotating part called the governor weight assembly, which consists of a Bakelite disk so supported that it is moved back and forth along the shaft by the operation of the governor weights. At standstill, the disk holds the cutout switch closed. If the disk does not hold the switch closed, motor cannot start. This may call for adjustment of the end-play washers. Dirty contact points may also keep the motor from starting. See that the contacts are clean. After the motor has accelerated to a predetermined speed, the disk is withdrawn from the switch, allowing it to open. With the load disconnected from the motor, close the starting switch. If the motor does not start, start it by hand and observe the operation of the governor as the motor speeds up, and also when the switch has been opened and the motor slows down. If the governor fails to operate, the governor weights may have become clogged. If it operates too soon or too late, the spring is too weak or too strong. Remove motor to service shop for adjustment. Governor weights are set to operate at about 75% of synchronous speed. Place rotor in balancing machine and, with a tachometer, determine if the governor operates at the correct speed.
(d) Open overload device	(d) If the motor is equipped with a built-in micro switch or similar overload device, remove the cover plate in the end bracket on which the switch is mounted and see if the switch contacts are closed. Do not attempt to adjust this switch or to test its operation with a match. Doing so may destroy it. If the switch is permanently open, remove the motor to the service shop for repairs.
(e) Grounded field	(e) If the motor overheats, produces shock when touched, or if idle watts are excessive, test for a field ground with a test lamp across the field leads and frame. If grounded, remove the motor to the service shop for repairs.

Symptom and Possible Cause	Possible Remedy
(f) Open-circuited field	(f) These motors have a main and a phase (starting) winding. Apply current to each winding separately with a test lamp. Do not leave the windings connected too long while rotor is stationary. If either winding is open, remove the motor to the repair shop for repairs.
(g) Short-circuited field	(g) If the motor draws excessive watts, and, at the same time lacks torque, overheats, or hums, a shorted field is indicated. Remove to the service shop for repairs.
(h) Incorrect end play	(h) Certain types of motors have steel-enclosed cork washers at each end to cushion the end thrust. Too great an end thrust, hammering on the shaft, or excessive heat may destroy the cork washers and interfere with the operation of the cutout switch mechanism. If necessary, install new end-thrust cushion bumper assemblies.End play should not exceed 0.01 in. (0.254 mm); if it does, install additional steel endplay washers. End play should be adjusted so that the cutout switch is closed at standstill and open when the motor is operating.
(i) Excessive load	(i) This may be approximately determined by checking the ampere input with the nameplate marking. Excessive load may prevent the motor from accelerating to the speed at which the governor acts and cause the phase winding to burn up.
(j) Tight bearings	(j) Test by turning armature by hand. If adding oil does not help, bearings must be replaced.
Motor overheats	
(a) Grounded field	(a) Test for a field ground with a test lamp between the field and motor frame. If grounded, remove the motor to the service shop for repair.
(b) Short-circuited field	(b) Test for excessive current draw, lack of torque, and presence of hum. Any of these symptoms indicates a shorted field. Remove the motor to the service shop for repair.
(c) Tight bearings	(c) Test by turning armature by hand. If oiling does not help, new bearings must be installed.
(d) Low voltage	(d) Measure voltage at motor terminals with switch closed. Voltage should be within 10% of nameplate voltage. Overloaded transformers or power circuits may cause low voltage. Check with power company. Overloaded building circuits can be found by comparing the voltage at the meter with the voltage at the motor terminals with the switch closed.
(e) Faulty cutout switch	(e) See Paragraph (c) under *Failure to Start.*
(f) Excessive load	(f) See Paragraph (i) under *Failure to Start.*
Motor does not come up to speed	
Same possible causes and possible remedies as under *Motor Overheats.*	
Excessive bearing wear	
(a) Belt too tight	(a) Adjust belt to tension recommended by manufacturer.
(b) Pulleys out of alignment	(b) Align pulleys correctly.
(c) Dirty, incorrect, or insufficient oil	(c) Use type of oil recommended by manufacturer.
(d) Dirty bearings	(d) Clean thoroughly. Replace worn berarings.
Excessive noise	
(a) Worn bearings	(a) See Paragraphs (a), (b), (c), and (d) under *Excessive Bearing Wear.*
(b) Excessive end play	(b) If necessary, add additional end-play washers.
(c) Loose parts	(c) Check for loose hold-down bolts, loose pulleys, etc.
(d) Misalignment	(d) Align pulleys correctly.
(e) Worn belts	(e) Replace belts.
(f) Bent shaft	(f) Straighten shaft, or replace armature or motor.
(g) Unbalanced rotor	(g) Balance rotor.
(h) Burrs on shaft	(h) Remove burrs.
Motor produces shock	
(a) Grounded field	(a) See Paragraph (e) under *Failure to Start.*
(b) Broken ground strap	(b) Replace ground strap.
(c) Poor ground connection	(c) Inspect and repair ground connection.
Rotor rubs stator	
(a) Dirt in motor	(a) Thoroughly clean motor.
(b) Burrs on rotor or stator	(b) Remove burrs.

Symptom and Possible Cause	Possible Remedy
(c) Worn bearings	(c) Replace bearings and inspect shaft for scoring.
(d) Bent shaft	(d) Repair and replace shaft or rotor.
Radio interference	
(a) Poor ground connection	(a) Check and repair any defective grounds.
(b) Loose contacts or connections	(b) Check and repair any loose contacts on switches or fuses, and loose connections on terminals.

Repulsion-Start Induction Brush-Lifting Motors

Symptom and Possible Cause	Possible Remedy
Failure to start	
(a) Fuses blown	(a) Check capacity of fuses. They should not be greater in ampere capacity than recommended by the manufacturer, and in no case smaller than the full-load ampere rating of the motor, and with a voltage capacity equal to or greater than the voltage of the supply circuit.
(b) No voltage or low voltage	(b) Measure voltage at motor terminals with switch closed. See that it is within 10% of the voltage stamped on the nameplate of the motor.
(c) Open-circuited field or armature	(c) Indicated by excessive sparking in starting, refusal to start at certain positions of the rotor, or by a humming sound when the switch is closed. Examine for broken wires, loose connections, or burned segments on the commutator at the point of loose or broken connections. Inspect the commutator for a foreign metallic substance that might cause a short between the commutator segments.
(d) Incorrect voltage or frequency	(d) Requires new motor built for operation on local power supply. Dc motors will not operate on ac circuit, or vice versa.
(e) Worn or sticking brushes	(e) When brushes are not making proper contact with the commutator, the motor will have a weak starting torque. This can be caused by worn brushes, brushes sticking in holders, weak brush springs, or a dirty commutator. The commutator should be polished with fine sandpaper (never use emery). The commutator should never be oiled or greased.
(f) Improper brush setting	(f) Unless a new armature has been installed, the brush holder or rocker arm should be opposite the index and locked in position. If a new armature has been installed, the position may be slightly off the original marking.
(g) Improper line connection	(g) See that the connections are made according to the connection diagram sent with the motor. The motor may, through error, be wired for a higher voltage.
(h) Excessive load	(h) If the motor starts with no load, and if all the foregoing conditions are satisfactory, then failure to start is most likely due to an excessive load.
(i) Shorted field	(i) Take separate current readings on each of the two halves of the stator winding. Unequal readings indicate a short. Shorted coil may also feel much hotter than the normal coil. An increase in hum may also be caused by a shorted winding.
(j) Shorted rotor	(j) Remove the brushes from the commutator and impress full voltage on the stator. If there is one or more points at which the rotor "hangs" (fails to revolve easily when turned), the rotor is shorted. Forcing the rotor to the position where it is most difficult to hold will cause the shorted coil to become hot. Do not hold in position too long or the coil will burn out.
Motor operates without lifting brushes	
(a) Dirty commutator	(a) Clean with fine sandpaper. (Do not use emery.)
(b) Governor mechanism or brushes sticking, or brushes worn too short for good contact	(b) See that brushes move freely in slots and that governor mechanism operates freely by hand. Replace worn brushes.
(c) Frequency of supply circuit incorrect	(c) Run motor idle. After brushes throw off, speed should be slightly in excess of full-load speed shown on nameplate. An idle speed varying more than 10% from nameplate speed indicates that motor is being used on an incorrect supply frequency. A different motor will be required.
(d) Low voltage	(d) See that voltage is within 10% of nameplate voltage with the switch closed.
(e) Line connection improperly or poorly made	(e) See that contacts are good and that connections correspond with diagram sent with motor.
(f) Incorrect brush setting	(f) Check to see that rocker-arm setting corresponds with index mark.
(g) Incorrect adjustment of governor	(g) The governor should operate and lift brushes at approximately 75% of speed stamped on nameplate. Below 65% or over 85% indicates incorrect spring tension.

Symptom and Possible Cause	*Possible Remedy*
(h) Excessive load	(h) An excessive load may be started but not carried to and held at full-load speed, which is beyond where the brushes lift. Tight motor bearings may contribute to overload. This is sometimes indicated by brushes lifting and returning to the commutator.
(i) Shorted field	(i) See Paragraph (i) under *Failure to Start.*
Excessive bearing wear	
(a) Belt too tight, or an unbalanced line coupling	(a) Correct the mechanical condition.
(b) Improper, dirty, or insufficient oil	(b) The lubrication system of most small motors provides for supplying the right amount of filtered oil to the bearings. It is necessary only for the user to keep the wool yarn saturated with a good grade of machine oil.
(c) Dirty bearnings	(c) When bearings become clogged with dirt, the motor may need protection from excessive dust. The application may be such that a specially constructed motor should be used.
Motor runs hot	
(a) Bearing trouble	(a) See Paragraphs (a), (b), and (c) under *Excessive Bearing Wear.*
(b) Short-circuited coils in stator	(b) Make separate wattmeter reading on each of the two halves of the stator winding. Sometimes the shorted coil may be located by the fact that one coil feels much hotter than the other. An increase over normal in the magnetic noise (hum) may also indicate a shorted stator.
(c) Rotor rubbing stator	(c) Extraneous matter may be between the rotor and the stator, or the bearings may be badly worn.
(d) Excessive load	(d) Be sure proper pulleys are on the motor and the machine. Driving the load at higher speeds requires more horsepower. Take an ammeter reading. If current draw exceeds the nameplate amperes for full load, the answer is evident.
(e) Low voltage	(e) Measure the voltage at the motor terminals with the switch closed. The reading should not vary more than 10% from the value stamped on the nameplate.
(f) High voltage	(f) See (e) above.
(g) Incorrect line connection to the motor	(g) Check the connection diagram sent with the motor.
Motor burns out	
(a) Frozen bearing	(a) See Paragraphs (a), (b), and (c) under *Excessive Bearing Wear.*
(b) Some condition of prolonged excessive overload	(b) Before replacing the burned-out motor, locate and remove the cause of the overload. Certain jobs that present a heavy load will, under unusual conditions of operation, apply prolonged overloads that may destroy a motor and be difficult to locate unless examined carefully. On jobs where it is assumed somewhat intermittent service will normally prevail, and that consequently are closely motored, the load cycle should be especially checked, as a change in this feature will easily produce excessive overload on the motor.
Motor is noisy	
(a) Unbalanced rotor	(a) When transportation handling has been so rough as to damage the heavy shipping case, it is well to test the motor for unbalanced conditions at once. It is even possible (though it rarely happens) that a shaft may be bent. In any event, the rotor should be rebalanced dynamically.
(b) Worn bearings	(b) See Paragraphs (a), (b), and (c) under *Excessive Bearing Wear.*
(c) Rough commutator, or brushes not seating properly	(c) Noise from this cause occurs only during the starting period, but conditions should be corrected to avoid consequent trouble.
(d) Excessive end play	(d) Proper end play is as follows: $\frac{1}{3}$ hp (248.7 W) and smaller—0.127 to 0.762 mm; $\frac{1}{2}$ (373 W) to 1 hp (0.746 kW)—0.254 to 1.905 mm. Washers supplied by the factory should be used. Be sure to tell factory all figures involved. Remember that too little end play is as bad as too much.
(e) Motor not properly aligned with the driven machine	(e) Correct the mechanical condition.
(f) Motor not firmly fastened to mounting base	(f) All small motors have steel bases so they can be firmly bolted to their mounting without fear of breaking. It is, of course, not to be expected that the base should be strained out of shape in order to make up for roughness in the mounting base.
(g) Loose accessories in motor	(g) Such parts as oil covers, guards (if any), end plates, etc., should be checked, especially

Symptom and Possible Cause	Possible Remedy
	if they have been removed for inspection of any sort. The conduit box should be tightened when the top is fitted after the connections are made.
(h) Air gap not uniform	(h) This results from a bent shaft or an unbalanced rotor. See Paragraph (a).
(i) Amplified motor noises	(i) When this condition is suspected, set the motor on a firm floor. If the motor is now quiet, then the mounting is acting as an amplifer to bring about certain noises in the motor. Frequently, the correction of slight details in the mounting will eliminate this, but rubber mounts are the surest cure.
Excessive brush wear	
(a) Dirty commutator	(a) Clean with fine sandpaper (never use emery).
(b) Poor contact with commutator	(b) See that the brushes are long enough to reach the commutator, that they move freely in the slots, and that the spring tension gives firm but not excessive pressure.
(c) Excessive load	(c) If the brush wear is due to overload, it can usually be checked by noting the time required for the brushes to lift from the commutator. The proper time is less than 10 seconds.
(d) Failure to lift promptly and stay off during the running period	(d) Examine for conditions listed under *Motor Operates Without Lifting Brushes.*
(e) High mica	(e) Examination will show this condition. Take a very light cut off the commutator face and polish with fine sandpaper. Undercut the mica.
(f) Rough commutator	(f) True up on lathe.
Brush-holder or rocker-arm wear	
(a) Failure to lift properly and stay off during the running period	(a) No noticeable wear of this part should occur during the life of the motor. Troublesome wear indicates faulty operation. See under *Motor Operates Without Lifting Brushes.*
Radio interference	
(a) Faulty ground	(a) Check for poor ground connections, and repair. Static electricity generated by the belts may cause radio noises if the motor frame is not thoroughly grounded. Check for loose connections or contacts in the switch, fuses, or starter.
	Capacitor-Start Induction Motors
Failure to start	
(a) Blown fuses or overload device tripped	(a) Examine motor bearings. Be sure that they are in good condition and properly lubricated. Be sure the motor and driven machine both turn freely. Check the circuit voltage at the motor terminals against the voltage stamped on the motor nameplate. Examine the overload protection of the motor. Overload relays operating on either magnetic or thermal principles (or a combination of the two) offer adequate protection to the motor. Ordinary fuses of sufficient size to permit the motor to start do not protect against burnout. A combination fuse and thermal relay, such as *Buss Fusetron,* protects the motor and is inexpensive. If the motor does not have overload protection, the fuses should be replaced with overload relays or *Buss Fusetrons.* After installing suitable fuses and resetting the overload relays, allow the machine to go through its operating cycle. If the protective devices again operate, check the load. If the motor is excessively overloaded, take the matter up with the manufacturer.
(b) No voltage or low voltage	(b) Measure the voltage at the motor terminals with the switch closed. See that it is within 10% of the voltage stamped on the motor nameplate.
(c) Open-circuited field	(c) Indicated by a humming sound when the switch is closed. Examine for broken wires and connections.
(d) Incorrect voltage or frequency	(d) Requires motor built for operation on power supply available. Ac motors will not operate on dc circuit, or vice versa.
(e) Cutout switch faulty	(e) The operation of the cutout switch may be observed by removing the inspection plate in the end bracket. If the governor disk does not hold the switch closed, the motor cannot start. This may call for additional end-play washers between the shaft shoulder and the bearing. Dirty or corroded contact points may also keep the motor from starting. See that the contacts are clean. With load disconnected from the motor, close the starting switch. If the motor does not start, start it by hand and listen for the characteristic click of the governor as the motor speeds up and also when the switch has been opened and the motor slows down. Absence of this click may indicate that the

Symptom and Possible Cause	Possible Remedy
	governor weights have become clogged, or that the spring is too strong. Continued operation under this condition may cause the phase winding to burn up. Remove the motor to the service shop for adjustment.
(f) Open field	(f) These motors have a main and phase winding in the stator. With the leads disconnected from the capacitor, apply current to the motor. If the main winding is all right, the motor will hum. If the main winding tests satisfactorily, connect a test lamp between the phase lead (the black lead) from the capacitor and the other capacitor lead. Close the starting switch.If the phase winding is all right, the lamp will glow and the motor may attempt to start. If either winding is open, remove the motor to the service shop for repairs.
(g) Faulty capacitor	(g) If the starting capacitor (electrolytic) is faulty, the motor starting torque will be weak and the motor may not start at all, but may run if started by hand. A capacitor can be tested for open circuit or short circuit as follows: Charge it with dc (if available), preferably through a resistance or test lamp. If no discharge is evident on immediate short circuit, an open or a short is indicated. If no dc is available, charge with ac. Try charging on ac several times to make certain that the capacitor has had a chance to become charged. If the capacitor is open, short-circuited, or weak, replace it. Replacement capacitors should not be of a lower capacity or voltage than the original. In soldering the connections, *do not use acid flux*. ***Note 1***—Electrolytic capacitors, if exposed to temperatures of 20°F (-6.7°C) and lower, may temporarily lose enough capacity so that the motor will not start and may cause the windings to burn up. The temperature of the capacitor should be raised by running the motor idle, or by other means. Capacitors should not be operated in temperatures exceeding 165°F. (74°C). ***Note 2***—The frequency of operation of electrolytic capacitors should not exceed two starts per minute of 3 seconds' acceleration each, or three to four starts per minute at less than 2 seconds' acceleration, provided that the total accelerating time (i.e., the time before the switch opens) does not exceed 1 to 2 minutes per hour. This may be approximately determined by checking the ampere input with the nameplate marking. Excessive load may prevent the motor from accelerating to the speed at which the governor acts, and thus cause the phase winding to burn up.
Radio interference	
(a) Faulty ground	(a) Check for poor ground connections. Static electricity generated by the belts may cause radio noises if the motor frame is not thoroughly grounded.
(b) Loose connections	(b) Check for loose connections or contacts in the switch, fuses, or starter. Capacitor motors ordinarily will not cause radio interference. Sometimes vibration may cause the capacitor to move so that it touches the metal container. This may cause radio interference. Open the container, move the capacitor, and replace the paper packing so that the capacitor cannot shift.

Source: Courtesy of Lincoln Electric Co.

Selection of Dual-Element Fuses for Motor-Running Overload Protection

Motors Marked With Not Less Than 1.15 S.F. Or Temp. Rise Not Over 40°C

Ampere Ratings		Ampere Ratings		Ampere Ratings	
Motor	**Fuse***	**Motor**	**Fuse***	**Motor**	**Fuse***
1.00 to 1.11	**1 1/4**	6.40 to 7.19	**8**	72.0 to 79.9	**90**
1.12 to 1.27	**1 4/10**	7.20 to 7.99	**9**	†80.0 to 87.9	**100**
1.28 to 1.43	**1 6/10**	8.00 to 9.59	**10**	88.0 to 99.9	**110**
1.44 to 1.59	**1 8/10**	9.60 to 11.9	**12**	100 to 119	**125**
1.60 to 1.79	**2**	12.0 to 13.9	**15**	120 to 139	**150**
1.80 to 1.99	**2 1/4**	14.0 to 15.9	**17 1/2**	140 to 159	**175**
2.00 to 2.23	**2 1/2**	16.0 to 19.9	**20**	†160 to 179	**200**
2.24 to 2.55	**2 8/10**	20.0 to 23.9	**25**	180 to 199	**225**
2.56 to 2.79	**3 2/10**	†24.0 to 27.9	**30**	200 to 239	**250**
2.80 to 3.19	**3 1/2**	28.0 to 31.9	**35**	240 to 279	**300**
3.20 to 3.59	**4**	32.0 to 35.9	**40**	280 to 319	**350**
3.60 to 3.99	**4 1/4**	36.0 to 39.9	**45**	†320 to 359	**400**
4.00 to 4.47	**5**	40.0 to 47.9	**50**	360 to 399	**450**
4.48 to 4.99	**5 6/10**	48.0 to 55.9	**60**	400 to 480	**500**
5.00 to 5.59	**6 1/4**	56.0 to 63.9	**70**	480 to 521	**600**
5.60 to 6.39	**7**	64.0 to 71.9	**80**		

†Disconnect switch must have an ampere rating at least 115% of motor ampere rating (430-110a). Next larger size switch with fuse reducers may be required.

*Use FUSETRON Fuses, FRN-R (250V) or FRS-R (600V); or LOW-PEAK fuses LPN-RK (250V) or LPS-RK (600V).

Abnormal Installations-May require FUSETRON or LOW-PEAK Fuses of a larger size than shown; will provide only short-circuit protection. These applications include:

(Courtesy of Bussmann)

All Other Motors (Less Than 1.15 S.F. Or Temp. Rise Greater Than 40°C)

Ampere Ratings		Ampere Ratings		Ampere Ratings	
Motor	**Fuse***	**Motor**	**Fuse***	**Motor**	**Fuse***
1.00 to 1.08	**1 1/8**	6.09 to 6.95	**7**	69.6 to 78.2	**80**
1.09 to 1.21	**1 1/4**	6.96 to 7.82	**8**	78.3 to 86.9	**90**
1.22 to 1.39	**1 4/10**	7.83 to 8.69	**9**	†87.0 to 95.6	**100**
1.40 to 1.56	**1 6/10**	8.70 to 10.0	**10**	95.7 to 108	**110**
1.57 to 1.73	**1 8/10**	10.5 to 12.0	**12**	109 to 125	**125**
1.74 to 1.95	**2**	13.1 to 15.0	**15**	131 to 150	**150**
1.96 to 2.17	**2 1/4**	15.3 to 17.3	**17 1/2**	153 to 173	**175**
2.18 to 2.43	**2 1/2**	17.4 to 20.0	**20**	†174 to 195	**200**
2.44 to 2.78	**2 8/10**	21.8 to 25.0	**25**	196 to 217	**225**
2.79 to 3.04	**3 2/10**	†26.1 to 30.0	**30**	218 to 250	**250**
3.05 to 3.47	**3 1/2**	30.5 to 34.7	**35**	261 to 300	**300**
3.48 to 3.91	**4**	34.8 to 39.1	**40**	305 to 347	**350**
3.92 to 4.34	**4 1/2**	39.2 to 43.4	**45**	†348 to 391	**400**
4.35 to 4.86	**5**	43.5 to 50.0	**50**	392 to 434	**450**
4.87 to 5.43	**5 6/10**	52.2 to 60.0	**60**	435 to 480	**500**
5.44 to 6.08	**6 1/4**	60.9 to 69.5	**70**		

(a) FUSETRON Fuses or LOW-PEAK Fuses in high ambient temperatures environments.

(b) A motor started frequently or rapidly reversed.
Motor is directly connected to a machine that cannot be brought up to full speed quickly (large fans, centrifugal machines such as extractors and pulverizers. Machines having large fly wheels such as large punch presses).

(c) Motor has a high Code Letter (or possibly no Code Letter) with full voltage start.

V

Tables and Formulas

The data contained in this section are provided for reference only.* Many formulas are for estimating purposes only because they cannot consider all factors in every machine application. Many formulas can assist the reader by demonstrating basic physical or electrical relationships needed to understand a more abstract concept in control or motor operation. Other data, such as conversion factors, are included for your convenience to provide a more comprehensive resource when working in an international design environment.

NOTE:
The following equations for calculating horsepower are meant to be used for estimating purposes only. These equations do not include any allowance for machine friction, windage, or other factors. These factors must be considered when selecting a drive for a machine application.

HORSEPOWER FORMULAS

Rotating Objects

$$\text{hp} = \frac{T \times N}{63{,}000}$$

where T = torque (lb − in.)
N = speed(rpm)

$$\text{hp} = \frac{T \times N}{5252}$$

where T = torque (lb − ft)
N = speed(rpm)

Objects in Linear Motion

$$\text{hp} = \frac{F \times V}{33{,}000}$$

where F = force (lb)
V = velocity(ft/min)

$$\text{hp} = \frac{F \times V}{396{,}000}$$

where F = force(lb)
V = velocity(in./min)

Pumps

$$\text{hp} = \frac{\text{gpm} \times \text{head} \times \text{specific gravity}}{3960 \times \text{efficiency of pump}}$$

$$\text{hp} = \frac{\text{gpm} \times \text{psi} \times \text{specific gravity}}{1713 \times \text{efficiency of pump}}$$

where gpm = gallons per minute
head = height of water (ft)
efficiency of pump = %/100
psi = pounds per square inch

Specific gravity of water = 1.0
1 cu ft per sec = 448 gpm
1 psi = a head of 2.309 ft — for water weighing 62.36 lb per cu ft at 62°F

*This information on formulas and tables is courtesy of Allen-Bradley Co.

Fans and Blowers

$$\text{hp} = \frac{\text{cfm} \times \text{psf}}{33{,}000 \times \text{efficiency of fan}}$$

$$\text{hp} = \frac{\text{cfm} \times \text{piw}}{6356 \times \text{efficiency of fan}}$$

$$\text{hp} = \frac{\text{cfm} \times \text{psi}}{229 \times \text{efficiency of fan}}$$

where cfm = cubic feet per minute
psf = pounds per square foot
piw = inches of water gage
psi = pounds per square inch
efficiency of fan = %/100

Conveyors

$$\text{hp (vertical)} = \frac{F \times V}{33{,}000}$$

$$\text{hp (horizontal)} = \frac{F \times V \times \text{coefficient of friction}}{33{,}000}$$

where F = force (lb)
V = velocity (ft/min)

Coefficient of friction:

Ball or roller slide = 0.02
Dovetail slide = 0.20
Hydrostatic ways = 0.01
Rectangle ways with gib = 0.1–0.25

TORQUE FORMULAS

$$T = \frac{\text{hp} \times 5252}{N}$$

where T = torque (lb-ft)
hp = horsepower
N = speed (rpm)

$$T = F \times R$$

where T = torque (lb – ft)
F = force (lb)
R = radius (ft)

$$T\text{(accelerating)} = \frac{\omega k^2 \times \Delta\text{rpm}}{308 \times t}$$

where T = torque (lb – ft)
ωk^2 = inertia reflected to the motor shaft (lb – ft^2)
Δrpm = change in speed
t = time to accelerate (sec)

Note:
To change lb-ft^2 to in.-lb-sec^2: Divide by 2.68
To change in.-lb-sec^2 to lb-ft^2: Multiply by 2.68

AC MOTOR FORMULAS

$$\text{sync speed} = \frac{\text{frequency} \times 120}{\text{number of poles}}$$

where sync speed = synchronous speed (rpm)
frequency = frequency (Hz)

$$\%\text{ slip} = \frac{(\text{sync speed} - \text{FL speed}) \times 100}{\text{sync speed}}$$

where FL speed = full-load speed (rpm)
sync speed = synchronous speed (rpm)

$$\text{reflected } \omega k^2 = \frac{\omega k^2 \text{ of load}}{(\text{reduction ratio})^2}$$

where ωk^2 = inertia (lb-ft^2)

ELECTRICAL FORMULAS

Ohm's Law

$$I = \frac{E}{R} \qquad R = \frac{E}{I} \qquad E = I \times R$$

where I = current (amperes)
E = EMF or voltage (volts)
R = resistance (ohms)

Power in DC Circuits

$$P = I \times E \qquad \text{hp} = \frac{I \times E}{746}$$

$$\text{kW} = \frac{I \times E}{1000} \qquad \text{kWh} = \frac{I \times E \times \text{hours}}{1000}$$

where P = power (watts)
I = current (amperes)
E = EMF or voltage (volts)
kW = kilowatts
kWh = kilowatthours

Power in AC Circuits

$$\text{kVA (one-phase)} = \frac{I \times E}{1000}$$

$$\text{kVA (three-phase)} = \frac{I \times E \times 1.73}{1000}$$

where kVA = kilovolt-amperes
I = current (amperes)
E = EMF or voltage (volts)

$$\text{kW (one-phase)} = \frac{I \times E \times \text{PF}}{1000}$$

$$\text{kW (two-phase)} = \frac{I \times E \times \text{PF} \times 1.42}{1000}$$

$$\text{kW (three-phase)} = \frac{I \times E \times \text{PF} \times 1.73}{1000}$$

$$\text{PF} = \frac{\text{W}}{\text{V} \times I} = \frac{\text{kW}}{\text{kVA}}$$

where kW = kilowatts
I = current (amperes)
E = EMF or voltage (volts)
PF = power factor
W = watts
V = volts
kVA = kilovolt-amperes

Calculating Motor Amperes

$$\text{motor amperes} = \frac{\text{hp} \times 746}{E \times 1.732 \times \text{Eff} \times \text{PF}}$$

$$\text{motor amperes} = \frac{\text{kVA} \times 1000}{1.73 \times E}$$

$$\text{motor amperes} = \frac{\text{kW} \times 1000}{1.73 \times E \times \text{PF}}$$

where hp = horsepower
E = EMF or voltage (volts)
Eff = efficiency of motor (%/100)
kVA = kilovolt-amperes
kW = kilowatts
PF = power factor

Calculating AC Motor Locked-Rotor Amperes

$$\text{LRA} = \frac{\text{hp} \times \left(\frac{\text{start kVA}}{\text{hp}}\right) \times 1000}{E \times 1.73}$$

where LRA = locked-rotor amperes
hp = horsepower
kVA = kilovolt-amperes
E = voltage (volts)

$$\text{LRA at freq. } X = \frac{\text{60-Hz LRA}}{\sqrt{\frac{60}{\text{freq. } X}}}$$

where 60 Hz LRA = locked-rotor amperes
freq. X = desired frequency (Hz)

ER FORMULAS

Calculating Accelerating Force for Linear Motion

$$F\text{(acceleration)} = \frac{W \times \Delta V}{1933 \times t}$$

where F = force (lb)
W = weight (lb)
ΔV = change of velocity (fpm)
t = time to accelerate weight (sec)

Calculating Minimum Accelerating Time of a Drive

$$t = \frac{\omega k^2 \times \Delta N}{308 \times T}$$

where t = Time required to accelerate load (sec)
ωk^2 = total inertia that the motor must accelerate (lb-ft^2; includes motor rotor, gear reducer, and load
ΔN = change in speed required (rpm)
T = accelerating torque (lb-ft)

Note:
To change lb-ft^2 to in.-lb-sec^2: Divide by 2.68
To change in.-lb-sec^2 to lb-ft^2: Multiply by 2.68

$$\text{rpm} = \frac{\text{fpm}}{0.262 \times D}$$

where rpm = revolutions per minute
fpm = feet per minute
D = diameter (ft)

$$\omega k^2 \text{ reflected to motor} = \text{load } \omega k^2 \times \left(\frac{\text{load rpm}}{\text{motor rpm}}\right)^2$$

where ωk^2 = inertia (lb-ft^2)
rpm = revolutions per minute

ENGINEERING CONSTANTS

Temperature

0°C = freezing point of water
32°F = freezing point of water = 0°C
100°C = boiling point of water at atmospheric pressure
212°F = boiling point of water at atmospheric pressure
1.8°F change = 1°C
0.252 kilocalorie = 1 Btu
−270°C = absolute zero
−459.6°F = absolute zero

Length

1760 yd = 1 mile
25.4 mm = 2.54 cm = 1 in.
3 ft = 1 yd
3.2808 ft = 1 m
39.37 in. = 1 m = 100 cm = 1000 mm
5280 ft = 1 mile
0.62137 mile = 1 km

Weight

16 oz = 1 lb
2.2046 lb = 1 kg
2.309 ft water at 62°F = 1 psi
28.35 g = 1 oz
59.76 lb = weight of 1 cu ft of water at 212°F
0.062428 lb per cu ft = 1 kg/cu m
62.355 lb = weight of 1 cu ft water at 62°F
8⅓ (8.32675) lb = weight 1 gal water at 62°F

Power

1.3410 hp = 1 kW
2.545 Btu per hr = 1 hp
33,000 ft-lb per min = 1 hp
550-ft-lb per sec = 1 hp
745.7 W = 1 hp

Area

10.764 sq ft = 1m^2
1,273,239 circular mils = 1 sq in.
144 sq in. = 1 sq ft
645 mm^2 = 1 sq in.
9 sq ft = 1 sq yd
0.0929 m^2 = 1 sq ft

Mathematic

1.4142 = square root of 2
1.7321 = square root of 3
3.1416 = π = ratio of circumference of circle to diameter = ratio of area of a circle to square of radius
57.296 degrees = 1 rad (angle)
0.7854 × diameter squared = area of a circle

Pressure

14.223 psi = 1 kg per cm^2 = 1 "metric atmosphere"
2.0355 in. Hg at 32°F = 1 psi
2.0416 in. Hg at 62°F = 1 psi
2116.3 psf = atmospheric pressure
27.71 in. water at 62°F = 1 psi
29.921 in. Hg at 32°F = atmospheric pressure
30 in. Hg at 62°F = atmospheric pressure (approximate)
33.974 ft water at 62°F = atmospheric pressure
0.433 psi = 1 ft of water at 62°F
5196 psf = 1 in. water at 62°F
760 mm Hg = atmospheric pressure at 0°C
0.07608 lb = weight 1 cu ft air at 62°F and 14.7 psi

Volume

1728 cu in. = 1 cu ft
231 cu in = 1 gal (U.S.)
277.274 cu in = 1 gal (British)
27 cu ft = 1 cu yd
31 gal (31.5 U.S. gal) = 1 barrel
35.314 cu ft = 1 m^3
3.785 liters = 1 gal
61.023 cu in. = 1 liter
7.4805 gal = 1 cu ft

Units of Pressure

kg per cm^2 = kilograms per square centimeter
Hg = symbol for mercury
psi = pounds per square inch
psf = pounds per square foot

Units of Volume

cu in. = cubic inch
gal = gallon
cu ft = cubic feet
ml = milliliter
fl oz = fluid ounce (U.S.)

CONVERSION FACTORS

Length

To Convert:	*To:*	*Multiply by:*
meters	feet	3.281
meters	inches	39.37
inches	meters	0.0254
feet	meters	0.3048
millimeters	inches	0.0394
inches	millimeters	25.4
threads/inch	millimeter pitch	Divide into 25.4
yards	meters	0.914

Example: 10 m × 3.281 = 32.81 ft

Area

To Convert:	*To:*	*Multiply by:*
circular mil	meter^2	0.50×10^{-9}
yard^2	meter^2	0.8361

Example: 100 circular mils × 0.5 × 10^{-9} = 0.5 × 10^{-7} m^2 (0.5 × 10^{-7} circular mil = 0.00000005 m^2)

Power

To Convert:	*To:*	*Multiply by:*
watts	hp	0.00134
ft-lb/min	hp	0.0000303
hp	kW	0.746

Example: 1500 W × 0.00134 = 2.01 hp

Rotation/Rate

To Convert:	To:	Multiply by:
rpm	deg/sec	6.00
rpm	rad/sec	0.1047
deg/sec	rpm	0.1667
rad/sec	rpm	9.549
fpm	m/s	0.00508
fps	m/s	0.3048
gal/min	cm^3/s	63.09
in./sec	m/s	0.0254
km/h	m/s	0.2778
mph	m/s	0.447
mph	km/h	1.609
rpm	rad/s	0.1047
yd^3/min	m^3/s	0.01274

Example: 1800 rpm × 6.00 = 10800 deg/sec

Moment of Inertia

To Convert:	To:	Multiply by:
newton-meters^2	lb-ft^2	2.42
oz-in.^{2}	lb-ft^2	0.000434
lb-in.^{2}	lb-ft^2	0.00694
slug-ft^2	lb-ft^2	32.17
oz-in.-sec^2	lb-ft^2	0.1675
in.-lb-sec^2	lb-ft^2	2.68

Example: 25 newton-meters^2 × 2.42 = 60.5 lb-ft^2

Mass/Weight

To Convert:	To:	Multiply by:
oz	g	31.1
kg	lb	2.205
lb	kg	0.4536
newtons	lb	0.2248

Example: 50 oz × 31.1 = 1555 g

Torque

To Convert:	To:	Multiply by:
newton-meters	lb-ft	0.7376
lb-ft	newton-meters	1.3558
lb-in.	lb-ft	0.0833
lb-ft	lb-in.	12.00

Example: 30 Newton-Meters × 0.7376 = 22.13 lb-ft

Volume

To Convert:	To:	Multiply by:
cm^3 (ml)	m^3	0.000001
fl oz	cm^3	29.57
ft^3 of water (39.2°F)	kg (or liter)	28.32
cfm	m^3/s	0.000472
liters	m^3	0.001
yd^3	m^3	0.7646

Example: 250 cm^3 × .000001 = .00025 m^3

Temperature

To Convert:	To:	Use the Formula:
°F	°C	$°C = \frac{°F - 32}{1.8}$
°C	°F	°F = (°C × 1.8) + 32

Example: $68°F = \frac{68 - 32}{1.8} = 20°C$

$20°C = (20 \times 1.8) + 32 = 68°F$

Fractional Inch to Equivalent Millimeters and Decimals

Inch	Equivalent		Inch	Equivalent		Inch	Equivalent		Inch	Equivalent	
	mm	Decimal		mm	Decimal		mm	Decimal		mm	Decimal
$\frac{1}{64}$	0.3969	0.0156	$\frac{17}{64}$	6.7469	0.2656	$\frac{33}{64}$	13.0969	0.5156	$\frac{49}{64}$	19.4469	0.7656
$\frac{1}{32}$	0.7938	0.0313	$\frac{9}{32}$	7.1438	0.2813	$\frac{17}{32}$	13.4938	0.5313	$\frac{25}{32}$	19.8438	0.7813
$\frac{3}{64}$	1.1906	0.0469	$\frac{19}{64}$	7.5406	0.2969	$\frac{35}{64}$	13.8906	0.5469	$\frac{51}{64}$	20.2406	0.7969
$\frac{1}{16}$	1.5875	0.0625	$\frac{5}{16}$	7.9375	0.3125	$\frac{9}{16}$	14.2875	0.5625	$\frac{13}{16}$	20.6375	0.8125
$\frac{5}{64}$	1.9844	0.0781	$\frac{21}{64}$	8.3344	0.3181	$\frac{37}{64}$	14.6844	0.5781	$\frac{53}{64}$	21.0344	0.8281
$\frac{3}{32}$	2.3813	0.0938	$\frac{11}{32}$	8.7313	0.3438	$\frac{19}{32}$	15.0813	0.5938	$\frac{27}{32}$	21.4313	0.8438

Inch	Equivalent mm	Equivalent Decimal	Inch	Equivalent mm	Equivalent Decimal	Inch	Equivalent mm	Equivalent Decimal	Inch	Equivalent mm	Equivalent Decimal
$\frac{7}{64}$	2.7781	0.1094	$\frac{23}{64}$	9.1281	0.3594	$\frac{39}{64}$	15.4781	0.6094	$\frac{55}{64}$	21.8281	0.8594
$\frac{1}{8}$	3.1750	0.1250	$\frac{3}{8}$	9.5250	0.3750	$\frac{5}{8}$	15.8750	0.6250	$\frac{7}{8}$	22.2250	0.8750
$\frac{9}{64}$	3.5719	0.1406	$\frac{25}{64}$	9.9219	0.3906	$\frac{41}{64}$	16.2719	0.6406	$\frac{57}{64}$	22.6219	0.8906
$\frac{5}{32}$	3.9688	0.1563	$\frac{13}{32}$	10.3188	0.4063	$\frac{21}{32}$	16.6688	0.6563	$\frac{29}{32}$	23.0188	0.9063
$\frac{11}{64}$	4.3656	0.1719	$\frac{27}{64}$	10.7156	0.4219	$\frac{43}{64}$	17.0656	0.6719	$\frac{59}{64}$	23.4156	0.9219
$\frac{3}{16}$	4.7625	0.1875	$\frac{7}{16}$	11.1125	0.4375	$\frac{11}{16}$	17.4625	0.6875	$\frac{15}{16}$	23.8125	0.9375
$\frac{13}{64}$	5.1594	0.2031	$\frac{29}{64}$	11.5094	0.4531	$\frac{45}{64}$	17.8594	0.7031	$\frac{61}{64}$	24.2094	0.9531
$\frac{7}{32}$	5.5563	0.2188	$\frac{15}{32}$	11.9063	0.4688	$\frac{23}{32}$	18.2563	0.7188	$\frac{31}{32}$	24.6063	0.9688
$\frac{15}{64}$	5.9531	0.2344	$\frac{31}{64}$	12.3031	0.4844	$\frac{47}{64}$	18.6531	0.7344	$\frac{63}{64}$	25.0031	0.9844
$\frac{1}{4}$	6.3500	0.2500	$\frac{1}{2}$	12.700	0.5000	$\frac{3}{4}$	19.0500	0.7500			

Inertia of a Solid Steel Shaft (lb-ft^2 per inch of length)

Diameter (in.)	ωk^2	Diameter (in.)	ωk^2	Diameter (in.)	ωk^2	Diameter (in.)	ωk^2	Diameter (in.)	ωk^2	Diameter (in.)	ωk^2
$\frac{3}{4}$	0.00006	8	0.791	$14\frac{1}{4}$	7.97	37	360.70	62	2844.3	87	11,028
1	0.00007	$8\frac{1}{4}$	0.895	$14\frac{1}{2}$	8.54	38	401.30	63	3032.3	88	11,544
$1\frac{1}{4}$	0.0005	$8\frac{1}{2}$	1.000	$14\frac{3}{4}$	9.15	39	445.30	64	3229.5	89	12,077
$1\frac{1}{2}$	0.001	$8\frac{3}{4}$	1.130	15	9.75	40	492.78	65	3436.1	90	12,629
$1\frac{3}{4}$	0.002	9	1.270	16	12.59	41	543.90	66	3652.5	91	13,200
2	0.003	$9\frac{1}{4}$	1.410	17	16.04	42	598.80	67	3879.0	92	13,790
$2\frac{1}{4}$	0.005	$9\frac{1}{2}$	1.550	18	20.16	43	658.10	68	4115.7	93	14,399
$2\frac{1}{2}$	0.008	$9\frac{3}{4}$	1.750	19	25.03	44	721.40	69	4363.2	94	15,029
$2\frac{3}{4}$	0.011	10	1.930	20	30.72	45	789.30	70	4621.7	95	15,679
3	0.016	$10\frac{1}{4}$	2.130	21	37.35	46	861.80	71	4891.5	96	16,349
$3\frac{1}{2}$	0.029	$10\frac{1}{2}$	2.350	22	44.99	47	939.30	72	5172	97	17,041
$3\frac{3}{4}$	0.038	$10\frac{3}{4}$	2.580	23	53.74	48	1021.80	73	5466	98	17,755
4	0.049	11	2.830	24	63.71	49	1109.60	74	5774	99	18,490
$4\frac{1}{4}$	0.063	$11\frac{1}{4}$	3.090	25	75.02	50	1203.07	75	6090	100	19,249
$4\frac{1}{2}$	0.079	$11\frac{1}{2}$	3.380	26	87.76	51	1302.2	76	6422		
5	0.120	$11\frac{3}{4}$	3.680	27	102.06	52	1407.4	77	6767	110	28,183
$5\frac{1}{2}$	0.177	12	4.000	28	118.04	53	1518.8	78	7125	120	39,914
6	0.250	$12\frac{1}{4}$	4.350	29	135.83	54	1636.7	79	7498	130	54,978
$6\frac{1}{4}$	0.296	$12\frac{1}{2}$	4.720	30	155.55	55	1761.4	80	7885	140	73,948
$6\frac{1}{2}$	0.345	$12\frac{3}{4}$	5.110	31	177.77	56	1893.1	81	8286	150	97,449
$6\frac{3}{4}$	0.402	13	5.58	32	201.80	57	2031.9	82	8703	160	126,152
7	0.464	$13\frac{1}{4}$	5.96	33	228.20	58	2178.3	83	9135	170	160,772
$7\frac{1}{4}$	0.535	$13\frac{1}{2}$	6.42	34	257.20	59	2332.5	84	9584	180	202,071
$7\frac{1}{2}$	0.611	$13\frac{3}{4}$	6.91	35	386.80	60	2494.7	85	10,048	190	250,858
$7\frac{3}{4}$	0.699	14	7.42	36	323.20	61	2665.2	86	10,529	200	307,988

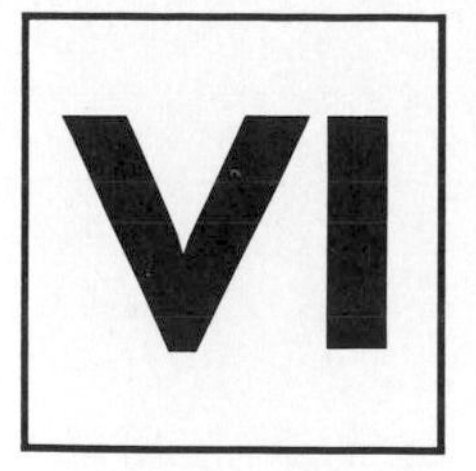

Full-Load Currents of AC and DC Motors

THREE-PHASE, 60-Hz AC INDUCTION MOTORS[a]

HP	RPM	Full Load Current					
		200 v	230 v	460 v	575 v	2200 v	4000 v
1/4	1800	1.10	0.96	0.48	0.38	....	....
	1200	1.33	1.16	0.58	0.46	....	....
	900	1.67	1.45	0.73	0.58	....	....
1/3	1800	1.33	1.16	0.58	0.47	....	....
	1200	1.64	1.43	0.72	0.58	....	....
	900	2.01	1.75	0.88	0.71	....	....
1/2	1800	1.93	1.68	0.84	0.67	....	....
	1200	2.38	2.07	1.04	0.83	....	....
	900	3.34	2.90	1.45	1.16	....	....
3/4	1800	2.68	2.33	1.17	0.93	....	....
	1200	3.28	2.85	1.43	1.14	....	....
	900	3.97	3.45	1.73	1.38	....	....
1	3600	3.16	2.75	1.38	1.10	....	....
	1800	3.51	3.05	1.53	1.22	....	....
	1200	4.07	3.54	1.77	1.42	....	....
	900	4.30	3.74	1.87	1.50	....	....
1 1/2	3600	4.80	4.17	2.09	1.67	....	....
	1800	4.92	4.28	2.14	1.71	....	....
	1200	5.58	4.85	2.43	1.94	....	....
	900	6.68	5.81	2.91	2.32	....	....
2	3600	6.39	5.56	2.78	2.22	....	....
	1800	6.62	5.76	2.88	2.30	....	....
	1200	7.30	6.35	3.18	2.54	....	....
	900	8.29	7.21	3.61	2.88	....	....
3	3600	9.05	7.87	3.94	3.14	....	....
	1800	9.53	8.29	4.14	3.32	....	....
	1200	10.3	8.92	4.46	3.56	....	....
	900	11.7	10.20	5.09	4.08	....	....
5	3600	14.6	12.7	6.34	5.08	....	....
	1800	15.2	13.2	6.60	5.28	....	....
	1200	16.2	14.1	7.05	5.64	....	....
	900	17.9	15.6	7.80	6.24	....	....
7 1/2	3600	22.1	19.2	9.6	7.68	....	....
	1800	22.2	19.3	9.7	7.72	....	....
	1200	23.3	20.3	10.2	8.12	....	....
	900	27.4	23.8	11.9	9.51	....	....
10	3600	28.2	24.5	12.3	9.8	....	....
	1800	29.0	25.2	12.6	10.1	....	....
	1200	30.6	26.6	13.3	10.6	....	....
	900	33.2	28.9	14.5	11.6	....	....
	600	38.9	33.8	16.9	13.5	....	....
15	3600	42.2	36.7	18.4	14.7	....	....
	1800	43.8	38.1	19.1	15.2	....	....
	1200	45.9	39.9	20.0	16.0	....	....
	900	48.2	41.9	21.0	16.8	....	....
	600	55.5	48.3	24.2	19.3	....	....
20	3600	56.4	49.0	24.5	19.6	5.2	2.9
	1800	58.1	50.5	25.3	20.2	5.3	3.0
	1200	59.5	51.7	25.9	20.6	5.4	3.1
	900	62.8	54.6	27.3	21.8	5.8	3.2
	600	70.7	61.5	30.8	24.6	6.4	3.5
25	3600	68.1	59.2	29.6	23.6	6.3	3.4
	1800	72.1	62.7	31.3	25.0	6.5	3.6
	1200	74.4	64.7	32.3	25.8	6.7	3.7
	900	77.5	67.4	33.7	27.0	6.9	3.8
	600	82.7	71.9	35.9	28.8	8.1	4.4

Source: Courtesy of Allen-Bradley.

[a]The full-load currents listed are "average values" for horsepower-rated motors of several manufacturers at the more common rated voltages and speeds. These "average values," along with the similar values listed in the NEC should be used only as a guide for selecting suitable components for the motor branch circuit. The rated full-load current, shown on the motor nameplate, may vary considerably from the value listed, depending on the specific motor design. The nameplate full-load current should always be used in determining the rating of the devices used for motor running overcurrent protection.

HP	RPM	Full Load Current					
		200 v	230 v	460 v	575 v	2200 v	4000 v
30	1800	83.7	72.8	36.4	29.2	7.8	4.3
	1200	88.7	77.1	38.6	30.8	8.0	4.4
	900	91.3	79.4	39.7	31.8	8.2	4.5
	600	101.1	87.9	43.9	35.2	9.3	5.0
40	1800	112.7	98	49.0	39.2	10.0	5.5
	1200	113.9	99	49.5	39.6	10.3	5.7
	900	119.6	104	52.0	41.6	10.6	5.8
	600	130.0	113	56.5	45.2	11.5	6.3
50	1800	139.2	121	60.5	48.4	12.3	6.8
	1200	140.3	122	61.0	48.8	12.4	6.8
	900	146.1	127	63.5	50.8	13.1	7.2
	600	158.7	138	69.0	55.2	14.2	7.8
60	1800	164.5	143	71.5	57.2	14.6	8.0
	1200	170.2	148	74.0	59.2	14.9	8.2
	900	173.7	151	75.5	60.4	15.4	8.5
	600	186.3	162	81.0	64.8	16.7	9.2
75	1800	204.7	178	89.0	71.2	18.0	9.9
	1200	208.2	181	90.5	72.4	18.2	10.0
	900	215.1	187	93.5	74.8	19.0	10.5
	600	228.9	199	99.5	79.6	21.0	11.6
100	1800	268.0	233	116	93.2	23.6	13.0
	1200	274.9	239	120	95.6	24.2	13.3
	900	281.8	245	123	98.0	24.8	13.6
	600	295.6	257	128	103.0	26.4	14.5
	450	333.5	290	145	116.0	29.8	16.4
125	1800	332.4	289	144	115	29.2	16.1
	1200	342.7	298	149	119	29.9	16.4
	900	350.8	305	153	122	30.9	17.0
	720	361.1	314	157	126	31.3	17.2
	600	368.0	320	160	128	32.8	18.0
	450	403.7	351	175	140	36.0	19.8
150	1800	397.9	346	173	138	34.8	19.1
	1200	402.5	350	175	140	35.5	19.5
	900	417.5	363	182	145	37.0	20.4
	720	432.4	376	188	150	37.0	20.4
	600	434.7	378	189	151	38.8	21.3
	450	480.7	418	209	166	42.0	23.1
200	1800	529.0	460	230	184	46.7	25.7
	1200	535.9	466	233	186	47.0	25.9
	900	563.5	490	245	196	49.4	27.2
	720	568.1	494	247	197	49.0	27.0
	600	572.7	498	249	199	50.9	28.0
	450	607.2	528	264	211	53.7	29.5
250	1800	657.8	572	286	229	57.5	31.6
	1200	667.0	580	290	232	58.5	32.2
	900	694.6	604	302	242	61.5	33.8
	720	718.8	625	312	250	61.5	33.8
	600	724.5	630	315	252	61.0	33.6
	450	724.5	630	315	252	65.3	35.9
	360	777.4	676	338	270	70.0	38.5
300	1800	787.8	685	342	274	69.0	38.0
	1200	800.4	696	348	278	70.0	38.5
	900	830.3	722	361	289	73.5	40.4
	600	830.3	722	361	289	72.3	39.8
	450	874.0	760	380	304	76.0	41.8
	360	954.5	830	415	332	82.8	45.5

Power Factor Correcting Capacitors

SUGGESTED MAXIMUM CAPACITOR RATING WHEN MOTOR AND CAPACITOR ARE SWITCHED AS UNIT

	Nominal Motor Speed (rpm):											
	3600		*1800*		*1200*		*900*		*720*		*600*	
Induction Motor Horse-Power Rating	*Capacitor Rating (kVAR)*	*Line Current Reduction (%)*	*Capacitor Rating (kVAR)*	*Line Current Reduction (%)*	*Capacitor Rating (kVAR)*	*Line Current Reduction (%)*	*Capacitor Rating (kVAR)*	*Line Current Reduction (%)*	*Capacitor Rating (kVAR)*	*Line Current Reduction (%)*	*Capacitor Rating (kVAR)*	*Line Current Reduction (%)*
3	1.5	14	1.5	15	1.5	20	2	27	2.5	35	3.5	41
5	2	12	2	13	2	17	3	25	4	32	4.5	37
$7\frac{1}{2}$	2.5	11	2.5	12	3	15	4	22	5.5	30	6	34
10	3	10	3	11	3.5	14	5	21	6.5	27	7.5	31
15	4	9	4	10	5	13	6.5	18	8	23	9.5	27
20	5	9	5	10	6.5	12	7.5	16	9	21	12	25
25	6	9	6	10	7.5	11	9	15	11	20	14	23
30	7	8	7	9	9	11	10	14	12	18	16	22
40	9	8	9	9	11	10	12	13	15	16	20	20
50	12	8	11	9	13	10	15	12	19	15	24	19
60	14	8	14	8	15	10	18	11	22	15	27	19
75	17	8	16	8	18	10	21	10	26	14	32.5	18
100	22	8	21	8	25	9	27	10	32.5	13	40	17
125	27	8	26	8	30	9	32.5	10	40	13	47.5	16
150	32.5	8	30	8	35	9	37.5	10	47.5	12	52.5	15
200	40	8	37.5	8	42.5	9	47.5	10	60	12	65	14

Induction Motor Horse-Power Rating	3600 Capacitor Rating (kVAR)	3600 Line Current Reduction (%)	1800 Capacitor Rating (kVAR)	1800 Line Current Reduction (%)	1200 Capacitor Rating (kVAR)	1200 Line Current Reduction (%)	900 Capacitor Rating (kVAR)	900 Line Current Reduction (%)	720 Capacitor Rating (kVAR)	720 Line Current Reduction (%)	600 Capacitor Rating (kVAR)	600 Line Current Reduction (%)
	Nominal Motor Speed (rpm):											
250	50	8	45	7	52.5	8	57.5	9	70	11	77.5	13
300	57.5	8	52.5	7	60	8	65	9	80	11	87.5	12
350	65	8	60	7	67.5	8	75	9	87.5	10	95	11
400	70	8	65	6	75	8	85	9	95	10	105	11
450	75	8	67.5	6	80	8	92.5		100	9	110	11
500	77.5	8	72.5	6	82.5	8	97.5	9	107.5	9	115	10

Note: For use with three-phase, 60-Hz NEMA class B motors to raise full-load power factor to approximately 95%.

KILOWATT MULTIPLIERS TO DETERMINE CAPACITOR KILOVAR REQUIRED FOR POWER FACTOR CORRECTION

Original Power Factor	0.80	0.81	0.82	0.83	0.84	0.85	0.86	0.87	0.88	0.89	0.90	0.91	0.92	0.93	0.94	0.95	0.96	0.97	0.98	0.99	1.0
	Corrected Power Factor																				
0.50	0.982	1.008	1.034	1.060	1.086	1.112	1.139	1.165	1.192	1.220	1.248	1.276	1.306	1.337	1.369	1.403	1.440	1.481	1.529	1.589	1.732
0.51	0.937	0.962	0.989	1.015	1.041	1.067	1.094	1.120	1.147	1.175	1.203	1.231	1.261	1.292	1.324	1.358	1.395	1.436	1.484	1.544	1.687
0.52	0.893	0.919	0.945	0.971	0.997	1.023	1.050	1.076	1.103	1.131	1.159	1.187	1.217	1.248	1.280	1.314	1.351	1.392	1.440	1.500	1.643
0.53	0.850	0.876	0.902	0.928	0.954	0.980	1.007	1.033	1.060	1.088	1.116	1.144	1.174	1.205	1.237	1.271	1.308	1.349	1.397	1.457	1.600
0.54	0.809	0.835	0.861	0.887	0.913	0.939	0.966	0.992	1.019	1.047	1.075	1.103	1.133	1.164	1.196	1.230	1.267	1.308	1.356	1.416	1.559
0.55	0.769	0.795	0.821	0.847	0.873	0.899	0.926	0.952	0.979	1.007	1.035	1.063	1.093	1.124	1.156	1.190	1.227	1.268	1.316	1.376	1.519
0.56	0.730	0.756	0.782	0.808	0.834	0.860	0.887	0.913	0.940	0.968	0.996	1.024	1.054	1.085	1.117	1.151	1.188	1.229	1.277	1.337	1.480
0.57	0.692	0.718	0.744	0.770	0.796	0.822	0.849	0.875	0.902	0.930	0.958	0.986	1.016	1.047	1.079	1.113	1.150	1.191	1.239	1.299	1.442
0.58	0.655	0.681	0.707	0.733	0.759	0.785	0.812	0.838	0.865	0.893	0.921	0.949	0.979	1.010	1.042	1.076	1.113	1.154	1.202	1.262	1.405
0.59	0.619	0.645	0.671	0.697	0.723	0.749	0.776	0.802	0.829	0.857	0.885	0.913	0.943	0.974	1.006	1.040	1.077	1.118	1.166	1.226	1.369
0.60	0.583	0.609	0.635	0.661	0.687	0.713	0.740	0.766	0.793	0.821	0.849	0.877	0.907	0.938	0.970	1.004	1.041	1.082	1.130	1.190	1.333
0.61	0.549	0.575	0.601	0.627	0.653	0.679	0.706	0.732	0.759	0.787	0.815	0.843	0.873	0.904	0.936	0.970	1.007	1.048	1.096	1.156	1.299
0.62	0.516	0.542	0.568	0.594	0.620	0.646	0.673	0.699	0.726	0.754	0.782	0.810	0.840	0.871	0.903	0.937	0.974	1.015	1.063	1.123	1.266
0.63	0.483	0.509	0.535	0.561	0.587	0.613	0.640	0.666	0.693	0.721	0.749	0.777	0.807	0.838	0.870	0.904	0.941	0.982	1.030	1.090	1.233
0.64	0.451	0.474	0.503	0.529	0.555	0.581	0.608	0.634	0.661	0.689	0.717	0.745	0.775	0.806	0.838	0.872	0.909	0.950	0.998	1.068	1.201
0.65	0.419	0.445	0.471	0.497	0.523	0.549	0.576	0.602	0.629	0.657	0.685	0.713	0.743	0.774	0.806	0.840	0.877	0.918	0.966	1.026	1.169
0.66	0.388	0.414	0.440	0.466	0.492	0.518	0.545	0.571	0.598	0.626	0.654	0.682	0.712	0.743	0.775	0.809	0.846	0.887	0.935	0.995	1.138
0.67	0.358	0.384	0.410	0.436	0.462	0.488	0.515	0.541	0.568	0.596	0.624	0.652	0.682	0.713	0.745	0.779	0.816	0.857	0.905	0.965	1.108
0.68	0.328	0.354	0.380	0.406	0.432	0.458	0.485	0.511	0.538	0.566	0.594	0.622	0.652	0.683	0.715	0.749	0.786	0.827	0.875	0.935	1.078
0.69	0.299	0.325	0.351	0.377	0.403	0.429	0.456	0.482	0.509	0.537	0.565	0.593	0.623	0.654	0.686	0.720	0.757	0.798	0.846	0.906	1.049
0.70	0.270	0.296	0.322	0.348	0.374	0.400	0.427	0.453	0.480	0.508	0.536	0.564	0.594	0.625	0.657	0.691	0.728	0.769	0.817	0.877	1.020
0.71	0.242	0.268	0.294	0.320	0.346	0.372	0.399	0.425	0.452	0.480	0.508	0.536	0.566	0.597	0.629	0.663	0.700	0.741	0.789	0.849	0.992
0.72	0.214	0.240	0.266	0.292	0.318	0.344	0.371	0.397	0.424	0.452	0.480	0.508	0.538	0.569	0.601	0.635	0.672	0.713	0.761	0.821	0.964
0.73	0.186	0.212	0.238	0.264	0.290	0.316	0.343	0.369	0.396	0.424	0.452	0.480	0.510	0.541	0.573	0.607	0.644	0.685	0.733	0.793	0.936
0.74	0.159	0.185	0.211	0.237	0.263	0.289	0.316	0.342	0.369	0.397	0.425	0.453	0.483	0.514	0.546	0.580	0.617	0.658	0.706	0.766	0.909
0.75	0.132	0.158	0.184	0.210	0.236	0.262	0.289	0.315	0.342	0.370	0.398	0.426	0.456	0.487	0.519	0.553	0.590	0.631	0.679	0.739	0.882

Original Power Factor	Corrected Power Factor																				
	0.80	*0.81*	*0.82*	*0.83*	*0.84*	*0.85*	*0.86*	*0.87*	*0.88*	*0.89*	*0.90*	*0.91*	*0.92*	*0.93*	*0.94*	*0.95*	*0.96*	*0.97*	*0.98*	*0.99*	*1.0*
0.76	0.105	0.131	0.157	0.183	0.209	0.235	0.262	0.288	0.315	0.343	0.371	0.399	0.429	0.460	0.492	0.526	0.563	0.604	0.652	0.712	0.855
0.77	0.079	0.105	0.131	0.157	0.183	0.209	0.236	0.262	0.289	0.317	0.345	0.373	0.403	0.434	0.466	0.500	0.537	0.578	0.626	0.686	0.829
0.78	0.052	0.078	0.104	0.130	0.156	0.182	0.209	0.235	0.262	0.290	0.318	0.346	0.376	0.407	0.439	0.473	0.510	0.551	0.599	0.659	0.802
0.79	0.026	0.052	0.078	0.104	0.130	0.156	0.183	0.209	0.236	0.264	0.292	0.320	0.350	0.381	0.413	0.447	0.484	0.525	0.573	0.633	0.776
0.80	0.000	0.026	0.052	0.078	0.104	0.130	0.157	0.183	0.210	0.238	0.266	0.294	0.324	0.355	0.387	0.421	0.458	0.499	0.547	0.609	0.750
0.81		0.000	0.026	0.052	0.078	0.104	0.131	0.157	0.184	0.212	0.240	0.268	0.298	0.329	0.361	0.395	0.432	0.473	0.521	0.581	0.724
0.82			0.000	0.026	0.052	0.078	0.105	0.131	0.158	0.186	0.214	0.242	0.272	0.303	0.335	0.369	0.406	0.447	0.495	0.555	0.698
0.83				0.000	0.026	0.052	0.079	0.105	0.132	0.160	0.188	0.216	0.246	0.277	0.309	0.343	0.380	0.421	0.469	0.529	0.672
0.84					0.000	0.026	0.053	0.079	0.106	0.134	0.162	0.190	0.220	0.251	0.283	0.317	0.354	0.395	0.443	0.503	0.646
0.85						0.000	0.027	0.053	0.080	0.108	0.136	0.164	0.194	0.225	0.257	0.291	0.328	0.369	0.417	0.477	0.620
0.86							0.000	0.026	0.053	0.081	0.109	0.137	0.167	0.198	0.230	0.264	0.301	0.342	0.390	0.450	0.593
0.87								0.000	0.027	0.055	0.083	0.111	0.141	0.172	0.204	0.238	0.275	0.316	0.364	0.424	0.567
0.88									0.000	0.028	0.056	0.084	0.114	0.145	0.177	0.211	0.248	0.289	0.337	0.397	0.540
0.89										0.000	0.028	0.056	0.086	0.117	0.149	0.183	0.220	0.261	0.309	0.369	0.512
0.90											0.000	0.028	0.058	0.089	0.121	0.155	0.192	0.233	0.281	0.341	0.484
0.91												0.000	0.030	0.061	0.093	0.127	0.164	0.205	0.253	0.313	0.456
0.92													0.000	0.031	0.063	0.097	0.134	0.175	0.223	0.283	0.426
0.93														0.000	0.032	0.066	0.103	0.144	0.192	0.252	0.395
0.94															0.000	0.034	0.071	0.112	0.160	0.220	0.363
0.95																0.000	0.037	0.079	0.126	0.186	0.329
0.96																	0.000	0.041	0.089	0.149	0.292
0.97																		0.000	0.048	0.108	0.251
0.98																			0.000	0.060	0.203
0.99																				0.000	0.143
																					0.000

Switch Symbols

COMMON SWITCH SYMBOLS

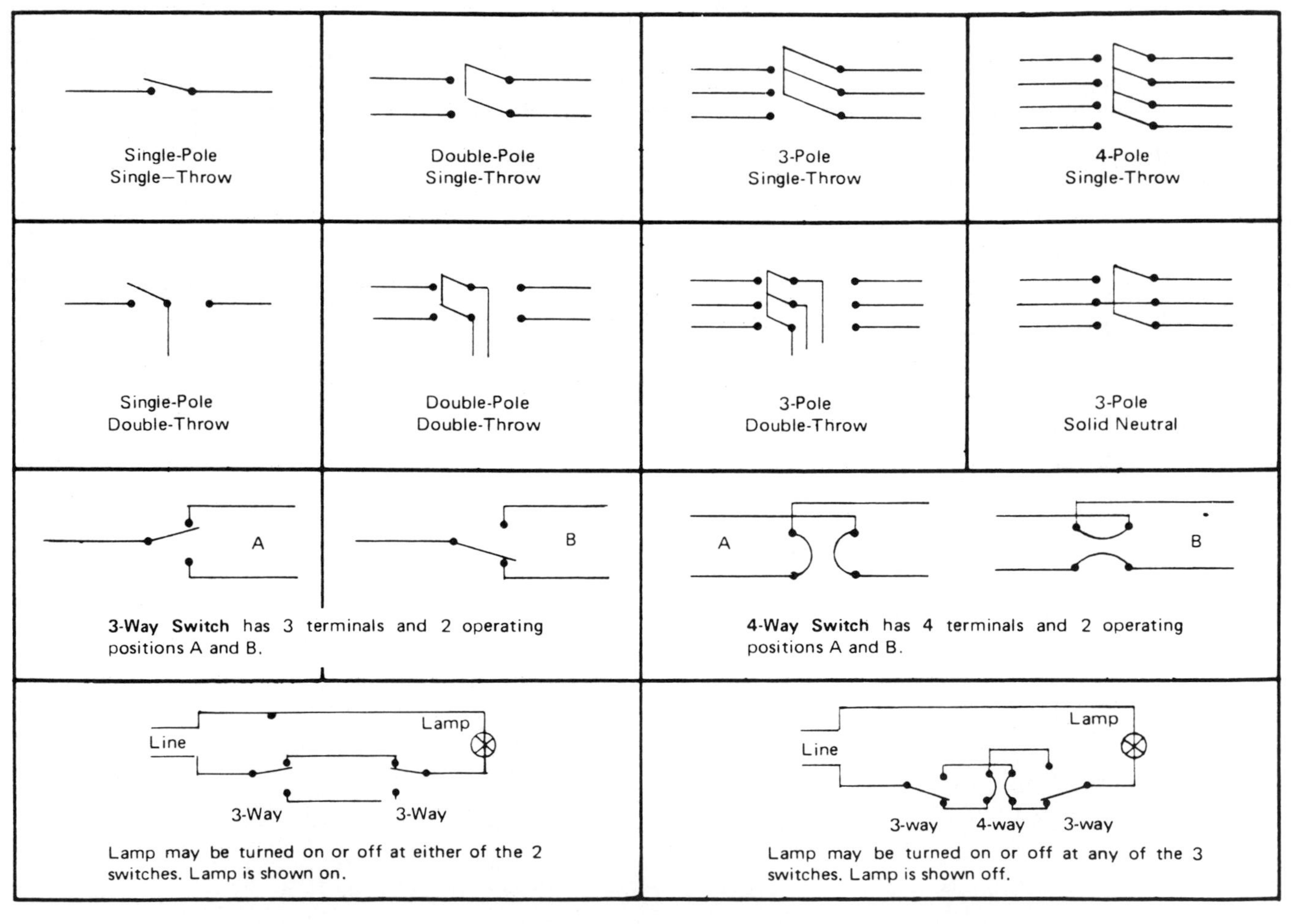

OTHER SWITCHES

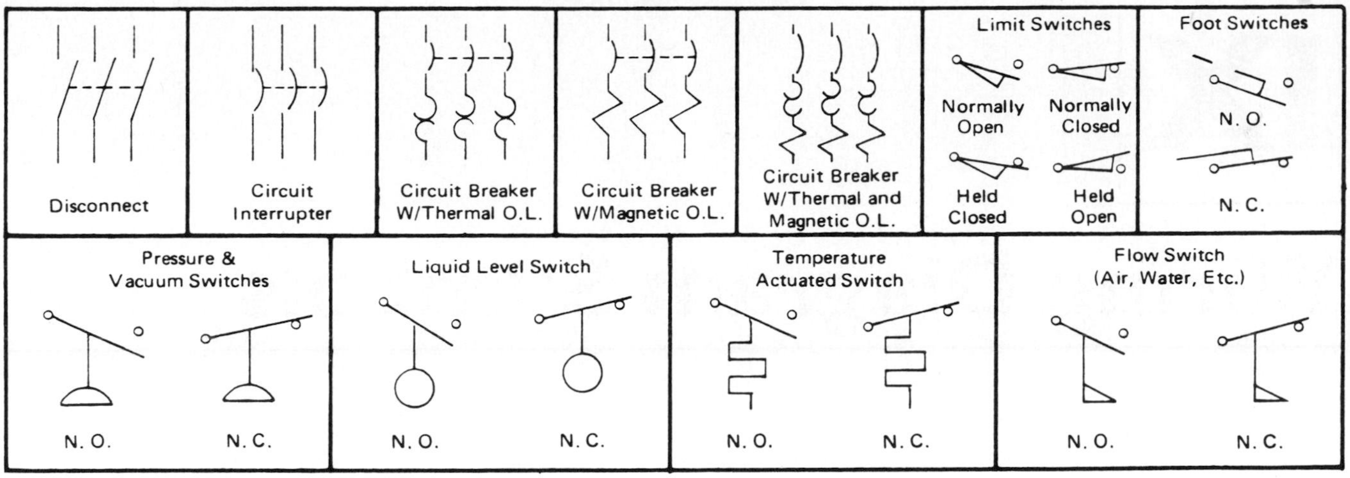

Disconnect	Circuit Interrupter	Circuit Breaker W/Thermal O.L.	Circuit Breaker W/Magnetic O.L.	Circuit Breaker W/Thermal and Magnetic O.L.	Limit Switches	Foot Switches
					Normally Open Normally Closed Held Closed Held Open	N. O. N. C.

Pressure & Vacuum Switches		Liquid Level Switch		Temperature Actuated Switch		Flow Switch (Air, Water, Etc.)	
N. O.	N. C.	N. O.	N. C.	N. O.	N. C.	N. O.	N. C.

Wiring Diagram Symbols

STANDARD ELEMENTARY WIRING DIAGRAM SYMBOLS

Speed (Plugging)

F F

R

Anti-Plug

F

R

SELECTOR

2 Position

A1	X	
A2		X
	Low	High

A1

A2

3 Position

A1	X		
A2			X
	Hand	Off	Auto

A1

A2

2 Pos. Sel. Push Button

A1	X			
A2		X	X	X
	Free	Depres'd	Free	Depres'd
	Jog		Run	

A1

A2

PUSH BUTTONS

Momentary Contact

Single Circuit: N. O. — N. C.

Double Circuit: N. O. — N. C.

Mushroom Head

Maintained Contact

One Double CKT — Two Single CKT

PILOT LIGHTS

Indicate Color By Letter

Non Push-To-Test (A) — Push-To-Test (R)

COILS

Shunt — Series

OVERLOAD RELAYS

Thermal — Magnetic

CONTACTS

Instant Operating				Timed Contacts – Contact Action Retarded After Coil Is			
With Blowout		Without Blowout		Energized		De-Energized	
N. O.	N. C.	N. O.	N. C.	N.O.T.C.	N.C.T.O.	N.O.T.O.	N.C.T.C.

INDUCTORS	TRANSFORMERS				
Iron Core	Auto	Iron Core	Air Core	Current	Dual Voltage
Air Core					

AC MOTORS				DC MOTORS			
Single Phase	3 Phase Squirrel Cage	2 Phase 4 Wire	Wound Rotor	Armature	Shunt Field	Series Field	Comm. or Compens. Field
					(Show 4 Loops)	(Show 3 Loops)	(Show 2 Loops)

Unit Prefixes

Prefix	*Symbol*	*Power of 10*	*Value*	*Name*
tera	T	10^{12}	1,000,000,000,000	trillion
		10^{11}	100,000,000,000	hundred-billion
		10^{10}	10,000,000,000	ten-billion
giga	G	10^{9}	1,000,000,000	billion
		10^{8}	100,000,000	hundred-million
		10^{7}	10,000,000	ten-million
mega	M	10^{6}	1,000,000	million
		10^{5}	100,000	hundred-thousand
myria	my	10^{4}	10,000	ten-thousand
kilo	k	10^{3}	1,000	thousand
hecto	h	10^{2}	100	hundred
deka	da	10^{1}	10	ten
		10^{0}	1	unit
deci	d	10^{-1}	.1	tenth
centi	c	10^{-2}	.01	hundredth
milli	m	10^{-3}	.001	thousandth
		10^{-4}	.000 1	ten-thousandth
		10^{-5}	.000 01	hundred-thousandth
micro	μ	10^{-6}	.000 001	millionth
		10^{-7}	.000 000 1	ten-millionth
		10^{-8}	.000 000 01	hundred-millionth
nano	n	10^{-9}	.000 000 001	billionth
		10^{-10}	.000 000 000 1	ten-billionth
		10^{-11}	.000 000 000 01	hundred-billionth
pico	p	10^{-12}	.000 000 000 001	trillionth
		10^{-13}	.000 000 000 000 1	ten-trillionth
		10^{-14}	.000 000 000 000 01	hundred-trillionth
femto	f	10^{-15}	.000 000 000 000 001	quadrillionth
		10^{-16}	.000 000 000 000 000 1	ten-quadrillionth
		10^{-17}	.000 000 000 000 000 01	hundred-quadrillionth
atto	a	10^{-18}	.000 000 000 000 000 001	quintillionth

Conversion Factors

Multiply:	*By:*	*To Obtain:*
British thermal units	778.3	foot-pounds
British thermal units	3.931×10^{-4}	horsepower-hours
British thermal units	1055	joules
centimeters	3.281×10^{-2}	feet
centimeters	0.3937	inches
centimeters per second	1.969	feet per minute
circular mils	5.067×10^{-6}	square centimeters
circular mils	7.854×10^{-7}	square inches
cubic centimeters	3.531×10^{-5}	cubic feet
cubic centimeters	6.102×10^{-2}	cubic inches
cubic feet	1728	cubic inches
cubic feet	0.02832	cubic meters
feet	30.48	centimeters
grams	0.3527	ounces
horsepower	550	foot pounds per second
horsepower	0.7457	kilowatts
inches	2.540	centimeters
kilometers	3281	feet
liters	0.2642	gallons
lumens per sq ft	1.0	foot-candles
meters	39.37	inches
miles per hour	88	feet per minute
pounds per sq ft	0.06804	atmospheres
radians	57.30	degrees
yards	91.44	centimeters

Decibel Table

dB	Power Ratio	Voltage or Current Ratio
0	1.00	1.00
0.5	1.12	1.06
1.0	1.26	1.12
1.5	1.41	1.19
2.0	1.58	1.26
3.0	2.00	1.41
4.0	2.51	1.58
5.0	3.16	1.78
6.0	3.98	2.00
7.0	5.01	2.24
8.0	6.31	2.51
9.0	7.94	2.82
10	10.0	3.16
15	31.6	5.62
20	100	10
25	316	17.8
30	1,000	31.6
40	10,000	100
50	10^5	316
60	10^6	1,000
70	10^7	3,162
80	10^8	10,000
90	10^9	31,620
100	10^{10}	10^5

Index

D

E

F

G

P

R

S

T

U

V

W